Springer Texts in Statistics

Springer Texts in Statistics (STS) includes advanced textbooks from 3rd- to 4th-year undergraduate courses to 1st- to 2nd-year graduate courses. Exercise sets should be included. The series editors are currently Genevera I. Allen, Richard D. De Veaux, and Rebecca Nugent. Stephen Fienberg, George Casella, and Ingram Olkin were editors of the series for many years.

More information about this series at http://www.springer.com/series/417

Ronald Christensen

Advanced Linear Modeling

Statistical Learning and Dependent Data

Third Edition

 Springer

Ronald Christensen
Department of Mathematics and Statistics
University of New Mexico
Albuquerque, NM, USA

ISSN 1431-875X   ISSN 2197-4136   (electronic)
Springer Texts in Statistics
ISBN 978-3-030-29166-2   ISBN 978-3-030-29164-8   (eBook)
https://doi.org/10.1007/978-3-030-29164-8

This Springer imprint is published by the registered company Springer Nature Switzerland AG.
The registered company address is: Gewerbestrasse 11, 6330 Cham, Switzerland

To Pete, Russ, and Scot,
my pals from high school;

also

Wes, Ed, and everyone from graduate school.

Preface to the Third Edition

This is the third edition of *Advanced Linear Modeling (ALM)*. It is roughly 50% longer than the previous edition. It discusses the extension of linear models into areas beyond those usually addressed in regression and analysis of variance. As in previous editions, its primary emphasis is on models in which the data display some form of dependence and many of the changes from the previous edition were made to systematize this emphasis on dependent data. Nonetheless, it begins with topics in modern regression analysis related to nonparametric regression and penalized estimation (regularization). R code for the analyses in the book is available at http://www.stat.unm.edu/~fletcher/R-ALMIII.pdf.

Mathematical background is contained in Appendix A on differentiation and Kronecker products. Also some notation used throughout the book is set in Sect. 1.1.

This edition has been written in conjunction with the fifth edition of Christensen (2011), often hereafter referred to as *PA*. Some discussions that previously appeared in *PA* have been moved here. Obviously, you cannot do advanced linear modeling without previously learning about linear modeling. I have tried to make this book readable to people who have studied linear model theory from sources other than *PA*, but I need to cite some source for basic results on linear models, so obviously I cite *PA*. In cases where I need to cite results for which the new version of *PA* is different from the previous edition(s), the citations are given as *PA-V*. I have rearranged the topics from the previous edition of *ALM* so that the material related to independent data comes first followed by the material on dependent data. The chapter on response surfaces has been dropped but is available in a new volume downloadable from my website: http://www.stat.unm.edu/~fletcher/TopicsInDesign. Some familiarity with inner products is assumed, especially in Chaps. 1 and 3. The required familiarity can be acquired from *PA*.

Chapter 1 expands the previous introduction to nonparametric regression. The discussion follows what is commonly known as the basis function approach, despite the fact that many of the techniques do not actually involve the use of basis functions per se. In fact, when dealing with spaces of functions the very idea of a basis is subject to competing definitions. Tim Hanson pointed out to me the obvious fact that if a group of functions are linearly independent, they always form a basis for the space that they span, but I think that in nonparametric regression the idea is to approximate wider collections of functions than just these spanning sets. Chapter 1 now also includes a short introduction to models involving an entire function of predictor variables.

Chapter 2 is an expanded version of the discussion of penalized regression from Christensen (2011). A new Chap. 3 extends this by introducing reproducing kernel Hilbert spaces.

Chapter 4 is new except for the last section. It gives results on an extremely general linear model for dependent or heteroscedastic data. It owes an obvious debt to Christensen (2011, Chapter 12). It contains several particularly useful exercises. In a standard course on linear model theory, the theory of estimation and testing for dependent data is typically introduced but not developed, see for example Christensen (2011, Sections 2.7 and 3.8). Section 4.1 of this book reviews, but does not re-prove, those results. This book then applies those fundamental results to develop theory for a wide variety of practical models.

I finally figured out how, without overwhelming the ideas in abstruse notation, to present MINQUE as linear modeling, so I have done that in Chap. 4. In a technical subsection, I give in to the abstruse notation so as to derive the MINQUE equations. Previously, I just referred the reader to Rao for the derivation.

Chapter 5 on mixed models originally appeared in *PA*. It has been shortened in places due of overlap with Chap. 4 but includes several new examples and exercises. It contains a new emphasis on linear covariance structures that leads not only to variance component models but the new Sect. 5.6 that examines a quite general longitudinal data model. The details of the recovery of interblock information for a balanced incomplete block design from *PA* no longer seem relevant, so they were relegated, along with the response surface material, to the volume on my website.

Chapters 6 and 7 introduce time series: first the frequency domain which uses models from Chap. 1 but with random effects as in Chap. 5 and then the time domain approach which can be viewed as applications of ideas from the frequency domain.

Chapter 8 on spatial data is little changed from the previous edition. Mostly, the references have been updated.

The former chapter on multivariate models has been split into three: Chap. 9 on general theory with a new section relating multivariate models to spatial and time series models and a new discussion of multiple comparisons, Chap. 10 on applications to specific models, and Chap. 11 with an expanded discussion of generalized multivariate linear models (also known as generalized multivariate analysis of variance (GMANOVA) and growth curve models).

Chapters 12 and 14 are updated versions of the previous chapters on discriminant analysis and principal components. Chapter 13 is a new chapter on binary regression and discrimination. Its raison d'être is that it devotes considerable attention to support vector machines. Chapter 14 contains a new section on classical multidimensional scaling.

From time to time, I mention the virtues of Bayesian approaches to problems discussed in the book. One place to look for more information is *BIDA*, i.e., Christensen, Johnson, Branscum, and Hanson (2010).

Thanks to my son Fletcher who is always the first person I ask when I have doubts. Joe Cavanaugh and Mohammad Hattab have been particularly helpful as have Tim Hanson, Wes Johnson, and Ed Bedrick. Finally, my thanks to Al Nosedal-Sanchez, Curt Storlie, and Thomas Lee for letting me modify our joint paper into Chap. 3.

As I have mentioned elsewhere, the large number of references to my other works is as much about sloth as it is ego. In some sense, with the exception of *BIDA*, all of my books are variations on a theme.

Albuquerque, NM, USA

Ronald Christensen

February 2019

Preface to the Second Edition

This is the second edition of *Linear Models for Multivariate, Time Series and Spatial Data*. It has a new title to indicate that it contains much new material. The primary changes are the addition of two new chapters: one on nonparametric regression and the other on response surface maximization. As before, the presentations focus on the linear model aspects of the subject. For example, in the nonparametric regression chapter there is very little about kernel regression estimation but quite a bit about series approximations, splines, and regression trees, all of which can be viewed as linear modeling.

The new edition also includes various smaller changes. Of particular note are a subsection in Chap. 1 on modeling longitudinal (repeated measures) data and a section in Chap. 6 on covariance structures for spatial lattice data. I would like to thank Dale Zimmerman for the suggestion of incorporating material on spatial lattices. Another change is that the subject index is now entirely alphabetical.

Albuquerque, NM, USA
May 9, 2000

Ronald Christensen

Preface to the First Edition

This is a companion volume to *Plane Answers to Complex Questions: The Theory of Linear Models*. It consists of six additional chapters written in the same spirit as the last six chapters of the earlier book. Brief introductions are given to topics related to linear model theory. No attempt is made to give a comprehensive treatment of the topics. Such an effort would be futile. Each chapter is on a topic so broad that an in-depth discussion would require a book-length treatment.

People need to impose structure on the world in order to understand it. There is a limit to the number of unrelated facts that anyone can remember. If ideas can be put within a broad, sophisticatedly simple structure, not only are they easier to remember but often new insights become available. In fact, sophisticatedly simple models of the world may be the only ones that work. I have often heard Arnold Zellner say that, to the best of his knowledge, this is true in econometrics. The process of modeling is fundamental to understanding the world.

In Statistics, the most widely used models revolve around linear structures. Often the linear structure is exploited in ways that are peculiar to the subject matter. Certainly, this is true of frequency domain time series and geostatistics. The purpose of this volume is to take three fundamental ideas from standard linear model theory and exploit their properties in examining multivariate, time series and spatial data. In decreasing order of importance to the presentation, the three ideas are: best linear prediction, projections, and Mahalanobis distance. (Actually, Mahalanobis distance is a fundamentally multivariate idea that has been appropriated for use in linear models.) Numerous references to results in *Plane Answers* are made. Nevertheless, I have tried to make this book as independent as possible. Typically, when a result from *Plane Answers* is needed not only is the reference given but also the result itself. Of course, for proofs of these results the reader will have to refer to the original source.

I want to reemphasize that this is a book about linear models. It is not traditional multivariate analysis, time series, or geostatistics. Multivariate linear models are viewed as linear models with a nondiagonal covariance matrix. Discriminant analysis is related to the Mahalanobis distance and multivariate analysis of variance. Principal components are best linear predictors. Frequency domain time series involves linear models with a peculiar design matrix. Time domain analysis involves models that are linear in the parameters but have random design matrices. Best linear predictors are used for forecasting time series; they are also fundamental to the estimation techniques used in time domain analysis. Spatial data analysis involves linear models in which the covariance matrix is modeled from the data; a primary objective in analyzing spatial data is making best linear unbiased predictions of future observables. While other approaches to these problems may yield different insights, there is value in having a unified approach to looking at these problems. Developing such a unified approach is the purpose of this book.

There are two well-known models with linear structure that are conspicuous by their absence in my two volumes on linear models. One is Cox's (1972) proportional hazards model. The other is the generalized linear model of Nelder and Wedderburn (1972). The proportional hazards methodology is a fundamentally nonparametric technique for dealing with censored data having linear structure. The emphasis on nonparametrics and censored data would make its inclusion here awkward. The interested reader can see Kalbfleisch and Prentice (1980). Generalized linear models allow the extension of linear model ideas to many situations that involve independent non-normally distributed observations. Beyond the presentation of basic linear model theory, these volumes focus on methods for analyzing correlated observations. While it is true that generalized linear models can be used for some types of correlated data, such applications do not flow from the essential theory. McCullagh and Nelder (1989) give a detailed exposition of generalized linear models, and Christensen (1997) contains a short introduction.

Acknowledgments

I would like to thank MINITAB[1] for providing me with a copy of release 6.1.1, BMDP with copies of their programs 4M, 1T, 2T, and 4V, and Dick Lund for providing me with a copy of MSUSTAT. Nearly all of the computations were performed with one of these programs. Many were performed with more than one.

I would not have tackled this project but for Larry Blackwood and Bob Shumway. Together Larry and I reconfirmed, in my mind anyway, that multivariate analysis is just the same old stuff. Bob's book put an end to a specter that has long haunted me: a career full of half-hearted attempts at figuring out basic time series analysis.

At my request, Ed Bedrick, Bert Koopmans, Wes Johnson, Bob Shumway, and Dale Zimmerman tried to turn me from the errors of my ways. I sincerely thank them for their valuable efforts. The reader must judge how successful they were with a recalcitrant subject. As always, I must thank my editors, Steve Fienberg and Ingram Olkin, for their suggestions. Jackie Damrau did an exceptional job in typing the first draft of the manuscript.

[1] MINITAB is a registered trademark of Minitab, Inc., 3081 Enterprise Drive, State College, PA 16801, telephone: (814) 238 3280.

Finally, I have to recognize the contribution of Magic Johnson. I was so upset when the 1987–88 Lakers won a second consecutive NBA title that I began writing this book in order to block the mental anguish. I am reminded of Woody Allen's dilemma: is the importance of life more accurately reflected in watching *The Sorrow and the Pity* or in watching the Knicks? (In my case, the Jazz and the Celtics.) It's a tough call. Perhaps life is about actually making movies and doing statistics.

Albuquerque, NM, USA
April 19, 1990

Ronald Christensen

References

Christensen, Ronald (1997). *Log-Linear Models and Logistic Regression*, Second Edition. Springer-Verlag, New York.

Christensen, R. (2011). *Plane answers to complex questions: The theory of linear models* (4th ed.). New York: Springer.

Christensen, R., Johnson, W., Branscum, A., & Hanson, T. E. (2010). *Bayesian ideas and data analysis: An introduction for scientists and statisticians*. Boca Raton, FL: Chapman and Hall/CRC Press.

Cox, D. R. (1972). Regression models and life tables (with discussion). *Journal of the Royal Statistical Society, Series B, 34*, 187–220.

Kalbfleisch, J. D., & Prentice, R. L. (1980). *The statistical analysis of failure time data*. New York: Wiley.

McCullagh, P., & Nelder, J. A. (1989). *Generalized linear models* (2nd Ed.). London: Chapman and Hall.

Nelder, J. A., & Wedderburn, R. W. M. (1972). Generalized linear models. *Journal of the Royal Statistical Society, Series A, 135*, 370–384.

Contents

Chapter 1
Nonparametric Regression

Abstract This chapter introduces nonparametric regression for a single predictor variable, discusses the curse of dimensionality that plagues nonparametric regression with multiple predictor variables, and discusses the kernel trick and related ideas as methods for overcoming the curse of dimensionality.

In the late 1990s, the orthogonal series approach to nonparametric regression became increasingly popular; see Hart (1997), Ogden (1997), and Efromovich (1999). In this approach, *orthogonal series* of functions are used to approximate the regression function. Later, the orthogonality was de-emphasized so that now a series of more general *basis functions* is often used to approximate the regression function. Basis functions, and other linear-approximation methods for which many of the series elements have small support, methods such as splines and wavelets, seem particularly useful. We discuss these approaches to nonparametric regression as fitting linear models.

Suppose we have a dependent variable y and a vector of predictor variables x. Regression is about estimating $E(y|x)$. In linear regression, we assume that $E(y|x) = x'\beta$ for some unknown parameter vector β. Recall that this includes fitting indicator variables and polynomials as special cases. In nonlinear regression we assume that $E(y|x) = f(x;\beta)$, where the function f is known but the vector β is unknown; see Sect. 7.4 and Christensen (1996, Chapter 18; 2015, Chapter 23). A special case of nonlinear regression involves linearizable models, including generalized linear models, that assume $E(y|x) = f(x'\beta)$ for f known, cf. Christensen (1997, Chapter 9). The key idea in nonlinear regression is using calculus to linearize the model. In *nonparametric regression*, we assume that $E(y|x) = f(x)$, where the function f is unknown. Note the absence of a vector of parameters β, hence the name nonparametric. Often, f is assumed to be continuous or to have some specified number of derivatives. In reality, nonparametric regression is exactly the opposite of what its name suggests. Nonparametric regression involves fitting far more parameters than either standard linear or nonlinear regression.

© Springer Nature Switzerland AG 2019
R. Christensen, *Advanced Linear Modeling*, Springer Texts in Statistics,
https://doi.org/10.1007/978-3-030-29164-8_1

EXAMPLE 1.0.1. Table 1.1 presents data from Montgomery and Peck (1982) and Eubank (1988) on voltage drops y over time t displayed by an electrical battery used in a guided missile. The 41 times go from 0 to 20. The variable x results from dividing t by 20, thus standardizing the times into the $[0,1]$ interval. The data comprise a time series as discussed in Chaps. 6 and 7, but the idea here is that the behavior over time is not a stationary stochastic process but rather a complicated regression function. An unusual feature of these data is that the t_i values are equally spaced (i.e., the t_is are ordered and $t_{i+1} - t_i$ is a constant). This typically occurs only when the data collection process is very well-controlled. However, when equal spacing does occur, it considerably simplifies data analysis. □

Table 1.1 Battery voltage drops versus time

Case	y	t	x	Case	y	t	x
1	8.33	0.0	0.000	22	14.92	10.5	0.525
2	8.23	0.5	0.025	23	14.37	11.0	0.550
3	7.17	1.0	0.050	24	14.63	11.5	0.575
4	7.14	1.5	0.075	25	15.18	12.0	0.600
5	7.31	2.0	0.100	26	14.51	12.5	0.625
6	7.60	2.5	0.125	27	14.34	13.0	0.650
7	7.94	3.0	0.150	28	13.81	13.5	0.675
8	8.30	3.5	0.175	29	13.79	14.0	0.700
9	8.76	4.0	0.200	30	13.05	14.5	0.725
10	8.71	4.5	0.225	31	13.04	15.0	0.750
11	9.71	5.0	0.250	32	12.06	15.5	0.775
12	10.26	5.5	0.275	33	12.05	16.0	0.800
13	10.91	6.0	0.300	34	11.15	16.5	0.825
14	11.67	6.5	0.325	35	11.15	17.0	0.850
15	11.76	7.0	0.350	36	10.14	17.5	0.875
16	12.81	7.5	0.375	37	10.08	18.0	0.900
17	13.30	8.0	0.400	38	9.78	18.5	0.925
18	13.88	8.5	0.425	39	9.80	19.0	0.950
19	14.59	9.0	0.450	40	9.95	19.5	0.975
20	14.05	9.5	0.475	41	9.51	20.0	1.000
21	14.48	10.0	0.500				

In Sect. 1.2, we discuss the basics of the linear-approximation approach. In Sect. 1.3, we examine its relationship to linear models. In Sect. 1.4, we discuss and illustrate least squares estimation and we discuss some of the estimation approaches specifically proposed for orthogonal series. In Sect. 1.5, we discuss variable selection and relate an orthogonal series proposal to the C_p statistic. Section 1.6 examines problems involving heteroscedastic variances. In Sect. 1.7, we discuss details of splines and introduce kernel estimation and other local polynomial regression techniques. Section 1.8 introduces nonparametric multiple regression. Section 1.9 examines testing lack of fit. Section 1.10 looks at regression trees. Finally, Sect. 1.11 makes a few comments on density estimation and Sect. 1.12 includes exercises.

1.1 Basic Notation

Before proceeding we set some notation that is used throughout the book, unless defined otherwise for particular purposes. A linear model has $Y = X\beta + e$ where Y is an $n \times 1$ vector of observable random variables, X is an $n \times p$ matrix of known values, β is a $p \times 1$ vector of fixed but unknown coefficients, and e is an $n \times 1$ vector of unobservable random errors. For this to be a linear model we need $E(e) = 0$ so that $E(Y) = X\beta$. A *standard linear model* assumes that an individual observation or error has variance σ^2 and that $Cov(Y) = Cov(e) = \sigma^2 I$. The assumption that the observations have a multivariate normal distribution can be written $Y \sim N(X\beta, \sigma^2 I)$. A partitioned linear model is written $Y = X\beta + Z\gamma + e$ where Z is also a matrix of known values and γ is also a vector of fixed, unknown coefficients. If Z has s columns, write

$$X = [X_1, \ldots, X_p] = \begin{bmatrix} x_1' \\ \vdots \\ x_n' \end{bmatrix}; \quad Z = [Z_1, \ldots, Z_s] = \begin{bmatrix} z_1' \\ \vdots \\ z_n' \end{bmatrix}.$$

For any vector v, $\|v\|^2 \equiv v'v$ is the squared (Euclidean) length of v. The Euclidean inner product between two vectors u and v is $u'v$. They are perpendicular (orthogonal), written $v \perp u$, if $v'u = 0$. A^- denotes the generalized inverse of the matrix A, $r(A)$ denotes its rank, and $tr(A)$ denotes its trace. M denotes the perpendicular projection operator (ppo) onto the column space of X. The column space of X is denoted $C(X)$. (With tongue slightly in cheek) the *Fundamental Theorem of Least Squares Estimation* is that in a linear model, $\hat{\beta}$ is a least squares estimate if and only if

$$X\hat{\beta} = MY.$$

More generally, M_A denotes the ppo onto $C(A)$. $C(A)^\perp$ denotes the orthogonal complement of $C(A)$, i.e. all the vectors that are orthogonal to $C(A)$. If $C(X) \subset C(A)$, $C(X)^\perp_{C(A)}$ denotes the orthogonal complement of $C(X)$ with respect to $C(A)$, i.e. all vectors in $C(A)$ that are orthogonal to $C(X)$. An $r \times c$ matrix of 1s is denoted J_r^c with $J_n \equiv J_n^1$ and $J \equiv J_n$.

This is all common notation and, except for the use of M and J, it is pretty much standard notation. (Some authors prefer P and $\mathbf{1}$.) It is important to understand the theory associated with this notation. For example, I expect the reader to know (or at least believe when I write) that the ppo onto $C(X)^\perp_{C(A)}$ is $M_A - M$. Such background can be found in many places including *PA*.

1.2 Linear Approximations

The key idea behind linear approximations is that a finite linear combination of some known functions can approximate a wide variety of functions on a closed bounded set, cf. the famous Stone-Weierstrass theorem. For convenience, we initially assume that f is defined on the interval $[0, 1]$ and is continuous. There are many ways to approximate f including polynomials, sines and cosines, step functions, and also by things similar to step functions called *wavelets*. Most often we assume that for some predictor variable x

$$f(x) = \sum_{j=0}^{\infty} \beta_j \phi_j(x),$$

where the ϕ_js are known functions that can be defined in many ways. Later we will use this characterization with x being a p vector instead of a scalar. In particular, with $p = 1$ and functions defined on the unit interval, we can take for $j = 0, 1, 2, \ldots$

$$\phi_j(x) = x^j, \tag{1.2.1}$$

or

$$\phi_j(x) = \cos(\pi j x), \tag{1.2.2}$$

or

$$\phi_{2j}(x) = \cos(\pi j x) \quad \phi_{2j+1}(x) = \sin(\pi j x). \tag{1.2.3}$$

When using (1.2.2), it should be noted that the derivative of every $\cos(\pi j x)$ function is 0 at $x = 0$, so the derivative of $f(x)$ should be 0 at $x = 0$.

   In practice we approximate f with a finite number of terms which determines a linear model in which only the β_js are unknown. We need to determine an appropriate finite approximation and estimate the corresponding β_js

   With a single predictor, another obvious approximation uses step functions but some care must be used. Let $\mathscr{I}_A$ be the *indicator function* for the set A, namely

$$\mathscr{I}_A(x) = \begin{cases} 1 & \text{if } x \in A \\ 0 & \text{otherwise.} \end{cases}$$

Obviously, if we define

$$\phi_j(x) = \mathscr{I}_{(\frac{j-1}{m}, \frac{j}{m}]}(x), \quad j = 0, 1, \ldots, m,$$

we can approximate any continuous function f, and as $m \to \infty$ we can approximate f arbitrarily well. Note that $\phi_0(x)$ is essentially $\mathscr{I}_{\{0\}}(x)$. Technically, rather than the infinite sum characterization, we are defining a *triangular array* of functions ϕ_{jm}, $j = 1, \ldots, m; m = 1, 2, 3, \ldots$ and assuming that

$$f(x) = \lim_{m \to \infty} \sum_{j=0}^{m} \beta_{jm} \phi_{jm}(x). \tag{1.2.4}$$

More generally, we could define the indicator functions using intervals between knots, $\tilde{x}_{-1,m} < 0 = \tilde{x}_{0,m} < \tilde{x}_{1,m} < \tilde{x}_{2,m} < \cdots < \tilde{x}_{m,m} = 1$ with the property that $\max_i\{\tilde{x}_{i+1,m} - \tilde{x}_{i,m}\}$ goes to zero as m goes to infinity.

Splines are more complicated than indicator functions. Choosing $m - 1$ knots in the interior of $[0,1]$ is fundamental to the use of splines. Rather than indicators, we can fit some low dimensional polynomial between the knots. In this context, indicator functions are 0 degree polynomials. For polynomials of degree greater than 0, traditional splines force the polynomials above and below each knot in $(0,1)$ to take the same value at the knot, thus forcing the splines to give a continuous function on $[0,1]$. *B-splines* use functions ϕ_{jm} that are nonzero only on small but overlapping subintervals with locations determined by (often centered around) a collection of knots. As with indicator functions, to get good approximations to an arbitrary regression function, the distances between consecutive knots must all get (asymptotically) small. As a practical matter, one tries to find one appropriate set of knots for the problem at hand. Technically, methods based on knots are not basis function methods because they do not provide a countable set of functions that are linearly independent and span the space of continuous functions. (B-spline is short for "basis spline" but that is something of a misnomer.)

As with basis function approaches based on an infinite sum, any triangular array satisfying Eq. (1.2.4) allows us to approximate f with a finite linear model in which only $\beta_{1m}, \ldots, \beta_{mm}$ are unknown. Triangular array approximations can also be used with vector inputs.

Rather than defining a triangular array of indicator functions, we can use the following device to define a single infinite series :

$$\phi_0(x) = 1, \quad \phi_1(x) = \mathscr{I}_{(0,.5]}(x), \quad \phi_2(x) = \mathscr{I}_{(.5,1]}(x),$$

$$\phi_3(x) = \mathscr{I}_{(0,.25]}(x), \quad \phi_4(x) = \mathscr{I}_{(.25,.5]}(x),$$

$$\phi_5(x) = \mathscr{I}_{(.5,.75]}(x), \quad \phi_6(x) = \mathscr{I}_{(.75,1]}(x),$$

$$\phi_7(x) = \mathscr{I}_{(0,2^{-3}]}(x), \ldots, \phi_{14}(x) = \mathscr{I}_{(\{2^3-1\}2^{-3},1]}(x),$$

$$\phi_{15}(x) = \mathscr{I}_{(0,2^{-4}]}(x), \ldots .$$

Technically, these ϕ_js constitute a spanning set of functions but are not basis functions. Except for approximating the point $f(0)$, including the function $\phi_0(x)$ is irrelevant once we include $\phi_1(x)$ and $\phi_2(x)$. Similarly, $\phi_1(x)$ and $\phi_2(x)$ are made irrelevant by $\phi_3(x), \ldots, \phi_6(x)$.

A sequence of basis functions, one that is equivalent to this spanning set of step functions, is the *Haar wavelet* collection

$$\phi_0(x) = 1, \quad \phi_1(x) = \mathscr{I}_{(0,.5]}(x) - \mathscr{I}_{(.5,1]}(x),$$

$$\phi_2(x) = \mathscr{I}_{(0,.25]}(x) - \mathscr{I}_{(.25,.5]}(x), \quad \phi_3(x) = \mathscr{I}_{(.5,.75]}(x) - \mathscr{I}_{(.75,1]}(x),$$

$$\phi_4(x) = \mathscr{I}_{(0,1/8]}(x) - \mathscr{I}_{(1/8,2/8]}(x), \ldots, \phi_7(x) = \mathscr{I}_{(6/8,7/8]}(x) - \mathscr{I}_{(7/8,1]}(x),$$

$$\phi_8(x) = \mathscr{I}_{(0,1/16]}(x) - \mathscr{I}_{(1/16,2/16]}(x), \ldots .$$

It is customary to call $\phi_0(x)$ the *father* wavelet function and $\phi_1(x)$ the *mother* function. Note that all of the subsequent functions are obtained from the mother function by changing the location and scale, for example, $\phi_3(x) = \phi_1(2x - 1)$, $\phi_7(x) = \phi_1(4x - 3)$, and, in general, if $j = 2^r + k$ for $k = 0, 1, \ldots, 2^r - 1$, then $\phi_j(x) = \phi_1(2^r x - k)$.

Actually, *this idea of changing location and scale can be applied to any mother function ϕ_1 that is 0 outside the unit interval and integrates to 0 over the unit interval*, hence generating different families of wavelets to be used as a basis series. (Rather than integrating to 0, theoretical developments often impose a stronger admissability condition on ϕ_1.) For simplicity we restrict ourselves to looking at Haar wavelets but my impression is that they are rarely used in practice. The *Mexican hat (Ricker) wavelet* seems to be quite popular.

Orthogonal series approximations use basis functions that are orthogonal in an appropriate inner product. Typically, the functions ϕ_j would be defined to be orthonormal in $\mathscr{L}^2$ space. $\mathscr{L}^2$ is the space of all functions that are square integrable, that is,

$$\int_0^1 f(x)^2 dx < \infty.$$

The inner product of two functions, say f and g, is

$$\langle f, g \rangle \equiv \int_0^1 f(x)g(x)dx,$$

so f and g are defined to be orthogonal if their inner product is 0. In other words, orthogonal functions ϕ_j are defined to have $\int_0^1 \phi_j(x)\phi_k(x)dx = 0$ for $j \neq k$. In particular, the polynomial functions given in (1.2.1) would have to be adjusted using the Gram–Schmidt theorem to make them orthogonal. The norm of a function f in $\mathscr{L}^2$ is

$$\|f\| \equiv \left[\int_0^1 f(x)^2 dx \right]^{1/2}.$$

Thus, for the ϕ_js to be orthonormal, they must be orthogonal and $\phi_j(x)$ needs to have $\int_0^1 [\phi_j(x)]^2 dx = 1$. Methods based on Legendre polynomials (Gram-Schmidt-ing the polynomials), cosines and/or sines, and many wavelets are designed as orthonormal series. Our definition of wavelets needs rescaling before they square integrate to 1, i.e., redefine $\phi_j(x) = 2^{r/2}\phi_1(2^r x - k)$ where ϕ_1 has norm 1. The choice of the mother function ϕ_1 determines whether the sequence is orthogonal.

Exercise 1.1.   Show that the Haar wavelets are orthogonal.

As we will see, for most regression problems $\mathscr{L}^2$ orthogonality is largely irrelevant, except perhaps for extremely large data sets.

A *technical note:* It is presumed that the reader has some familiarity with the concepts of random variables converging in distribution, in probability, with probability one (almost surely), and in $\mathscr{L}^2$. Recall that $\mathscr{L}^2$ convergence and almost sure convergence imply convergence in probability, which implies convergence in distribution but neither $\mathscr{L}^2$ nor almost sure convergence imply the other. Similar concepts apply to the functions f wherein almost sure convergence is analogous to convergence almost everywhere. In the relationships $f(x) = \sum_{j=0}^{\infty} \beta_j \phi_j(x) \equiv \lim_{m \to \infty} \sum_{j=0}^{m} \beta_j \phi_j(x)$ and $f(x) = \lim_{m \to \infty} \sum_{j=0}^{m} \beta_{jm} \phi_{jm}(x)$, the convergence involved might be for every single $x \in [0, 1]$ when f is continuous or it might be almost everywhere convergence or the convergence might be $\mathscr{L}^2$ convergence, i.e.,

$$\lim_{m \to \infty} \left\| f(x) - \sum_{j=0}^{m} \beta_{jm} \phi_{jm}(x) \right\| = 0,$$

for any $f \in \mathscr{L}^2$. Such issues have little practical effect on fitting approximating linear models to finite data.

If x is a vector instead of a scalar, alternative ϕ_j functions need to be used; see Sect. 1.8.

1.3 Simple Nonparametric Regression

The simple nonparametric regression model is

$$y_i = f(x_i) + \varepsilon_i, \quad \mathrm{E}(\varepsilon_i) = 0,$$

$i = 1, \ldots, n$, where y_i is a random variable, x_i is a known (scalar) constant, f is an unknown continuous function, and the ε_is are unobservable independent errors with $\mathrm{Var}(\varepsilon_i) = \sigma^2$. Traditionally, the errors are assumed independent, rather than just uncorrelated, to facilitate asymptotic results. In matrix form, write

$$\begin{bmatrix} y_1 \\ \vdots \\ y_n \end{bmatrix} = \begin{bmatrix} f(x_1) \\ \vdots \\ f(x_n) \end{bmatrix} + \begin{bmatrix} \varepsilon_1 \\ \vdots \\ \varepsilon_n \end{bmatrix}$$

or

$$Y = F(X) + e, \quad \mathrm{E}(e) = 0, \quad \mathrm{Cov}(e) = \sigma^2 I,$$

where $X \equiv (x_1, \ldots, x_n)'$ and $F(X) \equiv [f(x_1), \ldots, f(x_n)]'$. Again, for ease of exposition, we assume that $x_i \in [0, 1]$ for all i.

Using the infinite basis representation

$$f(x) = \sum_{j=0}^{\infty} \beta_j \phi_j(x),$$

the nonparametric regression model becomes an infinite linear model,

$$y_i = \sum_{j=0}^{\infty} \beta_j \phi_j(x_i) + \varepsilon_i.$$

This is not useful because it involves an infinite sum, so we use a finite linear model approximation,

$$y_i = \sum_{j=0}^{s-1} \beta_j \phi_j(x_i) + \varepsilon_i. \tag{1.3.1}$$

Essentially the same approximation results from a triangular array representation of f. If we define $\Phi_j \equiv [\phi_j(x_1), \ldots, \phi_j(x_n)]'$, in matrix terms model (1.3.1) becomes

$$Y = [\Phi_0, \Phi_1, \ldots, \Phi_{s-1}] \begin{bmatrix} \beta_0 \\ \beta_1 \\ \vdots \\ \beta_{s-1} \end{bmatrix} + e,$$

or, defining $\Phi \equiv [\Phi_0, \Phi_1, \ldots, \Phi_{s-1}]$, we get

$$Y = \Phi\beta + e.$$

The linear model (1.3.1) is only an approximation, so in reality the errors will be biased. For basis functions $E(\varepsilon_i) = \sum_{j=s}^{\infty} \beta_j \phi_j(x_i)$. It is important to know that for s large, these bias terms are small; see Efromovich (1999, Section 2.2).

Perhaps the two most important statistical questions are how to estimate the β_js and how to choose an appropriate value of s. These issues are addressed in the next two sections.

1.4 Estimation

Choose s so that, for all practical purposes,

$$Y = \Phi\beta + e, \quad E(e) = 0, \quad \text{Cov}(e) = \sigma^2 I. \tag{1.4.1}$$

Clearly, in this model, least squares estimates are BLUEs, so

$$\hat{\beta} = (\Phi'\Phi)^{-1}\Phi'Y.$$

To construct tests or confidence intervals, we would need to assume independent normal errors. The regression function is estimated by

$$\hat{f}(x) = \sum_{j=0}^{s-1} \hat{\beta}_j \phi_j(x).$$

This methodology requires $r(\Phi) \leq n$. Often the model will fit the data perfectly when $s = n$, but this would not occur if the Φ_js are linearly dependent (i.e., if $r(\Phi) < s$). In the next chapter we consider alternatives to least squares estimation.

For the voltage drop data we now examine the use of several methods of nonparametric regression: fitting polynomials, cosines, Haar wavelets, and cubic splines. We begin with the most familiar of these methodologies, fitting polynomials.

1.4.1 Polynomials

Fitting high-order polynomials becomes difficult numerically unless we do something toward orthogonalizing them. We will only fit a sixth degree polynomial, so for the battery data we can get by with simply subtracting the mean before defining the polynomials. The fitted sixth degree regression is

$$\hat{y} = 14.6 + 7.84(x - 0.5) - 66.3(x - 0.5)^2 - 28.7(x - 0.5)^3$$
$$+ 199(x - 0.5)^4 + 10.2(x - 0.5)^5 - 92(x - 0.5)^6$$

with $R^2 = 0.991$. The regression coefficients, ANOVA table, and sequential sums of squares are:

Table of coefficients: 6th degree polynomial

Predictor	$\hat{\beta}_k$	$SE(\hat{\beta}_k)$	t	P
Constant	14.6156	0.0901	162.24	0.000
$(x - 0.5)$	7.8385	0.6107	12.83	0.000
$(x - 0.5)^2$	−66.259	4.182	−15.84	0.000
$(x - 0.5)^3$	−28.692	9.190	−3.12	0.004
$(x - 0.5)^4$	199.03	43.87	4.54	0.000
$(x - 0.5)^5$	10.17	30.84	0.33	0.744
$(x - 0.5)^6$	−91.6	121.2	−0.76	0.455

Analysis of variance: 6th degree polynomial

Source	df	SS	MS	F	P
Regression	6	259.256	43.209	624.77	0.000
Error	34	2.351	0.069		
Total	40	261.608			

Source	df	Seq. SS
$(x - 0.5)$	1	47.081
$(x - 0.5)^2$	1	170.159
$(x - 0.5)^3$	1	11.155
$(x - 0.5)^4$	1	30.815
$(x - 0.5)^5$	1	0.008
$(x - 0.5)^6$	1	0.039

From the sequential sums of squares, the F test for dropping to a fourth degree polynomial is

$$F = \frac{[0.039 + 0.008]/2}{0.069} < 1,$$

so, refitting, we can get by with the regression equation

$$\hat{y} = 14.6 + 7.67(x - 0.5) - 63.4(x - 0.5)^2 - 25.7(x - 0.5)^3 + 166(x - 0.5)^4,$$

which still has $R^2 = 0.991$. The regression coefficients and ANOVA table are

Table of coefficients: 4th degree polynomial				
Predictor	$\hat{\beta}_k$	SE($\hat{\beta}_k$)	t	P
Constant	14.5804	192.64	0.0757	0.000
$(x - 0.5)$	7.6730	22.47	0.3414	0.000
$(x - 0.5)^2$	−63.424	−34.99	1.812	0.000
$(x - 0.5)^3$	−25.737	−12.94	1.989	0.000
$(x - 0.5)^4$	166.418	21.51	7.738	0.000

Analysis of variance: 4th degree polynomial					
Source	df	SS	MS	F	P
Regression	4	259.209	64.802	972.66	0.000
Error	36	2.398	0.0676		
Total	40	261.608			

Note that the estimated regression coefficients have changed with the dropping of the fifth and sixth degree terms. Figure 1.1 displays the data and the fitted curve.

Polynomials fit these data very well. Other linear approximations may fit the data better or worse. What fits well depends on the particular data being analyzed.

1.4.2 Cosines

For fitting cosines, define the variable $c_j \equiv \cos(\pi j x)$. I arbitrarily decided to fit cosines up to $j = 30$. The fitted regression equation is

$$\begin{aligned}
\hat{y} = {}& 11.4 - 1.63c_1 - 3.11c_2 + 0.457c_3 + 0.216c_4 + 0.185c_5 \\
& + 0.150c_6 + 0.0055c_7 + 0.0734c_8 + 0.0726c_9 + 0.141c_{10} \\
& + 0.0077c_{11} + 0.0603c_{12} + 0.125c_{13} + 0.120c_{14} + 0.0413c_{15} \\
& + 0.0184c_{16} + 0.0223c_{17} - 0.0320c_{18} + 0.0823c_{19} + 0.0409c_{20} \\
& - 0.0005c_{21} + 0.0017c_{22} + 0.0908c_{23} + 0.0036c_{24} - 0.0660c_{25} \\
& + 0.0104c_{26} + 0.0592c_{27} - 0.0726c_{28} - 0.0760c_{29} + 0.0134c_{30}
\end{aligned}$$

with $R^2 = 0.997$ and ANOVA table

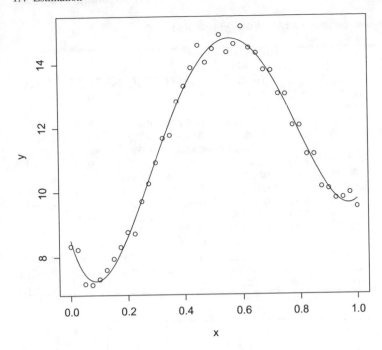

Fig. 1.1 Fourth-degree polynomial fit to battery data

Analysis of variance: 30 Cosines					
Source	df	SS	MS	F	P
Regression	30	260.7275	8.6909	98.75	0.000
Error	10	0.8801	0.0880		
Total	40	261.6076			

The table of regression coefficients is Table 1.2 and Fig. 1.2 displays the data and the fitted model. Note that most of the action in Table 1.2 takes place from $j = 0, \ldots, 6$ with no other terms having P values less than 0.05. However, these all are tests of effects fitted last and are not generally appropriate for deciding on the smallest level of j. In this case, the xs are equally spaced, so the c_js are very nearly orthogonal, so a model based on $j = 0, \ldots, 6$ will probably work well.

The regression equation based on only $j = 0, \ldots, 6$ is

$$\hat{y} = 11.4 - 1.61c_1 - 3.10c_2 + 0.473c_3 + 0.232c_4 + 0.201c_5 + 0.166c_6$$

with $MSE = 0.094 = 3.195/34$ and $R^2 = 98.8\%$. Notice the slight changes in the regression coefficients relative to the first 7 terms in Table 1.2 due to nonorthogonality. The correlation matrix of c_1 to c_6 is not quite the identity:

Table 1.2 Regression coefficients for fitting cosines with $s - 1 = 30$

j	$\hat{\beta}_k$	SE	t	P	j	$\hat{\beta}_k$	SE	t	P
0	11.3802	0.0466	244.34	0.000	16	0.01844	0.06539	0.28	0.784
1	−1.62549	0.06538	−24.86	0.000	17	0.02225	0.06538	0.34	0.741
2	−3.11216	0.06539	−47.59	0.000	18	−0.03197	0.06539	−0.49	0.635
3	0.45701	0.06538	6.99	0.000	19	0.08235	0.06538	1.26	0.236
4	0.21605	0.06539	3.30	0.008	20	0.04087	0.06539	0.62	0.546
5	0.18491	0.06538	2.83	0.018	21	−0.00048	0.06538	−0.01	0.994
6	0.14984	0.06539	2.29	0.045	22	0.00165	0.06539	0.03	0.980
7	0.00553	0.06538	0.08	0.934	23	0.09076	0.06538	1.39	0.195
8	0.07343	0.06539	1.12	0.288	24	0.00356	0.06539	0.05	0.958
9	0.07262	0.06538	1.11	0.293	25	−0.06597	0.06538	−1.01	0.337
10	0.14136	0.06539	2.16	0.056	26	0.01038	0.06539	0.16	0.877
11	0.00765	0.06538	0.12	0.909	27	0.05924	0.06538	0.91	0.386
12	0.06032	0.06539	0.92	0.378	28	−0.07257	0.06539	−1.11	0.293
13	0.12514	0.06538	1.91	0.085	29	−0.07600	0.06538	−1.16	0.272
14	0.11983	0.06539	1.83	0.097	30	0.01338	0.06539	0.20	0.842
15	0.04128	0.06538	0.63	0.542					

(Table of coefficients)

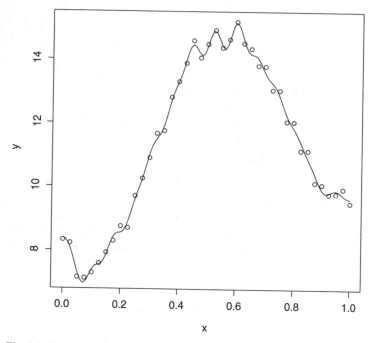

Fig. 1.2 Cosine fit with $s - 1 = 30$ for the battery data

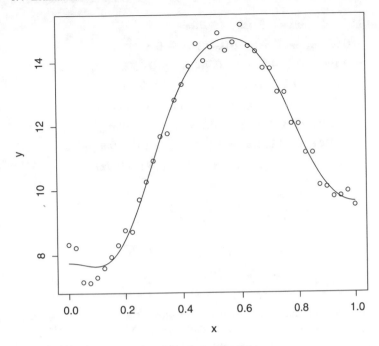

Fig. 1.3 Cosine fit with $s - 1 = 6$ for the battery data

	Correlations					
	c_1	c_2	c_3	c_4	c_5	c_6
c_1	1.00	0.00	0.05	0.00	0.05	0.00
c_2	0.00	1.00	0.00	0.05	0.00	0.05
c_3	0.05	0.00	1.00	0.00	0.05	0.00
c_4	0.00	0.05	0.00	1.00	0.00	0.05
c_5	0.05	0.00	0.05	0.00	0.00	0.00
c_6	0.00	0.05	0.00	0.05	0.00	1.00,

although the real issue is the correlations between these 6 and the other 24 variables. Figure 1.3 displays the data along with the fitted cosine curve for $s - 1 = 6$. Both the figures and the R^2 values establish that fitting the 5 parameters in a 4th degree polynomial fits these data better than an intercept and 6 cosine terms. That a polynomial fits better than cosines is a peculiarity of these data.

1.4.3 Haar Wavelets

Consider fitting Haar wavelets. We fit 32 functions, the father wavelet $\phi_0(x) \equiv p_0(x) \equiv 1$, the mother wavelet $\phi_1(x) \equiv m_0$, and then transformations of the mother wavelet, $\phi_j(x) \equiv m_{rk}$ where $j = 2^r + k - 1$, $k = 1, \ldots, 2^r$. The fitted regression equation is

$$\hat{y} = 11.3 - 1.05m_0 - 2.31m_{11} + 1.78m_{12}$$
$$- 0.527m_{21} - 1.36m_{22} + 0.472m_{23} + 0.814m_{24}$$
$$+ 0.190m_{31} - 0.444m_{32} - 0.708m_{33} - 0.430m_{34}$$
$$- 0.058m_{35} + 0.317m_{36} + 0.567m_{37} + 0.071m_{38}$$
$$+ 0.530m_{4,1} - 0.181m_{4,2} - 0.180m_{4,3} - 0.248m_{4,4}$$
$$- 0.325m_{4,5} - 0.331m_{4,6} - 0.290m_{4,7} + 0.139m_{4,8}$$
$$+ 0.275m_{4,9} - 0.131m_{4,10} + 0.265m_{4,11} + 0.349m_{4,12}$$
$$+ 0.005m_{4,13} + 0.229m_{4,14} + 0.150m_{4,15} + 0.012m_{4,16}$$

with $R^2 = 0.957$ and ANOVA table

Analysis of variance: 32 Haar wavelets					
Source	df	SS	MS	F	P
Regression	31	250.489	8.080	6.54	0.003
Error	9	11.118	1.235		
Total	40	261.6076			

Based on R^2, this fits the data much worse than either the fourth degree polynomial regression or the cosine regression model with $s - 1 = 6$. Table 1.3 gives the estimated regression coefficients. This table gives little indication that either the third- or fourth-order wavelets are contributing to the fit of the model, but again, the predictor variables are not orthogonal, so definite conclusions cannot be reached. For example, the $m_{4,k}$s are defined so that they are orthogonal to each other, but they are not orthogonal to all of the m_{3k}s. In particular, $m_{4,2}$ is not orthogonal to m_{31}. To see this, note that the first six entries of the 41 dimensional vector $m_{4,2}$ are $(0, 0, 0, 1, -1, -1)$, with the rest being 0's and the first six entries of m_{31} are $(0, 1, 1, -1, -1, -1)$, with the rest being 0's. Clearly, the two vectors are not orthogonal. The problem is that, even though the observations are equally spaced, the wavelets are based on powers of $1/2$, whereas the 41 observations occur at intervals of $1/40$.

Figure 1.4 displays the data along with the fitted Haar wavelets for $s = 32$. The figure displays a worse fit than the 4th degree polynomial and the 7 parameter cosine model despite fitting many more parameters. This is consistent with the relatively poor Haar wavelet R^2. Some kind of curved mother wavelet function would probably fit better than the Haar wavelets. Notice the curious behavior of the plot at $x = 0$. The way the Haar wavelets have been defined here, if the columns of Φ are orthogonal, the estimate of $f(0)$ will always be $\bar{y}$. Here, the columns of Φ are not orthogonal, but they are not ridiculously far from orthogonality, so the estimate of $f(0)$ is close to $\bar{y}$. If we had standardized the data in Table 1.1 so that $x_1 \neq 0$, this would not have been a problem. In particular, if we had defined $x_i = 2(t_i + .5)/42 = i/(n+1)$, we would not have $x_1 = 0$.

Table 1.3 Regression coefficients for fitting 32 Haar wavelets

Var.	$\hat{\beta}_k$	SE	t	P	Var.	$\hat{\beta}_k$	SE	t	P
p_0	11.3064	0.1813	62.35	0.000	$m_{4,1}$	0.5300	0.7859	0.67	0.517
m_0	−1.0525	0.1838	−5.73	0.000	$m_{4,2}$	−0.1808	0.6806	−0.27	0.797
m_{11}	−2.3097	0.2599	−8.89	0.000	$m_{4,3}$	0.1800	0.7859	−0.23	0.824
m_{12}	1.7784	0.2599	6.84	0.000	$m_{4,4}$	−0.2483	0.6806	−0.36	0.724
m_{21}	−0.5269	0.3676	−1.43	0.186	$m_{4,5}$	−0.3250	0.7859	−0.41	0.689
m_{22}	−1.3637	0.3676	−3.71	0.005	$m_{4,6}$	−0.3308	0.6806	−0.49	0.639
m_{23}	0.4725	0.3676	1.29	0.231	$m_{4,7}$	−0.2900	0.7859	−0.37	0.721
m_{24}	0.8144	0.3676	2.22	0.054	$m_{4,8}$	0.1392	0.6806	0.20	0.842
m_{31}	0.1896	0.5198	0.36	0.724	$m_{4,9}$	0.2750	0.7859	0.35	0.734
m_{32}	−0.4441	0.5198	−0.85	0.415	$m_{4,10}$	−0.1308	0.6806	−0.19	0.852
m_{33}	−0.7079	0.5198	−1.36	0.206	$m_{4,11}$	0.2650	0.7859	0.34	0.744
m_{34}	−0.4304	0.5198	−0.83	0.429	$m_{4,12}$	0.3492	0.6806	0.51	0.620
m_{35}	−0.0579	0.5198	−0.11	0.914	$m_{4,13}$	0.0050	0.7859	0.01	0.995
m_{36}	0.3171	0.5198	0.61	0.557	$m_{4,14}$	0.2292	0.6806	0.34	0.744
m_{37}	0.5671	0.5198	1.09	0.304	$m_{4,15}$	0.1500	0.7859	0.19	0.853
m_{38}	0.0709	0.5198	0.14	0.895	$m_{4,16}$	0.0117	0.6806	0.02	0.987

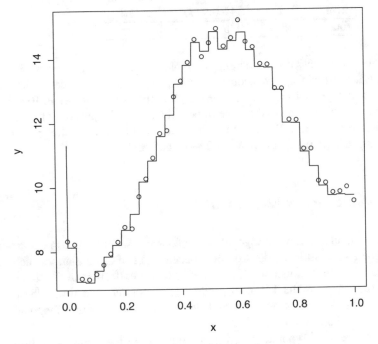

Fig. 1.4 Haar wavelet fit with $s = 32$ for the battery data

1.4.4 Cubic Splines

Splines are discussed in detail in Sect. 1.7. It turns out that a general cubic spline model with $m-1$ interior knots $\tilde{x}_j$ can be written

$$y_i = \beta_0 + \beta_1 x_i + \beta_2 x_i^2 + \beta_3 x^3 + \sum_{j=1}^{m-1} \beta_{j+3}[(x_i - \tilde{x}_j)_+]^3 + \varepsilon_i$$

where for any scalar a

$$(x-a)_+ \equiv \begin{cases} x-a & \text{if } x > a \\ 0 & \text{if } x \le a. \end{cases}$$

We began by fitting 30 equally spaced interior knots, to get

Analysis of variance: splines with 30 knots					
Source	df	SS	MS	F	P
Regression	33	261.064	7.9110	101.8	0.0000
Error	7		0.544	0.0777	
Total	40	261.6076			

with $R^2 = 0.998$ and regression coefficients in Table 1.4. The fitted curve is displayed in Fig. 1.5. Most of the regression coefficients are quite large because they are multiplying numbers that are quite small, i.e., the cube of a number in $[0,1]$. There is little useful information in the table of coefficients for the purpose of picking a smaller number of knots. We also fitted a model with only 4 equally spaced knots that fits surprisingly well, having $R^2 = 0.991$, see Fig. 1.6.

1.4.5 Orthonormal Series Estimation

Although restricting attention to orthogonal series does not seem to be as popular now as it was in the late 1990s, we discuss some peculiarities of estimation for such functions. Recall that an orthogonal series is defined as a collection of functions $\phi_j(x)$ that are orthonormal in the $\mathscr{L}^2$ inner product. (I do not understand why these are not called "orthornormal" series, just as I do not understand why orthogonal matrices are not called orthonormal matrices.)

From a linear models viewpoint, the primary weakness in the usual estimation procedures proposed for orthogonal series is that they ignore collinearity. Because the functions ϕ_j are orthonormal in $\mathscr{L}^2$, the usual methods perform estimation as though the vectors Φ_j were orthogonal. More specifically, *the usual orthogonal series estimation procedures act as if the vectors $\frac{1}{\sqrt{n}}\Phi_j$ are orthonormal. This orthonormality condition* is unlikely to occur in practice because orthogonality of the

Table 1.4 Regression coefficients for fitting splines with 30 interior knots

	Table of coefficients								
Var.	$\hat{\beta}_k$	SE	t	P	Var.	$\hat{\beta}_k$	SE	t	P
Const.	8.330	0.2788	29.881	0.0000	ϕ_{17}	78360	34500	2.271	0.0574
x	123.6	12.84	0.962	0.3680	ϕ_{18}	−91100	34620	−2.631	0.0338
x^2	−7530	7422	−1.015	0.3441	ϕ_{19}	85420	34620	2.467	0.0430
x^3	97080	103000	0.943	0.3773	ϕ_{20}	−62290	34500	−1.806	0.1140
ϕ_4	−116300	142700	−0.815	0.4419	ϕ_{21}	25890	34570	0.749	0.4783
ϕ_5	21860	64770	0.338	0.7456	ϕ_{22}	5164	34670	0.149	0.8858
ϕ_6	−7743	43780	−0.177	0.8646	ϕ_{23}	−13920	34560	−0.403	0.6991
ϕ_7	12430	37310	0.333	0.7487	ϕ_{24}	10190	34520	0.295	0.7765
ϕ_8	−21280	35460	−0.600	0.5674	ϕ_{25}	−9532	34640	−0.275	0.7911
ϕ_9	36210	34810	1.040	0.3329	ϕ_{26}	11840	34630	0.342	0.7425
ϕ_{10}	−45710	34560	−1.323	0.2275	ϕ_{27}	−9615	34560	−0.278	0.7889
ϕ_{11}	42350	34630	1.223	0.2608	ϕ_{28}	1079	34810	0.031	0.9761
ϕ_{12}	−39120	34640	−1.129	0.2960	ϕ_{29}	8318	35460	0.235	0.8213
ϕ_{13}	45090	34520	1.306	0.2328	ϕ_{30}	−10490	37310	−0.281	0.7868
ϕ_{14}	−51080	34560	−1.478	0.1829	ϕ_{31}	8146	43780	0.186	0.8577
ϕ_{15}	51290	34670	1.479	0.1826	ϕ_{32}	−8246	64770	−0.127	0.9023
ϕ_{16}	−58680	34570	−1.697	0.1334	ϕ_{33}	−10490	142700	−0.074	0.9434

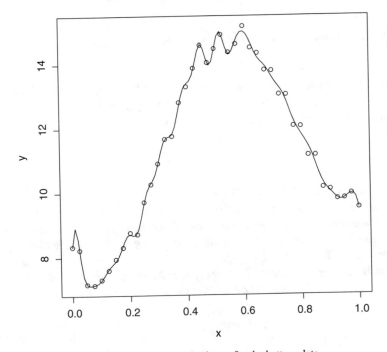

Fig. 1.5 Cubic spline fit with 30 interior knots for the battery data

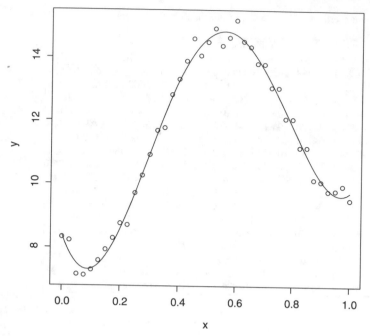

Fig. 1.6 Cubic spline fit with 4 interior knots for the battery data

Φ_j vectors is unlikely to occur unless the x_is are chosen to be equally spaced, and, as we have seen, orthonormality of the vectors may not happen even then.

A basic orthonormal series estimation method is to take

$$\hat{\beta}_k = \frac{1}{n}\Phi_k'Y = \frac{1}{n}\sum_{i=1}^{n} y_i\phi_k(x_i);$$

see Hart (1997, 165), Ogden (1997, p. 108), Efromovich (1999, p. 121). This only gives the least squares estimates under the orthonormality condition. If n is very large and the x_is are equally spaced, this is probably a cost effective approximation to the least squares estimate because you do not have to pay to find the inverse of $\Phi'\Phi$, which we know will be close to $(1/n)I$.

Another popular estimation method is to apply shrinkage estimators. Under the orthonormality assumption, Goldstein and Smith (1974) have shown that, for constants h_j, if

$$\tilde{\beta}_j = h_j\hat{\beta}_j$$

and

$$\frac{\beta_j^2}{\sigma^2/n} < \frac{1+h_j}{1-h_j}, \tag{1.4.2}$$

then $\tilde{\beta}_j$ is a better estimate in that

$$E(\tilde{\beta}_j - \beta_j)^2 \le E(\hat{\beta}_j - \beta_j)^2.$$

Efromovich (1999, p. 125) recommends adaptive shrinkage estimates $\tilde{\beta}_j$, where $h_j = [(F_j - 1)/F_j]_+$, the subscript $+$ indicates that h_j is taken to be 0 if the right-hand side is negative, and F_j is the F statistic for testing $H_0 : \beta_j = 0$, namely

$$F_j \equiv \frac{\hat{\beta}_j^2}{MSE/n}, \tag{1.4.3}$$

under orthonormality. This amounts to dropping the jth term if F_j is less than 1 and giving it progressively more weight up to a value of 1 as F_j increases. Although inequality (1.4.2) does not apply directly because F_j and thus h_j are random, this should work reasonably well. Ogden (1997, p. 124) discusses other methods of shrinkage. But again, this is all based on the assumption of orthonormality. Standard methods for shrinkage estimation in nonorthogonal linear models are principal component regression, generalized inverse regression, Bayesian regression, ridge regression, and lasso regression. The first three topics are discussed in PA, the last two topics are discussed in the next chapter.

Efromovich (1999, p. 128) proposes to deal with unequally spaced x_i data by using the estimator

$$\hat{\beta}_j = \frac{1}{n} \sum_{i=1}^{n} \frac{y_i \phi_j(x_i)}{h(x_i)} = \frac{1}{n} \Phi_j'[D(h(X))]^{-1}Y,$$

where $h(x)$ is the density for the randomly distributed x_is. Efromovich shows that this has a nice unbiased property when integrating over both Y and X; however, in model (1.4.1), the least squares estimates are superior in that they are conditionally unbiased given the x_is and therefore unconditionally unbiased, besides being BLUEs. Moreover, incorporating h does not seem to deal with the collinearity caused by unequal spacings, and it requires one to know, or at least estimate, h.

Another popular estimation method (see Efromovich 1999, p. 129 and Ogden 1997, pp. 43, 55) is

$$\hat{\beta}_j = \frac{1}{n} \sum_{i=1}^{n} y_i \tilde{\phi}_j(x_i),$$

where for some r

$$\tilde{\phi}_j(x_i) = \frac{1}{2r} \int_{x_i - r}^{x_i + r} \phi_j(x)dx.$$

Obviously, the idea is to smooth out the ϕ_j functions in the neighborhood of the observed x_i values. From a linear models viewpoint, all this does is change the model matrix Φ into a new matrix that we could call $\tilde{\Phi}$, but this substitution seems difficult to justify from the linear models viewpoint. If model (1.4.1) is appropriate, why would we want to replace the $\phi_j(x_i)$s with $\tilde{\phi}_j(x_i)$s?

1.5 Variable Selection

Variable selection is of key importance in these problems because the linear model is only an approximation. The problem is to select an appropriate value of s in model (1.3.1).

In the special case where the vectors $\frac{1}{\sqrt{n}}\Phi_j$ are orthonormal, the situation is analogous to identifying the important features in a 2^n factorial design; see Christensen (1996, Sections 17.3 and 17.4) or http://www.stat.unm.edu/~fletcher/TopicsInDesign. For example, we could begin by taking $s = n$ and construct a normal or half-normal plot of the $\hat{\beta}_j$s to identify the important ϕ_j functions. Similarly, we could construct a $\chi^2(1)$ plot for the sequential sums of squares. Here the orthonormality condition ensures that the estimates and sums of squares are independent under normality and that the sequencing of the sequential sums of squares is irrelevant. These ideas are applied in Chap. 6 on frequency domain time series analysis in which the model matrix *is* orthogonal. For the case against using such methods, see Lenth (2015).

Another method of choosing s is by cross-validation. For example, one can minimize the PRESS statistic; see *PA-V* Sect. 12.5 (Christensen 2011, Section 13.5) or Hart (1997, Section 4.2.1).

An alternative to variable selection is using a penalized estimation procedure as discussed in the next chapter.

In the remainder of this section we will assume that we have fitted a model with s predictors where s was chosen to be so large that it clearly gives a reasonable approximation. We want to find a reduced model with p predictors that does not over fit the model.

Hart (1997, Section 4.2.2) and Efromovich (1999, p. 125) suggest selecting p to maximize $A_p \equiv \sum_{j=0}^{p-1}(F_j - 2)$ when using the cosine ϕs but the same idea applies whenever the ϕ_j are ordered, e.g., polynomials. Here, the definition of F_j is based on (1.4.3) and orthonormality. If the ϕ_js are ordered, one can define F_j statistics more generally as sequential F tests and thus account for collinearity. In particular, redefine

$$F_j \equiv \frac{SSR(\Phi_j|\Phi_0,\ldots,\Phi_{j-1})}{MSE}. \tag{1.5.1}$$

Selecting p by maximizing A_p does not allow dropping lower-order terms if higher ones are included (i.e., it is similar, in polynomial regression, to not allowing x^2 to be eliminated if x^3 remains in the model). Efromovich suggests picking $s = 6\tilde{p}$, where $\tilde{p}$ is the smallest value of p for which

$$MSE < 2[1.48\,\text{median}|y_i - \hat{y}_i|]^2.$$

Based on Hart's discussion of Hurvich and Tsai (1995), another crude upper bound might be $s = \sqrt{n}$, although in practice this seems to give too small values of s. Based on the coefficient of variation for the variance estimator, *PA-V* Subsection 14.3.6 suggests that $n - s$ should be at least 8 with values of 18 or more preferable. In Example 1.5.1, s was chosen by the seat of my pants.

The A_p procedure is equivalent to the standard variable selection procedure based on the C_p statistic. (Note that in A_p, p is the number of mean parameters in the regression whereas in C_p the p is just part of the name. Denote the C_p statistic based on r parameters $C_p(r)$.) The C_p statistic for a model with p parameters can be rewritten as $C_p(p) = (s-p)(F-2) + s$, where F is the statistic for comparing the p parameter and s parameter models. Moreover, using (1.5.1), $C_p(p) - s = \sum_{j=p}^{s-1}(F_j - 2)$, where s is a constant, so $A_p + C_p(p) - s = \sum_{j=0}^{s-1}(F_j - 2)$, a constant, and the A_p procedure is equivalent to the standard regression procedure (i.e., pick the model in the sequence with the smallest $C_p(r)$ statistic).

EXAMPLE 1.5.1.   Using the battery data and fitting cosines with $s - 1 = 30$, Table 1.5 gives sequential sums of squares and values of $A_p - A_1$. The $(A_p - A_1)$s are easily computed from the sequential sums of squares as partial sums of the $(F_j - 2)$ statistics. For example, using the sequential sum of squares and the MSE, $F_5 = 0.8427/0.0880$ and $A_6 - A_1 = (F_1 - 2) + \cdots + (F_5 - 2)$. The C_p statistic is minimized when $A_p - A_1$ is maximized, so the best models from the sequence have $s - 1 = 6, 10, 13, 14$. If one were willing to consider models that do not include a contiguous set of j values, the problem becomes a traditional variable selection problem. Given the near orthogonality of the predictors in this example, it is fairly obvious from the sequential sums of squares alone that the most important predictors are $j = 1, \ldots, 6, 10, 13, 14$. With more collinear data, such a conclusion could not be made from the sequential sums of squares.

Table 1.5 Selection of s based on the C_p statistic

j	Seq SS	$A_{j+1} - A_1$	j	Seq SS	$A_{j+1} - A_1$
1	52.2633	591.90	16	0.0061	2922.20
2	198.6634	2847.44	17	0.0133	2920.35
3	4.8674	2900.75	18	0.0213	2918.59
4	1.2009	2912.40	19	0.1412	2918.19
5	0.8427	2919.97	20	0.0322	2916.56
6	0.5753	2924.51	21	0.0000	2914.56
7	0.0088	2922.61	22	0.0000	2912.56
8	0.1538	2922.36	23	0.1605	2912.38
9	0.1472	2922.03	24	0.0001	2910.39
10	0.4547	2925.20	25	0.0911	2909.42
11	0.0070	2923.28	26	0.0015	2907.44
12	0.0857	2922.25	27	0.0669	2906.20
13	0.3554	2924.29	28	0.1073	2905.42
14	0.2951	2925.64	29	0.1189	2904.77
15	0.0425	2924.13	30	0.0037	2902.81

Figure 1.7 gives fitted cosine curves for $s - 1 = 6, 10, 14, 30$. I suspect that, visually, $s - 1 = 14$ in the bottom left is the one that would most appeal to practitioners of nonparametric regression.   □

Often the ϕ functions are only partially ordered but they can be collected into groups that are ordered, so one could evaluate whether the entire group is required.

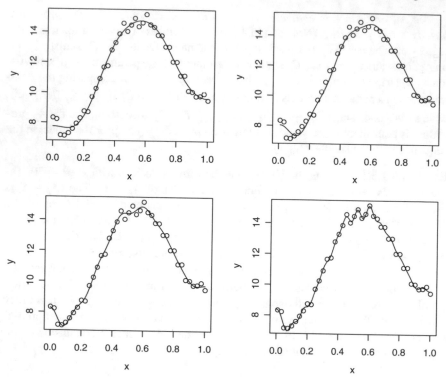

Fig. 1.7 Cosine fit with $s - 1 = 6, 10, 14, 30$ for the battery data. Read across and down

For the sine-cosine functions of (1.2.3), the ϕ_js occur in natural pairs having a common frequency of oscillation with the pairs ordered by their frequencies. One can easily use the C_p statistic to determine the largest frequency needed. If the orthonormality conditions holds, instead of finding the largest important frequency, we could identify all of the important frequencies. In Chap. 6 we will have orthogonality and periodograms are defined as mean squares for different sine-cosine pairs.

Similarly, wavelets are partially ordered, so one could decide on a level of partitioning needing in fitting wavelets. (Some regression functions are smoother in some places than others, so the required level of partitioning might vary with location.) To find an appropriate p for the Haar wavelets, we create ordered groups of 2^j predictors. For example, test the reduced model

$$Y = [\Phi_0, \Phi_1, \ldots, \Phi_7] \begin{bmatrix} \beta_0 \\ \beta_1 \\ \vdots \\ \beta_7 \end{bmatrix} + e$$

against the full model

$$Y = [\Phi_0, \Phi_1, \ldots, \Phi_{15}] \begin{bmatrix} \beta_0 \\ \beta_1 \\ \vdots \\ \beta_{15} \end{bmatrix} + e$$

to see if the $m_{3,k+1}(x) = \mathscr{I}_{(2k/16,(2k+1)/16]}(x) - \mathscr{I}_{((2k+1)/16,(2k+2)/16]}(x)$ terms, $k = 0, \ldots, 2^3 - 1$, are needed at all. Recall that when simultaneously testing for the effects of a large group of predictors, it is easy for some worthwhile predictors to get overlooked (averaged out) in a small F statistic.

EXAMPLE 1.5.2.   For fitting the Haar wavelets to the battery data, we have obvious groups of variables that occur in powers of 2. We can consider the highest-order group that we need, or we could consider including individual terms from any order group. In the first case, we would consider tests based on the ANOVA tables reported in Table 1.6.

Table 1.6 ANOVA tables for Haar wavelets

Analysis of variance: fitting p_0 to $m_{4,16}$.					
Source	df	SS	MS	F	P
Regression	31	250.489	8.080	6.54	0.003
Error	9	11.118	1.235		
Total	40	261.6076			

Analysis of variance: fitting p_0 to $m_{3,8}$.					
Source	df	SS	MS	F	P
Regression	15	248.040	16.536	30.47	0.000
Residual error	25	13.568	0.543		
Total	40	261.608			

Analysis of variance: fitting p_0 to $m_{2,4}$.					
Source	df	SS	MS	F	P
Regression	7	240.705	34.386	54.29	0.000
Residual error	33	20.902	0.633		
Total	40	261.608			

To test whether we can drop the $m_{4,k}$s, the test statistic is

$$F = \frac{[13.568 - 11.118]/16}{1.235} < 1.$$

To test whether we can drop the m_{3k}s, the test statistic is

$$F = \frac{[20.902 - 13.568]/8}{0.543} \doteq 2$$

or, using the MSE from the largest model fitted,

$$F = \frac{[20.902 - 13.568]/8}{1.235} < 1$$

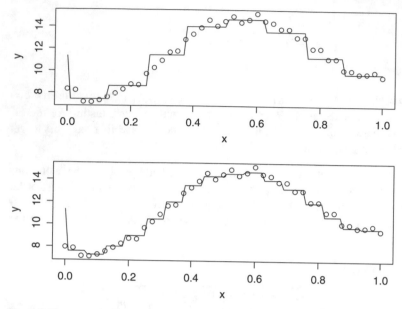

Fig. 1.8 Haar wavelet fit with $s = 8, 16$ for the battery data

If we allow elimination of individual variables from any group, the problem becomes a traditional variable selection problem. The number of wavelets needed is related to the smoothness of f, and the smoothness can change on different subsets of $[0,1]$. Figure 1.8 gives the fitted Haar wavelet curves for $s = 8$ and $s = 16$. Relative to the $s = 16$ fit, the $s = 8$ wavelets work pretty well from 0.5 to 0.625 and also from 0.875 to 1 but not very well anywhere else. □

One could even impose order on fitting splines by defining the knots as a sequence of refining partitions of $[0,1]$, i.e., just keep adding more knots to the existing ones. Perhaps a more appealing idea would be, if you have $m + 1$ knots, add another m knots with one inside each of the previous partition sets. That is essentially what wavelets are doing only wavelets specify that the partition sets are always of equal size.

1.6 Heteroscedastic Simple Nonparametric Regression

The heteroscedastic simple nonparametric regression model is

$$y_i = f(x_i) + \sigma(x_i)\varepsilon_i,$$

$i = 1, \dots, n$, where y_i is a random variable, x_i is a known (scalar) constant, f is an unknown continuous function, and the ε_is are unobservable independent errors with

$E(\varepsilon_i) = 0$ and $Var(\varepsilon_i) = 1$. The function $\sigma(x)$ is assumed to be nonnegative and is often assumed to be monotone. Note that $Var(y_i) = [\sigma(x_i)]^2$.

Treating $\sigma(\cdot)$ as known, this is simply a weighted least squares model. Let

$$\sigma^2 \equiv [\sigma(x_1)^2, \ldots, \sigma(x_n)^2]' \tag{1.6.1}$$

and let $D(\sigma^2)$ be a diagonal matrix with the elements of σ^2 along the diagonal. The approximate linear model is

$$Y = \Phi\beta + e, \quad E(e) = 0, \quad Cov(e) = D(\sigma^2).$$

Weighted least squares estimates are BLUEs, so

$$\hat{\beta} = [\Phi'D(\sigma^2)^{-1}\Phi]^{-1}\Phi'D(\sigma^2)^{-1}Y. \tag{1.6.2}$$

In orthogonal series estimation it seems to be standard practice to ignore the heteroscedastic variances in the estimation of β; see Efromovich (1999, Section 4.3).

Typically, the function $\sigma(x)$ will be unknown and must be estimated. Given an estimate of $\sigma(x)$, just plug the estimate into Eqs. (1.6.1) and (1.6.2) to get estimated regression coefficients. To estimate $\sigma(x)$, note that

$$E[y_i - f(x_i)]^2 = \sigma(x_i)^2,$$

so if f is known, estimating $\sigma(x_i)^2$ is just a (heteroscedastic) regression problem. (For normal data, $Var([y_i - f(x_i)]^2)$ is $3\sigma(x_i)^4$.) Of course, f is not known, but we can estimate it to get $\hat{y}_i \equiv \hat{f}(x_i)$. Without assuming structure on the form of $\sigma(x_i)^2$, we can use nonparametric regression methods on the pairs $(x_i, [y_i - \hat{y}_i]^2)$ to estimate $\sigma(x_i)^2$. Initially, we would use ordinary least squares to get the predicted values $\hat{y}_i$, but the estimate of $\sigma(x_i)^2$ will lead to a new weighted least squares estimate of β, which leads to new predicted values $\hat{y}_i$ and a new estimate of $\sigma(x_i)^2$. This process can be iterated once or iterated until the estimates of $\sigma(x_i)^2$ settle down. Carroll and Ruppert (1988) discuss how to handle heteroscedasticity in nonlinear regression models. Estimating heteroscedastic variances is a special case of estimating the covariance matrix which is the subject of Chap. 4. In particular, the results from Chap. 4 on how estimating the covariance matrix affects estimation of β apply.

1.7 Approximating-Functions with Small Support

Except for a finite number of 0s, the *support* of the polynomials, sines, and cosines is the entire interval $[0,1]$, i.e., the functions are nonzero except for a finite number of points. This can cause strange behavior when s gets close to n, especially when the x_is are unevenly spaced. As regards polynomials, this strange behavior is a well-known fact, but it is, perhaps, less well known as regards sines and cosines. Christensen (2015, Chapter 8) illustrates these phenomena. The fundamental idea of polynomial splines, b-splines, and wavelets is to fit functions that are nonzero only on small subsets of the domain space $[0,1]$.

As discussed in Sect. 1.2, splines and b-splines both involve knots. Earlier we needed a subscript m on the knots to indicate how many knots were being used. In this section, we will drop m from the subscript when m is not subject to change.

1.7.1 Polynomial Splines

The basic idea behind using polynomial splines is to connect the dots. Suppose we have data (x_i, y_i), $i = 1, \ldots, n$ in which the x_is are ordered from smallest to largest. Linear splines quite simply give, as a regression function, the function that fits a line segment between the consecutive pairs of points. Cubic splines fit a cubic polynomial between every pair of points rather than a line. Note that all of the action here has nothing to do with fitting a model to the data. The data are being fitted perfectly (at least in this simplest form of spline fitting). The key issue is how to model what goes on between data points.

As discussed earlier, in practice, splines often do not actually connect the dots, they create smooth functions between knots. Christensen (2015, Section 8.5) illustrates the use of a linear spline, whereas here we focus on cubic splines. We introduce the topic by discussing how to connect the dots from the data smoothly and then move on to fitting with knots.

The reason for using cubic splines is to make the curve look smooth. With cubic splines we require the fitted regression function to have continuous second derivatives. Recall that if we have a linear model $Y = X\beta + e$ and a constraint $\Lambda'\beta = 0$, that together these define another linear model; see PA Sect. 3.3. Depending on the nature of Λ, this can be either a reduced model or a reparameterization. In fitting cubic splines, we will define an overparameterized saturated model based on the cubic polynomials and then incorporate nonestimable constraints so as to create a particular parameterization which is the function fitted by cubic splines. Remember that, when connecting the dots, none of this has anything to do with fitting the data, only with what the model says about places where we have no data. In the next chapter we will introduce generalized ridge regression which would create some additional smoothing of the fitted regression function so that it does not fit every data point perfectly. Alternatively, introducing $m - 1$ arbitrarily chosen knots with $m < n$ will also further smooth the fitted regression.

The basic idea of fitting cubic splines to interpolate between data points is to fit the function

$$f(x) = a_i + b_i(x - x_i) + c_i(x - x_i)^2 + d_i(x - x_i)^3, \quad x \in [x_i, x_{i+1}].$$

Note that f is defined in two distinct ways at all x_i, except the first and last, that is, for $i = 2, \ldots, n - 1$

$$f(x_i) = a_i + b_i(x_i - x_i) + c_i(x_i - x_i)^2 + d_i(x_i - x_i)^3 = a_i$$
$$= a_{i-1} + b_{i-1}(x_i - x_{i-1}) + c_{i-1}(x_i - x_{i-1})^2 + d_{i-1}(x_i - x_{i-1})^3.$$

For f to be continuous, indeed for f to be a function, these must be equal, so we are led to the constraints

$$a_i = a_{i-1} + b_{i-1}(x_i - x_{i-1}) + c_{i-1}(x_i - x_{i-1})^2 + d_{i-1}(x_i - x_{i-1})^3,$$

$i = 2, \ldots, n-1$. Similarly, we want the right first and second derivatives of f to be equal to the left first and second derivatives of f at every interior point, which leads to the constraints

$$b_i = b_{i-1} + 2c_{i-1}(x_i - x_{i-1}) + 3d_{i-1}(x_i - x_{i-1})^2, \quad i = 2, \ldots, n-1$$

and

$$c_i = c_{i-1} + 3d_{i-1}(x_i - x_{i-1}), \quad i = 2, \ldots, n-1.$$

Altogether, for n data points, we have $n-1$ cubic polynomials being fitted, so we have $4(n-1)$ parameters being fitted to n data points. The model is somewhat overparameterized. The constraints for continuity and first and second derivatives give us $3(n-2)$ constraints, leaving us with $4(n-1) - 3(n-2) = n+2$ free parameters to fit to n data points. We need two more constraints if we are going to get a unique set of parameter estimates. These two constraints are that the second derivative should be 0 at x_1 and x_n—two points that are not involved in the previous continuity and derivative constraints. The additional constraints reduce to

$$0 = c_1 = c_{n-1} + d_{n-1}(x_n - x_{n-1}).$$

We will illustrate the linear modeling ideas for the case of $n = 4$. We can write a linear model $Y = X\beta + e$ for fitting the cubic splines by taking

$$\beta = (a_1, b_1, c_1, d_1, a_2, b_2, c_2, d_2, a_3, b_3, c_3, d_3)' \tag{1.7.1}$$

and the transpose of X as

$$X' = \begin{bmatrix} 1 & 1 & 0 & 0 \\ 0 & (x_2 - x_1) & 0 & 0 \\ 0 & (x_2 - x_1)^2 & 0 & 0 \\ 0 & (x_2 - x_1)^3 & 0 & 0 \\ 0 & 0 & 1 & 0 \\ 0 & 0 & (x_3 - x_2) & 0 \\ 0 & 0 & (x_3 - x_2)^2 & 0 \\ 0 & 0 & (x_3 - x_2)^3 & 0 \\ 0 & 0 & 0 & 1 \\ 0 & 0 & 0 & (x_4 - x_3) \\ 0 & 0 & 0 & (x_4 - x_3)^2 \\ 0 & 0 & 0 & (x_4 - x_3)^3 \end{bmatrix}.$$

Defining $t_2 \equiv (x_2 - x_1)$, $t_3 \equiv (x_3 - x_2)$, $t_4 \equiv (x_4 - x_3)$, and

$$\Lambda \equiv \begin{bmatrix} 1 & 0 & 0 & 0 & 0 & 0 & 0 & 0 \\ t_2 & 1 & 0 & 0 & 0 & 0 & 0 & 0 \\ t_2^2 & 2t_2 & 1 & 0 & 0 & 0 & 1 & 0 \\ t_2^3 & 3t_2^2 & 3t_2 & 0 & 0 & 0 & 0 & 0 \\ -1 & 0 & 0 & 1 & 0 & 0 & 0 & 0 \\ 0 & -1 & 0 & t_3 & 1 & 0 & 0 & 0 \\ 0 & 0 & -1 & t_3^2 & 2t_3 & 1 & 0 & 0 \\ 0 & 0 & 0 & t_3^3 & 3t_3^2 & 3t_3 & 0 & 0 \\ 0 & 0 & 0 & -1 & 0 & 0 & 0 & 0 \\ 0 & 0 & 0 & 0 & -1 & 0 & 0 & 0 \\ 0 & 0 & 0 & 0 & 0 & -1 & 0 & 1 \\ 0 & 0 & 0 & 0 & 0 & 0 & 0 & 3t_4 \end{bmatrix},$$

the constraints are $\Lambda'\beta = 0$.

If, as in *PA* Sect. 3.3, we find a matrix U such that $C(U) = C(\Lambda)^{\perp}$ and $X_0 = XU$, then the linear model $Y = X_0\gamma + e$ has the constraint $\Lambda'\beta = 0$ built into it. The constrained least squares estimate of β is $\hat{\beta} = U\hat{\gamma}$. Obviously, from the choice of U, $\Lambda'\hat{\beta} = \Lambda'U\hat{\gamma} = 0$. Because our spline model is saturated, X_0 will be nonsingular and we can obtain the estimates $\hat{\gamma} = (X_0'X_0)^{-1}X_0'Y = X_0^{-1}Y$ and then $\hat{\beta} = U\hat{\gamma} = U(XU)^{-1}Y$.

We have used cubic splines to *interpolate* between the observed data points. An alternative is that we could (a) have some fixed set of $m+1$ knots $0 = \tilde{x}_0 < \tilde{x}_1 < \tilde{x}_2 < \cdots < \tilde{x}_m = 1$ as endpoints for the cubic polynomials, (b) observe data (x_i, y_i), $i = 1, \ldots, n$, and (c) fit low order polynomials to the data between the knots. As such, each fitted polynomial has support only between the knots. As with interpolation, we require continuity of the fitted function and, depending on the order of the polynomial being fitted, perhaps require some continuous derivatives. The discussion so far has used $m - 1 = n$ and $\tilde{x}_k = x_k$ but has ignored fitting polynomials on $[0, x_1]$ and $[x_n, 1]$.

Using knots, consider fitting m cubic polynomials. For $j = 1 \ldots, m$

$$f(x) = a_j + b_j(x - \tilde{x}_{j-1}) + c_j(x - \tilde{x}_{j-1})^2 + d_j(x - \tilde{x}_{j-1})^3, \quad x \in [\tilde{x}_{j-1}, \tilde{x}_j]$$

so we have $4m$ parameters. The $3(m-1)$ constraints for continuity and equality of the right and left first and second derivatives at the $m - 1$ interior knots leave us $m + 3$ free parameters. We *do not* require that the second derivatives be zero at the boundaries $x = 0, 1$. To illustrate, for $m = 3$, the constraints are $\Lambda'\beta = 0$ where β is as in Eq. (1.7.1) and

$$\Lambda \equiv \begin{bmatrix} 1 & 0 & 0 & 0 & 0 & 0 \\ t_1 & 1 & 0 & 0 & 0 & 0 \\ t_1^2 & 2t_1 & 1 & 0 & 0 & 0 \\ t_1^3 & 3t_1^2 & 3t_1 & 0 & 0 & 0 \\ -1 & 0 & 0 & 1 & 0 & 0 \\ 0 & -1 & 0 & t_2 & 1 & 0 \\ 0 & 0 & -1 & t_2^2 & 2t_2 & 1 \\ 0 & 0 & 0 & t_2^3 & 3t_2^2 & 3t_2 \\ 0 & 0 & 0 & -1 & 0 & 0 \\ 0 & 0 & 0 & 0 & -1 & 0 \\ 0 & 0 & 0 & 0 & 0 & -1 \\ 0 & 0 & 0 & 0 & 0 & 0 \end{bmatrix},$$

where $t_j = \tilde{x}_j - \tilde{x}_{j-1}$. In general, the cubic spline regression model is

$$y_i = f(x_i) + \varepsilon_i,$$

where, if $x_i \in [\tilde{x}_{j-1}, \tilde{x}_j]$,

$$y_i = a_j + b_j(x_i - \tilde{x}_{j-1}) + c_j(x_i - \tilde{x}_{j-1})^2 + d_j(x_i - \tilde{x}_{j-1})^3 + \varepsilon_i.$$

For us to be able to estimate all of the parameters, we have a necessary condition of $n \geq m+3$. But even this is not sufficient for estimability of all parameters. It is not hard to see that if two adjacent intervals $[\tilde{x}_j, \tilde{x}_{j+1}]$ both contain no data, the spline function will not be estimable. In particular, if $r(X_0) < m+3$, we are in trouble. Moreover, we can get bad fitting models by having $m+3 \doteq n$. For example, fitting a cubic polynomial to only 4 data points can lead to strange results. Even with $m-1 = 30$ we see some undesirable behavior in Fig. 1.5 between $x_1 = 0$ and $x_2 = 1/40$. The behavior of the spline fit to the battery data gets worse as $m-1$ increases above 30.

In Sect. 1.4 we claimed that the cubic spline model can be conveniently written as

$$y_i = \beta_0 + \beta_1 x_i + \beta_2 x_i^2 + \beta_3 x^3 + \sum_{j=1}^{m-1} \beta_{j+3}[(x_i - \tilde{x}_j)_+]^3 + \varepsilon_i.$$

This has precisely the $m+3$ estimable parameters that we need. Christensen (2015, Exercise 8.7.8) and Sect. 1.7.1.2 both examine the relationship between this model and the constrained model discussed above. The equivalence relies on the fact that having separate polynomials between pairs of knots is equivalent to adding a new polynomial that begins at each knot, i.e.,

$$f(x) = \sum_{j=1}^{m} \left[\alpha_j + \eta_j(x - \tilde{x}_{j-1}) + \gamma_j(x - \tilde{x}_{j-1})^2 + \delta_j(x - \tilde{x}_{j-1})^3\right] \mathscr{I}_{[\tilde{x}_{j-1},1]}(x),$$

and showing that imposing the continuity and derivative constraints requires $0 = \alpha_j = \eta_j = \gamma_j$, $j = 2, \ldots, m$. Under these conditions f reduces to

$$f(x) = \alpha_1 + \eta_1 x + \gamma_1 x^2 + \delta_1 x^3 + \sum_{j=2}^{m} \delta_j(x - \tilde{x}_{j-1})^3 \mathscr{I}_{[\tilde{x}_{j-1},1]}(x),$$

which is equivalent to the regression function in the model written with β_js.

In general you can write a d order spline model as

$$y_i = \sum_{k=0}^{d} \beta_k x^k + \sum_{j=1}^{m-1} \beta_{j+d}[(x_i - \tilde{x}_j)_+]^d + \varepsilon_i. \tag{1.7.2}$$

1.7.1.1 B-Splines

B-splines are supposed to be *basis splines* but, as we have seen earlier, they do not actually define a meaningful basis in the space of functions. (Obviously they define a basis for their spanning space.) *B-splines provide the same fit as splines* but do so by defining a mother spline and then defining ϕ_j functions by rescaling and relocating the mother.

The mother spline is itself a low order polynomial spline. The mother spline of degree 2 is nonzero over $(0,3)$ and defined as

$$\Psi_2(x) = \frac{x^2}{2}\mathscr{I}_{[0,1]}(x) - \left\{[x-1.5]^2 - 0.75\right\}\mathscr{I}_{(1,2]}(x) + \frac{[3-x]^2}{2}\mathscr{I}_{(2,3]}(x).$$

This is a bell-shaped curve, similar to a normal density centered at 1.5, but it is 0 outside the interval $[0,3]$ while still being smooth in that it is differentiable everywhere. Ψ_2 is itself a quadratic spline, i.e., quadratics have been pasted together as a smooth function.

A mother spline Ψ_d of degree d has support on the interval $(0,d+1)$. It splices together $(d+1)$ different d-degree polynomials, each defined on a length 1 interval, so that the whole function is differentiable $d-1$ times and looks like a mean-shifted Gaussian density. Commonly d is either 2 or 3. For $d=3$, the cubic mother spline on $[0,4]$ is

$$\Psi_3(x) = \frac{x^3}{3}\mathscr{I}_{[0,1]}(x) + \left\{-x^3 + 4x^2 - 4x + \frac{4}{3}\right\}\mathscr{I}_{(1,2]}(x)$$

$$+ \left\{-[4-x]^3 + 4[4-x]^2 - 4[4-x] + \frac{4}{3}\right\}\mathscr{I}_{(2,3]}(x) + \frac{[4-x]^3}{3}\mathscr{I}_{(3,4]}(x).$$

Figure 1.9 shows these b-spline mother functions. Other than the domain on which they are defined, they look quite unremarkable. There is a body of theory associated with b-splines that includes defining the $d+1$ order mother spline recursively from the d order mother.

The approximating functions $\phi_j(x)$ are defined by rescaling and relocating the mother splines. For simplicity, consider $d=2$ with $m-1$ equally spaced interior knots. If the knots are equally spaced, the same rescaling of Ψ_2 works for all ϕ_j. Ψ_2 is defined on $[0,3]$ and pastes together 3 polynomials on three intervals of length one. To define ϕ_0 we rescale Ψ_2 to live on $[0,3/m]$ and then shift it to the left $2/m$ units so that only the polynomial originally defined on $[2,3]$ now overlaps the interval $[0,1/m]$ and ϕ_0 is 0 elsewhere in $[0,1]$. To define ϕ_1, again rescale Ψ_2 to live on $[0,3/m]$ but now shift it to the left only $1/m$ units so that the polynomial originally defined on $[2,3]$ now overlaps the interval $[1/m,2/m]$ and the polynomial originally defined on $[1,2]$ now overlaps the interval $[0/m,1/m]$. ϕ_1 is 0 elsewhere in $[0,1]$. ϕ_2 is just the rescaled version of Ψ_2. ϕ_3 is the rescaled Ψ_2 shifted to the *right* by $1/m$. More generally, ϕ_{2+j} is the rescaled Ψ_2 shifted to the *right* by j/m.

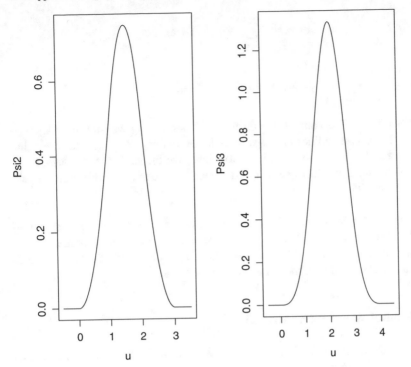

Fig. 1.9 B-spline mother functions for $d = 2, 3$

For arbitrary d, Ψ_d is rescaled so that its support is $(0, \{d+1\}/m)$ and ϕ_0 is the rescaled Ψ_d shifted to the left d/m units. Each successive ϕ_j is shifted to the right by an additional $1/m$ so that ϕ_d is the rescaled version of Ψ_d and ϕ_{d+j} is the rescaled Ψ_d shifted to the right by j/m.

For a quadratic spline model with $m - 1$ equally spaced interior knots,

$$y_i = f(x_i) + \varepsilon_i,$$

where, if $x_i \in [(j-1)/m, j/m]$,

$$y_i = a_j + b_j\{x_i - (j-1)/m\} + c_j\{x_i - (j-1)/m\}^2 + \varepsilon_i$$

More generally, splines fit a d dimensional polynomial between the knots with continuity of the function and of the $d - 1$ order derivatives at each interior knot. As mentioned earlier and shown later, this is equivalent to the model

$$y_i = \sum_{k=0}^{d} \beta_k x^k + \sum_{j=1}^{m-1} \beta_{j+d}[(x_i - \tilde{x}_j)_+]^d + \varepsilon_i$$

which involves $m + d$ parameters. The equivalent b-spline model is

$$y_i = \sum_{j=0}^{m+d-1} \gamma_j \, \phi_j(x_i) + \varepsilon_i$$

where

$$\phi_j(x) = \Psi_d \, (mx - j + d).$$

EXAMPLE 1.7.1.   The computer code for this book on my website provides programs for fitting both splines and b-splines to the battery data and illustrates that the two approaches are equivalent. Notably, they give the same fitted values, Error degrees of freedom, and Error sums of squares. The code is currently written only for using Ψ_2 and Ψ_3.   □

Pedagogically, I do not see why anyone would use b-splines rather than fitting model (1.7.2) but Eilers and Marx (1996) argue that B-splines have numerical advantages. Christensen, Johnson, Branscum, and Hanson (2010, Subsection 15.2) contains a discussion of b-splines in a Bayesian setting.

1.7.1.2 Equivalence of Spline Methods*

We now provide a more formal discussion of why our three different approaches to fitting splines are equivalent.

The idea of fitting splines is that the regression function f is a general d order polynomial between the knots but with the restrictions that the regression function be continuous at all interior knots, as are the derivatives of order up to $d - 1$. For the purpose of discussing equivalence we will break up f at the knots, i.e.,

$$f_j(x) = f(x) \mathscr{I}_{[\tilde{x}_{j-1}, \tilde{x}_j]}(x), \qquad j = 1, \ldots, m.$$

It is convenient to define f_j on closed intervals so we can discuss the value of the function at the end knots rather than the function's limit. The spline function is $f(x) = \sum_{j=1}^m f_j(x)$ except at the interior knots where $\sum_{j=1}^m f_j(\tilde{x}_k) = 2f(\tilde{x}_k)$, $k = 1, \ldots, m - 1$

We change our basic polynomial spline notation for f in this discussion. The basic definition of the spline function will be the arbitrary polynomial

$$f_j(x) = \left[\sum_{k=0}^d \beta_{k,j}(x - \tilde{x}_{j-1})^k \right] \mathscr{I}_{[\tilde{x}_{j-1}, \tilde{x}_j]}(x), \qquad j = 1, \ldots, m$$

but these polynomials are subject to the continuity and differentiability constraints

$$f_j(\tilde{x}_j) = f_{j+1}(\tilde{x}_j), \qquad j = 1, \ldots, m - 1$$

and

$$\mathbf{d}^k f_j(\tilde{x}_j) = \mathbf{d}^k f_{j+1}(\tilde{x}_j), \qquad k = 1, \ldots, d - 1, \quad j = 1, \ldots, m - 1$$

where $\mathbf{d}^k$ denotes the kth order left or right derivative as is appropriate, evaluated at the indicated point.

To establish equivalences between this and the other approaches to fitting splines, we provide alternative inductive definitions of f_j that satisfy the general criterion of fitting an arbitrary polynomial between the knots subject to constraints and establish how these lead to Eq. (1.7.2) and to b-splines. In these inductive definitions, f_1 always needs to be an arbitrary polynomial of order d, i.e.,

$$f_1(x) \equiv \left[\sum_{k=0}^{d} \beta_{k,1}(x - \tilde{x}_0)^k \right] \mathscr{I}_{[\tilde{x}_0, \tilde{x}_1]}(x).$$

Then, assuming that f_j is defined appropriately, we define f_{j+1} so that it is an arbitrary d order polynomial subject to the continuity and differentiation constraints and also takes the form of either Eq. (1.7.2) or b-splines.

We begin with establishing (1.7.2). The idea is to take $f_{j+1} = f_j + p_{j+1}$ where p_{j+1} is an arbitrary polynomial of order d and show that the continuity and differentiability constraints lead to adding a term like the one used in (1.7.2). The technicalities get a little more complex. What we really do is extend the polynomial from f_j onto the new domain and add a new arbitrary polynomial, i.e.,

$$f_{j+1}(x) = \mathscr{I}_{[\tilde{x}_j, \tilde{x}_{j+1}]}(x) \left\{ \left[\sum_{k=0}^{d} \beta_{k,j}(x - \tilde{x}_{j-1})^k \right] + \left[\sum_{k=0}^{d} \beta_{k,j+1}(x - \tilde{x}_j)^k \right] \right\}.$$

To get $f_j(\tilde{x}_j) = f_{j+1}(\tilde{x}_j)$, we need $\beta_{0,j+1} = 0$. Similarly, to get $\mathbf{d}^k f_j(\tilde{x}_j) = \mathbf{d}^k f_{j+1}(\tilde{x}_j)$, $k = 1, \ldots, d-1$, we need $\beta_{k,j+1} = 0$, $k = 1, \ldots, d-1$. So after incorporating the constraints,

$$f_{j+1}(x) = \mathscr{I}_{[\tilde{x}_j, \tilde{x}_{j+1}]}(x) \left\{ \left[\sum_{k=0}^{d} \beta_{k,j}(x - \tilde{x}_{j-1})^k \right] + \beta_{d,j+1}(x - \tilde{x}_j)^d \right\}.$$

It is now easy to see that if you define f from model (1.7.2) you get exactly the necessary structure. Model (1.7.2) defines some polynomial on $[\tilde{x}_{j-1}, \tilde{x}_j]$ and any polynomial can be written in the form indicated for f_j. Plus, model (1.7.2) extends the f_j polynomial to $[\tilde{x}_j, \tilde{x}_{j+1}]$ and makes the necessary addition.

Alas, the argument for b-splines is considerably more complicated. Before doing it for general d we examine the case $d = 2$. Write second degree polynomials $p_1(x) \equiv x^2/2$, $p_2(x) \equiv -[(x-.5)^2 - 0.75]$, and $p_3(x) \equiv (1-x)^2/2 = p_1(1-x)$. Our interest is in how these polynomials behave on $[0,1]$. Define the mother spline in terms of these polynomials.

$$\Psi_2(x) = p_1(x) \mathscr{I}_{[0,1]}(x) + p_2(x-1) \mathscr{I}_{(0,1]}(x-1) + p_3(x-2) \mathscr{I}_{(0,1]}(x-2).$$

Note that continuity of Ψ_2 derives from the facts that

$$p_1(1) = p_2(0); \quad p_2(1) = p_3(0).$$

Continuity of the derivatives derives from the facts that,

$$\mathbf{d}^1 p_1(1) = \mathbf{d}^1 p_2(0); \quad \mathbf{d}^1 p_2(1) = \mathbf{d}^1 p_3(0).$$

Another important fact is that $0 = p_3(1) = \mathbf{d}^1 p_3(1)$.

In b-splines, for $x \in [0, 1/m]$,

$$f(x) = f_1(x) \equiv \gamma_0 p_3(mx) + \gamma_1 p_2(mx) + \gamma_2 p_1(mx).$$

For $x \in [0, 1/m]$, $f_1(x)$ is an arbitrary quadratic polynomial as required because the γ_js are arbitrary and p_1, p_2, p_3 are linearly independent quadratics. More generally for $x \in [(j-1)/m, j/m]$,

$$f(x) = f_j(x) \equiv \gamma_{j-1} p_3(mx - j - 1) + \gamma_j p_2(mx - j - 1) + \gamma_{j+1} p_1(mx - j - 1).$$

Changing tacks just a little for notational convenience, we want to argue that, given f_{j-1}, which determines $\gamma_{j-2}, \gamma_{j-1}, \gamma_j$, then for $x \in [(j-1)/m, j/m]$, $f = f_j$ is again an arbitrary quadratic polynomial subject only to the constraints that the function is continuous and has continuous derivatives. For $x \in [1/m, 2/m]$,

$$f(x) = f_2(x) = \gamma_1 p_3(mx - 1) + \gamma_2 p_2(mx - 1) + \gamma_3 p_1(mx - 1).$$

One way to write an arbitrary polynomial on $[1/m, 2/m]$ would be to write

$$f_*(x) = \gamma_1 p_3(mx - 1) + \gamma_2 p_2(mx - 1) + \beta_{0,*} + \beta_{1,*}(x - 1/m) + \beta_{2,*}(x - 1/m)^2$$

where γ_1 and γ_2 are fixed to agree with their values on the previous interval but $\beta_{0,*}, \beta_{1,*}, \beta_{2,*}$ are arbitrary parameters. We want to show that imposing the continuity conditions on f_* at $1/m$ forces the function to be f_2. To get continuity we need $f_1(1/m) = f_*(1/m)$ and for continuous derivatives $\mathbf{d}^1 f_1(1/m) = \mathbf{d}^1 f_*(1/m)$. Observe that from the formula for $x \in [0, 1/m]$,

$$\begin{aligned}
f_1(1/m) &= \gamma_0 p_3(1) + \gamma_1 p_2(1) + \gamma_2 p_1(1) \\
&= 0 + \gamma_1 p_2(1) + \gamma_2 p_1(1) \\
&= \gamma_1 p_3(0) + \gamma_2 p_2(0).
\end{aligned}$$

To get continuity this must equal

$$\begin{aligned}
f_*(1/m) &= \gamma_1 p_3(0) + \gamma_2 p_2(0) + \beta_{0,*} + \beta_{1,*}(0) + \beta_{2,*}(0)^2 \\
&= \gamma_1 p_3(0) + \gamma_2 p_2(0) + \beta_{0,*}.
\end{aligned}$$

so we must have $\beta_{0,*} = 0$. To get continuous derivatives at $1/m$ we need

$$\begin{aligned}
\mathbf{d}^1 f_1(1/m) &= \gamma_0 \mathbf{d}^1 p_3(1) + \gamma_1 \mathbf{d}^1 p_2(1) + \gamma_2 \mathbf{d}^1 p_1(1) \\
&= 0 + \gamma_1 \mathbf{d}^1 p_2(1) + \gamma_2 \mathbf{d}^1 p_1(1) \\
&= \gamma_1 \mathbf{d}^1 p_3(0) + \gamma_2 \mathbf{d}^1 p_2(0)
\end{aligned}$$

to equal

$$\mathbf{d}^1 f_*(1/m) = \gamma_1 \mathbf{d}^1 p_3(0) + \gamma_2 \mathbf{d}^1 p_2(0) + \beta_{1,*} + 2\beta_{2,*}(0)$$
$$= \gamma_1 \mathbf{d}^1 p_3(0) + \gamma_2 \mathbf{d}^1 p_2(0) + \beta_{1,*},$$

so we must have $\beta_{1,*} = 0$. Thus, in the arbitrary polynomial the constraints imposed reduce it to $\beta_{2,*}(x - 1/m)^2$ where $\beta_{2,*}$ is an arbitrary parameter. Overall we get

$$f_*(x) = \gamma_1 p_3(mx - 1) + \gamma_2 p_2(mx - 1) + \beta_{2,*}(x - 1/m)^2$$
$$= \gamma_1 p_3(mx - 1) + \gamma_2 p_2(mx - 1) + \gamma_3 p_1(mx - 1)^2 = f_2(x)$$

after we relabel the free parameter $2\beta_{2,*}/m^2$ as γ_3.

The same ideas work in general. For a d order b-spline define the mother spline in terms of d order component polynomials $p_1, p_2, \ldots, p_{d+1}$ where $p_1(x) \propto x^d$, with $p_{d+1}(x) = p_1(1 - x)$ so that $0 = p_{d+1}(1) = \mathbf{d}^1 p_{d+1}(1) = \cdots = \mathbf{d}^{d-1} p_{d+1}(1)$, and the other p_js are defined so that the mother spline is

$$\Psi_d(x) = p_1(x)\mathscr{I}_{[0,1]}(x) + \sum_{k=2}^{d+1} p_k(x - k + 1)\mathscr{I}_{(0,1]}(x - k + 1).$$

It is an exercise to figure out how to write the p_js for $d = 3$.

A key feature is that Ψ_d is itself a spline so

$$p_j(1) = p_{j+1}(0), \qquad j = 1, \ldots, d$$

and

$$\mathbf{d}^k p_j(1) = \mathbf{d}^k p_{j+1}(0), \qquad k = 1, \ldots, d-1, \quad j = 1, \ldots, d.$$

In b-splines the regression function is

$$f(x) = \sum_{j=0}^{m+d-1} \gamma_j \phi_j(x)$$

where

$$\phi_j(x) = \Psi_d(mx - j + d).$$

In particular, between the first two knots

$$f_1(x) = \mathscr{I}_{[0,1]}(mx) \sum_{k=0}^{d} \gamma_k p_{d+1-k}(mx).$$

The p_ks are linearly independent polynomials so for arbitrary γ_js, f_1 is an arbitrary d order polynomial as we need it to be for our induction. According the b-spline model,

$$f_j(x) = \mathscr{I}_{[0,1]}(mx - j + 1) \sum_{k=0}^{d} \gamma_{k+j-1} p_{d+1-k}(mx - j + 1).$$

The idea is to take

$$f_*(x) = \mathscr{I}_{[0,1]}(mx - j)\left[p_*(x) + \sum_{k=0}^{d-1} \gamma_{k+j} p_{d+1-k}(mx - j)\right].$$

where p_* is an arbitrary polynomial of order d but the γs are determined by f_j and show that the continuity and differentiability constraints lead f_* to agree with the definition of f_{j+1} from the b-spline model. Write $p_*(x) = \sum_{k=0}^{d} \beta_{k,*}(mx - j)^k$ so that

$$f_*(x) =$$
$$\mathscr{I}_{[0,1]}(mx - j)\left\{\left[\sum_{k=0}^{d} \beta_{k,*}(mx - j)^k\right] + \left[\sum_{k=0}^{d-1} \gamma_{k+j} p_{d+1-k}(mx - j)\right]\right\}. \quad (1.7.3)$$

Continuity requires

$$f_j(j/m) = f_*(j/m).$$

Using the fact that $p_{d+1}(1) = 0$ and the spline properties of Ψ_d

$$f_j(j/m) = \sum_{k=0}^{d} \gamma_{k+j-1} p_{d+1-k}(1)$$
$$= 0 + \sum_{k=1}^{d} \gamma_{k+j-1} p_{d+1-k}(1)$$
$$= \sum_{k=0}^{d-1} \gamma_{k+j} p_{d-k}(1)$$
$$= \sum_{k=0}^{d-1} \gamma_{k+j} p_{d+1-k}(0).$$

From (1.7.3) it follows that

$$f_*(j/m) = \left[\beta_{0,*} + \sum_{k=1}^{d} \beta_{k,*} 0^k\right] + \left[\sum_{k=0}^{d-1} \gamma_{k+j} p_{d+1-k}(0)\right],$$

so to achieve continuity we must have $\beta_{0,*} = 0$. Similar arguments for the requirement of continuous derivatives force $0 = \beta_{1,*} = \cdots = \beta_{d-1,*}$. Incorporating these requirements, recalling that $p_1(x)$ is proportional to x^d, and replacing the arbitrary parameter $\beta_{d,*}$ with an arbitrary γ_{d+j} that incorporates the proportionality constant, we get

$$f_*(x) = \mathscr{I}_{[0,1]}(mx - j)\left\{\beta_{d,*}(mx - j)^d + \sum_{k=0}^{d-1} \gamma_{k+j} p_{d+1-k}(mx - j)\right\}$$
$$= \mathscr{I}_{[0,1]}(mx - j)\left\{\gamma_{d+j} p_1(mx - j) + \sum_{k=0}^{d-1} \gamma_{k+j} p_{d+1-k}(mx - j)\right\}$$

$$= \mathscr{I}_{[0,1]}(mx - j) \sum_{k=0}^{d} \gamma_{k+j} \, p_{d+1-k}(mx - j) = f_{j+1}(x).$$

Exercise 1.2. Show, for $d = 2$ b-splines, that $\sum_{k=0}^{m+1} \phi_k(x_i)$ is a constant. Hint: Recall that exactly three nonzero ϕ_j functions overlap on every subinterval and argue that it is enough to show that for $x \in [0,1]$ the function $x^2/2 - [(1+x) - 1.5]^2 + [3 - (2+x)]^2/2$ is a constant.

1.7.2 Fitting Local Functions

In discussing b-splines for $m+1$ equally spaced knots we carefully defined $\phi_j \equiv \phi_{jm}$ functions so that they were equivalent to fitting polynomial splines. But in another sense, we merely fitted a bunch of bell shaped curves that were rescaled to have small support and shifted so that the functions had centers that were spread over the entire unit interval.

Just about any function Ψ that goes to 0 as $|x| \to \infty$ can be used as a "mother" to define a triangular array ϕ_{jm} of approximating-functions with small (practical) support. These include indicator functions, mother splines, mother wavelets, normal densities, etc. Given a set of knots $\tilde{x}_{j,m}$, take $s - 1 = m$ with ϕ_{jm} a rescaled mother function with a location tied to (often centered at) $\tilde{x}_{j,m}$. The success of this enterprize will depend on the number and placement of the knots and how the mother function is rescaled. The process becomes a method based on approximating functions with small support when the mother function is rescaled in such a way that it becomes, for all practical purposes, 0 outside of a small interval.

1.7.3 Local Regression

Local (polynomial) regression, often referred to as *loess* or *lowess* (local weighted scatterplot smoothing) provides fitted values by fitting a separate low order polynomial for every prediction. It provides a collection of $(x, \hat{y})$ values that can be plotted, but it does not provide a formula for the estimated regression curve. As with splines, we assume that in the data (x_i, y_i), the x_is are ordered.

The key to local regression is that it uses weighted regression with weights determined by the distance between the actual data x_i and the location being fitted x. What makes this "local" regression is that the weights are either zero, or close to zero, outside a small region around x. The weights are determined by a *kernel* function (not to be confused with the reproducing kernels introduced in Chap. 3). (I jokingly once proposed such a function to my son who has always referred to it as Pop's corny kernel.)

Originally, this procedure was performed using 0 order polynomials and is known as *kernel smoothing*, see Green and Silverman (1994) or Efromovich (1999). The idea of kernel smoothing is to base estimation on the continuity of $f(x)$. The estimate $\hat{f}(x)$ is a weighted average of the y_i values in a small neighborhood of x. Less weight is given to a y_i for which the corresponding x_i is far from x. The weights are defined by a nonnegative kernel function $K(z)$ that gets small rapidly as z gets away from 0. The *Nadaraya–Watson kernel estimate* is

$$\hat{f}(x) = \sum_{i=1}^{n} y_i K[(x - x_i)] \Big/ \sum_{i=1}^{n} K[(x - x_i)],$$

which is just a weighted average, as advertised.

More generally, take a low order polynomial model, say,

$$Y = X\beta + 0, \quad E(e) = 0$$

on which we perform generalized least squares with a diagonal weighting matrix $D(w)$ having some vector of weights w down the diagonal. The definition of the weights is all-important. The ith element of w is

$$w_i \equiv K[(x_i - x)/h]$$

for some scalar tuning parameter h and kernel function K. From fitting this model we obtain only one thing, the fitted value $\hat{y}$ for the new data point x. You do this for a lot of xs and plot the result. Obviously, fitting a separate linear model for every fitted value requires modern computing power.

In loess the most commonly used weighting seems to be the *tri-weight* where the kernel function is

$$K(z) = \begin{cases} (1 - |z|^3)^3 & \text{if } |z| < 1 \\ 0 & \text{if } |z| \geq 1. \end{cases}$$

In R the default is to fit a quadratic polynomial.

For the battery data the default `loess` fit in R seems to me to oversmooth the data. It gives $R^2 = 0.962$. Code for fitting and plotting it is included with the other code on my website.

1.8 Nonparametric Multiple Regression

Nonparametric multiple regression involves using a p vector x as the argument for $\phi_j(\cdot)$ in an infinite sum or $\phi_{jm}(\cdot)$ in a triangular array. The difficulty is in choosing which ϕ functions to use. There are two common approaches. One is to construct the vector functions ϕ from the scalar ϕ functions already discussed. The other method uses the *kernel trick* to replace explicit consideration of the ϕ functions with evaluation of a *reproducing kernel* function.

1.8.1 Redefining ϕ and the Curse of Dimensionality

In nonparametric multiple regression, the scalars x_i are replaced by vectors $x_i = (x_{i1}, x_{i2}, \ldots, x_{ip})'$. Theoretically, the only real complication is that the ϕ_j functions have to be redefined as functions of vectors rather than scalars.

In practice, we often construct vector ϕ functions from scalar ϕ functions. The ideas become clear in the case of $p = 2$. For variables x_1 and x_2, define

$$\phi_{jk}(x_1, x_2) \equiv \phi_j(x_1)\phi_k(x_2),$$

and the regression function approximation is

$$f(x_1, x_2) \doteq \sum_{j=0}^{s_1-1} \sum_{k=0}^{s_2-1} \beta_{jk} \phi_{jk}(x_1, x_2). \tag{1.8.1}$$

In general, for $x = (x_1, \ldots, x_p)'$,

$$f(x) \doteq \sum_{k_1=0}^{s_1-1} \cdots \sum_{k_p=0}^{s_p-1} \beta_{k_1 \ldots k_p} \phi_{k_1}(x_1) \cdots \phi_{k_p}(x_p), \tag{1.8.2}$$

where most often $\phi_0 \equiv 1$.

Note that there are a lot of ϕ functions involved. For example, if we needed $s_1 = 10$ functions to approximate a function in x_1 and $s_2 = 8$ functions to approximate a function in x_2, it takes 80 functions to approximate a function in (x_1, x_2), and this is a very simple case. It is not uncommon to have $p = 5$ or more. If we need $s_* = 8$ for each dimension, we are talking about fitting $s = 8^5 = 32,768$ parameters for a very moderately sized problem. Clearly, this approach to nonparametric multiple regression is only practical for very large data sets if $p > 2$. However, nonparametric multiple regression seems to be a reasonable approach for $p = 2$ with moderately large amounts of data, such as are often found in problems such as two-dimensional image reconstruction and smoothing two-dimensional spatial data. Another way to think of the dimensionality problems is that, roughly, if we need n observations to do a good job of estimation with one predictor, we might expect to need n^2 observations to do a good job with two predictors and n^p observations to do a good job with p predictors. For example, if we needed 40 observations to get a good fit in one dimension, and we have $p = 5$ predictors, we need about 100 million observations. (An intercept can be included as either $\phi_0 \equiv 1$ or $x_{i1} \equiv 1$. In the latter case, s_*^{p-1} or n^{p-1} would be more appropriate.) This *curse of dimensionality* can easily make it impractical to fit nonparametric regression models.

One way to deal with having too many parameters is to use *generalized additive models*. Christensen (2015, Section 9.10) contains some additional details about writing out generalized additive models but the fundamental idea is an analogy to multifactor analysis of variance. Fitting the full model with $p = 5$ and, say, $s = 8^5$ parameters is analogous to fitting a 5 factor interaction term. If we fit the model with only the 10 three-factor interaction terms, we could get by with $10(8^3) = 5120$ parameters. If we fit the model with only the 10 two-factor interaction terms, we could get by with $10(8^2) = 640$ parameters. In particular, with the two-factor interactions, $f(x_1, \ldots, x_5)$ is modeled as the sum of 10 terms each looking like Eq. (1.8.1) but with each involving a different pair of predictor variables.

The all three-factor and all two-factor models still seem like a lot of parameters but the decrease is enormous compared to the five-factor model. The price for this decrease is the simplifying assumptions being made. And if we cannot fit the 5-factor interaction model, we cannot test the validity of those simplifying assumptions, e.g., whether it is alright to drop, say, all of the 4-factor interactions. Of course we don't have to restrict ourselves to the all 4-factor, all three-factor, all two-factor, and main-effects only models. We can create models with some three-factor interactions, some two-factors, and some main effects. Like ANOVA, we need to be concerned about creating linear dependencies in the model matrix Φ.

The most difficult part of computing least squares estimates is that they generally involve finding the inverse or generalized inverse of the $p \times p$ matrix $X'X$ (or some similarly sized computation). When p is large, the computation is difficult. When applying linear-approximation nonparametric methods the problem is finding the generalized inverse of the $s \times s$ matrix $\Phi'\Phi$, which typically has s much larger than p. This becomes particularly awkward when $s > n$. We now consider a device that gives us a model matrix that is always $n \times n$.

1.8.2 Reproducing Kernel Hilbert Space Regression

In Chap. 3 we introduce the theory of *reproducing kernel Hilbert spaces (RKHSs)* but for now we introduce a simple way to use them in nonparametric regression.

An RKHS transforms a p vector x_i into an s vector $\phi_i = [\phi_0(x_i), \ldots, \phi_{s-1}(x_i)]'$, where not infrequently $s = \infty$. Just as X has rows made up of the x_i's, Φ has rows made up of the ϕ_i's. Just as $XX' = [x_i'x_j]$ is an $n \times n$ matrix of inner products of the x_is, the whole point of RKHSs is that there exists a *reproducing kernel (r.k.)* function $R(\cdot, \cdot)$ with the property that

$$\tilde{R} \equiv [R(x_i, x_j)] = [\phi_i' D(\eta)\phi_j] = \Phi D(\eta)\Phi'$$

is an $n \times n$ inner product matrix of the ϕ_is where $D(\eta)$ is a positive definite diagonal matrix. Moreover, for s finite, $C[\Phi D(\eta)\Phi'] = C(\Phi)$ (see *PA* Section B.4), so fitting the r.k. model

$$Y = \tilde{R}\gamma + e$$

is equivalent to fitting the nonparametric model

$$Y = \Phi\beta + e.$$

The r.k. model is just a reparameterization with $\beta = D(\eta)\Phi'\gamma$. In particular, predictions are easy using the r.k. model,

$$\hat{y}(x) = [R(x,x_1),\ldots,R(x,x_n)]\hat{\gamma}.$$

This equivalence between fitting a linear structure with Φ and fitting one with the $n \times n$ matrix $\tilde{R}$ is sometimes known as the *kernel trick*.

A primary advantage of the kernel trick is simply that, for a known function $R(\cdot,\cdot)$, it is very easy to construct the matrix $\tilde{R}$. (It is time consuming to specify s different $\phi_j(\cdot)$ functions, as opposed to one $R(\cdot,\cdot)$ function.) Moreover, the $n \times s$ matrix Φ is awkward to use when s is large. $\tilde{R}$ is always $n \times n$, which limits how awkward it can become to use, but also prevents the simplifications that arise when $s < n$.

When $s \geq n$ and the x_is are distinct, it is to be expected that $\tilde{R}$ will be an $n \times n$ matrix of rank n, so it defines a saturated model. Least squares estimates (and, for generalized linear models, maximum likelihood estimates) will give fitted values that equal the observations and zero degrees of freedom for error. Nothing interesting will come of fitting a saturated model. We need to deal with this overfitting. Indeed, the kernel trick is typically used together with a penalized (regularized) estimation method such as those discussed in Chap. 2.

If the x_is are not all distinct, as in the discussion of Fisher's Lack-of-Fit Test from Chap. 6 of *PA*, the row structures of X, Φ, and $\tilde{R}$ (no longer nonsingular) are the same. Fitting any of $X\xi$, $\Phi\beta$, and $\tilde{R}\gamma$ by least squares would give exactly the same *pure error* sum of squares ($SSPE$) and degrees of freedom ($dfPE$). Moreover, fitting $\Phi\beta$ and $\tilde{R}\gamma$ would give exactly the same *lack-of-fit* sum of squares and degrees of freedom but, depending on the size of s, there is a good chance that fitting $\Phi\beta$ and $\tilde{R}\gamma$ would give $SSLF = 0$ on 0 $dfLF$. (This is the equivalent of fitting a saturated model when the x_is are not all distinct.)

Different choices of $R(\cdot,\cdot)$, if they have $s \geq n$, typically all give the same $C(\tilde{R})$, which defines either a saturated model or a model with no lack of fit. Thus different choices of $R(\cdot,\cdot)$ typically all give the same model, but they typically are reparameterizations of each other. They give the same least squares fits. But we will see in the next chapter that if you have two different parameterizations of the same model, and obtain estimates by penalizing parameters in the same way (i.e. use the same penalty function for every parameterization), that having the same penalties applied to different parameters leads to different fitted models. So, even though different $R(\cdot,\cdot)$ functions define essentially the same model, applying any standard penalty like ridge regression or lasso, will lead to different fitted values because the equivalent linear models have different parameters that are being shrunk in the same way. The process of shrinking is the same but the parameters are different, thus the end results are different. We saw that different ϕ_js work better or worse on the battery data and there is no way to tell ahead of time which collection will work best. Similarly, different $R(\cdot,\cdot)$s (with the same penalty) work better or worse on different data and there is no way to tell, ahead of time, which will work best.

If you know what ϕ_j functions you want to use, there is not much mathematical advantage to using r.k.s. But you can use R functions that are known to be r.k.s for which it is difficult or, in the case of $s = \infty$, impossible to write down all the ϕ_js. In Chap. 3 we will examine r.k.s that correspond to finite polynomial regression and to fitting splines. But there are a wide variety of potential r.k.s, many that correspond to $s = \infty$.

Table 1.7 gives some commonly used r.k.s. More generally, any function that can serve as a covariance function in the sense of Chap. 8 can serve as an r.k. Any r.k. that depends only on $\|u - v\|$ is a *radial basis function* r.k.

Table 1.7 Some common r.k. functions. b and c are scalars

Names	$R(u,v)$
Polynomial of degree d	$(1 + u'v)^d$
Polynomial of degree d	$b(c + u'v)^d$
Gaussian (Radial Basis)	$\exp(-b\|u - v\|^2)$
Sigmoid (Hyperbolic Tangent)	$\tanh(bu'v + c)$
Linear Spline (u, v scalars)	$\min(u,v)$
Cubic Spline (u, v scalars)	$\max(u,v)\min^2(u,v)/2 - \min^3(u,v)/6$
Thin Plate Spline (2 dimensions)	$\|u - v\|^2 \log(\|u - v\|)$

The hyperbolic tangent in Table 1.7 is not really an r.k. because it can give $\tilde{R}$ matrices that are not nonnegative definite. But any function $R(u,v)$ that is continuous in u can give plausible answers because it leads to fitting models of the form

$$m(x) = \sum_{j=1}^{n} \gamma_j R(x, x_j). \tag{1.8.3}$$

This idea can be viewed as extending the use of approximating functions with small support, cf. Sect. 1.7.2, from one to higher dimensions in a way that limits the curse of dimensionality. With local support methods, in one dimension you partition the line into say s_* sets and fit a separate one-dimensional wavelet, B spline, or other function for each partition set. The problem is that in p dimensions the number of partition sets (obtained by Cartesian products) quickly gets out of control, s_*^p. The key idea behind kernel methods is to fit a p-dimensional function, not for each partition set but for each observed data point. The number of functions being fitted is n, which is large but manageable, rather than s_*^p which rapidly becomes unmanageably large. The p-dimensional functions used in fitting can be defined as a product of p one-dimensional wavelet, spline, or other functions or they can be defined directly as p-dimensional functions via some kernel function. The tuning values b and c in Table 1.7 can be viewed as tools for getting the functions centered and scaled appropriately. Fitting n functions to n data points would typically result in overfitting, so penalizing the coefficients, as discussed in the next chapter, is appropriate. As mentioned earlier, when $\tilde{R}$ is a nonsingular matrix (or more generally has the column space associated with finding pure error), it does not matter what function you used to define $\tilde{R}$ because all such matrices are reparameterizations of each other and give the same least squares fitted values. But if you penalize the parameters in a fixed way, the parameterization penalty will have different effects on different parameterizations.

EXAMPLE 1.8.1.   I fitted the battery data with the R language's lm command using the polynomial functions $R(u,v) = (u'v)^4$, $R(u,v) = (1+u'v)^4$, $R(u,v) = 5(7+u'v)^4$ and the Gaussian functions $R(u,v) = \exp(-\|u-v\|^2)$ and $R(u,v) = \exp(-1000\|u-v\|^2)$. I defined x_i to include the intercept. The three polynomial functions gave fitted values $\hat{y}_i$ identical to those from fitting a fourth degree polynomial. (I fitted the fourth degree polynomial several ways including using $\Phi\Phi'$ as the model matrix.) The Gaussian r.k.s have $s = \infty$. The first Gaussian function gave an $\tilde{R}$ matrix that was computationally singular and gave fitted values that were (to me) unexplainable except as a convenient fitting device similar to the hyperbolic tangent discussed later. The last function gave an $\tilde{R}$ that was computationally invertible and hence gave fitted values with $\hat{y}_i = y_i$. This has overfit the model so penalizing the regression coefficients, as discussed in the next chapter, is advisable.

Figure 1.10 contains the fit of the hyperbolic tangent "kernel" to the battery data using $b = 1$ and $c = 0$. It turns out that (at least computationally) $\tilde{R}$ is a rank 8 matrix with R's lm command including only the 1st, 2nd, 3rd, 4th, 11th, 16th, 29th, and 41st columns of $\tilde{R}$. For an 8 parameter model this has a remarkably high value of $R^2 = 0.9996$. Incidentally, this $\tilde{R}$ has negative eigenvalues so is not nonnegative definite. With $b = 5$ and $c = 0$ lm uses columns 1, 2, 3, 6, 12, 22, 37 of $\tilde{R}$ and again has $R^2 = 0.9996$. With $b = 10$ and $c = 10$ lm fits only the first column of $\tilde{R}$ yet has $R^2 = 0.9526$. In none of these cases has the hyperbolic tangent led to serious overfitting (although it is quite clear from inspecting the R output that we could drop at least one of the columns used in each $c = 0$ example).   □

1.9 Testing Lack of Fit in Linear Models

Given a linear model
$$Y = X\beta + e, \qquad E(e) = 0, \qquad (1.9.1)$$
any form of fitting a nonparametric regression determines a potential lack-of-fit test procedure. When fitting nonparametric regression via the linear-approximation models discussed in this chapter, lack-of-fit tests are easy to specify. Because the procedure is based on having a linear-approximation model, essentially the same procedure works regardless of whether one is fitting polynomials, trigonometric functions, wavelets, or splines. (Local polynomials [lowess], because they do not fit a single linear model, do not seem to fit into this procedure.)

Suppose we have the linear model (1.9.1) based on predictor variables $x_1, \ldots, x_s$. Given enough data, it may be feasible to produce a nonparametric multiple regression model, say $Y = \Phi\gamma + e$. In practice, this may need to be some generalized additive model. If $C(X) \subset C(\Phi)$, we could just test the reduced model against the

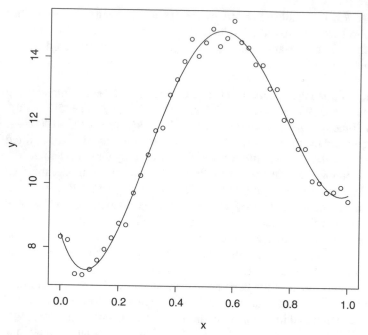

Fig. 1.10 Hyperbolic tangent fit to battery data

full model. Unless Φ is based on polynomials (including polynomial splines), more often than not $C(X) \not\subset C(\Phi)$. In that case we can test the reduced model (1.9.1) against the analysis of covariance full model

$$Y = X\beta + \Phi\gamma + e. \tag{1.9.2}$$

The F statistic is the standard

$$F = \frac{[SSE(1) - SSE(2)][dfE(1) - dfE(2)]}{MSE(2)}$$

and should be about 1 if the original model is correct. If $e \sim N(0, \sigma^2 I)$ is a good approximation, the test will have an approximate F distribution.

This procedure does not define a single lack-of-fit test. Every different method of picking Φ defines a different test. Which test is best? It seems pretty clear that no best test can possibly exist. If the lack of fit is due to the true model involving cosine curves that were not included in model (1.9.1), picking a Φ based on cosine curves should work better than picking a Φ based on polynomials or wavelets. If the lack of fit is due to the true model involving polynomial terms not included in model (1.9.1), picking Φ based on polynomials should work better than picking a Φ based on sines, cosines, or wavelets.

What works best will depend on the true nature of the lack of fit. Unfortunately, we do not know the true model. We won't know which of these tests will work best unless we have a very good idea about the true lack of fit and how to model it. If we had those good ideas, we probably wouldn't be thinking about doing a lack-of-fit test.

Incidentally, it is clearly impossible for such tests to be sensitive only to lack of fit in the mean structure. As will be discussed in Chap. 5, it is perfectly reasonable to assume that γ in (1.9.2) is a random vector with mean 0. In such a case, $E(Y) = X\beta$, so there is no lack of fit. However the F test will still be sensitive to seeing random values of γ that are very different from 0. Such values of γ will be the result of some combination of heteroscedasticity or dependence among the observations in Y. There is no way to tell from the test itself whether lack of fit or heteroscedasticity or dependence or some combination is causing a large F statistic.

The ACOVA lack-of-fit test examined in Christensen (1989, 1991) can be thought of as adding a nonparametric component to the original model that is simply a step function with steps defined for groups of near replicates among the predictor variables. Neill and Johnson (1989) and Christensen (1991) established that the ACOVA test tends to be worse than another test that is UMPI for orthogonal lack of fit between clusters of near replicates.

In fact, most traditional forms of lack-of-fit testing can be viewed through a lens of performing some kind of nonparametric regression on subsets of the data. Most traditional lack-of-fit tests rely on partitioning the data and fitting some kind of linear model within the partition sets. Fitting models on partitions is nothing more than fitting approximating-functions with small support. Lack-of-fit tests are reviewed in *PA* as well as Christensen (2015). Atwood and Ryan's idea for testing lack of fit is just fitting the original model on subsets of the data, so it is essentially fitting multivariate linear splines *without* the requirement that the fitted splines be continuous. Utts' method relies on fitting the original model on only a central group of points, so it implicitly puts each point not in the central group into a separate partition set and fits a separate parameter to each of those noncentral data points. (As alluded to earlier, the irony is that the more parameters you fit the more "nonparametric" your procedure.) Fisher's test fits the biggest model possible that maintains the row structure of the data, cf. *PA-V* Sect. 6.7.2, i.e., the data are partitioned into sets where the predictor variables are identical and a separate parameter if fitted to each set.

Clearly, this model based approach to performing lack-of-fit tests can be extended to testing lack of fit in logistic regression and other generalized linear models.

1.10 Regression Trees

Regression trees can be viewed as a form of linear modeling. In fact, they can be thought of as using forward selection to deal with the dimensionality problems of nonparametric multiple regression. But, unlike standard forward selection, the variables considered for inclusion in the model change with each step of the process.

There are a number of different algorithms available for constructing regression trees, cf. Loh (2011). We merely discuss their general motivation. Constructing trees is also known as *recursive partitioning*.

A simple approach to nonparametric regression is to turn the problem into a multifactor ANOVA. With p predictor variables, partition each predictor variable into s_* groups. In other words, define s_* indicator functions to partition each variable. Construct the predictor functions as in (1.8.2) by multiplying the indicator functions. This amounts to partitioning p dimensional space into s_*^p subsets. Fitting the regression function (1.8.2) amounts to fitting an ANOVA with p factors each at s_* levels, i.e., an s_*^p ANOVA. Fitting the ANOVA model that includes the p-factor interaction is equivalent to fitting a one-way ANOVA with s_*^p groups. If you want to make a prediction for a new point $x = (x_1, \ldots, x_p)'$, just figure out which of the s_*^p partition sets includes x and the prediction is the sample mean of the y observations that fell into that set. Of course this does nothing to help with the curse of dimensionality, but fitting generalized additive models is clearly nothing more than fitting a model that eliminates many of the interactions in the s_*^p ANOVA.

How do you pick the partition sets? The more partition sets you have, the more "nonparametric" the model will be. A reasonable rule of thumb might be to require that if a partition set includes any data at all, it has to include, say, 5 observations. The sets with no data we will ignore and never make predictions there. Five observations in a partition set is not a crazy small number of observations on which to base a prediction. Once the partition has be determined, we could use backward elimination to find partition sets that can be pooled together. (It would probably be wise to require that partition sets to be pooled must be contiguous in p dimensional space.)

Fitting regression trees is basically the same idea except that they are based on forward selection rather than backward elimination. By using forward selection, the procedure avoids the curse of dimensionality. Usually forward selection can easily miss important features. A nice feature of regression trees is that they pick the partition sets as well as deciding which partition sets need further dividing. In other words, they search through far more than a single set of s_*^p partition sets. In practice, regression trees are often used with bagging or random forests as discussed in *PA-V*, Chap. 14.

We now consider two examples. A very simple one to illustrate the ideas and a slightly more complicated one that examines the process.

EXAMPLE 1.10.1.   Consider a simple example with $n = 7$ observations and two predictor variables x_1, x_2, specifically

$$
Y = \begin{bmatrix} y_1 \\ y_2 \\ y_3 \\ y_4 \\ y_5 \\ y_6 \\ y_7 \end{bmatrix} \qquad [X_1, X_2] = \begin{bmatrix} 1 & 2 \\ 2 & 4 \\ 3 & 6 \\ 4 & 1 \\ 5 & 5 \\ 6 & 7 \\ 7 & 3 \end{bmatrix}.
$$

The first step is to split the data into two parts based on the size of X_1 or X_2. For instance, we can consider a split that consists of the smallest x_1 value and the six largest; or the two smallest x_1 values and the five largest; or the smallest three x_2 values and the largest four. We consider all such splits and posit an initial regression tree model $Y = \Phi^{(1)}\beta + e$, where

$$\Phi^{(1)} = \begin{bmatrix} 1 & 1 & 1 & 1 & 1 & 1 & 1 & 0 & 1 & 1 & 1 & 1 & 1 \\ 1 & 0 & 1 & 1 & 1 & 1 & 1 & 0 & 0 & 0 & 1 & 1 & 1 \\ 1 & 0 & 0 & 1 & 1 & 1 & 1 & 0 & 0 & 0 & 0 & 0 & 1 \\ 1 & 0 & 0 & 0 & 1 & 1 & 1 & 1 & 1 & 1 & 1 & 1 & 1 \\ 1 & 0 & 0 & 0 & 0 & 1 & 1 & 0 & 0 & 0 & 0 & 1 & 1 \\ 1 & 0 & 0 & 0 & 0 & 0 & 1 & 0 & 0 & 0 & 0 & 0 & 0 \\ 1 & 0 & 0 & 0 & 0 & 0 & 0 & 0 & 0 & 1 & 1 & 1 & 1 \end{bmatrix}.$$

The last 12 columns identify all of the possible splits. Columns 2 through 7 are the splits based on x_1 and columns 8 through 13 are the splits based on x_2, with, for example, the tenth column identifying the smallest three x_2 values and, by default since a column of 1's is included, the largest four. Obviously, this initial model is overparameterized; it has 13 predictor variables to explain 7 observations. The first (intercept) column is forced into the model and one other column is chosen by forward selection. Suppose that column is the fifth, so at the second stage we have the columns

$$\begin{bmatrix} 1 & 1 \\ 1 & 1 \\ 1 & 1 \\ 1 & 1 \\ 1 & 0 \\ 1 & 0 \\ 1 & 0 \end{bmatrix} \quad \text{or equivalently} \quad \begin{bmatrix} 1 & 0 \\ 1 & 0 \\ 1 & 0 \\ 1 & 0 \\ 0 & 1 \\ 0 & 1 \\ 0 & 1 \end{bmatrix}$$

forced into the second-stage model matrix. We now consider possible splits *within* the two groups that we have already identified. The first four observations can be split based on the sizes of either x_1 or x_2 and similarly for the last three. The second stage model is $Y = \Phi^{(2)}\beta + e$, where

$$\Phi^{(2)} = \begin{bmatrix} 1 & 0 & 1 & 1 & 1 & 0 & 1 & 1 & 0 & 0 & 0 & 0 \\ 1 & 0 & 0 & 1 & 1 & 0 & 0 & 1 & 0 & 0 & 0 & 0 \\ 1 & 0 & 0 & 0 & 1 & 0 & 0 & 0 & 0 & 0 & 0 & 0 \\ 1 & 0 & 0 & 0 & 0 & 1 & 1 & 1 & 0 & 0 & 0 & 0 \\ 0 & 1 & 0 & 0 & 0 & 0 & 0 & 0 & 1 & 1 & 0 & 1 \\ 0 & 1 & 0 & 0 & 0 & 0 & 0 & 0 & 0 & 1 & 0 & 0 \\ 0 & 1 & 0 & 0 & 0 & 0 & 0 & 0 & 0 & 0 & 1 & 1 \end{bmatrix}.$$

Here, columns 3, 4, and 5 are splits of the first group based on the size of x_1 and columns 6, 7, and 8 are splits of the first group based on the size of x_2. Columns 9 and 10 are splits of the second group based on x_1 and columns 11 and 12 are based on x_2. Again, the model is grossly overparameterized. Columns 1 and 2 are forced into the model, and one more column is chosen by forward selection. Suppose it is column 7, so at the third stage we have

$$
\begin{bmatrix}
1 & 0 & 1 \\
1 & 0 & 0 \\
1 & 0 & 0 \\
1 & 0 & 1 \\
0 & 1 & 0 \\
0 & 1 & 0 \\
0 & 1 & 0
\end{bmatrix}
\quad \text{or equivalently} \quad
\begin{bmatrix}
0 & 0 & 1 \\
1 & 0 & 0 \\
1 & 0 & 0 \\
0 & 0 & 1 \\
0 & 1 & 0 \\
0 & 1 & 0 \\
0 & 1 & 0
\end{bmatrix}
$$

forced into the model. We now have three groups, and again we consider splitting within groups. At the third stage, we have $Y = \Phi^{(3)}\beta + e$, where

$$
\Phi^{(3)} =
\begin{bmatrix}
0 & 0 & 1 & 0 & 0 & 0 & 0 & 0 & 0 \\
1 & 0 & 0 & 0 & 1 & 0 & 0 & 0 & 0 \\
1 & 0 & 0 & 0 & 0 & 0 & 0 & 0 & 0 \\
0 & 0 & 1 & 1 & 0 & 0 & 0 & 0 & 0 \\
0 & 1 & 0 & 0 & 0 & 1 & 1 & 0 & 1 \\
0 & 1 & 0 & 0 & 0 & 0 & 1 & 0 & 0 \\
0 & 1 & 0 & 0 & 0 & 0 & 0 & 1 & 1
\end{bmatrix}.
$$

Again, we add a column by forward selection. If no column can be added, we return to the model with the three forced variables,

$$
Y =
\begin{bmatrix}
0 & 0 & 1 \\
1 & 0 & 0 \\
1 & 0 & 0 \\
0 & 0 & 1 \\
0 & 1 & 0 \\
0 & 1 & 0 \\
0 & 1 & 0
\end{bmatrix}
\beta + e.
$$

Note that this is just a one-way ANOVA model, so the parameter estimates are group means. We can identify the groups as (1) $x_1 < 4.5$, $x_2 < 2.5$; (2) $x_1 > 4.5$; and (3) $x_1 < 4.5$, $x_2 > 2.5$. Predictions are based on identifying the appropriate group and use the group mean as a point prediction. Note that this is essentially fitting a step function to the data.

Going back to the original parameterization of the model (i.e., the original choices of columns), the model is

$$
Y =
\begin{bmatrix}
1 & 1 & 1 \\
1 & 1 & 0 \\
1 & 1 & 0 \\
1 & 1 & 1 \\
1 & 0 & 0 \\
1 & 0 & 0 \\
1 & 0 & 0
\end{bmatrix}
\beta + e.
$$

With these choices of the columns, the columns are ordered from left to right, and dropping columns successively from the right still gives a regression tree. □

As discussed in *PA*, forward selection defines a sequence of larger and larger models with various ways to determine which variable is added next and various ways to determine when to stop adding variables. Regression trees typically add variables based on minimizing the *SSE*, which is the traditional method employed in forward selection. Regression trees often employ an unusual stopping rule. Breiman et al. (1984, Section 8.5) suggest continuing the forward selection until each group has five or fewer observations. At that point, one can either accept the final model or pick a best model from the sequence using something like the C_p statistic (assuming that the final model gives a reasonable *MSE*).

EXAMPLE 1.10.2.   Mosteller and Tukey (1977) considered data from *The Cole-man Report*. The data are from schools in the New England and Mid-Atlantic states. The variables are y, the mean verbal test score for sixth graders; x_1, staff salaries per pupil; x_2, percentage of sixth graders whose fathers have white-collar jobs; x_3, a composite measure of socioeconomic status; x_4, the mean score of a verbal test given to the teachers; and x_5, the mean educational level of the sixth graders' mothers (one unit equals two school years). The data are given in Table 1.8. In this example, only two predictor variables are used: x_3 and x_5.

Table 1.8 *Coleman Report* data

School	y	x_1	x_2	x_3	x_4	x_5
1	37.01	3.83	28.87	7.20	26.60	6.19
2	26.51	2.89	20.10	−11.71	24.40	5.17
3	36.51	2.86	69.05	12.32	25.70	7.04
4	40.70	2.92	65.40	14.28	25.70	7.10
5	37.10	3.06	29.59	6.31	25.40	6.15
6	33.90	2.07	44.82	6.16	21.60	6.41
7	41.80	2.52	77.37	12.70	24.90	6.86
8	33.40	2.45	24.67	−0.17	25.01	5.78
9	41.01	3.13	65.01	9.85	26.60	6.51
10	37.20	2.44	9.99	−0.05	28.01	5.57
11	23.30	2.09	12.20	−12.86	23.51	5.62
12	35.20	2.52	22.55	0.92	23.60	5.34
13	34.90	2.22	14.30	4.77	24.51	5.80
14	33.10	2.67	31.79	−0.96	25.80	6.19
15	22.70	2.71	11.60	−16.04	25.20	5.62
16	39.70	3.14	68.47	10.62	25.01	6.94
17	31.80	3.54	42.64	2.66	25.01	6.33
18	31.70	2.52	16.70	−10.99	24.80	6.01
19	43.10	2.68	86.27	15.03	25.51	7.51
20	41.01	2.37	76.73	12.77	24.51	6.96

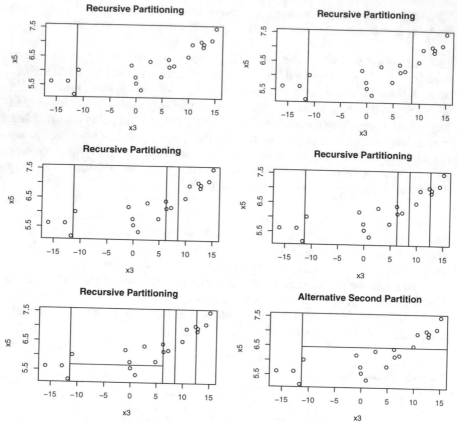

Fig. 1.11 Partition sets for x_3, x_5

We examine a partitioning created using the R package `rpart`. Details are given in the computing document on my website. For illustrative purposes, I required that there could be no fewer than two data points in any partition set. Typically one would do this on a bigger set of data and perhaps require more observations in every partition set. Figures 1.11 and 1.12 illustrate the recursive partitioning process. Figure 1.12 is the tree diagram produced by `rpart`. The algorithm begins by partitioning x_3 four times before it involves x_5. I set up `rpart` to keep running until it did a partition based on x_5.

Table 1.9 contains the statistics that determine the first partitioning of the data. The values $x_{3(i)}$ are the ordered values x_3 with $y_{3(i)}$ the corresponding y values (order statistics and induced order statistics). $x_{5(j)}$ and $y_{5(j)}$ are the corresponding values for x_5. For $i = 3$, the partition sets consist of the 3 smallest x_3 observations, and the 17 largest. For $i = 18$ the partition sets are the 18 smallest x_3 observations, and the 2 largest. For $j = 18$ the partition sets are the 18 smallest x_5 observations, and the 2 largest. Built in is the requirement that each partition set contain at least 2 observations. The SSEs are from fitting a one-way ANOVA on the two groups. Note

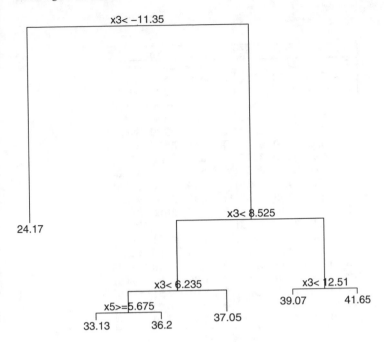

Fig. 1.12 Regression tree for x_3, x_5

that the smallest SSE corresponds to $i = 3$, so that is the first partition used. The first split illustrated in Figs. 1.11 and 1.12 is at $-11.35 = [x_{3(3)} + x_{3(4)}]/2$ so that the two partition sets include the 3 smallest x_3 observations, and the 17 largest.

For the second split we consider all the splits of each of the two partition sets from the first stage. Fortunately for our illustration, the partition set $x_3 < -11.35$ has only 3 observations, so we are not allowed to split it further because splitting it has to create a partition set with less than 2 observations. Thus we only need to consider all splits of the set $x_3 \geq -11.35$. In Table 1.10 I have blanked out the observations with $x_3 < -11.35$ but remember that these three observations are still included in the SSEs. The minimum SSE occurs when $i = 13$ so the next partition set goes from -11.35 to $8.525 = [x_{3(13)} + x_{3(14)}]/2$, as illustrated in Figs. 1.11 and 1.12.

While `rpart` created the partition just mentioned, in Table 1.10 the value $j = 13$ gives the same SSE as $i = 13$. The alternative partition of the (x_3, x_5) plane with $x_3 \geq -11.35$ and x_5 divided at $6.46 = [x_{5(13)} + x_{5(14)}]/2$ is given in the bottom right of Fig. 1.11. *It separates the data into exactly the same three groups* as the `rpart` partition. I have no idea why `rpart` chose the partition based on x_3 rather than the alternative partition based on x_5. It looks like, after incorporating the alternative partition, the subsequent partitions would continue to divide *the data* in the same way. However, the final partition sets would be different, which means that predictions could be different. There are 6 final partition sets, so there are only 6 distinct values that will be used to predict, and they will be the same 6 numbers for either partitioning. But the ranges of (x_3, x_5) values over which those 6 predictions are applied change with the different partitions.

Table 1.9 First tree split

i	$x_{3(i)}$	$y_{3(i)}$	SSE	j	$x_{5(j)}$	$y_{5(j)}$	SSE
1	−16.04	22.70		1	5.17	26.51	
2	−12.86	23.30	318.5	2	5.34	35.20	603.2
3	−11.71	26.51	222.6	3	5.57	37.20	627.2
4	−10.99	31.70	235.2	4	5.62	22.70	533.4
5	−0.96	33.10	255.8	5	5.62	23.30	394.8
6	−0.17	33.40	266.1	6	5.78	33.40	396.3
7	−0.05	37.20	331.1	7	5.80	34.90	412.7
8	0.92	35.20	349.2	8	6.01	31.70	376.7
9	2.66	31.80	306.1	9	6.15	37.10	413.0
10	4.77	34.90	306.5	10	6.19	33.10	387.8
11	6.16	33.90	283.2	11	6.19	37.01	412.3
12	6.31	37.10	306.6	12	6.33	31.80	356.6
13	7.20	37.01	321.3	13	6.41	33.90	321.3
14	9.85	41.01	394.1	14	6.51	41.01	394.1
15	10.62	39.70	438.2	15	6.86	41.80	468.1
16	12.32	36.51	427.1	16	6.94	39.70	505.2
17	12.70	41.80	492.8	17	6.96	41.01	554.0
18	12.77	41.01	539.6	18	7.04	36.51	539.6
19	14.28	40.70		19	7.10	40.70	
20	15.03	43.10		20	7.51	43.10	

Table 1.10 Second tree split

i	$x_{3(i)}$	$y_{3(i)}$	SSE	j	$x_{5(j)}$	$y_{5(j)}$	SSE
1				1			
2				2	5.34	35.20	
3				3	5.57	37.20	221.2
4	−10.99	31.70		4			
5	−0.96	33.10	174.5	5			
6	−0.17	33.40	156.1	6	5.78	33.40	211.6
7	−0.05	37.20	170.5	7	5.80	34.90	205.1
8	0.92	35.20	163.5	8	6.01	31.70	177.4
9	2.66	31.80	123.2	9	6.15	37.10	182.1
10	4.77	34.90	107.7	10	6.19	33.10	156.8
11	6.16	33.90	76.6	11	6.19	37.01	158.7
12	6.31	37.10	77.7	12	6.33	31.80	111.8
13	7.20	37.01	73.6	13	6.41	33.90	73.6
14	9.85	41.01	111.6	14	6.51	41.01	111.5
15	10.62	39.70	130.0	15	6.86	41.80	150.3
16	12.32	36.51	109.8	16	6.94	39.70	164.9
17	12.70	41.80	145.7	17	6.96	41.01	187.7
18	12.77	41.01	168.4	18	7.04	36.51	168.4
19	14.28	40.70		19	7.10	40.70	
20	15.03	43.10		20	7.51	43.10	

The set $x_3 < -11.35$ cannot be split further because of our requirement that all partition sets include two data points. But $-11.35 \leq x_3 < 8.525$ has 10 data points, so it can be split $14 = 2 \times (10 - 3)$ ways, and $x_3 \geq 8.525$ has 7 points, so can be split $8 = 2 \times (7 - 3)$ ways. That is another 22 ANOVAs to run from which we pick the one with the smallest SSE. The minimum occurs when splitting x_3 between $x_{3(11)}$ and $x_{3(12)}$, cf. Figs. 1.11 and 1.12. We discontinue the detailed illustration.

An advantage of doing one full s_*^p partition is that you can easily identify empty cells and cells with little data. Prediction variances will reflect that. With a forward selection partition, the algorithms typically create partition sets that restrict the minimum number of observations in a partition. However, looking a Fig. 1.11, an observation with, for example, a small value of x_3 and a large value of x_5 is far from the other data in its partition set, so it is unlikely to be predicted well by the mean of the observations in that set. It is not clear to me how one could identify that troublesome phenomenon when fitting a regression tree in higher dimensions □

Exercise 1.3. Without a computer, find the predictions for the point (15,6) from the two partitions. Hint: the `rpart` prediction is based on the average of four y values and the alternative partition prediction is based on two.

There is a strong tendency for regression trees to *overfit* the data, causing poor predictive performance from tree models. Random forests, bagging, and boosting were developed to improve the predictive performance of tree models. Those topics are discussed in the last section of the last chapter of *PA-V*.

1.11 Regression on Functional Predictors

For each dependent variable y_i, $i = 1, \ldots, n$, suppose we observe a function of predictor variables, say $\mathscr{X}_i(t)$, $t \in \mathscr{T} \subset \mathbf{R}^d$. The predictor function might be observed over time or might result from some sort of medical imaging. For some unknown function $\gamma(t)$ of regression coefficients, we assume the model

$$y_i = \alpha + \int_{\mathscr{T}} \mathscr{X}_i(t)\gamma(t)dt + e_i, \qquad \mathrm{E}(e_i) = 0.$$

As a practical matter, we incorporate a nonparametric characterization of the regression coefficient function using a standard spanning set of functions ϕ_j to get

$$\gamma(t) \doteq \sum_{j=0}^{s-1} \beta_j \phi_j(t) = \phi(t)'\beta.$$

where

$$\phi(t)' \equiv [\phi_0(t), \ldots, \phi_{s-1}(t)]', \qquad \beta \equiv (\beta_0, \ldots, \beta_{s-1})'.$$

This leads to a standard linear model for the observations

$$
\begin{aligned}
y_i &= \alpha + \int_{\mathscr{T}} \mathscr{X}_i(t)\gamma(t)dt + e_i \\
&\doteq \alpha + \int_{\mathscr{T}} \mathscr{X}_i(t)\phi(t)'\beta dt + e_i \\
&= \alpha + \left[\int_{\mathscr{T}} \mathscr{X}_i(t)\phi(t)'dt \right]\beta + e_i \\
&= \alpha + x_i'\beta + e_i
\end{aligned}
$$

where

$$
x_i' \equiv (x_{i0},\ldots,x_{i,s-1}), \quad \text{and} \quad x_{ij} \equiv \int_{\mathscr{T}} \mathscr{X}_i(t)\phi_j(t)dt.
$$

See Reiss, Goldsmith, Shang, and Ogden (2017) for additional discussion.

In reality, it is impossible to observe $\mathscr{X}_i(t)$ for every t in an infinite set of points $\mathscr{T}$. At most we can observe $\mathscr{X}_i(t)$ at t_{ik}, $k = 1,\ldots,N_i$ where we would expect N_i to be a very large number. In this case, we would want to use numerical approximations to the integrals, perhaps even something as simple as

$$
x_{ij} = \frac{1}{N_i}\sum_{k=1}^{N_i} \mathscr{X}_i(t_{ik})\phi_j(t_{ik}).
$$

1.12 Density Estimation

Let $y_1,\ldots,y_N$ be a random sample with density $f(y)$. Without loss of generality, we assume that the support of the density is the unit interval $[0,1]$. Partition the unit interval into S very small equal-sized intervals centered at $t_h = (2h-1)/2S$, $h = 1,\ldots,S$. Let $n \equiv (n_1,\ldots,n_S)'$ be the vector of counts in each interval, that is, the number of the y_i that fall into each interval. For S large, these should all be 0's and 1's.

The vector n has a multinomial distribution. The vector of expected values is $E(n) \equiv m$. The vector of probabilities for the cells is $(1/N)E(n) = p$. By assumption, $p_h \doteq f(t_h)/S$. Define $f \equiv [f(t_1),\ldots,f(t_S)]'$. We want to estimate the vector

$$
f \doteq Sp = \frac{S}{N}E(n) = \frac{S}{N}m.
$$

As in other aspects of nonparametric regression, we define the matrix Φ such that $\Phi = [\Phi_0,\ldots,\Phi_{s-1}]$ and $\Phi_j = [\phi_j(t_1),\ldots,\phi_j(t_S)]'$. Because the t_hs are equally spaced, there is little problem in taking $(1/\sqrt{S})\Phi$ to have orthonormal columns. Standard methods based on function series apparently use a linear model

$$
m = E(n) = \Phi\beta.
$$

Estimating via least squares gives

$$\hat{m} = \Phi\hat{\beta} = \frac{1}{S}\Phi\Phi'n$$

and

$$\hat{f} = \frac{S}{N}\hat{m} = \frac{S}{N}\Phi\hat{\beta} = \frac{1}{N}\Phi\Phi'n.$$

In particular, for $j = 0, \ldots, s-1$

$$\frac{S}{N}\hat{\beta}_j = \frac{1}{N}\Phi'_j n = \frac{1}{N}\sum_{h=1}^{S}\phi_j(t_h)n_h \doteq \frac{1}{N}\sum_{i=1}^{N}\phi_j(y_i).$$

The right-hand side is the estimator used in Efromovich (1999, Equation 3.1.4).

The complete density can be estimated by

$$\hat{f}(y) = \frac{S}{N}\sum_{j=0}^{s-1}\hat{\beta}_j\phi_j(y),$$

but this function need not integrate to 1 nor even be positive.

It would be more common to analyze count data using a log-linear model, say,

$$\log(m) = \log E(n) = \Phi\beta,$$

with estimate

$$\hat{f} = \frac{S}{N}\exp(\Phi\hat{\beta})$$

or more generally

$$\hat{f}(y) = \frac{S}{N}\exp\left[\sum_{j=0}^{s-1}\hat{\beta}_j\phi_j(y)\right].$$

One advantage of this is that the density estimates are forced to be positive, unlike the linear model estimates. Christensen (1997) discusses log-linear modeling. In particular, the log-linear model estimate of $\hat{\beta}$ is a solution

$$\Phi'[n - \exp(\Phi\hat{\beta})] = 0. \qquad (1.12.1)$$

With $\phi_0(x) \equiv 1$, a consequence of (1.12.1) is $(1/S)J'\hat{f} = 1$, so $\hat{f}(y)$ approximately integrates to 1. There is still no assurance that $\hat{f}(y)$ will integrate exactly to 1. In any case, it is a simple matter to standardize the estimate.

1.13 Exercises

The first six exercises reexamine *The Coleman Report* data of Table 1.8. The first 5 consider only two variables: y, the mean verbal test score for sixth graders, and x_3, the composite measure of socioeconomic status.

Exercise 1.13.1.   Rescale x_3 to make its values lie between 0 and 1. Plot the data. Using least squares, fit models with $p = 10$ using polynomials and cosines. Plot the regression lines along with the data. Which family works better on these data, cosines or polynomials?

Exercise 1.13.2.   Using $s = 8$, fit the *Coleman Report* data using Haar wavelets. How well does this do compared to the cosine and polynomial fits?

Exercise 1.13.3.   Based on the $s = 10$ polynomial and cosine models fitted in Exercise 1.13.1, use C_p to determine a best submodel for each fit. Plot the regression lines for the best submodels. Which family works better on these data, cosines or polynomials? Use C_p to determine the highest frequency needed when fitting sines and cosines.

Exercise 1.13.4.   Investigate whether there is a need to consider heteroscedastic variances with the *Coleman Report* data. If appropriate, refit the data.

Exercise 1.13.5.   Fit a cubic spline nonparametric regression to the *Coleman Report* data.

Exercise 1.13.6.   Fit a regression tree to the *Coleman Report* data using just variable x_4.

Exercise 1.13.7.   In Sect. 1.7 we set up the interpolating cubic spline problem as one of fitting a saturated linear model that is forced be continuous, have continuous first and second derivatives, and have 0 as the second derivative on the boundaries. In our discussion, the dots are only connected implicitly because a saturated model must fit every data point perfectly. Show that you can find the parameters of the fitted polynomials without using least squares by setting up a system of linear equations requiring the polynomials to connect the (x_i, y_i) dots along with satisfying the derivative conditions.

Exercise 1.13.8.   Fit a tree model to the battery data and compare the results to fitting Haar wavelets.

References

Breiman, L., Friedman, J. H., Olshen, R. A., & Stone, C. J. (1984). *Classification and regression trees*. Belmont, CA: Wadsworth.

Carroll, R. J., & Ruppert, D. (1988). *Transformations and weighting in regression*. New York: Chapman and Hall.

Christensen, R. (1989). Lack of fit tests based on near or exact replicates. *Annals of Statistics, 17*, 673–683.

Christensen, R. (1991). Small sample characterizations of near replicate lack of fit tests. *Journal of the American Statistical Association, 86*, 752–756.

Christensen, R. (1996). *Analysis of variance, design, and regression: Applied statistical methods*. London: Chapman and Hall.

Christensen, R. (1997). *Log-linear models and logistic regression* (2nd ed.). New York: Springer.

Christensen, R. (2011). *Plane answers to complex questions: The theory of linear models* (4th ed.). New York: Springer.

Christensen, R. (2015). *Analysis of variance, design, and regression: Linear modeling for unbalanced data* (2nd ed.). Boca Raton, FL: Chapman and Hall/CRC Press.

Christensen, R., Johnson, W., Branscum, A., & Hanson, T. E. (2010). *Bayesian ideas and data analysis: An introduction for scientists and statisticians*. Boca Raton, FL: Chapman and Hall/CRC Press.

Efromovich, S. (1999). *Nonparametric curve estimation: Methods, theory, and applications*. New York: Springer.

Eilers, P. H. C., & Marx, B. D. (1996). Flexible smoothing with b-splines and penalties (with discussion). *Statistical Science, 11*, 89–121.

Eubank, R. L. (1988). *Spline smoothing and nonparametric regression*. New York: Marcel Dekker.

Goldstein, M., & Smith, A. F. M. (1974). Ridge-type estimators for regression analysis. *Journal of the Royal Statistical Society, Series B, 26*, 284–291.

Green, P. J., & Silverman, B. W. (1994). *Nonparametric regression and generalized linear models: A roughness penalty approach*. London: Chapman and Hall.

Hart, J. D. (1997). *Nonparametric smoothing and lack-of-fit tests*. New York: Springer.

Hurvich, C. M., & Tsai, C.-L. (1995). Relative rates of convergence for efficient model selection criteria in linear regression. *Biometrika, 82*, 418–425.

Lenth, R. V. (2015). The case against normal plots of effects (with discussion). *Journal of Quality Technology, 47*, 91–97.

Loh, W.-Y. (2011). Classification and regression trees. *WIREs Data Mining and Knowledge Discovery, 1*, 14–23.

Montgomery, D. C., & Peck, E. A. (1982). *Introduction to linear regression analysis*. New York: Wiley.

Mosteller, F., & Tukey, J. W. (1977). *Data analysis and regression*. Reading, MA: Addison-Wesley.

Neill, J. W., & Johnson, D. E. (1989). A comparison of some lack of fit tests based on near replicates. *Communications in Statistics, Part A—Theory and Methods, 18*, 3533–3570.

Ogden, R. T. (1997). *Essential wavelets for statistical applications and data analysis*. Boston: Birkhäuser.

Reiss, P. T., Goldsmith, J., Shang, H. L., & Ogden, R. T. (2017). Methods for scalar-on-function regression. *International Statistical Review, 85*, 228–249.

Chapter 2
Penalized Estimation

Abstract The nonparametric methods of Chap. 1 are really highly parametric methods and suffer from issues of overfitting, i.e., fitting so many parameters to the data that the models lose their ability to make effective predictions. One way to stop overfitting is by using penalized estimation (regularization) methods. Penalized estimation provides an automated method of keeping the estimates from tracking the data more closely than is justified.

2.1 Introduction

In applications of linear model theory to situations where the number of model parameters is large relative to the sample size n, it is not uncommon to replace least squares estimates with estimates that incorporate a penalty on (some of) the regression coefficients. The nonparametric regression models of the previous chapter and *PA* Sect. 6.2.1 are germane examples. Penalty functions are often used to avoid *overfitting* a model. They can make it possible to use numbers of parameters that are similar to the number of observations without overfitting the model. As a tool to avoid overfitting, penalized estimation constitutes an alternative to variable selection, cf. *PA-V*, Chap. 14. Penalized estimation generally results in biased estimates (but not necessarily so biased as to be a bad thing). Chapter 1 and Christensen (2015, Chapter 8) contain some plots of overfitted models.

Penalized estimates are determined by adding some multiple of a nonnegative *penalty function* to the least squares criterion function and minimizing this new criterion function. For example, in the partitioned linear model, say,

$$Y = X\beta + Z\gamma + e, \quad \mathrm{E}(e) = 0, \quad \mathrm{Cov}(e) = \sigma^2 I, \tag{2.1.1}$$

if $\gamma_{s \times 1}$ is a high dimensional vector, one might minimize

$$(Y - X\beta - Z\gamma)'(Y - X\beta - Z\gamma) + k\,\mathscr{P}(\gamma), \tag{2.1.2}$$

© Springer Nature Switzerland AG 2019

R. Christensen, *Advanced Linear Modeling*, Springer Texts in Statistics,

https://doi.org/10.1007/978-3-030-29164-8_2

where $\mathscr{P}(\gamma)$ is a nonnegative penalty function and $k \geq 0$ is a *tuning parameter*. Obviously, if $k = 0$, the estimates are least squares estimates. Typical penalty functions are minimized at the vector $\gamma = 0$, so as k gets large, the penalty function dominates the minimization and the procedure, in some fashion, shrinks the least squares estimate of γ towards 0. Incorporating a penalty function is sometimes referred to as *regularization*.

For model (2.1.1) the estimates can be viewed as a combination of least squares estimation and penalized least squares estimation. As in *PA* Chap. 9, rewrite model (2.1.1) as

$$Y = X\delta + (I - M)Z\gamma + e$$

and rewrite the estimation criterion as

$$
\begin{aligned}
[Y - X\delta &- (I - M)Z\gamma]'[Y - X\delta - (I - M)Z\gamma] + k\mathscr{P}(\gamma) \\
&= [MY - X\delta]'[MY - X\delta] \\
&\quad + [(I - M)Y - (I - M)Z\gamma]'[(I - M)Y - (I - M)Z\gamma] + k\mathscr{P}(\gamma).
\end{aligned}
\tag{2.1.3}
$$

The parameter δ appears only in the first term of the second expression for the estimation criterion, so any minimizing estimate of δ has $X\hat{\delta} = MY$, i.e., $\hat{\delta}$ is least squares. The parameter γ only appears in the last two terms, so the penalized estimate of γ is, say, $\tilde{\gamma}_k$ that minimizes

$$[(I - M)Y - (I - M)Z\gamma]'[(I - M)Y - (I - M)Z\gamma] + k\mathscr{P}(\gamma).$$

If we first fit $Y = X\delta + e$ to obtain least squares residuals $\hat{e} \equiv (I - M)Y$, then $\tilde{\gamma}_k$ is a penalized estimate from fitting $\hat{e} = (I - M)Z\gamma + e$. Since $X\delta = X\beta + MZ\gamma$, the penalized estimate of $X\beta$ satisfies $X\hat{\delta} = X\tilde{\beta}_k + MZ\tilde{\gamma}_k$ or

$$X\tilde{\beta}_k = M(Y - Z\tilde{\gamma}_k),$$

which is analogous to a formula for least squares estimation in *PA* Chap. 9.

Rarely is there an obvious choice for k. Extending the idea of Hoerl and Kennard (1970), we can use a *trace* plot to pick k. Denote the penalized estimates for given k and the jth predictor variable as $\tilde{\gamma}_{kj}$. The trace plot is, for all j, a simultaneous plot of the curves defined by $(k, \tilde{\gamma}_{kj})$ as k varies. As mentioned, for $k = 0$ the $\tilde{\gamma}_{0j}$s are the least squares estimates and as k increases they typically all shrink towards 0. For the purpose of dealing with collinearity issues, Hoerl and Kennard suggested picking a small k for which the estimates settle down, i.e., stop varying wildly. Draper and van Nostrand (1979) conclude that for ridge regression the problems with picking k using trace plots outweigh their benefits. More modern methods of picking k include cross-validation and generalized cross-validation, cf. Hastie, Tibshirani, and Friedman (2016) or any number of other sources.

As discussed in *PA*, least squares is a geometric estimation criterion, not a statistical criterion, but least squares has many nice statistical properties. Penalized least squares is also a geometric criterion, not a statistical one. Unfortunately, it is harder to establish nice statistical properties for penalized estimates. That is *not* to say that they don't have any. Section 2.5 illustrates some geometry related to penalized least squares.

Many of the commonly used $\mathscr{P}$s penalize each variable the same amount. As a result, it is often suggested that the predictors in the model matrix Z should be standardized onto a common scale. If the height of my doghouse is a predictor variable, the appropriate regression coefficient depends a great deal on whether the height is measured in miles or microns. For a penalty function to be meaningful, it needs to be defined on an appropriate scale for each predictor variable.

Typically we assume that J, the vector of 1s, satisfies $J \in C(X)$, i.e., the $X\beta$ portion of the model contains an intercept or its equivalent. Frequently, Z is standardized in some way that does not change $C(J,Z)$ but so that $C(Z) \subset C(J)^\perp$, e.g., Z is replaced by $[I - (1/n)JJ']Z = Z - \bar{z}'J$, or more commonly Z is additionally modified so that $Z'Z$ also has 1's (or some other constant) down it's diagonal, cf. *PA-V* Sect. 13.1 (Christensen 2011, Section 15.1). The latter standardization puts all the regression variables in Z on the same scale, i.e., every column of Z has the same length, so it makes some sense to penalize their regression coefficients equally.

Any linear model can be written in the form (2.1.1) and any penalty function can be written as a function of γ alone. The partitioned model merely allows us to focus on penalizing some, but not all, of the coefficients. The same thing could be accomplished by defining a penalty function that nominally depends on both β and γ but actually only depends on γ. The most common form for (2.1.1) probably puts only the intercept in $X\beta$ and all other predictors in $Z\gamma$.

On occasion it may be simpler to think about fitting an unpartitioned model

$$Y = X\beta + e, \quad \mathrm{E}(e) = 0, \quad \mathrm{Cov}(e) = \sigma^2 I, \tag{2.1.4}$$

with estimation criterion

$$(Y - X\beta)'(Y - X\beta) + k\mathscr{P}(\beta). \tag{2.1.5}$$

One instance is the following argument.

If model (2.1.4) has multivariate normal errors, the least squares residuals will be independent of the penalized fitted values, so if (2.1.4) provides an adequate number of degrees of freedom for error, it may be advantageous to use the least squares residuals, rather than residuals based on the penalized estimates, to estimate the variance and to check model assumptions. To establish this, we merely need to show that the penalized estimates are a function of the least squares estimates. Using ideas similar to the proof of *PA* Theorem 2.2.1 with $\hat{\beta}$ a least squares estimate, write

$$\begin{aligned}
\|Y - X\beta\|^2 &= (Y - X\beta)'(Y - X\beta) \\
&= (Y - MY + MY - X\beta)'(Y - MY + MY - X\beta) \\
&= (Y - MY)'(Y - MY) + (MY - X\beta)'(MY - X\beta) \\
&= (Y - X\hat{\beta})'(Y - X\hat{\beta}) + (X\hat{\beta} - X\beta)'(X\hat{\beta} - X\beta) \\
&= (Y - X\hat{\beta})'(Y - X\hat{\beta}) + (\beta - \hat{\beta})'X'X(\beta - \hat{\beta}). \tag{2.1.6}
\end{aligned}$$

The estimation criterion (2.1.5) becomes

$$(Y - X\hat{\beta})'(Y - X\hat{\beta}) + (\beta - \hat{\beta})'X'X(\beta - \hat{\beta}) + k\mathscr{P}(\beta)$$

in which the first term $(Y - X\hat{\beta})'(Y - X\hat{\beta})$ does not involve β, so is irrelevant to the minimization, and the other terms depend on Y only through $\hat{\beta}$. Since the residuals $\hat{e} = (I - M)Y$ are independent of $\hat{\beta}$, the residuals are independent of the penalized estimate which must be a function of $\hat{\beta}$. Moreover, if the penalized estimate is a *linear* function of $\hat{\beta}$ (e.g. ridge regression), $\hat{e}$ is uncorrelated with the penalized estimate even without the assumption of multivariate normality. The decomposition of the squared distance in (2.1.6) will be used again in Sect. 2.5.1 to facilitate geometric interpretations.

2.1.1 Reparameterization and RKHS Regression: It's All About the Penalty

In traditional linear models it is well know that reparameterizations are irrelevant, i.e., two models for the same data, say, $Y = X_1\beta_1 + e$ and $Y = X_2\beta_2 + e$ are equivalent if $C(X_1) = C(X_2)$. In particular, least squares gives the same fitted values $\hat{Y}$ and residuals $\hat{e}$ for each model. Moreover, the least squares estimates for either β_1 or β_2 may not be uniquely defined, but we don't much care. In penalized least squares, if you use the same penalty function $\mathscr{P}(\cdot)$ for each of two equivalent models, you typically get different results, i.e., minimizing $\|Y - X_1\beta_1\|^2 + \mathscr{P}(\beta_1)$ does not give the same fitted values and residuals as minimizing $\|Y - X_2\beta_2\|^2 + \mathscr{P}(\beta_2)$. Moreover, incorporating the penalty function typically generates unique estimates, even when ordinary least squares does not. This reparameterization issue is a substantial one when using software that provides a default penalty function or even a menu of penalty functions.

To get equivalent results from equivalent models you need appropriate penalty functions. In particular, if X_1 and X_2 are both regression models so that $X_1 = X_2T$ for some invertible matrix T, then $\beta_2 = T\beta_1$ and minimizing $\|Y - X_1\beta_1\|^2 + \mathscr{P}(\beta_1)$ clearly gives the same fitted values and residuals as minimizing $\|Y - X_2\beta_2\|^2 + \mathscr{P}(T^{-1}\beta_2)$.

This discussion is particularly germane when applying the kernel trick as in Sect. 1.8.2. With two different reproducing kernels $R_1(\cdot, \cdot)$ and $R_2(\cdot, \cdot)$, for which $C(\tilde{R}_1) = C(\tilde{R}_2)$, the models $Y = \tilde{R}_1\gamma_1 + e$ and $Y = \tilde{R}_2\gamma_2 + e$ are reparameterizations of each other. Their least squares fits will be identical and with many kernels $Y = \hat{Y}_1 = \hat{Y}_2$ (when the x_i are distinct). If, to avoid overfitting, we use an off the shelf penalty function $\mathscr{P}(\cdot)$ to estimate the γ_is, that common penalty function will be entirely responsible for the differences between the fitted values $\tilde{Y}_1 \equiv \tilde{R}_1\tilde{\gamma}_1$ and $\tilde{Y}_2 \equiv \tilde{R}_2\tilde{\gamma}_2$ and the differences in other predictions made with the two models.

2.1.2 *Nonparametric Regression*

As discussed in the previous chapter, one approach to nonparametric regression of y on a scalar predictor x is to fit a linear model, say,

$$y_i = \beta_0 + \gamma_1 \phi_1(x_i) + \cdots + \gamma_s \phi_s(x_i) + \varepsilon_i$$

for known functions ϕ_j, e.g., polynomial regression. Penalized estimates are used in nonparametric regression to ensure smoothness. Penalized regression typically shrinks all regression estimates towards 0, some more than others when applied to nonparametric regression. Variable selection differs in that it shrinks the estimates of the eliminated variables to (not towards) 0 but lets least squares decide what happens to the estimates of the remaining variables.

The functions $\phi_j(x)$ in nonparametric regression are frequently subjected to some form of standardization (such as $\mathscr{L}^2$ normalization) when they are defined, thus obviating a strong need for further standardization of the vectors $\Phi_j \equiv [\phi_j(x_1), \cdots, \phi_j(x_n)]'$, especially when the x_is are equally spaced. For example, with x_is equally spaced from 0 to 1 and $\phi_j(x) = \cos(\pi j x)$, there is little need to standardize $Z \equiv [\Phi_1, \ldots, \Phi_s]$ further. When using simple polynomials $\phi_j(x) = x^j$, the model matrix should be standardized. When using the corresponding *Legendre polynomials* on equally spaced data, Z need not be.

In the context of nonparametric regression, not overfitting the model typically means ensuring appropriate smoothness. For example, with $\phi_j(x) = \cos(\pi j x)$, when j is large the cosine functions oscillate very rapidly, leading to *nonsmooth* or *noisy* behavior. Typically, with linear-approximation approaches to nonparametric regression, large j is indicative of more noisy behavior. We want to allow noisy behavior if the data require it, but we prefer smooth functions if they seem reasonable. It therefore makes sense to place larger penalties on the regression coefficients for large j. In other words, for large values of j we shrink the least squares estimate $\hat{\gamma}_j$ towards 0 more than when j is small.

2.2 Ridge Regression

Classical ridge regression provides one application of penalty functions. Write the multiple linear regression model $y_i = \beta_0 + \sum_{j=1}^{p-1} \beta_j x_{ij} + \varepsilon_i$ in matrix form as $Y = X\beta + e$ or, following *PA* Sect. 6.2, as

$$Y = J\beta_0 + Z\beta_* + e = [J, Z]\begin{bmatrix} \beta_0 \\ \beta_* \end{bmatrix} + e$$

where the elements of Z are x_{ij}, $i = 1, \ldots, n$, $j = 1, \ldots, p-1$. Relative to model (2.1.1), $J = X$, $\beta_0 = \beta$, and $\beta_* = \gamma$.

Classical ridge regression estimates β_0 and β_* by minimizing

$$[Y - J\beta_0 - Z\beta_*]'[Y - J\beta_0 - Z\beta_*] + k\beta_*'\beta_* \tag{2.2.1}$$

which amounts to using the penalty function

$$\mathscr{P}_R(\beta) \equiv \beta_*'\beta_* = \sum_{j=1}^{p-1} \beta_j^2.$$

Other than β_0, this penalizes each β_j the same amount, so it is important that the columns of Z be standardized to a common length or that they be defined in such a way that they are already nearly standardized. It is easy to see that the function (2.2.1) is the least squares criterion function for the model

$$\begin{bmatrix} Y \\ 0 \end{bmatrix} = \begin{bmatrix} J \\ 0 \end{bmatrix} \beta_0 + \begin{bmatrix} Z \\ \sqrt{k}I \end{bmatrix} \beta_* + \begin{bmatrix} e \\ \tilde{e} \end{bmatrix}.$$

Using ANCOVA to fit this augmented model by least squares yields the classical ridge regression estimates

$$\tilde{\beta}_{R*} = \{Z'[I - (1/n)JJ']Z + kI\}^{-1} Z'[I - (1/n)JJ']Y \quad \text{and} \quad \tilde{\beta}_{R0} = \bar{y}. - \bar{x}'.\tilde{\beta}_{R*}.$$

Exercise 2.1. Prove the formulae for $\tilde{\beta}_{R*}$ and $\tilde{\beta}_{R0}$.

The augmented regression model shows quite clearly that ridge regression is shrinking the regression parameters toward 0. The bottom part of the augmented model specifies

$$0 = \sqrt{k}\beta_* + \tilde{e},$$

so we are acting like 0 is an observation with mean vector $\sqrt{k}\beta_*$, which will shrink the estimate of β_* toward the 0 vector. Note that if $\sqrt{k}$ is already a very small number, then one expects $\sqrt{k}\beta_*$ to be small, so the shrinking effect of the artificial observations 0 will be small. If $\sqrt{k}$ is large, say 1, then we are acting like we have seen that β_* is near 0 and the shrinkage will be larger.

Like many sources, the discussion of classical ridge regression in *PA-V* Sect. 13.3 (Christensen 2011, 15.3) penalizes all the parameters and leaves any mean correction implicit, which results in a simpler ridge estimate. For model (2.1.4), minimizing (2.1.5) with $\mathscr{P}(\beta) = \beta'\beta$, the ridge estimate is $\tilde{\beta}_R = (X'X + kI)^{-1}X'Y$, which can be arrived at by least squares fitting of

$$\begin{bmatrix} Y \\ 0 \end{bmatrix} = \begin{bmatrix} X \\ \sqrt{k}I \end{bmatrix} \beta + \begin{bmatrix} e \\ \tilde{e} \end{bmatrix}.$$

2.2.1 Generalized Ridge Regression

Returning to the partitioned model (2.1.1), *generalized ridge regression* takes the form of a penalty

$$\mathscr{P}_{GR}(\gamma) \equiv \gamma'Q\gamma,$$

where Q is a nonnegative definite matrix. Most often Q is diagonal so that $\mathscr{P}_{GR}(\gamma) = \sum_{j=1}^{s} q_{jj}\gamma_j^2$. In *PA-V* Sect. 13.3 (Christensen 2011, 15.3), the generalized ridge estimates used $k_j \equiv kq_{jj}$.

We can minimize

$$(Y - X\beta - Z\gamma)'(Y - X\beta - Z\gamma) + k\gamma'Q\gamma \qquad (2.2.2)$$

using the least squares fit to the augmented linear model

$$\begin{bmatrix} Y \\ 0 \end{bmatrix} = \begin{bmatrix} X & Z \\ 0 & \sqrt{k}\tilde{Q} \end{bmatrix} \begin{bmatrix} \beta \\ \gamma \end{bmatrix} + \begin{bmatrix} e \\ \tilde{e} \end{bmatrix} \qquad (2.2.3)$$

where $Q = \tilde{Q}'\tilde{Q}$. Note that if the jth row and column of Q contain only 0s, γ_j is not penalized. That can only happen when the jth column of $\tilde{Q}$ is 0, in which case we could redefine γ_j as part of β and redefine Q eliminating the row and column of 0s.

The least squares estimates of model (2.2.3) minimize the quantity

$$\begin{bmatrix} Y - X\beta - Z\gamma \\ -\sqrt{k}\tilde{Q}\gamma \end{bmatrix}' \begin{bmatrix} Y - X\beta - Z\gamma \\ -\sqrt{k}\tilde{Q}\gamma \end{bmatrix} = (Y - X\beta - Z\gamma)(Y - X\beta - Z\gamma) + k\gamma'Q\gamma.$$

Using ideas from *PA* Chap. 9 with M the perpendicular projection operator onto $C(X)$, it is not difficult to show that

$$\tilde{\gamma} = [Z'(I - M)Z + kQ]^- Z'(I - M)Y; \qquad \tilde{\beta} = [X'X]^- X'(Y - Z\tilde{\gamma}) \qquad (2.2.4)$$

are least squares estimates for model (2.2.3) and thus are generalized ridge estimates. In most regression problems the generalized inverses in (2.2.4) can be replaced by true inverses.

Alternatively, when Q is nonsingular, we can minimize (2.2.2) using the generalized least squares fit to the augmented linear model

$$\begin{bmatrix} Y \\ 0 \end{bmatrix} = \begin{bmatrix} X & Z \\ 0 & I \end{bmatrix} \begin{bmatrix} \beta \\ \gamma \end{bmatrix} + \begin{bmatrix} e \\ \tilde{e} \end{bmatrix}, \; \mathrm{E}\begin{bmatrix} e \\ \tilde{e} \end{bmatrix} = \begin{bmatrix} 0 \\ 0 \end{bmatrix}, \; \mathrm{Cov}\begin{bmatrix} e \\ \tilde{e} \end{bmatrix} = \sigma^2 \begin{bmatrix} I & 0 \\ 0 & (1/k)Q^{-1} \end{bmatrix}. \qquad (2.2.5)$$

The generalized least squares estimates of model (2.2.5) minimize the quantity

$$\begin{bmatrix} Y - X\beta - Z\gamma \\ -\gamma \end{bmatrix}' \begin{bmatrix} I & 0 \\ 0 & kQ \end{bmatrix} \begin{bmatrix} Y - X\beta - Z\gamma \\ -\gamma \end{bmatrix}$$
$$= (Y - X\beta - Z\gamma)'(Y - X\beta - Z\gamma) + k\gamma'Q\gamma,$$

which is the generalized ridge regression criterion (2.2.2).

Green and Silverman (1994, Section 3.6) discuss different choices for Q. Those choices generally follow the pattern of more shrinkage for γ_js that incorporate noisier behavior into the model and are discussed in Sect. 2.2.1.

The generalized ridge augmented model (2.2.5) becomes a classical ridge augmented model when $Q = I$. It then involves fitting the simplest form of generalized least squares which is weighted least squares wherein the weights vector is

$w' = [J_n', kJ_s']$. Finding a computer program to fit generalized least squares can be difficult but almost all regression software fits weighted least squares. This idea will be useful in Chap. 13 when we generalize ridge regression to binomial generalized linear models.

Section 5.3 examines a relationship between ridge regression and fitting models with random effects.

2.2.2 Picking k

As alluded to in Sect. 2.1, to pick k in (2.2.1) for classical ridge regression, Hoerl and Kennard (1970) suggested using a *ridge trace*. If the classical ridge regression estimates are denoted $\tilde{\beta}_{kj}$ for the jth predictor variable and tuning parameter k, the ridge trace is, for all j, a simultaneous plot of the curves defined by $(k, \tilde{\beta}_{kj})$ as k varies. For $k = 0$ the $\tilde{\beta}_{0j}$s are the least squares estimates and as k increases they all shrink towards 0 (except β_{k0} which is not penalized). The idea is to pick k just big enough to stabilize the regression coefficients. Hoerl and Kennard's original idea was using ridge to deal with high collinearity in $[I - (1/n)JJ']Z$, rather than using shrinkage as an alternative to variable selection. As mentioned earlier, the trace idea applies to all penalized regression but Draper and van Nostrand (1979) found it lacking for ridge regression.

More recently, cross-validation and generalized cross-validation have been used to pick k, for example see Green and Silverman (1994, Sections 3.1 and 3.2).

2.2.3 Nonparametric Regression

Again we use the notation from Chap. 1 for simple nonparametric regression but the ideas extend immediately to multiple nonparametric regression.

Assuming that $f(x) = \sum_{j=0}^{p-1} \beta_j \phi_j(x)$, the generalized ridge regression estimate minimizes

$$(Y - \Phi\beta)'(Y - \Phi\beta) + k\beta'Q\beta,$$

where Q is a nonnegative definite matrix of penalties for unsmoothness and k is a tuning parameter which for many purposes is considered fixed but which ultimately is estimated. It seems most common not to penalize the intercept parameter when one exists, but that is optional.

In the special case in which $\frac{1}{\sqrt{n}}\Phi$ has orthonormal columns and Q is diagonal, it is not difficult to see that the generalized ridge estimate is

$$\tilde{\beta} = [nI + kD(q_{ii})]^{-1}\Phi'Y$$
$$= [D(n + kq_{ii})]^{-1}\Phi'Y$$
$$= D\left(\frac{n}{n + kq_{ii}}\right)\hat{\beta},$$

where $\hat{\beta}$ is the least squares estimate. By letting $\alpha = k/n$, we get

$$\tilde{\beta}_j = \frac{1}{1 + \alpha q_{ii}} \hat{\beta}_j,$$

which shows quite clearly the nature of the shrinkage.

A frequently used penalty matrix (that does not seem to require Φ to have columns of near equal length) has

$$Q = [q_{rs}], \qquad q_{rs} = \int_0^1 \phi_r^{(2)}(x) \phi_s^{(2)}(x) dx$$

with $\phi_r^{(2)}(x) \equiv \mathbf{d}^2 \phi_r(x)$ is the second derivative of $\phi_r(x)$. This penalty function does not depend on the data (X, Y). Whenever $\phi_0 \equiv 1$, $\phi_0^{(2)} \equiv 0$, so the first row and column of the second derivative Q will be 0. This places no penalty on the intercept and we could choose to think of penalizing a partitioned model.

Clearly, any constant multiple of the matrix Q works equivalently to Q if we make a corresponding change to k. If we use the cosine basis of (1.2.2), the second derivative matrix is proportional to

$$Q = \text{Diag}(0, 1^4, 2^4, \ldots, (s-1)^4).$$

For the sines and cosines of (1.2.3),

$$Q = \text{Diag}(0, 1^4, 1^4, 2^4, 2^4, \ldots).$$

In both cases, we have the diagonal matrix form of generalized ridge regression. Moreover, it is clear that the terms getting the greatest shrinkage are the terms with the largest values of j in (1.2.2) and (1.2.3), i.e., the highest frequency terms.

EXAMPLE 2.2.1.   For the voltage drop data of Chap. 1, using cosines with $j = 0, \ldots, 10$ and least squares, the estimated regression equation is

$$y = 11.4 - 1.61c_1 - 3.11c_2 + 0.468c_3 + 0.222c_4 + 0.196c_5$$
$$+ 0.156c_6 + 0.0170c_7 + 0.0799c_8 + 0.0841c_9 + 0.148c_{10}.$$

Using the generalized ridge regression augmented model

$$\begin{bmatrix} Y \\ 0 \end{bmatrix} = \begin{bmatrix} \Phi \\ \sqrt{k}\tilde{Q} \end{bmatrix} \beta + \begin{bmatrix} e \\ \tilde{e} \end{bmatrix}$$

with $\sqrt{k} = 0.2$ and $\tilde{Q} = \text{Diag}(0, 1, 4, 9, \ldots, 100)$, the estimated regression equation is

$$y = 11.4 - 1.60c_1 - 3.00c_2 + 0.413c_3 + 0.156c_4 + 0.0925c_5$$
$$+ 0.0473c_6 + 0.0049c_7 + 0.0102c_8 + 0.0068c_9 + 0.0077c_{10}.$$

Note the shrinkage towards 0 of the coefficients relative to least squares, with more shrinkage for higher values of j.

Figure 2.1 gives the data along with the generalized ridge regression fitted cosine curve using $j = 0, \ldots, 10$. With $k = 0.04$, the plot is very similar to the unpenalized cosine curve using $j = 0, \ldots, 6$ which is also plotted. Defining R^2 as the squared correlation between the observations and the fitted values, cf. *PA* Chap. 6, the ridge regression gives $R^2 = 0.985$ which is less than the value 0.988 from the least squares fit with 6 cosines.

$\square$

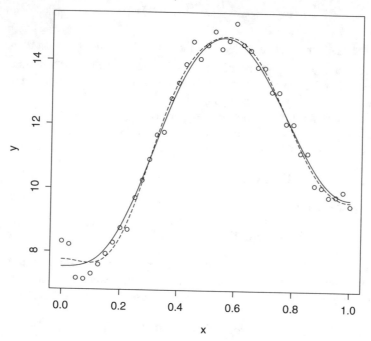

Fig. 2.1 Solid: Generalized ridge regression cosine fit with $k = 0.04$, $s - 1 = 10$ and second derivative weights for the battery data. Dashed: Least squares cosine fit with $s - 1 = 6$

Exercise 2.2. Perform a classical ridge regression with equal weights on the 10 term cosine model and compare the results to those from Example 2.2.1.

The second derivative penalty approach is worthless for Haar wavelets because it gives $Q = 0$ and thus the least squares estimates. Theoretically, one could use penalized least squares with other wavelets. The integral of the product of the second derivatives would be difficult to find for many wavelets. Fitting functions with small support inherently makes the fitted functions less smooth. Choosing how small to make the supports, e.g. choosing how many wavelets to fit, is already a choice of how smooth to make the fitted function. In such cases a smoothing penalty associated with ϕ_j should increase as the size (length, area, volume) of the support of ϕ_j gets smaller.

Exercise 2.3.   Find the second derivative penalty function matrix Q for the cubic spline model of the form (1.7.2) with 3 interior knots. Find Q for the corresponding linear spline model.

In Sect. 1.7.1 we discussed connecting the dots using splines by fitting a linear model $E(Y) = X\beta$ that incorporates linear constraints. We also discussed the reparameterized reduced model $E(Y) = X_0\gamma$ that incorporated the spline smoothness linear constraints into the original linear model. To introduce more smoothing (i.e., smoothing at the observed data), we can perform generalized ridge regression on the reduced model to obtain

$$\tilde{\gamma} = (X_0'X_0 + kQ)^{-1}X_0'Y.$$

Note that when we back transform to $\tilde{\beta} = U\tilde{\gamma}$, $\tilde{\beta}$ satisfies all of the constraints necessary to make the fitted spline function smooth. Green and Silverman (1994, Section 2.1) give an appropriate matrix Q based on a roughness penalty. ($Q = X_0'KX_0$, where they define K in their Eq. (1.4).)

If x is a vector, the second derivative of $\phi_j(x)$ is a square matrix as is the product of the second derivatives for different j. One might use something like the determinant of the integral of the matrix product to define Q.

2.3 Lasso Regression

Currently, a very popular penalty function is Tibshirani's (1996) *lasso (least absolute shrinkage and selection operator)*,

$$\mathscr{P}_L(\gamma) \equiv \sum_{j=1}^{s} |\gamma_j| \equiv \|\gamma\|_1. \tag{2.3.1}$$

The book by Hastie, Tibshirani, and Wainwright (2015) provides a wealth of information on this procedure. For applications with $p+s > n$ see Bühlmann and van de Geer (2011) or Bühlmann, Kalisch, and Meier (2014).

Because the lasso penalty function is not a quadratic form in γ, unlike ridge regression the estimate cannot be obtained by fitting an augmented linear model. Lasso estimates can be computed efficiently for a variety of values k using a modification of the *LARS* algorithm of Efron et al. (2004).

Less computationally efficient than LARS, but easier to understand, is an algorithm that involves obtaining estimates $\tilde{\beta}^{h+1}$, $\tilde{\gamma}^{h+1}$ for model (2.1.1) by repeatedly fitting

$$\begin{bmatrix} Y \\ 0 \end{bmatrix} = \begin{bmatrix} X & Z \\ 0 & I \end{bmatrix} \begin{bmatrix} \beta \\ \gamma \end{bmatrix} + \begin{bmatrix} e \\ \tilde{e} \end{bmatrix}, \ \mathrm{E}\begin{bmatrix} e \\ \tilde{e} \end{bmatrix} = \begin{bmatrix} 0 \\ 0 \end{bmatrix}, \ \mathrm{Cov}\begin{bmatrix} e \\ \tilde{e} \end{bmatrix} = \sigma^2 \begin{bmatrix} I & 0 \\ 0 & (1/k)Q_h^{-1} \end{bmatrix},$$

wherein Q_h is a diagonal matrix with elements $q_{jj} = 1/|\tilde{\gamma}_j^h|$. This has a penalty function

$$\mathscr{P}(\gamma) = \sum_{j=1}^{s} \frac{\gamma_j^2}{|\tilde{\gamma}_j^h|} \doteq \sum_{j=1}^{s} |\gamma_j|.$$

When $|\tilde{\gamma}_j^h|$ gets small, γ_j becomes very highly penalized, thus forcing $\tilde{\gamma}^{h+1}$ even closer to 0.

As the actual name (not the acronym) suggests, one of the benefits of the lasso penalty is that it automates variable selection. Rather than gradually shrinking all regression coefficients towards 0 like ridge regression, lasso can make some of the regression coefficients collapse to 0.

Just as *PA-V* Sect. 13.3 (Christensen 2011, 15.3) used canonical regression to explore ridge estimation, we can use canonical regression to explicate the behavior of lasso regression. Rather than analyzing model (2.1.1), consider the standard linear model $Y = X\beta + e$ and its canonical version

$$Y_* = \begin{bmatrix} L \\ 0 \end{bmatrix} \gamma + e,$$

where $L = D(\lambda_j)$ is positive definite. Applying the lasso criterion to canonical regression we need to minimize

$$\sum_{j=1}^{p} \left[(y_{*j} - \lambda_j \gamma_j)^2 + k|\gamma_j| \right].$$

Because of the simple structure of canonical regression, the lasso criterion acts independently on each coefficient. Without loss of generality, assume $y_{*j} > 0$. Clearly, $\gamma_j < 0$ will not minimize the criterion, because $\gamma_j = 0$ will be better. We therefore want to minimize $(y_{*j} - \lambda_j \gamma_j)^2 + k\gamma_j$ for $\gamma_j \geq 0$.

With $\hat{\gamma}_j = y_{*j}/\lambda_j$ being the least squares estimate, a little bit of work shows that the derivative of the criterion function with respect to γ_j is zero at $\gamma_j = \hat{\gamma}_j - k/2\lambda_j^2$. However, if $\hat{\gamma}_j < k/2\lambda_j^2$, the critical point is outside the domain of the function, so the minimum must occur at the boundary. Therefore, the lasso estimate is

$$\tilde{\gamma}_{Lj} = \begin{cases} \hat{\gamma}_j - k/2\lambda_j^2, & \text{if } \hat{\gamma}_j \geq k/2\lambda_j^2 \\ 0, & \text{if } |\hat{\gamma}_j| < k/2\lambda_j^2 \\ \hat{\gamma}_j + k/2\lambda_j^2, & \text{if } \hat{\gamma}_j \leq -k/2\lambda_j^2 . \end{cases} \tag{2.3.2}$$

Clearly, if the least squares estimate is too close to 0, the lasso estimate is zero and the variable is effectively removed from the model.

The lasso penalty (2.3.1) treats every coefficient the same. An obvious modification of lasso to penalize coefficients at different rates has

$$\mathscr{P}_{GL}(\gamma) = \sum_{j=0}^{s} q_{jj}|\gamma_j|$$

with q_{jj} often increasing in j.

Christensen (2015, Section 10.5) contains an example of the lasso applied to a 5 predictor regression problem. (The *Coleman Report* data.) Here we illustrate its use in nonparametric regression.

EXAMPLE 2.3.1.   For the battery data of Chap. 1, Fig. 2.2 shows the least squares cosine fits for $p - 1 = 6, 30$ and the R package lasso2's default fit except that, with equally spaced cosine predictors, the predictors were not standardized. (I also looked at the standardized version and it made little difference.) The default lasso fit has $k = 12.2133$, which is a lot of shrinkage. (The default is actually $\delta = 0.5\|\hat{\beta}_*\|_1$ where $\hat{\beta}_*$ is the $p - 1 = 30$ least squares estimate vector without the intercept, $\|\hat{\beta}_*\|_1 \equiv \sum_{j=1}^{30}|\hat{\beta}_j|$, and δ is defined in Sect. 2.5.)

The default is a shockingly bad fit. It gives $R^2 = 0.951$, which is poor for this problem. It has zeroed out too many of the cosine terms. A more reasonable lasso fit is given in Fig. 2.3. The fit in Fig. 2.3 has nonzero coefficients on precisely the first six cosine terms (and the constant) and it gives $R^2 = 0.981$, which cannot be greater than the R^2 provided by the least squares fit on the six cosine terms. Unlike our ridge regression example, in neither of the lasso fits have we put larger penalties on more noisy variables.   □

Exercise 2.4.   Fit the 30 cosine model using generalized lasso with the second derivative weights q_{jj} defined as in the ridge regression section.

2.4 Bayesian Connections

The augmented model (2.2.3) used to find ridge regression estimates is a special case of the augmented model used to fit Bayesian linear models in *PA-V* Section 2.10 (Christensen 2011, Section 2.9), see also Christensen, Johnson, Branscum, and Hanson (2010). Quite generally, penalized least squares estimates can be viewed as the mode of the posterior distribution of β when the penalty function is used to determine a specific prior distribution for β given σ^2. We begin by looking at a standard linear model and conclude with the partitioned model (2.1.1).

The likelihood function for $Y = X\beta + e$ with independent homoscedastic normal data is

$$L(\beta, \sigma^2) = (2\pi)^{-n/2}[\det(\sigma^2 I)]^{-1/2}\exp\left[-(Y - X\beta)'(Y - X\beta)/2\sigma^2\right].$$

We take a (possibly improper) prior density of the form $\pi(\beta, \sigma^2) \equiv \pi_1(\beta|\sigma^2)\pi_2(\sigma^2)$ where the conditional density of β given σ^2 is written as

$$\pi_1(\beta|\sigma^2) = h(\sigma^2)\exp\left[-k\,\mathscr{P}(\beta)/2\sigma^2\right],$$

with $\mathscr{P}(\beta)$ once again being the nonnegative penalty function. The posterior is proportional to the likelihood times the prior, so it has the form

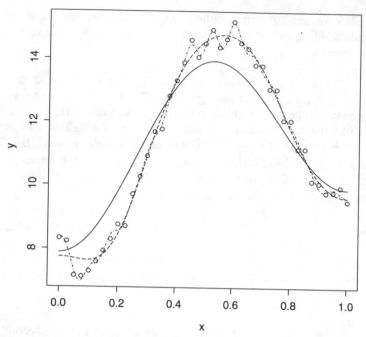

Fig. 2.2 Solid: Lasso cosine fit with $k = 12.2133$ ($\delta = 0.5\|\hat{\beta}_*\|_1$), $p - 1 = 30$. Dot-dash: Least squares cosine fit with $p - 1 = 30$. Dashed: Least squares cosine fit with $p - 1 = 6$

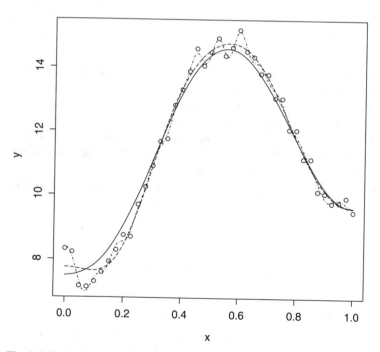

Fig. 2.3 Solid: Lasso cosine fit with $k = 3.045046$ ($\delta = 0.7\|\hat{\beta}_*\|_1$), $p - 1 = 30$. Dot-dash: Least squares cosine fit with $p - 1 = 30$. Dashed: Cosine least squares fit with $p - 1 = 6$

$$\pi(\beta, \sigma^2 | Y) \propto (\sigma^2)^{-n/2} \pi_2(\sigma^2) h(\sigma^2) \times$$

$$\exp\left\{-\frac{1}{2\sigma^2}\left[(Y - X\beta)'(Y - X\beta) + k\,\mathscr{P}(\beta)\right]\right\}.$$

The posterior mode consists of the values $\tilde{\beta}$, $\tilde{\sigma}^2$ that maximize the posterior. Similar to finding maximum likelihood estimates in a standard linear model, $\tilde{\beta}$ minimizes $[(Y - X\beta)'(Y - X\beta) + k\,\mathscr{P}(\beta)]$ regardless of the value of σ^2. Thus the posterior mode of β is also the penalized least squares estimate. A further prior can be placed on k.

Ridge regression amounts to placing a normal prior on β and using the one number that is the posterior mean, median, and mode as an estimate of β. In particular, the generalized ridge estimate devolves from the prior distribution

$$\beta | \sigma^2 \sim N\left(0, \frac{\sigma^2}{k} Q^{-1}\right).$$

When Q is diagonal, large penalties clearly correspond to small prior variances, i.e., strong prior beliefs that β_j is near the prior mean of 0. Classical ridge regression uses independent, homoscedastic, mean 0 normal priors on the β_js given σ, i.e., $\beta_j \sim N(0, \sigma^2 k)$, except possibly a flat prior on the intercept β_0.

Lasso-regression estimates are the posterior mode of β when placing independent, homoscedastic, mean 0 Laplace (double exponential) priors on the β_js given σ, except possibly a flat prior on the intercept β_0. Given the discontinuous nature of the lasso minimization problem, it is not surprising that technical difficulties can arise. Park and Casella (2008) provide a good discussion, but use a slightly different prior.

An alternative Bayesian method for avoiding overfitting is *thresholding*, see Smith and Kohn (1996), Clyde and George (2004), or Christensen et al. (2010, Section 15.2). The idea is to put positive prior probability on each regression coefficient being 0, so there will be positive, although perhaps very small, posterior probability of it being 0. For example, with a linear-approximation nonparametric regression model, a form of generalized ridge regression corresponds to independent normal priors on β_j with mean 0 and a variance σ^2/kq_{jj}, decreasing in j when high j indicates noisier behavior. Instead, a thresholding model prior might write

$$\beta_j \equiv \delta_j \beta_j^*$$

with $\beta_j^* \sim N(0, \sigma^2/k)$ independent of $\delta_j \sim \text{Bern}(p_j)$, i.e., $\Pr[\delta_j = 1] = p_j = 1 - \Pr[\delta_j = 0]$ where p_j decreases with j. This obviously makes it harder, but not impossible, for β_j to be nonzero as j increases (unless $p_j = 0$).

For the partitioned model (2.1.1) with multivariate normal data,

$$L(\beta, \gamma, \sigma^2) = (2\pi)^{-n/2} [\det(\sigma^2 I)]^{-1/2} \exp\left[-(Y - X\beta - Z\gamma)'(Y - X\beta - Z\gamma)/2\sigma^2\right].$$

We take a prior density of the form $\pi(\beta, \gamma, \sigma^2) \equiv \pi_1(\gamma|\sigma^2)\pi_2(\sigma^2)$ where the conditional density of γ given σ^2 is written as

$$\pi_1(\gamma|\sigma^2) = h(\sigma^2)\exp\left[-k\mathscr{P}(\gamma)/2\sigma^2\right],$$

with $\mathscr{P}(\gamma)$ once again being the nonnegative penalty function. Implicitly, this prior puts an improper flat prior on $\beta|\gamma, \sigma^2$, i.e., $\pi(\beta|\gamma, \sigma^2)$ is a constant. Again, the posterior mode equals the penalized least squares estimate.

2.5 Another Approach

For an unpartitioned linear model we can think about penalized least squares estimation as minimizing

$$\|Y - X\beta\|^2 + k\mathscr{P}(\beta) \tag{2.5.1}$$

for some tuning parameter $k \geq 0$. Alternatively, the procedure can be defined as choosing β to minimize the least squares criterion

$$\|Y - X\beta\|^2 \tag{2.5.2}$$

subject to a restriction on the regression coefficients,

$$\mathscr{P}(\beta) \leq \delta. \tag{2.5.3}$$

We want to establish the equivalence of these two procedures and explore the geometry of the alternative procedure. In penalized regression, we do not have a good reason for choosing any particular δ in (2.5.3), so we look at all possible values of δ or, more often and equivalently, all possible values of k in (2.5.1). In practice we will want to introduce the refinements associated with model (2.1.1) but for simplicity we examine the unpartitioned model.

2.5.1 Geometry

The restricted least squares problem of minimizing (2.5.2) subject to the inequality constraint (2.5.3) lends itself to a geometric interpretation. Our discussion is reasonably general but most illustrations are of the lasso in two dimensions. For simplicity, we examine a standard linear model $Y = X\beta + e$ but in practice penalized regression is always applied to some version of the partitioned model (2.1.1), even if the first term in the partition corresponds only to an intercept term.

To explore the geometry, we want to facilitate our ability to create contour maps of the least squares criterion surface as a function of β. Using the decomposition of $\|Y - X\beta\|^2$ given in (2.1.6), rewrite (2.5.2) in terms of a quadratic function of β minimized at the least squares estimate $\hat{\beta}$ plus a constant. The first term of the last

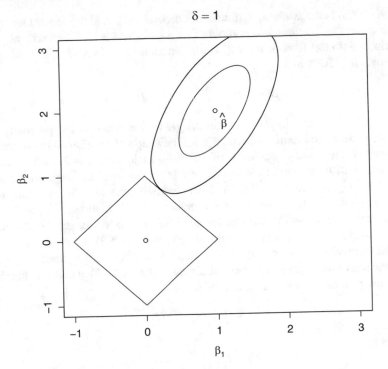

Fig. 2.4 Lasso shrinkage without variable selection

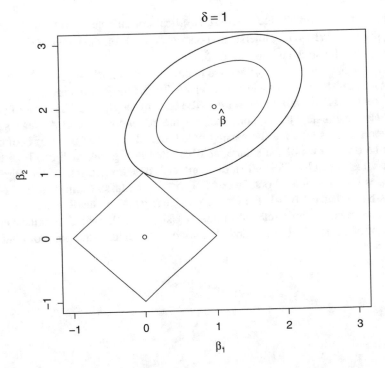

Fig. 2.5 Lasso shrinkage and variable selection

line in (2.1.6) is the constant that does not depend on β and the second term, since it is a quadratic function, has contours that are ellipsoids in β centered at $\hat{\beta}$. The function minimum is at $\hat{\beta}$, for which the function value is SSE. The contours are ellipsoids in β for which

$$SSE + (\beta - \hat{\beta})'X'X(\beta - \hat{\beta}) = D$$

for some D. As D gets larger, the contours get farther from $\hat{\beta}$. The geometry of ellipsoids is discussed more in Sect. 14.1.3. The shape of an ellipsoid is determined by the eigenvalues and eigenvectors of $X'X$—the major axis is in the direction of the eigenvector with the largest eigenvalue and is proportional in length to the square root of the eigenvalue. Note that, with multivariate normal data, each ellipsoid is also the confidence region for β corresponding to some confidence level. The least squares estimate subject to the constraint (2.5.3) is a β vector on the smallest elliptical contour that intersects the region defined by (2.5.3). Of course if the least squares estimates already satisfy (2.5.3), there is nothing more to find.

If the least squares estimate does not already satisfy (2.5.3), a multiple regression lasso that penalizes the intercept minimizes

$$SSE + (\beta - \hat{\beta})'X'X(\beta - \hat{\beta})$$

subject to

$$\sum_{j=0}^{p-1} |\beta_j| = \delta.$$

We need a β vector on the smallest elliptical contour that intersects the region $\sum_{j=0}^{p-1} |\beta_j| = \delta$. Where that intersection occurs depends on the value of $\hat{\beta}$, the orientation of the ellipsoid, and the size of δ.

For $y_i = \beta_1 x_{i1} + \beta_2 x_{i2} + \varepsilon_i$, the lasso penalty constraint $|\beta_1| + |\beta_2| \leq \delta$ is a square (diamond) centered at (0,0) with diameter 2δ. To find the lasso estimate, grow the ellipses centered at $\hat{\beta}$ until they just touch the edge of the square. The point of contact is the lasso estimate, i.e., the point that has the minimum value of the least squares criterion (2.5.2) subject to the penalty constraint (2.5.3). The point of contact can either be on the face of the square, as illustrated in Fig. 2.4, or it can be a corner of the square as in Fig. 2.5. When the contact is on a corner, one of the regression estimates has been zeroed out. In Fig. 2.5, $\delta = 1$, the lasso estimate of β_1 is 0 and the lasso estimate of β_2 is 1. For classical ridge regression, the diamonds in the two figures are replaced by circles of radius 1. Using a circle would definitely change the point of contact in Fig. 2.4 and almost certainly change the point of contact in Fig. 2.5.

2.5.1.1 More Lasso Geometry

In two dimensions the lasso estimate feels easy to find. With $\delta = 1$ and $\hat{\beta}$ in the first quadrant like it is in the two figures, the lasso estimate feels like it should be $(1,0)'$ or $(0,1)'$ or it should be the least squares estimate subject to the linear constraint $\beta_1 + \beta_2 = 1$. Finding the least squares estimate subject to a linear equality constraint is straightforward. It is precisely what one needs to do to find spline estimates as in the previous chapter. Unfortunately, things are not that simple. Figure 2.6 shows that it is possible for the lasso estimate of β_1 to be negative even when the least squares estimate is positive. And things get much more complicated in higher dimensions.

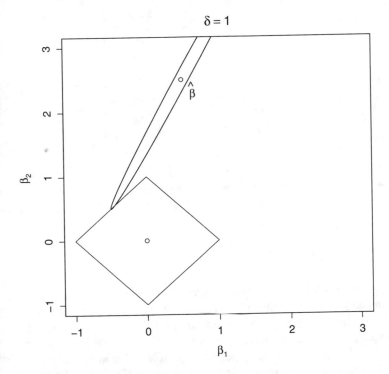

Fig. 2.6 Lasso sign change

Figures 2.7, 2.8 and 2.9 illustrate the geometry behind a trace plot. Remember that varying δ is equivalent to varying k, although the exact relationship is not simple. With both least squares $\hat{\beta}_j$s positive as in Fig. 2.7 and with δ large enough, the lasso estimate is just the least squares estimate constrained to be on $\beta_1 + \beta_2 = \delta$, unless δ is big enough that $\hat{\beta}_1 + \hat{\beta}_2 \leq \delta$ in which case least squares is lasso. Figure 2.8 has smaller δs than Fig. 2.7 but both plots in its top row have the same δ with the second plot being a closeup. The bottom row of Fig. 2.8 shows that as δ decreases, the penalized estimates remain on $\hat{\beta}_1 + \hat{\beta}_2 = \delta$ until β_1 becomes 0. The top row of

Fig. 2.9 has the lasso estimate of β_1 zeroed out for two additional δs. The bottom row shows the lasso estimate becoming negative.

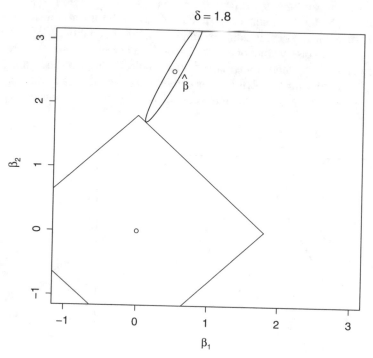

Fig. 2.7 Lasso trace $\delta = 1.8$

In three dimensions the lasso geometry is of throwing an American football (or a rugby ball) at an octohedron. Technically, the football should have someone sitting on it and, instead of throwing the football, we should blow it up until it hits the octohedron. The squashed football denotes the ellipsoids of the least squares criterion. The octohedron, see Fig. 2.10, is the lasso penalty region and should be centered at 0. The octohedron has 8 sides consisting of isosceles triangles, 12 edges between the sides, and 6 corners. The football can hit any of these 26 features. If we knew which of the 26 features the ellipsoid was hitting, it would be easy to find the restricted least squares estimate because it would be the least squares estimate subject to a set of linear constraints. The problem is knowing which of the 26 features the ellipsoid is hitting. If the ellipsoid hits a corner, two regression estimates are zeroed out and the third takes a value $\pm\delta$. If it hits an edge, one estimate is zeroed out and the other two are shrunk towards but not to 0. If it hits a surface, no estimates are zeroed but all are shrunk.

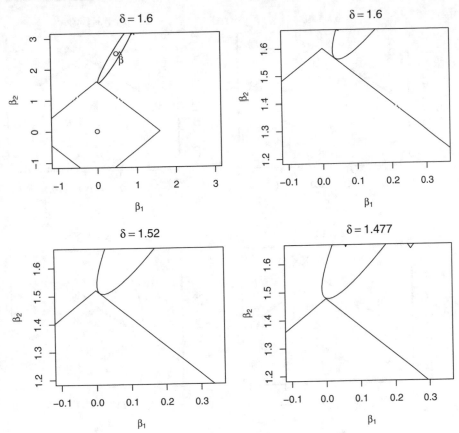

Fig. 2.8 Lasso trace

Typically, if you make δ big enough $\mathscr{P}(\hat{\beta}) \leq \delta$, so the penalty function has no effect on the least squares estimates. As soon as $\delta < \mathscr{P}(\hat{\beta})$, penalized least squares should be different from least squares. For the lasso, if all the elements of $\hat{\beta}$ are positive and δ is below, but sufficiently close to, $\sum_{j=0}^{p-1} |\hat{\beta}_j|$, the lasso estimate equals the least squares estimate subject to the linear constraint $\sum_{j=0}^{p-1} \beta_j = \delta$. More generally, the pattern of positive and negative values in $\hat{\beta}$ determines the pattern of positive and negative values in the linear constraint, i.e., $\sum_{j=0}^{p-1} \text{sign}(\hat{\beta}_j) \beta_j = \delta$. As you continue to decrease δ, the penalized least squares estimates gradually change, continuing to satisfy the constraint $\sum_{j=0}^{p-1} \text{sign}(\hat{\beta}_j) \beta_j = \delta$ until one of the estimates satisfies an additional linear constraint associated with $\sum_{j=0}^{p-1} \pm \hat{\beta}_j = \delta$, one that changes only one coefficient sign from the original constraint, so that together they cause the coefficient with the sign change to be zero. As δ further decreases, typically both linear constraints continue to hold for a while and then it is possible that the first linear constraint is supplanted by the second one.

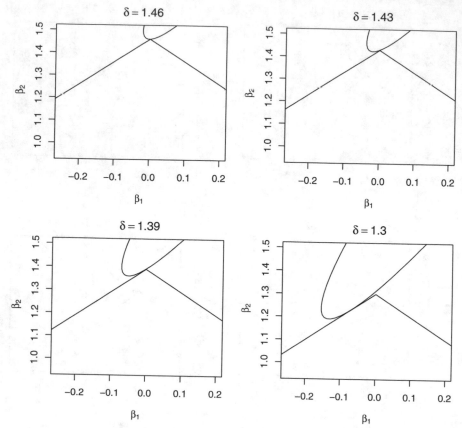

Fig. 2.9 Lasso trace

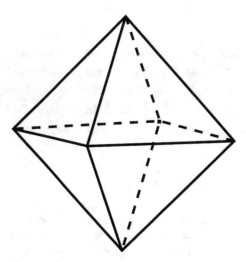

Fig. 2.10 Octohedron

It is convenient to think about the estimates moving along the surface of the penalty region (diamond, octahedron, etc.) as δ changes but that is not quite true because the surface itself changes with δ. Yet it is clear that features of the surface (diamond: 4 edges and 4 corners, octohedron: 26 features) are comparable for all δ.

2.5.2 Equivalence of Approaches

Suppose we want to minimize a continuous function $f(\beta)$ subject to $\mathscr{P}(\beta) \leq \delta$ where $\mathscr{P}(\beta)$ is a continuous, nonnegative "penalty" function. We assume that the set of βs with $\mathscr{P}(\beta) \leq \delta$ is a compact set. The minimum of a continuous function over a compact set will occur either in the interior of the set or on the boundary. The boundary is the set of βs that satisfy $\mathscr{P}(\beta) = \delta$. If the minimum occurs in the interior, we don't have to worry about the boundary, we just minimize $f(\beta)$. In other words, if the minimum of $f(\beta)$ occurs at $\hat{\beta}$ and if $\mathscr{P}(\hat{\beta}) \leq \delta$ then the restricted minimizer is just $\hat{\beta}$.

From the theory of Lagrange multipliers, to minimize the function $f(\beta)$ on the boundary, we minimize the continuous function

$$F_\delta(\beta, k) \equiv f(\beta) + k[\mathscr{P}(\beta) - \delta]$$

with respect to β and $k > 0$. The idea is that for βs on the boundary, $\mathscr{P}(\beta) = \delta$ so that minimizing $f(\beta)$ is equivalent to minimizing $f(\beta) + k[\mathscr{P}(\beta) - \delta]$ regardless of the value of k. The trick is that minimizers of $f(\beta) + k[\mathscr{P}(\beta) - \delta]$ are forced to be on the boundary. Note that the global minimum of $F_\delta(\beta, k)$ might occur when $k = 0$, which is precisely what happens when $\mathscr{P}(\hat{\beta}) \leq \delta$.

Setting equal to 0 the partial derivative with respect to k of $F_\delta(\beta, k)$ gives,

$$[\mathscr{P}(\beta) - \delta] = 0,$$

so any critical point will be on the boundary $\mathscr{P}(\beta) = \delta$. In particular, for any fixed β the minimizer of $F_\delta(\beta, k)$ must be on the boundary, hence the global minimizer of $F_\delta(\beta, k)$ must be on the boundary.

In practice, for fixed k find a minimizer for, what in applications would be the penalized least squares criterion function,

$$\tilde{\beta}(k) \equiv \arg\min_\beta [f(\beta) + k\mathscr{P}(\beta)],$$

and let us assume that there is a unique k_0 such that $\delta = \mathscr{P}(\tilde{\beta}(k_0))$. Think about minimizing the function $F_\delta(\beta, k)$. Clearly, for any β and k

$$f(\tilde{\beta}(k)) + k\mathscr{P}(\tilde{\beta}(k)) - k\delta \leq f(\beta) + k\mathscr{P}(\beta) - k\delta = F_\delta(\beta, k).$$

But we know that the global minimizer of $F_\delta(\beta, k)$ has to be on the boundary $\mathscr{P}(\beta) = \delta$, and by assumption the only k that puts $\tilde{\beta}(k)$ on the boundary is k_0, so

$$F_\delta(\tilde{\beta}(k_0), k_0) = \min_{\beta,k} F_\delta(\beta, k)$$

and $f(\tilde{\beta}(k_0))$ minimizes $f(\beta)$ subject to $\mathscr{P}(\beta) = \delta$.

All of this was for fixed δ with k a variable. Picking δ and picking k are interchangable. For every value of the tuning parameter k, say k_0, there exists a δ_0 for which $\tilde{\beta}(k_0)$ is the optimal estimate, namely $\delta_0 = \mathscr{P}(\tilde{\beta}(k_0))$. In this case $F_{\delta_0}(\tilde{\beta}(k_0), k_0)$ is the global minimum of $F_{\delta_0}(\beta, k)$ so that $f(\tilde{\beta}(k_0))$ minimizes $f(\beta)$ on the boundary $\delta_0 = \mathscr{P}(\beta)$.

Ideally, we would find $\tilde{\beta}(k)$ by setting equal to 0 the partial derivative with respect to β of $F_\delta(\beta, k)$. This requires a solution to

$$\mathbf{d}_\beta f(\beta) + k \mathbf{d}_\beta \mathscr{P}(\beta) = 0 \qquad (2.5.4)$$

for fixed k. Unfortunately, in applications like lasso regression, the function $\mathscr{P}$ is not differentiable everywhere, so we cannot rely on this to get us a lasso solution.

In practice, we look at all solutions $\tilde{\beta}(k)$.

Consider the simplest practical version of lasso. For a multiple regression model with an intercept take

$$\mathscr{P}_L(\beta) = \sum_{j=1}^{p-1} |\beta_j|.$$

Unless the least squares estimate $\hat{\beta}$ already has $\sum_{j=1}^{p-1} |\hat{\beta}_j| \le \delta$, the minimum will occur at a β vector with $\sum_{j=1}^{p-1} |\beta_j| = \delta$. As discussed, it is equivalent to minimize

$$\|Y - X\beta\|^2 + k \left[\sum_{j=1}^{p-1} |\beta_j| - \delta \right]$$

with respect to k and β and the lasso estimate $\tilde{\beta}(k)$ is found by minimizing

$$\|Y - X\beta\|^2 + k \sum_{j=1}^{p-1} |\beta_j|.$$

The squared error function $f(\beta) \equiv \|Y - X\beta\|^2$ is very well behaved but our lasso restricting function $\mathscr{P}_L(\beta) \equiv \sum_{j=1}^{p-1} |\beta_j|$, while continuous, has points where the derivatives are not defined, namely, anywhere that a $\beta_j = 0$, $j \ne 0$. Indeed, the whole point of lasso regression is that the minimums often occur at points where the derivatives do not exist. What this does is make the task of programming a solution $\tilde{\beta}(k)$ much more difficult because you have to check for minimums wherever the derivatives do not exist. And in lasso regression, that is where we like to find them, at k values where many of the $\tilde{\beta}_j(k)$s are 0.

Some computer programs (including `lasso2`) redefine the restriction

$$\mathscr{P}(\beta) \le \delta$$

as

$$\mathscr{P}(\beta) \leq \eta \, \mathscr{P}(\hat{\beta})$$

where $0 \leq \eta \leq 1$ and $\hat{\beta}$ is the least squares estimate. The beauty of this redefinition is that for $\eta < 1$, the minimizer of $f(\beta)$ cannot occur in the interior, it must occur on the boundary (provided $\mathscr{P}(\hat{\beta}) > 0$). For $\eta = 1$, the least squares estimate will be the penalized estimate (so $k = 0$). In particular, for lasso regression with an intercept,

$$\sum_{j=1}^{p-1} |\beta_j| \leq \delta$$

is redefined as

$$\sum_{j=1}^{p-1} |\beta_j| \leq \eta \sum_{j=1}^{p-1} |\hat{\beta}_j|$$

where $0 \leq \eta \leq 1$. For $\eta = 1$, the least squares estimates will be the lasso estimates. For $\eta = 0$, the estimates of $\beta_1, \ldots, \beta_{p-1}$ are all required to be 0, so the lasso estimate will be the least squares estimate for the intercept only model (so, essentially, $k = \infty$). lasso2's default is $\eta = 0.5$.

For generalized linear models, we could replace $f(\beta) = \|Y - X\beta\|^2$ with the negative log-likelihood, add a penalty function, and similar arguments continue to hold providing *maximum penalized likelihood estimates*.

2.6 Two Other Penalty Functions

For the partitioned model (2.1.1) recent approaches to regularization minimize

$$(Y - X\beta - Z\gamma)'(Y - X\beta - Z\gamma) + \mathscr{P}_\theta(\gamma),$$

where θ is a vector of tuning parameters involved in defining the penalty function $\mathscr{P}_\theta(\gamma)$.

The *Elastic Net* penalty combines the ridge and lasso penalties but incorporates another tuning parameter $\alpha \in [0, 1]$,

$$\mathscr{P}_{EN}(\gamma) \equiv \alpha \mathscr{P}_R(\gamma) + (1 - \alpha) \mathscr{P}_L(\gamma) = \alpha \sum_{j=1}^{s} \gamma_j^2 + (1 - \alpha) \sum_{j=1}^{s} |\gamma_j|.$$

Thus $\mathscr{P}_\theta(\gamma) = k \mathscr{P}_{EN}(\gamma)$ with $\theta = (k, \alpha)'$.

Fan and Li (2001) suggested the *SCAD (Smoothly Clipped Absolute Deviation)* penalty. The advantage of SCAD is that, like the LASSO, it shrinks small estimates to zero but unlike LASSO, it does not shrink large estimates at all. The penalty function is

$$\mathscr{P}_S(\gamma) \equiv \sum_{j=1}^{s} P_S(\gamma_j),$$

where for $a > 2$,

$$P_S(\gamma_j) \equiv \begin{cases} |\gamma_j| & \text{if } |\gamma_j| \le k, \\ -\left(\frac{|\gamma_j|^2 - 2ak|\gamma_j| + k^2}{2(a-1)k}\right) & \text{if } k < |\gamma_j| \le ak, \\ \frac{(a+1)k}{2} & |\gamma_j| > ak. \end{cases}$$

The SCAD penalty function depends not only on a new tuning parameter a but also on the original tuning parameter k, so $\mathscr{P}_\theta(\gamma) = k\mathscr{P}_S(\gamma)$ with $\theta = (k, a)'$. Fan and Li suggest that $a = 3.7$ often works well.

When X is vacuous and the columns of Z are orthonormal, it can be shown that SCAD results in the following modifications to the least squares estimates $\hat{\gamma}_j$,

$$\tilde{\gamma}_{Sj} = \begin{cases} 0, & \text{if } |\hat{\gamma}_j| \le k, \\ \hat{\gamma}_j - \text{sign}(\hat{\gamma}_j)\,k, & \text{if } k < |\hat{\gamma}_j| \le 2k, \\ \frac{(a-1)\hat{\gamma}_j - \text{sign}(\hat{\gamma}_j)ak}{a-2}, & \text{if } 2k < |\hat{\gamma}_j| \le ak, \\ \hat{\gamma}_j, & \text{if } |\hat{\gamma}_j| > ak. \end{cases}$$

A similar result for LASSO can be obtained from (2.3.2) by doubling the LASSO tuning parameter and taking all $\lambda_j = 1$,

$$\tilde{\gamma}_{Lj} = \begin{cases} 0, & \text{if } |\hat{\gamma}_j| < k, \\ \hat{\gamma}_j - \text{sign}(\hat{\gamma}_j)\,k, & \text{if } |\hat{\gamma}_j| \ge k. \end{cases}$$

The estimates agree for $|\hat{\gamma}_j| \le 2k$ but SCAD does less shrinkage on larger least squares estimates.

References

Bühlmann, P., & van de Geer, S. (2011). *Statistics for high-dimensional data: Methods, theory and applications.* Heidelberg: Springer.

Bühlmann, P., Kalisch, M., & Meier, L. (2014). High-dimensional statistics with a view toward applications in biology. *Annual Review of Statistics and Its Applications, 1,* 255–278.

Christensen, R. (2011). *Plane answers to complex questions: The theory of linear models* (4th ed.). New York, NY: Springer.

Christensen, R. (2015). *Analysis of variance, design, and regression: Linear modeling for unbalanced data* (2nd ed.). Boca Raton, FL: Chapman and Hall/CRC Press

Christensen, R., Johnson, W., Branscum, A., & Hanson, T. E. (2010). *Bayesian ideas and data analysis: An introduction for scientists and statisticians.* Boca Raton, FL: Chapman and Hall/CRC Press.

Clyde, M., & George, E. I. (2004). Model uncertainty. *Statistical Science, 19,* 81–94.

Draper, N. R., & Van Nostrand, R. C. (1979). Ridge regression and James–Stein estimation: Review and comments *Technometrics, 21,* 451–466.

Efron, B., Hastie, T., Johnstone, I., & Tibshirani, R. (2004). Least angle regression, with discussion. *The Annals of Statistics, 32*, 407–499.

Fan, J., & Li, R. (2001). Variable selection via nonconcave penalized likelihood and its oracle properties. *Journal of the American Statistical Association, 96*, 1348–1360.

Green, P. J., & Silverman, B. W. (1994). *Nonparametric regression and generalized linear models: A roughness penalty approach*. London: Chapman and Hall.

Hastie, T., Tibshirani, R., & Friedman, J. (2016). *The elements of statistical learning: Data mining, inference, and prediction* (2nd ed.). New York, NY: Springer.

Hastie, T., Tibshirani, R., & Wainwright, M. (2015). *Statistical learning with sparcity: The lasso and generalizations*. Boca Raton, FL: Chapman and Hall.

Hoerl, A. E., & Kennard, R. (1970). Ridge regression: Biased estimation for non-orthogonal problems. *Technometrics, 12*, 55–67.

Park, T., & Casella, G. (2008). The Bayesian lasso. *Journal of the American Statistical Association, 103*, 681–686.

Smith, M., & Kohn, R. (1996). Nonparametric regression using Bayesian variable selection. *Journal of Econometrics, 75*, 317–343.

Tibshirani, R. (1996). Regression shrinkage and selection via the lasso. *Journal of the Royal Statistical Society: Series B, 58*, 267–288.

Chapter 3
Reproducing Kernel Hilbert Spaces

Abstract This chapter introduces an elegant mathematical theory that has been developed for nonparametric regression with penalized estimation.

Chapter 1 included a brief description of using reproducing kernel Hilbert spaces in nonparametric regression. We now introduce their theory.

The ideas used in analyzing linear models extend naturally to much more general spaces than $\mathbf{R}^n$. One theoretical extension was introduced in *PA* Sect. 6.3.5. These extensions lead to interesting tools for examining penalized estimation in regression (Pearce & Wand 2006) and support vector machines in classification/discrimination problems (Moguerza & Muñoz 2006 and Zhu 2008), They are commonly used in areas such as functional data analysis (Ramsay & Silverman 2005), computer model analysis (Storlie, Swiler, Helton, & Sallaberry 2009), image processing (Berman 1994), and various applications of spatial statistics (Bivand, Pebesma, & Gómez-Rubio 2013 or Storlie, Bondell, & Reich 2010), to name a few. The flexibility and elegance of the methods are remarkable. Texts and survey articles on related subjects include Wahba (1990), Eubank (1999), Hastie, Tibshirani, and Friedman (2016), Gu (2002), Berlinet and Thomas-Agnan (2004), Bühlmann and van de Geer (2011), Heckman (2012), Bühlmann, Kalisch, and Meier (2014), and Wainwright (2014). These include a wealth of additional references as do the papers mentioned earlier.

In particular, the key ideas of linear models extend completely to *finite dimensional Hilbert spaces* whereas much of the newer theory has been developed for infinite dimensional *Reproducing Kernel Hilbert Spaces* (*RKHS*s). We provide an introduction to the mathematical ideas behind this work emphasizing its connections to linear model theory and two applications to problems that we have previously solved using linear model theory: ridge regression from Chap. 2 and (without a penalty function) spline regression from Chap. 1. We also provide an illustration of using reproducing kernels to test lack of fit in a linear model. Our development follows closely that of Nosedal-Sanchez, Storlie, Lee, and Christensen (2012).

Ridge regression and smoothing splines can both be viewed as solutions to minimization problems in a function space. If such a minimization problem is posed on an RKHS, the solution is guaranteed to exist and has a very simple form. We begin

© Springer Nature Switzerland AG 2019
R. Christensen, *Advanced Linear Modeling*, Springer Texts in Statistics,
https://doi.org/10.1007/978-3-030-29164-8_3

by solving some simple problems that relate to our broader goals. Section 3.2 introduces Banach spaces and Hilbert spaces. Section 3.3 provides basic ideas of RKHSs. Section 3.4 discusses the two distinct ways of using RKHS results and exploits the one not primarily used here to look at testing lack of fit in a linear model. Section 3.5 discusses penalized regression with RKHSs and the two specific examples of ridge regression and smoothing splines.

3.1 Introduction

For a known $n \times p$ matrix

$$X = [X_1, \cdots, X_p] = \begin{bmatrix} x_1' \\ \vdots \\ x_n' \end{bmatrix}$$

and vector Y, consider solving the system of n equations in p unknowns

$$X\beta = Y. \tag{3.1.1}$$

For a solution to exist, we must have $Y \in C(X)$. In standard linear model problems, we typically have $n > r(X)$, which typically causes $Y \notin C(X)$ and precludes us from finding a solution. That is precisely why we go to the trouble of finding (generalized) least squares estimates for β. In the current era of big data, people have become interested in solving such problems when $r(X) \geq n$ in which case solutions often exist.

We wish to find the solution that minimizes the norm $\|\beta\| = \sqrt{\beta'\beta}$ but we solve the problem using concepts that extend to RKHSs. As mentioned, a solution $\tilde{\beta}$ (not necessarily a minimum norm solution) exists whenever $Y \in C(X)$. Given one solution $\tilde{\beta}$, all solutions β must satisfy

$$X\tilde{\beta} = X\beta$$

or

$$X(\tilde{\beta} - \beta) = 0.$$

Write $\tilde{\beta}$ uniquely as $\tilde{\beta} = \beta_0 + \beta_1$ with $\beta_0 \in C(X') = \text{span}\{x_1, \ldots, x_n\}$ and $\beta_1 \in C(X')^\perp$, then β_0 is a solution because $X(\tilde{\beta} - \beta_0) = X\beta_1 = 0$.

In fact, β_0 is both the unique solution in $C(X')$ and the minimum norm solution. If β is any other solution in $C(X')$ then $X(\beta - \beta_0) = 0$ so we have both $(\beta - \beta_0) \in C(X')^\perp$ and $(\beta - \beta_0) \in C(X')$, two sets whose intersection is only the 0 vector. Thus $\beta - \beta_0 = 0$ and $\beta = \beta_0$. In other words, every solution $\tilde{\beta}$ has the same β_0 vector. Finally, β_0 is also the minimum norm solution because the arbitrary solution $\tilde{\beta}$ has

$$\beta_0'\beta_0 \leq \beta_0'\beta_0 + \beta_1'\beta_1 = \tilde{\beta}'\tilde{\beta}.$$

We have established the existence of a unique, minimum norm solution in $C(X')$ that can be written as

$$\beta_0 = X'\xi = \sum_{i=1}^{n} \xi_i x_i,$$

for some ξ_i, $i = 1, \ldots, n$. To find β_0 explicitly, write $\beta_0 = X'\xi$ and the defining Eq. (3.1.1) becomes

$$XX'\xi = Y, \tag{3.1.2}$$

which is again a system of linear equations. Even if there exist multiple solutions ξ, $X'\xi$ is unique.

EXAMPLE 3.1.1.   We use this framework to illustrate the "smallest" solution to the system of equations $\tilde{\beta}_1 + \tilde{\beta}_3 = 0$ and $\tilde{\beta}_2 = 1$. In the general framework (3.1.1), these become

$$X\beta \equiv \begin{bmatrix} 1 & 0 & 1 \\ 0 & 1 & 0 \end{bmatrix} \begin{bmatrix} \tilde{\beta}_1 \\ \tilde{\beta}_2 \\ \tilde{\beta}_3 \end{bmatrix} = \begin{bmatrix} 0 \\ 1 \end{bmatrix},$$

whereas (3.1.2) becomes

$$\begin{bmatrix} 2 & 0 \\ 0 & 1 \end{bmatrix} \begin{bmatrix} \xi_1 \\ \xi_2 \end{bmatrix} = \begin{bmatrix} 0 \\ 1 \end{bmatrix}.$$

The unique solution is $(\xi_1, \xi_2)' = (0, 1)'$ which implies that the solution to the original problem is $\beta_0 = X'\xi = 0x_1 + 1x_2 = (0, 1, 0)'$.   □

Virtually the same methods can be used to solve similar problems in any inner-product space $\mathcal{M}$. As discussed in *PA* Sect. 6.3.5, an inner product $\langle \cdot, \cdot \rangle$ assigns real numbers to pairs of vectors. There the notion was used to treat random variables as vectors whereas in most of this book and *PA*, vectors are elements of $\mathbf{R}^n$ or $\mathbf{R}^p$. In this chapter we often use functions from some set $\mathcal{E}$ into $\mathbf{R}$ as vectors. Frequently we take $\mathcal{E} \subset \mathbf{R}$ but $\mathcal{E}$ can be a subset of a more general vector space. Note that an element of $\mathbf{R}^n$ can be thought of as a function from the integers $\{1, 2, \ldots, n\}$ into $\mathbf{R}$.

We generalize the problem of solving a system of linear equations as follows. For n given vectors $x_i \in \mathcal{M}$ and numbers $y_i \in \mathbf{R}$, we might want to find the vector $\beta \in \mathcal{M}$ such that

$$\langle x_i, \beta \rangle = y_i, \qquad i = 1, 2, \ldots, n \tag{3.1.3}$$

for which the norm $\|\beta\| \equiv \sqrt{\langle \beta, \beta \rangle}$ is minimal. The solution (if one exists) has the form

$$\beta_0 = \sum_{i=1}^{n} \xi_i x_i, \tag{3.1.4}$$

with ξ_i satisfying the linear equations

$$\sum_{k=1}^{n} \langle x_i, x_k \rangle \xi_k = y_i, \qquad i = 1, \ldots, n \tag{3.1.5}$$

or, equivalently, we can solve the matrix equation

$$\tilde{R}\xi = Y$$

where the $n \times n$ matrix $\tilde{R}$ has elements $\tilde{r}_{ij}$ with

$$\tilde{r}_{ij} = \langle x_i, x_j \rangle.$$

For a formal proof see Máté (1990). From the matrix equation it is easy to check whether a solution exists. After a brief digression, we apply this result to the interpolating spline problem.

The illustrations in this chapter focus on three vector spaces: $\mathbf{R}^n$, because it is familiar;

$$\mathscr{L}^2[0,1] = \left\{ f : \int_0^1 [f(t)]^2 dt < \infty \right\},$$

because it relates to fitting splines; and the vector space of all functions from $\mathbf{R}^{p-1}$ into $\mathbf{R}$

$$\mathscr{F} = \left\{ f : f(x) \in \mathbf{R}, x \in \mathbf{R}^{p-1} \right\},$$

because it relates to fitting multiple regression models that include an intercept. We will examine two different inner products on $\mathbf{R}^n$. We will look at the standard inner product on $\mathscr{L}^2[0,1]$ and also at two subspaces with different inner products on each. For $\mathscr{F}$ we focus on linear multiple regression by focusing on subspaces of constant functions, linear functions, and affine functions along with the orthogonality properties that they display.

The first illustration looks at a subset of $\mathscr{L}^2[0,1]$. Throughout, $f^{(m)}(t)$ denotes the m-th derivative of f with $\dot{f} \equiv f^{(1)}$ and $\ddot{f} \equiv f^{(2)}$. From Appendix A we also have the notations $\dot{f}(t) \equiv \mathbf{d}_t f(t)$ and $\ddot{f}(t) \equiv \mathbf{d}_{tt}^2 f(t)$.

3.1.1 Interpolating Splines

Suppose we want to find a function $f(t)$ that interpolates between the points (t_i, y_i), $i = 0, 1, \ldots, n$, where $y_0 \equiv 0$ and $0 = t_0 < t_1 < t_2 < \cdots < t_n \leq 1$. We restrict attention to functions $f \in \mathscr{W}_0^1$ where

$$\mathscr{W}_0^1 = \left\{ f \in \mathscr{L}^2[0,1] : f(0) = 0, \int_0^1 [\dot{f}(t)]^2 dt < \infty \right\}$$

and it is understood that the derivatives exist because they are assumed to be square integrable. The restriction that $y_0 = f(0) = 0$ is not really necessary, but simplifies the presentation.

We want to find the smoothest function $f(t)$ that satisfies $f(t_i) = y_i$, $i = 1, \ldots, n$. Defining an inner product on $\mathscr{W}_0^1$ by

$$\langle f, g \rangle = \int_0^1 \dot{f}(x)\dot{g}(x)dx, \tag{3.1.6}$$

this determines a norm (length) $\|f\| \equiv \sqrt{\langle f, f \rangle}$ for $\mathscr{W}_0^1$ that is small for "smooth" functions. To address the interpolation problem, define the indicator function $\mathscr{I}_A(t)$ for a set A to be 1 if $t \in A$ and 0 if $t \notin A$. Note that the functions

$$R_i(s) \equiv \min(s, t_i) = s\mathscr{I}_{[0,t_i]}(s) + t_i \mathscr{I}_{(t_i,1]}(s), \quad i = 1, 2, \ldots, n,$$

have $R_i(0) = 0$, $\dot{R}_i(s) = \mathscr{I}_{[0,t_i]}(s)$ so that $R_i(\cdot) \in \mathscr{W}_0^1$, and have the property that $\langle R_i, f \rangle = f(t_i)$ because

$$\langle f, R_i \rangle = \int_0^1 \dot{f}(s)\dot{R}_i(s)ds$$
$$= \int_0^{t_i} \dot{f}(s)ds = f(t_i) - f(0) = f(t_i).$$

Thus, any interpolator f satisfies a system of equations like (3.1.3), namely

$$f(t_i) = \langle R_i, f \rangle = y_i, \qquad i = 1, \ldots, n. \tag{3.1.7}$$

and by (3.1.4), the smoothest function f (minimum norm) that satisfies the requirements has the form

$$\hat{f}(t) = \sum_{i=1}^n \xi_i R_i(t).$$

The ξ_j's are the solutions to the system of real linear equations obtained by substituting $\hat{f}$ into (3.1.7), that is

$$\sum_{j=1}^n \langle R_i, R_j \rangle \xi_j = y_i, \qquad i = 1, 2, \ldots, n$$

or

$$\tilde{R}\xi = Y.$$

Note that

$$\langle R_i, R_j \rangle = R_j(t_i) = R_i(t_j) = \min(t_i, t_j)$$

and we define the function

$$R(s, t) = \min(s, t)$$

that turns out to be a reproducing kernel.

EXAMPLE 3.1.2. Given points $f(t_i) = y_i$, say, $f(0) = 0$, $f(0.1) = 0.1$, $f(0.25) = 1$, $f(0.5) = 2$, $f(0.75) = 1.5$, and $f(1) = 1.75$, we can now find $f \in \mathscr{W}_0^1$ that satisfies these six conditions and minimizes the norm associated with (3.1.6). The vector $\xi \in \mathbf{R}^5$ that satisfies the system of equations

$$\begin{bmatrix} 0.1 & 0.1 & 0.1 & 0.1 & 0.1 \\ 0.1 & 0.25 & 0.25 & 0.25 & 0.25 \\ 0.1 & 0.25 & 0.5 & 0.5 & 0.5 \\ 0.1 & 0.25 & 0.5 & 0.75 & 0.75 \\ 0.1 & 0.25 & 0.5 & 0.75 & 1 \end{bmatrix} \begin{bmatrix} \xi_1 \\ \xi_2 \\ \xi_3 \\ \xi_4 \\ \xi_5 \end{bmatrix} = \begin{bmatrix} 0.1 \\ 1 \\ 2 \\ 1.5 \\ 1.75 \end{bmatrix}$$

is $\xi = (-5, 2, 6, -3, 1)'$, which implies that the smoothest interpolating function is

$$\hat{f}(t) = -5R_1(t) + 2R_2(t) + 6R_3(t) - 3R_4(t) + 1R_5(t) \tag{3.1.8}$$
$$= -5R(t, t_1) + 2R(t, t_2) + 6R(t, t_3) - 3R(t, t_4) + 1R(t, t_5)$$

or, adding together the slopes for $t > t_i$ and finding the intercepts,

$$\hat{f}(t) = \begin{cases} t & 0 \le t \le 0.1 \\ 6t - 0.5 & 0.1 \le t \le 0.25 \\ 4t & 0.25 \le t \le 0.5 \\ -2t + 3 & 0.5 \le t \le 0.75 \\ t + 0.75 & 0.75 \le t \le 1. \end{cases}$$

This is the linear interpolating spline as can be seen graphically in Fig. 3.1. □

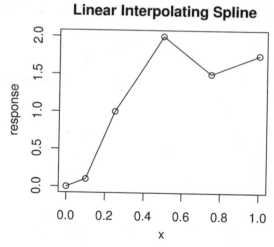

Fig. 3.1 Linear interpolating spline

For this illustration we restricted f so that $f(0) = 0$ but only for convenience of presentation. It can be shown that the form of the solution remains the same with any shift to the function, so that in general the solution takes the form $\hat{f}(t) = \beta_0 + \sum_{j=1}^{n} \xi_j R_j(t)$ where $\beta_0 = y_0$.

The two key points are (a) that the functions $R_j(t)$ allow us to express a function evaluated at a point as an inner-product constraint and (b) the restriction to functions in $\mathcal{W}_0^1$. $\mathcal{W}_0^1$ is a very special function space, a reproducing kernel Hilbert space, and R_j is determined by a reproducing kernel function R.

Ultimately, our goal is to address more complicated regression problems like

EXAMPLE 3.1.3. *Smoothing Splines.*
Consider simple regression data (x_i, y_i), $i = 1, \ldots, n$ and finding the function that minimizes

$$\frac{1}{n} \sum_{i=1}^{n} [y_i - f(x_i)]^2 + \lambda \int_0^1 \left[f^{(m)}(x) \right]^2 dx. \tag{3.1.9}$$

If $f(x)$ is restricted to be in an appropriate class of functions, minimizing only the first term gives least squares estimation within the class. If the class contains functions with $f(x_i) = y_i$ for all i, such functions minimize the first term but are typically very "unsmooth," i.e., have large second term. For example, an $n - 1$ order polynomial will always fit the data perfectly but is typically very unsmooth. The second "penalty" term is minimized whenever the mth derivative is 0 everywhere, but (at least for small m) that rarely has a small first term. For $m = 1$, $0 \le x \le 1$, and a suitable class of functions, as we will see later, the minimizer takes the form

$$\hat{f}(x) = \beta_0 + \sum_{i=1}^{n} \xi_i R_i(x),$$

where the R_is were given earlier and the coefficients are found by solving a slightly different system of linear equations. Choosing $m = 1, 2$ determines linear and cubic smoothing splines, respectively. □

If our goal is only to derive the solution to the linear smoothing spline problem with one predictor variable, RKHS theory is overkill. The value of RKHS theory lies in its generality. For example, the spline penalty can be replaced by many other penalties having associated inner products, and the x_i's can be vectors. Using RKHS results, we can solve the general problem of finding the minimizer of $\frac{1}{n} \sum_{i=1}^{n} [y_i - f(x_i)]^2 + \lambda Q(f)$ for quite general functions Q that correspond to a squared norm in a Hilbert subspace. See Wahba (1990) or Gu (2002) for a full treatment.

3.2 Banach and Hilbert Spaces

As discussed in *PA*'s Appendix A, a vector space is a set $\mathcal{M}$ that contains elements called *vectors* and supports two kinds of operations: addition of vectors and multiplication by scalars. The scalars are real numbers in this book and in *PA* but in general they can be from any *field*. We rely on context to distinguish between the vector $0 \in \mathcal{M}$ and the scalar $0 \in \mathbf{R}$. For vectors $u, v \in \mathcal{M}$, we also write $u + (-1 \times v)$ as $u - v$. Any subset of $\mathcal{M}$ that is closed under vector addition and scalar multiplication is a subspace of $\mathcal{M}$. The classic book on finite dimensional vector spaces is Halmos (1958). Harville (1997) also contains a wealth of information. For more on vector spaces and the other topics in this section, see, for example, Naylor and Sell (1982), Young (1988), Máté (1990) or Rustagi (1994).

3.2.1 Banach Spaces

A Banach space is a vector space that has some additional structure. First, a Banach space has a length measure, called a *norm*, associated with it and, second, a Banach space is *complete* under that norm. Banach spaces provide a convenient introduction to Hilbert spaces, who have another bit of structure, namely an inner product.

Definition 3.2.1.   A *norm* of a vector space $\mathcal{M}$, denoted $||\cdot||$, is a nonnegative real valued function satisfying the following properties for all $u, v \in \mathcal{M}$ and all $a \in \mathbf{R}$.

1. Non-negative: $||u|| \geq 0$.
2. Strictly positive: $||u|| = 0$ implies $u = 0$.
3. Homogeneous: $||au|| = |a|\, ||u||$.
4. Triangle inequality: $||u + v|| \leq ||u|| + ||v||$.

A vector space is called a *normed vector space* when a norm is defined on the space. The norm of a vector is also called its *length*. For vectors $u, v \in \mathcal{M}$, the *distance* between u and v is defined as $||u - v||$.

Definition 3.2.2.   A sequence $\{v_n\}$ in a normed vector space $\mathcal{M}$ is said to *converge* to $v_0 \in \mathcal{M}$ if

$$\lim_{n \to \infty} ||v_n - v_0|| = 0.$$

Definition 3.2.3.   A sequence $\{v_n\} \subset \mathcal{M}$ is called a *Cauchy sequence* if for any given $\varepsilon > 0$, there exists an integer N such that

$$||v_m - v_n|| < \varepsilon, \qquad \text{whenever} \quad m, n \geq N.$$

Convergence of sequences in normed vector spaces follows the same general idea

as sequences of real numbers except that the distance between two vectors of the space is measured by the norm of the difference between the two vectors.

Definition 3.2.4. (Banach Space).   A normed vector space $\mathcal{M}$ is *complete* if every Cauchy sequence in $\mathcal{M}$ converges to an element of $\mathcal{M}$. A complete normed vector space is a *Banach Space*.

EXAMPLE 3.2.5. $\mathcal{M} = \mathbf{R}$ with the absolute value norm $||x|| \equiv |x|$ is a complete, normed vector space over $\mathbf{R}$, and is thus a Banach space.   □

EXAMPLE 3.2.6.   For $\mathcal{M} = \mathbf{R}^n$, let $x = (x_1,\ldots,x_n)'$ be a vector. The $\mathscr{L}_p$ norm on $\mathbf{R}^n$ is defined by

$$\|x\|_p \equiv \left[\sum_{i=1}^{n} |x_i|^p\right]^{1/p} \qquad \text{for } 1 \le p < \infty.$$

One can verify properties 1–4 of Definition 3.2.1 for each p, validating that $\|x\|_p$ is a norm on $\mathbf{R}^n$. Under the $\mathscr{L}_p$ norm, $\mathbf{R}^n$ is complete and thus a Banach space. Euclidean distance on $\mathbf{R}^n$ corresponds to choosing $p = 2$.

Alternatively, if A is a positive definite matrix,

$$\|x\|_A \equiv \sqrt{x'Ax}$$

defines a norm and a Banach space on $\mathbf{R}^n$. Euclidean distance on $\mathbf{R}^n$ corresponds to choosing I for A.   □

3.2.2 Hilbert Spaces

A Hilbert Space is a Banach space in which the norm is defined by an inner-product (also called a dot-product) that maps any two vectors into a real number. Banach spaces incorporate concepts of length and distance; Hilbert spaces add the concept of orthogonality (perpendicularity).

We typically denote Hilbert spaces by $\mathscr{H}$. For elements $u, v \in \mathscr{H}$, write the inner product of u and v as either $\langle u, v \rangle_{\mathscr{H}}$ or, when it is clear from the context that the inner product is taking place in $\mathscr{H}$, as $\langle u, v \rangle$. An *inner product* must satisfy four properties for all $u, v, w \in \mathscr{H}$ and all $a \in \mathbf{R}$.

1. Associative: $\langle au, v \rangle = a\langle u, v \rangle$.
2. Commutative: $\langle u, v \rangle = \langle v, u \rangle$.
3. Distributive: $\langle u, v + w \rangle = \langle u, v \rangle + \langle u, w \rangle$.
4. Positive Definite: $\langle u, u \rangle \ge 0$ with equality holding only if $u = 0$.

Definition 3.2.7.   A vector space $\mathcal{M}$ with an inner product defined on it is called an *inner product space*. Vectors u and v are *orthogonal*, written $u \perp v$, if $\langle u, v \rangle = 0$. Two sets of vectors are said to be orthogonal if every vector in one set is orthogonal to every vector in the other. The set of all vectors orthogonal to a subspace $\mathscr{N}$ of $\mathcal{M}$ is called the *orthogonal complement* of $\mathscr{N}$ with respect to $\mathcal{M}$ and is written $\mathscr{N}_{\mathcal{M}}^{\perp}$, or just $\mathscr{N}^{\perp}$. The *norm* of u in an inner product space is $\|u\| \equiv \sqrt{\langle u, u \rangle}$. The angle θ between two vectors u and v is defined by

$$\cos(\theta) \equiv \frac{\langle x, y \rangle}{\|u\| \, \|v\|}.$$

A complete inner-product space is called a *Hilbert space*.

Most of the ideas related to orthogonality that we have exploited here and in *PA* extend immediately to finite dimensional Hilbert spaces. It is easy to see that an orthogonal complement is a subspace because it is closed under vector addition and scalar multiplication. Moreover, because only the 0 vector can be orthogonal to itself, $\mathcal{N} \cap \mathcal{N}_{\mathcal{M}}^{\perp} = \{0\}$, which means that for any vector x that can be written as $x = x_0 + x_1$ with $x_0 \in \mathcal{N}$ and $x_1 \in \mathcal{N}_{\mathcal{M}}^{\perp}$, the representation is unique. If $\mathcal{M} = \text{span}\{x_1, \cdots, x_n\}$, Gram–Schmidt applies so that we can find an orthonormal spanning set $\{o_1, \cdots, o_n\}$ in which all vectors are orthogonal to each other and each $\|o_j\|$ is 0 or 1 and $\text{span}\{x_1, \cdots, x_r\} = \text{span}\{o_1, \cdots, o_r\}$, $r = 1, \ldots, n$. Eliminating the 0 vectors from $\{o_1, \cdots, o_n\}$ gives an orthonormal basis for $\mathcal{M}$. The key idea in the inductive proof of Gram–Schmidt is to set

$$w_{s+1} = x_{s+1} - \sum_{j=1}^{s} o_j \langle o_j, x_{s+1} \rangle \quad \text{and} \quad o_{s+1} = \begin{cases} \frac{1}{\|w_{s+1}\|} w_{s+1} & \text{if } \|w_{s+1}\| > 0 \\ 0 & \text{if } \|w_{s+1}\| = 0 . \end{cases}$$

By taking spanning sets $\{x_1, \cdots, x_n\}$ for $\mathcal{M}$ and $\{v_1, \cdots, v_s\}$ for a subspace $\mathcal{N}$, we can Gram–Schmidt the spanning set $\{v_1, \cdots, v_s, x_1, \cdots, x_n\}$ of $\mathcal{M}$ to get orthonormal bases for $\mathcal{N}$ and $\mathcal{N}_{\mathcal{M}}^{\perp}$ that combine to give a basis for $\mathcal{M}$ thus establishing that any vector $x \in \mathcal{M}$ can be written uniquely as $x = x_0 + x_1$ with $x_0 \in \mathcal{N}$ and $x_1 \in \mathcal{N}_{\mathcal{M}}^{\perp}$ and allowing us to define x_0 as the perpendicular projection of x onto $\mathcal{N}$. Sometimes the perpendicular projection of x into a subspace $\mathcal{N}$ of $\mathcal{M}$ is defined as the unique vector $x_0 \in \mathcal{N}$ with the property that $\langle x - x_0, u \rangle = 0$ for any $u \in \mathcal{N}$.

EXAMPLE 3.2.8.   For $\mathbf{R}^n$ we can define a Hilbert space with the inner product

$$\langle u, v \rangle \equiv u'v = \sum_{i=1}^{n} u_i v_i$$

that conforms with Euclidean geometry. More generally, for any positive definite matrix A, we can define a Hilbert space with the inner product $\langle u, v \rangle \equiv u'Av$.   □

EXAMPLE 3.2.9.   For

$$\mathcal{L}^2[0, 1] = \left\{ f : \int_0^1 [f(t)]^2 dt < \infty \right\},$$

we can define a Hilbert space with the inner product

$$\langle f, g \rangle_{\mathcal{L}^2[0,1]} \equiv \int_0^1 f(t)g(t)dt.$$

The space is well-known to be complete, see de Barra (1981).

For the subspace

$$\mathscr{W}_0^1 = \left\{ f \in \mathscr{L}^2[0,1] : f(0) = 0, \int_0^1 [\dot{f}(t)]^2 dt < \infty \right\},$$

define the inner product

$$\langle f, g \rangle_{\mathscr{W}_0^1} = \int_0^1 \dot{f}(t)\dot{g}(t) dt. \qquad (3.2.1)$$

Note that $\langle f, f \rangle_{\mathscr{W}_0^1} = 0$ if and only if $f = 0$ because if $\langle f, f \rangle_{\mathscr{W}_0^1} = 0$, $\dot{f}(t) = 0$ for all t, so $f(t)$ must be a constant, however $f(0) = 0$, so $f(t) = 0$ for all t.

For the subspace

$$\mathscr{W}_0^2 = \left\{ f \in \mathscr{L}^2[0,1] : f(0) = \dot{f}(0) = 0, \int_0^1 [\ddot{f}(t)]^2 dt < \infty \right\},$$

define an inner product

$$\langle f, g \rangle_{\mathscr{W}_0^2} = \int_0^1 \ddot{f}(t)\ddot{g}(t) dt. \qquad (3.2.2)$$

Note that $\langle f, f \rangle_{\mathscr{W}_0^2} = 0$ if and only if $f = 0$ because if $\langle f, f \rangle_{\mathscr{W}_0^2} = 0$, $\ddot{f}(t) = 0$ for all t, so $\dot{f}(t)$ must be a constant, however $\dot{f}(0) = 0$, so $\dot{f}(t) = 0$ for all t, hence $f(t)$ must be a constant, however $f(0) = 0$, so $f(t) = 0$ for all t. $\square$

EXAMPLE 3.2.10. Consider the vector space of all functions from $\mathbf{R}^{p-1}$ to $\mathbf{R}$,

$$\mathscr{F} = \left\{ f : f(x) \in \mathbf{R}, x \in \mathbf{R}^{p-1} \right\}.$$

The subspace of all constant functions on $\mathbf{R}^{p-1}$ is

$$\mathscr{F}_0 = \{ f_a \in \mathscr{F} : f_a(x) = a, a \in \mathbf{R}, x \in \mathbf{R}^{p-1} \}$$

and define the inner product
$$\langle f_a, f_b \rangle_{\mathscr{F}_0} = ab.$$

Since $\mathbf{R}^{p-1}$ is a Hilbert Space, so is $\mathscr{F}_0$.

The subspace of all linear functions on $\mathbf{R}^{p-1}$ passing through the origin is

$$\mathscr{F}_1 = \{ f_\gamma \in \mathscr{F} : f_\gamma(x) = x'\gamma, \gamma \in \mathbf{R}^{p-1}, x \in \mathbf{R}^{p-1} \}$$

and define the inner product

$$\langle f_\eta, f_\gamma \rangle_{\mathscr{F}_1} = \eta'\gamma = \eta_1\gamma_1 + \eta_2\gamma_2 + \cdots + \eta_{p-1}\gamma_{p-1}.$$

Again, since $\mathbf{R}^{p-1}$ is a Hilbert Space, so is $\mathscr{F}_1$

Now consider the subspace of all affine (i.e., linear plus a constant) functions on $\mathbf{R}^{p-1}$,

$$\mathscr{F}_* = \{f_\beta \in \mathscr{F} : f_\beta(x) = \beta_0 + \beta_1 x_1 + \ldots + \beta_{p-1} x_{p-1},\ \beta \in \mathbf{R}^p,\ x \in \mathbf{R}^{p-1}\},$$

with the inner product

$$\langle f_\beta, f_\eta \rangle_{\mathscr{F}_*} = \beta_0 \eta_0 + \beta_1 \eta_1 + \ldots + \beta_{p-1} \eta_{p-1}.$$

This too is a Hilbert space.

The subspace of $\mathscr{F}_*$ that contains constant functions,

$$\mathscr{F}_0 = \{f_\beta \in \mathscr{F}_* : f_\beta(x) = \beta_0,\ 0 = \beta_1 = \cdots = \beta_p\}$$

has an orthogonal complement with respect to $\mathscr{F}_*$ of

$$\mathscr{F}_0^\perp = \{f_\beta \in \mathscr{F}_* : f_\beta(x) = \beta_1 x_1 + \ldots + \beta_{p-1} x_{p-1},\ 0 = \beta_0\} = \mathscr{F}_1.$$

For any $f_\beta, f_\eta \in \mathscr{F}_*$, write $\beta = [\beta_0, \beta'_*]$ and $\eta = [\eta_0, \eta'_*]$. We have the unique decompositions $f_\beta = f_{\beta_0} + f_{\beta_*}$ and $f_\eta = f_{\eta_0} + f_{\eta_*}$, and

$$\langle f_\beta, f_\eta \rangle_{\mathscr{F}_*} = \langle f_{\beta_0}, f_{\eta_0} \rangle_{\mathscr{F}_0} + \langle f_{\gamma_*}, f_{\eta_*} \rangle_{\mathscr{F}_1}. \qquad \square$$

3.3 Reproducing Kernel Hilbert Spaces

Hilbert spaces that display certain properties on certain linear operators are called reproducing kernel Hilbert spaces.

Definition 3.3.1. A function (operator) T mapping a vector space $\mathscr{X}$ into another vector space $\mathscr{Y}$ is called *linear* if $T(\lambda_1 x_1 + \lambda_2 x_2) = \lambda_1 T(x_1) + \lambda_2 T(x_2)$ for any $x_1, x_2 \in \mathscr{X}$ and any $\lambda_1, \lambda_2 \in \mathbf{R}$.

Any $p \times n$ matrix A maps vectors in $\mathbf{R}^n$ into vectors in $\mathbf{R}^p$ via $T_A(x) \equiv Ax$ and is linear.

Exercise 3.1. Consider a finite dimensional Hilbert space $\mathscr{H}$ (one that contains a finite basis) and a subspace $\mathscr{H}_0$. The operator M is a *perpendicular projection operator* onto $\mathscr{H}_0$ if $M(x) = x$ for any $x \in \mathscr{H}_0$ and $M(w) = 0$ for any $w \in \mathscr{H}_0^\perp$. Show that M must be both unique and linear. Let $\{o_1, \cdots, o_r\}$ be an orthonormal basis for $\mathscr{H}_0$ and show that

$$M(x) = \sum_{j=1}^r o_j \langle o_j, x \rangle.$$

Do these results hold for subspaces with infinite dimensions? In particular, if $\mathscr{H}$ has a countable basis and a subspace $\mathscr{H}_0$ is finite dimensional, can any $x \in \mathscr{H}$ be decomposed into $x = x_0 + x_1$ with $x_0 \in \mathscr{H}_0$ and $x_1 \in \mathscr{H}_0^\perp$?

Definition 3.3.2. The operator $T : \mathscr{X} \to \mathscr{Y}$ mapping a Banach space into a Banach space is *continuous* at $x_0 \in \mathscr{X}$ if and only if for every $\varepsilon > 0$ there exists $\delta = \delta(\varepsilon) > 0$ such that for every x with $||x - x_0||_{\mathscr{X}} < \delta$ we have $||T(x) - T(x_0)||_{\mathscr{Y}} < \varepsilon$.

Linear operators are continuous everywhere if they are continuous at 0.

We can write $x_m \to x_0$, if $||x_m - x_0||_{\mathscr{X}} \to 0$. Continuity occurs if $x_m \to x_0$ implies $T(x_m) \to T(x_0)$ in the sense that $||T(x_m) - T(x_0)||_{\mathscr{Y}} \to 0$.

Definition 3.3.3. A real valued function defined on a vector space is called a *functional*.

Any vector $a \in \mathbf{R}^n$ defines a linear functional on $\mathbf{R}^n$ via $\phi_a(x) \equiv a'x$.

EXAMPLE 3.3.4. Let $\mathscr{S}$ be the set of bounded real valued differentiable functions $\{f(x)\}$ defined on the real line. Then $\mathscr{S}$ is a vector space with the usual $+$ and $\times$ operations for functions. Some linear functionals on $\mathscr{S}$ are $\phi(f) = \int_a^b f(x)dx$ and $\phi_a(f) = \dot{f}(a)$ for some fixed a. A nonlinear functional is $\phi_{a,b}(f) = \sup_{x \in [a,b]} f(x)$. $\square$

A linear functional of particular importance is the evaluation functional.

Definition 3.3.5. Let $\mathscr{M}$ be a vector space of functions defined from $\mathscr{E}$ into $\mathbf{R}$. For any $t \in \mathscr{E}$, denote by e_t the *evaluation functional* at the point t, i.e., for $g \in \mathscr{M}$, the mapping is $e_t(g) = g(t)$.

For $\mathscr{M} = \mathbf{R}^n$, vectors can be viewed as functions from the set $\mathscr{E} = \{1, 2, \dots, n\}$ into $\mathbf{R}$. An evaluation functional is $e_i(x) = x_i$.

While it is simplest to take $\mathscr{E} \subset \mathbf{R}$ in Definition 3.3.5, we will need to consider $\mathscr{E} \subset \mathbf{R}^p$, and there is no reason not to use even more general vector spaces to define $\mathscr{E}$.

In a Hilbert space (or any normed vector space) of functions, the notion of pointwise convergence is related to the continuity of the evaluation functionals. The following are equivalent for a normed vector space $\mathscr{H}$ of real valued functions defined on $\mathscr{E}$.

(i) The evaluation functionals are continuous for all $t \in \mathscr{E}$.
(ii) If $f, f_1, f_2, \dots \in \mathscr{H}$ and $||f_n - f|| \to 0$ then $f_n(t) \to f(t)$ for every $t \in \mathscr{E}$.
(iii) For every $t \in \mathscr{E}$ there exists $K_t > 0$ such that $|f(t)| \le K_t||f||$ for all $f \in \mathscr{H}$.

Here (ii) is the definition of (i). See Máté (1990) for a proof of (iii).

To define a reproducing kernel, we need the famous *Riesz Representation Theorem*.

Theorem 3.3.6. Let $\mathscr{H}$ be a Hilbert space and let ϕ be a continuous linear functional on $\mathscr{H}$. Then there is one and only one vector $g \in \mathscr{H}$ such that

$$\phi(f) = \langle g, f \rangle, \qquad \text{for all } f \in \mathcal{H}.$$

The vector g is sometimes called the *representation* of ϕ. Nonetheless, ϕ and g are different objects: ϕ is a linear functional on $\mathcal{H}$ and g is a vector in $\mathcal{H}$. For a proof of this theorem, see Naylor and Sell (1982) or Máté (1990).

For $\mathcal{H} = \mathbf{R}^n$ with the Euclidean inner product, the representation theorem is well known because for $\phi(x)$ to be a linear functional there must exist a vector g such that

$$\phi(x) = g'x.$$

In particular, an evaluation functional is $e_i(x) = x_i$. The representation of this linear functional is the indicator vector $R_i \in \mathbf{R}^n$ that is 0 everywhere except has a 1 in the ith place because

$$e_i(x) = x_i = R_i'x.$$

In the future we will use e_i to denote both the indicator vector in $\mathbf{R}^n$ that is 0 everywhere except has a 1 in the ith place and the evaluation functional that the indicator vector represents.

An element of a set of functions, say f, from $\mathcal{E}$ into $\mathbf{R}$, is sometimes denoted $f(\cdot)$ to be explicit that the elements are functions, whereas $f(t)$ is the value of $f(\cdot)$ evaluated at $t \in \mathcal{E}$.

Applying the Riesz Representation Theorem to a Hilbert space $\mathcal{H}$ of real valued functions in which all evaluation functionals are continuous, for every $t \in \mathcal{E}$ there is a unique symmetric function $R : \mathcal{E} \times \mathcal{E} \to \mathbf{R}$ for which $R(\cdot, t) \in \mathcal{H}$ is the representation of the evaluation functional e_t, so that

$$f(t) = e_t(f) = \langle R(\cdot, t), f(\cdot) \rangle, \qquad f \in \mathcal{H}.$$

The function R is called a *reproducing kernel* (r.k.) and $f(t) = \langle R(\cdot, t), f(\cdot) \rangle$ is called the *reproducing property* of R. In particular, by the reproducing property

$$R(t, s) = \langle R(\cdot, t), R(\cdot, s) \rangle.$$

The fact that $\langle R(\cdot, t), R(\cdot, s) \rangle = \langle R(\cdot, s), R(\cdot, t) \rangle$ is why R must be a symmetric function.

Again, our use of t is suggestive of it being a real number but in general it can be a vector.

Definition 3.3.7. A Hilbert space $\mathcal{H}$ of functions defined on $\mathcal{E}$ into $\mathbf{R}$ is called a *reproducing kernel Hilbert space* if all evaluation functionals are continuous.

EXAMPLE 3.3.8. For $\mathbf{R}^n$ with inner product $\langle u, v \rangle \equiv u'Av$ where the $n \times n$ matrix A is positive definite, the r.k. $R(\cdot, \cdot)$ maps $s, t = 1, \ldots, n$ into $\mathbf{R}$, so it is really just a matrix. To see that $[R(s, t)] = A^{-1}$, note that $R(\cdot, t)$ is the tth column of $[R(s, t)]$ and we must have that $R(\cdot, t) = A^{-1}e_t$ because for any x

$$\langle R(\cdot,t),x \rangle \equiv x_t = e'_t x = e'_t A^{-1} A x = \langle A^{-1} e_t, x \rangle.$$

As earlier, e_t is the vector with 0s everywhere except a 1 in the tth place. For Euclidean $\mathbf{R}^n$, the r.k. is I. □

EXAMPLE 3.3.9. For

$$\mathscr{L}^2[0,1] = \left\{ f : \int_0^1 [f(t)]^2 dt < \infty \right\}$$

with the inner product

$$\langle f,g \rangle_{\mathscr{L}^2[0,1]} \equiv \int_0^1 f(t)g(t)dt,$$

the evaluation functionals are not continuous, so no r.k. exists. For example, if $t_n = t_0 - (1/n)$ and we define the $\mathscr{L}^2[0,1]$ functions $f(t) = \mathscr{I}_{[0,t_0]}(t)$ and $f_n(t) = \mathscr{I}_{[0,t_n)}(t)$ we have

$$\|f - f_n\| = \sqrt{\int_0^1 \mathscr{I}_{[t_n,t_0]}(t)dt} = \frac{1}{\sqrt{n}} \to 0$$

but

$$f_n(t_0) = 0 \nrightarrow 1 = f(t_0). \qquad \square$$

EXAMPLE 3.3.10. Consider the Hilbert space

$$\mathscr{W}_0^1 = \left\{ f \in \mathscr{L}^2[0,1] : f(0) = 0, \int_0^1 [\dot{f}(t)]^2 dt < \infty \right\}$$

with inner product

$$\langle f,g \rangle_{\mathscr{W}_0^1} = \int_0^1 \dot{f}(t)\dot{g}(t)dt. \tag{3.3.1}$$

In Sect. 3.1.1, we found the reproducing kernel to be $R(s,t) = \min(s,t)$. For fixed t, $R(\cdot,t)$ is an element of the function space $\mathscr{W}_0^1$, since $R(0,t) = 0$ and $\int_0^1 [\mathbf{d}_s R(s,t)]^2 ds < \infty$. Also, as shown earlier, $R(\cdot,\cdot)$ has the reproducing property. □

EXAMPLE 3.3.11. Consider the Hilbert space

$$\mathscr{W}_0^2 = \left\{ f \in \mathscr{L}^2[0,1] : f(0) = \dot{f}(0) = 0, \int_0^1 [\ddot{f}(t)]^2 dt < \infty \right\}$$

with inner product

$$\langle f,g \rangle_{\mathscr{W}_0^2} = \int_0^1 \ddot{f}(t)\ddot{g}(t)dt.$$

We begin by finding the second derivative of the representation of the evaluation functional. In particular, we show that

$$f(s) = \int_0^1 (s-u)_+ \ddot{f}(u)\,du, \tag{3.3.2}$$

where $(a)_+$ is a for $a > 0$ and 0 for $a \leq 0$.

Given any arbitrary and fixed $s \in [0,1]$,

$$\int_0^1 (s-u)_+ \ddot{f}(u)\,du = \int_0^s (s-u)\ddot{f}(u)\,du.$$

Integrating by parts

$$\int_0^s (s-u)\ddot{f}(u)\,du = (s-s)\dot{f}(s) - (s-0)\dot{f}(0) + \int_0^s \dot{f}(u)\,du = \int_0^s \dot{f}(u)\,du$$

and applying the Fundamental Theorem of Calculus to the last term,

$$\int_0^s (s-u)\ddot{f}(u)\,du = f(s) - f(0) = f(s).$$

Since the r.k. of the space $\mathscr{W}_2^0$ must satisfy $f(s) = \langle f(\cdot), R(\cdot,s)\rangle$, we see that $R(\cdot,s)$ is a function such that

$$\mathbf{d}_{uu}^2 R(u,s) = (s-u)_+.$$

We also know that $R(\cdot,s) \in \mathscr{W}_2^0$, so using $R(s,t) = \langle R(\cdot,t), R(\cdot,s)\rangle$

$$R(s,t) = \int_0^1 (t-u)_+(s-u)_+\,du = \frac{\max(s,t)\min^2(s,t)}{2} - \frac{\min^3(s,t)}{6}. \tag{3.3.3}$$

$\square$

Exercise 3.2. Do the calculus to establish Eq. (3.3.1).

EXAMPLE 3.3.12. Consider all constant functionals on $\mathbf{R}^{p-1}$,

$$\mathscr{F}_0 = \{f_a \in \mathscr{F} : f_a(x) = a,\ a \in \mathbf{R},\ x \in \mathbf{R}^{p-1}\},$$

with inner product

$$\langle f_a, f_b\rangle_{\mathscr{F}_0} = ab.$$

$\mathscr{F}_0$ has continuous evaluation functionals $e_x(f) = f(x)$, so it is an RKHS and has a unique reproducing kernel. To find the r.k., observe that $R(\cdot,x) \in \mathscr{F}_0$, so it is a constant for any x. Write $R(x) \equiv R(\cdot,x)$. By the representation theorem and the defined inner product

$$a = f_a(x) = \langle f_a(\cdot), R(\cdot,x)\rangle_{\mathscr{F}_0} = aR(x)$$

for any x and a. This implies that $R(x) \equiv 1$ so that $R(\cdot,x) \equiv 1$ and $R(\cdot,\cdot) \equiv 1$. $\square$

EXAMPLE 3.3.13.   Consider all linear functionals on $\mathbf{R}^{p-1}$ passing through the origin,

$$\mathscr{F}_1 = \{f_\gamma \in \mathscr{F} : f_\gamma(x) = x'\gamma,\ \gamma \in \mathbf{R}^{p-1},\ x \in \mathbf{R}^{p-1}\},$$

with inner product

$$\langle f_\eta, f_\gamma \rangle_{\mathscr{F}_1} = \eta'\gamma = \eta_1\gamma_1 + \eta_2\gamma_2 + \cdots + \eta_{p-1}\gamma_{p-1}.$$

The r.k. R must satisfy

$$f_\gamma(x) = \langle f_\gamma(\cdot), R(\cdot, x) \rangle_{\mathscr{F}_1}$$

for all γ and any x. Since $R(\cdot, x) \in \mathscr{F}_1$, $R(v, x) = v'u$ for some u that depends on x, i.e., $u(x)$ is the vector in $\mathbf{R}^{p-1}$ that determines the functional $R(\cdot, x) \in \mathscr{F}_1$, so $R(\cdot, x) = f_{u(x)}(\cdot)$ and $R(v, x) = v'u(x)$. By our definition of $\mathscr{F}_1$ we have

$$x'\gamma = f_\gamma(x) = \langle f_\gamma(\cdot), R(\cdot, x) \rangle_{\mathscr{F}_1} = \langle f_\gamma(\cdot), f_{u(x)}(\cdot) \rangle_{\mathscr{F}_1} = \gamma'u(x),$$

so we need $u(x)$ such that for any γ and x we have

$$x'\gamma = u(x)'\gamma.$$

It follows that $x = u(x)$. For example, taking γ to be the indicator vector e_i implies that $x_i = u_i(x)$ for every $i = 1, \ldots, p-1$. We now have $R(\cdot, x) = f_x(\cdot)$ so that

$$R(\tilde{x}, x) = \langle R(\cdot, \tilde{x}), R\cdot, x) \rangle_{\mathscr{F}_1} = \langle f_{\tilde{x}}, f_x \rangle_{\mathscr{F}_1} = \tilde{x}'x = x_1\tilde{x}_1 + x_2\tilde{x}_2 + \ldots + x_{p-1}\tilde{x}_{p-1}.\ \ \square$$

One point of these examples is that if you can characterize the evaluation functional $e_t(\cdot)$, then frequently you can find $R(s, t)$. For further examples of RKHSs with various inner products, see Berlinet and Thomas-Agnan (2004).

One further concept is useful in RKHS approaches to regression problems.

3.3.1 The Projection Principle for an RKHS

Consider the connection between the reproducing kernel R of the RKHS $\mathscr{H}$ and the reproducing kernel R_0 for a subspace $\mathscr{H}_0 \subset \mathscr{H}$. When any vector $f \in \mathscr{H}$ can be written uniquely as $f = f_0 + f_1$ with $f_0 \in \mathscr{H}_0$ and $f_1 \in \mathscr{H}_0^\perp$, more particularly, $R(\cdot, t) = R_0(\cdot, t) + R_1(\cdot, t)$ with $R_0(\cdot, t) \in \mathscr{H}_0$ and $R_1(\cdot, t) \in \mathscr{H}_0^\perp$ if and only if R_0 is the r.k. of $\mathscr{H}_0$ and R_1 is the r.k. of $\mathscr{H}_0^\perp$. For a proof see Gu (2002).

EXAMPLE 3.3.14.   Consider the affine functionals on $\mathbf{R}^{p-1}$,

$$\mathscr{F}_* = \{f_\beta \in \mathscr{F} : f_\beta(x) = \beta_0 + \beta_1 x_1 + \ldots + \beta_{p-1} x_{p-1},\ \beta \in \mathbf{R}^p,\ x \in \mathbf{R}^{p-1}\}$$

with inner product

$$\langle f_\beta, f_\eta \rangle_{\mathscr{F}_*} = \beta' \eta = \beta_0 \eta_0 + \beta_1 \eta_1 + \ldots + \beta_{p-1} \eta_{p-1}.$$

We have already derived the r.k.'s for the constant functionals $\mathscr{F}_0$ and linear functionals $\mathscr{F}_1 \equiv \mathscr{F}_0^\perp$ (call them R_0 and R_1, respectively) in Examples 3.3.12 and 3.3.13. Applying the projection principle, the r.k. for $\mathscr{F}_*$ is the sum of R_0 and R_1, i.e.,

$$R(\tilde{x}, x) = 1 + \tilde{x}'x. \qquad \qquad \square$$

For two subspaces $\mathscr{A}$ and $\mathscr{B}$ contained in a vector space $\mathscr{H}$, the *direct sum* is the space $\mathscr{D} = \{a + b : a \in \mathscr{A}, b \in \mathscr{B}\}$, written $\mathscr{D} = \mathscr{A} \oplus \mathscr{B}$ or just $\mathscr{D} = \mathscr{A} + \mathscr{B}$. Any elements $d_1, d_2 \in \mathscr{D}$ can be written as $d_1 = a_1 + b_1$ and $d_2 = a_2 + b_2$, respectively, for some $a_1, a_2 \in \mathscr{A}$ and $b_1, b_2 \in \mathscr{B}$. When the two subspaces have $\mathscr{A} \cap \mathscr{B} = \{0\}$, those decompositions are unique. If the vector space $\mathscr{H}$ is also a Hilbert space and $\mathscr{A} \perp \mathscr{B}$, the decomposition is unique and the inner product between d_1 and d_2 is $\langle d_1, d_2 \rangle = \langle a_1, a_2 \rangle + \langle b_1, b_2 \rangle$. In particular, for any finite dimensional subspace $\mathscr{H}_0$ of a Hilbert space $\mathscr{H}$, we have $\mathscr{H} = \mathscr{H}_0 \oplus \mathscr{H}_0^\perp$ and $\mathscr{H}_0 \perp \mathscr{H}_0^\perp$.

In something of a converse, suppose $\mathscr{A}$ and $\mathscr{B}$ are Hilbert space subspaces of a vector space $\mathscr{M}$ that has no norm, and suppose $\mathscr{A} \cap \mathscr{B} = \{0\}$. Then we can define a Hilbert space on $\mathscr{D} = \mathscr{A} \oplus \mathscr{B}$ by defining $\langle d_1, d_2 \rangle_{\mathscr{D}} \equiv \langle a_1, a_2 \rangle_{\mathscr{A}} + \langle b_1, b_2 \rangle_{\mathscr{B}}$. With respect to the Hilbert space $\mathscr{D}$, $\mathscr{A} \perp \mathscr{B}$. For more information about direct sum decompositions see, for example, Berlinet and Thomas-Agnan (2004) or Gu (2002).

3.4 Two Approaches

There seems to be at least two ways to use RKHS results. One is to define a penalty function for estimation that relates to an RKHS and figure out (or at least demonstrate) what the appropriate reproducing kernel is. Our applications up to this point and in the next section focus on that approach. The other approach is to choose an appropriate "kernel" function for vectors in $\mathscr{M}$, say, $R(u, v)$ and to build an RKHS around the kernel. The idea then is to work within this RKHS without really ever leaving $\mathscr{M}$. We now show how to construct such an RKHS. The next subsection illustrates the use of such an RKHS.

For an appropriate kernel function $R(\cdot, \cdot)$, one can find (but the point is that one need not find) a positive definite eigen-decomposition (*singular value decomposition*) of the function,

$$R(u, v) = \sum_{j=1}^{\infty} \eta_j \phi_j(u) \phi_j(v). \qquad (3.4.1)$$

The existence of such an eigen-decomposition implies that

$$R(\cdot, v) = \sum_{j=1}^{\infty} \eta_j \phi_j(\cdot) \phi_j(v).$$

Associated with the eigen-decomposition is a Hilbert space

$$\mathcal{H} = \left\{ f : f = \sum_{j=1}^{\infty} \alpha_j \phi_j \quad \text{with} \quad \sum_{j=1}^{\infty} \alpha_j^2 / \eta_j < \infty \right\}$$

having inner product

$$\left\langle \sum_{j=1}^{\infty} \alpha_j \phi_j, \sum_{j=1}^{\infty} \beta_j \phi_j \right\rangle \equiv \left\langle \sum_{j=1}^{\infty} \alpha_j \phi_j(\cdot), \sum_{j=1}^{\infty} \beta_j \phi_j(\cdot) \right\rangle \equiv \sum_{j=1}^{\infty} \alpha_j \beta_j / \eta_j.$$

Although the eigen-decomposition in (3.4.1) need not be unique because eigenvalues η_j may have multiplicities greater than 1, nonetheless every decomposition involves a set of basis functions ϕ_j that define a unique representation and all span the same space $\mathcal{H}$.

The chosen kernal function is the r.k. on $\mathcal{H}$ because $R(\cdot, v)$ represents the evaluation functional, i.e.,

$$\left\langle R(\cdot, v), \sum_{j=1}^{\infty} \alpha_j \phi_j(\cdot) \right\rangle = \left\langle \sum_{j=1}^{\infty} \eta_j \phi_j(\cdot) \phi_j(v), \sum_{j=1}^{\infty} \alpha_j \phi_j(\cdot) \right\rangle$$

$$= \sum_{j=1}^{\infty} \eta_j \phi_j(v) \alpha_j / \eta_j$$

$$= \sum_{j=1}^{\infty} \alpha_j \phi_j(v).$$

In this context, some commonly used kernels are, when $\mathcal{M}$ is a Hilbert space, the polynomial kernels for positive integers r

$$R(u, v) \equiv (1 + \langle u, v \rangle_{\mathcal{M}})^r$$

and, when $\mathcal{M}$ is a Banach space, the radial basis kernel

$$R(u, v) \equiv \exp(-\gamma \|u - v\|_{\mathcal{M}})$$

for some positive scalar γ. Most often the space $\mathcal{M}$ is Euclidean $\mathbf{R}^p$.

We now briefly apply these ideas, and the idea of never actually using the space $\mathcal{H}$, to the problem of testing lack of fit in a linear model.

3.4.1 Testing Lack of Fit

To translate these ideas back into linear regression theory, consider a linear model $Y = X\beta + e$ with X written in terms of its p dimensional row vectors as

$$X = \begin{bmatrix} x_1' \\ \vdots \\ x_n' \end{bmatrix}.$$

For a kernel function defined on $\mathbf{R}^p$, whenever the eigen-decomposition is finite, say,

$$R(u,v) = \sum_{j=1}^{s} \eta_j \phi_j(u) \phi_j(v)$$

we can define

$$\Phi_k = [\phi_k(x_1), \cdots, \phi_k(x_n)]', \quad \Phi = [\Phi_1, \cdots, \Phi_s],$$

and fit a new regression model

$$Y = \Phi\alpha + e, \quad \mathrm{E}(e) = 0. \tag{3.4.2}$$

This new model is appropriate for testing lack of fit because either we get $C(X) \subset C(\Phi)$ or, if not, we could use the model matrix $[X, \Phi]$ instead, with little change to our discussion.

Because the η_ks are positive, using *PA* Proposition B.51 we see that

$$C(\Phi) = C(\Phi D(\sqrt{\eta_i})) = C(\Phi D(\eta_i)\Phi').$$

The key fact is that

$$\Phi D(\eta_i)\Phi' = [R(x_h, x_i)]_{n \times n} \equiv \tilde{R},$$

so instead of fitting the linear model (3.4.2) we can fit the equivalent model

$$Y = \tilde{R}\xi + e$$

to obtain $\hat{\xi}$, fitted values, and residuals. Moreover, we can make a prediction for a new observed vector x_0 by using

$$\hat{y}_0 = [R(x_0, x_1), \cdots, R(x_0, x_n)]\hat{\xi}.$$

Thus we can execute the analysis without ever specifying the ϕ_ks or η_ks, but it would be useful to know s in order to find the MSE. The fact that the matrix Φ can be replaced by the matrix $\tilde{R}$ is known as the *kernel trick*.

Typically, one would also want to modify this discussion to deal with an intercept in the model and thus use vectors in $\mathbf{R}^{p-1}$ rather than $\mathbf{R}^p$. Moreover, there is almost nothing in this discussion that precludes the eigen-decomposition from being infinite. The only problem with an infinite decomposition is the likelihood of getting $C(\tilde{R}) = \mathbf{R}^n$ when the x_is are distinct and a not very interesting regression model (even when the x_is are not distinct). In fact, even for finite s we need to worry about overfitting the data.

EXAMPLE 3.4.1.   Consider a linear model $y_i = \beta_0 + x_i'\beta_* + \varepsilon_i$ or, in matrix form,

$$Y = X\beta + e = [J, Z] \begin{bmatrix} \beta_0 \\ \beta_* \end{bmatrix} + e$$

with

$$Z = \begin{bmatrix} x_1' \\ \vdots \\ x_n' \end{bmatrix}.$$

This model was used in Chap. 2 and in *PA* Sect. 6.2.
    We illustrate the use of

$$R(x_i, x_h) \equiv (1 + x_h'x_i)^2$$

when $p - 1 = 2$ so that $x_i' = (x_{i1}, x_{i2})$. Write

$$Z = [X_1, X_2], \quad X_1 = [x_{i1}], \quad X_2 = [x_{i2}]$$

and define the n dimensional vectors

$$X_1^2 \equiv [x_{i1}^2], \quad X_2^2 \equiv [x_{i2}^2], \quad X_1 X_2 \equiv [x_{i1} x_{i2}].$$

Note that the linear model

$$Y = W\gamma + e \equiv \gamma_{00} + \gamma_{10}X_1 + \gamma_{01}X_2 + \gamma_{20}X_1^2 + \gamma_{02}X_2^2 + \gamma_{11}X_1 X_2 + e$$

is a quadratic model in the two predictor variables. The W model contains the linear by linear interaction term $X_1 X_2$. It does not contain the analogously defined linear by quadratic terms $X_1 X_2^2$, $X_1^2 X_2$ or the quadratic by quadratic term $X_1^2 X_2^2$
    It is not difficult to see that in our example

$$R(x_i, x_h) \equiv (1 + x_h'x_i)^2 = 1 + 2x_{h1}x_{i1} + 2x_{h2}x_{i2} + x_{h1}^2 x_{i1}^2 + x_{h2}^2 x_{i2}^2 + 2x_{h1}x_{i1}x_{h2}x_{i2}$$

which leads to the hth column of $\tilde{R}$ being

$$\tilde{R}_h = J + 2x_{h1}X_1 + 2x_{h2}X_2 + x_{h1}^2 X_1^2 + x_{h2}^2 X_2^2 + 2x_{h1}x_{h2}X_1 X_2.$$

Clearly, $\tilde{R}_h \in C(J, X_1, X_2, X_1^2, X_2^2, X_1 X_2) = C(W)$, so $C(\tilde{R}) \subset C(W)$. Moreover, for regression data it is almost inconceivable that no 6 of the columns of the $n \times n$ matrix $\tilde{R}$ would be linearly independent, so in all likelihood $C(\tilde{R}) = C(W)$. In this case, fitting the model using the kernel trick is equivalent to fitting a quadratic model with linear by linear interaction to the data. We will come back to this with a slightly different and more precise argument shortly.

In this illustration, the point of the kernel trick is that instead of having to go to the trouble of defining W, one can use $R(x_i, x_h)$ to define $\tilde{R}$. Doesn't seem like much of an advantage in this problem, does it? In fact, it is hard to see where using the kernel trick with the polynomial kernels would ever provide much of an advantage. On the other hand, we saw in Chap. 1 that fitting cubic splines involved fitting a linear model with some pretty horrible linear constraints. (Admittedly, we found other ways to fit spline models without using the horrible linear constraints.) In the next section, an appropriate r.k. enables us to fit the cubic spline model without linear constraints.

A slightly more detailed argument makes the quadratic polynomial kernel relationship more precise. Redefine $x_1' = (1, a, b)$, $x_2' = (1, c, d)$. Then $x_1' x_2 = 1 + \{ac + bd\}$. Now consider transforming x_1 and x_2 into $\phi_1' \equiv [\phi_1(x_1), \ldots, \phi_6(x_1)] = (2, 2a, 2b, a^2, \sqrt{2}ab, b^2)$ and $\phi_2' = (2, 2c, 2d, c^2, \sqrt{2}cd, d^2)$. With the inner product defined by the second order polynomial r.k.,

$$
\begin{aligned}
R(x_1, x_2) &= \left(1 + x_1' x_2\right)^2 \\
&= 1 + 2x_1' x_2 + \left(x_1' x_2\right)^2 \\
&= 1 + 2\left(1 + \{ac + bd\}\right) + (1 + \{ac + bd\})^2 \\
&= 1 + 2 + 2\{ac + bd\} + \left[1 + 2\{ac + bd\} + (\{ac + bd\})^2\right] \\
&= 1 + 2 + 2\{ac + bd\} + 1 + 2\{ac + bd\} + \left[(ac)^2 + 2abcd + (bd)^2\right] \\
&= 4 + 4ac + 4bd + a^2 c^2 + 2abcd + b^2 d^2 \\
&= \phi_1' \phi_2.
\end{aligned}
$$

Now, with

$$
\Phi = \begin{bmatrix} \phi_1' \\ \vdots \\ \phi_n' \end{bmatrix},
$$

we have

$$
\tilde{R} = [R(x_i, x_j)] = [\phi_i' \phi_j] = \Phi\Phi',
$$

and we know from *PA* Appendix B that $C(\tilde{R}) = C(\Phi\Phi') = C(\Phi)$. It is quite obvious from the construction of Φ that $C(\Phi) = C(W)$. In this case, typically $r(\tilde{R}) = 6$. Again, this model only includes a linear by linear interaction term $X_1 X_2$ whereas the basic proposal for multivariate nonparametric regression in Chap. 1 also included the linear by quadratic term $X_1 X_2^2$, the quadratic by linear term $X_1^2 X_2$, and the quadratic by quadratic term $X_1^2 X_2^2$. □

Exercise 3.3. In terms of the notation for this section, the argument in Example 3.4.1 using $x_i' = (x_{i1}, x_{i2})$ involves $\mathcal{M} = \mathbf{R}^2$. What is $\mathcal{H}$?

Typically, with distinct x_is, the radial basis kernel leads to ϕ_i being an infinite vector and $r(\tilde{R}) = n$. Even the polynomial models quickly rise in rank when p gets to be of moderate size. In such cases, one should worry about overfitting. One way to address overfitting is penalized estimation.

3.5 Penalized Regression with RKHSs

The nonparametric regression model is

$$y_i = f(x_i) + \varepsilon_i, \quad i = 1, 2, \ldots, n,$$

where f is an unknown regression function and the ε_i are independent, mean 0 error terms. We start this section with three common examples of penalized regression: ridge regression, lasso regression, and smoothing splines.

3.5.1 Ridge and Lasso Regression

The classical linear regression setting is notationally identical to Example 3.4.1, i.e., $y_i = \beta_0 + x_i'\beta_* + \varepsilon_i$ with $\beta' = (\beta_0, \beta_*')$. The classical ridge regression estimator $\hat{\beta}_R$ proposed by Hoerl and Kennard (1970) minimizes

$$\sum_{i=1}^{n} \left(y_i - \beta_0 - \sum_{j=1}^{p-1} \beta_j x_{ij} \right)^2 + \lambda n \sum_{j=1}^{p-1} \beta_j^2 \tag{3.5.1}$$

where x_{ij} is the ith observation on the jth predictor variable. The resulting estimate is biased but can reduce the variance relative to least squares estimates. The tuning parameter $\lambda \geq 0$ is a constant that controls the trade-off between bias and variance in $\hat{\beta}_R$, and is often selected by some form of cross validation, see Sect. 3.6.

The lasso regression estimator $\hat{\beta}_L$ proposed by Tibshirani (1996) minimizes

$$\sum_{i=1}^{n} \left(y_i - \beta_0 - \sum_{j=1}^{p-1} \beta_j x_{ij} \right)^2 + \lambda n \sum_{j=1}^{p-1} |\beta_j|$$

The $\mathscr{L}_1$ norm for the penalty defines a Banach space but does not lend itself to an inner product, so RKHS results do not readily apply to lasso problems. However, Lin and Zhang (2006) and Storlie et al. (2010) used the RKHS framework to develop smoothing spline versions of the lasso and the adaptive lasso (cf. Zou 2006), respectively.

3.5.2 Smoothing Splines

Smoothing splines are among the most popular methods for the estimation of f due to their good empirical performance and sound theoretical support. We assume that the domain of f is $[0, 1]$. With $f^{(m)}$ the mth derivative of f, a smoothing spline estimate $\hat{f}$ is a minimizer of

$$\sum_{i=1}^{n} [y_i - f(x_i)]^2 + \lambda n \int \left[f^{(m)}(x) \right]^2 dx \tag{3.5.2}$$

The minimization of (3.5.2) is implicitly over functions with square integrable m-th derivatives. The first term in (3.5.2) encourages the fitted f to be close to the data, while the second term penalizes the roughness of f. The smoothing parameter λ controls the trade-off between the two conflicting goals. The special case of $m = 1$ is the linear smoothing spline problem.

In practice it is common to choose $m = 2$ in which case the minimizer $\hat{f}$ of (3.5.2) is called a cubic smoothing spline. As $\lambda \to \infty$, $\hat{f}$ approaches the least squares simple linear regression line, while as $\lambda \to 0$, $\hat{f}$ approaches the minimum curvature interpolant.

3.5.3 Solving the General Penalized Regression Problem

We now review a general framework that allows us to minimize (3.5.1) and (3.5.2) and many other similar problems, cf. O'Sullivan, Yandell, and Raynor (1986), Lin and Zhang (2006), Storlie et al. (2009, 2010), Gu (2002). The data model is

$$y_i = f(x_i) + \varepsilon_i, \quad E(\varepsilon_i) = 0, \quad i = 1, 2, \ldots, n, \tag{3.5.3}$$

where the ε_i are error terms and $f \in \mathcal{M}$, a given vector space of functions on $\mathcal{E}$.

Let $\mathcal{P}$ be a nonnegative penalty functional on $\mathcal{M}$ with restrictions that are discussed later. An estimate of f is obtained by minimizing

$$\frac{1}{n} \sum_{i=1}^{n} [y_i - f(x_i)]^2 + \lambda \mathcal{P}(f), \tag{3.5.4}$$

over $f \in \mathcal{S} \subset \mathcal{M}$ where $\mathcal{S}$ is chosen in a specific way.

We require that $\mathcal{P}$, in addition to being nonnegative, has a null set $\mathcal{N} = \{f \in \mathcal{M} : \mathcal{P}(f) = 0\}$ that is a subspace, that for $f_N \in \mathcal{N}$ and $f \in \mathcal{M}$, we have $\mathcal{P}(f_N + f) = \mathcal{P}(f)$, and that there exists an RKHS $\mathcal{H} \subset \mathcal{M}$ for which the inner product satisfies $\langle f, f \rangle = \mathcal{P}(f)$. This condition forces $\mathcal{N} \cap \mathcal{H} = \{0\}$. Finally, we consider a finite dimensional subspace of $\mathcal{N}$, say $\mathcal{N}_0$, with basis functions $\{\psi_1, \ldots, \psi_s\}$ and define

$$\mathcal{S} \equiv \mathcal{N}_0 \oplus \mathcal{H}.$$

In many applications $\mathcal{N}_0 = \mathcal{N}$.

EXAMPLE 3.5.1. *Linear Interpolating Splines.*
Using Example 3.3.10 we can take $\mathcal{M}$ as the subspace of $\mathcal{L}^2[0,1]$ with finite values of

$$\mathcal{P}(f) \equiv \int_0^1 \left[\dot{f}(t)\right]^2 dt.$$

This $\mathcal{P}(f)$ satisfies our four conditions with $\mathcal{H} = \mathcal{W}_0^1$. The constant functions $f(t) = a$ comprise $\mathcal{N} = \mathcal{N}_0$, so we take $\psi_1(x) \equiv 1$. Note that Eq. (3.1.6) defines an inner product on $\mathcal{W}_0^1$ but does not define an inner product on all of $\mathcal{M}$ because nonzero functions can have a zero inner product with themselves, hence the nontrivial nature of $\mathcal{N}$. □

Some nice things happen if $\mathcal{M}$ is a Hilbert space. In particular, for any two subspaces $\mathcal{N}$ and $\mathcal{H}$ with $\mathcal{N} \perp \mathcal{H}$, we have for any f in their direct sum a unique decomposition $f = f_0 + f_1$ with $f_0 \in \mathcal{N}$ and $f_1 \in \mathcal{H}$. This allows us to define $\mathcal{P}(f) \equiv \langle f_1, f_1 \rangle$ and have all our assumptions met. (Technically, we should define $\mathcal{P}$ for all functions $f \in \mathcal{M}$ but how we extend it beyond the direct sum is irrelevant.) The orthogonal decomposition of $\mathcal{S}$ is closely related to generalized additive models, cf. Wood (2006), and more generally to smoothing spline ANOVA models, cf. Gu (2002), which also include tensor product splines as a special case. Thin plate splines also fall nicely into the general RKHS framework, cf. Wahba (1990).

The key result, Wahba's Representation Theorem, also known as the *dual form* or *kernel trick*, cf. Pearce and Wand (2006), is that the minimizer of (3.5.4) is a finite linear combination of known basis functions and functions involving the reproducing kernel on $\mathcal{H}$. This allows us to find the coefficients of the linear combination by solving a quadratic minimization problem similar to those in standard linear models.

Theorem 3.5.2. *Representation Theorem.*
Any minimizer $\hat{f} \in \mathcal{S}$ of Eq. (3.5.4) has the form

$$\hat{f}(x) = \sum_{j=1}^s \beta_j \psi_j(x) + \sum_{i=1}^n \xi_i R(x_i, x), \qquad (3.5.5)$$

where $R(\cdot, \cdot)$ is the r.k. for $\mathcal{H}$.

PROOF: An informal proof is given. See Wahba (1990) or Gu (2002) for a formal proof.

Since we are working in $\mathcal{S}$, clearly, any minimizer $\hat{f}$ must have $\hat{f} = \hat{f}_0 + \hat{f}_1$ with $\hat{f}_0 \in \mathcal{N}_0$ and $\hat{f}_1 \in \mathcal{H}$. We want to establish that $\hat{f}_1(\cdot) = \sum_{i=1}^n \xi_i R(x_i, \cdot)$ for some ξ_is. Decompose $\mathcal{H}$ as $\mathcal{H} = \mathcal{H}_0 \oplus \mathcal{H}_0^\perp$ where $\mathcal{H}_0 = \text{span}\{R(x_i, \cdot), i = 1, \ldots, n\}$ so that

$$\hat{f}_1(\cdot) = \hat{f}_R(\cdot) + \eta(\cdot),$$

with

$$\hat{f}_R(\cdot) \equiv \sum_{i=1}^{n} \xi_i R(x_i, \cdot) \tag{3.5.6}$$

and $\eta(\cdot) \in \mathcal{H}_0^{\perp}$. By orthogonality and the reproducing property of the r.k.,

$$0 = \langle R(x_i, \cdot), \eta(\cdot) \rangle = \eta(x_i), \quad i = 1, \ldots, n.$$

We now establish the representation theorem. Using our assumptions about $\mathcal{P}$,

$$\frac{1}{n} \sum_{i=1}^{n} [y_i - \hat{f}(x_i)]^2 + \lambda \mathcal{P}(\hat{f}) = \frac{1}{n} \sum_{i=1}^{n} [y_i - \hat{f}_0(x_i) - \hat{f}_1(x_i)]^2 + \lambda \mathcal{P}(\hat{f}_0 + \hat{f}_1)$$

$$= \frac{1}{n} \sum_{i=1}^{n} [y_i - \hat{f}_0(x_i) - \hat{f}_1(x_i)]^2 + \lambda \mathcal{P}(\hat{f}_1)$$

$$= \frac{1}{n} \sum_{i=1}^{n} [y_i - \hat{f}_0(x_i) - \hat{f}_R(x_i) - \eta(x_i)]^2 + \lambda \mathcal{P}(\hat{f}_R + \eta).$$

Because $\eta(x_i) = 0$ and using orthogonality within $\mathcal{H}$

$$\frac{1}{n} \sum_{i=1}^{n} [y_i - \hat{f}(x_i)]^2 + \lambda \mathcal{P}(\hat{f}) = \frac{1}{n} \sum_{i=1}^{n} [y_i - \hat{f}_0(x_i) - \hat{f}_R(x_i)]^2 + \lambda \left[\mathcal{P}(\hat{f}_R) + \mathcal{P}(\eta) \right]$$

$$\geq \frac{1}{n} \sum_{i=1}^{n} [y_i - \hat{f}_0(x_i) - \hat{f}_R(x_i)]^2 + \lambda \mathcal{P}(\hat{f}_R).$$

Clearly, any $\eta \neq 0$ makes the inequality strict, so minimizers have $\eta = 0$ and $\hat{f} = \hat{f}_0 + \hat{f}_R$ with the last inequality an equality. $\qquad\square$

Corollary 3.5.3.   Among $f \in \mathcal{S}$, the least squares estimate with minimum penalty satisfies the relation (3.5.5).

PROOF: As in the proof of the Theorem 3.5.2, any $f \in \mathcal{S}$ can be written as $f_0 + f_R + \eta$ and

$$\frac{1}{n} \sum_{i=1}^{n} [y_i - f_0(x_i) - f_R(x_i) - \eta(x_i)]^2 = \frac{1}{n} \sum_{i=1}^{n} [y_i - f_0(x_i) - f_R(x_i)]^2.$$

However,

$$\mathcal{P}(f) = \mathcal{P}(f_0 + f_R + \eta) = \mathcal{P}(f_R + \eta) = \mathcal{P}(f_R) + \mathcal{P}(\eta) \geq \mathcal{P}(f_R) = \mathcal{P}(f_0 + f_R).$$

Thus for any $f \in \mathcal{S}$, there is a function of the form (3.5.5) that has the same squared error and at least as small a penalty. $\qquad\square$

A remarkable feature of the result in (3.5.5) is that the form of the minimizer is represented by a finite dimensional basis, regardless of the dimension of $\mathcal{H}$.

For example, $\mathscr{H} = \mathscr{W}_0^1$ requires an infinite expansion of basis functions to represent all functions in the space, yet the *solution* of the minimization can be represented by a finite basis!

Once we know that the minimizer takes the form (3.5.5), we can find the coefficients of the linear combination by solving a quadratic minimization problem similar to those in standard linear models. This occurs because we can write $\mathscr{P}(\hat{f}) = \mathscr{P}(\hat{f}_R)$ as a quadratic form in $\xi = (\xi_1, \ldots, \xi_n)'$. Define $\tilde{R}$ as the $n \times n$ matrix where the i, j entry is $\tilde{r}_{ij} = R(x_i, x_j)$. The matrix $\tilde{R}$ is commonly referred to as the *Gram* matrix. Now, using the reproducing property of R, write

$$\mathscr{P}(\hat{f}_R) = \left\langle \sum_{i=1}^n \xi_i R(x_i, \cdot), \sum_{j=1}^n \xi_j R(x_j, \cdot) \right\rangle = \sum_{i=1}^n \sum_{j=1}^n \xi_i \xi_j R(x_i, x_j) = \xi' \tilde{R} \xi.$$

Define the observation vector $Y = [y_1, \ldots, y_n]'$, and let W be the $n \times s$ matrix defined by $w_{ij} = \psi_j(x_i)$. With the usual norm for Euclidean $\mathbf{R}^n$, the minimization of (3.5.4) takes the form

$$\min_{\beta, \xi} \left\{ \frac{1}{n} \|Y - W\beta - \tilde{R}\xi\|^2 + \lambda \xi' \tilde{R} \xi \right\}. \tag{3.5.7}$$

The minimization in (3.5.7) is a special case of the generalized ridge regression minimization problem (2.2.2) and has the solutions of (2.2.4),

$$\hat{\xi} = [\tilde{R}(I - M_W)\tilde{R} + \lambda n \tilde{R}]^- \tilde{R}(I - M_W)Y \qquad \hat{\beta} = [W'W]^- W'(Y - \tilde{R}\hat{\xi}), \tag{3.5.8}$$

where M_W is the ppo onto $C(W)$.

Alternatively, the solution to the minimization of (3.5.4) can be found from the normal equations associated with the linear model (2.2.3) used for solving generalized ridge regression. In this application the model becomes

$$\begin{bmatrix} Y \\ 0 \end{bmatrix} = \begin{bmatrix} W & \tilde{R} \\ 0 & \sqrt{\lambda n}\tilde{S} \end{bmatrix} \begin{bmatrix} \beta \\ \xi \end{bmatrix} + \begin{bmatrix} e \\ \tilde{e} \end{bmatrix} \tag{3.5.9}$$

where $\tilde{R} = \tilde{S}'\tilde{S}$. The normal equations are

$$\begin{bmatrix} W'W & W'\tilde{R} \\ \tilde{R}W & \tilde{R}\tilde{R} + \lambda n \tilde{R} \end{bmatrix} \begin{bmatrix} \beta \\ \xi \end{bmatrix} = \begin{bmatrix} W'Y \\ \tilde{R}Y \end{bmatrix} \tag{3.5.10}$$

or they can be reduced to

$$\begin{bmatrix} W'W & 0 \\ 0 & \tilde{R}(I - M_W)\tilde{R} + \lambda n \tilde{R} \end{bmatrix} \begin{bmatrix} \delta \\ \xi \end{bmatrix} = \begin{bmatrix} W'Y \\ \tilde{R}(I - M_W)Y \end{bmatrix} \tag{3.5.11}$$

with $W\delta \equiv W\beta + M_W\tilde{R}\xi$. Both of these require solving a system of $s + n$ linear equations to find the estimates.

The simplest way to *program* these results may be to find

$$\begin{bmatrix} \hat{\beta} \\ \hat{\xi} \end{bmatrix} = \begin{bmatrix} W'W & W'\tilde{R} \\ \tilde{R}W & \tilde{R}\tilde{R}+\lambda n\tilde{R} \end{bmatrix}^{-} \begin{bmatrix} W'Y \\ \tilde{R}Y \end{bmatrix}$$

where it is easy to find the Moore–Penrose generalized inverse of a nonnegative definite matrix by using its eigen-decomposition as demonstrated in the proof of *PA* Theorem B.38. See the supplemental material to Nosedal-Sanchez et al. (2012) for examples of programs in the R language. However, it may be more efficient to find the two generalized inverses of dimensions s and n associated with solving (3.5.11) rather than one generalized inverse of dimension $s+n$ to solve (3.5.10).

For clarity, we have restricted our attention to minimizing (3.5.4), which incorporates squared error loss between the observations and the unknown function evaluations. The representation theorem holds for more general loss functions, e.g., those from logistic or Poisson regression, see Gu (2002).

EXAMPLE 3.5.4.   Reconsider fitting linear splines to the data of Example 3.1.2 but without the requirement that $f(0) = 0 = y_0$. The observed data are

$$Y \equiv \begin{bmatrix} y_1 \\ \vdots \\ y_n \end{bmatrix} = \begin{bmatrix} 0.1 \\ 1 \\ 2 \\ 1.5 \\ 1.75 \end{bmatrix}, \quad X \equiv \begin{bmatrix} x'_1 \\ \vdots \\ x'_n \end{bmatrix} \equiv \begin{bmatrix} t_1 \\ \vdots \\ t_n \end{bmatrix} = \begin{bmatrix} 0.1 \\ 0.25 \\ 0.5 \\ 0.75 \\ 1 \end{bmatrix}$$

We have

$$\mathscr{P}(f) \equiv \int_0^1 [\dot{f}(t)]^2 dt,$$

$\mathscr{N} = \mathscr{N}_0$, the one dimensional space spanned by $\psi(x) = 1$, and $\mathscr{H} = \mathscr{W}_0^1$, so $R(s,t) = \min(s,t)$. It follows that $W = J_n$ and for these data

$$\tilde{R} = \begin{bmatrix} 0.1 & 0.1 & 0.1 & 0.1 & 0.1 \\ 0.1 & 0.25 & 0.25 & 0.25 & 0.25 \\ 0.1 & 0.25 & 0.5 & 0.5 & 0.5 \\ 0.1 & 0.25 & 0.5 & 0.75 & 0.75 \\ 0.1 & 0.25 & 0.5 & 0.75 & 1 \end{bmatrix}.$$

A similar structure holds whenever $t_1 \le t_2 \le \cdots \le t_n$, which we henceforth assume.

If we set $\lambda = 0$ to get linear interpolating splines, we are just solving a least squares problem and the least squares problem is minimized by any function with $f(x_i) = y_i$, $i = 1,\ldots,n$. In particular, with five distinct x_i values, a quartic (fourth degree) polynomial will fit the data perfectly as would appropriate sine and cosine models as discussed in Chap. 1. One problem with these solutions is that they will not be particularly smooth as judged by having small values of $\mathscr{P}(f)$. In any case, Theorem 3.5.2 tells us that there is also a solution that can be found by fitting the model

$$Y = J\beta_0 + \tilde{R}\xi + e.$$

The first column of $\tilde{R}$ is $t_1 J$, so the estimates are not unique. We could fit this model using a standard regression package deleting the first column of $\tilde{R}$, which is equivalent to imposing a side condition of $\xi_1 = 0$. If we do so, we get $\hat{\beta}_0 = -0.5$ and $\hat{\xi} = (0, 2, 6, -3, 1)'$. A predicted value for a new t is

$$\hat{f}(t) = \hat{\beta}_0 + [R(t, t_1), \cdots, R(t, t_n)]\hat{\xi}$$
$$= -0.5 + 0R(t, t_1) + 2R(t, t_2) + 6R(t, t_3) - 3R(t, t_4) + 1R(t, t_5).$$

Or we could drop the "intercept" by imposing the side condition $\beta_0 = 0$ and fit only $Y = \tilde{R}\xi + e$. Now a predicted value is

$$\hat{f}(t) = \hat{\beta}_0 + [R(t, t_1), \cdots, R(t, t_n)]\hat{\xi}$$
$$= 0 - 5R(t, t_1) + 2R(t, t_2) + 6R(t, t_3) - 3R(t, t_4) + 1R(t, t_5).$$

which is the solution we found in Example 3.1.2. These two solutions are different for $0 \le t < t_1$ but agree for $t_1 \le t \le 1$. There is no unique solution. In fact, for these data we need only have $\hat{\beta}_0$ and $\hat{\xi}_1$ satisfying

$$-0.5 = \hat{\beta}_0 + \hat{\xi}_1 \times 0.1 = \hat{\beta}_0 + \hat{\xi}_1 R(t_1, t_1).$$

Moreover all solutions give a horizontal line from t_n to 1, an issue hidden in this example by the fact that $t_n = 1$. Actually, to fit interpolating splines, you do not need to do regression, you merely need to solve the system of linear equations $Y = J\beta_0 + \tilde{R}\xi$. □

We can find an interpolating function that uses the form of (3.5.11) but what if we wanted an interpolator that minimized $\mathscr{P}(f)$? Corollary 3.5.3 tells us the minimizer has this form, but not how to find the minimizer itself. Intuitively, the answer is quite clear. We need an interpolator that is a flat a possible. Which of the possible answers from the linear model will minimize the roughness penalty? Clearly, pick $\hat{\xi}_1$ to give a flat function on $[0, t_1]$, so the derivative is 0 and its contribution to $\mathscr{P}$ is minimized. Moreover, this shows that having a horizontal line from t_n to 1 is actually a feature and not a bug. We return to this question after the next example.

EXAMPLE 3.5.5.   Reconsider Example 3.5.1 but now including the point at $(t_0, y_0) = (0, 0)$. The observed data are

$$Y = \begin{bmatrix} y_0 \\ \vdots \\ y_{n-1} \end{bmatrix} = \begin{bmatrix} 0 \\ 0.1 \\ 1 \\ 2 \\ 1.5 \\ 1.75 \end{bmatrix}, \quad \begin{bmatrix} t_0 \\ \vdots \\ t_{n-1} \end{bmatrix} = \begin{bmatrix} 0 \\ 0.1 \\ 0.25 \\ 0.5 \\ 0.75 \\ 1 \end{bmatrix}$$

As before $W = J_n$ (but n is 6 rather than 5) and for these data

$$\tilde{R} = \begin{bmatrix} 0 & 0 & 0 & 0 & 0 & 0 \\ 0 & 0.1 & 0.1 & 0.1 & 0.1 & 0.1 \\ 0 & 0.1 & 0.25 & 0.25 & 0.25 & 0.25 \\ 0 & 0.1 & 0.25 & 0.5 & 0.5 & 0.5 \\ 0 & 0.1 & 0.25 & 0.5 & 0.75 & 0.75 \\ 0 & 0.1 & 0.25 & 0.5 & 0.75 & 1 \end{bmatrix}.$$

For $\lambda = 0$, as with the previous example, any interpolation function minimizes the squared errors. Similar to the previous example, fitting a fifth degree polynomal provides an interpolator, but not one that is necessarily very smooth. As before, fitting the linear model $Y = J\beta_0 + \tilde{R}\xi + e$ with $\xi = (\xi_0, \ldots, \xi_5)$ is one way to produce an interpolator. The estimate of ξ_0 can be anything. ξ_0 is irrelevant to fitting the data because it corresponds to a column of 0s. A predicted value is

$$\hat{f}(t) = \hat{\beta}_0 + [R(t,t_0), \cdots, R(t,t_n)]\hat{\xi}$$
$$= 0 - 5R(t,t_1) + 2R(t,t_2) + 6R(t,t_3) - 3R(t,t_4) + 1R(t,t_5)$$

because $R(t,t_0) = 0$, making the estimate of ξ_0 irrelevant. This is the solution we found in Example 3.1.2.

Now suppose $y_0 \neq 0$. We get a similar solution but $\hat{\beta}_0 = y_0$ with a different $\hat{\xi}_1$ (the other $\hat{\xi}_k$s remain the same). In particular, if $y_0 = 2$, we get $\hat{\xi}_1 = -25$ so that

$$y_1 = 0.1 = 2 - 25R(t_1,t_1) + 2R(t_1,t_2) + 6R(t_1,t_3) - 3R(t_1,t_4) + 1R(t_1,t_5)$$
$$= 2 + (-25 + 2 + 6 - 3 + 1)(0.1). \quad \square$$

Typically, for bivariate data (x_i, y_i), $i = 1, \ldots, n$, the predictor variable x_i is not between 0 and 1, although our procedure requires that it be. If we standardize the predictor so that

$$t_i = \frac{x_i - \min_k(x_k)}{\max_k(x_k) - \min_k(x_k)}$$

neither the nonuniqueness at the beginning of $\hat{f}(t)$ or the flatness of the end of $\hat{f}(t)$ remain issues in linear interpolation. In that case, as in Example 3.5.3, $t_1 = 0$, so the first column of $\tilde{R}$ is 0, the estimates of all parameters are unique except for ξ_1, but $R(t,t_1) = 0$ so ξ_1 is irrelevant to predictions. Of course, this standardization disallows any possibility of extrapolating the results beyond the observed data.

Proposition 3.5.6. For simple regression data (t_i, y_i), $i = 1, \ldots, n$ with strictly increasing t_i, there exists a minimum "linear spline" penalty interpolator for $f \in \mathscr{S}$ that has the form $\hat{f}(t) = \hat{\beta}_0 + \sum_{i=1}^n \hat{\xi}_i \min(t_i, t)$ with $\sum_{i=1}^n \hat{\xi}_i = 0$ and thus has $\hat{f}(t) = y_1$ for $0 \leq t \leq t_1$ and $\hat{f}(t) = y_n$ for $t_n \leq t \leq 1$. This minimizer is unique.

PROOF: We obtain interpolating coefficients (that minimize the sum of squared errors), say, $\hat{\beta}_0$ and $\hat{\xi}_i$s from solving $Y = J\beta_0 + \tilde{R}\xi$. Using the structure of $\tilde{R}$, if we set $\hat{\xi}_1 = 0$, there is a unique solution with, say, $\hat{\beta}_{00}$, but in general we merely have

$\hat{\beta}_{00} = \hat{\beta}_0 + \hat{\xi}_1 t_1$ with no other restrictions on the choices of $\hat{\beta}_0$ and $\hat{\xi}_1$. Choose $\hat{\xi}_1$ to minimize the penalty by writing

$$\mathscr{P}(\hat{f}) = \hat{\xi}'\tilde{R}\hat{\xi} = \begin{bmatrix} \hat{\xi}_1 & \hat{\xi}'_* \end{bmatrix} \begin{bmatrix} t_1 & t_1 J'_{n-1} \\ t_1 J_{n-1} & \tilde{R}_2 \end{bmatrix} \begin{bmatrix} \hat{\xi}_1 \\ \hat{\xi}_* \end{bmatrix}$$

or

$$\mathscr{P}(\hat{f}) = t_1 \hat{\xi}_1^2 + 2 t_1 \hat{\xi}_1 \hat{\xi}'_* J_{n-1} + \hat{\xi}'_* \tilde{R}_2 \hat{\xi}_*.$$

For $t_1 = 0$, $\hat{\xi}_1$ can be anything, so it can be chosen so that the proposition holds, noting that $\hat{f}$ is unique regardless of how $\hat{\xi}_1$ is chosen. For $t_1 > 0$, to minimize the penalty as a function of $\hat{\xi}_1$ set the derivative equal to 0 yielding $0 = \hat{\xi}_1 + \hat{\xi}'_* J_{n-1} = \sum_{i=1}^n \hat{\xi}_i$. Since $\hat{f}$ is an interpolator,

$$y_1 = \hat{f}(t_1) = \hat{\beta}_0 + \sum_{i=1}^n \hat{\xi}_i \min(t_i, t_1) = \hat{\beta}_0 + \sum_{i=1}^n \hat{\xi}_i t_1 = \hat{\beta}_0.$$

and for $0 \le t \le t_1$,

$$\hat{f}(t) = \hat{\beta}_0 + \sum_{i=1}^n \hat{\xi}_i \min(t_i, t) = \hat{\beta}_0 + \sum_{i=1}^n \hat{\xi}_i t = \hat{\beta}_0.$$

Finally, any interpolator of this form based on $\min(t_i, t)$ has, for $t_n \le t \le 1$, $y_n = \hat{f}(t_n) = \hat{f}(t)$. □

It is pretty clear that we do not want to use this procedure to extrapolate the data! In general, to fit linear smoothing splines with $\lambda > 0$ using a regression program, we need to incorporate

$$\tilde{S} = \begin{bmatrix} \sqrt{t_1} & \sqrt{t_1} & \sqrt{t_1} & \cdots & \sqrt{t_1} & \sqrt{t_1} \\ 0 & \sqrt{t_2 - t_1} & \sqrt{t_2 - t_1} & \cdots & \sqrt{t_2 - t_1} & \sqrt{t_2 - t_1} \\ 0 & 0 & \sqrt{t_3 - t_2} & \cdots & \sqrt{t_3 - t_2} & \sqrt{t_3 - t_2} \\ \vdots & \vdots & & \ddots & & \vdots \\ 0 & 0 & 0 & \cdots & 0 & \sqrt{t_n - t_{n-1}} \end{bmatrix}$$

into model (3.5.9).

Exercise 3.4.   Fit linear smoothing splines to the data of Example 3.5.2 using $\lambda = 0.001, 0.01, 0.1, 0.5, 1, 2$ and compare the results. Graph the results.

3.5.4 General Solution Applied to Ridge Regression

In Chap. 2, we used standard linear model ideas to solve the generalized ridge regression problem. In this section we used the generalized ridge regression results of

Chap. 2, along with RKHS theory, to solve the general penalized estimation problem. Now, somewhat circularly, we demonstrate how the general penalized estimation results apply to the classical ridge regression problem.

To solve the classical ridge problem we use the same notation as in Example 3.4.1, which is that of Chap. 2 and *PA* Sect. 6.2,

$$Y = X\beta + e = [J, Z]\begin{bmatrix} \beta_0 \\ \beta_* \end{bmatrix} + e$$

where the rows of Z are the predictor vectors x_i, $i = 1, \ldots, n$. In particular, the classical ridge regression problem estimates β_0 and β_* by minimizing

$$[Y - J\beta_0 - Z\beta_*]'[Y - J\beta_0 - Z\beta_*] + \lambda n\beta_*'\beta_*,$$

where $\lambda n = k$ from Chap. 2. We will see that RKHS theory essentially rewrites the model as $Y = [J, ZZ'][\beta_0, \xi] + e$ before solving the generalized ridge regression problem of minimizing

$$[Y - J\beta_0 - ZZ'\xi]'[Y - J\beta_0 - ZZ'\xi] + \lambda n\xi'ZZ'\xi$$

in which, quite clearly, $\beta_* = Z'\xi$.

To put the ridge regression problem in the framework of Theorem 3.5.2, reconsider Example 3.3.14. Take the affine functions

$$\mathcal{M} = \mathcal{F}_* = \left\{ f : f(x) = \beta_0 + \sum_{j=1}^{p-1} \beta_j x_j \right\}. \tag{3.5.12}$$

with the penalty function

$$\mathcal{P}(f) = \sum_{j=1}^{p-1} \beta_j^2.$$

Take $\mathcal{N}$ as $\mathcal{F}_0$ and $\mathcal{H}$ as $\mathcal{F}_0^\perp$ (which is $\mathcal{F}_1$ from Example 3.3.13). The dimension of $\mathcal{N} = \mathcal{N}_0$ is $s = 1$ with $\psi_1(x) = 1$. The r.k. on $\mathcal{H}$ comes from Example 3.3.13 and is

$$R(\tilde{x}, x) = x'\tilde{x}.$$

All of this leads us to

$$W = J, \qquad \tilde{R} = ZZ', \qquad \tilde{S} = Z'$$

in the minimization (3.5.7) and the model (3.5.9).

The estimates from (3.5.8) reduce to

$$\hat{\xi} = \left[ZZ'\left(I - \frac{1}{n}JJ'\right)ZZ' + \lambda nZZ' \right]^{-} ZZ'\left(I - \frac{1}{n}JJ'\right)Y,$$

$$\hat{\beta}_0 = (J'J)^{-1}J'(Y - ZZ'\hat{\xi})$$

or

$$\hat{\xi} = \left\{ Z \left[Z' \left(I - \frac{1}{n} JJ' \right) Z + \lambda nI \right] Z' \right\}^{-} ZZ' \left(I - \frac{1}{n} JJ' \right) Y, \quad \hat{\beta}_0 = \frac{1}{n} J' (Y - ZZ' \hat{\xi}).$$

It is easy to see by checking the definition of a generalized inverse that we can take

$$\left\{ Z \left[Z' \left(I - \frac{1}{n} JJ' \right) Z + \lambda nI \right] Z' \right\}^{-}$$

$$= Z(Z'Z)^{-1} \left[Z' \left(I - \frac{1}{n} JJ' \right) Z + \lambda nI \right]^{-1} (Z'Z)^{-1} Z'$$

so

$$\hat{\xi} = Z(Z'Z)^{-1} \left[Z' \left(I - \frac{1}{n} JJ' \right) Z + \lambda nI \right]^{-1} (Z'Z)^{-1} Z'ZZ' \left(I - \frac{1}{n} JJ' \right) Y$$

$$= Z(Z'Z)^{-1} \left[Z' \left(I - \frac{1}{n} JJ' \right) Z + \lambda nI \right]^{-1} Z' \left(I - \frac{1}{n} JJ' \right) Y$$

$$= Z(Z'Z)^{-1} \hat{\gamma}$$

where $\hat{\gamma}$ is the classical ridge estimator from (2.2.4) in which "classical" means that $Q = I$ in that formula. With $\beta_* = Z' \xi$,

$$\hat{\beta}_{*R} = Z' \hat{\xi} = Z'Z(Z'Z)^{-1} \hat{\gamma} = \hat{\gamma}.$$

This version of $\hat{\xi}$ leads to

$$\hat{\beta}_{0R} = \frac{1}{n} J'[Y - ZZ'Z(Z'Z)^{-1} \hat{\gamma}] = \frac{1}{n} J' (Y - Z\hat{\beta}_{*R}).$$

which also agrees with the classical ridge estimates from (2.2.4). Finally we have

$$\hat{Y} = \hat{\beta}_{0R} J + ZZ' \hat{\xi} = \hat{\beta}_{0R} J + ZZ' \left[Z(Z'Z)^{-1} \hat{\gamma} \right] = \hat{\beta}_{0R} J + Z\hat{\beta}_{*R}.$$

In particular, for prediction at a new vector x_0 we have

$$\hat{f}(x_0) = \hat{\beta}_{0R} + [R(x_0, x_1), \cdots, R(x_0, x_1)] \hat{\xi} = \hat{\beta}_{0R} + x_0' Z' \hat{\xi} = \hat{\beta}_{0R} + x_0' \hat{\beta}_{*R}.$$

3.5.5 General Solution Applied to Cubic Smoothing Splines

Consider again the bivariate regression problem $y_i = f(t_i) + \varepsilon_i$, $i = 1, 2, \ldots, n$ where $t_i \in [0, 1]$ and $E(\varepsilon_i) = 0$. Now consider the cubic smoothing spline estimates obtained by finding a function that minimizes

$$\sum_{i=1}^{n} [y_i - f(t_i)]^2 + \lambda n \int \ddot{f}(u)^2 du.$$

For the general solution we take $\mathcal{M}$ as the subset of $\mathscr{L}^2[0,1]$ with square integrable second derivatives. We take $\mathscr{P}(f)$ as in (3.2.2) which is also the inner product on $\mathscr{H} = \mathscr{W}_0^2$ from Example 3.3.11. The r.k. was given in Example 3.3.11 as Eq. (3.3) and is also given later. With this $\mathscr{P}$, the space $\mathscr{N}$ is all functions with $\ddot{f} = 0$. We take $\mathscr{N}_0$ to be linear functions $f_0(t) = \beta_0 + \beta_1 t$, so basis functions are $\psi_1(t) = 1$ and $\psi_2(t) = t$.

This structure leads to

$$W = \begin{bmatrix} 1 & t_1 \\ \vdots & \vdots \\ 1 & t_n \end{bmatrix}$$

and an $\tilde{R}$ that is ugly, but easily computed, in the minimization (3.5.7) and the model (3.5.9). Using

$$R(s,t) = \int_0^1 (t-u)_+(s-u)_+ du = \frac{\max(s,t)\min^2(s,t)}{2} - \frac{\min^3(s,t)}{6},$$

it is well known that the basis functions $R(t_h, t)$, that determine the columns $\tilde{R}_h$ of $\tilde{R}$ when evaluated at the data points t_i, form a natural cubic spline with knots at the distinct values of t_h. See Wahba (1990) for a justification, which just involves some algebra. The $\max(t_h, t)$ and $\min(t_h, t)$ in $R(t_h, t)$ combine in a way to produce knots at the t_h, while the degree of the polynomial spline is clearly three, since it is the highest power present in $R(s,t)$. This is the reason that the minimization problem in this section has been given the name *cubic smoothing spline*.

Nosedal-Sanchez et al. (2012) demonstrate with R code the fitting of cubic smoothing splines on some real data. The demonstration also includes searching for the best value of the tuning parameter λ which is briefly discussed in the next section.

3.6 Choosing the Degree of Smoothness

With the penalized regression procedures described above, the choice of the smoothing parameter λ is an important issue. There are many methods available for this task, e.g., visual inspection of the fit; m-fold cross-validation, Kohavi (1995); AIC/unbiased risk estimation; generalized maximum likelihood, Wahba (1990); generalized cross-validation (GCV), Craven and Wahba (1979); among others. Nosedal-Sanchez et al. (2012) used the GCV approach which works as follows. Suppose that for fixed λ the estimates determine fitted values $\hat{Y} = A(\lambda)Y$. The GCV choice of λ is the minimizer of

$$V(\lambda) = \frac{\frac{1}{n}\|I - A(\lambda)Y\|^2}{\left\{\frac{1}{n}\text{tr}[I - A(\lambda)]\right\}^2}.$$

For more details about GCV and other methods of finding λ see Golub, Heath, and Wahba (1979), Allen (1974), Wecker and Ansley (1983), and Wahba (1990).

References

Allen, D. (1974). The relationship between variable selection and data augmentation and a method for prediction. *Technometrics, 16*, 125–127.

Berman, M. (1994). Automated smoothing of image and other regularly spaced data. *IEEE Transactions on Pattern Analysis and Machine Intelligence, 16*, 460–468.

Berlinet, A., & Thomas-Agnan, C. (2004). *Reproducing kernel Hilbert spaces in probability and statistics*. Norwell, MA: Kluwer Academic Publishers.

Bivand, R., Pebesma, E., & Gómez-Rubio, V. (2013). *Applied spatial data analysis with R* (2nd ed.). New York, NY: Springer.

Bühlmann, P., & van de Geer, S. (2011). *Statistics for high-dimensional data: Methods, theory and applications*. Heidelberg, NY: Springer.

Bühlmann, P., Kalisch, M., & Meier, L. (2014). High-dimensional statistics with a view toward applications in biology. *Annual Review of Statistics and Its Applications, 1*, 255–278.

Craven, P., & Wahba, G. (1979). Smoothing noisy data with spline functions: Estimating the correct degree of smoothing by the method of generalized cross-validation. *Numerical Mathematics, 31*, 377–403.

de Barra, G. (1981). *Measure theory and integration*. West Sussex: Horwood Publishing.

Eubank, R. (1999). *Nonparametric regression and spline smoothing*. New York, NY: Marcel Dekker.

Golub, G., Heath, M., & Wahba, G. (1979). Generalized cross-validation as a method for choosing a good ridge parameter. *Technometrics, 21*, 215–223.

Gu, C. (2002). *Smoothing spline ANOVA models*. New York, NY: Springer.

Halmos, P. R. (1958). *Finite-dimensional vector spaces* (2nd ed.). Princeton, NJ: Van Nostrand.

Harville, D. A. (1997). *Matrix algebra from a statistician's perspective*. New York, NY: Springer.

Hastie, T., Tibshirani, R., & Friedman, J. (2016). *The elements of statistical learning: Data mining, inference, and prediction* (2nd ed.). New York, NY: Springer.

Heckman, N. (2012). The theory and application of penalized methods or Reproducing Kernel Hilbert Spaces made easy. *Statistics Surveys, 6*, 113–141.

Hoerl, A. E., & Kennard, R. (1970). Ridge regression: Biased estimation for non-orthogonal problems. *Technometrics, 12*, 55–67.

Kohavi, R. (1995). A study of cross-validation and bootstrap for accuracy estimation and model selection. *Proceedings of the Fourteenth International Joint Conference on Artificial Intelligence, 2*(12), 1137–1143.

Lin, Y., & Zhang, H. (2006). Component selection and smoothing in smoothing spline analysis of variance models. *Annals of Statistics, 34*, 2272–2297.

Máté, L. (1990). *Hilbert space methods in science and engineering*. London: Taylor & Francis.

Moguerza, J. M., & Muñoz, A. (2006). Support vector machines with applications. *Statistical Science, 21*, 322–336.

Naylor, A., & Sell, G. (1982). *Linear operator theory in engineering and science*. New York, NY: Springer.

Nosedal-Sanchez, A., Storlie, C.B., Lee, T. C. M., & Christensen, R. (2012). Reproducing kernel Hilbert spaces for penalized regression: A tutorial. *The American Statistician, 66*, 50–60

O'Sullivan, F., Yandell, B., & Raynor, W. (1986). Automatic smoothing of regression functions in generalized linear models. *Journal of the American Statistical Association, 81*, 96–103.

Pearce, N. D., & Wand, M. P. (2006). Penalized splines and reproducing kernel methods. *The American Statistician, 60*, 233–240.

Ramsay, J., & Silverman, B. (2005). *Functional data analysis*. New York, NY: Springer.

Rustagi, J. (1994). *Optimization techniques in statistics*. London: Academic Press.

Storlie, C., Bondell, H., & Reich, B. (2010). A locally adaptive penalty for estimation of functions with varying roughness. *Journal of Computational and Graphical Statistics, 19*, 569–589.

Storlie, C., Bondell, H., Reich, B., & Zhang, H. (2010). Surface estimation, variable selection, and the nonparametric oracle property. *Statistica Sinica, 21*, 679–705.

Storlie, C., Swiler, L., Helton, J., & Sallaberry, C. (2009). Implementation and evaluation of nonparametric regression procedures for sensitivity analysis of computationally demanding models. *Reliability Engineering and System Safety, 94*, 1735–1763.

Tibshirani, R. (1996). Regression shrinkage and selection via the lasso. *Journal of the Royal Statistical Society: Series B, 58*, 267–288.

Wahba, G. (1990). *Spline models for observational data* (Vol. 59). CBMS-NSF Regional Conference Series in Applied Mathematics. Philadelphia, PA: SIAM.

Wainwright, M. J. (2014). Structured regularizers for high-dimensional problems: Statistical and computational issues. *Annual Review of Statistics and Its Applications, 1*, 233–253.

Wecker, W., & Ansley, C. (1983). The signal extraction approach to nonlinear regression and spline smoothing. *Journal of the American Statistical Association, 78*, 81–89.

Wood, S. (2006). *Generalized additive models: An introduction with R*. Boca Raton, FL: CRC Press.

Young, N. (1988). *An introduction to Hilbert space*. Cambridge: Cambridge University Press.

Zhu, M. (2008). Kernels and ensembles. *The American Statistician, 62*, 97–109.

Zou, H. (2006). The adaptive lasso and its oracle properties. *Journal of the American Statistical Association, 101*, 1418–1429.

Chapter 4
Covariance Parameter Estimation

Abstract This chapter reviews fundamental ideas from linear model theory for dealing with dependent or heteroscedastic data when the nature of the dependence or heteroscedasticity is known. It then introduces general ideas for estimating dependence or heteroscedasticity when their exact natures are unknown. Most of the book, after this chapter, consists of applications of these ideas to specific models.

About three fourths of this book deals with dependent data in some sort of linear model. In particular, Chaps. 4 through 12 and 14 all examine dependent data to which some form of linear model theory is relevant. Many issues in these applications rely on results that are valid quite generally. Rather than treating these issues piecemeal, as they come up, we discuss them all at once in this chapter and refer back to this discussion as necessary. By far the most important of these issues are procedures for estimating parameters in the covariance matrix of a linear model. We also discuss the effect that estimation of covariance parameters has on estimates of mean parameters and on predictions. Parts of this chapter rely heavily on Appendix A. Some mathematical issues related to residual maximum likelihood estimation have been relegated to Appendix C.

4.1 Introduction and Review

Linear models have the form

$$Y = X\beta + e, \qquad \mathrm{E}(e) = 0.$$

In this chapter we look at linear models that have covariance matrices that are more complicated than those considered in *PA*. Primarily, *PA* considered $\mathrm{Cov}(e) = \sigma^2 I$ or, for V known and positive definite, $\mathrm{Cov}(e) = \sigma^2 V$. Chapter 11 of *PA* considered $\mathrm{Cov}(e) = \sigma_s^2 I + \sigma_w^2 X_1 X_1'$ for a known matrix of indicator variables X_1 but unknown variance parameters σ_s^2 and σ_w^2.

© Springer Nature Switzerland AG 2019
R. Christensen, *Advanced Linear Modeling*, Springer Texts in Statistics,
https://doi.org/10.1007/978-3-030-29164-8_4

We now consider normal theory general linear models wherein the covariance matrix depends on a vector of parameters θ, say,

$$Y = X\beta + e, \qquad e \sim N[0, V(\theta)]$$

or, equivalently,

$$Y \sim N[X\beta, V(\theta)].$$

To a lesser extent we examine models with the same mean and covariance but without the normality assumption. This chapter focuses on methods for estimating the vector of covariance parameters θ and on some consequences of that estimation. Here θ is taken to be functionally distinct from the unknown mean parameter β. Some authors, notably Carroll and Ruppert (1988), assume a diagonal covariance matrix but also allow the matrix to depend on β. We will not consider such models.

In subsequent chapters we look at special cases of this very general model. Chapter 5 looks at mixed linear models. Chapter 6 looks at the application of mixed models to time series data. Chapter 8 looks at spatial data models. Chapters 9 and 10 look at multivariate linear models. Chapter 11 examines generalized multivariate linear models and linear models for longitudinal data. Our main approaches to estimating covariance parameters are two versions of maximum likelihood estimation: standard maximum likelihood and another procedure, called residual maximum likelihood (or restricted maximum likelihood), that eliminates the effect of the unknown mean vector prior to maximizing the likelihood. Residual/restricted maximum likelihood is almost universally referred to as REML. To maximize the (restricted) likelihood, we take partial derivatives of the (restricted) log-likelihood and set them equal to zero, i.e., we solve the (restricted) likelihood equations. We also show that both the REML and likelihood methods can be recast as method of moments (estimating equation) estimators which suggests that the methods are at least reasonable even when the data are not normally distributed.

In this chapter we begin by reviewing results on estimation and prediction that apply when θ, and therefore $V(\theta)$, is known. The chapter then examines maximum likelihood estimation and REML estimation of θ. Next we present a useful linear model for the covariance matrix that leads to minimum norm quadratic unbiased estimation (MINQUE) of θ. The procedure misnamed as minimum variance quadratic unbiased estimation (MIVQUE) is addressed. Finally, we examine the effect that estimating covariance parameters has on the procedures that were developed assuming that the covariance parameters were known.

In *PA* we established a number of optimal estimation and prediction properties that hold when θ is known or, more generally, when $V(\theta)$ is known up to a scalar multiple. These are now reviewed. Typically we assume $V(\theta)$ is positive definite for all θ and, just to be clear, we write

$$V^{-1}(\theta) \equiv [V(\theta)]^{-1}.$$

4.1.1 Estimation of β

Following *PA* Sect. 2.7, by definition the generalized least squares estimates of β minimize $(Y - X\beta)'[V^{-1}(\theta)](Y - X\beta)$. These estimates are vectors $\hat{\beta}(\theta)$ that satisfy

$$X\hat{\beta}(\theta) = A(\theta)Y, \qquad (4.1.1)$$

where

$$A(\theta) \equiv X[X'V^{-1}(\theta)X]^{-}X'V^{-1}(\theta). \qquad (4.1.2)$$

$A(\theta)$ is an oblique projection operator onto the column space of X, $C(X)$. In particular, $A(\theta)X = X$ and $C[A(\theta)] = C(X)$. The generalized least squares estimates are also best linear unbiased estimates (BLUEs) without the assumption of normality, and are maximum likelihood estimates (MLEs) and uniformly minimum variance unbiased estimates (UMVUs) under normality. Alternatively, the generalized least squares estimates can be obtained from the normal equations

$$X'V^{-1}(\theta)X\beta = X'V^{-1}(\theta)Y.$$

For an estimable function $\lambda'\beta$ with $\lambda' = \rho'X$ for some ρ, the generalized least squares estimate is $\lambda'\hat{\beta}(\theta) = \rho'[A(\theta)]Y$. The estimate is unbiased and its variance is

$$\text{Var}[\lambda'\hat{\beta}(\theta)] = \rho'A(\theta)V(\theta)A'(\theta)\rho = \lambda'[X'V^{-1}(\theta)X]^{-}\lambda. \qquad (4.1.3)$$

Of course if θ is unknown, $\hat{\beta}(\theta)$ is not an estimate except in special cases, like those discussed in PA, *for which $\hat{\beta}(\theta)$ does not depend on θ.* If we have an estimate of θ, say $\tilde{\theta}$, we can define an (empirical or plug-in) estimate $\tilde{\beta} \equiv \hat{\beta}(\tilde{\theta})$. Under reasonable conditions discussed in Sect. 4.7, $\lambda'\tilde{\beta}$ is an unbiased estimate.

Knowing θ should be better than estimating it, so one would expect to have

$$\text{Var}[\lambda'\hat{\beta}(\theta)] \leq \text{Var}[\lambda'\hat{\beta}(\tilde{\theta})]$$

and Sect. 4.7 provides conditions under which this is true. Tarpey, Ogden, Petkova, and Christensen (2014, 2015) discuss an example where knowing the parameter is not always better than estimating it. Section 4.7 also provides conditions under which

$$\text{E}\{\lambda'[X'V^{-1}(\tilde{\theta})X]^{-}\lambda\} \leq \lambda'[X'V^{-1}(\theta)X]^{-}\lambda$$

so that

$$\text{E}\{\lambda'[X'V^{-1}(\tilde{\theta})X]^{-}\lambda\} \leq \text{Var}[\lambda'\hat{\beta}(\theta)] \leq \text{Var}[\lambda'\hat{\beta}(\tilde{\theta})].$$

Thus we see that using $\lambda'[X'V^{-1}(\tilde{\theta})X]^{-}\lambda$ as an estimate of $\text{Var}[\lambda'\hat{\beta}(\tilde{\theta})]$ can seriously underestimate the variability of the plug-in estimate. (Bayesian analyses do not suffer from these problems!)

Exercise 4.1.   Consider the linear model

$$\begin{bmatrix} Y_1 \\ Y_2 \end{bmatrix} = \begin{bmatrix} J_{N_1} \\ J_{N_2} \end{bmatrix} \mu + e, \quad E(e) = 0, \quad \text{Cov}\left(\begin{bmatrix} e_1 \\ e_2 \end{bmatrix}\right) = \begin{bmatrix} \sigma_1^2 I_{N_1} & 0 \\ 0 & \sigma_2^2 I_{N_2} \end{bmatrix}.$$

Find the BLUE $\hat{\mu}$ and $\text{Var}(\hat{\mu})$ assuming that the variance parameters are known. How does the result change if σ_1^2 is unknown but σ_2^2/σ_1^2 is known? (In general J_a^b denotes a matrix of 1s with a rows and b columns; $J_a \equiv J_a^1; J \equiv J_n$.)

4.1.2 Testing

Exact results for testing mean parameters are well known when V is either completely known or known up to a scalar multiple, cf. *PA* Sect. 3.8. A consistent estimate of θ is one that comes very close to θ whenever the sample sizes are large (whatever "large" means). If an estimate $\tilde{\theta}$ is sufficiently close to θ, the results for known θ should be nearly correct.

To test an estimable hypothesis $H_0 : \Lambda'\beta = d$, assuming that $\tilde{\theta}$ is a consistent estimate of θ, asymptotic theory suggests using a result similar to *PA* Theorem 3.8.3 (with $\sigma^2 = 1$),

$$(\Lambda'\tilde{\beta} - d)' \left[\Lambda'(X'\tilde{V}^{-1}X)^-\Lambda\right]^- (\Lambda'\tilde{\beta} - d) \sim \chi^2[r(\Lambda)], \quad (4.1.4)$$

wherein $\tilde{V} \equiv V(\tilde{\theta})$. Similarly, for testing a reduced model

$$Y = X_0\gamma + e, \quad C(X_0) \subset C(X),$$

with $\tilde{A} \equiv A(\tilde{\theta})$ and $\tilde{A}_0$ defined in the obvious way, one could base a test on the asymptotic result

$$Y'(\tilde{A} - \tilde{A}_0)'\tilde{V}^{-1}(\tilde{A} - \tilde{A}_0)Y \sim \chi^2[r(X) - r(X_0)]. \quad (4.1.5)$$

Note that $\tilde{V}, \tilde{\beta}, \tilde{A}$, and $\tilde{A}_0$ are all quite complicated functions of Y.

Many covariance models can be written to involve a scalar parameter times a matrix. Let $\theta' = [\theta_1, \theta_*']$ and suppose

$$V(\theta) \equiv \theta_1 V_*(\theta_*).$$

A test that is asymptotically equivalent to (4.1.4) is based on

$$\frac{(\Lambda'\tilde{\beta} - d)' \left[\Lambda'(X'\tilde{V}_*^{-1}X)^-\Lambda\right]^- (\Lambda'\tilde{\beta} - d)/r(\Lambda)}{MSE(\tilde{\theta}_*)} \sim F[r(\Lambda), n - r(X)] \quad (4.1.6)$$

where $\tilde{V}_* \equiv V_*(\tilde{\theta}_*)$,

$$MSE(\theta_*) \equiv \frac{Y'[I - A(\theta_*)]'\{V_*(\theta_*)\}^{-1}[I - A(\theta_*)]Y}{n - r(X)},$$

and

$$A(\theta_*) \equiv X[X'V_*^{-1}(\theta_*)X]^- X'V_*^{-1}(\theta_*).$$

There are several points worth noting:

- $A(\theta) = A(\theta_*)$, see Exercise 4.5, so that

$$MSE(\theta_*) = \frac{[Y - X\hat{\beta}(\theta)]'[V_*(\theta_*)]^{-1}[Y - X\hat{\beta}(\theta)]}{n - r(X)}$$

and

$$MSE(\tilde{\theta}_*) = \frac{(Y - X\tilde{\beta})'\tilde{V}_*^{-1}(Y - X\tilde{\beta})}{n - r(X)}.$$

- The distributions in (4.1.4) and (4.1.6) are exact for multivariate normal data with $\tilde{\theta} = \theta$ and $\tilde{\theta}_* = \theta_*$, respectively.
- $MSE(\tilde{\theta}_*)$ provides an estimate for θ_1 that may or may not be different from $\tilde{\theta}_1$, see Exercise 4.5.
- Since $V^{-1}(\theta) = (1/\theta_1)V_*^{-1}(\theta_*)$, if $MSE(\tilde{\theta}_*) = \tilde{\theta}_1$, the test statistic in (4.1.4), divided by the known constant $r(\Lambda)$, is the test statistic in (4.1.6).

Both (4.1.4) and (4.1.6) probably underrepresent the true variability of their test statistics, but (4.1.6), when applicable, probably does it a little less than (4.1.4). Similar conclusions apply to the test in (4.1.5) and the following test.

To test a reduced model, the F test corresponding to (4.1.5) uses the asymptotic approximation

$$\frac{Y'(\tilde{A} - \tilde{A}_0)'\tilde{V}_*^{-1}(\tilde{A} - \tilde{A}_0)Y/[r(X) - r(X_0)]}{MSE(\tilde{\theta}_*)} \sim F[r(X) - r(X_0), n - r(X)].$$

Generalized likelihood ratio tests are discussed in Sect. 4.2.1.

4.1.2.1 Improved Tests: Satterthwaite and Kenward-Roger

When testing for equal means in random samples from two normal distributions with possibly different variances, the usual t statistic does not have a t distribution, e.g. Christensen (2015, Section 4.3). It is common practice to use an approximate distribution proposed by Satterthwaite (1946). The same idea can be applied quite generally. In fact, the Satterthwaite approximation has been improved by Kenward and Roger (1997, 2009).

4.1.3 Prediction

Point prediction with independent observations is relatively easy, one just predicts with the estimated mean of the observation being predicted. The situation is more complicated with dependent observations. We now review the key ideas of

Best Linear Unbiased Prediction as presented in *PA-V* Sect. 6.6 (Christensen 2011, Sec. 12.2).

Consider a set of random variables y_i, $i = 0, 1, \ldots, n$. We want to use $y_1, \ldots, y_n$ to predict y_0. Let $Y = (y_1, \ldots, y_n)'$. In *PA* Sect. 6.3 it is shown that the best predictor (BP) of y_0 is $E(y_0|Y)$. Typically, the joint distribution of y_0 and Y is not available, so $E(y_0|Y)$ cannot be found. If the means and covariances of the y_is are available, then we can find the best linear predictor (BLP) of y_0. If the y_is have a multivariate normal joint distribution, the best predictor and the best linear predictor are identical.

Let $\mathrm{Cov}(Y) = V(\theta)$, $\mathrm{Cov}(Y, y_0) = V_{y0}(\theta)$, $E(y_i) = \mu_i$, $i = 0, 1, \ldots, n$, and $\mu = (\mu_1, \ldots, \mu_n)'$. From *PA* Sect. 6.3.4, the BLP of y_0 is

$$\hat{E}(y_0|Y) \equiv \mu_0 + [\delta_*(\theta)]'(Y - \mu), \tag{4.1.7}$$

where $\delta_*(\theta)$ satisfies $[V(\theta)]\delta_*(\theta) = V_{y0}(\theta)$. In our discussions of Principal Components (Chap. 14) and Time Domain (Chap. 7) analysis we will need additional properties of BLPs that are developed in Appendix B. Of course we cannot actually predict anything unless all of these parameters are known (or we can estimate them).

Following *PA-V* Sect. 6.6 (Christensen 2011, Section 12.2), we now weaken the assumptions that μ and μ_0 are known. Since the prediction is based on Y and there are as many unknown parameters in μ as there are observations in Y, we need to impose some structure on the mean vector μ before we can generalize the theory. This is done by specifying a linear model for Y. Since y_0 is being predicted and has not been observed, it is necessary either to know μ_0 or to know that μ_0 is related to μ in a specified manner. In the theory below, it is assumed that μ_0 is related to the linear model for Y.

Suppose that a vector of known concomitant variables $x_i' = (x_{i1}, \ldots, x_{ip})$ is associated with each random observation y_i, $i = 0, \ldots, n$. We impose structure on the μ_is by assuming that $\mu_i = x_i'\beta$ for some vector of unknown parameters β and all $i = 0, 1, \ldots, n$.

We can now reset our notation in terms of linear model theory. Let

$$X = \begin{bmatrix} x_1' \\ \vdots \\ x_n' \end{bmatrix}.$$

The observed vector Y satisfies the linear model

$$Y = X\beta + e, \quad E(e) = 0, \quad \mathrm{Cov}(e) = V(\theta).$$

With this additional structure, the BLP of y_0 becomes

$$\hat{E}(y_0|Y) = x_0'\beta + [\delta_*(\theta)]'(Y - X\beta), \tag{4.1.8}$$

where again $[V(\theta)]\delta_*(\theta) = V_{y0}(\theta)$.

The standard assumption that μ and μ_0 are known now amounts to the assumption that β is known. It is this assumption that we renounce. By weakening the assumptions, we consequently weaken the predictor. If β is known, we can find the best linear predictor. When β is unknown, we find the best linear unbiased predictor.

Before proceeding, a technical detail must be mentioned. To satisfy estimability conditions, we need to assume that $x_0' = \rho'X$ for some vector ρ. If the linear model is a regression model, choosing $\rho' = x_0'(X'X)^{-1}X'$ suffices. In applications with $E(y_0) = 0$, we take $x_0' = 0$, so $\rho - 0$ will suffice.

Definition 4.1.1. A *predictor* $f(Y)$ of y_0 is said to be *unbiased* if

$$E[f(Y)] = E(y_0).$$

Definition 4.1.2. $a_0 + a'Y$ is a *best linear unbiased predictor* of y_0 if $a_0 + a'Y$ is unbiased and if, for any other unbiased linear predictor $b_0 + b'Y$,

$$E[y_0 - a_0 - a'Y]^2 \leq E[y_0 - b_0 - b'Y]^2.$$

Theorem 4.1.3. The best linear unbiased predictor (BLUP) of y_0 is

$$\hat{y}_0(\theta) \equiv x_0'\hat{\beta}(\theta) + [\delta_*(\theta)]'[Y - X\hat{\beta}(\theta)],$$

where $[V(\theta)]\delta_*(\theta) = V_{y0}(\theta)$ and $X\hat{\beta}(\theta) = A(\theta)Y$.

Standard linear models have independent homoscedastic observations, so $\delta_*(\theta) = 0$ and $\hat{\beta}(\theta)$ reduces to a least squares estimate. Typically, both $\hat{\beta}(\theta)$ and $\delta_*(\theta)$ depend crucially on θ. When all inverses exist, $\hat{\beta}(\theta) = [X'V^{-1}(\theta)X]^{-1}X'[V^{-1}(\theta)]Y$ and $\delta_*(\theta) = [V^{-1}(\theta)]V_{y0}(\theta)$ so that

$$\hat{y}_0(\theta) = x_0'\hat{\beta}(\theta) + V_{y0}'(\theta)[V^{-1}(\theta)][Y - X\hat{\beta}(\theta)].$$

Given $\hat{\beta}(\theta)$ and $[V^{-1}(\theta)][Y - X\hat{\beta}(\theta)]$ any number of predictions can be made easily because the only things that change are the row vectors x_0' and $V_{y0}'(\theta)$ that premultiply these.

To present the prediction variance of the BLUP, let $\text{Var}(y_0) = \sigma_0^2(\theta)$. As shown in *PA*,

$$E\left\{y_0 - x_0'\hat{\beta}(\theta) - [\delta_*(\theta)]'[Y - X\hat{\beta}(\theta)]\right\}^2 = \sigma_0^2(\theta) - [V_{y0}(\theta)]'[V^-(\theta)][V_{y0}(\theta)]$$
$$+ [x_0 - X'\delta_*(\theta)]'[X'V^-(\theta)X]^-[x_0 - X'\delta_*(\theta)], \tag{4.1.9}$$

or, writing the BLUP as $[b(\theta)]'Y$, the prediction variance becomes

$$E\left\{y_0 - [b(\theta)]'Y\right\}^2 = \sigma_0^2(\theta) - 2[b(\theta)]'V_{y0}(\theta) + [b(\theta)]'V(\theta)b(\theta). \tag{4.1.10}$$

Similar to the discussion on estimation, when estimating θ with $\tilde{\theta}$, Sect. 4.7 establishes that under certain conditions

$$\mathrm{E}\left\{y_0 - [b(\theta)]'Y\right\}^2 \leq \mathrm{E}\left\{y_0 - [b(\tilde{\theta})]'Y\right\}^2,$$

so, not surprisingly, knowing θ gives better results than using the plug-in predictor. Section 4.7 also shows that under some conditions

$$\mathrm{E}\left\{\sigma_0^2(\tilde{\theta}) - 2[b(\tilde{\theta})]'V_{y0}(\tilde{\theta}) + [b(\tilde{\theta})]'V(\tilde{\theta})b(\tilde{\theta})\right\}$$
$$\leq \sigma_0^2(\theta) - 2[b(\theta)]'V_{y0}(\theta) + [b(\theta)]'V(\theta)b(\theta),$$

so, as with estimation, the plug-in variance estimate $\sigma_0^2(\tilde{\theta}) - 2[b(\tilde{\theta})]'V_{y0}(\tilde{\theta}) + [b(\tilde{\theta})]'V(\tilde{\theta})b(\tilde{\theta})$ can seriously underestimate the variability of the plug-in predictor.

4.1.4 Quadratic Estimation of θ

Consider the linear model

$$\mathrm{E}(Y) = X\beta; \quad \mathrm{Cov}(Y) = V(\theta).$$

We can base estimation of θ on the idea of setting quadratic forms equal to their expectations.

To estimate the p parameters in β and the s parameters in θ we can solve a system of $p+s$ equations in $p+s$ unknowns. Based on our previous theory, we will always start with the p normal equations

$$X'V^{-1}(\theta)X\beta = X'V^{-1}(\theta)Y$$

but we want to think about them in a little different way, as setting a linear function of the data equal to its expectation,

$$X'V^{-1}(\theta)Y = \mathrm{E}\left[X'V^{-1}(\theta)Y\right] = X'V^{-1}(\theta)X\beta.$$

To get s more equations pick any symmetric matrices $W_1(\theta), \ldots, W_s(\theta)$ to define quadratic forms. One way to proceed is to set those equal to their expectations

$$Y'W_j(\theta)Y = \mathrm{E}\left[Y'W_j(\theta)Y\right] = \mathrm{tr}\left[W_j(\theta)V(\theta)\right] + \beta'X'W_j(\theta)X\beta, \quad j = 1, \ldots, s.$$

A slightly different way to proceed is to set

$$(Y - X\beta)'W_j(\theta)(Y - X\beta) = \mathrm{E}\left[(Y - X\beta)'W_j(\theta)(Y - X\beta)\right] = \mathrm{tr}\left[W_j(\theta)V(\theta)\right],$$
$$j = 1, \ldots, s.$$

In either case, the equations all involve setting functions of observations equal to their expectations, so the equations are statistically reasonable and solving them should give reasonable estimates. Whether the solutions actually give good estimates depends on how one chooses the $W_j(\theta)$s. This is similar to the fact that least squares estimates are always reasonable for any linear model but, depending on the covariance structure, they may not always be good. We will see in the next section that maximum likelihood estimation fits the second paradigm with a particular choice of $W_j(\theta)$s, namely

$$W_j(\theta) = V^{-1}(\theta)[\mathbf{d}_{\theta_j}V(\theta)]V^{-1}(\theta).$$

REML estimation is discussed in Sect. 4.3 and fits into a slightly different paradigm that involves only s equations. Rather than picking just any old $W_j(\theta)$s, we now restrict ourselves to picking matrices $\tilde{W}_j(\theta)$ with the property that $\tilde{W}_j(\theta)X = 0$. The defining equations for estimating θ become

$$Y'\tilde{W}_j(\theta)Y = \mathrm{E}\left[Y'\tilde{W}_j(\theta)Y\right] = \mathrm{tr}\left[\tilde{W}_j(\theta)V(\theta)\right], \quad j = 1,\ldots,s.$$

Again, REML determines a particular choice of $\tilde{W}_j(\theta)$s that should be good for normal data but that should be reasonable even for nonnormal data. In particular, REML uses

$$\tilde{W}_j(\theta) = [I - A(\theta)]'V^{-1}(\theta)[\mathbf{d}_{\theta_j}V(\theta)]V^{-1}(\theta)[I - A(\theta)].$$

MINQUE from Sect. 4.5 and Henderson's Method 3 from Sect. 5.7 fit into this paradigm. They involve choosing matrices $\tilde{W}_j$ that do not depend on θ but have $\tilde{W}_jX = 0$. These methods apply to models with linear covariance structures like

$$V(\theta) \equiv \sum_{k=1}^{s} \theta_k V_k$$

for known nonnegative definite V_ks. The system of equations becomes

$$Y'\tilde{W}_jY = \mathrm{E}\left[Y'\tilde{W}_jY\right] = \mathrm{tr}\left[\tilde{W}_jV(\theta)\right] = \sum_{k=1}^{s} \theta_k \mathrm{tr}(\tilde{W}_jV_k), \quad j = 1,\ldots,s.$$

The $Y'\tilde{W}_jY$s are observable numbers, so this method involves merely solving a system of linear equations,

$$\begin{bmatrix} Y'\tilde{W}_1Y \\ \vdots \\ Y'\tilde{W}_sY \end{bmatrix} = \begin{bmatrix} \mathrm{tr}(\tilde{W}_1V_1) & \cdots & \mathrm{tr}(\tilde{W}_1V_s) \\ \vdots & \ddots & \vdots \\ \mathrm{tr}(\tilde{W}_sV_1) & \cdots & \mathrm{tr}(\tilde{W}_sV_s) \end{bmatrix}\theta,$$

and gives unbiased estimates. For a known vector of weights w, MINQUE uses

$$\tilde{W}_j = [I - A(w)]'V^{-1}(w)[\mathbf{d}_{\theta_j}V(\theta)|_{\theta=w}]V^{-1}(w)[I - A(w)].$$

Unfortunately, there is no compelling rationale (like maximum likelihood, REML, or MINQUE) for picking a particular set of $\tilde{W}_j$s for Henderson's Method 3. In Method 3 the $\tilde{W}_j$s are a convenient set of ppos.

4.2 Maximum Likelihood

Assume that for $\theta = (\theta_1, \ldots, \theta_s)'$,

$$Y \sim N[X\beta, V(\theta)]$$

and that $V(\theta)$ is nonsingular for all (allowable) θ, so that the density of Y exists. We write determinants as $|V| \equiv \det(V)$. (The notation $\det(V)$ becomes awkward.) The density/likelihood associated with Y is

$$(2\pi)^{-n/2}|V(\theta)|^{-1/2}\exp\left[-(Y-X\beta)'[V^{-1}(\theta)](Y-X\beta)/2\right].$$

The log-likelihood is

$$\ell(\beta, \theta) \equiv -\frac{n}{2}\log(2\pi) - \frac{1}{2}\log[|V(\theta)|] - \frac{1}{2}(Y-X\beta)'[V^{-1}(\theta)](Y-X\beta).$$

The (partial) derivative of the log-likelihood with respect to β uses Propositions A.1.1 and A.1.2 from Appendix A and the chain rule to get

$$\begin{aligned}
\mathbf{d}_\beta\ell(\beta, \theta) &= \frac{-1}{2}\mathbf{d}_\beta\left[(Y-X\beta)'[V^{-1}(\theta)](Y-X\beta)\right] \\
&= \frac{-1}{2}\left\{\mathbf{d}_{(Y-X\beta)}\left[(Y-X\beta)'[V^{-1}(\theta)](Y-X\beta)\right]\right\}\mathbf{d}_\beta(Y-X\beta) \\
&= \frac{-1}{2}\left\{2(Y-X\beta)'[V^{-1}(\theta)]\right\}(-X) \\
&= (Y-X\beta)'[V^{-1}(\theta)]X.
\end{aligned}$$

The (partial) derivatives of the log-likelihood with respect to θ_j, $j = 1, \ldots, s$, use Proposition A.1.2 parts (e), (c) and (b) to get

$$\mathbf{d}_{\theta_j}\ell(\beta, \theta) =$$
$$-\frac{1}{2}\mathrm{tr}\left\{V^{-1}(\theta)[\mathbf{d}_{\theta_j}V(\theta)]\right\} + \frac{1}{2}(Y-X\beta)'V^{-1}(\theta)[\mathbf{d}_{\theta_j}V(\theta)]V^{-1}(\theta)(Y-X\beta).$$

Setting the partial derivatives equal to zero leads to solving the likelihood equations $\mathbf{d}_\beta\ell(\beta, \theta) = 0$ or equivalently

$$X'V^{-1}(\theta)X\beta = X'V^{-1}(\theta)Y$$

and, for $j = 1, \ldots, s$, $\mathbf{d}_{\theta_j}\ell(\beta, \theta) = 0$ which is equivalent to

$$\mathrm{tr}\left\{V^{-1}(\theta)[\mathbf{d}_{\theta_j}V(\theta)]\right\} = (Y-X\beta)'V^{-1}(\theta)[\mathbf{d}_{\theta_j}V(\theta)]V^{-1}(\theta)(Y-X\beta). \quad (4.2.1)$$

Most of the time, some iterative computational technique is needed to solve the equations in (4.2.1). (SAS's "proc mixed" seems to use Newton–Raphson.) For the generalized split plot models of *PA* Chap. 11, relatively nice closed form solutions can be found. Even nicer solutions can be found for the multivariate linear models of Chaps. 9 and 10. If one uses structured covariances matrices in multivariate linear models, such as those inspired by time series models for longitudinal data, closed form estimates are no longer available. For the linear covariance models introduced in Sect. 4.4, relatively easy computational methods can be used, methods that merely involve iteratively solving systems of linear equations. Most of the covariance models discussed for mixed models in Chap. 5 and some of the covariance models discussed for spatial data in Chap. 8 have linear covariance structure. The covariance structures for most spatial models provide no obvious simplifications in solving these equations.

From *PA* Chap. 2, the likelihood is maximized for any θ by taking $\hat{\beta}(\theta)$ to satisfy

$$X\hat{\beta}(\theta) = A(\theta)Y,$$

where

$$A(\theta) = X[X'V^{-1}(\theta)X]^{-}X'V^{-1}(\theta).$$

This allows us to solve the equations in (4.2.1) without reference to the likelihood equations for β. Replacing β with its maximum likelihood estimate $\hat{\beta}(\theta)$ in (4.2.1), it follows that the MLE of θ satisfies

$$\begin{aligned}
\operatorname{tr}\left\{V^{-1}(\theta)[\mathbf{d}_{\theta_j}V(\theta)]\right\} &= [Y - X\hat{\beta}(\theta)]'V^{-1}(\theta)[\mathbf{d}_{\theta_j}V(\theta)]V^{-1}(\theta)[Y - X\hat{\beta}(\theta)] \\
&= Y'[I - A(\theta)]'V^{-1}(\theta)[\mathbf{d}_{\theta_j}V(\theta)]V^{-1}(\theta)[I - A(\theta)]Y
\end{aligned}$$

$$(4.2.2)$$

for $j = 1, \ldots, s$.

Some applications use parameterizations in terms of $V^{-1}(\theta)$ rather than $V(\theta)$ so that it is easier to find $\mathbf{d}_{\theta_j}V(\theta)^{-1}$ directly. In that case, using $\mathbf{d}_{\theta_j}V(\theta)^{-1} = V^{-1}(\theta)[\mathbf{d}_{\theta_j}V(\theta)]V^{-1}(\theta)$ gives the equations

$$\operatorname{tr}\left\{V(\theta)[\mathbf{d}_{\theta_j}V^{-1}(\theta)]\right\} = Y'[I - A(\theta)]'[\mathbf{d}_{\theta_j}V^{-1}(\theta)][I - A(\theta)]Y.$$

It is interesting to note that solving the full set of likelihood equations also gives method of moments (or estimating equation) estimates. The likelihood equations can be viewed as setting

$$E[X'V^{-1}(\theta)Y] = X'V^{-1}(\theta)Y$$

and for $j = 1, \ldots, s$

$$\begin{aligned}
E\{(Y - X\beta)'V^{-1}(\theta)[\mathbf{d}_{\theta_j}V(\theta)]V^{-1}(\theta)(Y - X\beta)\} \\
= (Y - X\beta)'V^{-1}(\theta)[\mathbf{d}_{\theta_j}V(\theta)]V^{-1}(\theta)(Y - X\beta).
\end{aligned}$$

The last equations follow because, similar to the proof of *PA* Theorem 1.3.2,

$$E\left\{(Y - X\beta)'V^{-1}(\theta)[\mathbf{d}_{\theta_j}V(\theta)]V^{-1}(\theta)(Y - X\beta)\right\}$$
$$= \mathrm{tr}\{V^{-1}(\theta)[\mathbf{d}_{\theta_j}V(\theta)]V^{-1}(\theta)V(\theta)\} = \mathrm{tr}\{V^{-1}(\theta)[\mathbf{d}_{\theta_j}V(\theta)]\}.$$

Note that the validity of this estimation procedure does not depend on the data being normal, although the motivation for using *these* estimating equations, as opposed to some others, was based on the data being normal.

Typically, there are constraints on the parameters in θ. Variances must be non-negative, correlations must be between -1 and 1, and $V(\theta)$ must be positive definite. *Solutions to the likelihood equations are not MLEs if they do not satisfy these constraints.* In models this general, it frequently occurs that the likelihood has more than one local maximum. Spatial data models are particularly notorious for this. Solutions to the likelihood equations can occur at local maxima, local minima, and at saddlepoints. Personally, I find the existence of multiple modes enough to call in question the entire idea of maximizing the likelihood. I personally think it makes far more sense in such problems to use Bayesian techniques that seek to explore the entire likelihood function, although they do so somewhat indirectly through the posterior distribution. These issues can be difficult even in simple models, cf. Lavine, Bray, and Hodges (2015).

4.2.1 Generalized Likelihood Ratio Tests

Consider testing the full model already considered against a reduced linear model with $C(X_0) \subset C(X)$, i.e.,

$$Y \sim N[X_0\gamma, V(\theta)].$$

For θ known, the generalized likelihood ratio test statistic reduces to the usual linear model test

$$[X_0\hat{\gamma}(\theta) - X\hat{\beta}(\theta)]'[V^{-1}(\theta)][X_0\hat{\gamma}(\theta) - X\hat{\beta}(\theta)]$$
$$\sim \chi^2[r(X) - r(X_0), (X_0\gamma - X\beta)'[V^{-1}(\theta)](X_0\gamma - X\beta)/2],$$

where $\chi^2(df, \pi)$ denotes a chi-squared distribution with df degrees of freedom and noncentrality parameter π.

For θ unknown but $\hat{\theta}$ and $\hat{\theta}_0$ denoting its MLEs under the full and reduced models respectively, we get the generalized likelihood ratio test statistic

$$\log[|V(\hat{\theta}_0)|] + [Y - X_0\hat{\gamma}(\hat{\theta}_0)]'[V^{-1}(\hat{\theta}_0)][Y - X_0\hat{\gamma}(\hat{\theta}_0)]$$
$$- \log[|V(\hat{\theta})|] - [Y - X\hat{\beta}(\hat{\theta})]'[V^{-1}(\hat{\theta})][Y - X\hat{\beta}(\hat{\theta})].$$

If standard likelihood theory applies, cf. Ferguson (1996), the test statistic should converge to $\chi^2[r(X) - r(X_0), 0]$ under the null (reduced) model.

Under the null model we expect both the full model MLE $\hat{\theta}$ and the reduced model MLE $\hat{\theta}_0$ to converge in probability to θ, so if $V(\theta)$ is well behaved,

$$\log[|V(\hat{\theta})|] - \log[|V(\hat{\theta}_0)|] \xrightarrow{P} 0$$

and we could drop the determinant terms without changing the asymptotic distribution of the test. Moreover, for any consistent estimate $\tilde{\theta}$ of θ, the generalized likelihood ratio test statistic should be asymptotically equivalent under the null model to the following, which I think is the most appealing test statistic,

$$[X_0\hat{\gamma}(\tilde{\theta}) - X\hat{\beta}(\tilde{\theta})]'[V^{-1}(\tilde{\theta})][X_0\hat{\gamma}(\tilde{\theta}) - X\hat{\beta}(\tilde{\theta})].$$

I would expect choosing $\tilde{\theta} = \hat{\theta}$ to have better power than $\tilde{\theta} = \hat{\theta}_0$. Note that this is the same test statistic as presented in relation (4.1.5).

We will see in Chap. 5 that merely letting the sample size n go to infinity is no guarantee for consistent estimation of θ.

This subsection contains a lot of hand waving because I have never wanted my linear models books to get into asymptotic theory. However, Christensen and Sun (2010) and Christensen and Lin (2015) contain some reasonably nice asymptotic theory for linear models.

4.3 Restricted Maximum Likelihood Estimation

The *residual (restricted) maximum likelihood (REML) estimation* procedure of Patterson and Thompson (1974) can be applied to linear models with general covariance structures. The method incorporates a matrix B that is a full column rank matrix with

$$C(B) = C(X)^{\perp}$$

and the method maximizes the likelihood associated with

$$B'Y \sim N[0, B'V(\theta)B]. \tag{4.3.1}$$

Fortunately, it turns out that the particular choice of B does not matter.

Incorporating such a matrix B certainly places a restriction on the distribution of Y and thus on the likelihood, but what has that got to do with residuals? It turns out that we can arrive at the same point by maximizing the likelihood associated with the residuals. Unfortunately, that is a more complicated matter. First of all, the residuals have a singular covariance matrix, so they do not have an n dimensional density. In Appendix C we examine maximum likelihood estimation for normal distributions with singular covariance matrices and we apply those results to finding the maximum likelihood estimates of covariance parameters using residuals. For now we focus on maximizing the likelihood of model (4.3.1).

Similar to the previous section the log-likelihood for $B'Y$ is

$$\ell_*(\theta) \equiv -\frac{n}{2}\log(2\pi) - \frac{1}{2}\log[|B'V(\theta)B|] - \frac{1}{2}(B'Y)'[B'V(\theta)B]^{-1}(B'Y).$$

Setting the partial derivatives of the log-likelihood function equal to zero leads to solving

$$\text{tr}\left\{[B'V(\theta)B]^{-1}[\mathbf{d}_{\theta_j}B'V(\theta)B]\right\} = Y'B[B'V(\theta)B]^{-1}[\mathbf{d}_{\theta_j}B'V(\theta)B][B'V(\theta)B]^{-1}B'Y,$$

for $j = 1,\ldots,s$. Noting that

$$\mathbf{d}_{\theta_j}B'V(\theta)B = B'[\mathbf{d}_{\theta_j}V(\theta)]B,$$

it is clearly equivalent to solve

$$\text{tr}\{[B'V(\theta)B]^{-1}B'[\mathbf{d}_{\theta_j}V(\theta)]B\} = Y'B(B'V(\theta)B)^{-1}B'[\mathbf{d}_{\theta_j}V(\theta)]B[B'V(\theta)B]^{-1}B'Y. \tag{4.3.2}$$

As demonstrated later in Lemmas 4.3.1 and 4.3.2,

$$[I - A(\theta)] = V(\theta)B(B'V(\theta)B)^{-1}B' \tag{4.3.3}$$

and thus

$$V^{-1}(\theta)[I - A(\theta)] = B(B'V(\theta)B)^{-1}B' = [I - A(\theta)]'V^{-1}(\theta). \tag{4.3.4}$$

Observing that

$$\text{tr}\{[B'V(\theta)B]^{-1}B'[\mathbf{d}_{\theta_j}V(\theta)]B\} = \text{tr}\{B[B'V(\theta)B]^{-1}B'[\mathbf{d}_{\theta_j}V(\theta)]\}$$

and using (4.3.4), Eq. (4.3.2) can be rewritten as

$$\text{tr}\left\{V^{-1}(\theta)[I - A(\theta)][\mathbf{d}_{\theta_j}V(\theta)]\right\}$$
$$= Y'[I - A(\theta)]'V^{-1}(\theta)[\mathbf{d}_{\theta_j}V(\theta)]V^{-1}(\theta)[I - A(\theta)]Y, \tag{4.3.5}$$

$j = 1,\ldots,s$, where $[I - A(\theta)]Y$ consists of the residuals from the BLUE of $X\beta$, i.e., $[I - A(\theta)]Y = Y - X\hat{\beta}(\theta)$.

Note that the equations in (4.3.5) do not depend on the particular choice of the matrix B, since B no longer appears in the equations. In PA we established that projections operators like $A(\theta)$ are unique. Note also that the only differences between the REML Eq. (4.3.5) and the corresponding MLE equations for θ in (4.2.2) is the existence of the term $[I - A(\theta)]$ in the trace in (4.3.5).

Some applications use parameterizations in terms of $V^{-1}(\theta)$ rather than $V(\theta)$ so that it is easier to find $\mathbf{d}_{\theta_j}V(\theta)^{-1}$ directly. In that case, using $\mathbf{d}_{\theta_j}V(\theta)^{-1} = V^{-1}(\theta)[\mathbf{d}_{\theta_j}V(\theta)]V^{-1}(\theta)$ on both sides of Eq. (4.3.5) gives

$$\text{tr}\left\{[I - A(\theta)]V(\theta)[\mathbf{d}_{\theta_j}V^{-1}(\theta)]\right\} = Y'[I - A(\theta)]'[\mathbf{d}_{\theta_j}V^{-1}(\theta)][I - A(\theta)]Y.$$

We now present two results that justify Eq. (4.3.3). Let $r(X) = r$ so that $r(B) = n - r$.

Lemma 4.3.1. Let $A = X(X'V^{-1}X)^- X'V^{-1}$. Then $I - A$ is the projection operator onto $C(VB)$ along $C(X)$.

PROOF. From *PA* Sect. 2.7 we know that A is an oblique projection operator onto $C(X)$. Since $C(B) = C(X)^\perp$ and V is nonsingular, $r(VB) = n - r = r(I - A)$. Also, $(I - A)VB = VB - X(X'V^{-1}X)^- X'B = VB$, so $I - A$ is a projection onto $C(VB)$. It is along $C(X)$ because $r(X) = r$ and $(I - A)X = X - X = 0$. $\qquad\square$

Lemma 4.3.2. $VB(B'VB)^{-1}B'$ is the projection operator onto $C(VB)$ along $C(X)$.

PROOF. $VB(B'VB)^{-1}B' = VB[(B'V)V^{-1}(VB)]^{-1}(B'V)V^{-1}$ is a projection operator onto $C(VB)$. Since $C(B) = C(X)^\perp$, $VB(B'VB)^{-1}B'X = 0$. Moreover, $r(VB) = n - r$ and $r(X) = r$. $\qquad\square$

Similar to maximum likelihood estimation, the REML Eq. (4.3.5) also give method of moments (estimating equation) estimates. The REML equations can be viewed as finding a solution to

$$E\{Y'[I - A(\theta)]'V^{-1}(\theta)[\mathbf{d}_{\theta_j}V(\theta)]V^{-1}(\theta)[I - A(\theta)]Y\}$$
$$= Y'[I - A(\theta)]'V^{-1}(\theta)[\mathbf{d}_{\theta_j}V(\theta)]V^{-1}(\theta)[I - A(\theta)]Y,$$

for $j = 1, \ldots, s$. The equivalence holds because $A(\theta)$ is a projection operator onto $C(X)$, $[I - A(\theta)]'V^{-1}(\theta) = V^{-1}(\theta)[I - A(\theta)]$ and, similar to the proof of *PA* Theorem 1.3.2,

$$E\{Y'[I - A(\theta)]'V^{-1}(\theta)[\mathbf{d}_{\theta_j}V(\theta)]V^{-1}(\theta)[I - A(\theta)]Y\}$$
$$= E\{(Y - X\beta)'[I - A(\theta)]'V^{-1}(\theta)[\mathbf{d}_{\theta_j}V(\theta)]V^{-1}(\theta)[I - A(\theta)](Y - X\beta)\}$$
$$= \mathrm{tr}\{[I - A(\theta)]'V^{-1}(\theta)[\mathbf{d}_{\theta_j}V(\theta)]V^{-1}(\theta)[I - A(\theta)]V(\theta)\}$$
$$= \mathrm{tr}\{V^{-1}(\theta)[I - A(\theta)]V(\theta)[I - A(\theta)]'V^{-1}(\theta)[\mathbf{d}_{\theta_j}V(\theta)]\}$$
$$= \mathrm{tr}\{V^{-1}(\theta)[I - A(\theta)]V(\theta)V^{-1}(\theta)[I - A(\theta)][\mathbf{d}_{\theta_j}V(\theta)]\}$$
$$= \mathrm{tr}\{V^{-1}(\theta)[I - A(\theta)][\mathbf{d}_{\theta_j}V(\theta)]\}.$$

As with the corresponding argument from the previous section, the validity of this estimation procedure does not depend on the assumption of normality even though the equations that define the procedure were determined using normality.

Exercise 4.2. Consider the model $Y = X\beta + e$, $e \sim N(0, \sigma^2 I)$. Show that the *MSE* is the REML estimate of σ^2.

Exercise 4.3. Consider the heteroscedastic two-sample problem, $y_{ij} = \mu_i + \varepsilon_{ij}$, $i = 1, 2$, $j = 1, \ldots, N_i$, with ε_{ij}s independent $N(0, \sigma_i^2)$. Write the linear model as

$$\begin{bmatrix} Y_1 \\ Y_2 \end{bmatrix} = \begin{bmatrix} J_{N_1} & 0 \\ 0 & J_{N_2} \end{bmatrix} \begin{bmatrix} \mu_1 \\ \mu_2 \end{bmatrix} + \begin{bmatrix} e_1 \\ e_2 \end{bmatrix}, \quad \mathrm{E}(e) = 0, \quad \mathrm{Cov}\left(\begin{bmatrix} e_1 \\ e_2 \end{bmatrix} \right) = \begin{bmatrix} \sigma_1^2 I_{N_1} & 0 \\ 0 & \sigma_2^2 I_{N_2} \end{bmatrix}.$$

(a) What is the vector θ? Show that the least squares estimates are always the BLUEs, i.e., show $C[V(\theta)X] \subset C(X)$ for any θ.

(b) Show that

$$[I - A(\theta)] = \mathrm{Blk\ diag}\left[I_{N_i} - \frac{1}{N_i} J_{N_i}^{N_i} \right].$$

Find $\mathrm{Cov}(\hat{\beta})$.

(c) Show that for $i = 1, 2$,

$$\mathrm{tr}\left\{ V^{-1}(\theta)[I - A(\theta)][\mathbf{d}_{\theta_i} V(\theta)] \right\} = (N_i - 1)\Big/ \sigma_i^2.$$

(d) Show that for $i = 1, 2$,

$$Y'[I - A(\theta)]'V^{-1}(\theta)[\mathbf{d}_{\theta_i} V(\theta)]V^{-1}(\theta)[I - A(\theta)]Y = (Y_i - \bar{y}_{i\cdot} J_{N_i})' \frac{1}{\sigma_i^4} (Y_i - \bar{y}_{i\cdot} J_{N_i}).$$

(e) Show that the REML estimate of σ_i^2 is the sample variance of the ith sample.

(f) Using the REML estimates and $\mathrm{Cov}(\hat{\beta})$, find a standard error for $\hat{\mu}_1 - \hat{\mu}_2$.

Exercise 4.4.   Generalizing the previous exercise, let

$$\begin{bmatrix} Y_1 \\ Y_2 \end{bmatrix} = \begin{bmatrix} X_1 & 0 \\ 0 & X_2 \end{bmatrix} \begin{bmatrix} \beta_1 \\ \beta_2 \end{bmatrix} + e, \quad \mathrm{E}(e) = 0, \quad \mathrm{Cov}\left(\begin{bmatrix} e_1 \\ e_2 \end{bmatrix} \right) = \begin{bmatrix} \sigma_1^2 I_{N_1} & 0 \\ 0 & \sigma_2^2 I_{N_2} \end{bmatrix}.$$

(a) Show that the REML estimate of σ_i^2 is $Y_i'(I - M_i)Y_i/[N_i - r(X_i)]$ where M_i is the perpendicular projection operator (ppo) onto $C(X_i)$.

(b) Show that the MLE of σ_i^2 is $Y_i'(I - M_i)Y_i/N_i$.

Exercise 4.5.   Consider the linear model

$$Y = X\beta + e, \qquad e \sim N[0, V(\theta)].$$

Partition θ as $\theta' = [\theta_1, \theta_*']$ and suppose

$$V(\theta) \equiv \theta_1 V_*(\theta_*),$$

where $\theta_1 > 0$ and $V_*(\theta_*)$ is positive definite for all allowable θ_*. Define

$$SSE(\theta) = Y'[I - A(\theta)]'V_*(\theta_*)^{-1}[I - A(\theta)]Y. \tag{4.3.6}$$

(a) Recalling its definition in (4.1.2), show that $A(\theta) = X[X'V_*(\theta_*)^{-1}X]^- X'V_*(\theta_*)^{-1}$.
    Show that $SSE(\theta)$ does not depend on θ_1.
(b) Show that if $\hat{\theta}$ is a solution to the REML equations, then

$$\hat{\theta}_1 = \frac{SSE(\hat{\theta})}{n - r(X)}.$$

(c) Show that if $\hat{\theta}$ is a solution to the likelihood equations, then

$$\hat{\theta}_1 = \frac{SSE(\hat{\theta})}{n}.$$

Hint: You need only look at the equations with $s = 1$.

The following exercise examines a quite general model in which solutions to the REML equations are available in closed form.

Exercise 4.6.   Consider a generalized split plot model as defined in *PA* Chap. 11, or the JSTOR-available (Christensen 1987). In particular you will need the definitions and properties of the perpendicular projections operators M_1, M_*, and M_2. The covariance matrix is

$$V(\theta) = \sigma_0^2 I + \sigma_1^2 X_1 X_1' = \sigma^2[(1-\rho)I + m\rho M_1]$$

where σ_0^2 is the subplot variance, σ_1^2 is the whole plot variance, X_1 is a matrix of indicator variables for the whole plots, m is the number of subplots in each whole plot, $\sigma^2 \equiv \sigma_0^2 + \sigma_1^2$, and $\rho \equiv \sigma_1^2/(\sigma_0^2 + \sigma_1^2)$. Reparameterize the problem as

$$\theta_1 = \sigma^2(1-\rho) = \sigma_0^2, \qquad \theta_2 = \frac{m\rho}{(1-\rho)}$$

so that

$$V(\theta) = \theta_1[I + \theta_2 M_1].$$

Note that $A(\theta) = M_* + M_2$ for any vector θ, that $C(M_*) \subset C(M_1)$, that $C(M_1) \perp C(M_2)$, and that V^{-1} is available from *PA-V* Proposition B.57 (or Christensen 2011, Proposition 12.11.1).

With this θ parameterization, the solution to the first REML equation, as discussed in Exercise 4.5, provides an estimate for θ_1. Show that the estimate of θ_1 reduces in this problem to the closed form $Y'(I - M_* - M_2)Y/r(I - M_* - M_2)$, which is the error estimate used in the subplot analysis of a generalized split plot model. Show that the second REML equation has a solution that involves estimating $\sigma^2[(1-\rho)+m\rho] = \sigma_0^2 + m\sigma_1^2$ by $Y'(M_1 - M_*)Y/r(M_1 - M_*)$, which is the error estimate used in the whole plot analysis of a generalized split plot model.

4.4 Linear Covariance Structures

Consider the linear model

$$Y = X\beta + e, \quad \mathrm{E}(e) = 0, \quad \mathrm{Cov}(e) = V(\theta),$$

wherein $V(\theta)$ has the linear structure

$$V(\theta) = V_0 + \theta_1 V_1 + \cdots + \theta_s V_s \tag{4.4.1}$$

for known matrices V_j that are most often symmetric but often fail to be nonnegative definite. The matrix V_0 allows some elements of the covariance matrix to be known as something other than 0. We continue to assume that $V(\theta)$ is nonsingular for all allowable θ. This covariance structure simplifies solving the likelihood and REML equations, it is fundamental to the development of MINQUE estimates in the next section, but it also arises naturally in mixed models (Chap. 5), spatial models (Chap. 8), and generalized multivariate linear models and their extensions to other longitudinal data (Chap. 11).

The remainder of this section is devoted to demonstrating that linear covariance structure allows one to solve the REML equations by iteratively solving systems of linear equations. This is also true for the likelihood equations but we only demonstrate it for the REML equations. In either case,

$$\mathbf{d}_{\theta_j} V(\theta) = V_j,$$

so, in particular, the REML equations reduce to

$$\mathrm{tr}\left\{ V^{-1}(\theta)[I - A(\theta)]V_j \right\} = Y'[I - A(\theta)]'V^{-1}(\theta)V_j V^{-1}(\theta)[I - A(\theta)]Y,$$

$j = 1, \ldots, s$. To simplify notation, henceforth we write $V(\theta)$ as V and $A(\theta)$ as A. Similar to the previous section when establishing that REML has an estimating equation justification,

$$\begin{aligned}
&\mathrm{tr}[V^{-1}(I - A)V_j] \\
&= \mathrm{tr}[V^{-1}(I - A)(I - A)V_j] \\
&= \mathrm{tr}[(I - A)V_j V^{-1}(I - A)] \\
&= \mathrm{tr}[VV^{-1}(I - A)V_j V^{-1}(I - A)] \\
&= \mathrm{tr}[V(I - A)'V^{-1}V_j V^{-1}(I - A)] \\
&= \mathrm{tr}\left[\left(V_0 + \sum_{k=1}^{s} \theta_k V_k \right) (I - A)'V^{-1}V_j V^{-1}(I - A) \right] \\
&= \sum_{k=1}^{s} \theta_k \mathrm{tr}\left[V_k(I - A)'V^{-1}V_j V^{-1}(I - A) \right] + \mathrm{tr}\left[V_0(I - A)'V^{-1}V_j V^{-1}(I - A) \right].
\end{aligned}$$

The equations for finding the REML estimates can now be written

$$\sum_{k=1}^{s} \theta_k \mathrm{tr}\left[V_k(I-A)'V^{-1}V_jV^{-1}(I-A)\right]$$
$$= Y'(I-A)'V^{-1}V_jV^{-1}(I-A)Y - \mathrm{tr}\left[V_0(I-A)'V^{-1}V_jV^{-1}(I-A)\right], \quad (1)$$

$j = 1,\ldots,s$. Since V is unknown, typically an initial guess for V will be made by making an initial guess for θ. With V and thus A fixed, estimates of the θ_js are computed as the solution to the system of linear equations. These estimates of the covariance parameters determine a new choice of V that can be used to get updated values of the θ_js. This iterative procedure is repeated until the θ_js converge. Since the equations are linear, solutions are easy to find. As mentioned, similar methods can be used to find unrestricted MLEs. In fact, doing so only involves removing the terms $(I-A)$ from the traces.

4.5 MINQUE

Before beginning on MINQUE, we quickly review generalized least squares estimation. A linear model has $E(Y) = X\beta$ and, for a positive definite weighting matrix W, the generalized least squares estimates are values of β that minimize the *norm*

$$(Y - X\beta)'W^{-1}(Y - X\beta).$$

Such estimates $\hat{\beta}_W$ satisfy the generalized normal equations $X'W^{-1}X\hat{\beta}_W = X'W^{-1}Y$. Recall that generalized least squares is a geometric optimality property, not a statistical optimality property. We will use these results to define MINQUE estimation.

Consider again the linear model

$$Y = X\beta + e, \quad E(e) = 0, \quad \mathrm{Cov}(e) = V(\theta),$$

with $V(\theta)$ having the linear structure

$$V(\theta) = V_0 + \theta_1 V_1 + \cdots + \theta_s V_s.$$

MINQUE provides a method for estimating the covariance parameter θ. MINQUE is short for *minimum norm quadratic unbiased (translation invariant) estimation*. Attention is restricted to estimates that are translation invariant unbiased quadratic forms in the observations. Our discussion follows Christensen (1993) which takes more care to discuss issues related to invariance.

Take B to be a full rank matrix with $C(B) = C(X)^{\perp}$. With such a choice, $E(B'Y) = 0$, so $\mathrm{Cov}(B'Y) = E[B'YY'B]$, and with linear covariance structure, the expected value of the observable random matrix $B'YY'B$ has

$$E[B'YY'B] = B'V(\theta)B = B'V_0B + \theta_1 B'V_1B + \cdots + \theta_s B'V_sB.$$

Using Vec operators and Kronecker products as reviewed in Appendix A.2, this becomes

$$
\begin{aligned}
\mathrm{E}[\mathrm{Vec}(B'YY'B)] \\
&= \mathrm{Vec}(B'V_0B) + \theta_1\mathrm{Vec}(B'V_1B) + \cdots + \theta_s\mathrm{Vec}(B'V_sB) \\
&= [B' \otimes B']\mathrm{Vec}(V_0) + \theta_1[B' \otimes B']\mathrm{Vec}(V_1) + \cdots + \theta_s[B' \otimes B']\mathrm{Vec}(V_s),
\end{aligned}
$$

which defines a linear model for the parameters θ. In particular, we write a linear model

$$
\mathrm{E}(\tilde{Y}) = \tilde{X}\theta + \tilde{v}
$$

with model matrix $\tilde{X}$ and offset vector $\tilde{v}$, wherein

$$
\tilde{Y} = \mathrm{Vec}(B'YY'B) = [B'Y \otimes B'Y], \qquad \tilde{v} = \mathrm{Vec}(B'V_0B) = [B' \otimes B']\mathrm{Vec}(V_0).
$$

$$
\tilde{X} = \{\tilde{X}_1, \ldots, \tilde{X}_s\} = [B' \otimes B'][\mathrm{Vec}(V_1), \ldots, \mathrm{Vec}(V_s)] \equiv [B' \otimes B']\tilde{U}.
$$

Equivalently, we can write the linear model as

$$
\mathrm{E}(\tilde{Y} - \tilde{v}) = \tilde{X}\theta \tag{4.5.1}
$$

with observable dependent variable vector $\tilde{Y} - \tilde{v}$.

MINQUE estimates of θ use generalized least squares, i.e., they minimize

$$
(\tilde{Y} - \tilde{v} - \tilde{X}\theta)'\tilde{W}^{-1}(\tilde{Y} - \tilde{v} - \tilde{X}\theta)'
$$

where, for some s dimensional weight vector w we define,

$$
\tilde{W} \equiv [B'V(w)B \otimes B'V(w)B].
$$

The weight vector w might consist of a vector of prior guesses for the parameters. MINQUE(1) takes the guess for θ as J_s. MINQUE(0) takes the guess for θ as $(1, 0, \ldots, 0)'$ but only when $V_1 = I_n$. In particular, estimates are typically obtained by solving the generalized normal equations

$$
\tilde{X}'\tilde{W}^{-1}\tilde{X}\theta = \tilde{X}'\tilde{W}^{-1}(\tilde{Y} - \tilde{v}).
$$

As seen below, for symmetric V_js, the generalized normal equations reduce to the well known MINQUE equations,

$$
\begin{aligned}
\sum_{k=1}^{s} \theta_k \mathrm{tr}&\left\{V_k[I - A(w)]'V^{-1}(w)V_jV^{-1}(w)[I - A(w)]\right\} \\
&= Y'[I - A(w)]'V^{-1}(w)V_jV^{-1}(w)[I - A(w)]Y \\
&\quad - \mathrm{tr}\left\{V_0[I - A(w)]'V^{-1}(w)V_jV^{-1}(w)[I - A(w)]\right\},
\end{aligned}
$$

Note the similarity to the REML equations of (4.4.1) wherein $V \equiv V(\theta)$ and $A \equiv A(\theta)$. The iterative process of solving the REML equations amounts to solving the

MINQUE equations once for an arbitrary w and then iteratively replacing w with the solutions to the previous set of equations. As a rule, this iterative MINQUE process provides neither minimum norm, nor quadratic, nor unbiased estimates.

Searle, Casella, and McCulloch (1992, Chapter 12), Christensen (1993), and Cressie (1993, Section 2.6) all discuss versions of this linear model approach. Justus Seely originally came up with the idea that quadratic forms could be recast as linear functions and Friedrich Pukelsheim importantly developed notation for working with them.

The remainder of this section is devoted to simplifying the generalized normal equations into the MINQUE equations.

4.5.1 Deriving the MINQUE Equations

The key to simplifying the generalized normal equations is that

$$[B \otimes B][B'V(w)B \otimes B'V(w)B]^{-1}[B' \otimes B']$$
$$= [V^{-1}(w)[I - A(w)] \otimes V^{-1}(w)[I - A(w)]] .$$

It suffices to show that

$$B[B'V(w)B]^{-1}B' = V^{-1}(w)[I - A(w)],$$

but this was established in Sect. 4.3 using Lemmas 4.3.1 and 4.3.2.

Consider the normal equations in pieces. To simplify notation, write $V \equiv V(w)$ and $A \equiv A(w)$.

$$\tilde{X}\tilde{W}^{-1}\tilde{Y}$$
$$= \tilde{U}'[B \otimes B]\left\{[B'VB]^{-1} \otimes [B'VB]^{-1}\right\}[B' \otimes B'][Y \otimes Y]$$
$$= \tilde{U}'\left\{V^{-1}(I - A) \otimes V^{-1}(I - A)\right\}[Y \otimes Y]$$
$$= [\text{Vec}(V_1), \dots, \text{Vec}(V_s)]'\left\{V^{-1}(I - A)Y \otimes V^{-1}(I - A)Y\right\}$$
$$= [\text{Vec}(V_1), \dots, \text{Vec}(V_s)]'\text{Vec}[V^{-1}(I - A)YY'(I - A)'V^{-1}].$$

The jth element of the vector is

$$\text{Vec}(V_j)'\text{Vec}[V^{-1}(I - A)YY'(I - A)'V^{-1}] = \text{tr}\left[V_j'V^{-1}(I - A)YY'(I - A)'V^{-1}\right]$$
$$= \text{tr}\left[Y'(I - A)'V^{-1}V_j'V^{-1}(I - A)Y\right]$$
$$= Y'(I - A)'V^{-1}V_j'V^{-1}(I - A)Y.$$

Now consider

$$\tilde{X}\tilde{W}^{-1}\tilde{v}$$
$$- \tilde{U}'[B \otimes B]\left\{[B'VB]^{-1} \otimes [B'VB]^{-1}\right\}[B' \otimes B']\text{Vec}(V_0)$$

$$= \tilde{U}'\left[V^{-1}(I-A)\otimes V^{-1}(I-A)\right]\mathrm{Vec}(V_0)$$
$$= [\mathrm{Vec}(V_1),\ldots,\mathrm{Vec}(V_s)]]'\,\mathrm{Vec}[V^{-1}(I-A)V_0(I-A)'V^{-1}].$$

The jth element of the vector is

$$\mathrm{Vec}(V_j)'\mathrm{Vec}[V^{-1}(I-A)V_0(I-A)'V^{-1}] = \mathrm{tr}\left[V_j'V^{-1}(I-A)V_0(I-A)'V^{-1}\right]$$
$$= \mathrm{tr}\left[V_0(I-A)'V^{-1}V_j'V^{-1}(I-A)\right].$$

Similarly,

$$\tilde{X}'\tilde{W}^{-1}\tilde{X}\theta$$
$$= \tilde{U}'[B\otimes B]\left\{[B'VB]^{-1}\otimes[B'VB]^{-1}\right\}[B'\otimes B']\tilde{U}\theta$$
$$= \tilde{U}'\left\{V^{-1}(I-A)\otimes V^{-1}(I-A)\right\}[\mathrm{Vec}(V_1),\ldots,\mathrm{Vec}(V_s)]\theta$$
$$= [\mathrm{Vec}(V_1),\ldots,\mathrm{Vec}(V_s)]'$$
$$\times\left\{\mathrm{Vec}[V^{-1}(I-A)V_1(I-A)'V^{-1}],\ldots,\mathrm{Vec}[V^{-1}(I-A)V_s(I-A)'V^{-1}]\right\}\theta.$$

The jth element of the vector is

$$\mathrm{Vec}(V_j)'\left\{\mathrm{Vec}[V^{-1}(I-A)V_1(I-A)'V^{-1}],\ldots,\mathrm{Vec}[V^{-1}(I-A)V_s(I-A)'V^{-1}]\right\}\theta$$
$$= \left\{\mathrm{tr}[V_j'V^{-1}(I-A)V_1(I-A)'V^{-1}],\ldots,\mathrm{tr}[V_j'V^{-1}(I-A)V_s(I-A)'V^{-1}]\right\}\theta$$
$$= \sum_{k=1}^{s}\theta_k\mathrm{tr}[V_j'V^{-1}(I-A)V_k(I-A)'V^{-1}]$$
$$= \sum_{k=1}^{s}\theta_k\mathrm{tr}[V_k(I-A)'V^{-1}V_j'V^{-1}(I-A)].$$

All together, the jth normal equation is

$$\sum_{k=1}^{s}\theta_k\mathrm{tr}[V_k(I-A)'V^{-1}V_j'V^{-1}(I-A)]$$
$$= Y'(I-A)'V^{-1}V_j'V^{-1}(I-A)Y - \mathrm{tr}\left[V_0(I-A)'V^{-1}V_j'V^{-1}(I-A)\right].$$

If $V_1,\ldots,V_s$ are symmetric, these are the MINQUE equations.

4.6 MIVQUE

Minimum variance quadratic unbiased translation invariant estimates (MIVQUEs) would be the best linear unbiased estimates (BLUEs) in model (4.5.1). Quadratic unbiased translation invariant estimates are precisely the linear unbiased estimates in this model. By definition, the best unbiased estimates are those with minimum variance. However, the model does not specify the covariance matrix of $\tilde{Y} = \mathrm{Vec}[B'YY'B]$, say $\check{V}$. The covariance matrix is needed to establish that something has minimum variance. Typically, $\check{V}$ is a singular matrix because the vector $[Y \otimes Y]$ contains both $y_i y_j$ and $y_j y_i$. While best linear unbiased estimation in linear models with an arbitrary singular covariance matrix is quite a complicated problem, a simple solution exists, cf. *PA* Chap. 10. For model (4.5.1), the BLUE of $\tilde{X}\theta$ is $\tilde{X}(\tilde{X}'\tilde{W}^-\tilde{X})^-\tilde{X}'\tilde{W}^-(\tilde{Y} - \tilde{v})$ where $\tilde{W} = \check{V} + \tilde{X}\tilde{X}'$ and it is assumed that $\tilde{v} \in C(\check{V}, \tilde{X})$ (which is usually true since $\tilde{v}$ is usually 0). However, finding the estimate requires us to know $\check{V}$, at least up to a scalar multiple. Even if we assume that Y is multivariate normal, the covariance matrix of $\tilde{Y}$ is unknown because it depends on the parameter vector θ. In particular, one can show that $B'YY'B$ has a central Wishart distribution in the sense of Chap. 9 so that $\check{V} = [B'V(\theta)B \otimes B'V(\theta)B](I+T)$ where T is the $[n - r(X)]^2 \times [n - r(X)]^2$ matrix that transforms $\mathrm{Vec}(Q)$ into $\mathrm{Vec}(Q')$. In short, there is no reason to think that anyone can find a MIVQUE estimate in anything but the most degenerate of models.

Typically, when people find "MIVQUE" estimates, they pretend that they know V, plug it into $\check{V} = [B'V(\theta)B \otimes B'V(\theta)B](I+T)$, and use linear model theory to get what would be the optimal estimate of θ if one really knew V. Then, because they know that they do not know V, they iterate the procedure. Modulo some results in Exercise 4.7, this is the same thing as iterated MINQUE, which is no excuse for the fact that the names MINQUE and MIVQUE are often used interchangeably. And like iterated MINQUE, these "MIVQUE" estimates are typically not minimum variance, nor quadratic, nor unbiased.

Theorem 4.6.1. Suppose $Y \sim N(\mu, V)$ and V is nonsingular. Then $\mathrm{Var}(Y'QY) = 2\mathrm{tr}[(QV)^2] + 4\mu'QVQ\mu$.

PROOF. See Searle (1971, Section 2.5). For $\mu = 0$, the result also follows from looking at $Y'QY = \mathrm{Vec}(Q)'\mathrm{Vec}(YY')$ where YY' is Wishart with the appropriate covariance matrix and $Q = Q'$. □

Exercise 4.7.

(a) Show that $[B'V(\theta)B \otimes B'V(\theta)B](I+T)$ is not positive definite. Hint: For a non-symmetric matrix Q, check the vector $\mathrm{Vec}(Q - Q')$.

(b) For any nonsymmetric matrix Q, show that $\mathrm{Vec}(Q)'\tilde{Y} = \mathrm{Vec}(Q+Q')\tilde{Y}/2$. Argue that in model (4.5.1) we can restrict attention to linear estimates $\mathrm{Vec}(Q)'\tilde{Y}$ where Q is symmetric. In particular, show that any MINQUE estimate of an estimable function $\lambda'\theta$ can be written as $\mathrm{Vec}(Q)'(\tilde{Y} - \tilde{v})$ where Q is symmetric.

(c) Show that if we restrict attention to symmetric linear combinations of $\tilde{Y}$, MIVQUE as discussed in this section is equivalent to MINQUE(θ).

(d) Consider the vector space of $p \times p$ matrices and let e_i be the ith column of I_p. Let the inner product between two matrices Q_1 and Q_2 be $\mathrm{Vec}(Q_1)'\mathrm{Vec}(Q_2) = \mathrm{tr}(Q_1'Q_2)$. Show that the matrices $e_i e_j'$ constitute an orthonormal basis for this p^2 dimensional space. Show that the vectors $e_i e_j' + e_j e_i'$, $i = 1, \ldots, p$, $j \leq i$ constitute an orthogonal basis for the $p(p+1)/2$ dimensional subspace of symmetric matrices. Show that the vectors $e_i e_j' - e_j e_i'$, $i = 1, \ldots, p$, $j < i$ constitute an orthogonal basis for the $p(p-1)/2$ dimensional subspace of matrices that are orthogonal to the symmetric matrices. Using the orthonormal basis, find the $p^2 \times p^2$ matrix T that maps $\text{Vec}(Q)$ into $\text{Vec}(Q')$.

(e) Show that $I = TT$ and that $(1/2)[I+T]$ is the ppo onto the space of symmetric matrices.

(f) Show that $(1/2)[I-T]$ is the ppo onto the orthogonal complement of the symmetric matrices.

(g) Show that if Q is in this orthogonal complement, then $q_{ij} = -q_{ji}$.

(h) Show that if the V_js in model (4.5.1) are symmetric, the entire linear model problem exists within in the subspace of $\mathbf{R}^{[n-r(X)]^2}$ that consists of vectors $\text{Vec}(Q)$ with $Q = Q'$.

(i) Show that

$$[B'V(\theta)B \otimes B'V(\theta)B](I+T) = 2\frac{1}{2}(I+T)[B'V(\theta)B \otimes B'V(\theta)B]\frac{1}{2}(I+T).$$

The second version is clearly nonnegative definite, unlike the first version. Hint: Pre- and post- multiply by $\text{Vec}(Q_1)'$ and $\text{Vec}(Q_2)$, respectively.

4.7 The Effect of Estimated Covariances

Typically, $V \equiv V(\theta)$ is not known, nor does it have a special form that allows a simple analysis. In practice, V is estimated with a function of Y, say $\tilde{V}(Y) = \tilde{V}$, that is nonnegative definite. More specifically, we estimate θ by $\tilde{\theta}$ and define $\tilde{V}(Y) = \tilde{V} = V(\tilde{\theta})$.

Estimation of β is performed by plugging $\tilde{V}$ in for V, that is, an estimate is $\tilde{\beta}$, where

$$X\tilde{\beta} = \tilde{A}Y$$

and $\tilde{A}$ is the random projection operator

$$\tilde{A} = X(X'\tilde{V}^{-1}X)^- X'\tilde{V}^{-1}.$$

$\tilde{A}$ is random because $\tilde{V}$ is a function of Y, hence random. For simplicity, in the discussion that follows, we will assume that V and $\tilde{V}$ are positive definite. $X\tilde{\beta}$ is called either an *empirical estimate* or a *plug-in estimate*.

For the prediction problem of Sect. 4.1.2, estimates $\tilde{V}(Y)$ and $\tilde{V}_{y0}(Y)$ are required. The empirical (plug-in) predictor is taken by analogy with Theorem 4.1.3 as

$$\tilde{y}_0 = x_0'\tilde{\beta} + \tilde{\delta}'(Y - X\tilde{\beta}),$$

where $\tilde{\beta}$ is defined as before,

$$\tilde{\delta} = \tilde{V}^{-1}\tilde{V}_{y0},$$

and $x_0'\beta$ is assumed to be estimable.

The empirical estimate of a function $\lambda'\beta$ or prediction of a point y_0 are complicated functions of Y. In this section we present general mathematical results originally due to Eaton (1985) and Harville (1985) showing that under reasonable conditions the empirical estimates and predictors are unbiased.

When $V(\theta)$ is known, there are formula given in Sect. 4.1 for the variances of BLUEs $\lambda'\hat\beta$ and BLUPs $\hat y_0$ that depend on $V(\theta)$. It is tempting to use these formulae with $V(\theta)$ replaced by $V(\tilde\theta)$ to provide standard errors for the empirical estimates and predictors. We also show that such standard errors systematically underestimate the true standard errors. It seems reasonable that the empirical estimates and predictors would be worse (have higher variance) than the BLUEs and BLUPs, since they involve estimating parameters that are taken as known in the BLUEs and BLUPs. We demonstrate, not only that the empirical values have larger variances, but that the naive standard error formulae that uses $\tilde\theta$ to replace θ tends to underestimate the true variability because it has an expected value bounded above by the formulae that use the true θ.

As discussed in Sect. 4.1.2, assume that $\theta' = [\theta_1, \theta_*']$ with

$$V(\theta) \equiv \theta_1 V_*(\theta_*).$$

One way to proceed is to find estimates $\tilde\theta_1$ and $\tilde\theta_*$, treat $V_*(\tilde\theta_*)$ as the true value $V_*(\theta_*)$, and construct standard errors and F statistics as usual for weighted least squares models. Based on the results of the next subsection, there should be a tendency for standard errors to be underestimated and tests to appear more significant than they are. The F statistics will not have exact F distributions. Part of the difficulty is that the theory for F tests involves $\tilde\theta_1$ being independent of the estimates of the fixed effects and having a distribution related to the χ^2 with some degrees of freedom. There is little reason to expect the independence and χ^2 properties to be true. In fact, to perform ad hoc F tests at all, we need to identify some number of degrees of freedom for $\tilde\theta_1$.

4.7.1 Mathematical Results*

Eaton (1985) and Harville (1985) have given conditions under which the empirical estimates and predictors are unbiased and have variances at least as great as the corresponding BLUEs and BLUPs. This is one of those cases in which two people have developed very similar ideas simultaneously and independently. We follow Eaton's development. To estimate the variance of a plug-in estimate or predictor, the variance formula for the BLUE or BLUP has frequently been used, with V replaced by $\tilde V$. Under mild conditions, when $\tilde V$ is unbiased for V, the expected value of this estimated variance is less than or equal to the variance of the BLUE or BLUP (which in turn is less than or equal to the true variance of the plug-in estimator or predictor). These results establish a theoretical basis for the often-observed phenomenon that these estimated variances for plug-in estimators (predictors) are often misleadingly small. Although Eaton's results do not explicitly use any parameterization for V, it is typically the case that the covariance matrix depends on a parameter vector θ, that is, $V = V(\theta)$, and that the estimate of $V(\theta)$ is $\tilde{V} = V(\tilde{\theta})$, where $\tilde{\theta}$ is an estimate of θ.

The first results on the unbiasedness of plug-in estimates and predictors are apparently due to Kackar and Harville (1981). Other results on improved variance estimation for plug-in predictors are given by Kackar and Harville (1984), Harville and Jeske (1992) and Zimmerman and Cressie (1992).

Definition 4.7.1. $\tilde{B}(Y)$ is a residual type statistic if

$$\tilde{B}(Y) = \tilde{B}(Y - X\beta) \qquad \text{for any} \quad \beta \tag{4.7.1}$$

and

$$\tilde{B}(Y) = \tilde{B}(-Y). \tag{4.7.2}$$

Note that any residual type statistic has $\tilde{B}(Y) = \tilde{B}(Y - X\hat{\beta}) = \tilde{B}((I-A)Y)$, so residual type statistics can be viewed as functions of the residual vector.

In the following discussion, we will assume that $\tilde{V}$ and $\tilde{V}_{y0}$ are residual type statistics. Most standard methods for estimating covariance matrices satisfy the conditions of Definition 4.7.1. Clearly, if $\tilde{V}$ and $\tilde{V}_{y0}$ are residual type statistics, any functions of them are also of the residual type. In particular, functions such as $\tilde{V}^{-1}$,

$$\tilde{A} = X(X'\tilde{V}^{-1}X)^{-}X'\tilde{V}^{-1},$$

and

$$\tilde{\delta} = \tilde{V}^{-1}\tilde{V}_{y0}$$

are residual type statistics.

Exercise 4.8.   Show that solving the likelihood or REML equations yield covariance matrix estimates that are residual type statistics.

The key result in establishing that plug-in estimators are unbiased is the following proposition.

Proposition 4.7.2.   If e and $-e$ have the same distribution and if $\tilde{B}(Y)$ is a residual type statistic, then

$$E[\tilde{B}(Y)Y] = E[\tilde{B}(Y)X\beta].$$

PROOF.

$$E[\tilde{B}(Y)Y] = E[\tilde{B}(Y)X\beta] + E[\tilde{B}(Y)e].$$

It suffices to show that $E[\tilde{B}(Y)e] = 0$. By Definition 4.7.1,

$$\tilde{B}(Y) = \tilde{B}(Y - X\beta) = \tilde{B}(e),$$

and by the symmetry property of e assumed in the proposition,

$$E[\tilde{B}(e)e] = -E[\tilde{B}(-e)e] = -E[\tilde{B}(e)e].$$

The only way a real vector can equal its negative is if the vector is zero, thus completing the proof. $\qquad\square$

Henceforth assume that e and $-e$ have the same distribution.

Proposition 4.7.2 leads immediately to two results on unbiased estimation.

Corollary 4.7.3. $\quad E[X\tilde{\beta}] = X\beta.$

PROOF. By definition, $X\tilde{\beta} = \tilde{A}Y$, where $\tilde{A}$ is a residual type statistic, so by Proposition 4.7.2, $E[X\tilde{\beta}] = E[\tilde{A}X\beta]$. Because $\tilde{A}$ is a projection operator onto $C(X)$ for any $\tilde{V}$, $E[\tilde{A}X\beta] = E[X\beta] = X\beta.$ $\qquad\square$

Corollary 4.7.4. $\quad$ If $\lambda'\beta$ is estimable, then

$$E[\lambda'\tilde{\beta}] = \lambda'\beta.$$

PROOF. $\quad$ By estimability, $\lambda' = \rho'X$ and

$$E[\lambda'\tilde{\beta}] = E[\rho'X\tilde{\beta}] = \rho'E[X\tilde{\beta}] = \rho'X\beta = \lambda'\beta.$$ $\qquad\square$

Now, consider the prediction problem. We seek to predict y_0, where $E(y_0) = x_0'\beta$ and $x_0'\beta$ is estimable. The plug-in predictor is

$$\tilde{y}_0 = x_0'\tilde{\beta} + \tilde{\delta}'(Y - X\tilde{\beta})$$
$$= x_0'\tilde{\beta} + \tilde{\delta}'(I - \tilde{A})Y,$$

where $\tilde{\delta} = \tilde{V}^{-1}\tilde{V}_{y0}$ is a residual type statistic. Before proving that the plug-in predictor is unbiased, we establish another result.

Lemma 4.7.5. $\quad E[\tilde{\delta}'(I - \tilde{A})Y] = 0.$

PROOF. $\quad$ Because $\tilde{\delta}$ and $(I - \tilde{A})$ are residual type statistics, $\tilde{\delta}'(I - \tilde{A})$ is also of residual type. Applying Proposition 4.7.2 and using the fact that for each Y, $\tilde{A}$ is a projection operator onto $C(X)$,

$$E[\tilde{\delta}'(I - \tilde{A})Y] = E[\tilde{\delta}'(I - \tilde{A})X\beta] = 0.$$ $\qquad\square$

The plug-in predictor is unbiased.

Proposition 4.7.6. $E[\tilde{y}_0] = x_0'\beta = E[y_0]$.

PROOF. By Corollary 4.7.4 and Lemma 4.7.5

$$E[\tilde{y}_0] = E[x_0'\tilde{\beta} + \tilde{\delta}'(I - \tilde{A})Y] = E[x_0'\tilde{\beta}] + E[\tilde{\delta}(I - \tilde{A})Y] = x_0'\beta. \qquad \square$$

The next two propositions establish conditions under which the variance of the plug-in estimate and the prediction variance of the plug-in predictor are known to be no less than the variance of the BLUE and the BLUP. After proving the results, a brief discussion of the conditions necessary for the results will be given.

Proposition 4.7.7. If $E[Ae|(I-A)e] = 0$ and $\lambda' = \rho'X$, then

$$\text{Var}[\lambda'\tilde{\beta}] = \text{Var}[\lambda'\hat{\beta}] + \text{Var}[\lambda'\tilde{\beta} - \lambda'\hat{\beta}].$$

PROOF.

$$\begin{aligned}
\text{Var}(\lambda'\tilde{\beta}) &= \text{Var}(\lambda'\tilde{\beta} - \lambda'\hat{\beta} + \lambda'\hat{\beta}) \\
&= \text{Var}(\lambda'\hat{\beta}) + \text{Var}(\lambda'\tilde{\beta} - \lambda'\hat{\beta}) + 2\text{Cov}\left(\lambda'\hat{\beta}, \lambda'(\tilde{\beta} - \hat{\beta})\right),
\end{aligned}$$

so it suffices to show that

$$\text{Cov}\left(\lambda'\hat{\beta}, \lambda'(\tilde{\beta} - \hat{\beta})\right) = 0.$$

Because $E[\lambda'\tilde{\beta} - \lambda'\hat{\beta}] = 0$, it is enough to show that

$$E[\lambda'\hat{\beta}\{\lambda'\tilde{\beta} - \lambda'\hat{\beta}\}] = 0. \tag{4.7.3}$$

Now, observe that because $AY \in C(X)$ and $\tilde{A}$ is a projection operator onto $C(X)$,

$$X\tilde{\beta} = \tilde{A}Y = \tilde{A}[AY + (I - A)Y] = AY + \tilde{A}(I - A)Y.$$

Hence,

$$\lambda'\tilde{\beta} = \rho'X\tilde{\beta} = \rho'AY + \rho'\tilde{A}(I - A)Y = \lambda'\hat{\beta} + \rho'\tilde{A}(I - A)Y$$

and

$$\lambda'\tilde{\beta} - \lambda'\hat{\beta} = \rho'\tilde{A}(I - A)Y.$$

Thus, (4.7.3) is equivalent to

$$E[\lambda'\hat{\beta}\{\rho'\tilde{A}(I - A)Y\}] = 0.$$

Before proceeding, note two things: first, $(I - A)Y = (I - A)e$ and second, because $\tilde{A}$ is a residual type function of Y, $\tilde{A}$ is also a function of $(I - A)Y$. Now, consider the conditional expectation

$$
\begin{aligned}
\mathrm{E}\left[\lambda'\hat{\beta}\{\rho'\tilde{A}(I - A)Y\}\big|(I - A)Y\right] \\
= \{\rho'\tilde{A}(I - A)Y\}\mathrm{E}\left[\lambda'\hat{\beta}\big|(I - A)Y\right] \\
= \{\rho'\tilde{A}(I - A)Y\}\mathrm{E}\left[\rho'X\beta + \rho'Ae\big|(I - A)Y\right] \\
= \{\rho'\tilde{A}(I - A)Y\}\rho'X\beta.
\end{aligned}
\tag{4.7.4}
$$

The last equality follows because $\rho'X\beta$ is a constant and $\mathrm{E}[\rho'Ae|(I - A)Y] = \rho'\mathrm{E}[Ae|(I - A)e] = 0$ by assumption. The statistic $\rho'\tilde{A}(I - A)$ is of residual type, so using (4.7.4) and Proposition 4.7.2,

$$
\begin{aligned}
\mathrm{E}\left[\lambda'\hat{\beta}\{\rho'\tilde{A}(I - A)Y\}\right] &= \mathrm{E}\left(\mathrm{E}\left[\lambda'\hat{\beta}\{\rho'\tilde{A}(I - A)Y\}\big|(I - A)Y\right]\right) \\
&= \mathrm{E}\left(\{\rho'\tilde{A}(I - A)Y\}\rho'X\beta\right) \\
&= \rho'X\beta\,\mathrm{E}\left(\rho'\tilde{A}(I - A)Y\right) \\
&= \rho'X\beta\,\mathrm{E}[\rho'\tilde{A}(I - A)X\beta] \\
&= 0,
\end{aligned}
$$

which proves the result. $\qquad\square$

Proposition 4.7.8. If $\mathrm{E}[y_0 - \hat{y}_0|(I - A)e] = 0$, then

$$
\mathrm{Var}(y_0 - \tilde{y}_0) = \mathrm{Var}(y_0 - \hat{y}_0) + \mathrm{Var}(\hat{y}_0 - \tilde{y}_0).
$$

PROOF. The proof is very similar to the proof of Proposition 4.7.7.

$$
\begin{aligned}
\mathrm{Var}(y_0 - \tilde{y}_0) &= \mathrm{Var}(y_0 - \hat{y}_0 + \hat{y}_0 - \tilde{y}_0) \\
&= \mathrm{Var}(y_0 - \hat{y}_0) + \mathrm{Var}(\hat{y}_0 - \tilde{y}_0) + 2\mathrm{Cov}(y_0 - \hat{y}_0, \hat{y}_0 - \tilde{y}_0).
\end{aligned}
$$

We show that $\mathrm{Cov}(y_0 - \hat{y}_0, \hat{y}_0 - \tilde{y}_0) = 0$, or equivalently, because $\mathrm{E}[\hat{y}_0 - \tilde{y}_0] = 0$, that

$$
\mathrm{E}[(y_0 - \hat{y}_0)(\hat{y}_0 - \tilde{y}_0)] = 0.
$$

Recall that $\tilde{b}$ satisfies

$$
\tilde{y}_0 = \tilde{b}'Y = \rho'\tilde{A}Y + \tilde{\delta}'(I - \tilde{A})Y.
\tag{4.7.5}
$$

Substituting AY for Y in (4.7.5) and using the facts that $AY \in C(X)$ and $\tilde{A}$ is a projection operator onto $C(X)$, we have $\tilde{b}'AY = \rho'\tilde{A}AY = \rho'AY = x_0'\hat{\beta}$. Moreover,

$$
\tilde{y}_0 = \tilde{b}'AY + \tilde{b}'(I - A)Y
$$

$$
\begin{aligned}
&= x_0'\hat{\beta} + \tilde{b}'(I-A)Y \\
&= x_0'\hat{\beta} + \delta'(I-A)Y - \delta'(I-A)Y + \tilde{b}'(I-A)Y \\
&= \hat{y}_0 + (\tilde{b} - \delta)'(I-A)Y,
\end{aligned}
$$

and

$$
\hat{y}_0 - \tilde{y}_0 = (\delta - \tilde{b})'(I-A)Y,
$$

which is a function of $(I-A)Y$ because $\tilde{b}$ is a residual type statistic.

As in the previous proof, evaluate the conditional expectation. This gives

$$
\mathrm{E}[(y_0 - \hat{y}_0)(\hat{y}_0 - \tilde{y}_0)|(I-A)Y] = (\hat{y}_0 - \tilde{y}_0)\mathrm{E}[(y_0 - \hat{y}_0)|(I-A)Y] = 0
$$

by assumption and the fact that $(I-A)Y = (I-A)e$. Because the conditional expectation is zero for all $(I-A)Y$, the unconditional expectation is zero and the result is proven.

$\square$

Proposition 4.7.7 shows that the variance of the plug-in estimator equals the BLUE variance plus a nonnegative quantity. Thus, the plug-in variance is at least as large as the BLUE variance. Proposition 4.7.8 gives a similar result for prediction.

Eaton (1985) discusses situations under which the conditions $\mathrm{E}[Ae|(I-A)e] = 0$ and $\mathrm{E}[y_0 - \hat{y}_0|(I-A)e] = 0$ hold. In particular, the first condition holds if e has an elliptical distribution, and the second condition holds if (e_0, e') has an elliptical distribution. Elliptical distributions can be generated as follows. Y has an elliptical distribution centered at 0 if Y has a density that is proportional to $\varphi(y'By)$, where B is a positive definite matrix and $\varphi(u)$ is a density on the nonnegative real numbers. From *PA* Sect. 1.3, we see that if $z_i \sim N(0,1)$, we get the multivariate normal distribution as a special case of elliptical distributions.

Eaton's last results involve concave functions and Jensen's inequality. A set ζ is convex if, for any $\alpha \in [0,1]$ and $s_1, s_2 \in \zeta$, the point $\alpha s_1 + (1-\alpha)s_2 \in \zeta$. A function $\Psi : \zeta \to \mathbf{R}$ is concave if $\Psi(\alpha s_1 + (1-\alpha)s_2) \geq \alpha\Psi(s_1) + (1-\alpha)\Psi(s_2)$. Finally, if s is random and defined on ζ, Jensen's inequality states that $\Psi(\mathrm{E}[s]) \geq \mathrm{E}[\Psi(s)]$. See Ferguson (1967) for a more complete discussion of convexity and a proof of Jensen's inequality.

Let ζ be the set of all positive definite matrices V, and observe that ζ is a convex set. Let $\mathscr{P}$ be the set of all matrices that are projection operators onto spaces that contain $C(X)$, namely

$$
\mathscr{P} = \{P | PP = P \text{ and } C(X) \subset C(P)\}.
$$

For any $\rho \in \mathbf{R}^n$ and each $P \in \mathscr{P}$ define

$$
\Psi_P(V) = \rho'PVP'\rho.
$$

Observe that $\Psi_P(V)$ is a concave function. To see this, note that

$$
\Psi_P(\alpha V_1 + (1-\alpha)V_2) = \rho'P\{\alpha V_1 + (1-\alpha)V_2\}P'\rho
$$

$$= \alpha \rho' PV_1 P' \rho + (1 - \alpha) \rho' PV_2 P' \rho$$
$$= \alpha \Psi_P(V_1) + (1 - \alpha) \Psi_P(V_2).$$

For any P, $\rho' PY$ is a linear unbiased estimate of $\rho' X\beta$, so the variance of the BLUE, $\rho' AY$, is at least as small as the variance of $\rho' PY$, that is,

$$\rho' AVA' \rho = \inf_{P \in \mathscr{P}} \rho' PVP' \rho.$$

Define

$$\Psi(V) \equiv \rho' AVA' \rho,$$

so we see that

$$\Psi(V) = \inf_{P \in \mathscr{P}} \Psi_P(V).$$

The infinum of a set of concave functions is also concave, so $\rho' AVA' \rho$ is concave. (Note that a direct proof of concavity is difficult because A is a function of V.)

Exercise 4.9. Show that if $f_\lambda(x)$ is concave for any $\lambda \in \Lambda$, then $f(x) = \inf_{\lambda \in \Lambda} f_\lambda(x)$ is also concave. Hint: By definition, for any point x_0 and any $\varepsilon > 0$, there exists λ such that $f(x_0) \geq f_\lambda(x_0) - \varepsilon$.

The estimated variance of the plug-in estimator is $\rho' \tilde{A} \tilde{V} \tilde{A}' \rho$. Define

$$\widetilde{\mathrm{Var}}(\rho' X\tilde{\beta}) \equiv \rho' \tilde{A} \tilde{V} \tilde{A}' \rho.$$

Recalling that

$$\mathrm{Var}(\rho' X\hat{\beta}) = \rho' AVA' \rho$$

and that by Proposition 4.7.7

$$\mathrm{Var}(\rho' X\hat{\beta}) \leq \mathrm{Var}(\rho' X\tilde{\beta}),$$

we can prove the following.

Proposition 4.7.9. If $\tilde{V}$ is unbiased for V and if Proposition 4.7.7 holds, then

$$E[\widetilde{\mathrm{Var}}(\rho' X\tilde{\beta})] \leq \mathrm{Var}(\rho' X\hat{\beta}) \leq \mathrm{Var}(\rho' X\tilde{\beta}).$$

PROOF. We need only prove that

$$E[\rho' \tilde{A} \tilde{V} \tilde{A}' \rho] \leq \rho' AVA' \rho.$$

Because $\Psi(V)$ is concave, Jensen's inequality gives

$$\rho' AVA' \rho = \Psi(V) = \Psi(E[\tilde{V}]) \geq E[\Psi(\tilde{V})] = E[\rho' \tilde{A} \tilde{V} \tilde{A}' \rho]. \qquad \square$$

Thus, for an unbiased covariance matrix estimate $\tilde{V}$, the expected value of the estimated variance of $\rho'X\tilde{\beta}$ is no greater than the variance of $\rho'X\hat{\beta}$ while the true variance of $\rho'X\tilde{\beta}$ is no less than the variance of $\rho'X\hat{\beta}$. This establishes that there is a tendency for the estimated variance of $\rho'X\tilde{\beta}$ to underestimate the true variance and illustrates how the underestimation could be very substantial.

The result for prediction follows similarly. Consider the set

$$\mathscr{D} = \{d \,|\, d'X = x_0'\}.$$

Then, any linear unbiased predictor can be written as $d'Y$ for some $d \in \mathscr{D}$. Let $V_{00} = \text{Var}(y_0)$,

$$V_{pred} = \begin{bmatrix} V_{00} & V_{0y} \\ V_{y0} & V \end{bmatrix},$$

and take ζ as the set of all positive definite V_{pred}. Define

$$\Psi_d(V_{pred}) \equiv \text{Var}(y_0 - d'Y) = V_{00} - 2d'V_{y0} + d'Vd.$$

It is easily seen that

$$\Psi_d(\alpha V_{pred1} + (1-\alpha)V_{pred2}) = \alpha\Psi_d(V_{pred1}) + (1-\alpha)\Psi_d(V_{pred2}),$$

so $\Psi_d(V_{pred})$ is concave. Define

$$\Psi(V_{pred}) \equiv \inf_{d \in \mathscr{D}} \Psi_d(V_{pred}).$$

Note that $\Psi(V_{pred})$ is concave and

$$\Psi(V_{pred}) = \text{Var}(y_0 - b'Y) = \text{Var}(y_0 - \hat{y}_0).$$

Once again, for $\tilde{V}_{pred}$ unbiased and writing (4.1.7) with $\tilde{V}_{pred}$ substituted for V_{pred} as $\widetilde{\text{Var}}(y_0 - \tilde{y}_0)$, Jensen's inequality gives

$$\text{E}[\widetilde{\text{Var}}(y_0 - \tilde{y}_0)] = \text{E}[\Psi(\tilde{V}_{pred})] \leq \Psi(\text{E}[\tilde{V}_{pred}]) = \Psi(V_{pred}) = \text{Var}(y_0 - \hat{y}_0).$$

We have proved the following proposition.

Proposition 4.7.10.   If $\tilde{V}_{pred}$ is unbiased for V_{pred} and if Proposition 4.7.8 holds, then

$$\text{E}[\widetilde{\text{Var}}(y_0 - \tilde{y}_0)] \leq \text{Var}(y_0 - \hat{y}_0) \leq \text{Var}(y_0 - \tilde{y}_0).$$

Note that these results depend on the assumption that the procedure used for estimating V_{pred} does not depend on the true value of β. Residual type estimators take on the same value when based on Y or on $e = Y - X\beta$. (The fact that e is unobservable is irrelevant to our argument.) Because the distribution of e does not depend on β, neither does the distribution of $\tilde{V}_{pred}(Y)$. For example, if $V_{pred}(\theta)$ is the covariance matrix for observations from a process with a stationary covariance function, then clearly, because the covariance function does not depend on β, residual type estimators for $V_{pred}(\theta)$ are reasonable. However, situations exist in which residual type estimators are not reasonable.

A common problem in regression analysis is the presence of heteroscedastic errors; see *PA-V* Sect. 12.4 (Christensen 2011, Section 13.4). To deal with this problem, one often assumes a model

$$V(\theta) = [V_{ij}(\theta)],$$

where

$$V_{ij}(\theta) = 0, \quad i \neq j,$$

and

$$V_{ii}(\theta) = h_i(\alpha, \beta).$$

Here,

$$\theta' = (\alpha', \beta').$$

Two common choices for the function h_i take α as a scalar and

$$h_i(\alpha, \beta) = \alpha(x_i'\beta)^2$$

or

$$h_i(\alpha, \beta) = \alpha x_i'\beta;$$

see Carroll and Ruppert (1988).

When the variance function depends on β, it is counter intuitive to use residual type estimation procedures. In particular, MLEs will not be residual type estimates for these variance functions. For normal errors, van Houwelingen (1988) has established that the variance of the optimal weighted least squares estimate $\hat{\beta}$ (based on known variances) is at least as great as the asymptotic variance of $\beta^* - \beta$, where β^* is the MLE of β. (Actually, the proper comparison is between $\sqrt{n}\hat{\beta}$ and $\sqrt{n}(\beta^* - \beta)$, so that both quantities have nontrivial asymptotic distributions.) This remarkable result occurs because there may be extra information to be gained about β from the variability of the observations. Moreover, van Houwelingen also established that for such variance models the MLE β^* *may* not even be consistent.

References

Carroll, R. J., & Ruppert, D. (1988). *Transformations and weighting in regression*. New York, NY: Chapman and Hall.

Christensen, R. (1987). The analysis of two-stage sampling data by ordinary least squares. *Journal of the American Statistical Association, 82*, 492–498.

Christensen, R. (1993). Quadratic covariance estimation and equivalence of predictions. *Mathematical Geology, 25*, 541–558.

Christensen, R. (2011). *Plane answers to complex questions: The theory of linear models* (4th ed.). New York, NY: Springer.

Christensen, R. (2015). *Analysis of variance, design, and regression: Linear modeling for unbalanced data* (2nd ed.). Boca Raton, FL: Chapman and Hall/CRC Press.

Christensen, R., & Sun, S. K. (2010). Alternative goodness-of-fit tests for linear models. *Journal of the American Statistical Association, 105*, 291–301.

Christensen, R., & Lin, Y. (2015). Lack-of-fit tests based on partial sums of residuals. *Communications in Statistics: Theory and Methods, 44*, 2862–2880.

Cressie, N. A. C. (1993). *Statistics for spatial data* (rev. ed.). New York, NY: Wiley.

Eaton, M. L. (1985). The Gauss–Markov theorem in multivariate analysis. In P. R. Krishnaiah *Multivariate analysis—VI*. Amsterdam: Elsevier.

Ferguson, T. S. (1967). *Mathematical statistics: A decision theoretic approach*. New York, NY: Academic Press.

Ferguson, T. S. (1996). *A Course in large sample theory*. Boca Raton, FL: Chapman & Hall/CRC.

Harville, D. A. (1985). Decomposition of prediction error. *Journal of the American Statistical Association, 80*, 132–138.

Harville, D. A., & Jeske, D. R. (1992). Mean squared error of estimation or prediction under a general linear model. *Journal of the American Statistical Association, 87*, 724–731.

Kackar, R. N., & Harville, D. A. (1981). Unbiasedness of two-stage estimation and prediction procedures for mixed linear models. *Communications in Statistics: Theory and Methods, A10*, 1249–1261.

Kackar, R. N., & Harville, D. A. (1984). Approximations for standard errors of estimators of fixed and random effects in mixed linear models. *Journal of the American Statistical Association, 79*, 853–862.

Kenward, M. G., & Roger, J. H. (1997). Small sample inference for fixed effects from restricted maximum likelihood, *Biometrics, 53*, 983–997.

Kenward, M. G., Roger, J. H. (2009). An improved approximation to the precision of fixed effects from restricted maximum likelihood. *Computational Statistics and Data Analysis, 53*, 2583–2595.

Lavine, M., Bray, A., Hodges, J., & Hodges, J. (2015). Approximately exact calculations for linear mixed models. *Electronic Journal of Statistics, 9*, 2293–2323.

Patterson, H. D., & Thompson, R. (1974). Maximum likelihood estimation of variance components. In *Proceedings of the 8th International Biometric Conference* (pp. 197–207).

Satterthwaite, F. E. (1946). An approximate distribution of estimates of variance components. *Biometrics, 2*, 110–114.

Searle, S. R. (1971). *Linear models*. New York, NY: Wiley.

Searle, S. R., Casella, G., & McCulloch, C. (1992). *Variance components*. New York, NY: Wiley.

Tarpey, T., Ogden, R. T., Petkova, E., & Christensen, R. (2014). A paradoxical result in estimating regression coefficients. *The American Statistician, 68*, 271–276.

Tarpey, T., Ogden, R., Petkova, E., & Christensen, R. (2015). Reply. *The American Statistician, 69*, 254–255.

van Houwelingen, J. C. (1988). Use and abuse of variance models in regression. *Biometrics, 44*, 1073–1081.

Zimmerman, D. L., & Cressie, N. (1992). Mean squared prediction error in the spatial linear model with estimated covariance parameters. *Annals of the Institute of Statistical Mathematics, 44*, 27–43.

Chapter 5
Mixed Models and Variance Components

Abstract This chapter particularizes the results of Chap. 4 for linear mixed models with special emphasis on variance component models and a particular longitudinal data model.

Traditionally, linear models have been divided into three categories: *fixed effects models*, *random effects models*, and *mixed models*. The categorization depends on whether the β vector in $Y = X\beta + e$ is fixed, random, or has both fixed and random elements. Random effects models always assume that there is a fixed overall mean for the observations, so random effects models are actually mixed.

Variance components are the variances of the random elements of β (particularly when the random elements are uncorrelated). Sections 5.1 and 5.2 discuss mixed models in general and prediction for mixed models. Section 5.3 introduces partitioning of the random effects vector and an associated variance component model as well as a particular longitudinal mixed model. Both of these special cases are examples in which the random effects have a linear covariance structure. A linear covariance structure for the random effects leads to a linear covariance structure for the data as discussed in Sect. 4.4. Traditionally, most mixed models have used linear covariance structures. The primary current exception is for some of the longitudinal models discussed in Chap. 11. Sections 5.4 and 5.5 discuss in more detail the variance component model and the longitudinal model. Section 5.6 introduces a method of unbiased estimation for variance components based on least squares fitting. Section 5.7 examines exact tests for variance components.

Searle, Casella, and McCulloch (1992) give an extensive discussion of variance component estimation. Khuri, Mathew, and Sinha (1998) give an extensive discussion of testing in mixed models. More recent books include Jiang (2007), McCulloch, Searle, and Neuhaus (2008), and Hodges (2014).

The methods considered in this chapter are presented in terms of fitting general linear models. In many special cases, considerable simplification results. For example, the RCB models of *PA* (11.1.5) and Exercise 11.4, the split plot model of *PA* (11.3.1), and the subsampling model of *PA* (11.4.2) are all mixed models with very special structures.

© Springer Nature Switzerland AG 2019

R. Christensen, *Advanced Linear Modeling*, Springer Texts in Statistics,

https://doi.org/10.1007/978-3-030-29164-8_5

5.1 Mixed Models

The *mixed model* is a linear model in which some of the parameters, instead of being fixed effects, are random. The model can be written

$$Y = X\beta + Z\gamma + e, \qquad \mathrm{E}(e) = 0, \tag{5.1.1}$$

where X and Z are known matrices with p and q columns respectively, β is an unobservable vector of fixed effects, and γ is an unobservable vector of random effects with $\mathrm{E}(\gamma) = 0$, $\mathrm{Cov}(\gamma) = D$, and $\mathrm{Cov}(\gamma, e) = 0$. Let $\mathrm{Cov}(e) = R$. Typically, D and R depend on some vector of parameters, say θ, and typically they depend on different subsets of the parameters.

Technically, mixed models are neither more nor less general than the models of Chap. 4. The mixed model is a special case of the Chap. 4 model in which the error term is all of $Z\gamma + e$. On the other hand, the Chap. 4 model is a special case of the mixed model with $\Pr[e = 0] = 1$, $Z = I_n$, and $D(\theta) = V(\theta)$. However, the mixed model lends itself to numerous useful applications.

EXAMPLE 5.1.1.   *One-Way ANOVA.*
Let $y_{ij} = \mu + \alpha_i + e_{ij}$, $i = 1, \ldots, a$, $j = 1, \ldots, N_i$, with the α_is independent $N(0, \sigma_1^2)$, the e_{ij}s independent $N(0, \sigma_0^2)$, and the α_is and e_{ij}s independent. Here $p = 1$, $\beta = \mu$, $q = a$, $\gamma = (\alpha_1, \ldots, \alpha_a)$, $D = \sigma_1^2 I_a$ and $R = \sigma_0^2 I_n$. Since the α_is are known to have mean 0, if we could observe the α_is, a natural estimate of σ_1^2 would be $\sum_{i=1}^{a} \alpha_i^2 / a$. Note that the only way you can get a good estimate of σ_1^2 is to have a large. No method, not maximum likelihood, not REML, not MINQUE; no method will produce good estimates of σ_1^2 if a is small.

□

Exercise 5.1.   Show that $\mathrm{E}\left(\sum_{i=1}^{a} \alpha_i^2 / a\right) = \sigma_1^2$.

EXAMPLE 5.1.2.   *Random Intercepts Model.*
Consider the one-way ACOVA model $y_{ij} = \alpha_i + x_{ij}'\beta + e_{ij}$, $i = 1, \ldots, a \equiv q$, $j = 1, \ldots, N_i$, but with intercepts α_i independent $N(0, \sigma_1^2)$. As usual, take the e_{ij}s independent $N(0, \sigma_0^2)$, but we also take the α_is and e_{ij}s independent. This model would be appropriate if we viewed each group i as a sample from some larger population rather than being of special importance in itself. The groups could be randomly chosen individuals on whom N_i observations are taken, randomly chosen classrooms from which N_i students are measured, or different lots of raw material for an industrial process. Typically, in this model the fixed effects term $x_{ij}'\beta$ would contain an intercept (grand mean).

Consider $a = 3$ and $(N_1, N_2, N_3) = (3, 2, 3)$. Write a model $Y = X\beta + Z\gamma + e$ with

$$Y = \begin{bmatrix} y_{11} \\ y_{12} \\ y_{13} \\ y_{21} \\ y_{22} \\ y_{31} \\ y_{32} \\ y_{33} \end{bmatrix}, \quad X = \begin{bmatrix} x'_{11} \\ x'_{12} \\ x'_{13} \\ x'_{21} \\ x'_{22} \\ x'_{31} \\ x'_{32} \\ x'_{33} \end{bmatrix}, \quad Z = \begin{bmatrix} 1 & 0 & 0 \\ 1 & 0 & 0 \\ 1 & 0 & 0 \\ 0 & 1 & 0 \\ 0 & 1 & 0 \\ 0 & 0 & 1 \\ 0 & 0 & 1 \\ 0 & 0 & 1 \end{bmatrix},$$

$$\gamma = (\alpha_1, \alpha_2, \alpha_3)',$$

$q = 3$, $D = \sigma_1^2 I_3$, $n = 8$, $R = \sigma_0^2 I_8$, and $\theta' = (\sigma_0^2, \sigma_1^2)$.

If the α_is are fixed effects, it does not matter whether an overall intercept term is included in the model, but with random α_is having mean 0, one typically wants a fixed intercept or, more generally, $J \in C(X)$. □

For estimation of the fixed effects β, the mixed model can be written as a general Gauss–Markov model. Write

$$V = \text{Cov}(Y) = \text{Cov}(Z\gamma + e) = ZDZ' + R$$

or, more accurately,

$$V(\theta) = \text{Cov}(Y) = \text{Cov}(Z\gamma + e) = ZD(\theta)Z' + R(\theta).$$

V is assumed to be nonsingular. Model (5.1.1) is equivalent to

$$Y = X\beta + \xi, \quad \text{E}(\xi) = 0, \quad \text{Cov}(\xi) = V.$$

The BLUE of $X\beta$ can be found using the theory for general Gauss-Markov models. Unfortunately, finding the BLUE requires knowledge (at least up to a constant multiple) of V. This is rarely available. The best procedure available for estimating $X\beta$ is to estimate V and then act as if the estimate is the real value of V. In other words, if V is estimated with $\tilde{V}$, then $X\beta$ is estimated with

$$X\tilde{\beta} = X \left[X'\tilde{V}^{-1}X \right]^- X'\tilde{V}^{-1}Y. \tag{5.1.2}$$

If $\tilde{V}$ is close to V, then the estimate $X\tilde{\beta}$ should be close to the BLUE, $X\hat{\beta}$. Corresponding standard errors tend to be underestimated, cf. Sects. 4.1 and 4.7.

It is important to be able to invert V in order to perform estimation and prediction. V can be hard to invert because its dimension n, the number of data points, is often large. Inverting V can be performed indirectly. Often R is chosen so that it is easy to invert and D, being $q \times q$ rather than $n \times n$, is easier to invert than V. Not infrequently, D is diagonal or block diagonal which makes it easier to invert. By *PA* Proposition B.56,

$$V^{-1} = R^{-1} - R^{-1}Z[D^{-1} + Z'R^{-1}Z]^{-1}Z'R^{-1}. \tag{5.1.3}$$

Note that a second $q \times q$ inverse is required in this formula. Frequently people assume $R = \sigma_0^2 I_n$, in which case

$$V^{-1} = \frac{1}{\sigma_0^2}\left\{I - Z[\sigma_0^2 D^{-1} + Z'Z]^{-1}Z'\right\}. \tag{5.1.4}$$

If γ were a fixed effect, we would be interested in estimating estimable functions like $\lambda'\gamma$. Since γ is random, we cannot estimate $\lambda'\gamma$, but we can predict $\lambda'\gamma$ from the observable vector Y. As discussed in Sect. 4.1.3, a best linear predictor requires knowledge of $\mathrm{E}(\lambda'\gamma)$, $\mathrm{Cov}(\lambda'\gamma, Y)$, $\mathrm{E}(Y)$, and $\mathrm{Cov}(Y)$. We know that $\mathrm{E}(\lambda'\gamma) = 0$, $\mathrm{Cov}(\lambda'\gamma, Y) = \mathrm{Cov}(\lambda'\gamma, Z\gamma + e) = \lambda'DZ'$, $\mathrm{E}(Y) = X\beta$, and $\mathrm{Cov}(Y) = V$. If we assume that D, R, and thus V are known,

$$\lambda'\hat{\gamma} = \lambda'\left[0 + DZ'V^{-1}(Y - X\beta)\right] \tag{5.1.5}$$

and our only problem in predicting $\lambda'\gamma$ is not knowing $\mathrm{E}(Y) = X\beta$. We could then replace β with a generalized least squares estimate $\hat{\beta}$ to get the *best linear unbiased predictor (BLUP)* of $\lambda'\gamma$. Alas, D and R are not known, but with good estimates of D and R, we can get empirical predictions of $\lambda'\gamma$ that are close to the BLUP.

Sometimes we would like to obtain an estimate of $X\beta$ without going to the trouble and expense of finding $\hat{V}^{-1}$, which is needed to apply Eq. (5.1.2). One simple estimate is the least squares estimate, MY. This gives an unbiased estimate of $X\beta$, but ignores the existence of the random effects. An alternative method is to fit the model

$$Y = X\beta + Z\delta + Z\gamma + e, \quad R = \sigma_0^2 I, \tag{5.1.6}$$

where δ is a vector of fixed effects corresponding to the vector of random effects γ. In model (5.1.6), there is no hope of estimating the covariance parameters in D because the fixed effects δ and the random effects γ are completely confounded, see Exercise 5.9 at the end of the chapter. However, it is easily seen that $C(V[X,Z]) \subset C(X,Z)$, so by *PA* Proposition 2.7.5 or Theorem 10.4.5, least squares estimates are BLUEs in model (5.1.6). To see that $C(V[X,Z]) \subset C(X,Z)$, observe that

$$\begin{aligned}
C(V[X,Z]) &= C(\{\sigma_0^2 I + ZDZ'\}[X,Z]) \\
&= C(\sigma_0^2 X + ZDZ'X, \sigma_0^2 Z + ZDZ'Z) \\
&\subset C(X,Z).
\end{aligned}$$

From *PA* Chap. 9, a least squares estimate of β is

$$\hat{\beta} = \left[X'(I - M_Z)X\right]^- X'(I - M_Z)Y, \tag{5.1.7}$$

where $M_Z = Z(Z'Z)^- Z'$. Although estimates obtained using (5.1.7) are not typically BLUEs for model (5.1.1), since model (5.1.6) is a larger model than model (5.1.1), the estimates should be reasonable. The only serious problem with using Eq. (5.1.7) is that it is not clear which functions of β are estimable in model (5.1.6).

5.2 Mixed Model Equations

We now develop the well-known *mixed model equations*. These equations are similar in spirit to normal equations, cf. *PA* Sect. 2.8. However, the mixed model equations simultaneously provide BLUEs and BLUPs, cf. Sect. 4.1.

Suppose that R is nonsingular. If γ were not random, the normal equations for model (5.1.1) would be

$$\begin{bmatrix} X' \\ Z' \end{bmatrix} R^{-1} [X, Z] \begin{bmatrix} \beta \\ \gamma \end{bmatrix} = \begin{bmatrix} X' \\ Z' \end{bmatrix} R^{-1} Y$$

or

$$\begin{bmatrix} X'R^{-1}X & X'R^{-1}Z \\ Z'R^{-1}X & Z'R^{-1}Z \end{bmatrix} \begin{bmatrix} \beta \\ \gamma \end{bmatrix} = \begin{bmatrix} X'R^{-1}Y \\ Z'R^{-1}Y \end{bmatrix}.$$

For D nonsingular, the mixed model equations are defined as

$$\begin{bmatrix} X'R^{-1}X & X'R^{-1}Z \\ Z'R^{-1}X & D^{-1}+Z'R^{-1}Z \end{bmatrix} \begin{bmatrix} \beta \\ \gamma \end{bmatrix} = \begin{bmatrix} X'R^{-1}Y \\ Z'R^{-1}Y \end{bmatrix}. \tag{5.2.1}$$

Theorem 5.2.1. If $[\hat{\beta}', \hat{\gamma}']$ is a solution to the mixed model equations, then $X\hat{\beta}$ is a BLUE of $X\beta$ and $\hat{\gamma}$ is a BLUP of γ.

PROOF. From *PA* Sect. 2.8, $X\hat{\beta}$ will be a BLUE of $X\beta$ if $\hat{\beta}$ is a solution to $X'V^{-1}X\beta = X'V^{-1}Y$. To use this equation we need a form for V^{-1} in terms of Z, D, and R, i.e., we need Eq. (5.1.3).

If $\hat{\beta}$ and $\hat{\gamma}$ are solutions, then the second row of the mixed model equations gives

$$Z'R^{-1}X\hat{\beta} + [D^{-1}+Z'R^{-1}Z] \hat{\gamma} = Z'R^{-1}Y$$

or

$$\hat{\gamma} = [D^{-1}+Z'R^{-1}Z]^{-1} Z'R^{-1}(Y-X\hat{\beta}). \tag{5.2.2}$$

The first row of the equations is

$$X'R^{-1}X\hat{\beta} + X'R^{-1}Z\hat{\gamma} = X'R^{-1}Y.$$

Substituting for $\hat{\gamma}$ gives

$$X'R^{-1}X\hat{\beta} + X'R^{-1}Z[D^{-1}+Z'R^{-1}Z]^{-1} Z'R^{-1}(Y-X\hat{\beta}) = X'R^{-1}Y$$

or, putting $\hat{\beta}$ on one side and Y on the other,

$$X'R^{-1}X\hat{\beta} - X'R^{-1}Z[D^{-1}+Z'R^{-1}Z]^{-1} Z'R^{-1}X\hat{\beta}$$
$$= X'R^{-1}Y - X'R^{-1}Z[D^{-1}+Z'R^{-1}Z]^{-1} Z'R^{-1}Y,$$

which, by Eq. (5.1.3), is $X'V^{-1}X\hat{\beta} = X'V^{-1}Y$. Thus, $\hat{\beta}$ is a generalized least squares solution and $X\hat{\beta}$ is a BLUE.

$\hat{\gamma}$ in (5.2.2) can be rewritten as

$$\hat{\gamma} = \left(D\left[D^{-1} + Z'R^{-1}Z\right] - DZ'R^{-1}Z\right)\left[D^{-1} + Z'R^{-1}Z\right]^{-1} Z'R^{-1}(Y - X\hat{\beta})$$
$$= \left(DZ'R^{-1} - DZ'R^{-1}Z\left[D^{-1} + Z'R^{-1}Z\right]^{-1} Z'R^{-1}\right)(Y - X\hat{\beta})$$
$$= DZ'V^{-1}(Y - X\hat{\beta}),$$

which is the BLUP of γ. Equation (5.1.5) suggests that this is the BLUP but, more formally, see *PA-V* Exercise 6.10 (Christensen 2011, Exercise 12.3) taking $Q = V^{-1}ZD$.   □

The mixed model equations' primary usefulness is that they are relatively easy to solve. Finding the solution to the generalized normal equations

$$X'V^{-1}X\beta = X'V^{-1}Y$$

requires inversion of the $n \times n$ matrix V. The mixed model equations

$$\begin{bmatrix} X'R^{-1}X & X'R^{-1}Z \\ Z'R^{-1}X & D^{-1} + Z'R^{-1}Z \end{bmatrix} \begin{bmatrix} \beta \\ \gamma \end{bmatrix} = \begin{bmatrix} X'R^{-1}Y \\ Z'R^{-1}Y \end{bmatrix}$$

require computation of two inverses, D^{-1}, which is q dimensional, and R^{-1}, which generally is taken to be a diagonal matrix. If there are many observations relative to the number of random effects, it is easier to solve the mixed model equations. Of course, using Eq. (5.1.3) to obtain V^{-1} for the generalized normal equations accomplishes the same thing.

An equivalent form of the mixed model equations that does not require D to be nonsingular is

$$\begin{bmatrix} X'R^{-1}X & X'R^{-1}ZD \\ Z'R^{-1}X & I + Z'R^{-1}ZD \end{bmatrix} \begin{bmatrix} \beta \\ \xi \end{bmatrix} = \begin{bmatrix} X'R^{-1}Y \\ Z'R^{-1}Y \end{bmatrix}. \tag{5.2.3}$$

Solutions $\hat{\beta}, \hat{\xi}$ have $X\hat{\beta}$ a BLUE of $X\beta$ and $D\hat{\xi} = \hat{\gamma}$ a BLUP of $\hat{\gamma}$.

Exercise 5.2.   Even when D is nonsingular, Eq. (5.2.3) has an advantage over Eq. (5.2.1) in that Eq. (5.2.3) does not require D^{-1}. Show that Eqs. (5.2.1) and (5.2.3) are equivalent when D is nonsingular.

The mixed model equations can also be arrived at from a Bayesian argument. Consider the model

$$Y = X\beta + Z\gamma + e, \quad e \sim N(0, R),$$

and, as discussed in *PA* Sect. 2.9, incorporate partial prior information in the form

$$\gamma \sim N(0, D),$$

where D is again assumed to be nonsingular. A minor generalization of *PA* Eq. (5.9.3) allows the data Y to have an arbitrary nonsingular covariance matrix, so the Bayesian analysis can be obtained from fitting the generalized least squares model

$$\begin{bmatrix} Y \\ 0 \end{bmatrix} = \begin{bmatrix} X & Z \\ 0 & I \end{bmatrix} \begin{bmatrix} \beta \\ \gamma \end{bmatrix} + \begin{bmatrix} e \\ \tilde{e} \end{bmatrix}, \quad \begin{bmatrix} e \\ \tilde{e} \end{bmatrix} \sim N\left(\begin{bmatrix} 0_{n \times 1} \\ 0_{r \times 1} \end{bmatrix}, \begin{bmatrix} R & 0 \\ 0 & D \end{bmatrix} \right).$$

The generalized least squares estimates from this model will be the "objective" posterior means of β and γ, respectively. However, the generalized least squares estimates can be obtained from the corresponding normal equations, and the normal equations are the mixed model equations (5.2.1). See also Sect. 5.3.

5.3 Equivalence of Random Effects and Ridge Regression

Models (2.1.1) and (5.1.1) are both partitioned linear models with, most commonly,

$$Y = X\beta + Z\gamma + e, \quad E(e) = 0, \quad \text{Cov}(e) = \sigma^2 I. \tag{5.3.1}$$

In both we were interested in estimating β but in (2.1.1) we were interested in penalizing the estimation of the fixed γ parameters whereas in (5.1.1) we were interested in predicting the random γ parameters. We now establish a numerical equivalence between obtaining $\hat{\beta}$ and $\hat{\gamma}$ using generalized ridge regression and obtaining $\hat{\beta}$ and $\hat{\gamma}$ by finding BLUEs and BLUPs. The fact of this equivalence was brought to my attention by Jim Hodges and Murry Clayton, see also Brumback et al. (1999), Pearce and Wand (2006), and Hodges (2014).

Consider again fitting the augmented linear model (2.2.3). We have already shown that the generalized least squares estimates for model (2.2.3) give the generalized ridge regression estimates for model (5.3.1). Now we show that they are also the BLUEs and BLUPs for a mixed model version of (5.3.1).

Consider model (5.3.1) as a mixed model with $E(\gamma) = 0$, $\text{Cov}(\gamma) = \sigma^2 (1/k) Q^{-1}$, $\text{Cov}(\gamma, e) = 0$. The mixed model equations of Sect. 5.2 for obtaining the BLUEs and BLUPs become

$$\begin{bmatrix} X'X & X'Z \\ Z'X & Z'Z + kQ \end{bmatrix} \begin{bmatrix} \beta \\ \gamma \end{bmatrix} = \begin{bmatrix} X'Y \\ Z'Y \end{bmatrix}.$$

It is easy to see that these mixed model equations are identical to the (generalized) normal equations, cf. *PA* Sect. 2.8, for model (2.2.3), namely

$$\begin{bmatrix} X & Z \\ 0 & I \end{bmatrix}' \begin{bmatrix} I & 0 \\ 0 & kQ \end{bmatrix} \begin{bmatrix} X & Z \\ 0 & I \end{bmatrix} \begin{bmatrix} \beta \\ \gamma \end{bmatrix} = \begin{bmatrix} X & Z \\ 0 & I \end{bmatrix}' \begin{bmatrix} I & 0 \\ 0 & kQ \end{bmatrix} \begin{bmatrix} Y \\ 0 \end{bmatrix}.$$

Thus generalized least squares estimates for model (2.2.3) are both the BLUEs and BLUPs for the mixed model version of (5.3.1) and the generalized ridge regression estimates for model (5.3.1).

5.4 Partitioning and Linear Covariance Structures

The basic mixed model is

$$Y = X\beta + Z\gamma + e,$$

where X is $n \times p$, Z is $n \times q$,

$$E(\gamma) = 0, \qquad E(e) = 0,$$

$$\mathrm{Cov}(\gamma) = D, \qquad \mathrm{Cov}(e) = R,$$

and

$$\mathrm{Cov}(\gamma, e) = 0.$$

When D and R are unknown, the positive definite matrix R can contain up to $n(n+1)/2$ unknown parameters and the positive definite matrix D can contain up to $q(q+1)/2$ unknown parameters. Obviously, we cannot do a good job of estimating that many parameters from the n observations in Y. We need to model R and D.

By far the most common model for R is

$$R = \sigma_0^2 I_n.$$

Henceforth we incorporate this assumption denoting $\theta_1 \equiv \sigma_0^2$ as needed. But our problem with estimation of D remains. We write an arbitrary parameterization of D as $D(\theta_*)$ where $\theta_* \equiv [\theta_2, \ldots, \theta_s]'$. As alluded to, $s-1$ can be as large as $q(q+1)/2$.

In estimating D, the best thing that could possibly happen is that we would observe γ. Under multivariate normality, if you observed both Y and γ, the likelihood would factor into two pieces based on $Y - Z\gamma \sim N(X\beta, \sigma_0^2 I)$; $\gamma \sim N[0, D(\theta_*)]$ with all the information on θ_* coming from γ alone. From our assumptions,

$$E(\gamma\gamma') = D,$$

so without further assumptions the best estimate we could possibly get is $\hat{D} = \gamma\gamma'$. This estimate is based on one observation of γ, so it is a horrible estimate. Moreover, this "best estimate" is not even available because we do not really observe γ. We have to make further assumptions about D to be productive.

Without loss of generality, partition Z and γ as

$$Z = [Z_1, \cdots, Z_r] \qquad \text{and} \qquad \gamma' = (\gamma_1', \cdots, \gamma_r').$$

where Z_k is $n \times q(k)$, γ_k is a $q(k)$ vector, and $q = \sum_{k=1}^r q(k)$. The linear model becomes

$$Y = X\beta + \sum_{k=1}^{r} Z_k \gamma_k + e.$$

Writing D in conformance with the partition gives

$$D = \begin{bmatrix} D_{11} & \cdots & D_{1r} \\ \vdots & \ddots & \vdots \\ D_{r1} & \cdots & D_{rr} \end{bmatrix}.$$

In particular, if we observed γ, we could estimate D_{jk} using $\gamma_j \gamma_k'$ since $E[\gamma_j \gamma_k'] = D_{jk}$, but the estimation is still horrible.

In terms of modeling D, the first thing we might do is assume the γ_ks are uncorrelated. This leads to

$$D = \begin{bmatrix} D_{11} & & 0 \\ & \ddots & \\ 0 & & D_{rr} \end{bmatrix},$$

but it does nothing to improve the quality of estimation for the terms that remain in the model. In the next two sections we examine models that do help with the quality of estimation.

In Sect. 5.5 we consider variance component models that have the form

$$D = \begin{bmatrix} \sigma_1^2 I_{q(1)} & & 0 \\ & \ddots & \\ 0 & & \sigma_r^2 I_{q(r)} \end{bmatrix}. \tag{5.4.1}$$

In the variance component model we have at least a hope of estimating the parameters well. Upon observing γ and using PA Proposition 1.3.2,

$$E\left[\gamma_k' \gamma_k\right] = \text{tr}\left[I_{q(k)} \sigma_k^2 I_{q(k)}\right] = \sigma_k^2 q(k).$$

Thus $\frac{1}{q(k)} \gamma_k' \gamma_k$ gives an unbiased estimate of σ_k^2 with $q(k)$ degrees of freedom. The larger $q(k)$, the better our estimate should be. Of course, this is the best we could hope to do. In reality, we do not observe γ, so our estimates will be worse.

In Sect. 5.6 we consider longitudinal data models that assume $q(1) = \cdots = q(r) \equiv \tilde{q}$ with

$$D = \begin{bmatrix} \Sigma & & 0 \\ & \ddots & \\ 0 & & \Sigma \end{bmatrix}, \tag{5.4.2}$$

where Σ is a positive definite covariance matrix of $\tilde{q}$ dimensions. Upon observing γ we have $E[\gamma_j \gamma_j'] = \Sigma$ and

$$E\left[\frac{1}{r} \sum_{j=1}^{r} \gamma_j \gamma_j'\right] = \Sigma.$$

Thus $\frac{1}{r}\sum_{j=1}^{r}\gamma_j\gamma_j'$ provides r degrees of freedom for estimating Σ. Without actually observing γ, we must expect to do less well.

The simplest example of both a variance component model and this longitudinal model is the random intercepts model of Example 5.1.2. In that model, $D = \sigma_1^2 I_q$, which fits form (5.4.1) with $r = 1$ and form (5.4.2) with $\tilde{q} = 1$.

The salient characteristic of models (5.4.1) and (5.4.2) is that they are both linear in their unknown parameters in the sense that for some s

$$D(\theta_*) = D_0 + \theta_2 D_2 + \cdots + \theta_s D_s, \tag{5.4.3}$$

where $D_0,\ldots,D_s$ are known matrices. Typically the D_js are symmetric but they may or may not be nonnegative definite. Note that D_j has no relationship to any D_{jk} and that D_0 has no relationship to σ_0^2. The variance component model (5.4.1) has

$$
\begin{bmatrix}
\sigma_1^2 I_{q(1)} & & & 0 \\
& \sigma_2^2 I_{q(2)} & & \\
& & \ddots & \\
0 & & & \sigma_r^2 I_{q(r)}
\end{bmatrix} =
$$

$$
\sigma_1^2
\begin{bmatrix}
I_{q(1)} & 0 & \cdots & 0 \\
0 & 0 & \cdots & 0 \\
\vdots & \vdots & \ddots & \vdots \\
0 & 0 & \cdots & 0
\end{bmatrix}
+ \cdots + \sigma_r^2
\begin{bmatrix}
0 & \cdots & 0 & 0 \\
\vdots & \ddots & \vdots & \vdots \\
0 & \cdots & 0 & 0 \\
0 & \cdots & 0 & I_{q(r)}
\end{bmatrix}.
$$

The longitudinal model (5.4.2), for $\tilde{q} = 2$, has

$$
\begin{bmatrix}
\sigma_{11} & \sigma_{12} & & 0 & 0 \\
\sigma_{12} & \sigma_{22} & & 0 & 0 \\
& & \ddots & & \\
0 & 0 & & \sigma_{11} & \sigma_{12} \\
0 & 0 & & \sigma_{12} & \sigma_{22}
\end{bmatrix}
= \sigma_{11}
\begin{bmatrix}
1 & 0 & & 0 & 0 \\
0 & 0 & & 0 & 0 \\
& & \ddots & & \\
0 & 0 & & 1 & 0 \\
0 & 0 & & 0 & 0
\end{bmatrix} \tag{5.4.4}
$$

$$
+ \sigma_{12}
\begin{bmatrix}
0 & 1 & & 0 & 0 \\
1 & 0 & & 0 & 0 \\
& & \ddots & & \\
0 & 0 & & 0 & 1 \\
0 & 0 & & 1 & 0
\end{bmatrix}
+ \sigma_{22}
\begin{bmatrix}
0 & 0 & & 0 & 0 \\
0 & 1 & & 0 & 0 \\
& & \ddots & & \\
0 & 0 & & 0 & 0 \\
0 & 0 & & 0 & 1
\end{bmatrix}.
$$

In (5.4.4) we have quite properly imposed the symmetry condition $\sigma_{12} = \sigma_{21}$. Note that in both examples $D_0 = 0$, all the D_js are symmetric, but in the longitudinal model (5.4.4) D_2 (corresponding to σ_{12}) is not nonnegative definite. It is pretty obvious how the decomposition changes if one does not impose symmetry.

The linear covariance structure imposed on $D(\theta_*)$ in (5.4.3) defines a linear covariance structure on $V(\theta)$ as discussed in Sect. 4.4. In particular, with $\theta_1 \equiv \sigma_0^2$,

$$V(\theta) = \theta_1 I_n + ZD(\theta_*)Z'$$

$$= \theta_1 I_n + Z(D_0 + \theta_2 D_2 + \cdots + \theta_s D_s)Z'$$
$$= ZD_0 Z' + \theta_1 I_n + \theta_2 Z D_2 Z' + \cdots + \theta_s Z D_s Z'$$
$$= V_0 + \theta_1 V_1 + \theta_2 V_2 + \cdots + \theta_s V_s.$$

Here $V_j = ZD_j Z'$ for $j \neq 1$ and $V_1 = I_n$.

5.5 Variance Component Models

In the old days, the terms "mixed model" and "variance component model" were essentially synonymous. That is no longer true. We describe traditional variance component models.

Variance component models assume a partitioned model

$$Y = X\beta + \sum_{k=1}^{r} Z_k \gamma_k + e,$$

$$E(e) = 0, \qquad E(\gamma) = 0,$$

$$\text{Cov}(e) = \sigma_0^2 I_n, \qquad \text{Cov}(\gamma, e) = 0,$$

and add the assumptions that $\text{Cov}(\gamma_k) = \sigma_k^2 I_{q(k)}$ and, for $j \neq k$, $\text{Cov}(\gamma_j, \gamma_k) = 0$. The covariance matrix of γ is given in Eq. (5.4.1), which, of course, is a diagonal matrix. With these conventions we can write

$$V(\theta) = \sigma_0^2 I_n + \sum_{k=1}^{r} \sigma_k^2 Z_k Z_k' = \sum_{k=0}^{r} \sigma_k^2 Z_k Z_k', \qquad (5.5.1)$$

where we take $Z_0 \equiv I_n$ and $\theta \equiv (\sigma_0^2, \cdots, \sigma_r^2)'$. Equation (5.5.1) constitutes a linear covariance structure with, relative to Sect. 4.4, $s = r+1$, $V_0 = 0$, and $V_{j+1} = Z_j Z_j'$, $j = 0, \ldots, r$. To justify the use of the asymptotic testing methods described in Chap. 4, we need n to get large but, as discussed in Sect. 5.4, we also need all of the $q(k)$s, $k = 1, \ldots, r$, to get large.

Traditionally, variance component models were used in analysis of variance situations in which the columns of Z_k consisted of 0-1 indicators of group membership. In such cases the diagonal elements of $Z_k Z_k'$ are all 1s and the variance of any individual observation is $\sum_{k=0}^{r} \sigma_k^2$, so every term σ_k^2 has a clear interpretation as a component of the overall variance. Even for general Z_ks, the variance of an individual observation is a linear combination of the variance components, the σ_k^2s.

EXAMPLE 5.5.1.   *Random Intercepts Model.*
Example 5.1.2 gives explicit versions of Y, X, Z, and γ for $n = 8$; $q = 3$. The partitioning in this model is degenerate in that we pick $r = 1$, so $Z = Z_1$, $q = q(1) = 3$, $\gamma = \gamma_1 = (\alpha_1, \alpha_2, \alpha_3)'$. With $\text{Cov}(\gamma_1) = \sigma_1^2 I_3 = D$, we get

$$V(\theta) = \sigma_0^2 I_n + \sigma_1^2 \begin{bmatrix} 1 & 1 & 1 & 0 & 0 & 0 & 0 & 0 \\ 1 & 1 & 1 & 0 & 0 & 0 & 0 & 0 \\ 1 & 1 & 1 & 0 & 0 & 0 & 0 & 0 \\ 0 & 0 & 0 & 1 & 1 & 0 & 0 & 0 \\ 0 & 0 & 0 & 1 & 1 & 0 & 0 & 0 \\ 0 & 0 & 0 & 0 & 0 & 1 & 1 & 1 \\ 0 & 0 & 0 & 0 & 0 & 1 & 1 & 1 \\ 0 & 0 & 0 & 0 & 0 & 1 & 1 & 1 \end{bmatrix}. \qquad \square$$

5.5.1 Variance Component Estimation

The variance component model is a general linear model so the likelihood theories of Sects. 4.2 and 4.3 apply with

$$\mathbf{d}_{\sigma_i^2} V(\theta) = Z_i Z_i'.$$

It follows that the likelihood equations for the variance components become

$$\text{tr}\{V^{-1}(\theta)Z_j Z_j'\} = Y'[I - A(\theta)]'V^{-1}(\theta)Z_j Z_j' V^{-1}(\theta)[I - A(\theta)]Y,$$

$j = 0, \ldots, r$, or, since $V(\theta)$ has the linear covariance structure of Sect. 4.4,

$$\sum_{k=0}^{r} \sigma_k^2 \text{tr}\{Z_k Z_k' V^{-1}(\theta)Z_j Z_j' V^{-1}(\theta)\} = Y'[I - A(\theta)]'V^{-1}(\theta)Z_j Z_j' V^{-1}(\theta)[I - A(\theta)]Y,$$

$j = 0, \ldots, r$.

The derivative also provides the REML equations for variance component models,

$$\text{tr}\{V^{-1}(\theta)[I - A(\theta)]Z_j Z_j'\} = Y'[I - A(\theta)]'V^{-1}(\theta)Z_j Z_j' V^{-1}(\theta)[I - A(\theta)]Y,$$

$j = 0, \ldots, r$. Again, since the covariance matrix is linear, the REML equations can be rewritten as

$$\sum_{k=0}^{r} \sigma_k^2 \text{tr}\{Z_k Z_k'[I - A(\theta)]'V^{-1}(\theta)Z_j Z_j' V^{-1}(\theta)[I - A(\theta)]\}$$

$$= Y'[I - A(\theta)]'V^{-1}(\theta)Z_j Z_j' V^{-1}(\theta)[I - A(\theta)]Y,$$

$j = 0, \ldots, r$.

There are many questions about using the likelihood or REML equations. The MLEs or REML estimates may be solutions to the equations with $\sigma_k^2 > 0$ for all k, or they may not be solutions, but rather be on a boundary of the parameter space.

There may be solutions other than the maximum. Such questions are beyond the scope of this book.

Exercise 5.3. In Exercise 4.6 we found closed form solutions to the REML equations for generalized split plot (GSP) models. The parameterization used in Exercise 4.6 was not the variance component parameterization. What we call Z in a mixed model is the matrix X_1 from a GSP. Show that the REML equations for the variance components of a GSP model lead to the same estimates as found in Exercise 4.6.

Following Sect. 4.5, the MINQUE equations for a variance component model can be written

$$\sum_{k=0}^{r} \sigma_k^2 \text{tr} \left\{ Z_k Z_k'[I - A(w)]' V^{-1}(w) Z_j Z_j' V^{-1}(w)[I - A(w)] \right\}$$

$$= Y'[I - A(w)]' V^{-1}(w) Z_j Z_j' V^{-1}(w)[I - A(w)] Y,$$

$j = 0, \ldots, r$.

EXAMPLE 5.5.2. *Balanced One-Way ANOVA.*
Let $y_{ij} = \mu + \alpha_i + e_{ij}$, $i = 1, \ldots, a$, $j = 1, \ldots, N$, with the α_is independent $N(0, \sigma_1^2)$, the e_{ij}s independent $N(0, \sigma_0^2)$, and the α_is and e_{ij}s independent.

The matrix $[X, Z]$ for the general mixed model is just the model matrix from *PA* Chap. 4, where now $X = J_n^1$, $q = a$, $r = 1$, $Z_1 = Z = [X_1, \ldots, X_a]$ and X_i is a 0–1 indicator vector for an observation being in group i. The covariance matrix is the same as that considered in *PA* Chap. 11 but in any case it is not difficult to see that

$$V = \sigma_0^2 I + \sigma_1^2 Z Z' = \sigma_0^2 I + N \sigma_1^2 M_Z,$$

where M_Z is the perpendicular projection matrix onto $C(Z)$. The inverse of V is easily seen to be

$$V^{-1} = \frac{1}{\sigma_0^2} \left[I - \frac{N \sigma_1^2}{\sigma_0^2 + N \sigma_1^2} M_Z \right].$$

This follows from Eq. (5.1.4) and simplifying but also follows from *PA-V* Proposition B.57 (Christensen 2011, Proposition 12.10.1).

We can now find the estimates. It is easily seen that $C(VX) \subset C(X)$, so, as discussed in *PA* Sects. 2.7 and 2.8, least squares estimates provide solutions to the normal equations $X'V^{-1}X\beta = X'V^{-1}Y$ which are also the likelihood equations for β. Simply put, $\hat{\mu} = \bar{y}_{..}$.

For $j = 0$, the likelihood equation is

$$\text{tr}(V^{-1}) = (Y - \hat{\mu}J)' V^{-2} (Y - \hat{\mu}J).$$

Observe that

$$V^{-1}(Y - \hat{\mu}J) = \frac{1}{\sigma_0^2}\left[I - \frac{N\sigma_1^2}{\sigma_0^2 + N\sigma_1^2}M_Z\right]\left(I - \frac{1}{n}J_n^n\right)Y$$

$$= \frac{1}{\sigma_0^2}\left[\left(I - \frac{1}{n}J_n^n\right)Y - \frac{N\sigma_1^2}{\sigma_0^2 + N\sigma_1^2}\left(M_Z - \frac{1}{n}J_n^n\right)Y\right]$$

$$= \frac{1}{\sigma_0^2}\left[(I - M_Z)Y - \left(\frac{N\sigma_1^2}{\sigma_0^2 + N\sigma_1^2} - 1\right)\left(M_Z - \frac{1}{n}J_n^n\right)Y\right]$$

$$= \frac{1}{\sigma_0^2}(I - M_Z)Y - \frac{1}{\sigma_0^2 + N\sigma_1^2}\left(M_Z - \frac{1}{n}J_n^n\right)Y.$$

Thus, evaluating $\mathrm{tr}(V^{-1})$ on the lefthand side of the likelihood equation and computing the squared length on the righthand side leads to the equation

$$\frac{Na[\sigma_0^2 + (N-1)\sigma_1^2]}{\sigma_0^2(\sigma_0^2 + N\sigma_1^2)} = \frac{SSE}{\sigma_0^4} + \frac{SSTrts}{(\sigma_0^2 + N\sigma_1^2)^2}. \tag{5.5.2}$$

For $j = 1$, the likelihood equation is

$$\mathrm{tr}(V^{-1}ZZ') = (Y - \hat{\mu}J)'V^{-1}ZZ'V^{-1}(Y - \hat{\mu}J). \tag{5.5.3}$$

Using $ZZ' = NM_Z$ and the characterization of $V^{-1}(Y - \hat{\mu}J)$, the righthand side of (5.5.3) can be written

$$(Y - \hat{\mu}J)'V^{-1}ZZ'V^{-1}(Y - \hat{\mu}J) = N(Y - \hat{\mu}J)'V^{-1}M_ZV^{-1}(Y - \hat{\mu}J)$$

$$= \frac{N}{(\sigma_0^2 + N\sigma_1^2)^2}Y'\left(M_Z - \frac{1}{n}J_n^n\right)Y$$

$$= \frac{N SSTrts}{(\sigma_0^2 + N\sigma_1^2)^2}.$$

To evaluate the lefthand side of (5.5.3), note that

$$\mathrm{tr}(V^{-1}ZZ') = \mathrm{tr}\left\{\frac{1}{\sigma_0^2}\left[I - \frac{N\sigma_1^2}{\sigma_0^2 + N\sigma_1^2}M_Z\right]NM_Z\right\}$$

$$= \frac{N}{\sigma_0^2}\mathrm{tr}\left[M_Z - \frac{N\sigma_1^2}{\sigma_0^2 + N\sigma_1^2}M_Z\right]$$

$$= \frac{N}{\sigma_0^2}\frac{\sigma_0^2}{\sigma_0^2 + N\sigma_1^2}\mathrm{tr}(M_Z)$$

$$= \frac{Na}{\sigma_0^2 + N\sigma_1^2}.$$

Equation (5.5.3) becomes

$$\frac{Na}{\sigma_0^2 + N\sigma_1^2} = \frac{N SSTrts}{(\sigma_0^2 + N\sigma_1^2)^2}$$

or
$$\sigma_0^2 + N\sigma_1^2 = SSTrts/a. \tag{5.5.4}$$

Substituting equation (5.5.4) into Eq. (5.5.2) and multiplying through by σ_0^4 gives

$$\frac{Na\sigma_0^2[\sigma_0^2 + (N-1)\sigma_1^2]}{(\sigma_0^2 + N\sigma_1^2)} - SSE + \frac{a\sigma_0^4}{(\sigma_0^2 + N\sigma_1^2)}$$

or, less obviously,
$$a(N-1)\sigma_0^2 = SSE.$$

The maximum likelihood estimates appear to be $\hat{\sigma}_0^2 = MSE$ and $\hat{\sigma}_1^2 = [(\hat{\sigma}_0^2 + N\hat{\sigma}_1^2) - \hat{\sigma}_0^2]/N = [SSTrts/a - MSE]/N$. However, this is true only if $SSTrts/a - MSE > 0$. Otherwise, the maximum is on a boundary, so $\hat{\sigma}_1^2 = 0$ and $\hat{\sigma}_0^2 = SSE/aN$. □

The maximum likelihood procedure tends to ignore the fact that mean parameters are being fitted. In the one-way ANOVA example, the estimate of $\sigma_0^2 + N\sigma_1^2$ was $SSTrts/a$ instead of the unbiased estimate $MSTrts$. No correction was included for fitting the parameter μ.

Note that the balanced one-way model of Example 5.5.2 is actually a special case of a generalized split plot model as discussed in Exercise 4.6.

5.6 A Longitudinal Model

Longitudinal models are characterized by having a number of individuals (often assumed to be independent), with multiple observations taken on each individual over time. In the random intercepts model of Example 5.1.2, the individuals would be identified as the groups i with the multiple observations then indexed by j. Many mixed models for longitudinal data have the block diagonal covariance structure of Eq. (5.4.2).

EXAMPLE 5.6.1.   *Random Intercepts Model.*
To view the random intercepts model of Examples 5.1.2 and 5.4.1 as having the covariance structure of (5.4.2), pick $r = q = a$ with the kth column of Z as Z_k, $q(k) = 1$, and $\gamma_k = \alpha_k$. □

EXAMPLE 5.6.2.   *Random Slopes and Intercepts Model.*
Starting with the random intercepts model and assuming an additional known predictor variable z, we incorporate random slopes η_i. For example, z_{ij} might be the jth time at which the ith individual was observed in which case $\alpha_i + \eta_i z_{ij}, j = 1, \ldots, N_i$, is a random linear time trend associated with the observations y_{ij} on the ith individual. Again, this is only appropriate when we have no particular interest in the individuals but rather think of them as a sample from a larger population. Let

$$y_{ij} = x'_{ij}\beta + \alpha_i + \eta_i z_{ij} + e_{ij},$$

$i = 1,\ldots,a$, $j = 1,\ldots,N_i$, with the $(\alpha_i,\eta_i)'$s independent $N(0,\Sigma)$, the e_{ij}s independent $N(0,\sigma_0^2)$, and the $(\alpha_i,\eta_i)'$s and e_{ij}s independent. Since α_i and η_i are both effects for the ith individual, it seems unlikely that they would be independent. Typically such models also include a fixed intercept and slope that do not depend on the individual i, but a fixed term $\beta_0 + \beta_1 z_{ij}$ can be incorporated into $x'_{ij}\beta$. If individuals are grouped according to some condition, typically there would be a fixed simple linear regression for each condition. Here $q = 2a$, $r = a$, and $q(k) = \tilde{q} = 2$.

Similar to the random intercepts model of Example 5.1.2, consider $a = 3$ and $(N_1,N_2,N_3) = (3,2,3)$. Write a model $Y = X\beta + Z\gamma + e$ with Y and X defined as in Example 5.1.2,

$$Z = [Z_1, Z_2, Z_3] = \begin{bmatrix} 1 & z_{11} & 0 & 0 & 0 & 0 \\ 1 & z_{12} & 0 & 0 & 0 & 0 \\ 1 & z_{13} & 0 & 0 & 0 & 0 \\ 0 & 0 & 1 & z_{21} & 0 & 0 \\ 0 & 0 & 1 & z_{22} & 0 & 0 \\ 0 & 0 & 0 & 0 & 1 & z_{31} \\ 0 & 0 & 0 & 0 & 1 & z_{32} \\ 0 & 0 & 0 & 0 & 1 & z_{33} \end{bmatrix},$$

and $\gamma_i = (\alpha_i,\eta_i)'$. For this model $r = 3$, $q = 6$, and $q(k) = 2 = \tilde{q}$. By assumption, γ has the covariance structure of (5.4.2). Writing the gh element of Σ as σ_{gh} leads to writing $\theta = (\sigma_0^2,\sigma_{11},\sigma_{12},\sigma_{22})'$ and

$$V(\theta) = \sigma_0^2 I_n + [Z_1,\ Z_2,\ Z_3]\begin{bmatrix} \Sigma & 0 & 0 \\ 0 & \Sigma & 0 \\ 0 & 0 & \Sigma \end{bmatrix}\begin{bmatrix} Z_1' \\ Z_2' \\ Z_3' \end{bmatrix} = \sigma_0^2 I_n + \sum_{k=1}^{3} Z_k \Sigma Z_k'.$$

While this is a nice looking way to write $V(\theta)$, it does not have the linear covariance structure of Sect. 4.4. However, as demonstrated in Eq. (5.4.4), these models do have linear covariance structure for $D(\theta_*)$ and therefore, as shown in Sect. 5.4, have the linear covariance structure of Sect. 4.4.

If we assume that the intercepts and slopes are independent, this longitudinal model also becomes a variance component model. If

$$\begin{bmatrix} \alpha_i \\ \eta_i \end{bmatrix} \sim N\left(\begin{bmatrix} 0 \\ 0 \end{bmatrix}, \begin{bmatrix} \sigma_1^2 & 0 \\ 0 & \sigma_2^2 \end{bmatrix}\right),$$

we could redefine Z with $r = 2$ and $q(k) = 3$ as

$$Z = \begin{bmatrix} 1 & 0 & 0 & z_{11} & 0 & 0 \\ 1 & 0 & 0 & z_{12} & 0 & 0 \\ 1 & 0 & 0 & z_{13} & 0 & 0 \\ 0 & 1 & 0 & 0 & z_{21} & 0 \\ 0 & 1 & 0 & 0 & z_{22} & 0 \\ 0 & 0 & 1 & 0 & 0 & z_{31} \\ 0 & 0 & 1 & 0 & 0 & z_{32} \\ 0 & 0 & 1 & 0 & 0 & z_{33} \end{bmatrix} = [Z_1, Z_2].$$

It follows that $\gamma_1 = (\alpha_1, \alpha_2, \alpha_3)'$ and $\gamma_2 = (\eta_1, \eta_2, \eta_3)'$ with $\mathrm{Cov}(\gamma_k) = \sigma_k^2 I_3$. Most importantly, and perhaps unrealistically, this model now assumes that $\mathrm{Cov}(\gamma_1, \gamma_2) = 0$. □

EXAMPLE 5.6.3. *Random Regression Models.*
More generally we can write longitudinal models that have

$$y_{ij} = x'_{ij}\beta + z'_{ij}\gamma_i + e_{ij},$$

$i = 1, \ldots, r$, $j = 1, \ldots, N_i$. Here β and the x_{ij}s are p vectors, the γ_is and z_{ij}s are $\tilde{q}$ vectors, and the γ_is are uncorrelated having $\mathrm{E}(\gamma_i) = 0$ and $\mathrm{Cov}(\gamma_i) = \Sigma$. As usual, the e_{ij}s are uncorrelated with each other as well as the γ_is, have $\mathrm{E}(e_{ij}) = 0$, and $\mathrm{Var}(e_{ij}) = \sigma_0^2$.

For the ith individual, we can write their N_i observations as a vector Y_i, with corresponding matrices X_i having rows consisting of the x'_{ij}s, and matrices Z_{ii} having rows consisting of the z'_{ij}s. The overall model becomes

$$\begin{bmatrix} Y_1 \\ \vdots \\ Y_r \end{bmatrix} = \begin{bmatrix} X_1 \\ \vdots \\ X_r \end{bmatrix} \beta + \begin{bmatrix} Z_{11} & & 0 \\ & \ddots & \\ 0 & & Z_{rr} \end{bmatrix} \begin{bmatrix} \gamma_1 \\ \vdots \\ \gamma_r \end{bmatrix} + e.$$

It is not difficult to see that this model gives the block diagonal covariance structure for γ of Eq. (5.4.2). Moreover,

$$V(\theta) = \begin{bmatrix} \sigma_0^2 I_{N_1} + Z_{11}\Sigma Z'_{11} & & 0 \\ & \ddots & \\ 0 & & \sigma_0^2 I_{N_r} + Z_{rr}\Sigma Z'_{rr} \end{bmatrix}.$$

This block diagonal form has the advantage of making $V^{-1}(\theta)$ easy to compute but does not seem to generate much additional simplification to the process of estimating parameters. In particular, using a result similar to (5.1.4),

$$V^{-1}(\theta) = \mathrm{Blk\ diag}\left[(\sigma_0^2 I_{N_j} + Z_{jj}\Sigma Z'_{jj})^{-1}\right]$$
$$= \mathrm{Blk\ diag}\left[\frac{1}{\sigma_0^2}\left\{I_{N_j} - Z_{jj}[\sigma_0^2\Sigma^{-1} + Z'_{jj}Z_{jj}]^{-1}Z'_{jj}\right\}\right].$$

In the last equality, the inverses are of $\tilde{q} \times \tilde{q}$ matrices and typically $\tilde{q}$ is quite small.

It is commonly assumed that

$$
C\left(\begin{bmatrix} Z_{11} \\ \vdots \\ Z_{rr} \end{bmatrix}\right) \subset C\left(\begin{bmatrix} X_1 \\ \vdots \\ X_r \end{bmatrix}\right).
$$

This specifies that there is a common fixed regression from which the individual random coefficient regressions deviate. If individuals are divided into categories (like treatment groups) there is typically a fixed regression for each category.

Relative to the partitioning notation of Sect. 5.3, the random regression model has $Z_j = Z_{jj} \otimes e_j$, where e_j is the jth column of I_r.   □

The model of *PA-V* Exercise 12.6b, used to illustrate that *Sandwich Estimators* can work, is a degenerate form of a random regressions model with $N_i \equiv N$, $Z_{ii} \equiv I_N$, and $\sigma_0^2 \equiv 0$.

Consider the partitioned model

$$
Y = X\beta + \sum_{j=1}^{r} Z_j \gamma_j + e
$$

where Z_j is $n \times \tilde{q}$, the γ_is are uncorrelated with $E(\gamma_i) = 0$ and $Cov(\gamma_i) = \Sigma$, e has $E(e) = 0$ and $Cov(e) = \sigma_0^2 I_n$, and $Cov(\gamma_j, e) = 0$ for all j. This gives the block diagonal covariance structure for $D(\theta_*)$ of (5.4.2) and

$$
V(\theta) = \sigma_0^2 I_n + \sum_{k=1}^{r} Z_k \Sigma Z_k'. \tag{5.6.1}
$$

It is not uncommon for people to refer to the covariance parameters in Σ as variance components, but that is infelicitous because variance components should provide a natural decomposition of the variance. While the linear structure of $V(\theta)$ in (5.6.1) is *not* the linear covariance structure of Sect. 4.4, similar to Sect. 5.3 we will see later that both $D(\theta_*)$ and $V(\theta)$ have linear covariance structure.

To use the asymptotic testing methods described in Chap. 4, we need consistent estimates. As discussed in Sect. 5.3, to get consistent estimation, we need r to be large. Often, that means a large number of individuals. Typically, if the number of individuals r is large, n automatically gets large in an appropriate way for consistent estimation. However, if $\tilde{q}$ is a reasonably large number, it may become difficult to get $\tilde{q}$ observations on each individual. For example, a random regression model having some values $N_i < \tilde{q}$ might cause problems. Such models share properties with the missing data models of Chap. 11.

We now address the issue of linear covariance structures for this model. Define the gh element of the $\tilde{q} \times \tilde{q}$ matrix Σ as σ_{gh} but recall that $\sigma_{gh} = \sigma_{hg}$, so we use a parameterization with $g \leq h$. The derivative of Σ with respect to σ_{gh} is a $\tilde{q}$ square matrix T_{gh} with elements t_{uv} wherein $t_{uv} = 1$ if $(u,v) = (g,h)$ or $(u,v) = (h,g)$ and is 0 otherwise. These derivatives are precisely the matrices we need to decompose Σ, and thus D, linearly. We can write

$$\Sigma = \sum_{g=1}^{\tilde{q}} \sum_{h \leq g} \sigma_{gh} T_{gh}$$

and thus

$$D(\theta_{\tau}) = \sum_{g=1}^{\tilde{q}} \sum_{h \leq g} \sigma_{gh} \text{Blk diag}(T_{gh}),$$

which is of the linear form (5.4.3). We also get

$$V(\theta) = \sigma_0^2 I_n + \sum_{g=1}^{\tilde{q}} \sum_{h \leq g} \sigma_{gh} \sum_{k=1}^{r} Z_k T_{gh} Z_k'$$

so that $V_0, V_1, \ldots, V_s$ from Sect. 4.4 are reindexed as $V_0 = 0$, $V_1 = I_n$, $V_{gh} = \sum_{k=1}^{r} Z_k T_{gh}$ $Z_k', g = 1, \ldots, \tilde{q}, h \leq g$. Some slight additional simplifications can be made by writing the partition matrices Z_k as

$$Z_k = \left[Z_{k1}, \cdots, Z_{k\tilde{q}} \right].$$

Then,

$$V_{gg} = \sum_{k=1}^{r} Z_{kg} Z_{kg}'$$

and for $g < h$,

$$V_{gh} = \sum_{k=1}^{r} \left[Z_{kg} Z_{kh}' + Z_{kh} Z_{kg}' \right].$$

With these identifications, the covariance parameter estimation methods are precisely as stated in Sects. 4.4 and 4.5.

5.7 Henderson's Method 3

Before the development of the estimation methods described in Chap. 4, Henderson (1953) presented a way of obtaining unbiased method of moment estimates for variance components using least squares computations. The method applies to variance component models but can also be applied to somewhat more general models. Consider a partitioned mixed linear model with

$$Y = X\beta + Z_* \gamma_* + Z_r \gamma_r + e, \tag{5.7.1}$$

where we have accumulated all but the last term of the partition into $Z_* = [Z_1, \ldots, Z_{r-1}]$ and $\gamma_* = [\gamma_1', \ldots, \gamma_{r-1}']$. We introduce a generalization of the variance component model by assuming

$$E(\gamma_*) = 0, \quad E(\gamma_r) = 0, \quad E(e) = 0,$$

$$\text{Cov}(\gamma_*) = D_*, \quad \text{Cov}(\gamma_r) = \sigma_r^2 I_{q(r)}, \quad \text{Cov}(e) = \sigma_0^2 I_n,$$

with all of these random vectors uncorrelated. Clearly, any variance component model can be written in this form.

As a general linear model we have

$$\text{E}(Y) = X\beta; \qquad \text{Cov}(Y) = V(\theta) = \sigma_0^2 I + Z_* D_* Z_*' + \sigma_r^2 Z_r Z_r'.$$

Henderson's method is based on least squares fitting and thus on perpendicular projection operators. For a general partitioned model define P_k as the ppo onto $C(X, Z_1, \ldots, Z_k)$ for $k = 1, \ldots, r$ and identify $P_0 \equiv M$. Obviously, $P_k X = X$ and, for $j \leq k$, $P_k Z_j = Z_j$. In particular, $P_r Z_* = P_{r-1} Z_* = Z_*$.

An unbiased estimate of σ_0^2 is

$$\hat{\sigma}_0^2 \equiv \frac{Y'(I - P_r)Y}{n - r(P_r)},$$

which is the MSE when treating the γs in the mixed model as fixed effects, rather than random effects. To see that this estimate is unbiased, observe that PA Proposition 1.3.2 gives

$$\begin{aligned}
\text{E}\left[Y'(I - P_r)Y\right] &= \text{tr}\left[(I - P_r)V(\theta)\right] + \beta' X'(I - P_r)X\beta \\
&= \text{tr}\left\{(I - P_r)\left[\sigma_0^2 I + Z_* D_* Z_*' + \sigma_r^2 Z_r Z_r'\right]\right\} + 0 \\
&= \text{tr}\left[\sigma_0^2(I - P_r)\right] + \text{tr}\left[(I - P_r)Z_* D_* Z_*'\right] + \sigma_r^2 \text{tr}\left[(I - P_r)Z_r Z_r'\right] \\
&= \sigma_0^2 \text{tr}[I - P_r].
\end{aligned}$$

To get an unbiased estimate of σ_r^2, consider $Y'(P_r - P_{r-1})Y$. Based on

$$\begin{aligned}
\text{E}\left[Y'(P_r - P_{r-1})Y\right] &= \text{tr}\left\{(P_r - P_{r-1})\left[\sigma_0^2 I + Z_* D_* Z_*' + \sigma_r^2 Z_r Z_r'\right]\right\} \\
&= \sigma_0^2 \text{tr}[(P_r - P_{r-1})] + \sigma_r^2 \text{tr}[(P_r - P_{r-1})Z_r Z_r'],
\end{aligned}$$

it is not hard to see that an unbiased estimate of σ_r^2 is

$$\hat{\sigma}_r^2 \equiv \frac{Y'(P_r - P_{r-1})Y - \hat{\sigma}_0^2 \text{tr}(P_r - P_{r-1})}{\text{tr}[Z_r'(P_r - P_{r-1})Z_r]}. \tag{5.7.2}$$

EXAMPLE 5.7.1. *Balanced One-Way ANOVA, continued from Example 5.4.2.* We relate the notation of this section to that of the earlier example:

$$r = 1, \qquad P_r = P_1 = M_Z, \qquad P_{r-1} = P_0 = (1/n)J_n^n,$$

$$Y'(I - P_1)Y = SSE, \qquad Y'(P_1 - P_0)Y = SSTrts,$$

$$\text{tr}(I - P_1) = a(N - 1), \qquad \text{tr}(P_1 - P_0) = a - 1,$$

$$Y'(I - P_1)Y / \text{tr}(I - P_1) = MSE = \hat{\sigma}_0^2.$$

Recall that $ZZ' = NM_Z = NP_1$, so

$$\begin{aligned}
\text{tr}\left[(P_1 - P_0)ZZ'\right] &= \text{tr}[(P_1 - P_0)NP_1] \\
&= N\text{tr}[(P_1 - P_0)] \\
&= N(a - 1).
\end{aligned}$$

From (5.7.2),

$$\hat{\sigma}_1^2 = \frac{SSTrts - MSE(a-1)}{N(a-1)} = \frac{MSTrts - MSE}{N}. \qquad \square$$

Exercise 5.4.   Consider the estimates of σ_0^2 and σ_1^2 in Example 5.7.1.

(a) Show that these are also the REML estimates.
(b) Show that the vector $(Y'Y, Y'M_ZY, J'Y)'$ is a complete sufficient statistic for the balanced one-way random effects model.
(c) Show that the Method 3 estimates are minimum variance unbiased.
(d) Find the distributions of $SSTrts$ and SSE.
(e) Find a generalized likelihood ratio test of level α for $H_0 : \sigma_1^2 = 0$.

Hint: Use the concepts of *PA* Sect. 2.5 and the methods of Example 5.4.2.

EXAMPLE 5.7.2.   *Unbalanced One-Way ANOVA.*
For unbalanced one-way ANOVA, the main things that change from the previous example are that $\text{tr}(I - P_1) = n - a$ and

$$\text{tr}\left[(P_1 - P_0)ZZ'\right] = \text{tr}\left[Z'(P_1 - P_0)Z\right] = \text{tr}\left[Z'\left(I - \frac{1}{n}J_n^n\right)Z\right] = n - \sum_{i=1}^{a}(N_i^2/n).$$

We still get $MSE = \hat{\sigma}_0^2$ but now

$$\hat{\sigma}_1^2 = \frac{SSTrts - MSE(a-1)}{n - \sum_{i=1}^{a}(N_i^2/n)}. \qquad \square$$

Henderson's Method 3 has no known (to me) optimality properties. Henderson himself recommended the use of other techniques. Method 3's greatest advantage is that it is easy to compute. It uses standard techniques from fitting linear models by least squares, except that it requires the computation of $\text{tr}[(P - P_{r-1})Z_rZ_r']$. Even this can be computed using standard techniques. Note that

$$\text{tr}\left[(P_r - P_{r-1})Z_rZ_r'\right] = \text{tr}\left[Z_r'(P_r - P_{r-1})Z_r\right] = \text{tr}\left[Z_r'(I - P_{r-1})Z_r\right].$$

Write $Z_r = [Z_{r1}, Z_{r2}, \ldots, Z_{rq(r)}]$, where each Z_{rj} is a vector. By fitting the model

$$Z_{rj} = X\beta + \sum_{i=1}^{r-1} Z_i\gamma_i + e$$

as if the γ_is are all fixed, we can obtain $(Z_{rj})'(I - P_{r-1})(Z_{rj})$ as the sum of squares error, and thus

$$\text{tr}\left[Z_r'(I - P_{r-1})Z_r\right] = \sum_{j=1}^{q(r)} (Z_{rj})'(I - P_{r-1})(Z_{rj}).$$

In other words, all of the numbers required for estimating σ_r^2 can be obtained from a standard least squares computer program.

5.7.1 Additional Estimates

Henderson's method is not limited to estimating σ_0^2 and σ_r^2. Consider a partitioned mixed linear model with

$$Y = X\beta + Z_\dagger \gamma_\dagger + Z_{r-1}\gamma_{r-1} + Z_r\gamma_r + e, \qquad (5.7.3)$$

where now $Z_\dagger = [Z_1, \ldots, Z_{r-2}]$, $\gamma_\dagger = [\gamma_1, \ldots, \gamma_{r-2}]$,

$$E(\gamma_\dagger) = 0, \quad E(\gamma_{r-1}) = 0, \quad E(\gamma_r) = 0, \quad E(e) = 0,$$

$$\text{Cov}(\gamma_\dagger) = D_\dagger, \quad \text{Cov}(\gamma_{r-1}) = \sigma_{r-1}^2 I_{q(r-1)}, \quad \text{Cov}(\gamma_r) = \sigma_r^2 I_{q(r)}, \quad \text{Cov}(e) = \sigma_0^2 I_n,$$

with all of the random vectors uncorrelated. Again, any variance component model can be written in this form. We now have

$$E(Y) = X\beta; \qquad \text{Cov}(Y) = V(\theta) = \sigma_0^2 I + Z_\dagger D_\dagger Z_\dagger' + \sigma_{r-1}^2 Z_{r-1} Z_{r-1}' + \sigma_r^2 Z_r Z_r'.$$

To get an unbiased estimate of σ_{r-1}^2, consider $Y'(P_{r-1} - P_{r-2})Y$. In particular

$$E\left[Y'(P_{r-1} - P_{r-2})Y\right]$$
$$= \text{tr}\left\{(P_{r-1} - P_{r-2})\left[\sigma_0^2 I + Z_\dagger D_\dagger Z_\dagger' + \sigma_{r-1}^2 Z_{r-1} Z_{r-1}' + \sigma_r^2 Z_r Z_r'\right]\right\}$$
$$= \sigma_0^2 \text{tr}[(P_{r-1} - P_{r-2})] + \sigma_r^2 \text{tr}[(P_{r-1} - P_{r-2})Z_r Z_r'] + \sigma_{r-1}^2 \text{tr}[(P_{r-1} - P_{r-2})Z_{r-1} Z_{r-1}'].$$

Write $t_0 \equiv \text{tr}[(P_{r-1} - P_{r-2})]$, $t_r \equiv \text{tr}[Z_r'(P_{r-1} - P_{r-2})Z_r]$, and $t_{r-1} \equiv \text{tr}[Z_{r-1}'(P_{r-1} - P_{r-2})Z_{r-1}]$, so that an unbiased estimate becomes

$$\hat{\sigma}_{r-1}^2 \equiv \frac{Y'(P_{r-1} - P_{r-2})Y - \hat{\sigma}_0^2 t_0 - \hat{\sigma}_r^2 t_r}{t_{r-1}}.$$

For a general variance component model, to get an unbiased estimate of σ_k^2, consider $Y'(P_k - P_{k-1})Y$.

$$E\left[Y'(P_k - P_{k-1})Y\right] = \text{tr}\left[(P_k - P_{k-1})\sum_{j=0}^{r}\sigma_j^2 Z_j Z_j'\right]$$

$$= \sigma_0^2 \text{tr}[P_k - P_{k-1}] + \sum_{j=k}^{r}\sigma_j^2 \text{tr}[(P_k - P_{k-1})Z_j Z_j'].$$

Define $s_j \equiv \text{tr}[Z_j'(P_k - P_{k-1})Z_j]$. It is not difficult to get an unbiased estimate using the previous unbiased estimates $\hat\sigma_0^2, \hat\sigma_r^2, \hat\sigma_{r-1}^2, \ldots, \hat\sigma_{k+1}^2$. In particular,

$$\hat\sigma_k^2 \equiv \frac{Y'(P_k - P_{k-1})Y - \hat\sigma_0^2 s_0 - \sum_{j=k+1}^{r}\hat\sigma_j^2 s_j}{s_k}.$$

For estimating more than one variance component, Method 3 is not, in general, well defined. We can permute the labels on the matrices $Z_1, \ldots, Z_r$ to get another set of estimates.

EXAMPLE 5.7.3.   *Two-Way ANOVA Without Interaction.*
The fixed effects analysis was discussed in *PA* Sect. 7.5 and we use similar notation, i.e.,

$$y_{ijk} = \mu + \alpha_i + \eta_j + e_{ijk},$$

$i = 1, \ldots, a$, $j = 1, \ldots, b$, $k = 1, \ldots, N_{ij}$. In the variance component model we assume that the main effects are random and in particular that the α_is are independent $N(0, \sigma_\alpha^2)$ and that the η_js are independent $N(0, \sigma_\eta^2)$.

We can identify σ_r^2 with σ_η^2 so that the sum of squares for fitting the η effects after the α effects, i.e. $R(\eta|\mu, \alpha)$, is identified with $Y'(P_r - P_{r-1})Y$. This leads to an estimate $\hat\sigma_{\eta,r}^2$. Alternatively, we could identify σ_r^2 with σ_α^2 so that the sum of squares for fitting the α effects after the η effects, i.e. $R(\alpha|\mu, \eta)$, is identified with $Y'(P_r - P_{r-1})Y$. This leads to an estimate $\hat\sigma_{\alpha,r}^2$.

We can also identify σ_{r-1}^2 with σ_η^2. Now we would identify $Y'(P_{r-1} - P_{r-2})Y$ with $R(\eta|\mu)$, the sum of squares for fitting the η effects ignoring the α effects, leading to an estimate $\hat\sigma_{\eta,r-1}^2$. Similarly, we can identify σ_{r-1}^2 with σ_α^2 and $Y'(P_{r-1} - P_{r-2})Y$ with $R(\alpha|\mu)$, the sum of squares for fitting the α effects ignoring the η effects, leading to an estimate $\hat\sigma_{\alpha,r-1}^2$.

Typically, $\hat\sigma_{\eta,r}^2 \neq \hat\sigma_{\eta,r-1}^2$ and $\hat\sigma_{\alpha,r}^2 \neq \hat\sigma_{\alpha,r-1}^2$. When the ANOVA is balanced or has proportional numbers, the estimates will be the same.   □

For *nested models*, the order in which variance components can be estimated is restricted. If $C(Z_{r-1}) \subset C(Z_r)$, we say that γ_r is nested within γ_{r-1}. This definition of nested effects is somewhat nonstandard. To many people, interaction effects are not considered nested. As used here, interaction effects are nested in more than one effect. Interaction effects are always nested within their main effects and within any lower order interaction effects. In Henderson's method, a nested effect has to have its variance components estimated before that of any effect it is nested within. In general, with γ_r is nested within γ_{r-1}, we can estimate σ_r^2 first and then use it to estimate σ_{r-1}^2. The alternative order is not possible. If we desire to estimate σ_{r-1}^2 first, we require the perpendicular projection operator onto the orthogonal complement of $C(X, Z_1, \ldots, Z_{r-2}, Z_r)$ with respect to $C(X, Z_1, \ldots, Z_r)$. Because $C(Z_{r-1}) \subset C(Z_r)$, we have $C(X, Z_1, \ldots, Z_{r-2}, Z_r) = C(X, Z_1, \ldots, Z_r)$. The orthogonal complement is the zero space, and the projection matrix is the zero matrix which makes estimation impossible. Thus an interaction effect's variance component must be estimated before any lower order variance components.

EXAMPLE 5.7.2.   *Two-Way ANOVA with Interaction.*
The fixed effects analysis was discussed in *PA* Sect. 7.5 and we use similar notation, i.e.,

$$y_{ijk} = \mu + \alpha_i + \eta_j + \xi_{ij} + e_{ijk},$$

$i = 1, \ldots, a, j = 1, \ldots, b, k = 1, \ldots, N_{ij}$. In the variance component model we assume that the main effects and interaction are random and in particular that the α_is are independent $N(0, \sigma_\alpha^2)$, that the η_js are independent $N(0, \sigma_\eta^2)$, and that the ξ_{ij}s are independent $N(0, \sigma_\xi^2)$. In the variance component version of this model we have $X = J$ and $Z = [Z_\alpha, Z_\eta, Z_\xi]$. In this model, $C(X, Z) = C(Z_\xi)$.

We can identify σ_r^2 with σ_ξ^2 so that the sum of squares for fitting the "interaction" ξ effects after the α and η effects, i.e. $R(\xi | \eta, \mu, \alpha)$, is identified with $Y'(P_r - P_{r-1})Y$. This leads to an estimate of the interaction variance $\hat{\sigma}_\xi^2$. In this model, we cannot permute the order of the Z_js in order to estimate either σ_α^2 or σ_η^2 because, due to nesting, that would involve identifying $Y'(P_r - P_{r-1})Y$ with either $R(\alpha | \mu, \eta, \xi) \equiv 0$ or $R(\eta | \mu, \alpha, \xi) \equiv 0$.

However, we can identify σ_{r-1}^2 with σ_η^2 so that the sum of squares for fitting the η effects after the α effects, i.e. $R(\eta | \mu, \alpha)$, is identified with $Y'(P_{r-1} - P_{r-2})Y$. This leads to an estimate $\hat{\sigma}_{\eta, r-1}^2$. Similarly, we can identify σ_{r-1}^2 with σ_α^2. Now we would identify $Y'(P_{r-1} - P_{r-2})Y$ with $R(\alpha | \mu, \eta)$, the sum of squares for fitting the α effects after the η effects, leading to an estimate $\hat{\sigma}_{\alpha, r-1}^2$. Moreover, we can identify σ_{r-2}^2 with σ_η^2 and $Y'(P_{r-2} - P_{r-3})Y$ with $R(\eta | \mu)$, the sum of squares for fitting the η effects ignoring the α effects, leading to an estimate $\hat{\sigma}_{\eta, r-2}^2$. Similarly, we can identify σ_{r-2}^2 with σ_α^2. Now we would identify $Y'(P_{r-2} - P_{r-3})Y$ with $R(\alpha | \mu)$, the sum of squares for fitting the α effects ignoring the η effects, leading to an estimate $\hat{\sigma}_{\alpha, r-2}^2$. Typically, $\hat{\sigma}_{\eta, r-1}^2 \neq \hat{\sigma}_{\eta, r-2}^2$ and $\hat{\sigma}_{\alpha, r-1}^2 \neq \hat{\sigma}_{\alpha, r-2}^2$. When the ANOVA is balanced or has proportional numbers, the estimates will be the same.   □

Another situation in which no estimates exist is if you fit a random main effect with a fixed interaction. One usually argues that it is philosophically nonsensical to have a fixed interaction that corresponds to a random main effect, but the mathematics also degenerate. If $C(Z_k) \subset C(X)$, we have $P_k = P_{k-1}$. Even the ML and REML equations degenerate because terms that involve $[I - A]'V^{-1}Z_k$ can be rewritten as $V^{-1}[I - A]Z_k = 0$.

For balanced ANOVA models, Henderson's Method 3 gives unique answers, because all of the effects are either orthogonal (e.g., main effects) or nested (e.g., a two-factor interaction is nested within both of the main effects).

Exercise 5.5. Data were generated according to the model

$$y_{ijk} = \mu + \alpha_i + \eta_j + \gamma_{ij} + e_{ijk},$$

$i = 1, 2$, $j = 1, 2, 3$, $k = 1, \ldots, N_{ij}$, where $E(\eta_j) = E(\gamma_{ij}) = E(e_{ijk}) = 0$, $\text{Var}(e_{ijk}) = \sigma_0^2 = 64$, $\text{Var}(\eta_j) = \sigma_1^2 = 784$, and $\text{Var}(\gamma_{ij}) = \sigma_2^2 = 25$. All of the random effects were taken independently and normally distributed. The fixed effects were taken as $\mu + \alpha_1 = 200$ and $\mu + \alpha_2 = 150$. The data are

| | | j | | | | j | |
i	1	2	3	i	1	2	3
1	250	211	199	2	195	153	131
	262	198	184		187	150	133
	251	199	200		203	135	135
		198	187		192		
		184	184		209		
					184		

Estimate the variance components using MINQUE(0), MINQUE(1), and Henderson's Method 3. Estimate the variance components using each of these three sets of estimates as starting values for REML and as starting values for maximum likelihood. For each method of estimation, find estimates of the fixed effects. Compare all the estimates with each other and with the true values. What tentative conclusions can you draw about the relative merits of the estimation procedures?

5.8 Exact F Tests for Variance Components

In this section we examine two procedures for testing whether a variance component is zero. The first test method is closely related to Henderson's estimation method. In fact, as discussed by Christensen and Bedrick (1999), it is actually quite easy to come up with exact tests of a single variance component.

5.8.1 Wald's Test

Seely and El-Bassiouni (1983) considered extensions of Wald's variance component test. They examined the mixed linear model

$$Y = X\beta + Z_*\gamma_* + Z_r\gamma_r + e \tag{5.8.1}$$

which is similar to model (5.7.1). Here, Z_* and Z_r are, respectively, $n \times q_*$ and $n \times q(r)$ matrices of known quantities. γ_*, γ_r, and e are independent random vectors with

$$\gamma_* \sim N(0, D_*), \quad \gamma_r \sim N(0, \sigma_r^2 I_{q(r)}), \quad e \sim N(0, \sigma_0^2 I_n).$$

The null hypothesis $H_0 : \sigma_r^2 = 0$ can be tested by using ordinary least squares calculations treating the γs as fixed effects. Let $SSE(1)$ be the sum of squares error from fitting model (5.8.1). The degrees of freedom error are $dfE(1)$. Also let $SSE(2)$ be the sum of squares error from the least squares fit of

$$Y = X\beta + Z_*\gamma_* + e \tag{5.8.2}$$

with degrees of freedom error $dfE(2)$. The Wald test is simply based on the fact that under H_0

$$\frac{[SSE(2) - SSE(1)]/[dfE(2) - dfE(1)]}{SSE(1)/dfE(1)} \sim F[dfE(2) - dfE(1), dfE(1)].$$

Mohammad Hattab has pointed out to me that when γ_* and γ_r have a joint multivariate normal distribution, even if they are not independent, the proof given below continues to hold. In longitudinal models it is not uncommon to assume that different individuals are independent but that each individual's results involve a linear time effect with random slope and intercept wherein the slope and intercept are correlated. This observation allows one to perform a Wald test of whether the variance of the slope or intercept is 0. Extensions to other longitudinal models with independent subjects but multiple correlated random effects within subjects are obvious.

Mohammad also pointed out that this statistic can be used to test $D_r = 0$ whenever $\gamma_r \sim N(0, D_r)$, although it would probably be better to view it as a test of whether $\text{tr}[Z_r'(P_r - P_*)Z_r D_r] = 0$, since $E[Y'(P_r - P_*)Y] = \sigma_0^2 \text{tr}(P_r - P_*) + \text{tr}[Z_r'(P_r - P_*)Z_r D_r]$.

Of course, if the two model matrix column spaces are the same, that is, if $C(X, Z_*, Z_r) = C(X, Z_*)$, then there is no test because both the numerator sum of squares and degrees of freedom are zero. For example, in a two-way ANOVA with interaction and both main effects random (so that the interaction must also be random), this provides only a test of whether the interaction variance component is zero. Neither of the main effect variance components could be tested this way. On the other hand, in a two-way ANOVA without interaction and both main effects random, the procedure can be applied in different ways to test whether each main effect variance component is zero. The next subsection addresses a very general way of testing whether a main effect variance component is zero when the interaction variance component is not.

We now verify the distribution given above. Note that under the mixed model,

$$\text{Cov}(Y) \equiv V(\theta) = \sigma_0^2 I + Z_* D_* Z_*' + \sigma_r^2 Z_r Z_r'.$$

Let $P_* = P_{r-1}$ be the perpendicular projection operator onto the column space $C(X, Z_*)$, and let P_r be the perpendicular projection operator onto $C(X, Z_*, Z_r)$. It follows that $SSE(2) - SSE(1) = Y'(P_r - P_*)Y$, $dfE(2) - dfE(1) = r(P_r - P_*)$, $SSE(1) = Y'(I - P_r)Y$, and $dfE(1) = r(I - P_r)$. Note that $P_r Z_* = Z_*$ and $P_r P_* = P_*$. We need to show

(1) that $Y'(I - P_r)Y/\sigma_0^2 \sim \chi^2[r(I - P_r)]$,
(2) that $Y'(I - P_r)Y$ and $Y'(P_r - P_*)Y$ are independent, and
(3) that, under H_0, $Y'(P_r - P_*)Y/\sigma_0^2 \sim \chi^2[r(P_r - P_*)]$.

Using results from PA Sect. 1.3, we need only show that $\sigma_0^{-4}(I - P_r)V(I - P_r) = \sigma_0^{-2}(I - P_r)$, that $(I - P_r)V(P_r - P_*) = 0$, and that, when $\sigma_r^2 = 0$, $\sigma_0^{-4}(P_r - P_*)V(P_r - P_*) = \sigma_0^{-2}(P_r - P_*)$. Verifying these results involves only straightforward linear algebra along with properties of projection operators. In general, the distribution of $Y'(P_r - P_*)Y$ seems to be intractable without the assumption that $\sigma_r^2 = 0$, but see Sect. 5.3.

To facilitate extensions of this test in the next subsection, we use a simple generalization of the Seely and El-Bassiouni results. The mixed model considered in (5.8.1) can also be written as

$$Y \sim N\left(X\beta, \sigma_0^2 I + \sigma_r^2 Z_r Z_r' + Z_* D_* Z_*'\right),$$

so the test applies for any data with a distribution of this form. If we let Σ be any known nonnegative definite matrix, the method also applies to

$$Y \sim N\left(X\beta, \sigma_0^2 \Sigma + \sigma_r^2 Z_r Z_r' + Z_* D_* Z_*'\right). \tag{5.8.3}$$

Simply write $\Sigma = QQ'$. Then there exists a matrix T such that $TQ = I$; so

$$TY \sim N\left(TX\beta, \sigma_0^2 I + \sigma_r^2(TZ_r)(TZ_r)' + (TZ_*)D_*(TZ_*)'\right),$$

and the method applies to TY. Obviously, for Σ positive definite, $T = Q^{-1}$. The test based on (5.8.3) is simply the standard generalized least squares test of model (5.8.2) versus (5.8.1) when $\text{Cov}(Y) = \sigma_0^2 \Sigma$, see PA Sect. 3.8.

5.8.2 Öfversten's Second Method

Consider again model (5.7.3),

$$Y = X\beta + Z_\dagger \gamma_\dagger + Z_{r-1}\gamma_{r-1} + Z_r\gamma_r + e,$$

where now

$$\gamma_\dagger \sim N(0, D_\dagger), \quad \gamma_{r-1} \sim N\left(0, \sigma_{r-1}^2 I_{q(r-1)}\right), \quad \gamma_r \sim N\left(0, \sigma_r^2 I_{q(r)}\right), \quad e \sim N\left(0, \sigma_0^2 I_n\right),$$

with all of the random vectors independent. The object of the second method is to obtain an exact F test for $H_0 : \sigma_{r-1}^2 = 0$.

EXAMPLE 5.8.1. *Two-Way ANOVA with Interaction.*
The fixed effects analysis was discussed in PA Sect. 7.5 and we use similar notation, i.e.,

$$y_{ijk} = \mu + \alpha_i + \eta_j + \xi_{ij} + e_{ijk},$$

$i = 1, \ldots, a$, $j = 1, \ldots, b$, $k = 1, \ldots, N_{ij}$. In this variance component model we assume that the α main effects are fixed, but that the η_js and ξ_{ij}s are independent and random and in particular that the η_js are independent $N(0, \sigma_\eta^2)$, and that the ξ_{ij}s are independent $N(0, \sigma_\xi^2)$. We now have $X = [J, X_\alpha]$ and $Z = [Z_\eta, Z_\xi]$ where X_α, Z_η, and Z_ξ consist of columns of indicators for the subscripted effect. We want to test $H_0 : \sigma_\eta^2 = 0$. In a balanced ANOVA it turns out that an F test statistic is just $MS(\eta)/MS(\xi)$ where the terms are defined as for a fixed effects model. It is not clear how to do a test if the data are unbalanced. $\qquad\square$

The notation for developing this test in a general partitioned model gets pretty hairy. To simplify, let's look at the model

$$Y = X\beta + Z_1\gamma_1 + Z_2\gamma_2 + e, \tag{5.8.4}$$

where now

$$\gamma_1 \sim N\left(0, \sigma_1^2 I_{q(1)}\right), \quad \gamma_2 \sim N\left(0, \sigma_2^2 I_{q(2)}\right), \quad e \sim N\left(0, \sigma_0^2 I_n\right).$$

It will be easy to see that the test development works just as well when β contains some random effects. In this model we are interested in testing whether $\sigma_1^2 = 0$.

Generalizing the ANOVA example, the test is of primary interest when $C(X, Z_1) \subset C(X, Z_2)$, e.g., main effects γ_1 are nested within interaction γ_2. If $C(X, Z_1) \not\subset C(X, Z_2)$, a Wald test of $H_0 : \sigma_1^2 = 0$ is available from the previous subsection by simply interchanging the roles of $Z_1\gamma_1$ and $Z_2\gamma_2$. If $C(X, Z_1) \subset C(X, Z_2)$, this interchange does not provide a test because $C(X, Z_2) = C(X, Z_1, Z_2)$. As developed here, if Öfversten's second method provides a test, that test is valid regardless of the relationship of $C(X, Z_1)$ and $C(X, Z_2)$.

Öfversten's (1993) second method as presented in Christensen (1996) involves using an orthonormal basis. For example, use Gram–Schmidt on the columns of $[X, Z_2, Z_1, I_n]$ to obtain an orthonormal basis for $\mathbf{R}^n$, say $c_1, \ldots, c_n$. Write these as columns of a matrix $C = [c_1, \ldots, c_n]$. Partition C as $C = [C_1, C_2, C_3, C_4]$, where the columns of C_1 are an orthonormal basis for $C(X)$, the columns of C_2 are an orthonormal basis for the orthogonal complement of $C(X)$ with respect to $C(X, Z_2)$, the columns of C_3 are an orthonormal basis for the orthogonal complement of $C(X, Z_2)$ with respect to $C(X, Z_2, Z_1)$, and the columns of C_4 are an orthonormal basis for the orthogonal complement of $C(X, Z_2, Z_1)$. Note that if $C(X, Z_1) \subset C(X, Z_2)$, then C_3 is vacuous.

Note that $M = C_1 C_1'$. We define $M_2 \equiv M + C_2 C_2'$ to be the ppo onto $C(X, Z_2)$ and $P \equiv P_2 = M_2 + C_3 C_3'$ is the ppo onto $C(X, Z_1, Z_2)$. Finally, $C_4 C_4' = I - P$.

The basic idea of this method is to choose a matrix K so that the extended Wald's test for model (5.8.3) can be applied to $C_2'Y + KC_4'Y$. In executing the test of $H_0 : \sigma_1^2 = 0$, σ_1^2 plays the role assigned to σ_r^2 in (5.8.3), a function of σ_0^2 and σ_2^2 plays the role assigned to σ_0^2 in (5.8.3), and the role of D_1 in (5.8.3) is vacuous. In particular, for some number λ and some matrix K, we want to have

$$C_2'Y + KC_4'Y \sim N[0, (\sigma_2^2 + \sigma_0^2/\lambda)C_2'Z_2Z_2'C_2 + \sigma_1^2 C_2'Z_1 Z_1'C_2]. \tag{5.8.5}$$

This is of the form (5.8.3). As shown in Exercise 5.6, $C_2'Z_2Z_2'C_2$ is a positive definite matrix, so the test follows immediately from generalized least squares. The test cannot be performed if $C_2'Z_1 = 0$, which occurs, for example, if $C(Z_1) \subset C(X)$. Note that $C_4C_4'Y$ is the vector of residuals from treating the random γ effects as fixed. Thus, in using $KC_4'Y = KC_4'C_4C_4'Y$ we are using some of the residual variability to construct the test.

To get the degrees of freedom for the test, we identify correspondences between (5.8.3) and (5.8.5). There are $r(C_2)$ "observations" available in (5.8.5). In (5.8.3), the numerator degrees of freedom for the test are $r(X, Z_1, Z_2) - r(X, Z_1)$. With mean zero in (5.8.5) there is no linear mean structure, i.e, nothing corresponding to X in (5.8.3), Z_1 from (5.8.3) is also vacuous in (5.8.5), and $C_2'Z_1$ is playing the role of Z_2 in (5.8.3). Thus the numerator degrees of freedom for the test are $r(C_2'Z_1)$ and the denominator degrees of freedom are $r(C_2) - r(C_2'Z_1)$. In model (5.8.5), $r(C_2) = r(X, Z_2) - r(X)$. If $C(Z_1) \subset C(X, Z_2)$, it is shown in Exercise 5.6 that $r(C_2'Z_1) = r(X, Z_1) - r(X)$ and the degrees of freedom for the test are $r(X, Z_1) - r(X)$ and $r(X, Z_1, Z_2) - r(X, Z_1)$, respectively. In the two-way example, these numbers are $(a + b - 1) - a = b - 1$ and $ab - (a + b - 1) = (a - 1)(b - 1)$ as they are in the balanced ANOVA F test.

Observe that

$$C_2'Y \sim N(0, \sigma_0^2 I + \sigma_2^2 C_2'Z_2Z_2'C_2 + \sigma_1^2 C_2'Z_1 Z_1'C_2).$$

In many interesting cases, $C_2'Z_2Z_2'C_2 = \lambda I$; in which case an ordinary least squares Wald test can be applied immediately without any use of $C_4'Y$ as long as $C_2'Z_1 \neq 0$. In the two-way ANOVA example, $C(Z_2) = C(X, Z_1, Z_2)$, so $C(C_2) \subset C(Z_2)$. In the balanced case, $N_{ij} \equiv N$, $Z_2Z_2' = NP$, so $C_2'Z_2Z_2'C_2 = NI$.

It is not difficult to see that in balanced ANOVA problems either $C_2'Z_2Z_2'C_2 = \lambda I$ when a standard balanced ANOVA test is available, or $C_2'Z_1 = 0$ when such a test is not available. For example, consider (5.8.4) as modeling a balanced two-way ANOVA without interaction $y_{ijk} = \mu + \gamma_{1i} + \gamma_{2j} + e_{ijk}$, $i = 1, \ldots, q(1)$, $j = 1, \ldots, q(2)$, $k = 1, \ldots, N$, with $X\beta$ being the vector $J\mu$. The γ_1 and γ_2 treatments are often said to be orthogonal. Letting M be the perpendicular projection operator onto $C(X)$, this orthogonality means precisely that $C[(I - M)Z_1]$ and $C[(I - M)Z_2]$ are orthogonal, i.e., $Z_2'(I - M)Z_1 = 0$, cf. PA Sect. 7.1. Now, by the definition of C_2, $C(C_2) = C[(I - M)Z_2]$, so $C_2'Z_1 = 0$ iff $Z_2'(I - M)Z_1 = 0$, which we know is true from orthogonality. Hence, no test of $H_0 : \sigma_1^2 = 0$ is available *from this method*. Of course the standard Wald test of the previous subsection does provide a test here.

Now consider a balanced nested model $y_{ijk} = \mu + \gamma_{1i} + \gamma_{2ij} + e_{ijk}$ with i, j, and k as above. In this model, $C(X, Z_1) \subset C(Z_2)$ and $\frac{1}{N}Z_2Z_2' \equiv P$ is the perpendicular projection operator onto $C(Z_2)$. Observing that $C(C_2) \subset C(Z_2)$ and using the orthonormality of the columns of C_2,

$$C_2'Z_2Z_2'C_2 = N C_2'PC_2 = N C_2'C_2 = NI.$$

Thus an ordinary least squares Wald test based on $C_2'Y$ is available. Given that a Wald test simply compares models in the usual way, for $H_0 : \sigma_1^2 = 0$ this test is simply the standard balanced ANOVA test for no fixed γ_{1i} effects when γ_2 is random, i.e, $MS(\gamma_1)/MS(\gamma_2)$. Similar orthogonality and containment results hold in more general balanced ANOVAs. For the special case of $C(Z_1) \subset C(X, Z_2)$ with $C_2'Z_2Z_2'C_2 = \lambda I$, a general explicit form for the test statistic is given in (5.8.8).

In general, $C_2'Z_2Z_2'C_2 \neq \lambda I$, so the test requires $C_4'Y$. If $r(X, Z_2, Z_1) \equiv t$,

$$C_4'Y \sim N\left(0, \sigma_0^2 I_{n-t}\right).$$

It is easy to see that $C_2'VC_4 = 0$, so $C_2'Y$ and $C_4'Y$ are independent. To obtain (5.8.5), simply pick K so that

$$KC_4'Y \sim N\left(0, \sigma_0^2\left[\lambda^{-1}C_2'Z_2Z_2'C_2 - I\right]\right).$$

Obviously one can do this provided that $\lambda^{-1}C_2'Z_2Z_2'C_2 - I$ is a nonnegative definite matrix. λ is chosen to ensure that the matrix is nonnegative definite. By the choice of C_2, $C_2'Z_2Z_2'C_2$ is a positive definite matrix and λ is taken as its smallest eigenvalue. Then if we use the eigenvalue-eigenvector decomposition $C_2'Z_2Z_2'C_2 = WD(\lambda_i)W'$ with W orthonormal,

$$\lambda^{-1}C_2'Z_2Z_2'C_2 - I = WD\left(\frac{\lambda_i}{\lambda} - 1\right)W',$$

which is clearly nonnegative definite.

Note that this development makes it obvious why λ needs to be the smallest eigenvalue. Actually, the test would still work if λ were chosen to be any positive number less than the smallest eigenvalue, but we want $KC_4'Y$ to increase the variability of $C_2'Y$ as little as possible, and this is accomplished by taking λ as large as possible. In particular, choosing λ as the smallest eigenvalue gives $KC_4'Y$ a singular covariance matrix and thus no variability in at least one direction. Other valid choices of λ can only increase variability.

Also note that $KC_4'Y = 0$ a.s. if the eigenvalues of $C_2'Z_2Z_2'C_2$ all happen to be the same. In this case, $\lambda^{-1}C_2'Z_2Z_2'C_2 - I = 0$ and $C_2'Z_2Z_2'C_2 = \lambda I$, so we get the simpler Wald test alluded to earlier.

The difficulty with the second method is that K is not unique, and typically the results of the test depend on the choice of K. In particular, K is a $w \times (n-t)$ matrix, where typically $w \equiv r(X, Z_2) - r(X) < (n-t)$, while $\lambda^{-1}C_2'Z_2Z_2'C_2 - I$ is a $w \times w$ matrix. Thus, we can take $K = \left[WD\left(\sqrt{\frac{\lambda_i}{\lambda} - 1}\right), 0\right]$ or $K = \left[0, WD\left(\sqrt{\frac{\lambda_i}{\lambda} - 1}\right)\right]$ or any number of other matrices. Modifying Öfversten, a reasonable procedure might be just to pick one of these convenient K matrices, but first randomly permute the rows of $C_4'Y$.

This method applies quite generally. A proof consists of observing that the test of $H_0 : \sigma_1^2 = 0$ remains valid when $X = [X_1, X_2]$, $\beta = (\beta_1', \beta_2')'$ with $\beta_2 \sim N(0, D_2)$, and β_2 independent of γ_1 and γ_2. Relative to the model at the beginning of the subsection, make the identifications $X = [X_1, X_2] \leftarrow [X, Z_\dagger]$, $\beta = (\beta_1', \beta_2')' \leftarrow (\beta', \gamma_r')'$ with $\gamma_1 \leftarrow$ γ_{r-1} and $\gamma_2 \leftarrow \gamma_r$ for a test of $H_0 : \sigma_{r-1}^2 = 0$. This model allows for interaction between two random factors and arbitrary numbers of factors.

To repeat, this method will be most useful when $C(X, Z_1) \subset C(X, Z_2)$; if this is not the case, the simpler Wald test is available. Whenever $C_2' Z_1 = 0$, no test is available. For example, this will occur whenever $C(Z_1) \subset C(X)$, which is precisely what happens when one tries to test the variance component of a random main effect in a two-way analysis of variance with a fixed interaction.

5.8.3 Comparison of Tests

When $C(Z_1) \not\subset C(X, Z_2)$ in model (5.8.4), we have two tests of $H_0 : \sigma_1^2 = 0$ available. Let M, M_2, P, and P_1 be perpendicular projection matrices onto $C(X)$, $C(X, Z_2)$, $C(X, Z_2, Z_1)$, and $C(X, Z_1)$, respectively. The simple Wald test has the F statistic

$$F = \frac{Y'(P - M_2)Y / r(P - M_2)}{Y'(I - P)Y / r(I - P)}.$$

See Lin and Harville (1991) and Christensen and Bedrick (1999) for some alternatives to Wald's test other than the second method just developed.

It can be of interest to examine the power (probability of rejecting the test) under some alternative to the null model, e.g., the model when the null hypothesis is false. The power of this test is quite complicated, but for given values of the parameters the power can be computed as in Davies (1980). Software is available through STATLIB. See also Christensen and Bedrick (1997).

Intuitively, the power depends in part (and only in part) on the degrees of freedom, $r(X, Z_2, Z_1) - r(X, Z_2)$, $n - r(X, Z_2, Z_1)$ and the ratio of the expected mean squares,

$$1 + \frac{\sigma_1^2}{\sigma_0^2} \frac{\operatorname{tr}[Z_1'(P - M_2)Z_1]}{r(X, Z_2, Z_1) - r(X, Z_2)}. \tag{5.8.6}$$

The basic idea behind F tests is that under the null hypothesis the test statistic is the ratio of two estimates of a common variance. Obviously, since the two are estimating the same thing under H_0, the ratio should be about 1. The F distribution quantifies the null variability about 1 for this ratio of estimates. If the numerator and denominator are actually estimates of very different things, the ratio should deviate substantially from the target value of 1. In fixed effects models, the power of an F test under the full model is simply a function of the ratio of expected values of the two estimates and the degrees of freedom of the estimates. In mixed models, the power is generally much more complicated, but the ratio of expected values can still provide some insight into the behavior of the tests. The ratio in (5.8.6) is strictly greater than one whenever the test exists and $\sigma_1^2 > 0$, thus indicating that larger values of the test statistic can be expected under the alternative. The power of the test should tend to increase as this ratio increases but in fact the power is quite complicated. Note that $Y'(P - M_2)Y = Y'C_3C_3'Y$ and $Y'(I - P)Y = Y'C_4C_4'Y$; so this test uses only $C_3'Y$ and $C_4'Y$.

The second test is based on $C_2'Y + KC_4'Y$. Again, exact powers can be computed as in Davies (1980). As shown in Exercise 5.6, the ratio of the expected mean squares for the second test is

$$1 + \frac{\sigma_1^2}{\sigma_2^2 + \sigma_0^2/\lambda} \frac{\text{tr}\left[(C_2'Z_2Z_2'C_2)^{-1}C_2'Z_1Z_1'C_2\right]}{r(C_2'Z_1)}. \tag{5.8.7}$$

Again, this is strictly greater than one whenever the test exists and $\sigma_1^2 > 0$.

The degrees of freedom for the second test were given earlier. To compare the degrees of freedom for the two tests, observe that

$$C(X) \subset C(X, \{M_2 - M\}Z_1) = C(X, M_2Z_1) \subset C(X, Z_2) \subset C(X, Z_2, Z_1).$$

The degrees of freedom for the second test are, respectively, the ranks of the orthogonal complement of $C(X)$ with respect to $C(X, \{M_2 - M\}Z_1)$ and the orthogonal complement of $C(X, \{M_2 - M\}Z_1)$ with respect to $C(X, Z_2)$. (The first orthogonal complement is $C(C_2C_2'Z_1)$ with the same rank as $C_2'Z_1$, and the second orthogonal complement has rank $r(X, Z_2) - [r(X) + r(C_2'Z_1)]$.) The degrees of freedom for the simple Wald test are, respectively, the ranks of the orthogonal complement of $C(X, Z_2)$ with respect to $C(X, Z_2, Z_1)$ and the orthogonal complement of $C(X, Z_2, Z_1)$. In practice, the simple Wald test would typically have an advantage in having larger denominator degrees of freedom, but that could be outweighed by other factors in a given situation. We also see that, in some sense, the second test is being constructed inside $C(X, Z_2)$; it focuses on the overlap of $C(X, Z_2)$ and $C(Z_1)$. On the other hand, the simple Wald test is constructed from the overlap of $C(Z_1)$ with the orthogonal complement of $C(X, Z_2)$.

In the special case of $C(Z_1) \subset C(X, Z_2)$ with $C_2'Z_2Z_2'C_2 = \lambda I$, the second method gives the test statistic

$$F = \frac{Y'(P_1 - M)Y/r(P_1 - M)}{Y'(M_2 - P_1)Y/r(M_2 - P_1)}. \tag{5.8.8}$$

See Exercise 5.6 for a proof. For example, in a two-way ANOVA, X can indicate the grand mean and a fixed main effect, Z_1 can indicate the random main effect to be tested, and Z_2 can indicate the interaction. When the two-way is balanced, $C_2'Z_2Z_2'C_2 = \lambda I$ and we have the traditional test, i.e., the mean square for the random main effect divided by the mean square for interaction.

It should be noted that under $H_0 : \sigma_1^2 = 0$, $C_3'Y$ also has a $N(0, \sigma_0^2 I)$ distribution so it could also be used, along with $C_4'Y$, to adjust the distribution of $C_2'Y$ and still maintain a valid F test. However, this would be likely to have a deleterious effect on the power since then both the expected numerator mean square and the expected denominator mean square would involve positive multiples of σ_1^2 under the alternative.

The material in this section is closely related to Christensen (1996). The near replicate lack of fit tests discussed in *PA* Sect. 6.7.2 can also be used to construct exact F tests for variance components. In fact, when used as a variance component test, Christensen's (1989) test is identical to Wald's test. See Christensen and Bedrick (1999) for an examination of these procedures.

Exercise 5.6.

(a) Prove that $r(C_2'Z_1) = r(X, Z_1) - r(X)$ when $C(Z_1) \subset C(X, Z_2)$.
(b) Prove that $C_2'Z_2Z_2'C_2$ is positive definite.
(c) Prove (7).
(d) Prove (8).

Exercise 5.7. Use the data and model of Exercise 5.5 to test $H_0 : \sigma_1^2 = 0$ and $H_0 : \sigma_2^2 = 0$.

Exercise 5.8. Show that all of the variance component estimation procedures in this chapter yield covariance matrix estimates that are residual type statistics as defined in Sect. 4.7.

Exercise 5.9. Consider model (5.1.6) with

$$\text{Cov}(Y) = V(\theta) = \theta_1 I_n + ZD(\theta_*)Z',$$

wherein $\theta_1 \equiv \sigma_0^2$ and $\theta' = [\theta_1, \theta_*']$. Show that $[I - A_{X,Z}(\theta)][d_{\theta_i}V(\theta)] = 0$, $i = 2, \ldots, s$. Here $A_{X,Z}(\theta)$ is the oblique projection operator onto $C(X, Z)$ along $C[V^{-1}(X, Z)]^{\perp}$. Use this result to show that the likelihood, REML, and MINQUE equations all degenerate for $i = 2, \ldots, s$.

References

Brumback, B. A., Ruppert, D., & Wand, M. P. (1999). Comment on Shively, Kohn, and Wood. *Journal of the American Statistical Association, 94,* 794–797.

Christensen, R. (1989). Lack of fit tests based on near or exact replicates. *Annals of Statistics, 17,* 673–683.

Christensen, R. (1996). Exact tests for variance components. *Biometrics, 52,* 309–315.

Christensen, R. (2011). *Plane answers to complex questions: The theory of linear models* (4th ed.). New York: Springer.

Christensen, R., & Bedrick, E. J. (1997). Testing the independence assumption in linear models. *Journal of the American Statistical Association, 92,* 1006–1016.

Christensen, R., & Bedrick, E. J. (1999). A survey of some new alternatives to Wald's variance component test. *Tatra Mountains Mathematical Publications, 17*, 91–102.

Davies, R. B. (1980). The distribution of linear combinations of χ^2 random variables. *Applied Statistics, 29*, 323–333.

Henderson, C. R. (1953). Estimation of variance and covariance components. *Biometrics, 9*, 226–252.

Hodges, J. S. (2014). *Richly parameterized linear models: Additive, time series, and spatial models using random effects.* Boca Raton, FL: Chapman and Hall/CRC.

Jiang, J. (2007). *Linear and generalized linear mixed models and their applications.* New York: Springer.

Khuri, A. I., Mathew, T., & Sinha, B. K. (1998). *Statistical tests for mixed linear models.* New York: Wiley.

Lin, T.-H., & Harville, D. A. (1991). Some alternatives to Wald's confidence interval and test. *Journal of the American Statistical Association, 86*, 179–187.

McCulloch, C. E., Searle, S. R., & Neuhaus, J. M. (2008). *Generalized, linear, and mixed models* (2nd ed.). New York: Wiley.

Öfversten, J. (1993). Exact tests for variance components in unbalanced linear models. *Biometrics, 49*, 45–57.

Pearce, N. D., & Wand, M. P. (2006). Penalized splines and reproducing kernel methods. *The American Statistician, 60*, 233–240.

Searle, S. R., Casella, G., & McCulloch, C. (1992). *Variance components.* New York: Wiley.

Seely, J. F. & El-Bassiouni, Y. (1983). Applying Wald's variance component test. *The Annals of Statistics, 11*, 197–201.

Chapter 6
Frequency Analysis of Time Series

Abstract This chapter examines the linear mixed models from Chap. 5 that have traditionally been used to analyze time series data. It also examines spectral distributions/densities and linear filtering of time series.

Consider a sequence of observations $y_1, y_2, \ldots, y_n$ taken at equally spaced time intervals. Some of the many sources of such data are production from industrial units, national economic data, and regularly recorded biological data (e.g., blood pressures taken at regular intervals). The distinguishing characteristic of time series data is that, because data are being taken on the same object over time, the individual observations are correlated.

EXAMPLE 6.0.1.   Tukey (1977) reported data, extracted from *The World Almanac*, on the yearly production of bituminous coal in the United States between 1920 and 1968. In 1969, the method of reporting the production figures changed. Bituminous coal was subdivided into bituminous, sub-bituminous, and lignite. These three figures were combined to yield the data in Table 6.1 and Fig. 6.1 Coal production was high during the economic boom of the 1920s, low during the great depression, high again during World War II, dropped after the war, and grew from the 1960s to 1980.   □

A key feature of time series data is that they are often *cyclical* in nature. For example, retail sales go up every year at Christmas. The number of people who vacation in Montana goes up every summer, down in the fall and spring, and up during the ski season. Time series analysis is designed to model cyclical elements.

There are two main schools of time series analysis: *frequency domain* analysis and *time domain* analysis. The frequency domain can be viewed as regression on independent variables that isolate the frequencies of the cyclical behavior. The regression predictor variables are cosines and sines evaluated at known frequencies and times. Most properly, the regression coefficients are taken as random variables, so the appropriate linear model is a mixed model as in Chap. 5. The justification for this approach is based on a very powerful result in probability theory called the Spectral Representation Theorem. Frequency domain analysis is the subject of this chapter.

© Springer Nature Switzerland AG 2019
R. Christensen, *Advanced Linear Modeling*, Springer Texts in Statistics,
https://doi.org/10.1007/978-3-030-29164-8_6

195

Table 6.1 Coal production data

	1920	1930	1940	1950	1960	1970	1980
0	569	468	461	516	416	602.9	823.7
1	416	382	511	534	403	552.2	
2	422	310	583	467	422	595.3	
3	565	334	590	457	459	591.7	
4	484	359	620	392	467	603.4	
5	520	372	578	467	512	648.4	
6	573	439	534	500	534	678.7	
7	518	446	631	493	552	691.3	
8	501	349	600	410	545	665.2	
9	505	395	438	412	560.5	776.3	
	Values are in millions of short tons.						

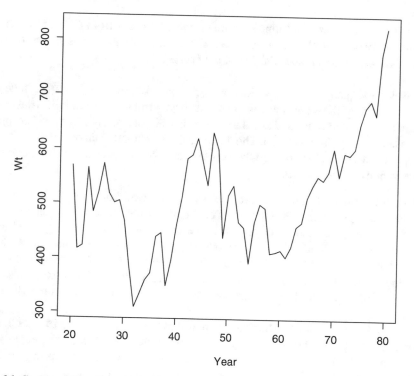

Fig. 6.1 Coal production data: 1920–1980

Chapter 7 discusses time domain analysis. Time domain analysis involves the modeling of observed time series as processes generated by a series of random errors. An important idea in the time domain is that of an autoregressive model. For example, the autoregressive model of order 2 is

$$y_t = \beta_1 y_{t-1} + \beta_2 y_{t-2} + e_t, \tag{6.0.1}$$

where the e_ts are uncorrelated errors. The current observation y_t is being regressed on the previous two observations. Because the model matrix for model (6.0.1) consists of random observations, it does not satisfy the standard linear model assumption that the model matrix is fixed. Recall from Chap. 4 that regression is closely related to best linear prediction. Prediction theory is based on having random predictors such as in model (6.0.1). Best linear prediction is an important tool in time domain analysis.

In addition to cyclical elements, there may also be *trends* in the data. For example, from year to year, the retail sales at Christmas time may display a relatively steady increase. The blood pressure of an overweight person on a diet may display a steady tendency to decrease. This is an important aspect of the analysis of time series. One way to handle trend is to remove it prior to analyzing the cyclical behavior of the time series. Another way of viewing, say, an increasing trend is to consider it as a cycle for which the downturn is nowhere in sight. In this view, trend is just a very slowly oscillating cycle. One can also fit linear models with both cycles and trends but that requires the more general ideas of Chap. 8.

Time series analysis is a large and important subject and there is a huge literature available. The books by Shumway and Stoffer (2011), Brockwell and Davis (1991, 2002) and Fuller (1996) discuss both the frequency and time domains. Prado and West (2010) devote roughly 100 pages to these topics and then get into more advanced models including state space models. Bloomfield (1976) and Koopmans (1974) are devoted to the frequency domain. At a more advanced level are the frequency domain books by Brillinger (1981) and Hannan (1970). Many other books are also available that examine frequency domain ideas.

It is common these days to combine time series analysis and spatial data analysis. One notable book is Cressie and Wikle (2011).

We begin with an introduction to stationary processes. This is followed in Sect. 6.2 with some basic ideas on analyzing data to identify important frequencies. The methods and models of Sect. 6.2 are related to stochastic processes in Sect. 6.3. The relationship suggests that random effects models are more appropriate than fixed effects models for stationary time series. Two random effects models are examined in Sects. 6.4 and 6.5: one with and one without uncorrelated residual errors.

A traditional data-analytic tool for time series is linear filtering. In Sect. 6.6, various types of linear filters are defined and their frequency properties are examined. The linear filters discussed in Sect. 6.6 are the basis for the time domain models of Chap. 7. The chapter closes with some ideas on the relationship between two different time series and a discussion of the commonly used (by other people) Fourier series notation for data analysis.

6.1 Stationary Processes

One fruitful approach to modeling time series data is through the use of stationary processes. A stationary random process is just a group of random variables that exhibit a property called stationarity. In a sense, they remain the same, i.e., stationary.

Typically, the groups of random variables are large: either countably or uncountably infinite. In this chapter and the next, we consider countable sequences of random variables that begin at a fixed time, say, $y_1, y_2, y_3, \ldots$ or, more generally, doubly infinite sequences $\ldots, y_{-1}, y_0, y_1, \ldots$.

In Chap. 8, we consider random variables $y(u)$ for arbitrary vectors u in a subset of $\mathbf{R}^d$ and in Sect. 8.1.1 we define stationarity for these more general processes. Random variables observable in time can be characterized as $y(t)$ for $t \in \mathbf{R}$. Here we assume that this continuous time process will only be observed at equally spaced times $t_1, t_2, \ldots$. (In other words, $t_k - t_{k-1}$ is the same for all k.) The time series we are concerned with is $y_1 = y(t_1)$, $y_2 = y(t_2)$, $\ldots$. If the $y(t)$ process is stationary as defined in Chap. 8, then the sequence $y_1, y_2, \ldots$ is also stationary here. Of course, we cannot actually observe more than a finite number of these random variables.

The property of stationarity is simply that the mechanisms generating the process do not vary—they remain stationary. The sense in which this occurs is that no matter where you start to examine the process, the distribution of the process looks the same. A formal definition of a (strictly) stationary sequence can be based on examining an arbitrary number of random variables, say k, both at the start of the process and at any other time t. Let $C_1, \ldots, C_k$ be arbitrary (Borel) sets, then $\ldots, y_1, y_2, \ldots$ is a *discrete time stationary process* if, for any t,

$$\Pr[y_1 \in C_1, \ldots, y_k \in C_k] = \Pr[y_{t+1} \in C_1, \ldots, y_{t+k} \in C_k]. \qquad (6.1.1)$$

Thus, the random vectors $(y_1, \ldots, y_k)'$ and $(y_{t+1}, \ldots, y_{t+k})'$ have the same distribution for any values t and k. In particular, if the expectation exists,

$$E(y_t) = \mu \qquad (6.1.2)$$

for any t and some scalar μ. If second moments exist, then for any t and k

$$\mathrm{Cov}(y_t, y_{t+k}) = \sigma(k) \qquad (6.1.3)$$

for some *covariance (autocorrelate) function* $\sigma(\cdot)$ that does not depend on t. Note that $\sigma(0)$ is the variance of y_t.

A concept related to stationarity is that of *second-order stationarity*, also known as *weak stationarity, covariance stationarity*, and *stationarity in the wide sense*. A process is said to be second-order stationary if it satisfies conditions (6.1.2) and (6.1.3). The name derives from the fact that conditions (6.1.2) and (6.1.3) only involve second-order moments of the process. As mentioned in the previous paragraph, any process with second moments that satisfies the stationarity condition (6.1.1) also satisfies (6.1.2) and (6.1.3), so any stationary process with second moments is second-order stationary. The converse, of course, does not hold. That random variables have the same first and second moments does not imply that they have the same distributions.

Interestingly, there is an important subclass of stationary processes for which stationarity and second-order stationarity are equivalent. If $(y_{t+1}, \ldots, y_{t+k})$ has a multivariate normal distribution for all t and k, the process is called a *Gaussian* process. Because multivariate normal distributions are completely determined by their means and covariances, conditions (6.1.2) and (6.1.3) imply that a Gaussian process is stationary.

In applying linear models to observations from stochastic processes with a known covariance function, the assumption of second-order stationarity leads to BLUEs. To obtain tests and confidence intervals or maximum likelihood estimates, the data need to result from a stationary Gaussian process. In practice the covariance function is unknown and we are in a situation similar to that discussed in Chap. 4.

6.2 Basic Data Analysis

Frequency domain analysis of an observed time series is based on identifying the frequencies associated with cycles displayed by the data. The most familiar mathematical functions that display cyclical behavior are the sine and cosine functions. The frequency domain analysis of time series can be viewed as doing regression on sines and cosines. In particular, if n is an odd number, we can fit the regression model for $t = 1, \ldots, n$,

$$y_t = \alpha_0 + \sum_{k=1}^{\frac{n-1}{2}} \left[\alpha_k \cos\left(2\pi \frac{k}{n} t\right) + \beta_k \sin\left(2\pi \frac{k}{n} t\right) \right]. \qquad (6.2.1)$$

Here the αs and βs are unknown regression parameters. The predictor variables are the sine and cosine functions evaluated as indicated. The independent variables are grouped by their frequency of oscillation. If $\cos\left(2\pi \frac{k}{n} t\right)$ is graphed on $[0, n]$, the function will complete k cycles. Thus, the frequency, the number of cycles in one unit of time, is k/n. Similarly, the frequency of $\sin\left(2\pi \frac{k}{n} t\right)$ is k/n. Notice that the model has $1 + 2\left(\frac{n-1}{2}\right) = n$ independent variables, so the model is saturated (fits the data perfectly). For this reason, we have not included an error term e_t. Figure 6.2 illustrates some of these predictor variables as functions of time for $n = 61$, the number of observations in the coal data. The figure illustrates the lowest frequency, the middle frequency, and the highest frequency. Solid diamonds indicated the actual data points with the underlying function also graphed for the years of the time series. (If you decide to count the number of cycles in the graphs, remember that they are essentially being graphed on $[1, n]$ rather than on $[0, n]$.)

If the number of observations n is even, a slightly different model is used: for $t = 1, \ldots, n$,

$$y_t = \alpha_0 + \sum_{k=1}^{\frac{n}{2}-1} \left[\alpha_k \cos\left(2\pi \frac{k}{n} t\right) + \beta_k \sin\left(2\pi \frac{k}{n} t\right) \right] + \alpha_{\frac{n}{2}} (-1)^t. \qquad (6.2.2)$$

Again, there are n predictors in the model, one for each observation, so again the data are fitted perfectly. Note that the upper limit of the sums in (6.2.1) and (6.2.2) can both be written as $\lfloor \frac{n-1}{2} \rfloor$, where $\lfloor \frac{n-1}{2} \rfloor$ is the greatest integer contained in (floor of) $\frac{n-1}{2}$. Considering that a saturated model is always fitted to the data, it may be

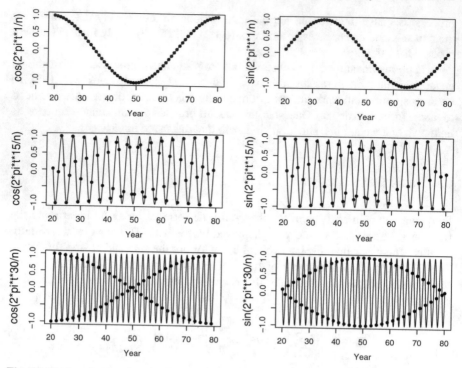

Fig. 6.2 Coal production data: predictor variables for $k/n = 1/61, 15/61, 30/61$

more appropriate to view these models as data-analytic tools rather than as realistic models for the data.

It is convenient to write these models in matrix form. Let

$$Y = (y_1, \ldots, y_n)',$$

$$C_k = \left[\cos\left(2\pi\frac{k}{n}1 \right), \cos\left(2\pi\frac{k}{n}2 \right), \ldots, \cos\left(2\pi\frac{k}{n}n \right) \right]',$$

$$S_k = \left[\sin\left(2\pi\frac{k}{n}1 \right), \sin\left(2\pi\frac{k}{n}2 \right), \ldots, \sin\left(2\pi\frac{k}{n}n \right) \right]',$$

and

$$Z_k = [C_k, S_k].$$

Also, let $\gamma_k = [\alpha_k, \beta_k]'$. The model for both n even and n odd can be written

$$Y = J\alpha_0 + \sum_{k=1}^{\lfloor \frac{n-1}{2} \rfloor} Z_k \gamma_k + \delta_{\lfloor \frac{n}{2} \rfloor \frac{n}{2}} C_{\frac{n}{2}} \alpha_{\frac{n}{2}}, \tag{6.2.3}$$

where J is a column of 1s, $\lfloor \frac{n}{2} \rfloor$ is the greatest integer in (floor of) $\frac{n}{2}$ (i.e., for n even $\lfloor \frac{n}{2} \rfloor = \frac{n}{2}$ and for n odd $\lfloor \frac{n}{2} \rfloor = \frac{n-1}{2}$), and

$$\delta_{ab} = \begin{cases} 1 & a=b \\ 0 & a \neq b. \end{cases}$$

Note that, for n even, $C_{\frac{n}{2}} = (-1,1,-1,1,\ldots,-1,1)'$. Also observe that $C_0 = J$.

One reason that these particular models are used is that the independent variable vectors in (6.2.3) are orthogonal. In particular,

$$C_i'C_j = \begin{cases} n & \text{if } i=j \in \{0,n/2\} \\ n/2 & \text{if } i=j \notin \{0,n/2\} \\ 0 & \text{if } i \neq j \end{cases}, \tag{6.2.4}$$

$$S_i'S_j = \begin{cases} n/2 & \text{if } i=j \notin \{0,n/2\} \\ 0 & \text{otherwise} \end{cases}, \tag{6.2.5}$$

$$C_i'S_j = 0 \quad \text{any } i \text{ and } j. \tag{6.2.6}$$

See Exercise 6.9.15 for a proof of these relationships. Because the vectors are orthogonal, the sum of squares associated with each independent variable does not depend on the other variables that may or may not be included in any submodel. The order of fitting of variables is also irrelevant. Denote the sum of squares associated with C_k as $SS(C_k)$ and the sum of squares for S_k as $SS(S_k)$. The total sum of squares associated with the frequency k/n is $SS(C_k)+SS(S_k)$.

The *periodogram* is a function $P(v)$ of the frequencies that indicates how important a particular frequency is in the time series. The periodogram is defined only for $v = k/n, k = 1, \ldots, \lfloor \frac{n-1}{2} \rfloor$. The periodogram is

$$P(k/n) \equiv \{SS(C_k)+SS(S_k)\}/2.$$

This is precisely *the mean square associated with the frequency k/n*. Clearly, if the mean square for the frequency k/n is large relative to the mean squares for the other frequencies, then the frequency k/n is important in explaining the time series.

We will not define $P(0.5)$ but for n even, some people define

$$P(0.5) \equiv Y'C_{\frac{n}{2}}C_{\frac{n}{2}}'Y/n = SS(C_{\frac{n}{2}}).$$

This is still the mean square associated with the frequency because $S_{\frac{n}{2}} = 0$. A couple of reasons for not defining $P(0.5)$ are that an oscillation that fast would rarely be interesting and that it is inconvenient to have 2 degrees of freedom associated with every periodogram value except this one. The details of a data analysis can depend on whether $P(0.5)$ is defined. Since the programming language R defines it, some of our output will use it.

Exercise 6.1. Show that for $k = 1, \ldots, \lfloor \frac{n-1}{2} \rfloor$, $P(k/n) = \{(C_k'Y)^2 + (S_k'Y)^2\}/n$. Show that for n even, $P(0.5) = \{(C_{\frac{n}{2}}'Y)^2 + (S_{\frac{n}{2}}'Y)^2\}/n$ so that the same formula applies even though this has 1 degree of freedom, not 2.

As always, our linear model is only an approximation to reality. The true frequencies associated with a time series are unlikely to be among the values k/n. If the true frequency is, say, $v \in \left[\frac{k-1}{n}, \frac{k}{n}\right]$, then we could expect the effect of this frequency to show up in $SS(C_{k-1})$, $SS(C_k)$, $SS(S_{k-1})$, and $SS(S_k)$. If we want a measure of the importance of all the frequencies in a neighborhood of $\frac{k}{n}$, it makes sense to compute the mean square for a group of frequencies near $\frac{k}{n}$. This idea is used to define a smoothed version of the periodogram called an *estimate of the spectral density* or more simply a *spectral estimator*. For r odd, define

$$\hat{f}_r(k/n) = \frac{1}{r} \sum_{i=-(r-1)/2}^{(r-1)/2} P\left(\frac{k+i}{n}\right), \qquad (6.2.7)$$

which is *the mean square for the $2r$ variables* $C_{k+\ell}, S_{k+\ell} : \ell = -(r-1)/2, \ldots, (r-1)/2$. This process is sometimes referred to as applying a *Daniell kernel* of order r to the periodogram. The frequencies $(k+\ell)/n$ for $\ell = -(r-1)/2, \ldots, (r-1)/2$ will be called the r *neighborhood (Daniell spectral window)* of k/n. Picking an r neighborhood is equivalent to picking frequencies in a band having *bandwidth* r/n. Choosing $r = 3$ (i.e., examining the mean square for the frequencies $\frac{k-1}{n}, \frac{k}{n}, \frac{k+1}{n}$, seems particularly appealing to the author but may not be particularly common in applications. (For consistent estimation of the spectral density that is defined later, r must be an increasing function of n.)

Strictly speaking, $\hat{f}_r(k/n)$ is only defined for $k = 1 + (r-1)/2, \ldots, \lfloor\frac{n-1}{2}\rfloor - (r-1)/2$. To get values for all $k = 1, \ldots, \lfloor\frac{n-1}{2}\rfloor$, it is common practice to replace any values of $(k+i)/n$ in Eq. (6.2.7) that are not defined with the frequency that is closest to it for which the periodogram actually exists. So, if $(k+i)/n \le 0$, use $1/n$. For high frequencies, the adjustment depends on whether you have defined $P(0.5)$.

Rather than using the simple average of the periodograms in the r neighborhood, the spectral density estimate is often taken as a weighted average of the periodogram values. The weights can be defined by evaluating a *weighting (kernel) function* at appropriate points. One common choice for a weighting function is the cosine. Within this context, the simple average corresponds to a rectangular weighting function. An indirect method of changing the weights is to apply the Daniell kernal first to the periodogram and then to the resulting spectral estimate.

Although the spectral density function estimator $\hat{f}_r(\cdot)$ is certainly a reasonable thing to examine, the name of the function must seem totally bizarre to anyone without a previous knowledge of time series analysis. The genesis of this ghostly (ghastly?) name will be discussed in the next section.

EXAMPLE 6.2.1.   Table 6.2 contains three versions of the periodogram for the coal production data of Example 6.0.1. As will be discussed in Sect. 6.8, the periodogram can be written as the squared absolute value of the discrete Fourier transform of the time series divided by the sample size. As such, something called the *fast Fourier transform (FFT)* is often used to compute the periodogram. We begin by discussing the fact that different computer software can give different periodograms.

Table 6.2 Coal production periodograms

Definition		BMDP		Default R	
v	$P(v)$	v	$P(v)$	v	$P(v)$
0.01639344	56528.08478	0.0159	0.612E+05	0.015625	28935.8824
0.03278689	149752.69159	0.0317	0.137E+06	0.031250	79666.0952
0.04918033	6943.81845	0.0476	0.878E+04	0.046875	79587.5992
0.06557377	11140.48878	0.0635	0.216E+05	0.062500	99027.9791
0.08196721	15541.26244	0.0794	0.183E+05	0.078125	43073.8576
0.09836066	18097.97961	0.0952	0.181E+05	0.093750	67434.1605
0.11475410	8261.15057	0.1111	0.373E+04	0.109375	22521.6558
0.13114754	7103.20125	0.1270	0.117E+04	0.125000	22321.6843
0.14754098	1562.75966	0.1429	0.541E+04	0.140625	56389.5988
0.16393443	5318.25725	0.1587	0.371E+04	0.156250	20906.5014
0.18032787	9111.07990	0.1746	0.164E+04	0.171875	14773.1099
0.19672131	722.57093	0.1905	0.156E+04	0.187500	30294.4949
0.21311475	8155.94896	0.2063	0.262E+04	0.203125	19647.4293
0.22950820	370.46263	0.2222	0.376E+04	0.218750	28577.5781
0.24590164	6134.12844	0.2381	0.340E+03	0.234375	8548.4000
0.26229508	731.47281	0.2540	0.432E+04	0.250000	20429.3934
0.27868852	8728.98983	0.2698	0.255E+02	0.265625	3951.1664
0.29508197	3504.96303	0.2857	0.667E+04	0.281250	16924.2177
0.31147541	740.04737	0.3016	0.890E+04	0.296875	10982.5091
0.32786885	769.01729	0.3175	0.403E+04	0.312500	1571.0365
0.34426230	704.15309	0.3333	0.306E+04	0.328125	987.0066
0.36065574	2006.14370	0.3492	0.247E+04	0.343750	833.8779
0.37704918	1007.92177	0.3651	0.516E+04	0.359375	1329.7808
0.39344262	317.64999	0.3810	0.421E+02	0.375000	6192.8272
0.40983607	1451.82227	0.3968	0.141E+04	0.390625	2775.2283
0.42622951	404.59414	0.4127	0.390E+03	0.406250	10371.4089
0.44262295	623.50094	0.4286	0.521E+02	0.421875	7015.8474
0.45901639	44.88373	0.4444	0.400E+03	0.437500	7090.1223
0.47540984	469.74898	0.4603	0.806E+02	0.453125	5031.5630
0.49180328	262.64055	0.4762	0.426E+03	0.468750	7146.6194
		0.4921	0.239E+03	0.484375	8210.8865
				0.500000	5897.7056

Once upon a time, discrete Fourier transforms were hard to compute. They still can be, when analyzing huge amounts of data. The FFT was a major computational improvement that was based on factorizations of the sample size n. If, say, $n = n_1 \times n_2 \times n_3$ you can rewrite the Fourier transform in terms of three easier to compute (because they have smaller size) transforms of sizes n_1, n_2, and n_3. The ultimate convenience was to have $n = 2^j$. To make optimal use of this fact, when computing transforms, computer software often adds additional observations to the end of the series.

The three versions of the periodogram in Table 6.2 come from different software. When I first analyzed the coal data decades ago, the software BMDP added two observations, both equal to the mean of the series, to make the sample size $63 = 7 \times 3^2$. I recently reanalyzed these data in R. The command spec.pgram by default adds

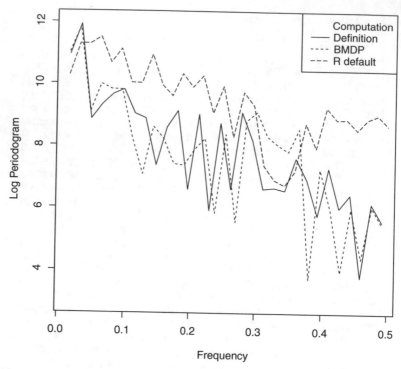

Fig. 6.3 Coal production data: three versions of $P(v)$

three observations to the time series before producing the periodogram to make the sample size $64 = 2^6$. The R documentation indicates that these added observations are 0s, but adding three 0s (or any other numbers I could think of) to the data file does not reproduce the numbers given by `spec.pgram`. You can tell how many observations have been added because the frequencies reported are k divided by the number of observations used. I also computed the periodogram directly from the definitions given here by (6.2.1) essentially computing the orthogonal ppos, as well as (6.2.2) by using R's `fft` command, and by (6.2.3) telling `spec.pgram` not to add observations. These numbers all agreed. R's `fft` does not add observations; it merely exploits any factorization available for n. R's `spec.pgram`, by default, speeds up the computations by modifying the sample size to some highly factorable number but specifying `fast=FALSE` stops that. Incidentally, setting `spec.pgram` to `fast=FALSE` is not the only default that has to be reset to get the periodogram.

Figure 6.3 plots the three versions of the periodogram on the natural log scale, i.e., it plots $(v, \log[P(v)])$. I did this with natural logs but it is more commonly plotted with base 10 logs. You can see substantial differences, especially with the default `spec.pgram` results. Figure 6.4 plots the three corresponding versions of $\log[\hat{f}_5(v)]$. Table 6.2 repeats the definitional periodogram and gives the corresponding spectral density estimates based on $r = 3$ and $r = 5$.

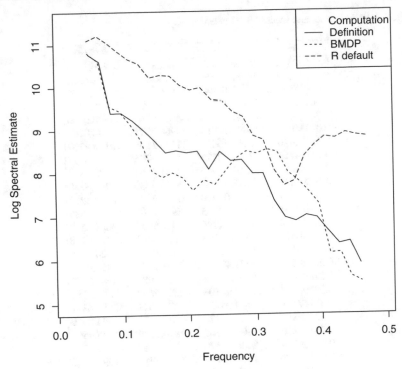

Fig. 6.4 Coal production data: three versions of $\hat{f}_5(v)$

As noted earlier, the frequencies v reported in Table 6.1 for BMDP are not the Fourier frequencies $k/61$ but rather $k/63$. In the interest of computational efficiency, BMDP has extended the time series from 61 to 63 observations by including artificial values $y_{62} = y_{63} = \bar{y}.$. Clearly, this procedure will not affect the estimate of the mean. All effects for positive frequencies are orthogonal to the mean, so working with $y_i - \bar{y}.$ is equivalent to working with the original data. After correcting for the mean, the artificial observations are zero; thus the artificial observations do not increase the sums of squares for positive frequencies. I assume that this is also what spec.pgram intends to produce, but the documentation I read was not clear, nor could I reproduce spec.pgram's results. In any case, it cannot be denied that the different methods provide solutions to slightly different problems than we set out to solve and Fig. 6.3 illustrates how different the solutions are.

Figure 6.5 contain plots, on a logarithmic scale, of the periodogram and the spectral density estimates. In Example 6.5.1 we will discuss confidence intervals for the true periodogram values, i.e., for the σ_k^2s. Perhaps the two most noteworthy aspects of these plots are that there seem to be substantial effects for low frequencies and small effects for high frequencies. Low frequencies are consistent with trend in the data. To eliminate the effect of a trend, one can perform the frequency analysis on the residuals from a simple linear regression on time. Figure 6.6 gives $\hat{f}_5$ with the

Table 6.3 Periodogram and spectral density estimates

v	$P(k/n)$	$\hat{f}_3(k/n)$	$\hat{f}_5(k/n)$
0.01639344	56528.08478	87602.9537	65256.1529
0.03278689	149752.69159	71074.8649	56178.6337
0.04918033	6943.81845	55945.6663	47981.2692
0.06557377	11140.48878	11208.5232	40295.2482
0.08196721	15541.26244	14926.5769	11996.9400
0.09836066	18097.97961	13966.7975	12028.8165
0.11475410	8261.15057	11154.1105	10113.2707
0.13114754	7103.20125	5642.3705	8068.6697
0.14754098	1562.75966	4661.4061	6271.2897
0.16393443	5318.25725	5330.6989	4763.5738
0.18032787	9111.07990	5050.6360	4974.1233
0.19672131	722.57093	5996.5333	4735.6639
0.21311475	8155.94896	3082.9942	4898.8382
0.22950820	370.46263	4886.8467	3222.9168
0.24590164	6134.12844	2412.0213	4824.2005
0.26229508	731.47281	5198.1970	3894.0033
0.27868852	8728.98983	4321.8086	3967.9203
0.29508197	3504.96303	4324.6667	2894.8981
0.31147541	740.04737	1671.3426	2889.4341
0.32786885	769.01729	737.7393	1544.8649
0.34426230	704.15309	1159.7714	1045.4566
0.36065574	2006.14370	1239.4062	960.9772
0.37704918	1007.92177	1110.5718	1097.5382
0.39344262	317.64999	925.7980	1037.6264
0.40983607	1451.82227	724.6888	761.0978
0.42622951	404.59414	826.6391	568.4902
0.44262295	623.50094	357.6596	598.9100
0.45901639	44.88373	379.3779	361.0737
0.47540984	469.74898	259.0911	332.6829
0.49180328	262.64055	331.6767	301.9326

trend removed. For comparison, the regular $\hat{f}_5$ is also plotted. The plots are not markedly different for the coal data.   □

6.3 Spectral Approximation of Stationary Time Series

We have used model (6.2.3) as a tool for identifying important frequencies in the data. We now justify the model as an approximate model for observations for any second-order stationary process. The argument is based on the *Spectral Representation Theorem*. This result can be found in a variety of advanced books on probability. The discussion in Doob (1953, Sections X.3 and X.4) is particularly germane, but quite sophisticated mathematically. The mathematics in Breiman (1968, Section

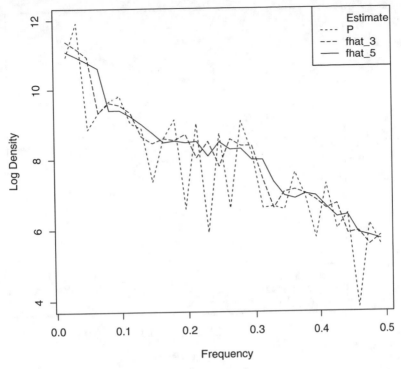

Fig. 6.5 Coal production data: plots of $P(v)$, $\hat{f}_3(v)$, and $\hat{f}_5(v)$

11.6) is probably easier to follow, but the discussion is less clearly applicable to the problem at hand.

In this section, applications of various results in probability are discussed. The results themselves are given without proof. In particular, we will use the fact that any second-order stationary process can be approximated by model (6.2.3) with the regression coefficients taken as random variables. This defines a random effects model as in Chap. 5. The variance components of the mixed model are related to the autocorrelate function through something called the *spectral distribution function*. Estimation of the variance components is via least squares, thus mimicking Henderson's method 3, which was discussed in Sect. 5.6. It is also equivalent to doing REML, cf. Exercise 6.3.

Let $\dots y_{-2}, y_{-1}, y_0, y_1, y_2, \dots$ be a second-order stationary process with $E(y_t) = \mu$ and covariance (autocorrelate) function $\sigma(k) = \text{Cov}(y_t, y_{t+k})$. As Doob (1953, p. 486) points out, the Spectral Representation Theorem implies that the process $y_t - \mu$ can be approximated arbitrarily closely by a process based on sines and cosines. In particular, for n large and some α_ks and β_ks,

$$y_t - \mu \doteq z_t, \tag{6.3.1}$$

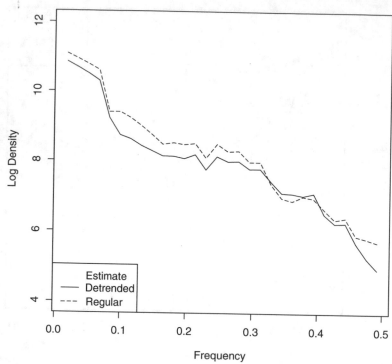

Fig. 6.6 Coal production data: plots of $\hat{f}_5(v)$ from detrended data and from original data

where

$$z_t = \sum_{k=0}^{\lfloor \frac{n-1}{2} \rfloor} \left[\alpha_k \cos\left(2\pi\frac{k}{n}t\right) + \beta_k \sin\left(2\pi\frac{k}{n}t\right) \right] + \delta_{\lfloor \frac{n}{2} \rfloor \frac{n}{2}} \cos(\pi t)\alpha_{\frac{n}{2}}$$

with, for all k and k', $E(\alpha_k) = E(\beta_k) = 0$, $\mathrm{Var}(\alpha_k) = \mathrm{Var}(\beta_k) = \sigma_k^2$, $\mathrm{Cov}(\alpha_k, \beta_{k'}) = 0$, and, for $k \neq k'$, $\mathrm{Cov}(\alpha_k, \alpha_{k'}) = \mathrm{Cov}(\beta_k, \beta_{k'}) = 0$. Although it has been suppressed in the notation, σ_k^2 also depends on n. Note that, for $k = 0$, $\sin\left(2\pi\frac{k}{n}t\right) = 0$ and $\cos\left(2\pi\frac{k}{n}t\right) = 1$ for all t. Thus, the right-hand side of (6.3.1) is identical to the right-hand sides of models (6.2.1) and (6.2.2), except that we had not previously assumed random regression coefficients.

In our later data analysis, the random effects α_0 and β_0 will not be considered. This requires some justification. First, β_0 is multiplied by $0 \equiv \sin(2\pi 0/nt)$, so β_0 has no effect. The random effect α_0 cannot be analyzed because it is statistically indistinguishable from the parameter μ. Because $1 \equiv \cos(2\pi 0/nt)$, α_0 is added to *every* observation y_t. Similarly, μ is added to every y_t. On the basis of one realization of the series, the two effects are hopelessly confounded. Fortunately, our interest is often in predicting the future of this realization or in simply explaining the behavior of these observations. For either purpose, it is reasonable to consider α_0 as fixed.

With α_0 fixed, the mean of y_t is $\mu + \alpha_0$. Because this involves two parameters for one object, we will suppress μ in the notation and use α_0 to denote the mean.

The spectral approximation (6.3.1) along with the discussion in the previous paragraph is the justification for using model (6.2.3) as an approximate model for stationary time series. Based on the approximation, in model (6.2.3) we assume that

$$E(\gamma_k) = 0,$$
$$Cov(\gamma_k) = \sigma_k^2 I_2,$$
$$Cov(\gamma_k, \gamma_{k'}) = 0 \qquad k \neq k',$$

α_0 is a fixed effect, and for n even $\alpha_{\frac{n}{2}}$ is a random effect with zero mean, variance $\sigma_{\frac{n}{2}}^2$, and zero covariance with the other random effects. Under these assumptions, model (6.2.3) is a variance component as in Chap. 5.

The key parameters in model (6.2.3) are the variance components, the σ_k^2s. These are closely tied to additional aspects of the Spectral Representation Theorem. To illustrate these aspects, assume that $\mu = 0$ and α_0 is a random effect. The Spectral Representation Theorem implies that there is a *unique* right continuous function F called the *spectral distribution function* that is (a) defined on $[-1/2, 1/2]$, (b) symmetric about zero [i.e., $F(v-) - F(0) = F(0) - F(-v)$, where $F(v-) = \lim_{\eta \nearrow v} F(\eta)$], and (c) satisfies

$$\sigma(k) = \int_{-1/2}^{1/2} e^{2\pi i v k} dF(v)$$

$$= \int_{-1/2}^{1/2} \cos(2\pi v k) dF(v). \qquad (6.3.2)$$

By definition, $e^{2\pi i v k} = \cos(2\pi v k) + i \sin(2\pi v k)$. The integrals are Riemann–Stieltjes integrals. They are similar to the standard Riemann integrals. If $\mathscr{P}_n = (v_{0n}, \ldots, v_{nn})$ defines a sequence of partitions of $[-1/2, 1/2]$ with $v_{in} - v_{i-1,n}$ approaching zero for all i and n, then

$$\int_{-1/2}^{1/2} \cos(2\pi v k) dF(v) = \lim_{n \to \infty} \sum_{i=0}^{n-1} \cos(2\pi v_{in} k)[F(v_{i+1,n}) - F(v_{in})].$$

The second equality in (6.3.2) follows from the symmetry of F and the fact that $\sin(2\pi v k)$ is an odd function in v.

EXAMPLE 6.3.1. Consider the stochastic process defined by

$$z_t = \sum_{j=0}^{\frac{n-1}{2}} \left[\alpha_j \cos\left(2\pi \frac{j}{n} t\right) + \beta_j \sin\left(2\pi \frac{j}{n} t\right) \right],$$

where n is odd and the αs and βs are random effects that satisfy the assumptions made along with Eq. (6.3.1). Note that this process is just the right-hand side of (6.3.1). In this example, we examine properties of the process that is used to approximate second-order stationary time series. Using the assumptions about the random effects and applying the formula $\cos(a - b) = \cos(a)\cos(b) + \sin(a)\sin(b)$ to $\cos(2\pi\frac{j}{n}k) = \cos(-2\pi\frac{j}{n}k) = \cos(\{2\pi(j/n)(t+k)\} - \{2\pi(j/n)t\})$, we get

$$\sigma(k) = \text{Cov}(z_t, z_{t+k}) = \text{E}(z_t z_{t+k})$$

$$= \sum_{j=0}^{\frac{n-1}{2}} \sigma_j^2 \cos\left(2\pi\frac{j}{n}t\right) \cos\left(2\pi\frac{j}{n}(t+k)\right)$$

$$+ \sum_{j=0}^{\frac{n-1}{2}} \sigma_j^2 \sin\left(2\pi\frac{j}{n}t\right) \sin\left(2\pi\frac{j}{n}(t+k)\right)$$

$$= \sum_{j=0}^{\frac{n-1}{2}} \sigma_j^2 \cos\left(2\pi\frac{j}{n}k\right). \tag{6.3.3}$$

Noticing that $\cos(2\pi(j/n)k) = \cos(-2\pi(j/n)k)$, the spectral distribution function F must be a step function that is zero at $v = -\frac{1}{2}$, has a jump of σ_0^2 at $v = 0$, and jumps of $\sigma_j^2/2$ at $v = \pm j/n$, $j = 1, \ldots, (n-1)/2$. [Computing (6.3.2) for this function F is just like computing the "expected value" of $\cos(2\pi v k)$, where v is a random variable that takes on the values $\pm j/n$ with "probability" $\sigma_j^2/2$ and the value 0 with "probability" σ_0^2. The only difference is that the "probabilities" do not add up to one.]

This random effects process is used to approximate an arbitrary time series process. The fact that it makes a good approximation to the series does not imply that it makes a realistic model for the time series. Realistic models for time series should probably have continuous spectral distributions. This process has a discrete spectral distribution. It may be reasonable to approximate a continuous distribution with a discrete distribution; it is less reasonable to *model* a continuous distribution as a discrete distribution.

Most realistic models for time series do not "remember" forever what has happened in the past. More technically, they have a covariance function $\sigma(k)$ that converges to zero as $k \to \infty$. From Eq. (6.3.3), that does not occur with the approximation process. Nonetheless, the approximation process provides a good tool for introducing basic concepts of frequency domain analysis to people with a background in linear models. □

A second-order stationary process y_t has a spectral distribution function F defined by (6.3.2). Typically, F will not be a discrete distribution. The important point is that the distribution F determines the spectral distribution of the approximation process z_t and thus the variance components of the approximation model. In particular, the Spectral Representation Theorem implies that the approximation in (6.3.1) has

$$\sigma_0^2 = F\left(\frac{1}{n}-\right) - F\left(-\frac{1}{n}\right),$$

$$\sigma_k^2 = \left[F\left(\frac{k+1}{n}-\right) - F\left(\frac{k}{n}-\right)\right] + \left[F\left(-\frac{k}{n}\right) - F\left(-\frac{k+1}{n}\right)\right] \quad (6.3.4)$$

$$= 2\left[F\left(\frac{k+1}{n}-\right) - F\left(\frac{k}{n}-\right)\right].$$

Much of standard frequency domain time series relates to the spectral density function. Assuming that

$$\sum_{k=0}^{\infty} |\sigma(k)| < \infty,$$

the spectral distribution function $F(v)$ has a derivative $f(v)$ and

$$\sigma(k) = \int_{-1/2}^{1/2} e^{2\pi i v k} f(v) dv \quad (6.3.5)$$

$$= \int_{-1/2}^{1/2} \cos(2\pi v k) f(v) dv.$$

The function $f(v)$ is called the *spectral density*. From (6.3.4), we see that for $k \neq 0$

$$\sigma_k^2 \doteq 2f\left(\frac{k}{n}\right)\left[\frac{k+1}{n} - \frac{k}{n}\right]$$

and

$$\frac{n}{2}\sigma_k^2 \doteq f\left(\frac{k}{n}\right). \quad (6.3.6)$$

We have assumed that the process has gone on in the infinite past. Thus, by second-order stationarity, the random variables $\ldots, y_{-2}, y_{-1}, y_0, y_1, y_2, \ldots$ have the property that

$$\sigma(k) = \sigma(-k).$$

Equation (6.3.5) leads to a well-known inverse relation,

$$f(v) = \sum_{k=-\infty}^{\infty} \sigma(k) e^{-2\pi i v k}$$

$$= \sum_{k=-\infty}^{\infty} \sigma(k) \cos(2\pi v k). \quad (6.3.7)$$

The last equality follows from the symmetry of $\sigma(\cdot)$.

White noise is a term used to describe one of the simplest yet most important examples of a second-order stationary process. A process e_t is said to be *white noise* if

$$E(e_t) = 0$$

and

$$\sigma(k) = \begin{cases} \sigma^2 & \text{if } k = 0 \\ 0 & \text{if } k \neq 0 \end{cases}.$$

Such processes are of particular importance in defining time domain models in Chap. 7. Note that random sampling (with replacement) from any population with mean zero and finite variance generates white noise.

Exercise 6.2. Let e_t be a white noise process. Show that the spectral density of e_t is

$$f_e(v) = \sigma^2.$$

6.4 The Random Effects Model

Consider now the matrix form of the spectral approximation random effects model (6.3.1) in all its gory detail. The model is

$$Y = J\mu + \sum_{k=0}^{\lfloor \frac{n-1}{2} \rfloor} Z_k \gamma_k + \delta_{\lfloor \frac{n}{2} \rfloor \frac{n}{2}} C_{\frac{n}{2}} \alpha_{\frac{n}{2}} , \qquad (6.4.1)$$

$$E(\gamma_k) = 0, \qquad \text{Cov}(\gamma_k) = \sigma_k^2 I_2 ,$$
$$E(\alpha_{\frac{n}{2}}) = 0, \qquad \text{Var}(\alpha_{\frac{n}{2}}) = \sigma_{\frac{n}{2}}^2 ,$$
$$\text{Cov}(\gamma_k, \gamma_{k'}) = 0, \qquad k \neq k' ,$$
$$\text{Cov}(\gamma_k, \alpha_{\frac{n}{2}}) = 0.$$

Let $\text{Cov}(Y) = V$; then,

$$V = \sum_{j=0}^{\lfloor \frac{n-1}{2} \rfloor} \sigma_j^2 Z_j Z_j' + \delta_{\lfloor \frac{n}{2} \rfloor \frac{n}{2}} \sigma_{\frac{n}{2}}^2 C_{\frac{n}{2}} C_{\frac{n}{2}}' .$$

Model (6.4.1) can be rewritten as

$$Y = J\mu + \xi, \ E(\xi) = 0, \ \text{Cov}(\xi) = V.$$

Typically V will be nonsingular and $C(VJ) \subset C(J)$ so the least squares estimate of μ, i.e. $\bar{y}.$, is the BLUE, cf. Exercise 6.3.

This random effects model is exactly correct for observations generated by the approximation process z_t of Sect. 6.3. It is approximately correct for other second-order stationary processes. As discussed in Example 6.3.1, the process z_t based on sines and cosines approximates the behavior of the true process but in broader terms may not be a very realistic model for the true process. In this section, we rely only on the quality of the approximation. We derive results assuming that model (6.4.1) is correct and mention the appropriate interpretation when model (6.4.1) is only an approximation.

Let P_k be the perpendicular projection operator onto the column space $C(C_k, S_k)$. Note that by the choice of the C_js and S_js, for $k = 1, \ldots, \lfloor \frac{n-1}{2} \rfloor$,

$$P_k = \frac{2}{n}[C_k C_k' + S_k S_k'] = \frac{2}{n} Z_k Z_k'.$$

The periodogram is

$$P(k/n) = Y' P_k Y / 2.$$

Proposition 6.4.1. For $k = 1, \ldots, \lfloor \frac{n-1}{2} \rfloor$,

$$E[P(k/n)] = \frac{n}{2} \sigma_k^2 \doteq f\left(\frac{k}{n}\right).$$

PROOF.   By Theorem 1.3.2 in *PA*,

$$
\begin{aligned}
E(Y' P_k Y) &= \text{tr}[P_k V] + (J\mu)' P_k (J\mu) \\
&= \sum_j \sigma_j^2 \text{tr}[P_k Z_j Z_j'] + 0 \\
&= \sigma_k^2 \text{tr}[P_k Z_k Z_k'] \\
&= \sigma_k^2 \text{tr}[Z_k' P_k Z_k] \\
&= \sigma_k^2 \text{tr}[C_k' C_k + S_k' S_k] \\
&= \sigma_k^2 \left[\frac{n}{2} + \frac{n}{2}\right] \\
&= n\sigma_k^2.
\end{aligned}
$$

Dividing by two and recalling (6.3.6) gives the result. □

Proposition 6.4.2.   If the data have a multivariate normal distribution and model (6.4.1) is correct, then for $k = 1, \ldots, \lfloor \frac{n-1}{2} \rfloor$,

$$\frac{2P\left(\frac{k}{n}\right)}{\frac{n}{2}\sigma_k^2} \sim \chi^2(2).$$

PROOF.   This follows from checking the conditions of *PA* Theorem 1.3.6. Note that

$$2P\left(\frac{k}{n}\right) \Big/ \frac{n}{2}\sigma_k^2 = Y' P_k Y \Big/ \frac{n}{2}\sigma_k^2 = Y' \left[\frac{2}{n\sigma_k^2} P_k\right] Y.$$

Because the only fixed effect is $J\alpha_0$, and because $P_k J = 0$, it suffices to show that $V \frac{2}{n\sigma_k^2} P_k V \frac{2}{n\sigma_k^2} P_k V = V \frac{2}{n\sigma_k^2} P_k V$. In fact, it suffices to show that

$$\left(\frac{2}{n\sigma_k^2}\right)^2 P_k V P_k V = \left(\frac{2}{n\sigma_k^2}\right) P_k V.$$

Note that

$$
\begin{aligned}
P_k V &= P_k \sum_j \sigma_j^2 Z_j Z_j' \\
&= \sigma_k^2 P_k [Z_k Z_k'] \\
&= \sigma_k^2 P_k \left[\frac{n}{2} P_k\right] \\
&= \frac{n}{2} \sigma_k^2 P_k.
\end{aligned}
$$

Clearly, $\left(\frac{2}{n\sigma_k^2}\right)^2 P_k V P_k V = \left(\frac{2}{n\sigma_k^2}\right) P_k V.$ □

Of course, in practice, model (6.4.1) will not be true. However, if the y_t process is a stationary Gaussian process and n is large, then by (6.3.1), model (6.4.1) is approximately correct, so $P\left(\frac{k}{n}\right) / \frac{n}{2} \sigma_k^2$ is approximately $\chi^2(2)$.

We can now derive confidence intervals for the $\frac{n}{2}\sigma_k^2$s. A $(1-\alpha)100\%$ confidence interval for $\frac{n}{2}\sigma_k^2$ is based on

$$
\begin{aligned}
1 - \alpha &= \Pr\left[\chi^2\left(\frac{\alpha}{2},2\right) \le \frac{2P\left(\frac{k}{n}\right)}{\frac{n}{2}\sigma_k^2} \le \chi^2\left(1-\frac{\alpha}{2},2\right)\right] \\
&= \Pr\left[\frac{2P\left(\frac{k}{n}\right)}{\chi^2\left(1-\frac{\alpha}{2},2\right)} \le \frac{n}{2}\sigma_k^2 \le \frac{2P\left(\frac{k}{n}\right)}{\chi\left(\frac{\alpha}{2},2\right)}\right].
\end{aligned}
$$

The confidence interval is

$$\left(\frac{2P\left(\frac{k}{n}\right)}{\chi^2\left(1-\frac{\alpha}{2},2\right)}, \frac{2P\left(\frac{k}{n}\right)}{\chi^2\left(\frac{\alpha}{2},2\right)}\right).$$

From (6.3.6), this is also an approximate confidence interval for $f\left(\frac{k}{n}\right)$. If you define $P(0.5)$, the confidence interval for $f(0.5)$ needs to be based on 1 degree of freedom rather than 2.

Note that, as n increases, the periodogram estimates the spectral density at more points, but the quality of the individual estimates does not improve. We are observing one realization of the process. Without combining information from nearby frequencies, to get improved periodogram estimates of the spectral density requires independent replication of the *series*, not additional observations on the realization at hand.

The spectral density estimate $\hat{f}_r\left(\frac{k}{n}\right)$ was defined in (6.2.7). Arguments similar to those just given establish that under model (6.4.1)

$$E\left[\hat{f}_r\left(\frac{k}{n}\right)\right] = \frac{1}{r}\sum_{i=-(r-1)/2}^{(r-1)/2}\frac{n}{2}\sigma^2_{k+i}$$

$$\doteq \frac{1}{r}\sum_{i=-(r-1)/2}^{(r-1)/2}f\left(\frac{k+i}{n}\right), \qquad (6.4.2)$$

which is the average of $f(v)$ in the r neighborhood of $\frac{k}{n}$. If $f(v)$ is continuous and n is large, all of these values should be similar so

$$E\left[\hat{f}_r\left(\frac{k}{n}\right)\right] \doteq f\left(\frac{k}{n}\right).$$

In fact, if we let r be an increasing function of n, under reasonable conditions on r and the process, it is possible to achieve consistent estimation of the spectral density from only one realization of the process.

If all the σ^2_{k+i}s are *equal* in the r neighborhood,

$$\frac{2r\hat{f}_r\left(\frac{k}{n}\right)}{\frac{n}{2}\sigma^2_k} \sim \chi^2(2r). \qquad (6.4.3)$$

Again, if $f(v)$ is continuous and n is large, the distribution should hold approximately. This yields a confidence interval for $\frac{n}{2}\sigma^2_k \doteq f(k/n)$ of

$$\left(\frac{2r\hat{f}_r\left(\frac{k}{n}\right)}{\chi^2\left(1-\frac{\alpha}{2},2r\right)}, \frac{2r\hat{f}_r\left(\frac{k}{n}\right)}{\chi^2\left(\frac{\alpha}{2},2r\right)}\right).$$

This is only really appropriate for frequencies for which $\hat{f}_r$ is the average of distinct periodogram values that each have 2 degrees of freedom. In other words, it is only appropriate for k/n with $k = 1 + (r-1)/2, \ldots, \lfloor\frac{n-1}{2}\rfloor - (r-1)/2$.

It is often convenient to have the length of the confidence intervals independent of the frequency. This can be accomplished by reporting the confidence intervals based on $\log\hat{f}_r\left(\frac{k}{n}\right)$ rather than $\hat{f}_r\left(\frac{k}{n}\right)$. The confidence interval for $\log\left(\frac{n}{2}\sigma^2_k\right)$ is

$$\left(\log\hat{f}_r\left(\frac{k}{n}\right) - \log\left[\frac{\chi^2\left(1-\frac{\alpha}{2},2r\right)}{2r}\right], \log\hat{f}_r\left(\frac{k}{n}\right) - \log\left[\frac{\chi^2\left(\frac{\alpha}{2},2r\right)}{2r}\right]\right).$$

As discussed earlier, $\hat{f}_r\left(\frac{k}{n}\right)$ is the mean square for the r neighborhood of $\frac{k}{n}$ and is a good indicator of the relative importance of frequencies near $\frac{k}{n}$. Confidence intervals allow more rigorous comparisons of the relative importance of the various frequencies.

Exercise 6.3.

(a) In model (6.4.1), $Z_0 = [J,0]$. Show that $C(VJ) \subset C(J)$.

(b) In model (6.4.1), show that for $j = 1, \ldots, \lfloor \frac{n-1}{2} \rfloor$, the REML equations amount to setting $P(j/n) = \frac{n}{2}\sigma_j^2$. When n is even, a similar result holds for $j/n = 0.5$.

(c) Show that the matrix

$$\sum_{j=1}^{\lfloor \frac{n-1}{2} \rfloor} \sigma_j^2 Z_j Z_j' + \delta_{\lfloor \frac{n}{2} \rfloor \frac{n}{2}} \sigma_{\frac{n}{2}}^2 C_{\frac{n}{2}} C_{\frac{n}{2}}'$$

is singular.

(d) In model (6.4.1), $C(Z_0) = C(J)$ with ppo $(1/n)JJ'$. Find $E[Y'(1/n)JJ'Y]$ and discuss the problems with estimating both μ and σ_0^2.

(e) Assuming that the random effects model (6.4.1) is Gaussian, prove that (6.4.2) and (6.4.3) hold.

6.5 The White Noise Model

It would be nice if we could arrive at some justification for looking at reduced models. Predictions based on saturated models are notoriously poor. Unfortunately, reduced models are not possible in model (6.4.1) because the estimates for all of the $\frac{n}{2}\sigma_k^2$s will be positive and there is no reason to conclude that any can be zero. In using model (6.4.1), we have overlooked the nearly ubiquitous fact that multiple measurements on the same unit differ. It seems reasonable to model the observations as the sum of a simple stationary process and individual uncorrelated measurement errors, i.e., white noise. (Although the measurement errors can be incorporated into (6.4.1), it is convenient to isolate them.) Dropping Z_0 from the random effects to eliminate the confounding of μ and σ_0^2, this suggests the model

$$Y = J\alpha_0 + \sum_{k-1}^{\lfloor \frac{n-1}{2} \rfloor} Z_k \gamma_k + \delta_{\lfloor \frac{n}{2} \rfloor \frac{n}{2}} C_{\frac{n}{2}} \alpha_{\frac{n}{2}} + e, \qquad (6.5.1)$$

$$\mathrm{Cov}(e) = \sigma^2 I_n, \quad \mathrm{Cov}(e, \gamma_k) = 0, \quad \mathrm{Cov}(e, \alpha_{\frac{n}{2}}) = 0,$$

$$\mathrm{Cov}(\gamma_k) = \sigma_k^2 I_2, \quad \mathrm{Var}(\alpha_{\frac{n}{2}}) = \sigma_{\frac{n}{2}}^2, \quad \mathrm{Cov}(\gamma_k, \gamma_{k'}) = 0, \; k \neq k', \quad \mathrm{Cov}(\gamma_k, \alpha_{\frac{n}{2}}) = 0.$$

The analysis of this model is similar to that of model (6.4.1). For $k = 1 + (r - 1)/2, \ldots, \lfloor \frac{n-1}{2} \rfloor - (r-1)/2$, when all variances in the r neighborhood of $\frac{k}{n}$ are equal,

$$E\left[\hat{f}_r\left(\frac{k}{n}\right)\right] = \sigma^2 + \frac{n}{2}\sigma_k^2 \qquad (6.5.2)$$

and

$$\frac{2r\hat{f}_r\left(\frac{k}{n}\right)}{\sigma^2 + \frac{n}{2}\sigma_k^2} \sim \chi^2(2r). \qquad (6.5.3)$$

This leads to confidence intervals for the values

$$\sigma^2 + \frac{n}{2}\sigma_k^2.$$

If all of the σ_k^2s are zero, the various confidence intervals are estimates of the same thing, σ^2. Confidence intervals containing distinctly larger values suggest the existence of a nonzero variance σ_k^2.

Another way of identifying important frequencies is to modify an approach used for identifying important effects in saturated models for designed experiments, cf. *PA-V* Example 12.2.4 (Christensen 2011, Example 13.2.4). Let $s = \lfloor \frac{n-1}{2} \rfloor$ and let $w_1, \ldots, w_s$ be i.i.d. $\chi^2(2r)$. Construct the order statistics $w_{(1)} \leq \cdots \leq w_{(s)}$, and compute the expected order statistics $E[w_{(1)}], \ldots, E[w_{(s)}]$. If all the σ_k^2s are zero, $\hat{f}_r\left(\frac{1}{n}\right), \hat{f}_r\left(\frac{2}{n}\right), \ldots, \hat{f}_r\left(\frac{s}{n}\right)$ are each σ^2 times a $\chi^2(2r)$ random variable, and the plot of $\left(E[w_{(k)}], \hat{f}_r\left(\frac{k}{n}\right)\right)$ should form an approximate straight line. In any case, the unimportant frequencies should form a straight line. Values $\hat{f}_r\left(\frac{k}{n}\right)$ that are so large as to be inconsistent with the straight line indicate important frequencies.

There is one obvious problem with this method. The w_ks were assumed to be independent, but the $\hat{f}_r\left(\frac{k}{n}\right)$s are only independent when $r = 1$. If $r \geq 3$ (recall that r is odd), then the neighborhoods overlap so, for example, $\hat{f}_r\left(\frac{k}{n}\right)$ and $\hat{f}_r\left(\frac{k-1}{n}\right)$ both involve $P\left(\frac{k}{n}\right)$ and $P\left(\frac{k-1}{n}\right)$, hence the $\hat{f}_r\left(\frac{k}{n}\right)$s are not independent. It is interesting to recall that normal plots are also plagued by a correlation problem when the residuals are used to check for normality. However, the order of magnitude of the correlation problem is very different in the two cases, see Example 6.5.1. Simulations provide an obvious method for dealing with such dependence.

Exercise 6.4.

(a) Assuming that model (6.5.1) is Gaussian with, for convenience, n odd, prove that (6.5.2) and (6.5.3) hold.
(b) Why is it impossible to learn what σ^2 is?
(c) How does this procedure differ from testing that all the σ_k^2s are equal?

If a group of s frequencies, say $k_1/n, \ldots, k_s/n$, have been identified as important, the random effects model becomes

$$Y = J\alpha_0 + \sum_{i=1}^{s} Z_{k_i}\gamma_{k_i} + e,$$

so best linear unbiased prediction methods can be used to obtain predictions of both current and future observations. As always, our approach here has been based on linear models. Brockwell and Davis (1991, Section 5.6) discuss a prediction method for the frequency domain that is founded on Fourier analysis.

EXAMPLE 6.5.1.   Consider again the coal production data of Example 6.2.1. Tables 6.4, 6.5, and 6.6 give 95% confidence intervals for the spectral density based on

Table 6.4 95% confidence intervals based on the periodogram

v	$P(v)$	Lower Limit	Upper Limit
0.01639344	56528.08478	15323.91760	2232740.086
0.03278689	149752.69159	40595.71299	5914915.370
0.04918033	6943.81845	1882.36524	274266.179
0.06557377	11140.48878	3020.01974	440025.803
0.08196721	15541.26244	4213.00361	613847.078
0.09836066	18097.97961	4906.09136	714832.012
0.11475410	8261.15057	2239.47426	326298.018
0.13114754	7103.20125	1925.57153	280561.463
0.14754098	1562.75966	423.64075	61725.710
0.16393443	5318.25725	1441.69993	210059.941
0.18032787	9111.07990	2469.87737	359868.433
0.19672131	722.57093	195.87816	28540.027
0.21311475	8155.94896	2210.95567	322142.777
0.22950820	370.46263	100.42687	14632.492
0.24590164	6134.12844	1662.87040	242285.132
0.26229508	731.47281	198.29133	28891.633
0.27868852	8728.98983	2366.29848	344776.682
0.29508197	3504.96303	950.14301	138438.645
0.31147541	740.04737	200.61577	29230.310
0.32786885	769.01729	208.46907	30374.560
0.34426230	704.15309	190.88536	27812.562
0.36065574	2006.14370	543.83553	79238.443
0.37704918	1007.92177	273.23250	39810.784
0.39344262	317.64999	86.11016	12546.505
0.40983607	1451.82227	393.56728	57343.917
0.42622951	404.59414	109.67941	15980.615
0.44262295	623.50094	169.02177	24626.971
0.45901639	44.88373	12.16731	1772.813
0.47540984	469.74898	127.34192	18554.094
0.49180328	262.64055	71.19792	10373.747

$P(k)$, $\hat{f}_3$, and $\hat{f}_5$ respectively. The confidence intervals confirm the impression that there are three levels to $f(v)$: an area of low frequencies with large contributions, an area of moderate frequencies with moderate contributions, and an area of high frequencies that account for little of the variability in the data.

Figures 6.7 and 6.8 contain chi-square probability plots for the values of $\hat{f}_5(v)$. The chi-square scores are $G^{-1}[i/(n+1)]$, where $G(\cdot)$ is the cumulative distribution function for the $\chi^2(10)$ distribution. Figure 6.7 contains the plot for the 30 frequencies. The four largest $\hat{f}_5$ values stand out as distinct from the rest. The frequencies associated with these $\hat{f}_5$ values are, in order, the four smallest frequencies. To further investigate the frequencies, the four that are clearly disparate were dropped and a chi-square plot for the remaining 26 frequencies was created. This plot appears as Fig. 6.8. There does not seem to be any particularly important frequencies in Fig. 6.8. (My analysis of these plots has changed considerably since the previous edition. That is partly due to using a different periodogram. This analysis is based

Table 6.5 95% confidence intervals based on $\hat{f}_3$

ν	$\hat{f}_3(\nu)$	Lower Limit	Upper Limit
0.01639344	87602.9537		
0.03278689	71074.8649	29513.3305	344648.784
0.04918033	55945.6663	23231.0387	271285.860
0.06557377	11208.5232	4654.2593	54351.196
0.08196721	14926.5769	6198.1546	72380.392
0.09836066	13966.7975	5799.6130	67726.330
0.11475410	11154.1105	4631.6648	54087.343
0.13114754	5642.3705	2342.9541	27360.391
0.14754098	4661.4061	1935.6156	22603.602
0.16393443	5330.6989	2213.5347	25849.066
0.18032787	5050.6360	2097.2406	24491.015
0.19672131	5996.5333	2490.0176	29077.760
0.21311475	3082.9942	1280.1913	14949.732
0.22950820	4886.8467	2029.2282	23696.785
0.24590164	2412.0213	1001.5746	11696.121
0.26229508	5198.1970	2158.5142	25206.552
0.27868852	4321.8086	1794.6002	20956.861
0.29508197	4324.6667	1795.7870	20970.721
0.31147541	1671.3426	694.0131	8104.499
0.32786885	737.7393	306.3410	3577.368
0.34426230	1159.7714	481.5868	5623.842
0.36065574	1239.4062	514.6546	6009.999
0.37704918	1110.5718	461.1570	5385.268
0.39344262	925.7980	384.4310	4489.283
0.40983607	724.6888	300.9219	3514.085
0.42622951	826.6391	343.2560	4008.452
0.44262295	357.6596	148.5156	1734.325
0.45901639	379.3779	157.5340	1839.639
0.47540984	259.0911	107.5857	1256.357
0.49180328	331.6767		

on the true periodogram and not the approximation used by BMDP to perform the fast Fourier transform. But I think the bigger change is that I have become more demanding of chi-square plots. I tend to discount anything except the most extreme effects.)

Figure 6.7 involves 30 $\hat{f}_5$ values, each with ten degrees of freedom. Supposedly there are 300 degrees of freedom involved in the plot but there are only 61 observations. The extra degrees of freedom are generated by the averaging involved in computing $\hat{f}_5$ so that contiguous frequencies use four of the same periodogram values. It is no accident that the four frequencies identified as important form a contiguous group. The plot is further complicated for the two largest and two smallest frequencies by their alternative definitions. And that does not address the complications that would result if one added extra observations to speed up the fast Fourier transform.

Table 6.6 95% confidence intervals based on $\hat{f}_5$

v	$\hat{f}_5(v)$	Lower Limit	Upper Limit
0.01639344	65256.1529		
0.03278689	56178.6337		
0.04918033	47981.2692	23424.7199	147772.3173
0.06557377	40295.2482	19672.3621	124100.9731
0.08196721	11996.9400	5856.9722	36948.0768
0.09836066	12028.8165	5872.5345	37046.2500
0.11475410	10113.2707	4937.3545	31146.7677
0.13114754	8068.6697	3939.1690	24849.8223
0.14754098	6271.2897	3061.6782	19314.2664
0.16393443	4763.5738	2325.6030	14670.8153
0.18032787	4974.1233	2428.3944	15319.2641
0.19672131	4735.6639	2311.9772	14584.8587
0.21311475	4898.8382	2391.6398	15087.4014
0.22950820	3222.9168	1573.4457	9925.9125
0.24590164	4824.2005	2355.2013	14857.5330
0.26229508	3894.0033	1901.0739	11992.7194
0.27868852	3967.9203	1937.1605	12220.3682
0.29508197	2894.8981	1413.3052	8915.6832
0.31147541	2889.4341	1410.6377	8898.8554
0.32786885	1544.8649	754.2116	4757.8622
0.34426230	1045.4566	510.3977	3219.7888
0.36065574	960.9772	469.1543	2959.6096
0.37704918	1097.5382	535.8242	3380.1890
0.39344262	1037.6264	506.5749	3195.6732
0.40983607	761.0978	371.5721	2344.0228
0.42622951	568.4902	277.5401	1750.8315
0.44262295	598.9100	292.3912	1844.5181
0.45901639	361.0737	176.2782	1112.0317
0.47540984	332.6829		
0.49180328	301.9326		

This correlation problem can be eliminated by plotting the periodogram $\hat{f}_1$, see Fig. 6.9. But the main benefit probably comes just from plotting the values systematically, rather than having a truly valid chi-squared plot.

In the periodogram, there are two frequencies that are clearly not white noise, $v = 0.03278689 = 2/61$ and $v = 1/61$. Figure 6.10 is a $\chi^2(2)$ plot after deleting the two periodogram values that are not considered white noise. It is definitely not straight, but there are also no definitely important frequencies. The three largest are $v = 6/61, 5/61, 4/61$. We present predictions based on all five only because it has a simple but not too simple structure and involves nonconsecutive frequencies.   □

We now proceed to predicting seven additional years of the coal production data (for which we actually have the data). First we need to estimate the variance components. We present a semiquick, semidirty analysis.

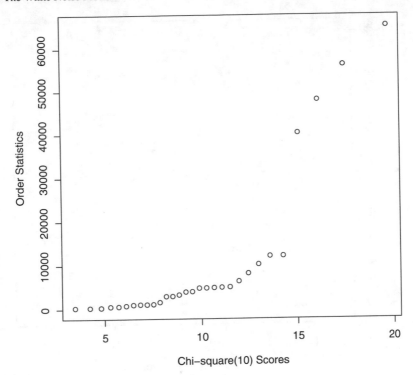

Fig. 6.7 $\chi^2(10)$ plot for 30 $\hat{f}_5$ observations

6.5.1 The Reduced Model: Estimation

If coal production is a stationary process, we probably would not have seen data like that observed from 1970 to 1980, cf. Fig. 6.1. If stationary, the data seem overdue for a downturn. During the entire time period, coal production should be tied to general economic level, since the period probably predates serious efforts to reduce coal consumption. But only the last 10 years of the data seem like clear evidence of nonstationarity. Our model comes to the same conclusion and predicts a downturn. Moreover, the actual data from 1981 to 1987 are also unwilling to cooperate with our model. Analyses in previous versions of this book were based on the BMDP periodogram and reflected adjustments for how that was computed. The current analysis is based on the definitional periodogram.

If there were an overall trend, we could have removed it using regression and performed frequency analysis on the residuals (probably ignoring the complications involved with the residuals not being the actual errors, the errors being what might reasonably be assumed stationary). Incidentally, in frequency analysis, trends are often picked up as very low frequency effects. Or perhaps it is just too much to hope that an economy affected by depression, world war, and a microelectronics

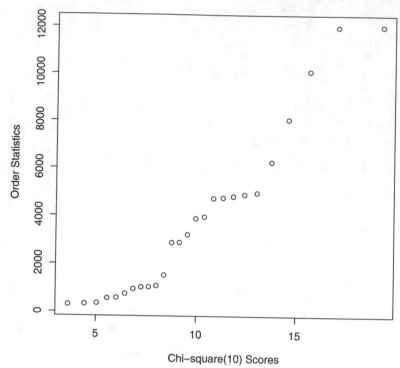

Fig. 6.8 $\chi^2(10)$ plot for 26 smallest $\hat{f}_5$ observations

revolution could provide stationary data. In any case, our analysis won't really be quick and it certainly won't be clean.

We begin with a reduced version of the white noise random effects model (6.5.1),

$$Y = J\mu + Z_1\gamma_1 + Z_2\gamma_2 + Z_4\gamma_4 + Z_5\gamma_5 + Z_6\gamma_6 + e, \tag{6.5.4}$$

with

$$\mathrm{E}(e) = 0, \qquad \mathrm{Cov}(e) = \sigma^2 I,$$

where the rows of Z_k are

$$z'_{kt} = \left(\cos\left(2\pi\frac{k}{61}t \right), \sin\left(2\pi\frac{k}{61}t \right) \right),$$

$t = 1,\ldots,61$. We also write $Z \equiv [Z_1, Z_2, Z_4, Z_5, Z_6]$ and $\gamma' \equiv [\gamma_1', \gamma_2', \gamma_4', \gamma_5', \gamma_6']$ so as to rewrite the model as a standard mixed model

$$Y = J\mu + Z\gamma + e.$$

Getting even dirtier, for illustrative purposes I decided to incorporate the following assumptions about the variance components:

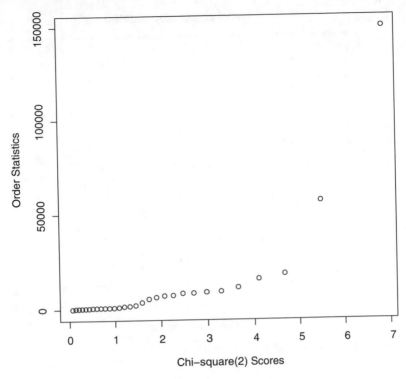

Fig. 6.9 $\chi^2(2)$ plot for 30 periodogram observations

$$\text{Cov}(\gamma_1) = \text{Cov}(\gamma_2) = \sigma_a^2 I$$

and

$$\text{Cov}(\gamma_4) = \text{Cov}(\gamma_5) = \text{Cov}(\gamma_6) = \sigma_b^2 I.$$

The disadvantage of these assumptions is that they may not be true. The advantage is that, if they approximate the truth, we should get better estimates of the variance components. (This is not dissimilar to averaging periodograms to define spectral density estimates.)

The entire mixed model can be fitted from the periodogram values and the mean of the observations, $\bar{y}. = 511.8131$. In particular, the white noise variance estimate is the usual *MSE* for the model, which in this case is

$$\hat{\sigma}^2 = \left[P(3/61) + \sum_{j=7}^{30} P(j/61) \right] \bigg/ [61 - 11]$$

$$= Y'(I - P)Y \big/ [61 - 11] = 150902/50 = 3018,$$

wherein $P \equiv (1/n)JJ' + P_1 + P_2 + P_4 + P_5 + P_6$.

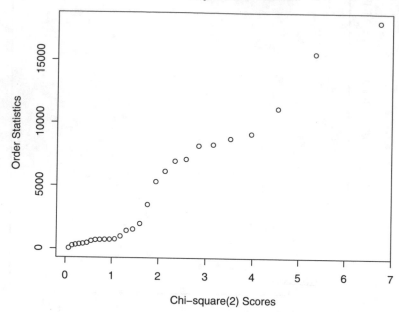

Fig. 6.10 $\chi^2(2)$ plot for 28 periodogram observations

The periodogram average $[P(1/61)+P(2/61)]/2$ provides an estimate of $\sigma^2 + \frac{61}{2}\sigma_a^2$, i.e.,

$$\hat{\sigma}^2 + \frac{61}{2}\hat{\sigma}_a^2 = [P(1/61)+P(2/61)]/2$$
$$= (56,528.08478 + 149,752.69159)/2 = 103,140.388185.$$

It follows that

$$\hat{\sigma}_a^2 = \left([P(1/61)+P(2/61)]/2 - \hat{\sigma}^2\right)\left(\frac{2}{61}\right)$$
$$= (103,140.38 - 3018)\left(\frac{2}{61}\right) = 3282.7.$$

The average of the periodogram values for $k = 4,5,6$ provides an estimate of $\sigma^2 + \frac{61}{2}\sigma_b^2$,

$$\hat{\sigma}^2 + \frac{61}{2}\hat{\sigma}_b^2 = \frac{P(4/61)+P(5/61)+P(6/61)}{3}$$
$$= (11,140.48878 + 15,541.26244 + 18,097.97961)/3 = 14,926.6.$$

It follows that an estimate of σ_b^2 is

$$\hat{\sigma}_b^2 = \left([P(4/61) + P(5/61) + P(6/61)]/3 - \hat{\sigma}^2\right)\left(\frac{2}{61}\right)$$

$$= (14{,}926.6 - 3018)\left(\frac{2}{61}\right) = 390.4.$$

These are a direct applications of Henderson's method 3. Generally, Henderson's method does not provide unique estimates of variance components; they depend on the order in which the components are estimated. However, due to the orthogonality of the columns involved, the estimates based on the periodogram are unique.

6.5.2 The Reduced Model: Prediction

We are now in a position to estimate the best linear unbiased predictors. We have to estimate the predictors because we do not know the variance components. Prediction is performed as in Sect. 5.1, see also Sect. 4.1.3. It is easily seen that the best linear unbiased predictor (BLUP) for an observation at time t is

$$\hat{y}_t = \hat{\mu} + \tilde{z}_t'\hat{\gamma},$$

where

$$\hat{\mu} = \bar{y}. = 511.8,$$

$$\tilde{z}_t = \begin{bmatrix} \cos\left(2\pi(\frac{1}{61})t\right) \\ \sin\left(2\pi(\frac{1}{61})t\right) \\ \cos\left(2\pi(\frac{2}{61})t\right) \\ \sin\left(2\pi(\frac{2}{61})t\right) \\ \cos\left(2\pi(\frac{4}{61})t\right) \\ \sin\left(2\pi(\frac{4}{61})t\right) \\ \cos\left(2\pi(\frac{5}{61})t\right) \\ \sin\left(2\pi(\frac{5}{61})t\right) \\ \cos\left(2\pi(\frac{6}{61})t\right) \\ \sin\left(2\pi(\frac{6}{61})t\right) \end{bmatrix}, \qquad \gamma = \begin{bmatrix} \gamma_1 \\ \gamma_2 \\ \gamma_4 \\ \gamma_5 \\ \gamma_6 \end{bmatrix},$$

and $\hat{\gamma}$ is the BLUP of γ.

Let

$$\tilde{Y} \equiv \begin{bmatrix} y_{62} \\ \vdots \\ y_{68} \end{bmatrix}, \qquad \tilde{Z} \equiv \begin{bmatrix} z_{1,62}', & z_{2,62}', & z_{4,62}', & z_{5,62}', & z_{6,62}' \\ \vdots & \vdots & \vdots & \vdots & \vdots \\ z_{1,68}', & z_{2,68}', & z_{4,68}', & z_{5,68}', & z_{6,68}' \end{bmatrix} \equiv [\tilde{Z}_1, \tilde{Z}_2, \tilde{Z}_4, \tilde{Z}_5, \tilde{Z}_6].$$

We want to predict a $\tilde{Y}$ that follows the model

$$\tilde{Y} = J_7\mu + \tilde{Z}\gamma + \tilde{e},$$

where $\tilde{e}$ is a white noise vector uncorrelated with e and γ. In other words, we want to estimate the BLP

$$\hat{E}(\tilde{Y}|Y) = J_{12}\mu + \mathrm{Cov}(\tilde{Y},Y)[\mathrm{Cov}(Y)]^{-1}(Y - J_{61}\mu).$$

From model (4) we have

$$
\begin{aligned}
V &\equiv \mathrm{Cov}(Y) = \mathrm{Cov}(e) \\
&= \sigma^2 I + \sigma_a^2 Z_1 Z_1' + \sigma_a^2 Z_2 Z_2' + \sigma_b^2 Z_4 Z_4' + \sigma_b^2 Z_5 Z_5' + \sigma_b^2 Z_6 Z_6' \\
&= \sigma^2 I + \sigma_a^2 (Z_1 Z_1' + Z_2 Z_2') + \sigma_b^2 (Z_4 Z_4' + Z_5 Z_5' + Z_6 Z_6') \\
&= \sigma^2 I + \sigma_a^2 \frac{n}{2}(P_1 + P_2) + \sigma_b^2 \frac{n}{2}(P_4 + P_5 + P_6) \\
&= \sigma^2 (I - \tilde{P}) + \left(\sigma^2 + \frac{n}{2}\sigma_a^2\right)(P_1 + P_2) + \left(\sigma^2 + \frac{n}{2}\sigma_b^2\right)(P_4 + P_5 + P_6),
\end{aligned}
$$

where $\tilde{P} \equiv P_1 + P_2 + P_4 + P_5 + P_6$. We need to estimate

$$V^{-1} = \frac{1}{\sigma^2}(I - \tilde{P}) + \frac{1}{\left(\sigma^2 + \frac{n}{2}\sigma_a^2\right)}(P_1 + P_2) + \frac{1}{\left(\sigma^2 + \frac{n}{2}\sigma_b^2\right)}(P_4 + P_5 + P_6).$$

We also need

$$
\begin{aligned}
\mathrm{Cov}(\tilde{Y},Y) &= \mathrm{Cov}(\tilde{Z}\gamma + \tilde{e}, Z\gamma + e) \\
&= \mathrm{Cov}(\tilde{Z}\gamma, Z\gamma) = \tilde{Z}\mathrm{Cov}(\gamma)Z' \\
&= \tilde{Z}\begin{bmatrix} \sigma_a^2 I_4 & 0 \\ 0 & \sigma_b^2 I_6 \end{bmatrix} Z' \\
&= \sigma_a^2(\tilde{Z}_1 Z_1' + \tilde{Z}_2 Z_2') + \sigma_b^2(\tilde{Z}_4 Z_4' + \tilde{Z}_5 Z_5' + \tilde{Z}_6 Z_6').
\end{aligned}
$$

This leads to

$$\mathrm{Cov}(\tilde{Y},Y)V^{-1} = \frac{\sigma_a^2}{\left(\sigma^2 + \frac{n}{2}\sigma_a^2\right)}(\tilde{Z}_1 Z_1' + \tilde{Z}_2 Z_2') + \frac{\sigma_b^2}{\left(\sigma^2 + \frac{n}{2}\sigma_b^2\right)}(\tilde{Z}_4 Z_4' + \tilde{Z}_5 Z_5' + \tilde{Z}_6 Z_6')$$

and the BLUP of $\tilde{Y}$ is $\bar{y}.J_7$ plus

$$
\begin{aligned}
&\mathrm{Cov}(\tilde{Y},Y)V^{-1}(Y - J_{61}\bar{y}.) \\
&= \frac{\sigma_a^2}{\left(\sigma^2 + \frac{n}{2}\sigma_a^2\right)}(\tilde{Z}_1 Z_1' Y + \tilde{Z}_2 Z_2' Y) + \frac{\sigma_b^2}{\left(\sigma^2 + \frac{n}{2}\sigma_b^2\right)}(\tilde{Z}_4 Z_4' + \tilde{Z}_5 Z_5' + \tilde{Z}_6 Z_6').
\end{aligned}
$$

$$= [\tilde{Z}_1, \tilde{Z}_2, \tilde{Z}_4, \tilde{Z}_5, \tilde{Z}_6] \begin{bmatrix} \frac{\sigma_a^2}{(\sigma^2 + \frac{n}{2}\sigma_a^2)} Z_1' Y \\ \frac{\sigma_a^2}{(\sigma^2 + \frac{n}{2}\sigma_a^2)} Z_2' Y \\ \frac{\sigma_b^2}{(\sigma^2 + \frac{n}{2}\sigma_b^2)} Z_4' Y \\ \frac{\sigma_b^2}{(\sigma^2 + \frac{n}{2}\sigma_b^2)} Z_5' Y \\ \frac{\sigma_b^2}{(\sigma^2 + \frac{n}{2}\sigma_b^2)} Z_6' Y \end{bmatrix}$$

$$= \tilde{Z}\hat{\gamma}$$

where $\hat{\gamma}$ is the BLUP (and the BLP) of γ. Note that if $\sigma^2 = 0$, $\hat{\gamma}$ becomes the least squares estimate of γ, so in general the BLUP is a shrinkage estimate. Of course we have to estimate the variance components as indicated earlier to obtain an empirical BLUP $\tilde{\gamma}$.

Figure 6.11 gives all the data from 1920 to 1987 along with the mixed model predicted values and least squares predicted values from just doing a regression on the 10 sine and cosine terms. As suggested earlier, the models, which were fitted only to the data from 1920 to 1980, turn down after 1980. In point of fact, coal production continued to increase during the 1980s, so the model and its predictions are not very

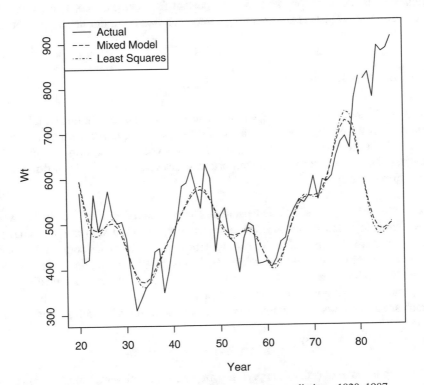

Fig. 6.11 Coal production with mixed model and least squares predictions: 1920–1987

good after 1980. Perhaps the data are simply not regular enough to allow predictions based on only 61 observations, or perhaps the process generating the data changed around 1960. The behavior of the plotted data certainly looks different in the last years than it did up to 1975. Another possibility is that, given the discussion of Example 6.3.1, the very act of prediction may involve taking the random effects model more seriously than is appropriate.

For a time domain analysis of these data, see Example 7.6.1.

6.5.3 Summary of Sects. 6.2, 6.3, 6.4, and 6.5

We began Sect. 6.2 by showing that the least squares fit of a model based on sines and cosines yields useful information about the frequencies that are important in describing the cyclical behavior of a time series. We then explained that any second-order stationary process can be approximated by a random effects model based on sines and cosines. The variances of the random effects were related to the spectral distribution function F, which is implicitly defined by the covariance function $\sigma(k)$. Estimates of $\frac{n}{2}$ times the variance components were obtained and related to the spectral density function $f(v) \equiv \mathbf{d}_v F(v)$. Confidence intervals and a χ^2 plot were suggested as methods for identifying important frequencies. Best linear unbiased prediction was suggested for obtaining forecasts based on important frequencies.

6.6 Linear Filtering

Many traditional ways of dealing with time series consist of computing simple linear functions of the observations to create a new series that is, in some sense, representative of the original series but is also better behaved than the original series. By "better behaved" we mean that the important structure of the series is clarified in the transformed series.

EXAMPLE 6.6.1.   A series that oscillates very quickly and erratically has very high frequencies that contribute substantially to the spectral approximation. (High frequencies are those near 1/2.) To examine the structure of such a series, it might be wise to try to eliminate the high frequencies while retaining the relative importance of the low and moderate-size frequencies. A traditional method of attenuating high frequencies is taking *moving averages*. For instance, a running average of order 5 (also called a centered 5-term moving average or applying a Daniell kernel of order 2) is

$$w_t^{(5)} = \frac{1}{5}[y_{t-2} + y_{t-1} + y_t + y_{t+1} + y_{t+2}].$$

A running average of order 6 (applying a modified Daniell kernel of order 3) is

$$w_t^{(6)} = \frac{1}{6}\left[\frac{1}{2}y_{t-3} + y_{t-2} + y_{t-1} + y_t + y_{t+1} + y_{t+2} + \frac{1}{2}y_{t+3}\right].$$ □

EXAMPLE 6.6.2.   Consider a nonstationary process

$$y_t = \gamma_0 + \gamma_1 t + y_t^{(1)},$$

where γ_0 and γ_1 are fixed and $y_t^{(1)}$ is second-order stationary. This is a process with a linear time trend. The difference series

$$w_t \equiv y_t - y_{t-1}$$

is second-order stationary. To see this, observe that

$$\begin{aligned}
E(w_t) &= E(y_i - y_{t-1}) \\
&= \gamma_0 + \gamma_1 t + E(y_t^{(1)}) \\
&\quad - \gamma_0 - \gamma_1(t-1) - E(y_{t-1}^{(1)}) \\
&= \gamma_1
\end{aligned}$$

and

$$\begin{aligned}
\mathrm{Cov}(w_t, w_{t+k}) &= \mathrm{Cov}(y_t^{(1)} - y_{t-1}^{(1)}, y_{t+k}^{(1)} - y_{t+k-1}^{(1)}) \\
&= \mathrm{Cov}(y_t^{(1)}, y_{t+k}^{(1)}) - \mathrm{Cov}(y_t^{(1)}, y_{t+k-1}^{(1)}) \\
&\quad - \mathrm{Cov}(y_{t-1}^{(1)}, y_{t+k}^{(1)}) + \mathrm{Cov}(y_{t-1}^{(1)}, y_{t+k-1}^{(1)}) \\
&= 2\sigma^{(1)}(k) - \sigma^{(1)}(k-1) - \sigma^{(1)}(k+1),
\end{aligned}$$

where $\sigma^{(1)}(\cdot)$ is the covariance function for $y_t^{(1)}$. Note that the covariance depends only on k. Often, the *first difference operator* is denoted ∇, so that w_t is

$$\nabla y_t \equiv y_t - y_{t-1}.$$

   The idea that trends can be thought of as low-frequency cyclical effects was mentioned earlier. In this example, we can think of the frequency generated by $\gamma_0 + \gamma_1 t$ as being so low that the oscillation will never be observed. The first difference process eliminates low frequencies from the spectral approximation.   □

Exercise 6.5.   Consider the process $y_t = \gamma_0 + \gamma_1 t + \gamma_2 t^2 + y_t^{(1)}$, where $y_t^{(1)}$ is second-order stationary. Let $w_t = \nabla y_t$ and $z_t = \nabla w_t$. In other words,

$$\begin{aligned}
z_t &= \nabla(\nabla y_t) \\
&= \nabla^2 y_t.
\end{aligned}$$

Show that z_t is second-order stationary.

These examples are special cases of the *general linear filter*

$$w_t = \sum_{s=-\infty}^{\infty} a_s y_{t-s}.$$ (6.6.1)

The special case

$$w_t = \sum_{s=0}^{\infty} a_s y_{t-s}$$

in which w_t depends only on the current and previous values of y_t will be referred to as a *causal* linear filter. The process y_t is viewed as causing w_t.

EXAMPLE 6.6.1 CONTINUED.   The process $w_t^{(5)}$ has $a_{-2} = a_{-1} = a_0 = a_1 = a_2 = \frac{1}{5}$ and $a_s = 0$ for all other s. For $w_t^{(6)}$, $a_{-3} = a_3 = \frac{1}{12}$, $a_{-2} = a_{-1} = a_0 = a_1 = a_2 = \frac{1}{6}$, and $a_s = 0$ for all others. Neither filter is causal.   □

EXAMPLE 6.6.2 CONTINUED.   The process ∇y_t has $a_0 = 1$, $a_1 = -1$, and $a_s = 0$ for all other s. The filter is causal.   □

Collectively, the series $\ldots, a_{-1}, a_0, a_1, \ldots$ is called the *impulse response function*; it is a function from the integers to the reals. If $\sum_{k=-\infty}^{\infty} |a_k| < \infty$, we can draw an analogy between a_k and $\sigma(k)$, thus defining a function similar to the spectral density as given in (6.3.7), say

$$A(v) = \sum_{k=-\infty}^{\infty} a_k e^{-2\pi i v k}.$$ (6.6.2)

(Because we need not have $a_k = a_{-k}$, we cannot, in general, reduce $A(v)$ to a sum involving only cosines.) The complex valued function $A(v)$ is called the *frequency response* function. As will be seen later, $A(v)$ identifies the effect that the filter has on different frequencies in the spectral approximation.

The behavior of the spectral approximation (6.3.1) is determined by the spectral density of y_t, say $f_y(v)$. Similarly, the process w_t determined by (6.6.1) has a spectral density $f_w(v)$. We wish to show that

$$f_w(v) = |A(v)|^2 f_y(v),$$ (6.6.3)

where $|A(v)|^2 = A(v)\overline{A(v)}$ and $\overline{A(v)}$ is the *complex conjugate* of $A(v)$. (The conjugate of the complex number $a + ib$ with a and b real is $\overline{a+ib} \equiv a - ib$.) If, for example,

$$|A(v)|^2 = \begin{cases} 1 & v \in [-0.4, 0.4] \\ 0 & \text{otherwise} \end{cases},$$

then in the spectral approximation to w_t, the variance components σ_k^2 corresponding to frequencies $\frac{k}{n} \in [-0.4, 0.4]$ are identical to the corresponding variance components in the spectral approximation to y_t. At the same time, the spectral approximation to w_t has $\sigma_k^2 = 0$ for $|\frac{k}{n}| > 0.4$ regardless of the size of the corresponding variance components in the approximation to y_t. A linear filter with $A(v)$ as given earlier perfectly eliminates high frequencies (greater than 0.4) while it leaves the contributions of the lower frequencies unchanged. In fact, a function

$$|A(v)|^2 = \begin{cases} 7 & v \in [-0.4, 0.4] \\ 0 & \text{otherwise} \end{cases}$$

would be equally effective because it retains the same relative contributions of the frequencies below 0.4.

A filter that eliminates high frequencies but does little to low frequencies is called a *low-pass* filter. Symmetric moving averages such as those defined in Example 6.6.1 are low-pass filters; see Exercise 6.9.7d. Filters that eliminate low frequencies but have little effect on high frequencies are called *high-pass* filters. The first difference filter of Example 6.6.2 is such a filter; see Exercise 6.9.7a.

It is by no means clear that an impulse response function (i.e., a sequence $\ldots, a_{-1}, a_0, a_1, \ldots$) exists that generates either of the functions $|A(v)|^2$ given earlier but the examples illustrate the potential usefulness of the frequency response function in interpreting the result of applying a linear filter to a process.

Of course, to discuss $f_w(v)$ presupposes that w_t is second-order stationary. If $\sum_{s=-\infty}^{\infty} |a_s| < \infty$,

$$E(w_t) = \sum_{s=-\infty}^{\infty} a_s E(y_{t-s})$$

$$= \sum_{s=-\infty}^{\infty} a_s \mu$$

$$= \mu \sum_{s=-\infty}^{\infty} a_s,$$

which is a constant. Covariances for the w_t process depend only on k because

$$\text{Cov}(w_t, w_{t+k}) = \text{Cov}\left(\sum_s a_s y_{t-s}, \sum_{s'} a_{s'} y_{t+k-s'} \right)$$

$$= \sum_s \sum_{s'} a_s a_{s'} \text{Cov}(y_{t-s}, y_{t+k-s'})$$

$$= \sum_{s=-\infty}^{\infty} \sum_{s'=-\infty}^{\infty} a_s a_{s'} \sigma_y(k - s' + s)$$

$$\equiv \sigma_w(k). \tag{6.6.4}$$

Thus, w_t is second-order stationary.

To establish (6.6.3), we use (6.6.4) and the representation of $\sigma_y(k)$ from (6.3.5).

$$\sigma_w(k) = \sum_s \sum_{s'} a_s a_{s'} \sigma_y(k - s + s')$$

$$= \sum_s \sum_{s'} a_s a_{s'} \int_{-1/2}^{1/2} f_y(v) e^{2\pi i v(k-s+s')} dv$$

$$= \int_{-1/2}^{1/2} f_y(v) e^{2\pi i v k} \left[\sum_s \sum_{s'} a_s a_{s'} e^{2\pi i v(s'-s)} \right] dv$$

$$= \int_{-1/2}^{1/2} f_y(v) e^{2\pi i v k} \sum_s a_s e^{-2\pi i v s} \sum_{s'} a_{s'} e^{2\pi i v s'} dv \qquad (6.6.5)$$

$$= \int_{-1/2}^{1/2} f_y(v) e^{2\pi i v k} A(v) \overline{A(v)} dv$$

$$= \int_{-1/2}^{1/2} |A(v)|^2 f_y(v) e^{2\pi i v k} dv.$$

By the spectral representation theorem, the spectral distribution function and hence the spectral density are unique. Thus, $f_w(v)$ is the unique function with

$$\sigma_w(k) = \int_{-1/2}^{1/2} f_w(v) e^{2\pi i v k} dv.$$

By uniqueness and (6.6.5),

$$f_w(v) = |A(v)|^2 f_y(v).$$

EXAMPLE 6.6.3.   In this example, we simply present the results of applying low-pass and high-pass filters to the coal production data of Example 6.0.1. The top of Fig. 6.12 displays the results of a low-pass filter. The behavior of the filtered process is extremely regular. The bottom of Fig. 6.12 shows the results of a high-pass filter. The early filtered observations do not show much pattern. The behavior after observation 40 seems a bit more regular.   □

6.6.1 Recursive Filters

A *general recursive filter* defines a current value w_t using previous values of the w process along with a linear filter in the current and previous values of the y_t process. A general recursive filter is written

$$w_t = \sum_{s=1}^{p} a_s w_{t-s} + y_t - \sum_{s=1}^{q} b_s y_{t-s}.$$

If the w process is stationary, we can determine the frequency properties of the recursive filter. Let

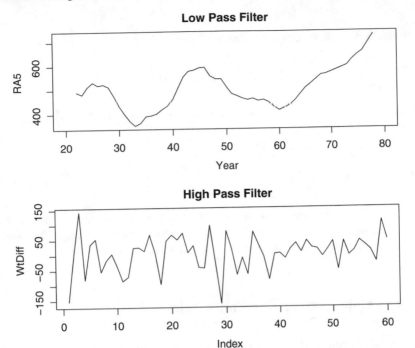

Fig. 6.12 Coal production data: Low pass filter is a running average of 5. High pass filter is the lag 1 difference

$$w_t - \sum_{s=1}^{b} a_s w_{t-s} = z_t = y_t - \sum_{s=1}^{q} b_s y_{t-s}.$$

The process z_t is the result of a linear filter in the w process, so

$$f_z(v) = |A(v)|^2 f_w(v).$$

It is also a linear filter in the y process so

$$f_z(v) = |B(v)|^2 f_y(v).$$

Clearly,

$$f_w(v) = \frac{|B(v)|^2}{|A(v)|^2} f_y(v),$$

thus $|B(v)|^2/|A(v)|^2$ is the frequency response function for the recursive filter.

A *simple recursive filter*, also called an *autoregressive filter*, is the special case

$$w_t = \sum_{s=1}^{p} a_s w_{t-s} + y_t.$$

Exercise 6.6. Show that for a simple recursive filter

$$f_w(v) = \frac{1}{|A(v)|^2} f_y(v)$$

and thus that the frequency response function of the w process is $1/|A(v)|^2$.

6.6.2 Summary

Properties of linear filters are of interest because they are commonly used in traditional data analysis to clarify the significant aspects of the time series. *Time domain analysis consists of the application of recursive filters to white noise.*

6.7 The Coherence of Two Time Series

Suppose we have two time series $y_1 = (y_{11}, y_{12}, \ldots)'$ and $y_2 = (y_{21}, y_{22}, \ldots)'$ with observations $Y_1 = (y_{11}, \ldots, y_{1n})'$ and $Y_2 = (y_{21}, \ldots, y_{2n})$. We wish to measure the correlation between y_1 and y_2 relative to the frequency k/n. One way to do that is to look at the sample partial correlation between Y_1 and Y_2 after eliminating all of the frequencies other than k/n; see Appendix B and Example 9.3.1. Because the vectors $C_0, \ldots, C_{\lfloor \frac{n}{2} \rfloor}, S_1, \ldots, S_{\lfloor \frac{n-1}{2} \rfloor}$ form an orthogonal basis for $\mathbf{R}^n$, eliminating all of the frequencies other than k/n amounts to projecting onto the column space $C(C_k, S_k)$. In particular, the squared sample partial correlation is

$$\frac{Y_1'[P_k Y_2 (Y_2' P_k Y_2)^{-1} Y_2' P_k] Y_1}{Y_1' P_k Y_1},$$

or equivalently

$$(Y_1' P_k Y_2)^2 \Big/ (Y_1' P_k Y_1)(Y_2' P_k Y_2).$$

Recall that $Y_1' P_k Y_1 = 2P_1(k/n)$, twice the periodogram for Y_1 evaluated at k/n, and $Y_2' P_k Y_2 = 2P_2(k/n)$. The other term, $Y_1' P_k Y_2/2$ also has strong connections too the periodogram.

As in Sect. 6.2, it may be desired to pool results over an r neighborhood of the frequency k/n. Let

$$M_k = \sum_{\ell=-(r-1)/2}^{(r-1)/2} P_{k+\ell}.$$

The sample partial correlation between Y_1 and Y_2 eliminating all frequencies except those in the r neighborhood of k/n is called the *squared sample real coherence function*. It is

$$\hat{\gamma}_{12}^{2(R)}(k/n) \equiv \frac{Y_1'[M_kY_2(Y_2'M_kY_2)^{-1}Y_2'M_k]Y_1}{Y_1'M_kY_1}$$

$$= (Y_1'M_kY_2)^2 \Big/ (Y_1'M_kY_1)(Y_2'M_kY_2).$$

The superscript (R) in $\hat{\gamma}_{12}^{2(R)}$ stands for "real" and will be explained in Sect. 6.8.

Consider the geometric interpretation of $\hat{\gamma}_{12}^{(R)}(k/n)$. M_k is the perpendicular projection operator onto the space spanned by $\{C_{k+\ell}, S_{k+\ell} : \ell = -(r-1)/2, \ldots, (r-1)/2\}$. This space has $2r$ degrees of freedom. The perpendicular projection operator $[M_kY_2(Y_2'M_kY_2)^{-1}Y_2'M_k]$ projects onto the one-dimensional subspace $C(M_kY_2) \subset C(M_k)$. M_kY_2 is the projection of Y_2 into the space associated with the r neighborhood of k/n. The value $\hat{\gamma}_{12}^{(R)}(k/n)$ is the squared length of the projection of Y_1 into $C(M_kY_2)$ divided by the squared length of Y_1 projected into $C(M_k)$. $Y_1'M_kY_1$ is the fraction of $Y_1'Y_1$ that is associated with the frequency k/n. $\hat{\gamma}_{12}^{2(R)}(k/n)$ is the proportion of the squared length of the vector Y_1 that can be attributed to the association between Y_1 and Y_2 (actually Y_1 and M_kY_2). In particular, if $Y_1 = aY_2$ for some scalar a, it is easily seen that $\hat{\gamma}_{12}^{(R)}(k/n) = 1$ for any k.

It is interesting to consider the special case $Y_1 = C_{k+\ell}$ and $Y_2 = C_{k+j}$ for $j, \ell = -(r-1)/2, \ldots, (r-\ell)/2$. If $j = \ell$, as indicated earlier, $\hat{\gamma}_{12}^{2(R)}(k/n) = 1$, but if $j \neq \ell$, $\hat{\gamma}_{12}^{2(R)}(k/n) = 0$. The same results hold for $Y_1 = S_{k+\ell}$ and $Y_2 = S_{k+j}$. If the two series have the same frequency and that frequency is in the r neighborhood, $\hat{\gamma}_{12}^{2(R)}(k/n) = 1$. However, if there are two different frequencies (even though both are in the r neighborhood), $\hat{\gamma}_{12}^{2(R)}(k/n)$ shows no relationship.

There is one problem with $\hat{\gamma}_{12}^{2(R)}(k/n)$ as a measure of the correlation between y_1 and y_2 relative to the frequency k/n. Suppose $Y_1 = C_k$ and $Y_2 = S_k$. Then, both series are completely determined by the frequency k/n. More to the point, the relationship between these series is completely determined by k/n. However, it is easily seen that $\hat{\gamma}_{12}^{2(R)}(k/n) = 0$. Simply observe that $M_kY_2 = S_k$ and thus $Y_1'M_kY_2 = C_k'S_k = 0$. Obviously, we need another measure that can pick up relationships that are in the same frequency but orthogonal to one another. We begin by considering just the frequency k/n. The discussion is then extended to r neighborhoods.

Within $C(C_k, S_k)$, we began by looking at

$$Y_1'[P_kY_2(Y_2'P_kY_2)^{-1}Y_2'P_k]Y_1 \Big/ Y_1'P_kY_1.$$

To detect an orthogonal relationship between Y_1 and Y_2 within $C(P_k)$, we should rotate P_kY_2 by $90°$. This rotation is well-defined because $C(P_k)$ is a two-dimensional space.

Let $c_k = C_k/\sqrt{n/2}$ and let $s_k = S_k/\sqrt{n/2}$. Thus, $c_k'c_k = 1$, $s_k's_k = 1$, and

$$P_k = c_kc_k' + s_ks_k'.$$

A $90°$ rotation of P_kY_2 is F_kY_2, where

$$F_k = (s_k c_k' - c_k s_k').$$

To see this we must establish that $(P_kY_2)'(F_kY_2) = 0$ and $(P_kY_2)'(P_kY_2) = (F_kY_2)'(F_kY_2)$. To see the first of these, note that

$$(P_kY_2)'(F_kY_2) = Y_2' P_k F_k Y_2 = Y_2' F_k Y_2$$

and that for any vector v

$$v' F_k v = (v' s_k)(c_k' v) - (v' c_k)(s_k' v) = 0.$$

To see that $(P_kY_2)'(P_kY_2) = (F_kY_2)'(F_kY_2)$, observe that

$$
\begin{aligned}
F_k' F_k &= (c_k s_k' - s_k c_k')(s_k c_k' - c_k s_k') \\
&= c_k s_k' s_k c_k' - s_k c_k' s_k c_k' - c_k s_k' c_k s_k + s_k c_k' c_k s_k' \\
&= c_k c_k' - 0 - 0 + s_k s_k' \\
&= P_k.
\end{aligned}
$$

We can now measure the orthogonal relationship between Y_1 and Y_2 in the space $C(C_k, S_k)$ by projecting Y_1 into $C(F_kY_2)$ and comparing the squared length to $Y_1' P_k Y_1$, namely,

$$\frac{Y_1'[F_kY_2(Y_2' F_k' F_k Y_2)^{-1} Y_2' F_k]Y_1}{Y_1' P_k Y_1}.$$

Because $F_k' F_k = P_k$, this can be written as

$$(Y_1' F_k Y_2)^2 \big/ (Y_1' P_k Y_1)(Y_2' P_k Y_2).$$

We have projected Y_1 onto the space $C(P_kY_2) \subset C(P_k)$ and onto $C(F_kY_2) \subset C(P_k)$, where $C(P_kY_2) \perp C(F_kY_2)$. Because $r(P_k) = 2$,

$$Y_1' P_k Y_1 = Y_1'[P_kY_2(Y_2' P_k Y_2)^{-1} Y_2' P_k]Y_1 + Y_1'[F_kY_2(Y_2' P_k Y_2)^{-1} Y_2' F_k]Y_1.$$

Thus, all of the variability of P_kY_1 can be accounted for in one of the two directions. In particular, the two correlation measures add up to 1, that is,

$$\frac{(Y_1' P_k Y_2)^2}{(Y_1' P_k Y_1)(Y_2' P_k Y_2)} + \frac{(Y_1' F_k Y_2)^2}{(Y_1' P_k Y_1)(Y_2' P_k Y_2)} = 1.$$

In a two-dimensional space, the partial correlation and the orthogonal partial correlation must add up to one.

The situation is not quite so degenerate when dealing with r neighborhoods. Let

$$G_k = \sum_{\ell=-(r-1)/2}^{(r-1)/2} F_{k+\ell} \, .$$

The partial correlation involves projecting Y_1 into $C(M_k Y_2)$. The orthogonal partial correlation is obtained by projecting Y_1 into $C(G_k Y_2)$. It is easily seen that $F'_{k+\ell} F_{k+j} = 0$ for any $\ell \neq j$. It follows that

$$G'_k G_k = \sum_{\ell} \sum_{j} F'_{k+\ell} F_{k+j} = \sum_{\ell} F'_{k+\ell} F_{k+\ell}$$
$$= \sum_{\ell} P_k = M_k \, .$$

Also, for any vector v,

$$v' G_k v = \sum_{\ell} v' F_{k+\ell} v = 0 \, .$$

From these facts, it is easily established that $(M_k Y_2)'(G_k Y_2) = 0$ and $(M_k Y_2)'(M_k Y_2) = (G_k Y_2)'(G_k Y_2)$. Thus, $G_k Y_2$ is an orthogonal rotation of $M_k Y_2$. In particular, $G_k Y_2$ is the sum of the orthogonal rotations in the spaces $C(P_{k+\ell})$.

As a measure of the orthogonal correlation between Y_1 and Y_2 in the frequency k/n, use the *squared sample imaginary coherence function*

$$\hat{\gamma}_{12}^{2(I)}(k/n) \equiv \frac{Y'_1 [G_k Y_2 (Y'_2 M_k Y_2)^{-1} Y'_2 G_k] Y_1}{Y'_1 M_k Y_1}$$
$$= (Y'_1 G_k Y_2)^2 \big/ (Y'_1 M_k Y_1)(Y'_2 M_k Y_2) \, .$$

The superscript (I) in $\hat{\gamma}_{12}^{2(I)}$ stands for "imaginary" and will be discussed in Sect. 6.8. The value $\hat{\gamma}_{12}^{2(I)}(k/n)$ is the fraction of $Y'_1 M_k Y_1$ that is associated with projecting Y_1 into $C(G_k Y_2)$.

To see how the imaginary sample coherence works, suppose $Y_1 = C_{k+\ell}$ and $Y_2 = S_{k+j}$. For $\ell \neq j$, $\hat{\gamma}_{12}^{2(I)}(k/n) = 0$ and for $\ell = j$, $\hat{\gamma}_{12}^{2(I)}(k/n) = 1$. Of course, the same results hold for $Y_1 = S_{k+\ell}$ and $Y_2 = C_{k+j}$. The squared sample imaginary coherence identifies a relationship only if the frequency is the same but one process is a cosine while the other is a sine. These are analogous to the results for $\hat{\gamma}_{12}^{2(R)}(k/n)$ with $Y_1 = C_{k+\ell}$, $Y_2 = C_{k+j}$, and $Y_1 = S_{k+\ell}$, $Y_2 = S_{k+j}$.

As a total measure of the correlation between Y_1 and Y_2 relative to the frequency k/n, define the *squared sample coherence function*

$$\hat{\gamma}_{12}^2(k/n) = \hat{\gamma}_{12}^{2(R)}(k/n) + \hat{\gamma}_{12}^{2(I)}(k/n) \, .$$

Exercise 6.7. Let $Y_1 = a_1 C_{k+\ell} + b_1 S_{k+\ell}$ and let $Y_2 = a_2 C_{k+j} + b_2 S_{k+j}$. Show that for $j \neq \ell$, $\hat{\gamma}_{12}^2 = 0$ and for $j = \ell$, $\hat{\gamma}_{12}^2(k/n) = 1$. What is $\hat{\gamma}_{12}^{2(I)}(k/n)$ when $b_1 = b_2 = 0$? What is $\hat{\gamma}_{12}^{2(I)}(k/n)$ when $a_1 = a_2 = 0$?

We now provide a test for the coherence. Let

$$Q_k = M_k Y_2 (Y_2' M_k Y_2)^{-1} Y_2' M_k + G_k Y_2 (Y_2' M_k Y_2)^{-1} Y_2' G_k.$$

The matrix Q_k is a perpendicular projection operator with rank two and $C(Q_k) \subset C(M_k)$. Note that

$$\hat{\gamma}_{12}^2(k/n) = \frac{Y_1' Q_k Y_1}{Y_1' M_k Y_1},$$

$$\frac{\hat{\gamma}_{12}^2(k/n)}{1 - \hat{\gamma}_{12}^2(k/n)} = \frac{Y_1' Q_k Y_1}{Y_1'(M_k - Q_k)Y_1},$$

and

$$\frac{2(r-1)}{2} \frac{\hat{\gamma}_{12}(k/n)}{1 - \hat{\gamma}_{12}(k/n)} = \frac{Y_1' Q_k Y_1 / 2}{Y_1'(M_k - Q_k)Y_1 / 2(r-1)}. \tag{6.7.1}$$

Let X be the model matrix for a linear model including all frequencies except those in the r neighborhood of k/n (i.e., the column space of X contains C_i, S_i for all i other than those near k). The test statistic (6.7.1) is appropriate for testing

$$Y_1 = X\beta + e$$

against

$$Y_1 = X\beta + (M_k Y_2)\gamma_1 + (G_k Y_2)\gamma_2 + e.$$

Under a variety of conditions, if there is no relationship between Y_1 and Y_2 in the k/n frequencies, then (6.7.1) has either an $F(2, 2(r-1), 0)$ distribution or has this approximate distribution for large samples.

6.8 Fourier Analysis

Our discussion of the frequency domain has been based on linear models. The traditional way of developing this material is via Fourier analysis. We now present the basic terminology used in Fourier analysis.

Consider observations on two time series $Y_1 = (y_{11}, \cdots, y_{1n})'$ and $Y_2 = (y_{21}, \cdots, y_{2n})'$. Define the *discrete Fourier transform* of Y_1 as

$$Y_1(k) \equiv \frac{1}{\sqrt{n}}[Y_1' C_k - iY_1' S_k] = \frac{1}{\sqrt{n}} \sum_{t=1}^{n} y_{1t} e^{-2\pi i \frac{k}{n} t}$$

and define $Y_2(k)$ similarly. It is easily shown that the periodogram of Y_1 is

$$P_1(k/n) = Y_1(k)\overline{Y_1(k)},$$

where $\overline{Y_1(k)}$ is the complex conjugate of $Y_1(k)$.

Continuing to use the notation of Sect. 6.7, define the *crossperiodogram* as

$$P_{12}(k/n) \equiv Y_1(k)\overline{Y_2(k)} = Y_1'P_kY_2 + iY_1'F_kY_2.$$

This is, in general, a complex valued function. Define the smoothed cross-spectral estimator from the r neighborhood of k/n as

$$\hat{f}_{12}(k/n) = \frac{1}{r} \sum_{\ell=-(r-1)/2}^{(r-1)/2} P_{12}\left(\frac{k+\ell}{n}\right).$$

The sample squared coherence equals

$$\hat{\gamma}_{12}^2(k/n) = \frac{|\hat{f}_{12}(k/n)|^2}{\hat{f}_1(k/n)\hat{f}_2(k/n)},$$

where $\hat{f}_1(k/n) = \frac{1}{2r}Y_1'M_kY_1$, $\hat{f}_2(k/n) = \frac{1}{2r}Y_2'M_kY_2$, and

$$|\hat{f}_{12}(k/n)|^2 = \hat{f}_{12}(k/n)\overline{\hat{f}_{12}(k/n)}.$$

Write the complex function $\hat{f}_{12}(k/n)$ as

$$\hat{f}_{12}(k/n) = \hat{f}_{12}^{(R)}(k/n) + i\hat{f}_{12}^{(I)}(k/n).$$

Then,

$$\hat{\gamma}_{12}^{2(R)}(k/n) = \frac{[\hat{f}_{12}^{(R)}(k/n)]^2}{\hat{f}_1(k/n)\hat{f}_2(k/n)}$$

and

$$\hat{\gamma}_{12}^{(I)}(k/n) = \frac{[\hat{f}_{12}^{(I)}(k/n)]^2}{\hat{f}_1(k/n)\hat{f}_2(k/n)}.$$

As discussed earlier, $\hat{f}_1(k/n)$ and $\hat{f}_2(k/n)$ are estimates of the corresponding theoretical spectral densities $f_1(v)$ and $f_2(v)$ of the two processes. If the covariances between the two processes are stationary (i.e., depend only on the time difference between the observations), define the crosscovariance function

$$\sigma_{12}(k) = \text{Cov}(y_{1,t+k}, y_{2,t}).$$

The bivariate process (y_1, y_2) is defined to be second-order stationary if each marginal process is second-order stationary and if the crosscovariances are stationary. The spectral representation of $\sigma_{12}(k)$ for a bivariate second-order stationary process is

$$\sigma_{12}(k) = \int_{-1/2}^{1/2} e^{2\pi ivk} f_{12}(v)dv,$$

which can be inverted as

$$f_{12}(v) = \sum_{k=-\infty}^{\infty} \sigma_{12}(k)e^{-2\pi ivk}.$$

The statistic $\hat{f}_{12}(k/n)$ can be viewed as an estimate of $f_{12}(k/n)$, and $\hat{\gamma}_{12}^2(k/n)$ can be viewed as an estimate of

$$\gamma_{12}^2(v) = \frac{|f_{12}(v)|^2}{f_1(v)f_2(v)}.$$

6.9 Additional Exercises

Exercise 6.9.1. Show that the following functions are nonnegative definite.
(a)

$$\sigma(k) = \begin{cases} 1 & \text{if } k = 0 \\ \frac{14}{27} & \text{if } k = \pm 1 \\ \frac{4}{27} & \text{if } k = \pm 2 \\ 0 & \text{other } k . \end{cases}$$

(b)

$$\sigma(k) = \begin{cases} 1 & \text{if } k = 0 \\ \rho & \text{if } k = \pm 1 \\ \rho^2 & \text{if } k = \pm 2 \\ 0 & \text{other } k . \end{cases}$$

The function $\sigma(k)$ is nonnegative definite if, for any n, the $n \times n$ matrix $\Sigma \equiv [\sigma(i - j)]$ is nonnegative definite.

Exercise 6.9.2. Let α and β be random variables with $E(\alpha) = E(\beta) = 0$, $\text{Var}(\alpha) = \text{Var}(\beta) = \sigma^2$, and $\text{Cov}(\alpha, \beta) = 0$. Show that

$$y_t = \alpha \cos(2\pi v t) + \beta \sin(2\pi v t)$$

is a second-order stationary process.

Exercise 6.9.3. Suppose e_t is second-order stationary. Which of the following processes are second-order stationary?
(a) $y_t = \exp[e_t]$.
(b) $y_t = y_{t-1} + e_t$.
(c) $y_t = x_t e_t$, where x_t is another second-order stationary process independent of e_t.

Exercise 6.9.4. Consider the simple linear regression model $y_t = \alpha + \beta t + e_t$, where e_t is a white noise process. Let w_t be the symmetric moving average of order 5 introduced in Example 6.6.1 as applied to the y_t process. Find $E(w_t)$ and $\text{Cov}(w_t, w_{t+k})$. Is the w_t process stationary? Why or why not?

Exercise 6.9.5. Shumway (1988) reports data from Waldmeier (1960-1978, 1961) on the number of sunspots from 1748 to 1978. The data are collected monthly, and a symmetric moving average of length 12 has been applied. The data in Table 6.7 are the moving averages corresponding to June and December (read across rows). Do a frequency analysis of these data including appropriate confidence intervals and plots.

Exercise 6.9.6. Box, Jenkins, and Reinsel (1994, p. 597) report data on international air travel. The values in Table 6.8 are the number of passengers in thousands. Do a frequency analysis of these data including appropriate transformations of the data, confidence intervals, and plots.

Exercise 6.9.7. Let e_t be a white noise process. The spectral density of e_t is $f_e(v) = \sigma^2$; see Exercise 6.2. Let $|\phi_1| < 1$ and $|\theta_1| < 1$.

(a) Find the spectral density of

$$y_t = e_t - \theta_1 e_{t-1}.$$

Sketch the graph of the spectral density for $\theta_1 = .5$ and $\theta_1 = 1$. Take $\sigma^2 = 1$. Which frequencies are most important in y_t?

(b) Show that the spectral density of

$$y_t = \phi_1 y_{t-1} + e_t$$

is

$$f_y(v) = \frac{\sigma^2}{1 - 2\phi_1 \cos(2\pi v) + \phi_1^2}.$$

Sketch the graph of the spectral density for $\phi_1 = .5$ and $\sigma^2 = 1$. Which frequencies are most important in y_t?

(c) Find the spectral density of

$$y_t = \phi_1 y_{t-1} + e_t - \theta_1 e_{t-1}.$$

Sketch the graph of the spectral density for $\phi_1 = .5$, $\theta_1 = .5$, and $\sigma^2 = 1$. Which frequencies are most important in y_t?

(d) Find the spectral density of the symmetric moving average of order 5,

$$w_t = \frac{1}{5}(e_{t-2} + e_{t-1} + e_t + e_{t+1} + e_{t+2}).$$

Sketch the graph of the spectral density for $\sigma^2 = 1$. Which frequencies are most important in w_t?

Table 6.7 Sunspot data

	1	2	3	4	5	6	7	8	9	10
000	89	84	70	49	47	48	41	32	17	12
010	11	9	11	11	17	31	45	47	46	53
020	62	63	72	87	72	60	53	46	49	38
030	26	22	18	11	19	36	50	67	78	106
040	112	101	94	82	80	68	53	36	39	32
050	11	7	11	19	41	88	139	157	142	126
060	107	87	80	69	51	39	31	24	14	10
070	12	24	46	79	108	130	138	133	127	118
080	109	93	76	67	63	61	55	47	41	41
090	33	23	20	17	10	7	5	4	7	7
100	6	13	24	34	41	45	43	43	45	47
110	48	43	35	29	20	11	7	8	7	3
120	0	0	0	1	3	5	7	12	15	14
130	20	34	48	47	43	41	36	31	24	24
140	23	17	11	7	6	4	1	1	6	9
150	10	16	23	34	47	51	60	63	61	67
160	71	70	62	50	41	28	14	9	7	12
170	24	55	93	116	139	142	126	105	82	87
180	82	64	50	37	29	25	19	11	12	15
190	28	38	47	61	65	97	123	122	121	99
200	78	69	67	64	60	55	45	40	30	21
210	16	7	3	4	9	22	36	54	76	93
220	95	95	91	78	68	61	55	44	43	48
230	41	31	24	17	7	6	17	36	57	73
240	106	138	135	113	98	102	92	68	52	45
250	33	18	13	11	13	13	7	3	2	6
260	16	31	44	54	62	60	55	62	74	65
270	55	53	41	26	14	13	11	7	5	6
280	6	7	17	35	56	71	77	84	87	79
290	71	64	53	43	35	27	26	28	20	12
300	11	10	5	3	3	5	11	23	33	42
310	51	63	62	53	60	63	51	49	51	43
320	33	21	13	6	3	3	3	2	4	9
330	24	47	59	56	69	101	98	84	78	65
340	51	38	31	27	23	14	7	6	8	16
350	24	41	61	65	72	71	69	77	68	63
360	57	39	28	22	15	11	9	6	5	8
370	15	34	57	77	101	116	110	109	103	91
380	76	67	58	49	48	31	21	17	9	9
390	19	33	56	89	126	152	145	135	139	136
400	118	87	72	70	47	32	26	15	7	4
410	12	35	81	137	164	188	200	187	181	161
420	131	114	84	56	49	38	30	28	21	10
430	11	15	25	45	73	91	101	107	110	106
440	105	105	84	67	69	71	51	39	32	35
450	25	16	16	12	15	26	57	89	110	

Table 6.8 Air passenger data

Year	Jan.	Feb.	March	April	May	June	July	Aug.	Sept.	Oct.	Nov.	Dec.
1949	112	118	132	129	121	135	148	148	136	119	104	118
1950	115	126	141	135	125	149	170	170	158	133	114	140
1951	145	150	178	163	172	178	199	199	184	162	146	166
1952	171	180	193	181	183	218	230	242	209	191	172	194
1953	196	196	236	235	229	243	264	272	237	211	180	201
1954	204	188	235	227	234	264	302	293	259	229	203	229
1955	242	233	267	269	270	315	364	347	312	274	237	278
1956	284	277	317	313	318	374	413	405	355	306	271	306
1957	315	301	356	348	355	422	465	467	404	347	305	336
1958	340	318	362	348	363	435	491	505	404	359	310	337
1959	360	342	406	396	420	472	548	559	463	407	362	405
1960	417	391	419	461	472	535	622	606	508	461	390	432

Exercise 6.9.8. Let e_t and ε_t be uncorrelated white noise processes. Let

$$y_t = \phi_1 y_{t-1} + e_t$$

and

$$w_t = \phi_1 w_{t-1} + y_t + \varepsilon_t.$$

Find the spectral density of w_t.

Exercise 6.9.9. Let e_t be second-order stationary and define

$$y_t = \sum_{s=-\infty}^{\infty} a_s e_{t-s}$$

and

$$w_t = \sum_{r=-\infty}^{\infty} b_r y_{t-r},$$

where $\sum_{s=-\infty}^{\infty} |a_s| < \infty$ and $\sum_{r=-\infty}^{\infty} |b_r| < \infty$.
(a) Show that

$$f_w(v) = |A(v)|^2 |B(v)|^2 f_e(v).$$

(b) Show that

$$w_t = \sum_{s=-\infty}^{\infty} c_s e_{t-s},$$

where

$$c_s = \sum_{r=-\infty}^{\infty} a_{s-r} b_r.$$

Exercise 6.9.10.   Let y_t and w_t be two second-order stationary processes and suppose that $f_w(v) \le f_y(v)$ for all $v \in [-\pi, \pi]$. Show that

$$\Sigma_{yy} - \Sigma_{ww}$$

is nonnegative definite, where Σ_{yy} and Σ_{ww} are the covariance matrices of $(y_1, \cdots, y_n)'$ and $(w_1, \cdots, w_n)'$, respectively.

Exercise 6.9.11.   Show that the spectral density

$$f(v) = \frac{\pi - |v|}{\pi^2}$$

determines the covariance function

$$\sigma(k) = \begin{cases} \sigma^2 & \text{if } k = 0 \\ 4\sigma^2/(\pi|k|)^2 & \text{if } k \text{ is odd} \\ 0 & \text{otherwise.} \end{cases}$$

Exercise 6.9.12.   If y_t and w_t are independent second-order stationary processes with spectral densities f_y and f_w, find the spectral distribution of the stationary process

$$z_t = y_t + w_t.$$

Exercise 6.9.13.   Suppose y_t is second-order stationary and $f_y(v)$ is nonnegative, bounded, and $f_y(1/2) \ne 0$. Let

$$w_t = y_t - y_{t-1}.$$

Find $f_w(v)$ in terms of $f_y(v)$.

Exercise 6.9.14.   For $v \in [0, \pi]$, define

$$f_y(v) = \begin{cases} 50 & v \in [\frac{1}{4} - .01, \frac{1}{4} + .01] \\ 0 & \text{otherwise} \end{cases}$$

and $f_y(-v) = f_y(v)$. Find $\sigma(0), \sigma(1), \sigma(2)$.

Exercise 6.9.15.
(a) Use the relationships

$$\sin(a+b) = \sin(a)\cos(b) + \cos(a)\sin(b)$$

and

$$\cos(a+b) = \cos(a)\cos(b) - \sin(a)\sin(b)$$

to show the following.

$$\cos(a)\cos(b) = \frac{1}{2}\{\cos(a+b) + \cos(a-b)\}.$$

$$\cos(a)\sin(b) = \frac{1}{2}\{\sin(a+b) - \sin(a-b)\}.$$

$$\sin(a)\sin(b) = \frac{1}{2}\{\cos(a-b) - \cos(a+b)\}.$$

(b) Recall that for complex numbers x with $|x| < 1$,

$$\sum_{t=1}^{n} x^t = \frac{x - x^{n+1}}{1 - x}.$$

Apply this fact to $\exp\left(2\pi i \frac{j}{n}\right)$ to show that, for any $j = 1, \ldots, n-1$,

$$\sum_{t=1}^{n} \cos\left(2\pi \frac{j}{n} t\right) = 0 = \sum_{t=1}^{n} \sin\left(2\pi \frac{j}{n} t\right).$$

(c) Prove Eqs. (6.2.4), (6.2.5), and (6.2.6).

Exercise 6.9.16.   Apply the relationship in Exercise 6.9.15a on the sine of a sum to the process in Exercise 6.9.2 to show that the process can be rewritten as

$$y_t = A\sin(2\pi v t + \phi),$$

where A is the random amplitude and ϕ is the random phase of the sine curve.

Find the relationship between (A, ϕ) and (α, β). How does this relate to the basic spectral approximation in (6.3.1)? If, instead, we write $y_t = A_* \cos(2\pi v t + \phi_*)$, how do (A_*, ϕ_*) differ from (A, ϕ)?

References

Bloomfield, P. (1976). *Fourier analysis of time series: An introduction*. New York: Wiley.

Box, G. E. P., Jenkins, G. M., & Reinsel, G. C. (1994). *Time series analysis: Forecasting and control* (3rd ed.). New York: Wiley.

Breiman, L. (1968). *Probability*. Reading: Addison-Wesley.

Brillinger, D. R. (1981). *Time series: Data analysis and theory* (2nd ed.). San Francisco: Holden Day.

Brockwell, P. J., & Davis, R. A. (1991). *Time series: Theory and methods* (2nd ed.). New York: Springer.

Brockwell, P. J., & Davis, R. A. (2002). *Introduction to time series and forecasting* (2nd ed.). New York: Springer.

Christensen, R. (2011). *Plane answers to complex questions: The theory of linear models* (4th edn.). New York: Springer.

Cressie, N. A.C., & Wikle, C. K. (2011). *Statistics for spatio-temporal data*. New York: Wiley.

Doob, J. L. (1953). *Stochastic processes*. New York: Wiley.

Fuller, W. A. (1996). *Introduction to statistical time series* (2nd ed.). New York: Wiley.

Hannan, E. J. (1970). *Multiple time series*. New York: Wiley.

Koopmans, L. H. (1974). *The spectral analysis of time series*. New York: Academic.

Prado, R., & West, M. (2010). *Time series: Modeling, computation, and inference*. Boca Raton: CRC Press.

Shumway, R. H. (1988). *Applied statistical time series analysis*. Englewood Cliffs: Prentice-Hall.

Shumway, R. H., & Stoffer, D. S. (2011). *Time series analysis and its applications: With R examples* (3rd ed.). New York: Springer.

Tukey, J. W. (1977). *Exploratory data analysis*. Reading: Addison-Wesley.

Waldmeier, M. (1960-1978). *Monthly sunspot bulletin*. Zurich: Swiss Federal Observatory.

Waldmeier, M. (1961). *The sunspot activity in the years 1610–1960*. Zurich: Swiss Federal Observatory.

Chapter 7
Time Domain Analysis

Abstract This chapter develops Box-Jenkins models. These involve applying the linear filters of Chap. 6 to white noise. It also introduces state-space models and the Kalman filter.

In the previous chapter frequency domain analysis was introduced using a spectral approximation for an arbitrary second-order stationary time series $\ldots, y_{-1}, y_0, y_1, \ldots$. In the traditional approach, the variance components in the spectral approximation will be nonzero, thus there is little chance of developing a parsimonious model. (The model with measurement error is more promising in that regard.)

The time domain approach to time series analysis does not apply to arbitrary stationary processes. The time domain assumes that the process can be modeled as a simple recursive filter, a causal linear filter, or a general recursive filter of an uncorrelated stationary error process. Linear filters were discussed in Sect. 6.6. The uncorrelated stationary error process is often referred to as *white noise*. It has the same properties as measurement error. When the filtered process is stationary, a general recursive filter, and hence a simple recursive filter, can also be modeled as a causal linear filter. The particular choice of a model is determined by selecting a filter that does a good job of explaining the observed series and is relatively parsimonious (i.e., has few parameters).

In time domain analysis, the covariance function is again very important. In addition, partial correlations between observations are important. In this chapter, correlations and general prediction are discussed in Sect. 7.1. Section 7.2 introduces the various time domain models. These sections are followed by discussions of prediction (forecasting), estimation, model selection, and seasonal adjustment. The chapter closes with consideration of dynamic linear models also known as the state-space model. State-space methodology is closely tied to the Kalman filter. The reader needs to be familiar with best linear prediction, which is discussed thoroughly in *PA* Sects. 6.3–6.5 and is reviewed in Appendix B. Example 9.3.1 relates partial correlations to multivariate linear models and best linear prediction is implicit in the review of best linear unbiased prediction in Sect. 4.1.

© Springer Nature Switzerland AG 2019
R. Christensen, *Advanced Linear Modeling*, Springer Texts in Statistics,
https://doi.org/10.1007/978-3-030-29164-8_7

The time domain analysis of time series is discussed in many books and research articles. The classic text on the subject is Box, Jenkins, Reinsel, and Ljung (2015). Shorter discussions can be found in general texts on time series, e.g., Shumway and Stoffer (2011), Brockwell and Davis (1991, 2002), Prado and West (2010), Chatfield (2003), Diggle (1990), Fuller (1996).

7.1 Correlations and Prediction

We have seen that the (auto)covariance function $\sigma(k)$ determines the spectral distribution and the variance components of the spectral approximation. Although time domain analysis is very different in spirit from frequency analysis, the covariance function again plays an important role. Along with the covariance function, time domain analysis uses the correlation function, the partial covariance function, and the partial correlation function. The purpose of this section is to define these three additional functions and to relate partial correlation to best linear prediction. Assume a second-order stationary process $\cdots, y_{-1}, y_0, y_1, \cdots$ with mean μ and covariance function $\sigma(k)$.

The *correlation function* is simply

$$
\begin{aligned}
\rho(k) &\equiv \mathrm{Cov}(y_t, y_{t+k}) \Big/ \sqrt{\mathrm{Var}(y_t)\mathrm{Var}(y_{t+k})} \\
&= \sigma(k) \Big/ \sqrt{\sigma(0)\sigma(0)} \\
&= \sigma(k)/\sigma(0).
\end{aligned}
$$

The *partial correlation function* is defined to be the partial correlation between y_t and y_{t+k} given $y_{t+1}, \ldots, y_{t+k-1}$, cf. *PA* Sect. 6.5. Of course, this only makes sense for $k \geq 2$. Using notation similar to that in Appendix B, let $y = (y_t, y_{t+k})'$ and $x = (y_{t+1}, \ldots, y_{t+k-1})'$. The *partial covariance* is the off-diagonal element of

$$
V_{yy} - V_{yx}V_{xx}^{-1}V_{xy},
$$

where

$$
V_{yy} = \begin{bmatrix} \sigma(0) & \sigma(k) \\ \sigma(k) & \sigma(0) \end{bmatrix},
$$

$$
V_{yx} \equiv V_{xy}' \equiv \begin{bmatrix} \sigma_1' \\ \sigma_2' \end{bmatrix} = \begin{bmatrix} \sigma(1) & \sigma(2) & \cdots & \sigma(k-1) \\ \sigma(k-1) & \sigma(k-2) & \cdots & \sigma(1) \end{bmatrix}, \tag{7.1.1}
$$

and

$$
V_{xx} = \begin{bmatrix} \sigma(0) & \sigma(1) & \sigma(2) & \cdots & \sigma(k-2) \\ \sigma(1) & \sigma(0) & \sigma(1) & \cdots & \sigma(k-3) \\ \sigma(2) & \sigma(1) & \sigma(0) & \cdots & \sigma(k-4) \\ \vdots & \vdots & \vdots & & \vdots \\ \sigma(k-2) & \sigma(k-3) & \sigma(k-4) & \cdots & \sigma(0) \end{bmatrix}. \tag{7.1.2}
$$

Note that none of these matrices depend on t; all are functions of k alone.

The partial correlation function is denoted $\phi(k)$. For $k = 1$, define $\phi(1) \equiv \rho(1)$. For $k \geq 2$, $\phi(k)$ is the ratio of the partial covariance of y_t and y_{t+k} to the product of the partial standard deviations of y_t and y_{t+k}. In particular, the partial covariance is

$$\sigma(k) - \sigma_1' V_{xx}^{-1} \sigma_2 .$$

The partial variances for y_t and y_{t+k} are, respectively,

$$\sigma(0) - \sigma_1' V_{xx}^{-1} \sigma_1$$

and

$$\sigma(0) - \sigma_2' V_{xx}^{-1} \sigma_2 .$$

Thus,

$$\phi(k) \equiv \{\sigma(k) - \sigma_1' V_{xx}^{-1} \sigma_2\} \bigg/ \sqrt{\{\sigma(0) - \sigma_1' V_{xx}^{-1} \sigma_1\}\{\sigma(0) - \sigma_2' V_{xx}^{-1} \sigma_2\}} .$$

Partial correlations can also be defined recursively which may be a better way to compute them (see *PA* Exercise 6.9.9).

Exercise 7.1. Let $H = [h_{ij}]$ be a $(k-1) \times (k-1)$ matrix, where

$$h_{ij} = \begin{cases} 1 & \text{if } i+j = k \\ 0 & \text{otherwise.} \end{cases}$$

H is of the form

$$H = \begin{bmatrix} 0 & \cdots & 0 & 0 & 1 \\ 0 & \cdots & 0 & 1 & 0 \\ 0 & \cdots & 1 & 0 & 0 \\ \vdots & & \vdots & \vdots & \vdots \\ 1 & \cdots & 0 & 0 & 0 \end{bmatrix} .$$

Show that $\sigma_1' H = \sigma_2'$, $\sigma_2' H = \sigma_1'$, $HV_{xx}H = V_{xx}$ and $HH = I$.

Using the results of Exercise 7.1, we can show that

$$\sigma(0) - \sigma_1' V_{xx}^{-1} \sigma_1 = \sigma(0) - \sigma_2' V_{xx}^{-1} \sigma_2$$

(i.e., the partial variances are the same for y_t and y_{t+k}). To see this, observe that

$$\begin{aligned} \sigma_2' V_{xx}^{-1} \sigma_2 &= \sigma_2' HHV_{xx}^{-1} HH\sigma_2 \\ &= \sigma_1' HV_{xx}^{-1} H\sigma_1 \\ &= \sigma_1' (H^{-1} V_{xx} H^{-1})^{-1} \sigma_1 \\ &= \sigma_1' (HV_{xx}H)^{-1} \sigma_1 \\ &= \sigma_1' V_{xx}^{-1} \sigma_1 . \end{aligned}$$

It follows that

$$\phi(k) = \{\sigma(k) - \sigma_1' V_{xx}^{-1} \sigma_2\} / \{\sigma(0) - \sigma_1' V_{xx}^{-1} \sigma_1\}.$$

7.1.1 Partial Correlation and Best Linear Prediction

We now investigate the relationship between the coefficient of y_t in the best linear predictor of y_{t+k} based on $y_t, y_{t+1}, \ldots, y_{t+k-1}$ and the partial correlation between y_{t+k} and y_t. In particular, we will show that these are identical. From Appendix B, the best linear predictor is

$$\hat{E}(y_{t+k}|y_t, x) = \mu + \delta' \left[\binom{y_t}{x} - J\mu \right],$$

where

$$\delta' = [\sigma(k), \sigma_2'] \begin{bmatrix} \sigma(0) & \sigma_1' \\ \sigma_1 & V_{xx} \end{bmatrix}^{-1}. \tag{7.1.3}$$

The coefficient of y_t is δ_k in $\delta' = (\delta_k, \ldots, \delta_1)$. Let $K = [\sigma(0) - \sigma_1' V_{xx}^{-1} \sigma_1]^{-1}$, from the standard result on the inverse of a partitioned matrix, see Exercise B.4,

$$\begin{bmatrix} \sigma(0) & \sigma_1' \\ \sigma_1 & V_{xx} \end{bmatrix}^{-1} = \begin{bmatrix} K & -K\sigma_1' V_{xx}^{-1} \\ -V_{xx}^{-1}\sigma_1 K & V_{xx}^{-1} + V_{xx}^{-1}\sigma_1 \sigma_1' V_{xx}^{-1} K \end{bmatrix},$$

so

$$\begin{aligned}
\delta_k &= [\sigma(k), \sigma_2'] \begin{bmatrix} K \\ -V_{xx}\sigma_1 K \end{bmatrix} \\
&= (\sigma(k) - \sigma_2' V_{xx}^{-1}\sigma_1) K \\
&= \{\sigma(k) - \sigma_1' V_{xx}^{-1}\sigma_2\} / \{\sigma(0) - \sigma_1' V_{xx}^{-1}\sigma_1\} \\
&= \phi(k).
\end{aligned}$$

Because this is a second-order property of the process y_t (i.e., involves only means, variances, and covariances), and because y_t is second-order stationary, it follows immediately that $\phi(k)$ is also the coefficient of y_{t-k} in the best linear predictor of y_t from $y_{t-1}, \ldots, y_{t-k}$.

Recursive methods exist for finding BLPs based on $Y = (y_1, \ldots, y_n)'$. The difficulty of computing the predictors directly is that finding best linear predictors requires solving a system of equations based on an $n \times n$ matrix. The obvious method of solution involves taking the inverse of the $n \times n$ matrix. To be computationally practical for large sample sizes, more sophisticated methods of finding a solution are needed. We discuss two methods for finding $\hat{E}(y_{n+1}|Y)$. A more detailed discussion of prediction methods based on Y is given by Brockwell and Davis (1991, Section 5.2).

7.1.2 The Durbin–Levinson Algorithm

The *Durbin–Levinson algorithm* is a recursive procedure for predicting the future in a stationary process. Having observed Y, to find the BLP of y_{n+1} Durbin–Levinson applies of the results in the previous subsection with $t = 1$ and $k = n$, repeats them for $t = 2$, $k = n - 1$, then for $t = 3$, $k = n - 2$, etc.,

$$\hat{E}(y_{n+1}|Y) = \mu + \delta'(Y - \mu J) = \mu + [\sigma(n), \sigma_2'] \begin{bmatrix} \sigma(0) & \sigma_1' \\ \sigma_1 & V_{xx} \end{bmatrix}^{-1} (Y - \mu J).$$

The BLP without including y_1 is

$$\hat{E}(y_{n+1}|y_2, \ldots, y_n) = \mu + \sigma_2' V_{xx}^{-1} \begin{bmatrix} y_2 - \mu \\ \vdots \\ y_n - \mu \end{bmatrix}.$$

The Durbin–Levinson algorithm establishes a simple formula for the BLP based on Y in terms of $\sigma_2' V_{xx}^{-1}$, which is the only difficult thing to find in $\hat{E}(y_{n+1}|y_2, \ldots, y_n)$. But finding $\sigma_2' V_{xx}^{-1}$ is easier than finding δ, because δ requires the inversion of an $n \times n$ matrix but V_{xx} is only $(n-1) \times (n-1)$. The idea then is to apply the partitioning idea to $\hat{E}(y_{n+1}|y_2, \ldots, y_n)$ and find it by inverting an $(n-2) \times (n-2)$ matrix and continue the process to where you only need to invert a 1×1 matrix.

Writing $\delta = (\delta_n, \ldots, \delta_1)$, Sect. 7.1.1 shows that

$$\delta_n = (\sigma(n) - \sigma_2' V_{xx}^{-1} \sigma_1) K,$$

where

$$K = [\sigma(0) - \sigma_1' V_{xx}^{-1} \sigma_1]^{-1}.$$

Write the other elements of δ as $\delta_* = (\delta_{n-1}, \ldots, \delta_1)$. Using the results of Sect. 7.1, it is easily seen that

$$\delta_* = \sigma_2' V_{xx}^{-1} - \delta_n \sigma_1' V_{xx}^{-1}$$
$$= \sigma_2' V_{xx}^{-1} - \delta_n \sigma_2' V_{xx}^{-1} H.$$

Recall that the vectors σ_1 and σ_2 have the same entries in reverse order. Thus, solving the reduced data problem to find $\sigma_2' V_{xx}^{-1}$ leads to simple formulae for the full data values K, δ_n, and δ_*.

7.1.3 Innovations Algorithm

The *innovations algorithm* is based on repeated application of Proposition B.1.8 to obtain $\hat{E}(y_{n+1}|Y)$ as a linear combination of the prediction errors, say, $\varepsilon_1 \equiv y_1$ and, for $t > 1$, $\varepsilon_t \equiv e(y_t|y_{t-1}, \ldots, y_1) \equiv y_t - \hat{E}(y_t|y_{t-1}, \ldots, y_1)$; see Exercise 7.2. Write the best linear predictor as

$$\hat{E}(y_{n+1}|Y) = \sum_{t=1}^{n} \eta_t \varepsilon_t. \tag{7.1.4}$$

To use this equation, one needs to know the η_ts and the ε_ts. The η_ts are determined by the repeated application of Proposition B.1.8. Note that the η_ts depend on n. The ε_ts are found recursively. The value of $\varepsilon_1 = y_1$ is known. For $t = 2, \ldots, n$, $\hat{E}(y_t|y_{t-1}, \ldots y_1)$, and thus ε_t, can be found using exactly the same procedure as used for $\hat{E}(y_{n+1}|Y)$.

Exercise 7.2.

(a) Using Proposition B.1.8, find $\hat{E}(y_4|y_3, y_2, y_1)$ in terms of $\sigma(\cdot)$, y_1, $e(y_2|y_1)$, and $e(y_3|y_2, y_1)$.
(b) Use induction to show that $y_1, e(y_2|y_1), \cdots, e(y_n|y_{n-1}, \cdots, y_1)$ are uncorrelated.

To obtain $\hat{E}(y_{n+k}|Y)$, first obtain $\hat{E}(y_{n+k}|Y, y_{n+1}, \ldots, y_{n+k-1})$ as a linear combination of prediction error terms. From Exercise B.5 and the equivalent of Eq. (7.1.4) for predicting y_{n+k},

$$\hat{E}(y_{n+k}|Y) = \hat{E}[\hat{E}(y_{n+k}|Y, y_{n+1}, \ldots, y_{n+k-1})|Y]$$
$$= \hat{E}\left[\sum_{t=1}^{n+k} \eta_t \varepsilon_t \Big| Y\right].$$

By Proposition B.1.4, Exercise 7.2, Proposition B.1.5, and Corollary B.1.3,

$$\hat{E}\left[\sum_{t=1}^{n+k} \eta_t \varepsilon_t \Big| Y\right] = \hat{E}\left[\sum_{t=1}^{n+k} \eta_t \varepsilon_t \Big| \varepsilon_n, \ldots, \varepsilon_1\right]$$
$$= \hat{E}\left[\sum_{t=1}^{n} \eta_t \varepsilon_t \Big| \varepsilon_n, \ldots, \varepsilon_1\right]$$
$$= \sum_{t=1}^{n} \eta_t \varepsilon_t.$$

Note that the η_ts depend on $n + k$.

Methods related to the Durbin–Levinson and innovations algorithms can also be used to obtain the exact prediction variance.

7.2 Time Domain Models

We now consider the various models used for analyzing time series data in the time domain. Generally, these are stationary models, so their frequency properties can be studied using the methods of Chap. 6. Time domain models are just linear filters of the white noise process. White noise is the name used for the uncorrelated error process e_t, where

$$E(e_t) = 0; \qquad \text{Var}(e_t) = \sigma^2; \qquad \text{Cov}(e_t, e_{t'}) = 0, \quad t \neq t'.$$

In particular, because time domain models are linear filters, their frequency properties are easily derived from Sect. 6.6 together with the fact, established in Exercise 6.5, that the spectral density of white noise is $f_e(v) = \sigma^2$.

7.2.1 Autoregressive Models: AR(p)s

Let $\ldots, e_{-1}, e_0, e_1, \ldots$ be a stationary process of uncorrelated errors (i.e., white noise). An autoregressive model of order p, denoted $AR(p)$, is a model that states that the observable time series is a simple recursive filter of the error process involving p terms, namely

$$y_t = \sum_{s=1}^{p} \phi_s y_{t-s} + e_t. \tag{7.2.1}$$

Because the $AR(p)$ model is just a simple recursive filter, the results of Exercises 6.2 and 6.6 (in Sects. 6.3 and 6.6 respectively) fully determine which frequencies are important in a stationary autoregressive process.

Whether an autoregressive process is stationary depends on another way of looking at the process. Let B be the *backshift operator*, namely

$$By_t \equiv y_{t-1}.$$

This is a very useful tool in writing models. For example, the ∇ operator of Example 6.6.2 is

$$\nabla y_t = (1 - B)y_t,$$

and

$$B^2 y_t = B(By_t) = By_{t-1} = y_{t-2}.$$

To examine the $AR(p)$ model, let

$$\Phi(B) \equiv 1 - \sum_{s=1}^{p} \phi_s B^s.$$

The $AR(p)$ model (7.2.1) can be rewritten as

$$y_t - \sum_{s=1}^{p} \phi_s y_{t-s} = e_t$$

or

$$\Phi(B)y_t = e_t. \tag{7.2.2}$$

Note that $\Phi(B)$ is a polynomial in the backshift operator. If we substitute a scalar variable x for B, we get a standard polynomial $\Phi(x)$. The roots of $\Phi(x)$ (i.e., the solutions to $\Phi(x) = 0$) are generally complex numbers. Let x_0 be an arbitrary root. If all of the roots satisfy $|x_0|^2 > 1$, then the rational polynomial $1/\Phi(x)$ can be written as an infinite polynomial. This follows from doing a Taylor expansion of the function $1/\Phi(x)$ about 0; see Exercise 7.9.14. In particular, there exists

$$\Psi(B) \equiv 1 + \sum_{s=1}^{\infty} \psi_s B^s$$

such that

$$\sum_{s=1}^{\infty} |\psi_s| < \infty \tag{7.2.3}$$

and

$$\Psi(B)\Phi(B) = 1. \tag{7.2.4}$$

Equation (7.2.4) states that the product of the polynomials $\Psi(B)$ and $\Phi(B)$ is the constant polynomial that takes only the value 1.

Applying $\Psi(B)$ to (7.2.2) yields

$$y_t = \Psi(B)\Phi(B)y_t = \Psi(B)e_t = e_t + \sum_{s=1}^{\infty} \psi_s e_{t-s}, \tag{7.2.5}$$

so the $AR(p)$ model (7.2.1) can be written as a causal linear filter of the error process. Condition (7.2.3) implies that y_t is a mean zero second-order stationary process (see Sect. 6.6). The restriction that all roots x_0 have $|x_0|^2 > 1$ is needed to obtain stationarity.

The coefficients in $\Psi(B)$ can be identified by solving an infinite system of equations. Note that (7.2.4) will occur if and only if the polynomials in the scalar x satisfy $\Psi(x)\Phi(x) = 1$. Because the constant terms in both $\Psi(x)$ and $\Phi(x)$ are 1, $\Psi(x)\Phi(x) = 1$ if and only if the coefficient of x^k in $\Psi(x)\Phi(x)$ equals zero for every $k \geq 1$.

Exercise 7.3. Show that in an $AR(1)$ model with $|\phi_1| < 1$,

$$\psi_k = \phi_1^k.$$

The forms (7.2.2) and (7.2.5) are useful in finding the covariance function $\sigma_y(\cdot)$ of the process y_t. Let k be nonnegative and note that

$$\text{Cov}(\Phi(B)y_t, y_{t-k}) = \text{Cov}(e_t, y_{t-k}). \tag{7.2.6}$$

Using the fact that $\sigma_y(k) = \sigma_y(-k)$, the left-hand side is

$$\text{Cov}(\Phi(B)y_t, y_{t-k}) = \text{Cov}\left(y_t - \sum_{s=1}^{p} \phi_s y_{t-s}, y_{t-k}\right)$$

$$- \sigma_y(-k) \sum_{s=1}^{p} \phi_s \sigma_y(-k+s)$$

$$= \sigma_y(k) - \sum_{s=1}^{p} \phi_s \sigma_y(k-s).$$

The right-hand side of (7.2.6) is

$$\text{Cov}(e_t, y_{t-k}) = \text{Cov}(e_t, \Psi(B)e_{t-k})$$

$$= \text{Cov}(e_t, e_{t-k}) + \sum_{s=1}^{\infty} \psi_s \text{Cov}(e_t, e_{t-k-s})$$

$$= \begin{cases} \sigma^2 & \text{if } k = 0 \\ 0 & \text{if } k \neq 0. \end{cases}$$

Setting equal the reexpressions of the left- and right-sides of (7.2.6) gives

$$\sigma_y(0) - \sum_{s=1}^{p} \phi_s \sigma_y(s) = \sigma^2$$

and for $k \geq 1$

$$\sigma_y(k) - \sum_{s=1}^{p} \phi_s \sigma_y(k-s) = 0.$$

These are known as the *Yule–Walker equations* and for given ϕ_ss are solved to find $\sigma_y(k)$ for $k = 0, 1, \ldots$. The symmetry of $\sigma_y(\cdot)$ about zero completes its characterization.

The covariance function can also be characterized in terms of the ψs. Details are given in Eq. (7.2.12) for autoregressive moving average (*ARMA*) models. Because autoregressive models are also *ARMA* models, the later discussion applies to the current case. In particular, the covariance function for an *AR*(1) process is given later in Example 7.2.3.

We now establish that the partial correlation function has the properties that

$$\phi(p) = \phi_p$$

and

$$\phi(k) = 0 \qquad k = p+1, p+2, \ldots.$$

This is a key feature in identifying the order p of an *AR* model. One can estimate the partial correlations and if they obviously drop off after some point, that should indicate p. The result follows from the fact that

$$\hat{E}(y_t|y_{t-1},\dots,y_{t-k}) = \sum_{s=1}^{p} \phi_s y_{t-s} \qquad k \geq p. \tag{7.2.7}$$

In Sect. 7.1 we established that the coefficient of y_{t-k} in the BLP is $\phi(k)$, so Eq. (7.2.7) immediately provides the result. If $k = p$, $\phi(p) = \phi_p$. If $k > p$, the BLP coefficient of y_{t-k} is 0, so $\phi(k) = 0$. We now establish Eq. (7.2.7).

First note that in model (7.2.1), y_t depends on all of $e_t, e_{t-1}, \dots$ but it depends on e_t only though the additive term; the e_{t-k}s, $k \geq 1$ only involve themselves in the y_{t-k}s, $k \geq 1$. This is even more clear when (7.2.5) applies. As a result, $\mathrm{Cov}(e_t, y_{t-k}) = 0$, $k \geq 1$. Now, using results from Appendix B and the fact that $k \geq p$,

$$\hat{E}(y_t|y_{t-1},\dots,y_{t-k}) = \hat{E}(\sum_{s=1}^{p} \phi_s y_{t-s} + e_t|y_{t-1},\dots,y_{t-k})$$

$$= \sum_{s=1}^{p} \phi_s \hat{E}(y_{t-s}|y_{t-1},\dots,y_{t-k}) + \hat{E}(e_t|y_{t-1},\dots,y_{t-k})$$

$$= \sum_{s=1}^{p} \phi_s y_{t-s} + 0,$$

where the last equality follows because the BLP of one of the predictors is just that predictor and the BLP of a random variable uncorrelated with the predictors is the unconditional mean of the random variable.

There is a danger of getting the partial correlation function $\phi(k)$ confused with the coefficients $\phi_1, \dots, \phi_p$ in the $AR(p)$ model. However, the equality $\phi(p) = \phi_p$ is the reason ϕ is used to denote both quantities.

EXAMPLE 7.2.1.   For $AR(1)$ and $AR(2)$ processes, the partial correlation functions have very simple forms. For an $AR(1)$ process, $y_t = \phi_1 y_{t-1} + e_t$,

$$\phi(1) = \phi_1 = \rho(1)$$

and

$$\phi(k) = 0 \qquad k > 1.$$

For an $AR(2)$ process, $y_t = \phi_1 y_{t-1} + \phi_2 y_{t-2} + e_t$

$$\phi(1) = \rho(1),$$
$$\phi(2) = \phi_2,$$

and

$$\phi(k) = 0 \qquad k > 2.$$

For an $AR(3)$ process, $\phi(1) = \rho(1)$, $\phi(3) = \phi_3$, and $\phi(k) = 0$ for $k > 3$, but $\phi(2)$ is a more complicated function of the ϕ_is and $\sigma(i)$s.   □

To deal with time series that have a nonzero mean, write $E(y_t) = \mu$ and use the $AR(p)$ model

$$\Phi(B)[y_t - \mu] = e_t.$$

Equivalently, we can write

$$\Phi(B)y_t = \Phi(1)\mu + e_t$$

$$= \left(1 - \sum_{s=1}^{p} \phi_s\right)\mu + e_t$$

$$= \alpha + e_t,$$

where $\alpha \equiv \left(1 - \sum_{s=1}^{p} \phi_s\right)\mu$. Another equivalent expression is

$$y_t = \alpha + \sum_{s=1}^{p} \phi_s y_{t-s} + e_t.$$

Note that

$$E(y_t) \equiv \mu = \alpha \left/ \left(1 - \sum_{s=1}^{p} \phi_s\right).\right.$$

7.2.2 Moving Average Models: MA(q)s

A moving average model of order q, denoted $MA(q)$, is just a causal linear filter of the error (white noise) process. In particular,

$$y_t = e_t - \sum_{s=1}^{q} \theta_s e_{t-s}. \tag{7.2.8}$$

The covariance function and spectral densities for linear filters were discussed in Sect. 6.6. Applying (6.6.4) gives

$$\sigma_y(k) = \begin{cases} \sigma^2\left(1 + \sum_{s=1}^{q} \theta_s^2\right) & k = 0 \\ \sigma^2\left(-\theta_k + \sum_{s=1}^{q-k} \theta_{s+k}\theta_s\right) & k = 1,\ldots,q-1 \\ -\sigma^2\theta_q & k = q \\ 0 & k > q. \end{cases}$$

For model selection, this result is analogous to the fact that an $AR(p)$ model has $\phi(k) = 0$ for $k > p$. If we estimate $\sigma_y(k)$ or equivalently the correlation $\rho_y(k)$, and if the estimated correlations are negligible for, say, $k > 5$, then an $MA(5)$ model is suggested.

Using the backshift operator, (7.2.8) can be written as

$$y_t = \Theta(B)e_t, \tag{7.2.9}$$

where

$$\Theta(B) \equiv 1 - \sum_{s=1}^{q} \theta_s B^s.$$

Note that with only a finite number of nonzero coefficients in the filter, the process y_t is a mean zero second-order stationary process. If the roots of $\Theta(x)$ are all outside the unit circle in the complex plane, the process defined by (7.2.9) is said to be *invertible*. In such a case, $1/\Theta(B)$ can be written as an infinite polynomial in B and e_t can be written as a causal linear filter of y_t.

Processes with nonzero mean are written

$$y_t - \mu = \Theta(B)e_t.$$

EXAMPLE 7.2.2.　The $MA(1)$ process, $y_t = \mu + e_t - \theta_1 e_{t-1}$, has

$$\begin{aligned}
E(y_t) &= \mu, \\
\sigma(0) &= \sigma^2(1 + \theta_1^2), \\
\sigma(1) &= -\sigma^2 \theta_1, \\
\sigma(k) &= 0, \qquad k > 1,
\end{aligned}$$

and

$$\rho(1) = \frac{-\theta_1}{1 + \theta_1^2}. \qquad \qquad \square$$

7.2.3 Autoregressive Moving Average Models: ARMA(p,q)s

In many ways, the most important time domain models are the autoregressive moving average models. These are general recursive filters of the error process, namely

$$y_t = \sum_{s=1}^{p} \phi_s y_{t-s} + e_t - \sum_{s=1}^{q} \theta_s e_{t-s}. \tag{7.2.10}$$

This is an $AR(p)$ with the addition of q moving average terms. It is also an $MA(q)$ with the addition of p autoregressive terms. Note that both the $AR(p)$ and $MA(q)$ models are special cases of $ARMA$ models. Model (7.2.10) can be rewritten as

$$y_t - \sum_{s=1}^{p} \phi_s y_{t-s} = e_t - \sum_{s=1}^{q} \theta_s e_{t-s},$$

or, using the backshift operator,

$$\Phi(B)y_t = \Theta(B)e_t. \tag{7.2.11}$$

Here, we assume that the roots of both $\Phi(x)$ and $\Theta(x)$ are outside the unit circle in the complex plane. This ensures that the process is both stationary and invertible. We also assume that the two polynomials have no roots in common. If the polynomials have s roots in common, then s common terms can be factored out of both sides of (7.2.11), thus creating an $ARMA(p-s, q-s)$ model. To eliminate duplication, we impose the condition of no common roots. The spectral density of model (7.2.11) can be obtained using the results in Sect. 6.6.

To compute the covariance function for an $ARMA(p,q)$, use the model in the form (7.2.11). Assuming the roots of the polynomial are outside the unit circle, the rational polynomial $1/\Phi(B)$ is itself an infinite polynomial. Write

$$\Psi(B) = \Theta(B)/\Phi(B);$$

thus, dividing (7.2.11) by $\Phi(B)$ gives

$$y_t = \Psi(B)e_t.$$

Computing the covariance function directly for $k \geq 0$,

$$
\begin{aligned}
\sigma(k) = \mathrm{Cov}(y_t, y_{t+k}) &= \mathrm{Cov}(\Psi(B)e_t, \Psi(B)e_{t+k}) \\
&= \mathrm{Cov}\left(\sum_{s=0}^{\infty} \psi_s e_{t-s}, \sum_{s'=0}^{\infty} \psi_{s'} e_{t+k-s'}\right) \\
&= \sum_{s=0}^{\infty}\sum_{s'=0}^{\infty} \psi_s \psi_{s'} \mathrm{Cov}(e_{t-s}, e_{t+k-s'}) \\
&= \sum_{s=0}^{\infty} \psi_s \psi_{s+k} \mathrm{Cov}(e_{t-s}, e_{t-s}) \\
&= \sigma^2 \sum_{s=0}^{\infty} \psi_s \psi_{s+k}.
\end{aligned}
\tag{7.2.12}
$$

Equation (7.2.12) is just a special case of (6.6.4). Note that it can also be used to find the covariance function for an $AR(p)$ process, where we consider an $AR(p)$ as an $ARMA(p, 0)$ model.

As usual, to model a process y_t with $E(y_t) = \mu$, use

$$\Phi(B)[y_t - \mu] = \Theta(B)e_t.$$

An equivalent form is

$$y_t = \alpha + \sum_{s=1}^{p} \phi_s y_{t-s} + e_t - \sum_{s=1}^{q} \theta_s e_{t-s},$$

where $\alpha \equiv \left(1 - \sum_{s=1}^{p} \phi_s\right)\mu = \Phi(1)\mu$. Covariance properties are not affected by this change.

It is interesting to note that $ARMA$ models can be used to approximate the covariance structure of any stationary process with $\sigma(k) \to 0$ as $k \to \infty$. In particular, for such a covariance function and any integer $K \geq 0$, there exists an $ARMA(p,q)$ process y_t with covariance function $\sigma_y(k)$ such that

$$\sigma_y(k) = \sigma(k) \text{ for } k = 0, 1, \ldots, K.$$

EXAMPLE 7.2.3. *The Covariance Function for an* AR(1).
Identify the $AR(1)$ process with an $ARMA(1,0)$; thus,

$$\Theta(B) = 1.$$

Writing

$$\Phi(B) = 1 - \phi_1 B,$$

it is easily seen that if $|\phi_1| < 1$,

$$\Psi(B) = 1/\Phi(B) = \sum_{s=0}^{\infty} \phi_1^s B^s.$$

Applying (7.2.12) and using the fact that $\sum_{s=0}^{\infty} v^s = \frac{1}{1-v}$ for $v \in (0,1)$ yields

$$\sigma(k) = \sigma^2 \sum_{s=0}^{\infty} \phi_1^{2s+k}$$

$$= \sigma^2 \phi_1^k \sum_{s=0}^{\infty} (\phi_1^2)^s$$

$$= \sigma^2 \phi_1^k \Big/ (1 - \phi_1^2)$$

for $k \geq 0$. From the symmetry of the covariance function, the correlation function is

$$\rho(k) = \phi_1^{|k|}. \qquad \qquad \square$$

EXAMPLE 7.2.4. The $ARMA(1,1)$ model is

$$[1 - \phi_1(B)](y_t - \mu) = e_t - \theta_1 e_{t-1}$$

or

$$y_t = (1 - \phi_1)\mu + \phi_1 y_{t-1} + e_t - \theta_1 e_{t-1}. \qquad (7.2.13)$$

The process is stationary if $|\phi_1| < 1$,

$$E(y_t) = (1 - \phi_1)\mu + \phi_1 E(y_{t-1}) + 0$$
$$= (1 - \phi_1)\mu + \phi_1 \mu$$
$$= \mu.$$

The covariance function $\sigma_y(\cdot)$ can be computed via application of equality (7.2.12). The autoregressive transformation is

$$\Phi(B) = 1 - \phi_1 B,$$

so

$$1/\Phi(B) = 1 + \phi_1 B + \phi_1^2 B^2 + \phi_1^3 B^3 + \cdots$$
$$= \sum_{s=0}^{\infty} \phi_1^s B^s.$$

The moving average polynomial is

$$\Theta(B) = 1 - \theta_1 B;$$

thus,

$$\Psi(B) = \Theta(B)/\Phi(B)$$
$$= \sum_{s=0}^{\infty} \phi_1^s B^s - \sum_{s=0}^{\infty} \theta_1 \phi_1^s B^{s+1}$$
$$= 1 + \sum_{s=1}^{\infty} (\phi_1^s - \theta_1 \phi_1^{s-1}) B^s$$
$$= 1 + \sum_{s=1}^{\infty} \phi_1^{s-1} (\phi_1 - \theta_1) B^s. \tag{7.2.14}$$

Applying (7.2.12),

$$\sigma(0) = \sigma^2 \left[1 + \sum_{s=1}^{\infty} \phi_1^{2(s-1)} (\phi_1 - \theta_1)^2 \right]$$
$$= \sigma^2 [1 + (\phi_1 - \theta_1)^2/(1 - \phi_1^2)]$$

and, for $k > 0$,

$$\sigma(k) = \sigma^2 \left[\phi_1^{k-1} (\phi_1 - \theta_1) + \sum_{s=1}^{\infty} \phi_1^{2s-2+k} (\phi_1 - \theta_1)^2 \right]$$
$$= \sigma^2 \{ (\phi_1 - \theta_1) [\phi_1^{k-1} + \phi_1^k (\phi_1 - \theta_1) / (1 - \phi_1^2)] \}$$
$$= \sigma^2 (\phi_1 - \theta_1) \phi_1^{k-1} [1 + \phi_1 (\phi_1 - \theta_1) / (1 - \phi_1^2)].$$

Note that

$$\sigma(k) = \phi_1^{k-1} \sigma(1). \qquad \qquad \square$$

7.2.4 Autoregressive Integrated Moving Average Models: ARIMA$(p,d,q)s$

The $ARMA(p,q)$ model and its special cases, the $AR(p)$ and $MA(q)$ models, are used to model second-order stationary time series. Using $ARIMA(p,d,q)$ models is the time domain method for dealing with nonstationarity. Recall that in Example 6.6.2 and Exercise 6.5, the difference operator was used to transform series that were nonstationary into stationary processes. The $ARIMA(p,d,q)$ model assumes that the dth difference

$$\nabla^d y_t \equiv (1-B)^d y_t$$

is a stationary $ARMA(p,q)$ process. The $ARIMA(p,d,q)$ model is written

$$\Phi(B)\nabla^d y_t = \Theta(B)e_t.$$

EXAMPLE 7.2.5.   The $ARIMA(1,1,1)$ model can be rewritten as

$$[1-\phi_1(B)](y_t - y_{t-1}) = e_t - \theta_1 e_{t-1}$$

or

$$(y_t - y_{t-1}) - \phi_1(y_{t-1} - y_{t-2}) = e_t - \theta_1 e_{t-1}$$

or

$$y_t = (1+\phi_1)y_{t-1} - \phi_1 y_{t-2} + e_t - \theta_1 e_{t-1}.$$ □

7.3 Time Domain Prediction

One of the prime motivations in analyzing time series is to be able to predict (forecast) the future of the series. The spectral approximation has limited value as a forecasting tool. The spectral model fits the data perfectly, so the prediction of the future is that the past will reoccur. In particular, if n is even, the predictions for the next n observations will be precisely $y_1,\ldots,y_n$. If measurement error is included, reduced models can be fitted and more interesting predictions result.

These problems do not occur in the time domain. Time domain models are particularly well-suited for making predictions. We wish to examine the best linear predictor of, say, y_{n+k}, based on the observations actually in hand, $y_1,\ldots,y_n$. For processes that are Gaussian, the best linear predictor is also the best predictor. The BLP methods of Sect. 7.1 continue to apply but some interesting twists arise.

Let $Y = (y_1,\ldots,y_n)'$. We begin our discussion of prediction with examples.

EXAMPLE 7.3.1.   *Prediction for an AR*(1) *Model.*
Because the model

$$y_t = \alpha + \phi_1 y_{t-1} + e_t$$

is linear and $\hat{E}(\cdot|Y)$ is a linear operator, we can develop, recursively, a formula for the BLP. For $k = 1, 2, \ldots$,

$$\begin{aligned} \hat{E}(y_{n+k}|Y) &= \hat{E}(\alpha + \phi_1 y_{n+k-1} + e_{n+k}|Y) \\ &= \alpha + \phi_1 \hat{E}(y_{n+k-1}|Y) + \hat{E}(e_{n+k}|Y) \\ &= \alpha + \phi_1 \hat{E}(y_{n+k-1}|Y). \end{aligned}$$

The last equality holds by Proposition B.1.5 because $E(e_{n+k}) = 0$ and $\text{Cov}(e_{n+k}, Y) = 0$. Similarly, for $k \geq 2$,

$$\hat{E}(y_{n+k-1}|Y) = \alpha + \phi_1 \hat{E}(y_{n+k-2}|Y),$$

so

$$\begin{aligned} \hat{E}(y_{n+k}|Y) &= \alpha + \phi_1[\alpha + \phi_1 \hat{E}(y_{n+k-2}|Y)] \\ &= \alpha(1 + \phi_1) + \phi_1^2 \hat{E}(y_{n+k-2}|Y). \end{aligned}$$

Continuing this procedure gives

$$\begin{aligned} \hat{E}(y_{n+k}|Y) &= \alpha \sum_{s=0}^{k-1} \phi_1^s + \phi_1^k \hat{E}(y_n|Y) \\ &= \alpha \sum_{s=0}^{k-1} \phi_1^s + \phi_1^k y_n, \end{aligned}$$

where the last equality holds by Proposition B.1.2 because we are predicting y_n from a vector of observations that includes y_n. $\qquad\square$

EXAMPLE 7.3.2. *Prediction for an MA(1) Model.*
The model is

$$y_t = \mu + e_t - \theta_1 e_{t-1}.$$

The BLP for $k = 2, 3, \ldots$ is

$$\begin{aligned} \hat{E}(y_{n+k}|Y) &= \hat{E}(\mu + e_{n+k} - \theta_1 e_{n+k-1}|Y) \\ &= \mu + \hat{E}(e_{n+k}|Y) - \theta_1 \hat{E}(e_{n+k-1}|Y) \\ &= \mu, \end{aligned} \qquad (7.3.1)$$

where the last equality is a result of

$$E(e_{n+k}) = E(e_{n+k-1}) = 0,$$

$$\text{Cov}(e_{n+k}, Y) = \text{Cov}(e_{n+k-1}, Y) = 0,$$

and Proposition B.1.5.

For $k = 1$, $\text{Cov}(e_{n+k-1}, Y) = \text{Cov}(e_n, Y) \neq 0$, so (1) does not hold. For $k = 1$, note that $\text{Cov}(y_{n+1}, y_{n-s}) = \sigma(s+1) = 0$, $s = 1, \ldots, n$, so $\text{Cov}(Y, y_{n+1}) \equiv V_{Yy} = (0, 0, \ldots, 0, \sigma(1))'$. The covariance matrix of Y is of the form (7.1.2) with $k - 1 = n$. This can also be written

$$\text{Cov}(Y) = V_{YY} = \begin{bmatrix} V_{22} & \sigma_1 \\ \sigma_1' & \sigma(0) \end{bmatrix},$$

where V_{22} is defined by (7.1.2) with $k = n$ and σ_1' is defined by (7.1.1) with $k = n$. Using Exercise B.4 on inverses of partitioned matrices, we can show that

$$V_{YY}^{-1} V_{Yy} = \begin{bmatrix} -\sigma(1) V_{22}^{-1} \sigma_1 / [\sigma(0) - \sigma_1' V_{22}^{-1} \sigma_1] \\ \sigma(1) / [\sigma(0) - \sigma_1' V_{22}^{-1} \sigma_1] \end{bmatrix}.$$

Thus,

$$\hat{E}(y_{n+1}|Y) = \mu + [Y - \mu J]' V_{YY}^{-1} V_{Yy}, \tag{7.3.2}$$

where $V_{YY}^{-1} V_{Yy}$ is characterized as before. This is quite complicated and likely to get more so for an $MA(q)$ with $q > 1$. It is also difficult to compute V_{22}^{-1} for large n. □

7.3.1 Conditioning on Y_∞

It is my impression that modern software is more likely to bite the bullet and perform the matrix computations necessary for making predictions. But in dealing with moving average processes, it is computationally convenient (although mathematically less precise) to condition not only on Y but on $Y_\infty = (y_n, \ldots, y_1, y_0, y_{-1}, y_{-2}, \ldots)'$. The use of Y_∞ allows us to develop a simple recursive prediction method similar to that used for the $AR(1)$ process in Example 7.3.1. For an invertible $MA(q)$ process

$$[y_t - \mu] = \Theta(B) e_t$$

with $\Psi(B) \Theta(B) = 1$, we can write

$$\Psi(B)[y_t - \mu] = e_t,$$

where

$$\Psi(B)[y_t - \mu] = \sum_{s=0}^{\infty} \psi_s [y_{t-s} - \mu] = \sum_{s=0}^{\infty} \psi_s y_{t-s} - \mu \sum_{s=0}^{\infty} \psi_s.$$

Thus, there is an invertible linear transformation between $(y_t, y_{t-1}, \ldots)'$ and $(e_t, e_{t-1}, \ldots)'$. In other words, there exists a nonsingular transformation between the two vectors for any value of t. Let $e_\infty = (e_n, e_{n-1}, \ldots)'$. By Proposition B.1.4, for any random variable w,

$$\hat{E}(w|Y_\infty) = \hat{E}(w|e_\infty). \tag{7.3.3}$$

EXAMPLE 7.3.3.   *MA*(1) *Prediction Using Y_∞.*
For $k = 2, 3, \ldots$, just as before,

$$\hat{E}(y_{n+k}|Y_\infty) = \mu.$$

However, for $k = 1$,

$$
\begin{aligned}
\hat{E}(y_{n+1}|Y_\infty) &= \mu + \hat{E}(e_{n+1}|Y_\infty) - \theta_1 \hat{E}(e_n|Y_\infty) \\
&= \mu + 0 - \theta_1 \hat{E}(e_n|Y_\infty) \\
&= \mu - \theta_1 e_n,
\end{aligned}
$$

where the last equality follows from either Corollary B.1.3 or Propositions B.1.4 and B.1.2. □

Comparing the preceeding result with (7.3.2), it becomes clear why predicting with Y_∞ is mathematically more convenient than predicting with Y. In general, (7.3.3) and Proposition B.1.2 imply that

$$\hat{E}(e_{n+k}|Y_\infty) = \begin{cases} 0 & k \geq 1 \\ e_{n+k} & k \leq 0. \end{cases} \tag{7.3.4}$$

This characterization is very useful in dealing with moving averages. □

Exercise 7.4.   Consider an $AR(p)$ model with $p < n$.

(a) Show for $k = 0, 1, 2, \ldots$ that $\hat{E}(y_{n+k}|Y)$ depends only on $y_n, y_{n-1}, \ldots, y_{n-p+1}$.
(b) Show for $k = 0, 1, \ldots$ that $\hat{E}(y_{n+k}|Y) = \hat{E}(y_{n+k}|Y_\infty)$.

Hint: For (a), use induction on k.

We now consider prediction for an $ARMA(p,q)$ model. Predictions for *ARIMA* models are found in a similar fashion as will be illustrated later. The forecasting procedure consists of building the predictions recursively. We wish to find $\hat{E}(y_{n+k}|Y_\infty)$. The value depends on all of p, q, and k. Moreover, to illustrate the recursive nature, assume that $p > 2$ and $q \geq 2$. The $ARMA(p,q)$ model is

$$y_t = \alpha + \sum_{s=1}^{p} \phi_s y_{t-s} + e_t - \sum_{s=1}^{q} \theta_s e_{t-s},$$

so

$$
\begin{aligned}
\hat{E}(y_{n+1}|Y_\infty) &= \alpha + \sum_{s=1}^{p} \phi_s \hat{E}(y_{n+1-s}|Y_\infty) \\
&\quad + \hat{E}(e_{n+1}|Y_\infty) - \sum_{s=1}^{q} \theta_s \hat{E}(e_{n+1-s}|Y_\infty) \\
&= \alpha + \sum_{s=0}^{p-1} \phi_{s+1} y_{n-s} + 0 - \sum_{s=0}^{q-1} \theta_s e_{n-s}.
\end{aligned}
$$

For $k = 2$,

$$\hat{E}(y_{n+2}|Y_\infty) = \alpha + \sum_{s=1}^{p} \phi_s \hat{E}(y_{n+2-s}|Y_\infty)$$

$$+ \hat{E}(e_{n+2}|Y_\infty) - \sum_{s=1}^{q} \theta_s \hat{E}(e_{n+2-s}|Y_\infty)$$

$$= \alpha + \phi_1 \hat{E}(y_{n+1}|Y_\infty)$$

$$+ \sum_{s=0}^{p-2} \phi_{s+2} y_{n-s} + 0 - 0 - \sum_{s=0}^{q-2} \theta_{s+2} e_{n-s},$$

where $\hat{E}(y_{n+1}|Y_\infty)$ was found earlier.

The general recursive pattern is clear, for $k = 1, 2, \ldots$,

$$\hat{E}(y_{n+k}|Y_\infty) = \alpha + \sum_{s=1}^{p} \phi_s \hat{E}(y_{n+k-s}|Y_\infty) + 0 - \sum_{s=1}^{q} \theta_s \hat{E}(e_{n+k-s}|Y_\infty), \qquad (7.3.5)$$

where $\hat{E}(y_t|Y_\infty) = y_t$ if $t \leq n$ and is found recursively for $t > n$ and where $\hat{E}(e_t|Y_\infty) = e_t$ for $t \leq n$ and $\hat{E}(e_t|Y_\infty) = 0$ for $t > n$.

There are two problems in using this result. First, we do not know the values of α, the ϕ_js, and the θ_js. Second, we have not actually observed Y_∞. In particular, because we do not know Y_∞, we also do not know $e_n, e_{n-1}, \ldots, e_{n-q}$, and these error terms are explicitly involved in $\hat{E}(y_{n+k}|Y_\infty)$.

The first problem is generally handled by substituting estimates of α, the ϕ_js, and the θ_js for the actual parameters. Parameter estimation is discussed in Sect. 7.5.

One approach to the second problem is to assume that $e_t = 0$ for $t = 0, -1, -2, \ldots$. Note that, by the invertability of $\Phi(\cdot)$, this also implies that $y_t = \mu = \alpha / \left(1 - \sum_{s=1}^{p} \phi_s\right)$ for $t = 0, -1, -2, \ldots$. With this assumption, we can simply solve for $e_1, \ldots, e_n$. The $ARMA(p, q)$ model can be rearranged as

$$e_t = \sum_{s=1}^{q} \theta_s e_{t-s} + (y_t - \mu) - \sum_{s=1}^{p} \phi_s(y_{t-s} - \mu). \qquad (7.3.6)$$

For $t = 1$, the right-hand side involves only "observed" random variables, so

$$e_1 = 0 + (y_1 - \mu) - 0$$

$$= y_1 - \mu$$

$$= y_1 - \left\{\alpha \Big/ \left(1 - \sum_{s=1}^{p} \phi_s\right)\right\}.$$

Similarly,

$$e_2 = \theta_1 e_1 + (y_2 - \mu) - \phi_1(y_1 - \mu).$$

Equation (7.3.6) can be used recursively to obtain $e_3, e_4, \ldots, e_n$. In practice, the parameters α, ϕ_s, $s = 1, \ldots, p$, and θ_s, $s = 1, \ldots, q$ must be estimated, so estimated errors $\hat{e}_t$, $t = n - q, \ldots, n$ are used in the BLP, Eq. (7.3.5).

Another, less appealing, approach to dealing with the problem that Y_∞ is not observed is to assume that $e_1 = \cdots = e_q = 0$ and to otherwise ignore the problem. With this assumption about the errors, $e_n, \ldots, e_{n-q}$ can be computed using (7.3.6) and then substituted into (7.3.5) to give predictions. If n is much greater than p and q, the two approaches should give similar results. Yet another approach to dealing with the "prehistoric" values in Y_∞ is a method called *backcasting*, which is discussed in Sect. 7.5.

EXAMPLE 7.3.4. *Prediction for an ARMA(1,1).*
Using Eq. (7.3.5),

$$\hat{E}(y_{n+1}|Y_\infty) = \alpha + \phi_1 y_n - \theta_1 e_n,$$

$$\hat{E}(y_{n+2}|Y_\infty) = \alpha + \phi_1 \hat{E}(y_{n+1}|Y_\infty)$$
$$= \alpha + \phi_1[\alpha + \phi_1 y_n - \theta_1 e_n],$$

$$\hat{E}(y_{n+3}|Y_\infty) = \alpha + \phi_1 \hat{E}(y_{n+2}|Y_\infty)$$
$$= \alpha + \phi_1\{\alpha + \phi_1[\alpha + \phi_1 y_n - \theta_1 e_n]\}$$
$$= \alpha + \phi_1 \alpha + \phi_1^2 \alpha + \phi_1^3 y_n - \phi_1^2 \theta_1 e_n,$$

and in general, for $k \geq 1$,

$$\hat{E}(y_{n+k}|Y_\infty) = \alpha \sum_{s=0}^{k-1} \phi_1^s + \phi_1^k y_n - \phi_1^{k-1} \theta_1 e_n. \qquad \square$$

Exercise 7.5.

(a) Show that with $e_t = 0$ for $t = 0, -1, -2, \ldots$, an *ARMA(1,1)* model has

$$e_t = (y_t - \mu) + \sum_{s=1}^{t-1} (\theta_1 - \phi_1)\theta_1^{t-1-s}(y_s - \mu)$$

for $t = 1, \ldots, n$.

(b) Show that the assumption $e_1 = 0$ leads to

$$e_t = (y_t - \mu) + \left[\sum_{s=1}^{t-1} (\theta_1 - \phi_1)\theta_1^{t-1-s}(y_s - \mu) \right] - \theta_1^{t-1}(y_1 - \mu).$$

We complete our discussion of Y_∞ prediction in an $ARMA(p,q)$ model by finding the mean squared error of prediction,

$$E[y_{n+k} - \hat{E}(y_{n+k}|Y_\infty)]^2.$$

Because the BLP is an unbiased estimate of y_{n+k}, this is also called the *prediction variance*.

By definition,

$$\Phi(B)[y_{n+k} - \mu] = \Theta(B)e_{n+k}. \tag{7.3.7}$$

Also, (7.3.5) can be restated as

$$\Phi(B)[\hat{E}(y_{n+k}|Y_\infty) - \mu] = \Theta(B)\hat{E}(e_{n+k}|Y_\infty). \tag{7.3.8}$$

Subtracting (7.3.8) from (7.3.7) gives

$$\Phi(B)\{y_{n+k} - \hat{E}(y_{n+k}|Y_\infty)\} = \Theta(B)\{e_{n+k} - \hat{E}(e_{n+k}|Y_\infty)\}. \tag{7.3.9}$$

By assumption, we can write

$$\Psi(B) = \Theta(B)/\Phi(B).$$

Multiplying (7.3.9) by $1/\Phi(B)$ gives

$$y_{n+k} - \hat{E}(y_{n+k}|Y_\infty) = \Psi(B)\{e_{n+k} - \hat{E}(e_{n+k}|Y_\infty)\}. \tag{7.3.10}$$

Recall that

$$\hat{E}(e_t|Y_\infty) = \begin{cases} e_t & t \le n \\ 0 & t > n \end{cases},$$

so

$$e_t - \hat{E}(e_t|Y_\infty) = \begin{cases} 0 & t \le n \\ e_t & t > n \end{cases}.$$

Substituting into (7.3.10) gives

$$y_{n+k} - \hat{E}(y_{n+k}|Y_\infty) = \sum_{s=0}^{k-1} \psi_s e_{n+k-s}. \tag{7.3.11}$$

From (7.3.11), the prediction variance is easily computed:

$$\begin{aligned} E[y_{n+k} - \hat{E}(y_{n+k}|Y_\infty)]^2 &= \text{Var}\left(\sum_{s=0}^{k-1} \psi_s e_{n+k-s}\right) \\ &= \sum_{s=0}^{k-1} \psi_s^2 \text{Var}(e_{n+k-s}) \\ &= \sigma^2 \sum_{s=0}^{k-1} \psi_s^2. \end{aligned}$$

With estimated parameters, this probably underestimates the true prediction error. For a Gaussian process, we get a $(1 - \alpha)100\%$ prediction interval for y_{n+k} with endpoints

$$\hat{E}(y_{n+k}|Y_\infty) \pm z\left(1 - \frac{\alpha}{2}\right)\sqrt{\sigma^2 \sum_{s=0}^{k-1} \psi_s^2}.$$

EXAMPLE 7.3.5. *Prediction Variance in an ARMA(1,1).*
In (7.2.14), the polynomial transformation $\Psi(B)$ was given for an $ARMA(1,1)$. Applying this gives

$$E[y_{n+1} - \hat{E}(y_{n+1}|Y_\infty)]^2 = \sigma^2$$

and, for $k \geq 2$,

$$
\begin{aligned}
& E[y_{n+k} - \hat{E}(y_{n+k}|Y_\infty)]^2 \\
&= \sigma^2\left[1 + \sum_{s=1}^{k-1} \phi_1^{2(s-1)}(\phi_1 - \theta_1)^2\right] \\
&= \sigma^2\left[1 + (\phi_1 - \theta_1)^2\left\{\sum_{s=0}^{\infty} \phi_1^{2s} - \phi_1^{2(k-1)}\sum_{s=0}^{\infty}\phi_1^{2s}\right\}\right] \\
&= \sigma^2\left[1 + (\phi_1 - \theta_1)^2\{(1 - \phi_1^{2(k-1)})/(1 - \phi_1^2)\}\right] \\
&= \sigma^2\left[1 + (\phi_1 - \theta_1)^2(1 - \phi_1^{2(k-1)})/(1 - \phi_1^2)\right].
\end{aligned}
$$

$\square$

To predict for an $ARIMA(p,d,q)$ model, let

$$\Phi(B)\nabla^d y_t = \Theta(B)e_t$$

and define

$$z_t = \nabla^d y_t,$$

so that z_t is an $ARMA(p,q)$. A new problem in prediction is that $z_1, \ldots, z_d$ are at most partially observed. We have to assume "prehistoric" (i.e., $t = 0, -1, -2, \ldots$) values not only for the e_ts but also for the y_ts. With some reasonable assumption about the prehistoric y_ts (e.g., $y_t = \alpha + \beta t$) and again assuming that $e_t = 0, t = 0, -1, -2, \ldots$, predictions

$$\hat{E}(z_{n+k}|Z_\infty)$$

can be made for any $k > 0$. Here, $Z_\infty = (z_n, z_{n-1}, \ldots)'$.
To predict y_{n+k}, write

$$
\begin{aligned}
\hat{E}(z_{n+k}|Z_\infty) &= \hat{E}(\nabla^d y_{n+k}|Z_\infty) \\
&= \hat{E}[(1 - B)^d y_{n+k}|Z_\infty]
\end{aligned}
$$

$$= \hat{E}\left[\sum_{s=0}^{d} \binom{d}{s}(-1)^d y_{n+k-s} \middle| Z_\infty\right]$$

$$= \sum_{s=0}^{d} \binom{d}{s}(-1)^d \hat{E}(y_{n+k-s}|Z_\infty). \qquad (7.3.12)$$

Using the approximation

$$\hat{E}(y_t|Z_\infty) \doteq y_t \qquad (7.3.13)$$

for $t \leq n$, Eq. (7.3.13) can be solved recursively to obtain $\hat{E}(y_{n+1}|Z_\infty), \hat{E}(y_{n+2}|Z_\infty), \ldots$. If n is much greater than p, d, and q, the assumptions about the prehistory should not have much effect on the predictions.

EXAMPLE 7.3.6. *Prediction for an ARIMA$(p,1,q)$ model.*
From Eq. (7.3.12),

$$\hat{E}(z_{n+k}|Z_\infty) = \hat{E}(y_{n+k}|Z_\infty) - \hat{E}(y_{n+k-1}|Z_\infty)$$

so

$$\hat{E}(y_{n+k}|Z_\infty) = \hat{E}(z_{n+k}|Z_\infty) + \hat{E}(y_{n+k-1}|Z_\infty).$$

In particular,

$$\hat{E}(y_{n+1}|Z_\infty) = \hat{E}(z_{n+1}|Z_\infty) + y_n.$$

The values $\hat{E}(z_{n+k}|Z_\infty)$ are found by applying the ARMA(p,q) prediction results to the first difference process z_t. Values of $\hat{E}(y_{n+k}|Z_\infty)$ for $k > 1$ are found in a straightforward recursive manner. $\qquad \square$

7.4 Nonlinear Least Squares

The idea of using estimates that minimize the sum of squared errors is a geometric idea, not a statistical idea; it does not depend on the statistical properties of the observations. As was seen in *PA* Chap. 2, for linear models the least squares estimates are not only intuitively appealing but are also BLUEs, MLEs, and minimum variance unbiased estimates under appropriate assumptions. These other properties of the estimates do depend on the statistical model.

The idea of least squares estimation can be extended to very general nonlinear situations. In particular, the idea can be used to obtain estimates for time domain models. In many such extensions, least squares estimates are also maximum likelihood estimates for normal data. In this section we discuss least squares in general and discuss the Gauss–Newton algorithm for computing the estimates. The section closes with a discussion of nonlinear regression. Although nonlinear regression is not directly applicable to time series analysis, it is important in its own right and illustrates one application of nonlinear least squares methodology. Estimation for time domain models is discussed in Sect. 7.5.

Consider a vector of observations $v = (v_1, v_2, \ldots, v_n)$ and a parameter $\xi \in \mathbf{R}^p$. For known functions $f_i(\cdot)$, we can write a model

$$v_i = f_i(\xi) + e_i.$$

Here the e_is are errors. The least squares estimates are values of ξ that minimize the sum of squared errors

$$\mathrm{SSE}(\xi) = \sum_{i=1}^{n} [v_i - f_i(\xi)]^2.$$

In matrix, notation write

$$F(\xi) = \begin{bmatrix} f_1(\xi) \\ \vdots \\ f_n(\xi) \end{bmatrix}$$

and $e = (e_1, \ldots, e_n)'$, so

$$v = F(\xi) + e.$$

Note that F is a function from $\mathbf{R}^p$ to $\mathbf{R}^n$. The sum of squared errors can be rewritten as

$$\mathrm{SSE}(\xi) = [v - F(\xi)]'[v - F(\xi)]. \tag{7.4.1}$$

The functions f_i can be very general. They can even depend on the v_js, a situation that occurs in the time domain models discussed in Sect. 7.5. We now illustrate the application of this model to standard univariate linear models.

EXAMPLE 7.4.1.   Consider the linear function of ξ, $F(\xi) = Z\xi$ for a fixed $n \times p$ matrix Z of rank p. The model is $v = Z\xi + e$ and the criterion to be minimized is

$$[v - Z\xi]'[v - Z\xi].$$

The minimizing value gives least squares estimates of ξ in the linear model.   □

7.4.1 The Gauss–Newton Algorithm

The *Gauss–Newton algorithm* is a method for finding least squares estimates in nonlinear problems. The algorithm consists of obtaining a sequence of linear least squares estimates that converge to the least squares estimate in the nonlinear problem.

The Gauss–Newton algorithm requires multivariate calculus. We use the notation set in Appendix A. The procedure also requires an initial guess (estimate) for ξ, say ξ_0, and defines a series of estimates ξ_r that converge to the least squares estimate $\hat{\xi}$.

Given ξ_r, we define ξ_{r+1}. Using the first-order Taylor's approximation, for ξ in a neighborhood of ξ_r,

$$F(\xi) \doteq F(\xi_r) + d_\xi F(\xi_r)(\xi - \xi_r), \tag{7.4.2}$$

where, because ξ_r is known, $F(\xi_r)$ and $\mathbf{d}_\xi F(\xi_r)$ are known. The derivative $\mathbf{d}_\xi F(\xi_r)$ is the $n \times p$ matrix of partial derivatives evaluated at ξ_r. We assume that $\mathbf{d}_\xi F(\xi_r)$ has full column rank.

Define

$$\text{SSE}_r(\xi) \equiv [v - F(\xi_r) - \mathbf{d}_\xi F(\xi_r)(\xi - \xi_r)]'[v - F(\xi_r) - \mathbf{d}_\xi F(\xi_r)(\xi - \xi_r)]. \quad (7.4.3)$$

Substituting the approximation (7.4.2) into Eq. (7.1.1), we see that $\text{SSE}_r(\xi) \doteq \text{SSE}(\xi)$ when ξ is near ξ_r. If ξ_r is near the least squares estimate $\hat{\xi}$, the minimum of $\text{SSE}_r(\xi)$ should be close to the minimum of $\text{SSE}(\xi)$. Now, make the following identifications:

$$Y = v - F(\xi_r),$$
$$X = \mathbf{d}_\xi F(\xi_r),$$
$$\beta = (\xi - \xi_r).$$

With these identifications, minimizing $\text{SSE}_r(\xi)$ is equivalent to minimizing

$$[Y - X\beta]'[Y - X\beta].$$

From standard linear model theory, this is minimized by

$$\beta_{r+1} = (X'X)^{-1}X'Y$$
$$= ([\mathbf{d}_\xi F(\xi_r)]'[\mathbf{d}_\xi F(\xi_r)])^{-1}[\mathbf{d}_\xi F(\xi_r)]'[v - F(\xi_r)].$$

Now, $\beta = \xi - \xi_r$, so

$$\xi_{r+1} \equiv \xi_r + \beta_{r+1}$$

minimizes $\text{SSE}_r(\xi)$. Although ξ_{r+1} minimizes $\text{SSE}_r(\xi)$ exactly, ξ_{r+1} is only an approximation to the value $\hat{\xi}$ that minimizes $\text{SSE}(\xi)$. However, as ξ_r converges to $\hat{\xi}$, the approximation (7.4.2) becomes increasingly better.

EXAMPLE 7.4.2. Again consider the linear function $F(\xi) = Z\xi$. From standard linear model theory, we know that $\hat{\xi} = (Z'Z)^{-1}Z'v$. Because $\mathbf{d}_\xi F(\xi) = Z$, given any ξ_0,

$$\beta_1 = (Z'Z)^{-1}Z'(v - Z\xi_0)$$
$$= (Z'Z)^{-1}Z'v - \xi_0$$

and

$$\xi_1 = \xi_0 + \beta_1 = (Z'Z)^{-1}Z'v = \hat{\xi}.$$

Thus, for a linear least squares problem, the Gauss–Newton method converges to $\hat{\xi}$ in only one iteration. □

An alternative to the Gauss–Newton method of finding least squares estimates is the method of steepest descent, see Draper and Smith (1981). Marquardt (1963) has presented a method that is a compromise between Gauss–Newton and the method of steepest descent. Marquardt's compromise is often used to improve the convergence properties of Gauss–Newton.

7.4.2 Nonlinear Regression

We present a very short account of the *nonlinear regression model* and the corresponding least squares estimates. Christensen (2015, Chapter 23) and Draper and Smith (1981, Chapter 10) give more extensive introductions. Seber and Wild (1989, 2003) give a complete account.

Consider a situation in which there are observations $y_1, \ldots, y_n$ taken on a dependent variable and corresponding observations on independent variables denoted by the row vectors $x_1', \ldots, x_n'$. A linear model is

$$y_i = x_i' \beta + e_i.$$

Nonlinear regression is the model

$$y_i = f(x_i; \beta) + e_i, \tag{7.4.4}$$

where $f(\cdot; \cdot)$ is a known function. This is a special case of the general nonparametric model introduced in Chap. 1. Write $Y = (y_1, \ldots, y_n)$ and $e = (e_1, \ldots, e_n)$. In the notation of our discussion of the Gauss–Newton algorithm,

$$v = Y,$$
$$\xi = \beta,$$
$$f_i(\cdot) = f(x_i; \cdot),$$

and

$$F(\beta) = \begin{bmatrix} f(x_1; \beta) \\ \vdots \\ f(x_n; \beta) \end{bmatrix}.$$

In this problem only one derivative vector needs to be determined, $\mathbf{d}_\beta f(x; \beta)$. However, the vector is evaluated at n different x_i values. Note that x need not be a p vector. We can now apply Gauss–Newton to find least squares estimates.

Nonlinear regression is a problem in which least squares estimates are MLEs under suitable conditions. Assume

$$e \sim N(0, \sigma^2 I);$$

then, because $F(\beta)$ is a fixed vector,

$$Y \sim N[F(\beta), \sigma^2 I]$$

with density

$$f(Y|\beta,\sigma^2) = (2\pi)^{-n/2}\sigma^{-n/2}\exp\left\{-[Y-F(\beta)]'[Y-F(\beta)]/2\sigma^2\right\}.$$

The likelihood is the density taken as a function of β and σ^2 for fixed Y. The MLEs maximize the likelihood. For any σ^2, the likelihood is maximized by minimizing $[Y-F(\beta)]'[Y-F(\beta)]$. Thus, least squares estimates $\hat{\beta}$ are also MLEs. The MLE of σ^2 can be found by maximizing

$$L(\sigma^2) = (2\pi)^{-n/2}(\sigma^2)^{-n}\exp\left\{-\left[Y-F(\hat{\beta})\right]'\left[Y-F(\hat{\beta})\right]/2\sigma^2\right\}.$$

Differentiation leads to

$$\hat{\sigma}^2 = [Y-F(\hat{\beta})]'[Y-F(\hat{\beta})]/n.$$

Asymptotic results for maximum likelihood estimation allow statistical inferences to be made for nonlinear regression models.

Some nonlinear relationships can be transformed into linear relationships. The nonlinear regression model (7.4.4) indicates that

$$y_i \doteq f(x_i;\beta).$$

If $f(\cdot;\cdot)$ can be written as

$$f(x_i;\beta) = f(x_i'\beta)$$

and f is invertible, the relationship is said to be linearizable. In that case,

$$f^{-1}(y_i) \doteq x_i'\beta.$$

An alternative to fitting model (4) is to fit

$$f^{-1}(y_i) = x_i'\beta + \varepsilon_i. \tag{7.4.5}$$

The choice between analyzing the nonlinear model (7.4.4) and the linear model (7.4.5) is typically based on which model better approximates the assumption of independent identically distributed normal errors. Just as in linear regression, the nonlinear least squares fit of β in model (7.4.4) generates residuals

$$\hat{e}_i = y_i - \hat{y}_i = y_i - f(x_i;\hat{\beta}).$$

These can be plotted to examine normality and heteroscedasticity.

7.5 Estimation

In this section, we discuss empirical estimates of correlations and partial correlations. These are empirical in the sense that they do not depend on any particular time domain model for the second-order stationary time series. We also discuss least squares and maximum likelihood estimates for the parameters of time domain models.

7.5.1 Correlations

The empirical estimate of the variance of a stationary process is similar to the sample variance of a simple random sample:

$$s(0) \equiv \frac{1}{n} \sum_{i=1}^{n} (y_i - \bar{y}_.)^2.$$

This differs from the unbiased estimate for a random sample in that it uses the multiplier $1/n$ instead of $1/(n-1)$. An empirical estimate of $\sigma(k) = \text{Cov}(y_t, y_{t+k})$ can be based on the sample covariance of $(y_1, \ldots, y_{n-k})$ and $(y_{k+1}, \ldots, y_n)$. An estimate of $\sigma(k)$ is

$$s(k) = \frac{1}{n-k} \sum_{i=1}^{n-k} (y_i - \bar{y}_.)(y_{i+k} - \bar{y}_.).$$

This differs from the estimate for simple random samples in that (a) the multiplier is $1/(n-k)$ rather than $1/(n-k-1)$ and (b) $\bar{y}_.$ is used to estimate the mean of both samples rather than using $(y_1 + \cdots + y_{n-k})/(n-k)$ and $(y_{k+1} + \cdots + y_n)/(n-k)$, respectively.

Sometimes the alternative estimates

$$\hat{\sigma}(k) = \frac{n-k}{n} s(k) = \frac{1}{n} \sum_{i=1}^{n-k} (y_i - \bar{y}_.)(y_{i+k} - \bar{y}_.)$$

are used. The fact that the $\hat{\sigma}(k)$s all use the same multiplier is convenient for some purposes. In particular, it ensures that estimates of covariance matrices such as (7.1.2) are nonnegative definite; see Brockwell and Davis (1991, Section 7.2). Note that $\hat{\sigma}(0) = s(0)$.

The correlation function $\rho(k) = \sigma(k)/\sigma(0)$ is estimated with either

$$r(k) = s(k)/s(0)$$

or

$$\hat{\rho}(k) = \hat{\sigma}(k)/\hat{\sigma}(0).$$

Partial covariances and partial correlations as defined in Sect. 7.1 are functions of $\sigma(k)$. Either estimate of $\sigma(k)$ can be used to define estimated partial covariances and correlations. For example, substituting $\hat{\sigma}(k)$ into (7.1.1), (7.1.2), and (7.1.3) leads to an estimated partial correlation

$$\hat{\phi}(k) = \hat{\delta}_k,$$

where $\hat{\delta}_k$ is the first component of $\hat{\delta} = (\hat{\delta}_k, \ldots, \hat{\delta}_1)$ and $\hat{\delta}$ is defined by replacing $\sigma(\cdot)$ with $\hat{\sigma}(\cdot)$ in (7.1.4). An alternative to using matrix operations for finding $\hat{\phi}(k)$ is to use the results of Exercise 7.9.16.

Box, Jenkins, and Reinsel (1994, p. 323) suggest that estimated covariances are only useful when $n \geq 50$ and $k \leq \frac{n}{4}$. Inferences for correlations can be based on asymptotic normality.

7.5.2 Conditional Estimation for AR(p) Models

Various aspects of estimation in an $ARMA(p,q)$ model simplify when $q = 0$. This simplification can be quite useful, so we consider the special case first. In particular, three estimation methods are examined: empirical estimation, least squares, and maximum likelihood. The $AR(p)$ model is

$$y_t = \alpha + \sum_{s=1}^{p} \phi_s y_{t-s} + e_t,$$

where $E(e_t) = 0$, $Var(e_t) = \sigma^2$. The parameters to be estimated are $\alpha, \phi_1, \phi_2, \ldots, \phi_p$ and σ^2.

As established in Sect. 7.2,

$$\hat{E}(y_t | y_{t-1}, \ldots, y_{t-p}) = \mu + \sum_{s=1}^{p} \phi_s (y_{t-s} - \mu) = \alpha + \sum_{s=1}^{p} \phi_s y_{t-s}.$$

Thus, modifying (7.2.7), we get estimates $\hat{\phi}_1, \ldots, \hat{\phi}_p$ that satisfy

$$\hat{\sigma}_y(k) = \sum_{s=1}^{p} \hat{\phi}_s \hat{\sigma}(s - k)$$

for $k = 1, \ldots, p$. In particular,

$$
\begin{bmatrix} \hat{\phi}_1 \\ \vdots \\ \hat{\phi}_p \end{bmatrix} =
\begin{bmatrix}
\hat{\sigma}(0) & \hat{\sigma}(1) & \cdots & \hat{\sigma}(p-1) \\
\hat{\sigma}(1) & \hat{\sigma}(0) & \cdots & \hat{\sigma}(p-2) \\
\vdots & \vdots & & \vdots \\
\hat{\sigma}(p-1) & \hat{\sigma}(p-2) & \cdots & \hat{\sigma}(0)
\end{bmatrix}^{-1}
\begin{bmatrix} \hat{\sigma}(1) \\ \hat{\sigma}(2) \\ \vdots \\ \hat{\sigma}(p) \end{bmatrix}.
$$

Note that from our earlier discussion

$$\hat{\phi}(p) = \hat{\phi}_p.$$

Moreover, taking $\hat{\mu} = \bar{y}.$ gives

$$\hat{\alpha} = \left(1 - \sum_{s=1}^{p} \hat{\phi}_s\right)\hat{\mu}.$$

The second method of estimating $\alpha, \phi_1, \ldots, \phi_p$ is least squares. The least squares estimates are the values that minimize $\sum_{t=p+1}^{n} e_t^2$ or equivalently

$$\sum_{t=p+1}^{n} \left[y_t - \alpha - \sum_{s=1}^{p} \phi_s y_{t-s}\right]^2.$$

The sum is from $t = p+1$ to n because not all of the y_{t-s}s are observed for $t \le p$. Conditional methods use precisely the cases for which we have complete data.

In applying the Gauss–Newton method to obtain least squares estimates, let $v = (y_n, \ldots, y_{p+1})'$, $\xi = (\alpha, \phi_1, \ldots, \phi_p)'$, $z_t' = (1, y_{t-1}, \ldots, y_{t-p})$, and for $t = n, \ldots, p+1$

$$\alpha + \sum_{s=1}^{p} \phi_s y_{t-s} = f_t(\xi) = z_t'\xi.$$

Note that each $f_t(\xi)$ is a linear function of ξ, so $F(\xi)$ is a linear function of ξ. In particular,

$$Z = \begin{bmatrix} z_n' \\ \vdots \\ z_{p+1}' \end{bmatrix}$$

and $F(\xi) = Z\xi$. Example 7.4.2 applies, giving

$$\begin{bmatrix} \hat{\alpha} \\ \hat{\phi}_1 \\ \vdots \\ \hat{\phi}_p \end{bmatrix} = \hat{\xi} = (Z'Z)^{-1}Z'v,$$

which is obtained from fitting the linear model

$$v = Z\xi + e, \ E(e) = 0, \ \text{Cov}(e) = \sigma^2 I.$$

Thus, conditional least squares estimates for an $AR(p)$ model are particularly easy to find.

The conditional least squares estimates are also conditional maximum likelihood estimates when the process is Gaussian. The $AR(p)$ model is a special case of the

$ARMA(p,q)$ model and conditional least squares estimates are conditional MLEs in the more general model. The equivalence will be established later.

Finally, to estimate σ^2, one can use either

$$\hat{\sigma}^2 = \frac{1}{n-p} \sum_{t=p+1}^{n} \left[y_t - \hat{\alpha} - \sum_{s=1}^{p} \hat{\phi}_s y_{t-s} \right]^2$$

or

$$\text{MSE} = \frac{1}{n-2p-1} \sum_{t=p+1}^{n} \left[y_t - \hat{\alpha} - \sum_{s=1}^{p} \hat{\phi}_s y_{t-s} \right]^2 .$$

7.5.3 Conditional Least Squares for $\text{ARMA}(p,q)s$

We now consider conditional least squares estimates for the model

$$y_t = \alpha + \sum_{s=1}^{p} \phi_s y_{t-s} + e_t - \sum_{s=1}^{q} \theta_s e_{t-s} .$$

Let $\xi = (\alpha, \phi_1, \ldots, \phi_p, \theta_1, \ldots, \theta_q)'$. Conditional least squares estimates $\hat{\xi} = (\hat{\alpha}, \hat{\phi}_1, \ldots, \hat{\phi}_p, \hat{\theta}_1, \ldots, \hat{\theta}_q)'$ minimize $\sum_{t=p+1}^{n} e_t^2$ or equivalently

$$\text{SSE}_C(\xi) \equiv \sum_{t=p+1}^{n} \left(y_t - \left[\alpha + \sum_{s=1}^{p} \phi_s y_{t-s} - \sum_{s=1}^{q} \theta_s e_{t-s} \right] \right)^2 .$$

The difficulty in minimizing this is that the e_ts depend on the other parameters.

To apply the Gauss–Newton method, let $v = (y_n, \ldots, y_{p+1})'$ and

$$f_t(\xi) = \alpha + \sum_{s=1}^{p} \phi_s y_{t-s} - \sum_{s=1}^{q} \theta_s e_{t-s}(\xi) .$$

Differentiating,

$$\mathbf{d}_\xi f_t(\xi) = \left[\mathbf{d}_\alpha f_t(\xi), \mathbf{d}_{\phi_1} f_t(\xi), \ldots, \mathbf{d}_{\phi_p} f_t(\xi), \mathbf{d}_{\theta_1} f_t(\xi), \ldots, \mathbf{d}_{\theta_q} f_t(\xi) \right],$$

where

$$\mathbf{d}_\alpha f_t(\xi) = 1,$$

$$\mathbf{d}_{\phi_s} f_t(\xi) = y_{t-s} - \sum_{s=1}^{q} \theta_s \mathbf{d}_{\phi_s} e_{t-s}(\xi),$$

and

$$\mathbf{d}_{\theta_s} f_t(\xi) = -\sum_{s=1}^{q} [e_{t-s}(\xi) + \theta_s \mathbf{d}_{\theta_s} e_{t-s}(\xi)].$$

Given a current estimate $\xi_r = (\alpha^{(r)}, \phi_1^{(r)}, \dots, \phi_p^{(r)}, \theta_1^{(r)}, \dots, \theta_q^{(r)})'$, we need to be able to evaluate $f_t(\xi_r)$ and $\mathbf{d}_\xi f_t(\xi_r)$ for $t = n, \dots, p+1$. In particular, we need to be able to evaluate $e_t(\xi_r)$, $\mathbf{d}_{\psi_s} e_t(\xi_r)$, and $\mathbf{d}_{\theta_s} e_t(\xi_r)$. To do this, we repeat the assumption made in Sect. 7.3 that

$$e_t = 0, \qquad t = 0, -1, -2, \dots, \tag{7.5.1}$$

where now we think of these statements as implying that $e_t(\xi) = 0$ for all ξ. Thus, we have conditioned on the values of e_t, $t = 0, -1, -2, \dots$. Because the functions are constant the derivatives are zero; thus,

$$\mathbf{d}_\alpha e_t(\xi) = \mathbf{d}_{\phi_s} e_t(\xi) = \mathbf{d}_{\theta_s} e_t(\xi) = 0 \qquad t = 0, -1, -2, \dots. \tag{7.5.2}$$

Recalling that

$$e_t(\xi) = \sum_{j=1}^{q} \theta_j e_{t-j}(\xi) + y_t - \sum_{j=1}^{p} \phi_j y_{t-j}, \tag{7.5.3}$$

we have

$$\mathbf{d}_{\phi_s} e_t(\xi) = -y_{t-s} + \sum_{j=1}^{q} \theta_j \mathbf{d}_{\phi_s} e_{t-j}(\xi) \tag{7.5.4}$$

and

$$\mathbf{d}_{\theta_s} e_t(\xi) = e_{t-s}(\xi) + \sum_{j=1}^{q} \theta_j \mathbf{d}_{\theta_s} e_{t-j}(\xi). \tag{7.5.5}$$

Assumptions (7.5.1) and (7.5.2) along with (7.5.3), (7.5.4), and (7.5.5) allow us to compute $e_t(\xi_r)$ and the necessary partial derivatives. Computation of the $e_t(\xi_r)$s assuming (7.5.1) was discussed in Sect. 7.3. Our discussion there assumed that $\xi = (\alpha, \phi_1, \dots, \phi_p, \theta_1, \dots, \theta_q)'$ was known. We simply use those results assuming that $\xi = \xi_r$.

Computation of the partial derivatives is a bit more involved. We illustrate the method for $\mathbf{d}_{\theta_s} e_t(\xi)$ with $q = 2$. Assume that for $t = 1, \dots, n$, $e_t = e_t(\xi_r)$ has already been computed. We simplify notation by writing $\theta_s^{(r)} = \theta_s$ and $\mathbf{d}_{\theta_s} e_t(\xi)$ as $\mathbf{d}_{\theta_s} e_t$. The method makes repeated use of (7.5.1), (7.5.2), and (7.5.5).

For $t = 1$,

$$\begin{aligned}
\mathbf{d}_{\theta_1} e_1 &= e_0 + \theta_1 \mathbf{d}_{\theta_1} e_0 + \theta_2 \mathbf{d}_{\theta_1} e_{-1} \\
&= 0 + \theta_1 0 + \theta_2 0 \\
&= 0,
\end{aligned}$$

$$\begin{aligned}
\mathbf{d}_{\theta_2} e_1 &= e_{-1} + \theta_1 \mathbf{d}_{\theta_2} e_0 + \theta_2 \mathbf{d}_{\theta_2} e_{-1} \\
&= 0 + \theta_1 0 + \theta_2 0 \\
&= 0.
\end{aligned}$$

For $t = 2$,

$$\mathbf{d}_{\theta_1} e_2 = e_1 + \theta_1 \mathbf{d}_{\theta_1} e_1 + \theta_2 \mathbf{d}_{\theta_1} e_0$$
$$= e_1 + \theta_1 0 + \theta_2 0$$
$$= e_1,$$

$$\mathbf{d}_{\theta_2} e_2 = e_0 + \theta_1 \mathbf{d}_{\theta_2} e_1 + \theta_2 \mathbf{d}_{\theta_2} e_0$$
$$= 0 + \theta_1 0 + \theta_2 0$$
$$= 0.$$

For $t = 3$,

$$\mathbf{d}_{\theta_1} e_3 = e_2 + \theta_1 \mathbf{d}_{\theta_1} e_2 + \theta_2 \mathbf{d}_{\theta_1} e_1$$
$$= e_2 + \theta_1 e_1,$$

$$\mathbf{d}_{\theta_2} e_3 = e_1 + \theta_1 \mathbf{d}_{\theta_2} e_2 + \theta_2 \mathbf{d}_{\theta_2} e_1$$
$$= e_1 + \theta_1 0 + \theta_2 0$$
$$= e_1.$$

For $t = 4$,

$$\mathbf{d}_{\theta_1} e_4 = e_3 + \theta_1 \mathbf{d}_{\theta_1} e_3 + \theta_2 \mathbf{d}_{\theta_1} e_2$$
$$= e_3 + \theta_1 [e_2 + \theta_1 e_1] + e_1$$
$$= e_3 + \theta_1 e_2 + (1 + \theta_1^2) e_1,$$

$$\mathbf{d}_{\theta_2} e_4 = e_2 + \theta_1 \mathbf{d}_{\theta_2} e_3 + \theta_2 \mathbf{d}_{\theta_2} e_2$$
$$= e_2 + \theta_1 e_1 + \theta_2 0$$
$$= e_2 + \theta_1 e_1.$$

This procedure goes on recursively, thus allowing computation of $\mathbf{d}_{\theta_s} f_t(\xi_r)$, which is necessary to execute the Gauss–Newton algorithm.

7.5.4 Conditional MLEs for ARMA(p, q)s

It remains to establish the equivalence of conditional least squares and conditional maximum likelihood for Gaussian processes. To do this, we need to find the joint distribution of $y_n, \ldots, y_{p+1}$. The joint distribution is conditional on the unknown parameters ξ. We also condition on $e_p, e_{p-1}, \ldots$ or equivalently $y_p, y_{p-1}, \ldots$. Writing the stationary invertible ARMA(p, q) process as

$$\Phi(B)[y_t - \mu] = \Theta(B)e_t$$

and letting

$$\Psi(B) = \Theta(B)/\Phi(B)$$

we have

$$[y_t - \mu] = \Psi(B)e_t. \tag{7.5.6}$$

From (7.5.6), with $e_p, e_{p-1}, \dots$ fixed, each $y_t - \mu, t > p$ is a linear function of the random variables $e_t, \dots, e_{p+1}$. In particular, write

$$\begin{pmatrix} y_n \\ \vdots \\ y_{p+1} \end{pmatrix} = A \begin{pmatrix} e_n \\ \vdots \\ e_{p+1} \end{pmatrix} + \eta.$$

The fixed vector η is $\eta = (\eta_n, \dots, \eta_{p+1})'$ with $\eta_t = \mu + \sum_{s=0}^{\infty} \psi_{t-p+s} e_{p-s}$. Moreover,

$$A = \begin{bmatrix} a_n' \\ \vdots \\ a_{p+1}' \end{bmatrix}$$

and $a_t' = (0, \dots, 0, \psi_0, \psi_1, \dots, \psi_{t-p-1})$ with $\psi_0 = 1$. ($\psi_0 = 1$ because the first coefficients in both $\Phi(B)$ and $\Theta(B)$ are 1.) Observe that A is a nonsingular, upper triangular matrix with 1s down the diagonal. The distribution of $y = (y_n, \dots, y_{p+1})'$ is

$$y \sim N(\eta, \sigma^2 AA'),$$

and the density is

$$\begin{aligned} f(y) &= (2\pi)^{-\left(\frac{n-p}{2}\right)} |\sigma^2 AA'|^{-\frac{1}{2}} \exp[-(y-\eta)'(AA')^{-1}(y-\eta)/2\sigma^2] \\ &= (2\pi)^{-\left(\frac{n-p}{2}\right)} (\sigma^2)^{-\frac{n-p}{2}} |A|^{-1} \\ &\quad \times \exp[-\{A^{-1}(y-\eta)\}'\{A^{-1}(y-\eta)\}/2\sigma^2] \\ &= (2\pi)^{-\left(\frac{n-p}{2}\right)} (\sigma^2)^{-\frac{n-p}{2}} \exp[-\{A^{-1}(y-\eta)\}'\{A^{-1}(y-\eta)\}/2\sigma^2], \end{aligned} \tag{7.5.7}$$

where the last equality follows from the fact that A is upper triangular with 1s on the diagonal, so $|A| = 1$.

The problem with (7.5.7) is that it is a complicated function of our original parameters: ξ and σ^2. Both A and η depend on the coefficients of $\Psi(B)$. Note, however, that

$$A^{-1}(y - \eta) = (e_n, \dots, e_{p+1})',$$

where each e_t is a function of both y and the parameters ξ. Writing the $ARMA(p,q)$ model as

$$e_t(y, \xi) = \sum_{s=1}^{q} \theta_s e_{t-s}(y, \xi) + y_t - \sum_{s=1}^{p} \phi_s y_{t-s} - \alpha,$$

we see that

$$\{A^{-1}(y - \eta)\}' \{A^{-1}(y - \eta)\}$$

$$= \sum_{t=p+1}^{n} \left[y_t - \alpha - \sum_{s=1}^{p} \phi_s y_{t-s} + \sum_{s=1}^{q} \theta_s e_{t-s}(y, \xi) \right]^2 \qquad (7.5.8)$$

$$= \mathrm{SSE}_C(\xi),$$

and the density can be rewritten as

$$f(y) = (2\pi)^{-\frac{(n-p)}{2}} (\sigma^2)^{-\frac{(n-p)}{2}} \exp\left\{ -\frac{1}{2\sigma^2} \mathrm{SSE}_C(\xi) \right\}.$$

Thinking of this as a function of σ^2 and ξ with y fixed, we have the likelihood

$$L(\xi, \sigma^2) = (2\pi)^{-\frac{(n-p)}{2}} (\sigma^2)^{-\frac{(n-p)}{2}} \exp\left\{ -\frac{1}{2\sigma^2} \mathrm{SSE}_C(\xi) \right\}. \qquad (7.5.9)$$

Clearly, for any σ^2, the value of ξ that minimizes (7.5.8) will maximize the likelihood. However, by definition, the minimizing value of (7.5.8) is given by the least squares estimates. Given the least squares estimate $\hat{\xi}$, the MLE of σ^2 can then be found by solving $\mathbf{d}_{\sigma^2} L(\hat{\xi}, \sigma^2) = 0$ to get,

$$\hat{\sigma}^2 = \frac{1}{n-p} \sum_{t=p+1}^{n} \left[y_t - \hat{\alpha} - \sum_{s=1}^{p} \hat{\phi}_s y_{t-s} + \sum_{s=1}^{q} \hat{\theta}_s e_{t-s}(y, \hat{\xi}) \right]^2. \qquad (7.5.10)$$

7.5.5 *Unconditional Estimation for* **ARMA**(p, q) *Models*

We begin with maximum likelihood estimation. This discussion will lead naturally into least squares estimation. For unconditional MLEs, we need the density of $Y = (y_n, \cdots, y_1)'$. Again consider the equality (7.5.6); however, we now redefine A as an $n \times \infty$ matrix

$$A = \begin{bmatrix} a'_n \\ \vdots \\ a'_1 \end{bmatrix},$$

where

$$a'_t = (0, \cdots, 0, \psi_0, \psi_1, \psi_2, \cdots),$$

with the first $n - t$ columns equal to 0 and $\psi_0 = 1$. Writing the infinite vector

$$e = (e_n, e_{n-1}, e_{n-2}, \cdots)',$$

Eq. (7.5.6) becomes

$$Y = Ae + \mu J, \tag{7.5.11}$$

where J is an n vector of 1s. Recalling that the e_ts are i.i.d. $N(0, \sigma^2)$, we see that

$$Y \sim N(\mu J, \sigma^2 AA'),$$

so the unconditional likelihood function is

$$f(Y|\xi, \sigma^2) = (2\pi)^{-\frac{n}{2}} (\sigma^2)^{-\frac{n}{2}} |AA'|^{-\frac{1}{2}}$$
$$\times \exp\left[-(Y - \mu J)'(AA')^{-1}(Y - \mu J)\Big/2\sigma^2\right]. \tag{7.5.12}$$

It is important to note that both μ and A depend on the parameters ξ. When convenient, we write $A(\xi)$ for A. Equation (7.2.12) can be written in a somewhat simpler form. Note that

$$\hat{E}(e|Y, \xi) \equiv \hat{E}(e|Y) = A'(AA')^{-1}(Y - \mu J).$$

Clearly,

$$(Y - \mu J)'(AA')^{-1}(Y - \mu J) = \{\hat{E}(e|Y, \xi)\}'\{\hat{E}(e|Y, \xi)\}$$
$$= \sum_{t=-\infty}^{n} \{\hat{E}(e_t|Y, \xi)\}^2,$$

so substituting into (7.5.12) gives the likelihood

$$L(\xi, \sigma^2) = (2\pi)^{-\frac{n}{2}} (\sigma^2)^{-\frac{n}{2}} |A(\xi)A'(\xi)|^{-\frac{1}{2}}$$
$$\times \exp\left\{-\hat{E}(e|Y, \xi)'\hat{E}(e|Y, \xi)\Big/2\sigma^2\right\}. \tag{7.5.13}$$

Maximum likelihood estimates maximize the function (7.5.13). Least squares estimates minimize

$$\text{SSE}(\xi) \equiv (Y - \mu J)'(AA')^{-1}(Y - \mu J)$$
$$= \hat{E}(e|Y, \xi)'\hat{E}(e|Y, \xi)$$
$$= \sum_{t=-\infty}^{n} \{\hat{E}(e_t|Y, \xi)\}^2.$$

In unconditional estimation, the least squares estimates need not equal the maximum likelihood estimates. However, for moderate to large sample sizes n and parameter values that are not near their boundaries, the least squares estimate of ξ gives a very good approximation to the MLE; see Box et al. (1994, Section 7.1.4) and McLeod (1977). This phenomenon occurs because the determinant of the covariance matrix usually plays a very minor role in determining the maximum of the likelihood.

To actually find estimates, an iterative procedure is required. The Gauss–Newton method can be used to find least squares estimates. Because the determinant $|A(\xi)A'(\xi)|$ depends on ξ, some other method (e.g., Newton–Raphson) must be used to obtain MLEs. (Newton–Raphson was discussed in Christensen (1997, Section 12.4).) In addition, the values $\hat{E}(e_t|Y,\xi)$ need to be evaluated. As in Sect. 7.3, we could use

$$\hat{E}(e_t|Y,\xi) \doteq \hat{E}(e_t|Y_\infty,\xi),$$

where

$$\hat{E}(e_t|Y_\infty,\xi) = \begin{cases} 0 & t > n \\ e_t & t \le n \end{cases}, \tag{7.5.14}$$

along with the assumption that $e_t = 0$, $t \le 0$. This assumption implies that $y_t = \mu$, $t \le 0$, and thus we can compute e_t for $t = 1,\dots,n$ using

$$e_t = \sum_{s=1}^q \theta_s e_{t-s} + y_t - \sum_{s=1}^p \phi_s y_{t-s} - \alpha.$$

These very strong assumptions can be improved upon in practice by incorporating Box and Jenkins' method of *back forecasting (backcasting)*. Backcasting is used to obtain values for $\hat{E}(e_t|Y,\xi)$. It is based on the observation that if w_t and e_t are both white noise processes, the mean and covariance function of a stationary process defined by

$$\Phi(B)(y_t - \mu) = \Theta(B)e_t$$

must be the same as the mean and covariance function of

$$\Phi(F)(y_t - \mu) = \Theta(F)w_t, \tag{7.5.15}$$

where F is the *forward* shift operator (i.e., $F(w_t) = w_{t+1}$). Thus, model (7.5.15) is every bit as good a model for the data as the standard $ARMA(p,q)$ model. Now, rather than assuming $e_t = 0$, $t = 0, -1, \dots$ we assume $w_t = 0$ for $t = n+1, n+2, \dots$. Rather than using (7.5.14), we use

$$\hat{E}(w_t|Y^\infty,\xi) = \begin{cases} w_t & t \ge 1 \\ 0 & t < n \end{cases}, \tag{7.5.16}$$

where $Y^\infty = (y_1,\cdots,y_n,\mu,\mu,\mu,\cdots)'$. Note that the assumption $w_t = 0$ for $t \ge n+1$ implies that $y_t = \mu$ for $t \ge n+1$. Rewriting (7.5.15) as

$$w_t = \sum_{s=1}^q \theta_s w_{t+s} + y_t - \sum_{s=1}^p \phi_s y_{t+s} - \alpha, \tag{7.5.17}$$

we can compute $w_n, w_{n-1}, \cdots, w_1, \cdots$ recursively. This in turn allows us to backcast values $\hat{E}(y_t|Y^\infty,\xi)$ for $t < 1$, namely

$$\hat{E}(y_t|Y^\infty,\xi) = \alpha + \sum_{s=1}^p \phi_s \hat{E}(y_{t+s}|Y^\infty,\xi)$$

$$+ \hat{E}(w_t|Y^\infty, \xi) - \sum_{s=1}^{q} \theta_s \hat{E}(w_{t+s}|Y^\infty, \xi),$$

where the terms $\hat{E}(w_t|Y^\infty, \xi)$ are computed using (7.5.16) and (7.5.17) and the values $\hat{E}(y_t|Y^\infty, \xi)$ are computed recursively for $t < 1$ and for $t \geq 1$, $\hat{E}(y_t|Y^\infty, \xi) = y_t$.

The backcasting values $\hat{E}(y_t|Y^\infty, \xi)$, $t < 1$ can be used in the standard ARMA model to improve on the assumption $e_t = 0$, $t < 1$ and its consequence $y_t = \mu$, $t < 1$. Instead of these assumptions, we choose a large value Q and assume that $\hat{E}(y_t|Y, \xi) = \hat{E}(y_t|Y^\infty, \xi)$ for $t = 0, 1, \ldots, 1 - Q$ and that $e_t = 0$ for $t < 1 - Q$. As before, the assumption on e_t implies that $y_t = \mu$ for $t < 1 - Q$.

The functions $L(\xi, \sigma^2)$ and SSE(ξ) are now evaluated using

$$\hat{E}(e_t|Y, \xi) = \sum_{s=1}^{q} \theta_s \hat{E}(e_{t-s}|Y, \xi) + \hat{E}(y_t|Y, \xi) - \sum_{s=1}^{p} \phi_s \hat{E}(y_{t-s}|Y, \xi) - \alpha,$$

where

$$\hat{E}(y_t|Y, \xi) \doteq \begin{cases} y_t & t = 1, \ldots, n \\ \hat{E}(y_t|Y^\infty, \xi) & t = 1 - Q, \ldots, 0 \\ \mu & t \leq -Q, \end{cases}$$

$$\hat{E}(e_t|Y, \xi) \doteq 0 \qquad t \leq -Q,$$

and $\hat{E}(e_t|Y, \xi)$ is computed recursively for $t > -Q$.

One problem is that a specific choice of Q must be made. Without getting into details, we mention a basic consideration in the selection of Q. For a stationary ARMA process, the covariance function is given in (7.2.12) as $\sigma(k) = \sigma^2 \sum_{s=0}^{\infty} \psi_s \psi_{s+k}$. Because $\sum_{s=0}^{\infty} \psi_s$ is finite, $\lim_{s \to \infty} \psi_s = 0$. It is intuitively obvious (and not hard to show) that if the correlation between y_{n+k} and the observed data is approaching zero, then $\hat{E}(y_{n+k}|Y)$ must approach μ, the mean of the stationary process. If the correlation between the future and the present is zero, there is no basis for predicting the future as anything other than the mean of the process.

The same phenomenon occurs in back forecasting. For a stationary process, the back forecasts should settle down around $\hat{\mu}$ for times t that are large and negative. Moreover, when the backcasts have approached μ, the predictions of e_t must be near zero. For a given Q, by checking whether these phenomena occur for t near $-Q$, one can decide whether Q is sufficiently large.

Brockwell and Davis (1991, Chapter 8) present methods for unconditional maximum likelihood and least squares estimation based on the innovations algorithm; see Exercises 7.2 and 7.6. These use the exact value of $\hat{E}(e_t|Y, \xi)$ rather than the approximation $\hat{E}(e_t|Y_\infty, \xi)$. Brockwell and Davis also present asymptotic distributional and inferential results.

Exercise 7.6. Let $e(y_t|y_{t-1}, \cdots, y_1; \xi) = y_t - \hat{E}(y_t|y_{t-1}, \cdots, y_1, \xi)$ from an ARMA(p, q) model with parameters (ξ, σ^2), and let $p_t(\xi, \sigma^2)$ be the corresponding prediction variance. Use the results of Exercise 7.2 to show that

$$L(\xi,\sigma^2) = (2\pi)^{-\frac{n}{2}} \left[\prod_{t=1}^{n} p_t(\xi,\sigma^2)\right]^{-\frac{1}{2}}$$

$$\times \exp\left\{-y_1^2/2\sigma(0)^2 - \sum_{t=1}^{n} [e(y_t|y_{t-1},\cdots,y_1;\xi)]^2 \Big/ 2p_t(\xi,\sigma^2)\right\}.$$

Hint: Show that there is a nonsingular linear transformation between the data vector Y and the vector of sequential prediction errors.

7.5.6 Estimation for ARIMA(p,d,q) Models

As in Sect. 7.3, $ARIMA(p,d,q)$ models are analyzed by considering the corresponding $ARMA(p,q)$ model for

$$z_t = \nabla^d y_t.$$

7.6 Model Selection

Box et al. (1994) suggest performing model selection by examination of the empirical correlations and partial correlations. Alternatively, general model selection criteria such as the Akaike Information Criterion (AIC) and Schwarz's asymptotic Bayesian modification of the AIC can be applied to time domain models; see Akaike (1973) and Schwarz (1978).

7.6.1 Box–Jenkins

The Box–Jenkins approach is based on two facts. First, for an $AR(p)$ process, $\phi(k) = 0$ for $k > p$. Second, for a $MA(q)$ process, $\rho(k) = 0$ for $k > q$. Empirical estimates $\hat{\phi}(k)$ and $\hat{\rho}(k)$ are obtained as in the previous section.

Recall that the stationary invertible $ARMA(p,q)$ model

$$\Phi(B)[y_t - \mu] = \Theta(B)e_t$$

can be written as an infinite moving average

$$y_t - \mu = \frac{\Theta(B)}{\Phi(B)}e_t$$

and also as an infinite autoregressive process

$$\frac{\Phi(B)}{\Theta(B)}[y_t - \mu] = e_t.$$

Moreover, the infinite sum $\Theta(1)/\Phi(1) = 1/[\Phi(1)/\Theta(1)]$ is a finite number, so the terms in each sequence converge to zero fairly quickly.

If $q = 0$, $\hat{\rho}(k)$ should gradually approach zero while $\hat{\phi}(k)$ drops off precipitously after $k = p$. Similarly if $p = 0$, $\hat{\phi}(k)$ should gradually approach zero while $\hat{\rho}(k)$ drops off after $k = p$. If both p and q are positive, then the first p terms of $\Phi(B)/\Theta(B)$ will tend to dominate, so $\hat{\phi}(k)$ should drop off substantially after $k = p$. Similarly, the first q terms of $[\Phi(B)]^{-1}\Theta(B)$ should dominate, leading $\hat{\rho}(k)$ to drop off substantially after $k = q$.

For example, if $|\hat{\phi}(k)|$ drops precipitously to near zero after $k = 3$, and if $|\hat{\rho}(k)|$ decreases gradually to zero, an $AR(3)$ model is suggested. If $|\hat{\rho}(k)|$ drops quickly after $k = 2$ and $|\hat{\phi}(k)|$ decreases gradually, a $MA(2)$ model is suggested. If $|\hat{\phi}(k)|$ drops after $k = 3$ and $|\hat{\rho}(k)|$ drops after $k = 2$, an $ARMA(3,2)$ model is suggested.

The standard errors of $\hat{\rho}(k)$ and $\hat{\phi}(k)$ can be used to help decide which correlations are important. For white noise (i.e., $\rho(k) = 0$ for all $k \geq 1$), the large sample standard deviation of $\hat{\rho}(k)$ is estimated by

$$\mathrm{SE}(\hat{\rho}(k)) = \frac{1}{\sqrt{n}}\left[1 + 2\sum_{j=1}^{k-1}\hat{\rho}(j)^2\right]^{1/2};$$

see Bartlett (1946). Sometimes people take $\mathrm{SE}(\hat{\rho}(k)) = 1/\sqrt{n}$ which comes from substituting $\rho(j) = 0$ for $\hat{\rho}(j)$. Similarly, Quenille (1949) has shown that for large samples from an $AR(p)$ the standard error of $\hat{\phi}(k)$ for $k > p$ is

$$\mathrm{SE}(\hat{\phi}(k)) \doteq 1/\sqrt{n}.$$

To identify important correlations, it is useful to plot $(k, \hat{\rho}(k))$ and $(k, \hat{\phi}(k))$. Values with absolute values greater than, say, twice the standard error are likely to be nonzero.

Nonstationarity can sometimes be identified by graphing the time series. A stationary process should display constant variability about its mean value μ. If the variability in the plot seems to increase with time, a log transformation may alleviate the problem. If a trend appears in the plot, differencing (i.e., considering $\nabla^d y_t$) would be in order. Once d has been determined so that the plot looks stationary, p and q can be identified as earlier, yielding an $ARIMA(p,d,q)$ model.

As a supplement to visual inspection of the graph, there are several statistics that give suggestions of nonstationarity. First, a value of $|\hat{\phi}(1)|$ near 1 suggests possible nonstationarity. Also, $\hat{\rho}(k)$ approaching zero very slowly suggests nonstationarity. Finally, nonstationarity is suggested in the frequency domain by the existence of one or more small frequencies that are very important. In practice, the differencing parameter d is rarely taken to be greater than 2.

Having identified a time domain model, we can attempt to check its appropriateness. In any time domain model, e_t is a white noise process. Assume the most general model, an $ARIMA(p,d,q)$. The e_t process can be estimated by

$$\hat{e}_t = \sum_{s=1}^{q} \hat{\theta}_s \hat{e}_{t-s} + \nabla^d (y_t - \hat{\mu}) - \sum_{s=1}^{p} \hat{\phi}_s \nabla^d (y_{t-s} - \hat{\mu}),$$

where it is assumed that $\hat{e}_t = 0$, $t = 0, -1, -2, \dots$. Brockwell and Davis (1991, Section 9.4) suggest the use of standardized residuals based only on previous data,

$$r_t = \left[y_t - \hat{E}(y_t | y_{t-1}, \cdots, y_1; \hat{\xi}) \right] \Big/ \sqrt{p_t(\hat{\xi}, \hat{\sigma}^2)},$$

where $p_t(\xi, \sigma^2)$ is the prediction variance as found in Sect. 7.3. These residuals arise naturally when using the innovations algorithm to obtain unconditional maximum likelihood estimates (see Exercise 7.6) and appear to be an improvement over the unstandardized residuals $\hat{e}_t$.

The residuals can be plotted against time to detect evidence of nonstationarity. For example, the residuals may display a trend or increasing variability.

The correlation and partial correlation functions for white noise are

$$\rho(0) \equiv \phi(0) = 1,$$
$$\rho(k) = \phi(k) = 0 \qquad k = 1, 2, \dots .$$

The estimated correlations and partial correlations for the residual process should be similar if the model is correct.

The spectral density of white noise is $f(v) = \sigma^2$. If the model is correct, the residual process should not have any important frequencies.

Finally, to check whether the white noise is Gaussian, a normal plot of the residuals can be used; see *PA-V* Sect. 12.2 (Christensen 2011, Section 13.2).

7.6.1.1 Testing Model Fit

The *Ljung-Box* modification of the *Box-Pierce* portmanteau test is used to evaluate whether the estimated autocorrelation function is consistent with the process being white noise. When evaluating the fit of an $ARMA(p,q)$ model, the autocorrelation function should be computed from the estimated errors (residuals). Specifically, compare the test statistic

$$Q(h) \equiv n(n+2) \sum_{k=1}^{h} \frac{\hat{\rho}(k)^2}{n-k}$$

to a $\chi^2(h - df)$ distribution where, excluding σ^2, $df = p + q + 1$ is the number of parameters in an $ARMA(p,q)$ model with nonzero mean. Frequently, this test is performed for several values of h. Minitab uses $h = 12, 24, 36, 48$. At https://robjhyndman.com/hyndsight/ljung-box-test/ (Last accessed on 13 August 2019) Rob Hyndman recommended $h = \min(10, n/5)$ and that you should not be too fussy about choosing h.

7.6.2 Model Selection Criteria

If one has identified several candidate models, a formal model selection criterion can be used to identify the best of these. The criteria to be considered require a maximum likelihood fit of the model, so computing the criteria for large numbers of models is expensive. (Using approximate fits for large numbers of models may be a reasonable practical procedure.)

The AIC and Schwarz's large sample Bayesian modification, the BIC, are discussed in Clayton, Geisser, and Jennings (1985). As applied to Gaussian ARMA models that include a mean and an unknown error variance, so that the number of fitted parameters is $k = p + q + 2$, and with a log-likelihood denoted $\ell(\xi, \sigma^2)$, these criteria can be estimated by

$$\text{AIC} = -2\ell(\hat{\xi}, \hat{\sigma}^2) + 2k$$

and

$$\text{BIC} = -2\ell(\hat{\xi}, \hat{\sigma}^2) + k\log(n)$$

where the estimates are MLEs.

The model that minimizes the criterion is the best fitting model. In practice, one should use these criteria to identify a small group of best models. For each of these models, the residual process should be evaluated to determine whether a Gaussian white noise assumption appears reasonable. For multivariate autoregressions, a simulation by Lütkepohl (1985) suggests that the BIC outperforms the AIC and many other criteria in terms of correct model identification and minimization of prediction errors. Clayton et al. (1985) examine simulations for data that are not time series. Hurvich and Tsai (1989) discuss correcting the AIC for bias which leads to

$$\text{AICc} = AIC + 2k(k+1)/[n-k-1]$$

PA-V gives a brief discussion of these measures for linear models and some additional references.

The AIC is not asymptotically consistent in the sense that a consistent criterion asymptotically chooses the correct model. The BIC criterion is a modification of the AIC specifically developed to achieve consistency. It is not surprising that the BIC outperforms criteria that are not consistent for large samples when the true model is one of the models being considered. In practice, however, our models are only approximations to reality. The appropriate question is not "Which criterion selects the correct model most often?" but "Which criterion selects the best approximation most often?"

7.6.3 An Example

EXAMPLE 7.6.1. Consider again the coal production data from Example 6.0.1. The data are given in Table 6.1 and are displayed in Fig. 6.1. Figure 7.1 gives the estimated correlation and partial correlation functions. Note that the correlation plot from R's acf function includes $\hat{\rho}(0) \equiv 1$ but the partial correlation (pacf) plot does not include $\hat{\phi}(0)$. (The forecast library's Acf and Pacf functions seem to be consistent.) Standard errors for the correlations can be based on Bartlett and Quenille's asymptotic formulae; the resulting 95% rejection regions for testing that the correlations are zero are displayed as dashed lines in the figures. Note that multiple tests of the correlations are being performed without controlling the overall error rate; thus, any marginally significant correlations are still questionable. The autocorrelations are dying out quite slowly, and the dominant partial autocorrelation is $\hat{\phi}(1) = 0.769$. These traits are consistent with a nonstationary process.

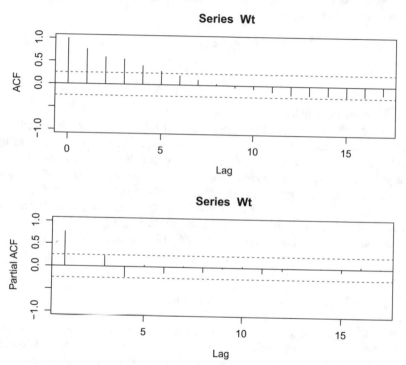

Fig. 7.1 Estimated autocorrelations and partial autocorrelations for the coal production data. The label "Wt" refers to the weight of coal produced

Figure 7.2 gives an old fashioned plot of the estimated correlations. It has two advantages. First, it actually gives the numerical correlation estimates $\hat{\rho}(k)$. Second, one can more readily apply the "ink test." The fact that the plot contains a great deal of ink suggests that the process in nonstationary. It is a little more work to see that in Fig. 7.1.

In an attempt to eliminate nonstationarity, we consider the first difference process $y_{i+1} - y_i$, $i = 1, \ldots, 60$. The data are plotted in Fig. 7.3. The estimated autocorrela-

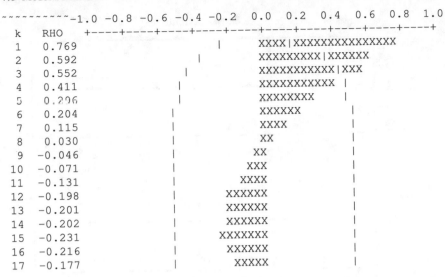

```
~~~~~~~~~~-1.0 -0.8 -0.6 -0.4 -0.2  0.0  0.2  0.4  0.6  0.8  1.0
   k   RHO    +----+----+----+----+----+----+----+----+----+----+
   1  0.769                       |     XXXX | XXXXXXXXXXXXXXXX
   2  0.592                   |         XXXXXXXXX | XXXXXX
   3  0.552                 |           XXXXXXXXXXX | XXX
   4  0.411             |               XXXXXXXXXXX  |
   5  0.296             |               XXXXXXXX      |
   6  0.204           |                 XXXXXX        |
   7  0.115           |                 XXXX          |
   8  0.030           |                 XX            |
   9 -0.046           |                 XX            |
  10 -0.071           |                 XXX           |
  11 -0.131           |                XXXX           |
  12 -0.198           |              XXXXXX            |
  13 -0.201           |              XXXXXX            |
  14 -0.202           |              XXXXXX            |
  15 -0.231           |             XXXXXXX            |
  16 -0.216           |              XXXXXX            |
  17 -0.177           |               XXXXX            |
```

Fig. 7.2 Estimated autocorrelations for the coal production data

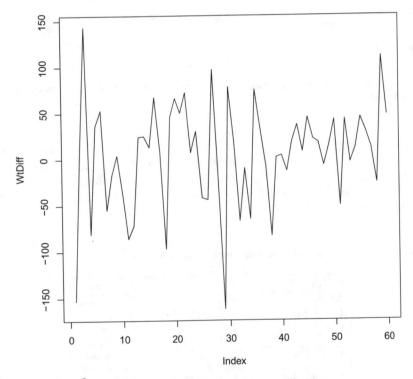

Fig. 7.3 First difference of the coal production data

tion and partial autocorrelation functions are given in Fig. 7.4. Neither plot gives any strong indication of nonstationarity, so we proceed to model the first differences.

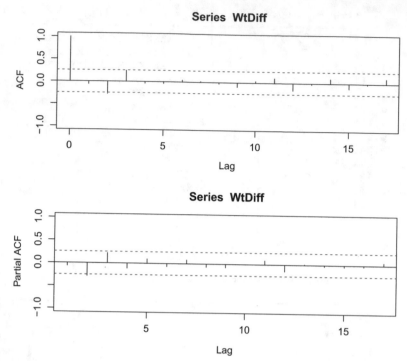

Fig. 7.4 Estimated autocorrelations and partial autocorrelations for the first difference of the coal production data

The small correlation and partial correlation values at $k = 1$ and the relatively large values at $k = 2, 3$ suggest that the *ARIMA* models $(0, 1, 0)$, $(2, 1, 2)$, and $(3, 1, 3)$ should be examined. Summary statistics for the three models along with $(1, 1, 1)$ and several others are contained in Table 7.1. All of the models in Table 7.1 were fitted by maximum likelihood.

Table 7.1 Coal production ARIMA model selection statistics

MODEL	$\hat{\sigma}^2$	$\ell(\hat{\xi}, \hat{\sigma}^2)$	AIC	AICc	BIC
$(0, 1, 0)$	3394	−328.52	661.04	661.25	665.23
$(1, 1, 1)$	3426	−327.79	663.59	664.31	671.96
$(2, 1, 2)$	2925	−322.31	656.62	658.20	669.18
$(3, 1, 3)$	3024	−322.21	660.43	663.25	677.18
$(2, 1, 1)$	2873	−322.32	654.63	655.74	665.11
$(1, 1, 2)$	2952	−323.09	656.18	657.29	666.65
$(2, 1, 1)*$	2857	−322.67	653.35	654.08	661.73
$(1, 1, 2)*$	2934	−323.43	654.86	655.59	663.24
* Indicates a model fitted without a constant					

Note that relatively small changes in the AIC and BIC criteria correspond to substantial changes in $\hat{\sigma}^2$. The importance of small changes in the information criteria should not be discounted. Of the four models that were considered with $p = q$, $(2,1,2)$ fits the best as judged by $\hat{\sigma}^2$, AIC, and AICc and it fits second best as judged by the BIC. The best BIC model is the rather uninteresting $(0,1,0)$.

The estimates of the parameters in the $ARIMA(2,1,2)$ model follow.

Table of coefficients: $ARIMA(2,1,2)$							
Unconditional least squares				Maximum likelihood			
Parameter	Estimate	SE	t	Parameter	Estimate	SE	t
ϕ_1	−0.8029	0.3075	−2.61	ϕ_1	−0.7828	0.3147	−2.49
ϕ_2	−0.4551	0.2576	−1.77	ϕ_2	−0.4416	0.3182	−1.39
θ_1	−0.8029	0.3400	−2.36	θ_1	−0.7840	0.3268	−2.40
θ_2	−0.0414	0.3301	−0.13	θ_2	−0.0443	0.3575	−0.12
α	10.61	12.71	0.83	α	4.6751	5.5226	0.85
σ^2	2848			σ^2	2925		

The two methods for fitting the model, Unconditional Least Squares (from Minitab) and Maximum Likelihood (from R), give comparable results.

I dropped the highly insignificant θ_2 parameter and fit the $ARIMA(2,1,1)$.

Table of coefficients: $ARIMA(2,1,1)$							
Unconditional least squares				Maximum likelihood			
Parameter	Estimate	SE	t	Parameter	Estimate	SE	t
ϕ_1	0.840	0.158	5.33	ϕ_1	−0.7514	0.1916	−3.92
ϕ_2	−0.011	0.143	−0.08	ϕ_2	−0.4055	0.1428	−2.84
θ_1	0.954	0.114	8.38	θ_1	−0.7513	0.1860	4.04
α	0.608	0.370	1.65	α	4.6741	5.4577	0.86
σ^2	3393			σ^2	2873		

Many of these estimates are not even close. Moreover, based on the estimates of σ^2, the Minitab fit indicates that $ARIMA(2,1,1)$ is a substantially worse model than $ARIMA(2,1,2)$, which is counterintuitive since the smaller model just drops θ_2, a parameter that both $ARIMA(2,1,2)$ fits agree has extremely little significance. Part of the problem may be associated with the fact that the parameter estimates are highly correlated. The asymptotic correlation matrix for the unconditional least squares fit of the $(2,1,2)$ model is given below.

	$\hat{\phi}_1$	$\hat{\phi}_2$	$\hat{\theta}_1$	$\hat{\theta}_2$	$\hat{\alpha}$
$\hat{\phi}_1$	1.000	0.801	0.918	0.867	−0.028
$\hat{\phi}_2$	0.801	1.000	0.731	0.863	−0.018
$\hat{\theta}_1$	0.918	0.731	1.000	0.920	−0.030
$\hat{\theta}_2$	0.867	0.863	0.920	1.000	−0.027
$\hat{\alpha}$	−0.028	−0.018	−0.030	−0.027	1.000

Unlike previous editions of this book, I'm going to believe the maximum likelihood fit (R was not available at the time of the previous edition) and proceed with that analysis.

From the statistics in Table 7.1, the models $(2,1,1)^*$, and $(2,1,1)$ appear reasonable. The model $(2,1,1)^*$ is an *ARIMA* model in which the constant is assumed to be zero and looks better than $(2,1,1)$. The R library forecast has a procedure for automatically finding the best model. In this case auto.arima arrived at the same model: *ARIMA*$(2,1,1)$ with no constant. (forecast also has testing procedures for determining how many differences are needed.)

Having had the benefit of looking at the 7 years of additional data displayed in Fig. 6.11, I am inclined to (cheat and) believe that the series has an increasing trend, so my choice for further consideration is $(2,1,1)$. Recall that a linear trend corresponds to a nonzero mean in the first difference process.

The asymptotic correlation matrix for the estimated parameters is

	$\hat{\phi}_1$	$\hat{\phi}_2$	$\hat{\theta}_2$	$\hat{\alpha}$
$\hat{\phi}_1$	1.000	0.042	-0.759	-0.022
$\hat{\phi}_2$	0.042	1.000	0.390	-0.022
$\hat{\theta}_2$	-0.759	0.390	1.000	0.008
$\hat{\alpha}$	-0.022	-0.022	0.008	1.000

The residuals from this fitted model should be evaluated to see if they are consistent with a white noise error process. The correlation and partial correlation functions are given in Fig. 7.5. Neither seems unreasonable. In fact, the P values for the Ljung-Box test at $h = 10, 15, 20$ are almost suspiciously high; all exceeding 0.99.

In results not shown, a rankit (normal scores) plot looks reasonably linear, and a frequency analysis of the residuals gives spectral estimates that are consistent with the error process being white noise. The primary problem with the residuals is that the variability seems to decrease after 1960. The plot of residuals versus year is given in Fig. 7.6.

Assuming that the possible heteroscedasticity is not a serious problem, what have we accomplished? Not a great deal. The model $(2,1,1)$ does not really explain the data very well. Fuller (1976, p. 364) suggests using F statistics for approximate tests of model adequacy. These are certainly reasonable statistics for model comparisons. The problem with their use is in determining an appropriate reference distribution. From the R output of a maximum likelihood fit to an *ARIMA*(p,d,q) with a constant, the sum of squares error for the model is

$$SSE = \hat{\sigma}^2 \times (n - p - d - q - 1)$$

Testing $(2,1,1)$ against $(0,1,0)$ gives

$$F = \frac{(200222.1 - 160901.9)/3}{2873.249} = 4.56.$$

If we compare the test statistic to an $F(3,56)$ distribution, we obtain a rough P *value* of 0.01. For comparing a model to the model of no *ARMA* structure, this is not particularly impressive. The R^2 type measure

$$\frac{200222.1 - 160901.9}{200222.1} = 0.196$$

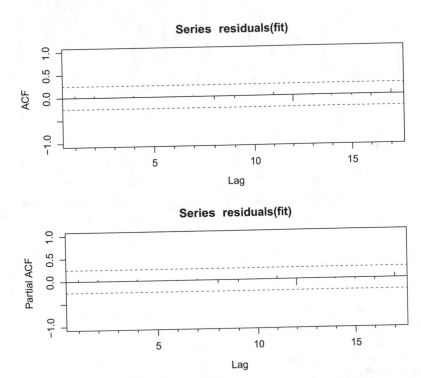

Fig. 7.5 Estimated autocorrelations and partial autocorrelations for the residuals of the *ARIMA*(2,1,1) model

indicates that the $(2,1,1)$ model accounts for only 20% of the variability in the data. This, by itself, does not mean that the fitted model is a poor one; we could have the perfect model but a process that is subject to a great deal of variability. However, whether the model is appropriate or not, we cannot expect to obtain accurate fore-casts from it. The squared correlation between the data y_t and the fitted values $\hat{y}_t$ is a respectable 75% but the justification of looking at this squared correlation when the y_ts are dependent is dubious.

One of the most important uses of time domain models is to predict the future. Table 7.2 contains the actual coal production figures for 1981 through 1987 along with the forecasts based on the $(2,1,1)$ model fitted both with and without a nonzero

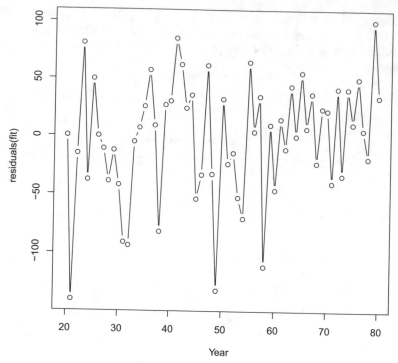

Fig. 7.6 Time plot for the residuals of the $(2, 1, 1)$ model

mean. Ninety-five percent prediction intervals are given for both models. While neither model gives very accurate predictions, the model that includes a trend does a
bit better. The model without a trend gives predictions that are settling down around
792. For both models, the prediction intervals are so large that, in spite of the poor
point predictions, the actual coal production values fall inside them. This results
from the large estimates of σ^2 associated with the models.

Table 7.2 *ARIMA* model forecasts

Year	Actual	*ARIMA* $(2,1,1)$	95% Limits Lower	Upper	*ARIMA* $(2,1,1)*$	95% Limits Lower	Upper
81	818.4	779.18	674.12	884.24	774.76	669.99	879.53
82	833.5	803.50	654.93	952.06	792.71	643.66	941.77
83	777.9	813.36	652.20	974.51	798.48	636.24	960.72
84	891.8	806.17	619.37	992.97	787.15	599.20	975.10
85	879.0	817.65	611.30	1024.01	793.34	585.68	1001.01
86	886.1	822.02	600.77	1043.27	793.16	570.20	1016.12
87	912.7	824.16	585.88	1062.45	790.87	550.78	1030.96
The model $(2,1,1)*$ is fitted with a mean of zero							

The data from 1921 to 1987 are plotted in Fig. 7.7 along with the predicted values from $(2,1,1)$. For the data from 1921 to 1980, which were used in estimation, the model consistently predicts the next observation to be similar to the last one. In other words, the predicted values are very similar to the data except that they are shifted to the right by 1 year. This phenomenon makes Fig. 7.7 appear misleadingly good. Visually, people tend to evaluate the proximity of two curves by their orthogonal distance. In Fig. 7.7, this is quite small because of the shifting phenomenon described earlier. In fact, for the purpose of prediction, we need to evaluate the vertical distance between the two curves at each time point. Thus, although it is less directly relevant to the prediction problem, the plot of the residuals, Fig. 7.7, gives a more accurate picture of the quality of the forecasts. The 1 year shift phenomenon in Fig. 7.7 is a direct result of the poor-fitting $(2,1,1)$ model. We are actually fitting an $ARMA(2,1)$ model to $z_t = y_t - y_{t-1}, t = 2, \ldots, 61$. The predicted differences are, say, $\hat{z}_t$, and the predicted values are $\hat{y}_t = \hat{z}_t + y_{t-1}$. As compared to the y_{t-1} values, the predicted differences are relatively close to zero. Thus, the prediction plot is essentially the data plot lagged by one.

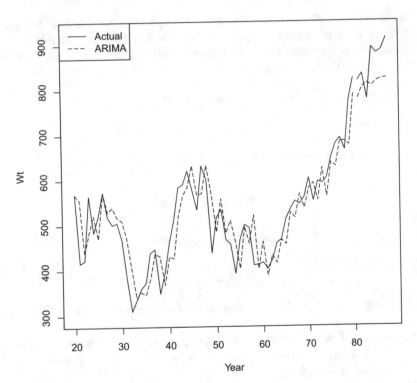

Fig. 7.7 Coal production (solid line) and predicted values (dashed line) for 1921 to 1987. $ARIMA(2,1,1)$ with constant

The quality of the predictions goes down markedly after 1980. It is always harder to predict the future than the past, and our model is not particularly effective. Actually, both Figs. 7.6 and 7.7 suggest that the behavior of the series since 1960 may be inconsistent with the previous data. Thus, it is not surprising that the predictions for 1981 to 1987 are not terribly good. Perhaps a better approach would be to analyze only the data from 1960 forward. Unfortunately, 21, or even 28 observations make a rather inadequate database for time series analysis.

In general, our only hope for prediction is that the future will behave like the relevant past. If there is no relevant past or if we include the irrelevant past, as we may have in this example, we are left to swim in the oceans of uncertainty, or worse, to follow random paths in the deserts of decision making that lead nowhere but give the illusion of progress. (Ok, ok, I'll take my tongue out of my cheek now.)   □

7.7 Seasonal Adjustment

Consider a time series consisting of monthly flypaper sales. (If you don't know what flypaper is, think of bug zappers.) It is reasonable that sales may be related to the previous couple of months, but sales may very well be related to the corresponding values in the previous year. An appropriate autoregressive model might be

$$y_t = \alpha + \phi_1 y_{t-1} + \phi_2 y_{t-2} + \phi_{1,1} y_{t-12} - \phi_1 \phi_{1,1} y_{t-13} - \phi_2 \phi_{1,1} y_{t-14} + e_t,$$

or equivalently,

$$y_t = \alpha + \phi_1 y_{t-1} + \phi_2 y_{t-2} + \phi_{1,1} [y_{t-12} - \phi_1 y_{t-13} - \phi_2 y_{t-14}] + e_t. \tag{7.7.1}$$

Write

$$\Phi(B) = 1 - \phi_1 B - \phi_2 B^2$$

and

$$\Phi_1(B^{12}) = 1 - \phi_{1,1} B^{12};$$

then, model (7.7.1) is

$$\Phi_1(B^{12})\Phi(B)[y_t - \mu] = e_t.$$

This is referred to as a *multiplicative seasonal autoregressive* model and is written $AR(2) \times (1)_{12}$.

In general, if the seasonal effects occur every T time units, we can define a *multiplicative seasonal autoregressive integrated moving average* model $ARIMA(p,d,q) \times (P,D,Q)_T$,

$$\Phi_P(B^T)\Phi(B)\nabla_T^D\nabla^d[y_t - \mu] = \Theta_Q(B^T)\Theta(B)e_t,$$

where

$$\Phi_P(B^T) = 1 - \sum_{s=1}^{P} \phi_{s,P} B^{Ts},$$

$$\nabla_T^D = (1 - B^T)^D,$$

and

$$\Theta_Q(B^T) = 1 - \sum_{s=1}^{Q} \theta_{s,Q} B^{Ts}.$$

These models are discussed in detail in Box et al. (2015).

EXAMPLE 7.7.1.   Table 7.3 consists of data from a daily census made of the in-patients at the University of Wisconsin Hospital between April 21 and July 29, 1974. These data were previously presented by Pandit and Wu (1983, Appendix A). Figure 7.8 presents the data graphically. There is a clear periodicity in the data. While this suggests that a frequency analysis of the data may be particularly enlightening, our object is to illustrate techniques used for identifying multiplicative seasonal ARIMA models. Figure 7.9 presents the estimated correlation function and the estimated partial correlation function for the series. Note that the correlation function displays periodic behavior with a period of 7. This makes sense in terms of a weekly cycle. Moreover, the correlations are dying out very slowly. Together, these suggest the appropriateness of a seasonal difference of order 7.

It should be noted that the need for seasonal differencing is often accompanied by the need for regular differencing. Moreover, the need for seasonal differencing is often hidden until the regular differencing is completed.

Table 7.3 University of Wisconsin hospital data

| Week | Day | | | | | | |
	Su	M	Tu	W	Th	F	Sa
I	397	462	486	483	477	438	407
II	421	480	484	486	479	415	400
III	419	477	510	503	500	435	408
IV	417	478	497	500	512	450	421
V	423	471	496	478	463	413	396
VI	366	375	444	469	480	439	402
VII	442	492	507	518	493	439	399
VIII	428	476	499	488	460	419	380
IX	406	472	502	495	490	443	398
X	417	490	505	499	484	430	384
XI	392	452	455	426	414	405	379
XII	410	485	514	525	511	461	436
XIII	444	488	494	510	493	429	392
XIV	420	466	476	494	484	423	388
XV	411	472					

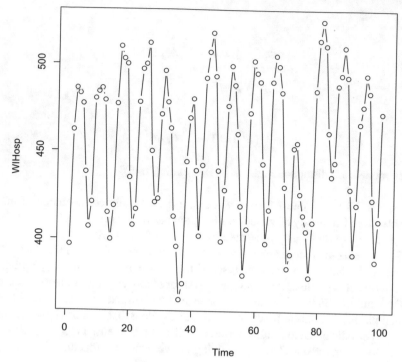

Fig. 7.8 Plot of hospital data

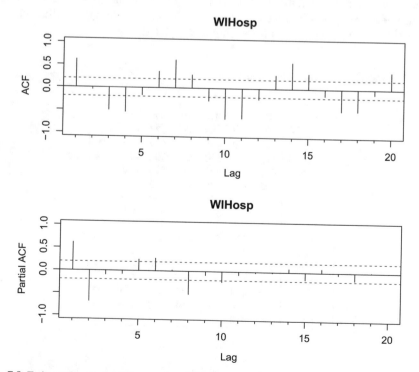

Fig. 7.9 Estimated autocorrelations and partial autocorrelations for the hospital data

Figure 7.10 gives the correlation function and partial correlation function for the series generated by taking differences of order 7. Figure 7.11 gives the seasonally differenced series. The correlations display the classic pattern of an autoregressive process with $p > 1$. The only problem is that they are dying out very slowly. The partial correlations show an extremely large value for ϕ_1. These conclusions are consistent with the need for a regular difference.

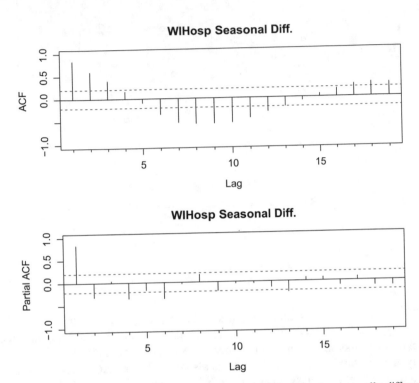

Fig. 7.10 Estimated autocorrelations and partial autocorrelations for the seasonally differenced hospital data

Figure 7.12 gives the correlation function and partial correlation function for the series generated by taking a seasonal difference of order 7 and a regular difference of order 1. The correlations suggest the need for a moving average term of order 7. Thus, the suggested moving average portion of the model is $\Theta(B) = 1$ and $\Theta_1(B^7) = 1 - \theta_{1,1}B^7$. The partial correlations are large for $k = 6, 7, 13$ and not small for $k = 1$. One possible way to model this would be to take $\Phi(B) = 1 - \phi_1 B - \phi_6 B^6$ and $\Phi_1(B^7) = 1 - \phi_{1,1}B^7$. Altogether, the model is an $ARIMA([1,6],1,0) \times (1,1,1)_7$, where the notation $[1,6]$ is used to indicate the peculiar nature of $\Phi(B)$. Although this is a model that involves only four parameters in addition to σ^2, it is really quite complex. The model for y_t involves y_{t-k} for $k = 1, 2, 6, 7, 8, 9, 13, 14, 15, 16, 20,$ and 21.

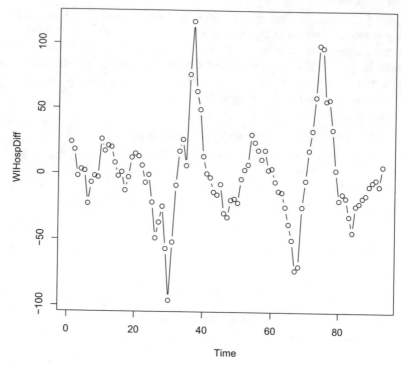

Fig. 7.11 Seasonally differenced hospital data

While we halt this example now, it should be recognized that this is only the beginning of an analysis of these data. The model needs to be fitted, the residuals checked, and other models need to be investigated. The plot in Fig. 7.13, of the regularly and seasonally differenced series, is also quite interesting. There is a strange increase in variability that occurs around 35 days and again around 70 days. This may just be an oddity of the data, or it may be an indication of some structure in the hospital that generates a 5 week seasonal pattern. More investigation of hospital practices or more data would be needed to verify a 5 week cycle. Of course, the odd behavior of the series could just mean that differencing has not achieved stationarity.   □

7.8 The Multivariate State-Space Model and the Kalman Filter

The *state-space model* provides a quite general paradigm for modeling multivariate time series. At each time t, a $q_t \times 1$ vector of observations Y_t is observed. We assume that Y_t satisfies a linear model

$$Y_t = X_t \beta_t + e_t \tag{7.8.1}$$

WIHosp Reg. and Seasonal Diff.

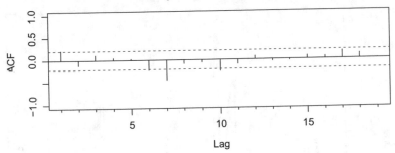

WIHosp Reg. and Seasonal Diff.

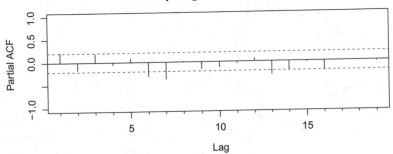

Fig. 7.12 Estimated autocorrelations and partial autocorrelations for the regularly and seasonally differenced hospital data

in which all of the components are allowed to vary with time but X_t always has p columns. The error vector e_t is assumed to satisfy

$$E(e_t) = 0; \qquad Cov(e_t) \equiv V_t.$$

The linear model (7.8.1) is often called the *observation equation*.

Dependencies between times are modeled through the $p \times 1$ vector β_t. Rather than assuming that β_t is fixed, we assume that β_t satisfies a multivariate autoregressive model. Let Φ be a $p \times p$ matrix. We assume that

$$\beta_t = \Phi \beta_{t-1} + \varepsilon_t, \qquad (7.8.2)$$

where ε_t is a $p \times 1$ error vector with

$$E(\varepsilon_t) = 0; \qquad Cov(\varepsilon_t) \equiv \Sigma_t.$$

The autoregressive model (7.8.2) is referred to as the *state* equation.

Finally, for $i = 1, \ldots, t$ and $j = 1, \ldots, t$, assume that

$$Cov(e_i, e_j) = 0 \quad i \neq j,$$

WIHosp Reg. and Seasonal Diff.

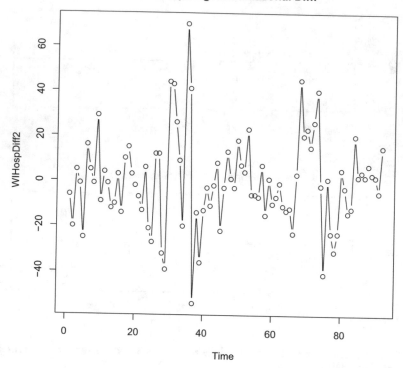

Fig. 7.13 Regularly and Seasonally differenced hospital data

$$\mathrm{Cov}(\varepsilon_i, \varepsilon_j) = 0 \quad i \neq j,$$

and

$$\mathrm{Cov}(e_i, \varepsilon_j) = 0 \qquad \text{all } i, j.$$

To start the sequence off, we assume that β_0 has

$$\mathrm{E}(\beta_0) \equiv \tilde{\beta}_0,$$
$$\mathrm{Cov}(\beta_0) \equiv P_0,$$
$$\mathrm{Cov}(\beta_0, e_i) = 0 \qquad i = 1, \dots, t,$$
$$\mathrm{Cov}(\beta_0, \varepsilon_i) = 0 \qquad i = 1, \dots, t.$$

Traditionally, prediction of the unobservable vector β_t has been viewed as the primary goal of state-space model analysis. The *Kalman filter* is a recursive procedure for predicting β_t on the basis of $Y_t, Y_{t-1}, \dots, Y_1$.

At first glance, Eq. (7.8.2) may seem rather restrictive. It appears to be a matrix version of a first-order autoregressive process. The fact that it appears to be first order may seem restrictive. In fact, because (7.8.2) involves matrices, there is nothing intrinsically first-order about it.

EXAMPLE 7.8.1.   A very basic model in many applications is that observations are the sum of a signal μ_t plus some noise e_t, namely

$$y_t = \mu_t + e_t.$$

If the signal is the result of an autoregressive process, a state-space model is appropriate. Suppose μ_t is an $AR(3)$ process, namely

$$\mu_t = \phi_1 \mu_{t-1} + \phi_2 \mu_{t-2} + \phi_3 \mu_{t-3} + \varepsilon_t.$$

The state equation is

$$\begin{pmatrix} \mu_t \\ \mu_{t-1} \\ \mu_{t-2} \end{pmatrix} = \begin{bmatrix} \phi_1 & \phi_2 & \phi_3 \\ 1 & 0 & 0 \\ 0 & 1 & 0 \end{bmatrix} \begin{pmatrix} \mu_{t-1} \\ \mu_{t-2} \\ \mu_{t-3} \end{pmatrix} + \begin{pmatrix} \varepsilon_t \\ 0 \\ 0 \end{pmatrix}.$$

The observation equation is

$$y_t = (1,0,0) \begin{pmatrix} \mu_t \\ \mu_{t-1} \\ \mu_{t-2} \end{pmatrix} + e_t.$$

Thus, $X_t = (1,0,0)$ and $\beta_t = (\mu_t, \mu_{t-1}, \mu_{t-2})'$. Note that, in practice, it would be unusual for the values ϕ_1, ϕ_2, and ϕ_3 to be known. In the development of the Kalman filter, these are assumed to be known. Typically, they will be estimated from the data, and the estimates will be substituted for the true parameters to make predictions. □

The state-space model was originally introduced to track missiles using satellite observations of their positions. The satellite observations Y_t are subject to error. Based on a first-order differential equation, the actual position of the missile is modeled by a first-order autoregressive process.

The following example illustrates a model in which the matrix Φ is completely known.

EXAMPLE 7.8.2.   Phadke (1981) and Meinhold and Singpurwalla (1983) present a model useful in quality control. The number of defective items in a process is transformed into a value y_t. (The transformation is used to make the data distribution approximate a normal distribution.) The transformed number of defectives is modeled as a signal plus noise,

$$y_t = \mu_t + e_t.$$

However, the signal is generated in an unusual fashion. The signal is subject to a drift. The underlying signal for defectives is a parameter θ_t determined by

$$\theta_t = \theta_{t-1} + w_{t,1},$$

where $w_{t,1}$ is an error term that is uncorrelated with other error terms. However, the signal for any particular observation is subject to additional error, namely

$$\mu_t = \theta_t + w_{t,2},$$

where $w_{t,2}$ is an uncorrelated error term. Upon observing that

$$\mu_t = \theta_{t-1} + w_{t,1} + w_{t,2},$$

we see that the state equation is

$$\begin{pmatrix} \theta_t \\ \mu_t \end{pmatrix} = \begin{pmatrix} 1 & 0 \\ 1 & 0 \end{pmatrix} \begin{pmatrix} \theta_{t-1} \\ \mu_{t-1} \end{pmatrix} + \begin{pmatrix} w_{t,1} \\ w_{t,1} + w_{t,2} \end{pmatrix}$$

and the observation equation is

$$y_t = (0,1) \begin{pmatrix} \theta_t \\ \mu_t \end{pmatrix} + e_t.$$

The matrix Φ in the state equation is completely known.   □

EXAMPLE 7.8.3.   We now present an $ARMA(p,q)$ model as a state-space model. Begin with an $ARMA(2,1)$,

$$y_t = \phi_1 y_{t-1} + \phi_2 y_{t-2} + \eta_t - \theta_1 \eta_{t-1},$$

where η_t is a white noise process. To write this as a state-space model let $q_t = 1$, $Y_t = y_t$, $X_t = [1,0]$, $e_t \equiv 0$,

$$\beta_t = \left[\begin{array}{c} y_t \\ \phi_2 y_{t-1} - \theta_1 \eta_t \end{array} \right], \quad \Phi = \left[\begin{array}{cc} \phi_1 & 1 \\ \phi_2 & 0 \end{array} \right], \quad \varepsilon_t = \left[\begin{array}{c} 1 \\ -\theta_1 \end{array} \right] \eta_t.$$

Note that

$$\beta_{t+1} = \Phi \beta_t + \varepsilon_{t+1}$$

becomes

$$\left[\begin{array}{c} y_{t+1} \\ \phi_2 y_t - \theta_1 \eta_{t+1} \end{array} \right] = \left[\begin{array}{cc} \phi_1 & 1 \\ \phi_2 & 0 \end{array} \right] \left[\begin{array}{c} y_t \\ \phi_2 y_{t-1} - \theta_1 \eta_t \end{array} \right] + \left[\begin{array}{c} 1 \\ -\theta_1 \end{array} \right] \eta_{t+1}$$

or

$$\left[\begin{array}{c} y_{t+1} \\ \phi_2 y_t - \theta_1 \eta_{t+1} \end{array} \right] = \left[\begin{array}{c} \phi_1 y_t + \phi_2 y_{t-1} - \theta_1 \eta_t \\ \phi_2 y_t \end{array} \right] + \left[\begin{array}{c} \eta_{t+1} \\ -\theta_1 \eta_{t+1} \end{array} \right]$$

$$= \left[\begin{array}{c} \phi_1 y_t + \phi_2 y_{t-1} + \eta_{t+1} - \theta_1 \eta_t \\ \phi_2 y_t - \theta_1 \eta_{t+1} \end{array} \right].$$

The first row of the equation is the ARMA(2,1) model and the second row is tautological.

Now consider an $ARMA(r, r-1)$ model,

$$y_t = \sum_{j=1}^{r} \phi_j y_{t-j} + \eta_t - \sum_{k=1}^{r-1} \theta_k \eta_{t-k}.$$

To write this as a state-space model, let

$$\beta_t = \begin{bmatrix} y_t \\ \sum_{j=1}^{r-1} \phi_{j+1} y_{t-j} - \sum_{k=1}^{r-1} \theta_k \eta_{t+1-k} \\ \sum_{j=2}^{r-1} \phi_{j+1} y_{t+1-j} - \sum_{k=2}^{r-1} \theta_k \eta_{t+2-k} \\ \vdots \\ \phi_r y_{t-1} - \theta_{r-1} \eta_t \end{bmatrix}$$

$$\Phi = \begin{bmatrix} \phi_1 & 1 & & 0 \\ \phi_2 & & \ddots & \\ \vdots & 0 & & 1 \\ \phi_r & 0 & \cdots & 0 \end{bmatrix}, \quad \varepsilon_t = \begin{bmatrix} 1 \\ -\theta_1 \\ -\theta_2 \\ \vdots \\ -\theta_{r-1} \end{bmatrix} \eta_t$$

To write a general $ARMA(p,q)$ model as a state-space model, take $r = \max(p, q+1)$ and set appropriate parameters equal to 0. □

Durbin and Koopman (2012) discuss how to write arbitrary ARIMA models as state-space models.

The state-space model and the Kalman filter were originally introduced by Kalman (1960) and Kalman and Bucy (1961). This was done in the engineering literature. Harrison and Stevens (1971, 1976) first presented the method as a Bayesian procedure based on Φ being known and a multivariate normal distribution for $(\beta_0', e_1', \ldots, e_t', \varepsilon_1', \ldots, \varepsilon_t')'$. Meinhold and Singpurwalla (1983) give a nice exposition of this approach. The approach taken here is based on linear expectations. Because linear expectations are the conditional expectations for multivariate normals, the derivation given here is also valid for multivariate normals. Diderrich (1985) has examined the relationship between the Kalman filter and Goldberger–Theil estimators. Wegman (1982) relates the Kalman filter to stochastic differential equations. Shumway and Stoffer (2011) give a number of references to various applications of the Kalman filter.

Since the previous edition of this work, state-space models have become a more standard part of time series, cf. Prado and West (2010) and Durbin and Koopman (2012).

7.8.1 The Kalman Filter

The Kalman filter is a procedure for predicting β_t based on the data $Y_1, \ldots, Y_t$. It is a recursive procedure in that the prediction of β_t is based on modifying the predictor of β_{t-1}. Let $x' = (Y'_{t-1}, Y'_{t-2}, \ldots, Y'_1)$. We wish to find $\hat{E}(\beta_t | Y_t, x)$. By Proposition B.1.8,

$$\hat{E}(\beta_t | Y_t, x) = \hat{E}(\beta_t | x)$$
$$+ \text{Cov}\left(\beta_t, Y_t - \hat{E}(Y_t | x)\right) \left[\text{Cov}(Y_t - \hat{E}(Y_t | x))\right]^{-1} [Y_t - \hat{E}(Y_t | x)] \quad (7.8.3)$$

We now proceed to identify the various parts of this equation.

First,

$$\hat{E}(\beta_t | x) = \hat{E}(\Phi \beta_{t-1} + \varepsilon_t | x)$$
$$= \Phi \hat{E}(\beta_{t-1} | x) + \hat{E}(\varepsilon_t | x).$$

However, ε_t is uncorrelated with the earlier errors and β_0 whereas x is a linear function of the earlier errors and β_0; thus, $\text{Cov}(\varepsilon_t, x) = 0$. By Proposition B.1.5, $\hat{E}(\varepsilon_t | x) = 0$ and

$$\hat{E}(\beta_t | x) = \Phi \hat{E}(\beta_{t-1} | x). \quad (7.8.4)$$

Next, examine $Y_t - \hat{E}(Y_t | x)$.

$$\hat{E}(Y_t | x) = \hat{E}(X_t \beta_t + e_t | x)$$
$$= X_t \hat{E}(\beta_t | x) + \hat{E}(e_t | x)$$
$$= X_t \Phi \hat{E}(\beta_{t-1} | x),$$

where we have used (7.8.4) and the fact that $\text{Cov}(e_t, x) = 0$. Thus,

$$Y_t - \hat{E}(Y_t | x) = Y_t - X_t \Phi \hat{E}(\beta_{t-1} | x). \quad (7.8.5)$$

The covariance matrix of this prediction error can be computed as follows. Using (7.8.1) and (7.8.2) and the fact that errors are uncorrelated with previous events

$$\text{Cov}[Y_t - \hat{E}(Y_t | x)] = \text{Cov}[Y_t - X_t \Phi \hat{E}(\beta_{t-1} | x)]$$
$$= \text{Cov}[(X_t \beta_t + e_t) - X_t \Phi \hat{E}(\beta_{t-1} | x)]$$
$$= \text{Cov}[(X_t [\Phi \beta_{t-1} + \varepsilon_t] + e_t) - X_t \Phi \hat{E}(\beta_{t-1} | x)]$$
$$= \text{Cov}[X_t \Phi (\beta_{t-1} - \hat{E}(\beta_{t-1} | x)) + X_t \varepsilon_t + e_t]$$
$$= X_t \Phi \text{Cov}[\beta_{t-1} - \hat{E}(\beta_{t-1} | x)] \Phi' X'_t + X_t \Sigma_t X'_t + V_t$$
$$= X_t [\Phi P_{t-1} \Phi' + \Sigma_t] X'_t + V_t, \quad (7.8.6)$$

where

$$P_{t-1} \equiv \text{Cov}[\beta_{t-1} - \hat{E}(\beta_{t-1} | x)]. \quad (7.8.7)$$

Finally, we need to compute $\text{Cov}[\beta_t, Y_t - \hat{E}(Y_t | x)]$. We begin the computation by mentioning four facts. First, β_t is a linear function of β_0, ε_t, and the errors previous to time t; thus,

$$\text{Cov}(\beta_t, e_t) = 0.$$

Second, β_{t-1} is a linear function of β_0 and errors previous to time t, so

$$\text{Cov}(\beta_{t-1}, \varepsilon_t) = 0.$$

Third, by Proposition B.1.10,

$$\text{Cov}(\beta_{t-1}, \beta_{t-1} - \hat{E}(\beta_{t-1}|x)) = P_{t-1}.$$

Fourth, because $\text{Cov}(\varepsilon_t, x) = 0$ and $\hat{E}(\beta_{t-1}|x)$ is a linear function of x,

$$\text{Cov}(\varepsilon_t, \beta_{t-1} - \hat{E}(\beta_{t-1}|x)) = 0.$$

The computation goes as follows.

$$
\begin{aligned}
&\text{Cov}[\beta_t, Y_t - \hat{E}(Y_t|x)]\\
&= \text{Cov}[\beta_t, X_t \Phi(\beta_{t-1} - \hat{E}(\beta_{t-1}|x)) + X_t \varepsilon_t + e_t]\\
&= \text{Cov}[\beta_t, X_t \Phi(\beta_{t-1} - \hat{E}(\beta_{t-1}|x))] + \text{Cov}[\beta_t, X_t \varepsilon_t]\\
&= \text{Cov}[\Phi\beta_{t-1} + \varepsilon_t, X_t \Phi(\beta_{t-1} - \hat{E}(\beta_{t-1}|x))]\\
&\quad + \text{Cov}[\Phi\beta_{t-1} + \varepsilon_t, X_t \varepsilon_t]\\
&= \text{Cov}[\Phi\beta_{t-1}, X_t \Phi(\beta_{t-1} - \hat{E}(\beta_{t-1}|x))] + \text{Cov}[\varepsilon_t, X_t \varepsilon_t]\\
&= \Phi\text{Cov}[\beta_{t-1}, \beta_{t-1} - \hat{E}(\beta_{t-1}|x)]\Phi'X_t' + \Sigma_t X_t'\\
&= (\Phi P_{t-1}\Phi' + \Sigma_t)X_t'.
\end{aligned}
\tag{7.8.8}
$$

Substituting (7.8.4), (7.8.5), (7.8.6) and (7.8.8) into (7.8.3) gives the standard form for the Kalman filter. Let

$$R_t \equiv \Phi P_{t-1}\Phi' + \Sigma_t; \tag{7.8.9}$$

then,

$$
\begin{aligned}
\hat{E}(\beta_t|Y_t, x) =\\
\Phi\hat{E}(\beta_{t-1}|x) + R_t X_t'[X_t R_t X_t' + V_t]^{-1}[Y_t - X_t \Phi\hat{E}(\beta_{t-1}|x)].
\end{aligned}
\tag{7.8.10}
$$

Equation (7.8.10) is a formula for predicting β_t given Y_t, the prediction of β_{t-1}, and the matrix P_{t-1} defined in (7.8.7). P_{t-1} enters through Eq. (7.8.9). Note that the only matrix assumed to be nonsingular is $X_t R_t X_t' + V_t$. Neither V_t nor Σ_t are assumed nonsingular. This is important in some applications (e.g., Example 7.8.1).

To use (7.8.10) recursively, we need a recursive formula for P_t. From Proposition B.1.9,

$$
\begin{aligned}
P_t &= \text{Cov}[\beta_t - \hat{E}(\beta_t|Y_t, x)]\\
&= \text{Cov}[\beta_t - \hat{E}(\beta_t|x)] - \text{Cov}[\beta_t, Y_t - \hat{E}(Y_t|x)]
\end{aligned}
$$

$$\times \, [\mathrm{Cov}(Y_t - \hat{E}(Y_t|x))]^{-1} \mathrm{Cov}[Y_t - \hat{E}(Y_t|x), \beta_t].$$

Parts of this have already been computed in (7.8.6) and (7.8.8). The only new computation is

$$\begin{aligned}
\mathrm{Cov}[\beta_t - \hat{E}(\beta_t|x)] &= \mathrm{Cov}[\Phi(\beta_{t-1} - \hat{E}(\beta_{t-1}|x)) + \varepsilon_t] \\
&= \mathrm{Cov}[\Phi(\beta_{t-1} - \hat{E}(\beta_{t-1}|x))] + \mathrm{Cov}[\varepsilon_t] \\
&= \Phi P_{t-1} \Phi' + \Sigma_t.
\end{aligned}$$

Thus, using (7.8.9), we get the recursive formula

$$P_t = R_t - R_t X_t' [X_t R_t X_t' + V_t]^{-1} X_t R_t. \tag{7.8.11}$$

Note that the recursive nature of (7.8.11) lies in the fact that R_t is a function of P_{t-1}. The matrix P_t is useful in updating predictions but also provides standard errors of prediction for linear combinations of β_t.

To start the recursive process based on (7.8.9), (7.8.10), and (7.8.11), we need an initial prediction for β_0 and the covariance matrix for the error of that prediction. The initial prediction is based on no data, so $\hat{E}(\beta_0) = E(\beta_0) = \tilde{\beta}_0$ and $\mathrm{Cov}(\beta_0 - \hat{E}(\beta_0)) = \mathrm{Cov}(\beta_0) = P_0$.

Having developed a procedure for obtaining $\hat{E}(\beta_t|Y_t, x)$, prediction of the future given $Y_1, \ldots, Y_t$ is easily performed:

$$\begin{aligned}
\hat{E}(\beta_{t+1}|Y_t, x) &= \hat{E}(\Phi \beta_t + \varepsilon_{t+1}|Y_t, x) \\
&= \Phi \hat{E}(\beta_t|Y_t, x).
\end{aligned}$$

The prediction covariance is

$$\begin{aligned}
\mathrm{Cov}[\beta_{t+1} - \hat{E}(\beta_{t+1}|Y_t, x)] &= \mathrm{Cov}[\Phi\{\beta_t - \hat{E}(\beta_t|Y_t, x)\} + \varepsilon_{t+1}] \\
&= \Phi P_t \Phi' + \Sigma_{t+1} \\
&= R_{t+1}.
\end{aligned}$$

Similarly,

$$\begin{aligned}
\hat{E}(Y_{t+1}|Y_t, x) &= \hat{E}(X_{t+1} \beta_{t+1} + e_{t+1}|Y_t, x) \\
&= X_{t+1} \hat{E}(\beta_{t+1}|Y_t, x) \\
&= X_{t+1} \Phi \hat{E}(\beta_t|Y_t, x),
\end{aligned}$$

with prediction covariance matrix

$$\begin{aligned}
\mathrm{Cov}[Y_{t+1} - \hat{E}(Y_{t+1}|Y_t, x)] &= \mathrm{Cov}[X_{t+1}\{\beta_{t+1} - \hat{E}(\beta_{t+1}|Y_t, x)\} + e_{t+1}] \\
&= X_{t+1} \mathrm{Cov}[\beta_{t+1} - \hat{E}(\beta_{t+1}|Y_t, x)] X_{t+1}' + V_{t+1} \\
&= X_{t+1} R_{t+1} X_{t+1}' + V_{t+1}.
\end{aligned}$$

To predict events further in the future,

$$\hat{E}(\beta_{t+r}|Y_t,x) = \Phi'\hat{E}(\beta_t|Y_t,x),$$

with the prediction covariance matrix obtained recursively using

$$\text{Cov}[\beta_{t+r} - \hat{E}(\beta_{t+r}|Y_t,x)] = \Phi\text{Cov}[\beta_{t+r-1} - \hat{E}(\beta_{t+r-1}|Y_t,x)]\Phi' + \Sigma_{t+r}.$$

The covariance matrix depends only on P_t, Φ, and $\Sigma_{t+1},\dots,\Sigma_{t+r}$. To predict future observations,

$$\begin{aligned}\hat{E}(Y_{t+r}|Y_t,x) &= X_{t+r}\hat{E}(\beta_{t+r}|Y_t,x) \\ &= X_{t+r}\Phi'\hat{E}(\beta_t|Y_t,x).\end{aligned}$$

Again, the prediction covariance matrix is obtained recursively:

$$\text{Cov}[Y_{t+r} - \hat{E}(Y_{t+r}|Y_t,x)] = X_{t+r}\text{Cov}[\beta_{t+r} - \hat{E}(\beta_{t+r}|Y_t,x)]X'_{t+r} + V_{t+r}.$$

Note that this covariance matrix does not involve any of $X_{t+1},\dots,X_{t+r-1}$.

If β_0, $e_1,\dots,e_t$, and $\varepsilon_1,\dots,\varepsilon_t$ are taken to have a joint normal distribution, then the posterior distribution of β_t given the data (Y_t,x) is multivariate normal with mean $\hat{E}(\beta_t|Y_t,x)$ and covariance matrix P_t. The subjective Bayesian aspect of this analysis lies entirely in the a priori assumption that

$$\beta_0 \sim N(\tilde{\beta}_0, P_0).$$

7.8.2 Parameter Estimation

For the purposes of prediction, the matrices Φ and X_i, V_i, Σ_i, $i = 1,\dots,t$ were assumed to be known. This is similar to the approach taken for prediction in time domain models. In practice, as with time domain models, many of these parameters must be estimated. In particular, the matrix Φ defining the multivariate autoregressive process and the covariances matrices V_i and Σ_i generally need to be estimated. The design matrices X_i are typically known. The estimation of the covariance matrices presents special problems in that they include many more parameters than there are observations. The covariance matrices must be modeled in some way if estimates are to be obtained. Suppose the data $Y_1,\dots,Y_n$ have been observed. One commonly used and particularly simple model is that, for some unknown positive definite matrices V and Σ,

$$V_1 = V_2 = \cdots = V_n = V$$

and

$$\Sigma_1 = \Sigma_2 = \cdots = \Sigma_n = \Sigma.$$

The assumption that the observation covariance matrices V_i are equal implies that the number of observations available at each time is the same. Although model (7.8.2) implies that the dimensions of β_t and ε_t remain constant, in general there is no such restriction on Y_t, X_t, and e_t.

The standard estimation method seems to be maximum likelihood based on the assumption that $(\beta_0, e_1, \ldots, e_n, \varepsilon_1, \ldots, \varepsilon_n)'$ has a multivariate normal distribution. It follows that the data $Y_1, Y_2, \ldots, Y_t$ have a multivariate normal distribution. The joint density can be written as the product of conditional densities, that is,

$$f(Y_1, \ldots, Y_t) = f(Y_1)f(Y_2|Y_1)f(Y_3|Y_1, Y_2) \cdots f(Y_t|Y_1, Y_2, \ldots, Y_{t-1}).$$

In fact, assuming that β_0 is fixed, all of the conditional distributions are normals with means and covariances given by the Kalman filter. For example,

$$Y_t|Y_1, \ldots, Y_{t-1} \sim N(\hat{E}(Y_t|x), \text{Cov}[Y_t - \hat{E}(Y_t|x)]),$$

where $\hat{E}(Y_t|x)$ is given before (7.8.5) and $\text{Cov}[Y_t - \hat{E}(Y_t|x)]$ is given by (7.8.6). Write

$$\hat{e}_t \equiv Y_t - \hat{E}(Y_t|Y_1, \ldots, Y_{t-1})$$

and

$$Q_t \equiv \text{Cov}[Y_t - \hat{E}(Y_t|Y_1, \ldots, Y_{t-1})].$$

The likelihood function is the product of the conditional normal densities, that is,

$$L(\Phi, V_1, \ldots, V_n, \Sigma_1, \ldots, \Sigma_n) =$$

$$\left(\prod_{t=1}^{n}(2\pi)^{-q_t/2}\right)\left(\prod_{t=1}^{n}|Q_t|^{-1/2}\right)\exp\left[-\frac{1}{2}\sum_{t=1}^{n}\hat{e}_t'Q_t^{-1}\hat{e}_t\right].$$

As usual, it is easier to maximize the log-likelihood,

$$\ell(\Phi, V_1, \ldots, V_n, \Sigma_1, \ldots, \Sigma_n) = -\log(2\pi)\sum_{t=1}^{n}\frac{q_t}{2} - \frac{1}{2}\sum_{t=1}^{n}\log(|Q_t|) - \frac{1}{2}\sum_{t=1}^{n}\hat{e}_t'Q_t\hat{e}_t.$$

Remembering that parametric models for $V_1, \ldots, V_n$ and $\Sigma_1, \ldots, \Sigma_n$ are mandatory, it is important to note that this log-likelihood function is valid for any such models. If $V_i = V_i(\eta)$ and $\Sigma_i = \Sigma_i(\zeta)$, $i = 1, \ldots, n$, then

$$\ell(\Phi, \eta, \zeta) = \ell(\Phi, V_1(\eta), \ldots, V_n(\eta), \Sigma_1(\zeta), \ldots, \Sigma_n(\zeta)).$$

Similarly, some or all of the components of Φ can be known without materially changing the likelihood function.

In practice, with almost any parameterization, the log-likelihood will be a very complicated nonquadratic function of the parameters. (Quadratic functions are relatively easy to maximize.) Some sort of iterative maximization technique is generally required. Much of the work on this maximization problem has used the covariance model that assumes equal covariance matrices. Shumway and Stoffer (1982, 2011) discuss maximization of $\ell(\Phi, V, \Sigma)$ using the EM algorithm. The EM algorithm is presented in Dempster, Laird, and Rubin (1977). The Newton–Raphson algorithm has also been used to maximize $\ell(\Phi, V, \Sigma)$. This approach is discussed by Gupta and Mehra (1974), Jones (1980), and Ansley and Kohn (1984).

With a likelihood this complex, it is probably safer to use Bayesian methods rather than relying on a single point that maximizes the likelihood, cf. West and Harrison (1997), Prado and West (2010), Cressie and Wikle (2011).

7.8.3 Missing Values

The assumption of equal covariance matrices seems to be the standard way of modeling the V_ts and Σ_ts. As mentioned earlier, equal V_ts imply that q_t, the dimension of Y_t, is the same for all t. This does not seem like a particularly harsh requirement. The vector Y_t will often consist of observations on q_t dependent variables. The assumption that, at each time t, the same number of variables are available to be observed is not very restrictive. The only problem is that some observations may be missing at some times. The generality of the state-space model allows missing values to be handled easily.

Suppose the $q \times 1$ vector Z_t is to be observed at time t. Assume an observation equation

$$Z_t = W_t \beta_t + \xi_t,$$

where

$$\mathrm{Cov}(\xi_t) = V.$$

However, the actual observation is Y_t, which consists of some of the components of Z_t. If no components are missing, $Y_t = Z_t$, $X_t = W_t$, and $e_t = \xi_t$. If some components are lost, write

$$Z_t = \begin{bmatrix} Y_t \\ Y_{tM} \end{bmatrix},$$

$$W_t = \begin{bmatrix} X_t \\ X_{tM} \end{bmatrix},$$

$$\xi_t = \begin{bmatrix} e_t \\ e_{tM} \end{bmatrix},$$

and

$$\mathrm{Cov}(\xi_t) = \begin{bmatrix} V_t & V_{tM} \\ V_{Mt} & V_{MM} \end{bmatrix}.$$

There is no real loss of generality in assuming that the last components of Z_t are the ones that were not observed. The observation equation for Z_t implies an observation equation for the actual observations,

$$Y_t = X_t \beta_t + e_t,$$

for which

$$\mathrm{Cov}(e_t) = V_t.$$

In estimating the parameters Φ, V, and Σ, the likelihood allows V_t to be a function of the parameters. Because V_t is a submatrix of V, V_t is clearly a function of the parameter matrix V. In fact, the likelihood function is not much more complicated than if none of the components of Z_t were lost.

Of course, the loss of components has no effect on the Kalman filter. The filter is based on Y_t and treats V_t as known. The lost components are simply ignored. The references given earlier for maximizing the likelihood also discuss the missing value problem.

7.9 Additional Exercises

Exercise 7.9.1. Which of the following processes are stationary? Which are invertible?

(a) $(1 - 1.5B + 0.54B^2)y_t = (1 - 0.5B)e_t$.
(b) $(1 - \frac{5}{8}B)y_t = (1 + 0.1B - 0.56B^2)e_t$.
(c) $(1 - 1.6B + 0.55B^2)y_t = (1 - 0.6B)e_t$.
(d) $(1 - B + 0.2475B^2)y_t = (1 - B - \frac{7}{36}B^2)e_t$.
(e) $(1 - 0.7B - 0.66B^2 + 0.432B^3)y_t = (1 - 0.8B + 0.07B^2)e_t$.
(f) $(1 - 0.8B + 0.07B^2)y_t = (1 - 0.1B)e_t$.

Exercise 7.9.2. Consider the $AR(1)$ model

$$y_t = 10 - 0.7y_{t-1} + e_t, \ \sigma^2 = 3.$$

(a) Find μ, $\sigma(0)$, $\sigma(1)$, $\sigma(2)$, and $\sigma(3)$.
(b) If $y_4 = 12$, will y_5 tend to be less than or greater than μ?
(c) Is the process stationary?
(d) Give $\hat{E}(y_{4+k}|Y)$ and the prediction variance for $k = 1,2,3$ and $Y = (12,11,11,12)'$.

Exercise 7.9.3. Consider the $MA(1)$ model

$$y_t = 10 + e_t - 0.4e_{t-1}, \ \sigma^2 = 3.$$

(a) Find μ, $\sigma(0)$, $\sigma(1)$, $\sigma(2)$, and $\sigma(3)$.
(b) If $y_4 = 12$, will y_5 tend to be less than or greater than μ?
(c) Is the process stationary?

(d) Give $\hat{E}(y_{4+k}|Y_\infty)$ and the prediction variance for $k = 1, 2, 3$ and $Y = (12, 11, 11, 12)'$.

Exercise 7.9.4. Consider the $ARMA(1, 1)$ model

$$y_t = 10 - 0.7y_{t-1} + e_t - 0.4e_{t-1}, \ \sigma^2 = 3$$

(a) Find μ, $\sigma(0)$, $\sigma(1)$, $\sigma(2)$, and $\sigma(3)$.
(b) If $y_4 = 12$, will y_5 tend to be less than or greater than μ?
(c) Is the process stationary?
(d) Give $\hat{E}(y_{4+k}|Y_\infty)$ and the prediction variance for $k = 1, 2, 3$ and $Y = (12, 11, 11, 12)'$.

Exercise 7.9.5. Do a time domain analysis of the sunspot data in Exercise 6.9.5. This should include estimation, model fitting, and checking of assumptions.

Exercise 7.9.6. Use a multiplicative seasonal model to analyze the international air passenger data of Exercise 6.9.6. This should include estimation, model fitting, and checking of assumptions.

Exercise 7.9.7. For an $MA(1)$ process, find an estimate of θ_1 in terms of $\hat{\rho}(1)$. Are any restrictions on $\hat{\rho}(1)$ needed?

Exercise 7.9.8. For an $AR(2)$ process, use the Yule–Walker equations, with $\hat{\sigma}_y(\cdot)$ replacing $\sigma_y(\cdot)$ to obtain estimates

$$\hat{\phi}_1 = \frac{\hat{\rho}(1)(1 - \hat{\rho}(1))}{1 - \hat{\rho}(1)^2}$$

and

$$\hat{\phi}_2 = \frac{\hat{\rho}(2) - \hat{\rho}(1)^2}{1 - \hat{\rho}(1)^2}.$$

Exercise 7.9.9. Show that the variance for predicting k steps ahead in an $AR(1)$ process is

$$\sigma^2 \frac{1 - \phi_1^{2k}}{1 - \phi_1^2}.$$

Exercise 7.9.10. Show, for an $MA(1)$ process with parameter θ_1, that

$$\hat{E}(y_{n+1}|Y_\infty) = -\sum_{s=0}^{\infty} \theta_1^s y_{n-s}.$$

Exercise 7.9.11. Let e_t be a second-order stationary process. Show that

$$y_t = e_t - \theta_1 e_{t-1}$$

and

$$w_t = e_t - \frac{1}{\theta_1} e_{t-1}$$

have the same correlation function.

Exercise 7.9.12. What are the largest possible values of $|\rho(1)|$ and $|\rho(2)|$ for an $MA(2)$ process?

Exercise 7.9.13. Find the ψ_is, μ, and $\sigma(k)$ for the following processes.

(a) The $AR(3)$ process

$$(1 - 0.9B)(1 + 0.8B)(1 - 0.6B)y_t = e_t.$$

(b) The $ARMA(2,1)$ process

$$(1 - 0.9B)(1 - 0.6B)y_t = (1 - 0.5)e_t.$$

Exercise 7.9.14. Show that if all the roots x_0 of $\Phi(x)$ have $|x_0| > 1$, then there exists a polynomial

$$\Psi(x) = \sum_{i=0}^{\infty} \psi_i x^i$$

with

$$\sum_{i=0}^{\infty} |\psi_i| < \infty$$

and, for $x \in [-1, 1]$,

$$[\Psi(x)][\Phi(x)] = 1.$$

Hint: Do a Taylor expansion of $1/\Phi(x)$ about 0.

Exercise 7.9.15. Generate 25 realizations of an $ARMA(1,1)$ process for $\phi_1 = -0.8, -0.2, 0, .2, .8$ and $\theta_1 = -0.8, -0.2, 0, .2, .8$.

Exercise 7.9.16. Prove *PA-V*'s Exercise 6.9.9, that

(a) $\rho_{12 \cdot 3} = \dfrac{\rho_{12} - \rho_{13}\rho_{23}}{\sqrt{1 - \rho_{13}^2}\sqrt{1 - \rho_{23}^2}}$

(b) $\rho_{12 \cdot 34} = \dfrac{\rho_{12 \cdot 4} - \rho_{13 \cdot 4}\rho_{23 \cdot 4}}{\sqrt{1 - \rho_{13 \cdot 4}^2}\sqrt{1 - \rho_{23 \cdot 4}^2}}.$

References

Akaike, H. (1973). Information theory and an extension of the maximum likelihood principle. In B.N. Petrov & F. Czaki (Eds.), *Proceedings of the 2nd International Symposium on Information*. Budapest: Akademiai Kiado.

Ansley, C. F., & Kohn, R. (1984). On the estimation of ARIMA models with missing values. In E. Parzen (Ed.) *Time series analysis of irregularly observed data*. New York: Springer.

Bartlett, M. S. (1946). On the theoretical specification of sampling properties of autocorrelated time series. *Journal of the Royal Statistical Society, Supplement, 8*, 27–41.

Box, G. E. P., Jenkins, G. M., & Reinsel, G. C. (1994). *Time series analysis: Forecasting and control* (3rd ed.). New York: Wiley.

Box, G. E. P., Jenkins, G. M., Reinsel, G. C., & Ljung, G. M. (2015). *Time series analysis: Forecasting and control* (5th ed.). New York: Wiley.

Brockwell, P. J., & Davis, R. A. (1991). *Time series: Theory and methods* (2nd ed.). New York: Springer.

Brockwell, P. J., & Davis, R. A. (2002). *Introduction to time series and forecasting* (2nd ed.). New York: Springer.

Chatfield, C. (2003). *The analysis of time series: An introduction* (6th ed.). New York: Chapman and Hall.

Christensen, R. (1997). *Log-linear models and logistic regression* (2nd ed.). New York: Springer.

Christensen, R. (2011). *Plane answers to complex questions: The theory of linear models* (4th ed.). New York: Springer.

Christensen, R. (2015). *Analysis of variance, design, and regression: Linear modeling for unbalanced data* (2nd ed.). Boca Raton, FL: Chapman and Hall/CRC Press.

Clayton, M. K., Geisser, S., & Jennings, D. E. (1985). A comparison of several model selection procedures. In P. Goel & A. Zellner (Eds.). *Bayesian inference and decision techniques*. Amsterdam: Elsevier Science Publishers B.V.

Cressie, N. A. C., & Wikle, C. K. (2011). *Statistics for spatio-temporal data*. New York: Wiley.

Dempster, A. P., Laird, N. M., & Rubin, D. B. (1977). Maximum likelihood from incomplete data via the EM algorithm. *Journal of the Royal Statistical Society, Series B, 39*, 1–38.

Diderrich, G. T. (1985). The Kalman filter from the perspective of Goldberger-Theil estimators. *The American Statistician, 39*, 193–198.

Diggle, P. J. (1990). *Time series: A biostatistical introduction*. New York: Oxford University Press.

Draper, N., & Smith, H. (1981). *Applied regression analysis* (2nd ed.). New York: Wiley.

Durbin, J., & Koopman, S. J. (2012). *Time series analysis by state space methods* (2nd ed.). Oxford: Oxford University Press.

Fuller, W. A. (1976). *Introduction to statistical time series*. New York: Wiley.

Fuller, W. A. (1996). *Introduction to statistical time series* (2nd ed.). New York: Wiley.

Gupta, N. K., & Mehra, R. K. (1974). Computational aspects of maximum likelihood estimation and reduction in sensitivity function calculations. *IEEE Transactions on Automatic Control, AC-19*, 774–783.

Harrison, P. J., & Stevens, C. F. (1971). A Bayesian approach to short-term forecasting. *Operations Research Quarterly, 22*, 341–362.

Harrison, P. J., & Stevens, C. F. (1976). Bayesian forecasting. *Journal of the Royal Statistical Society, Series B, 38*, 205–247.

Hurvich, C. M., & Tsai, C.-L. (1989). Regression and time series model selection in small samples. *Biometrika, 76*, 297–308.

Jones, R. H. (1980). Maximum likelihood fitting of ARMA models to time series with missing observations. *Technometrics, 22*, 389–396.

Kalman, R. E. (1960). A new approach to linear filtering and prediction problems. *Journal of Basic Engineering, 82*, 34–45.

Kalman, R. E., & Bucy, R. S. (1961). New results in linear filtering and prediction theory. *Journal of Basic Engineering, 83*, 95–108.

Lütkepohl, H. (1985). Comparison of criteria for estimating the order of a vector autoregressive process. *Journal of Time Series Analysis, 65*, 297–303.

Marquardt, D. W. (1963). An algorithm for least squares estimation of non-linear parameters. *Journal of the Society for Industrial and Applied Mathematics, 2*, 431–441.

McLeod, A. I. (1977). Improved Box–Jenkins estimators. *Biometrika, 64*, 531–534.

Meinhold, R. J., & Singpurwalla, N. D. (1983). Understanding the Kalman filter. *The American Statistician, 37*, 123–127.

Pandit, S. M., & Wu, S. M. (1983). *Time series and system analysis with applications*. New York: Wiley.

Phadke, M. S. (1981). Quality audit using adaptive Kalman filtering. *ASQC Quality Congress Transactions – San Francisco*, 1045–1052.

Prado, R., & West, M. (2010). *Time series: Modeling, computation, and inference*. Boca Raton, FL: CRC Press.

Quenille, M. H. (1949). Approximate tests of correlation in time-series. *Journal of the Royal Statistical Society, Series B, 11*, 68–84.

Schwarz, G. (1978). Estimating the dimension of a model. *Annals of Statistics, 16*, 461–464.

Seber, G. A. F., & Wild, C. J. (1989, 2003). *Nonlinear Regression*. New York: Wiley (The 2003 version seems to be a reprint of the 1989 version.).

Shumway, R. H., & Stoffer, D. S. (1982). An approach to time-series smoothing and forecasting using the EM algorithm. *Journal of Time Series Analysis, 3*, 253–264.

Shumway, R. H., & Stoffer, D. S. (2011). *Time Series Analysis and Its Applications: With R Examples* (3rd ed.). New York: Springer.

Wegman, E. J. (1982). Kalman filtering. In N. Johnson & S. Kotz (Eds.) *Encyclopedia of statistics*. New York: Wiley.

West, M., & Harrison, P. J. (1997). *Bayesian forecasting and dynamic models* (2nd ed.). New York: Springer.

Chapter 8
Linear Models for Spatial Data: Kriging

Abstract This chapter addresses linear models for spatial data. A key aspect is the introduction of models for the covariance between data points separated in space. The same ideas can be used to model time series but, unlike the methods in the previous two chapters, time is not required to be observed at regular intervals.

There are innumerable situations in which data are collected at various locations in space and thus innumerable potential applications for methods of analysis for spatial data. Data collected at known locations in space are often correlated. For example, deposits of high-quality copper are more likely to occur near other high-quality deposits. The levels of lead contamination in the soil around a smelter are likely to be correlated. The prevalences of AIDS (or any other communicable disease) viewed geographically are correlated. We make no attempt to cover the entire array of procedures developed for spatial data. We only examine the relation of linear models to spatial prediction.

One branch of statistics concerned with spatial prediction is known as *geostatistics*. The practical application of geostatistics was originally developed in relative isolation from the mainstream of statistics. Not surprisingly, it uses some terminology that is unfamiliar to classically trained statisticians. David (1977) and Journel and Hüijbregts (1978) give details of the geostatistical approach using geostatistical terminology. Ripley (1981) takes a point of view that is probably more familiar to most statisticians. He uses ideas of prediction for stochastic processes that are closely related to time series methods. Cressie (1993) gives an excellent presentation of both the theory and application of statistics for spatial data. Stein (1999) gives an excellent account of the theory. Isaaks and Srivastava (1989) give a relatively elementary introduction; see also Cliff and Ord (1981) and Kitanidis (1997). In this chapter, we present both traditional and geostatistical terminologies.

Most often, spatial data are sampled at irregular locations. Occasionally, data are collected at all the locations on a uniformly spaced grid. Such data are called *regular lattice data* and allow some models that are not available with irregular locations.

Contamination around a smelter is largely a two-dimensional issue. Mining minerals and metals is clearly a three-dimensional issue. To a large extent, time series

© Springer Nature Switzerland AG 2019
R. Christensen, *Advanced Linear Modeling*, Springer Texts in Statistics,
https://doi.org/10.1007/978-3-030-29164-8_8

can be viewed as a special case of spatial data in which the dimension of the space is $d = 1$. The time series models of the previous chapters assume that observations are taken at equally spaced time intervals. These observations are taken on a one-dimensional regular lattice and special models have been developed for such data. When time series observations are not equally spaced, you might as well view them as arbitrary one-dimensional spatial data.

The state-space model and associated Kalman filter of Sect. 7.8 provide a natural method for examining how spatial linear models change over time. Early work on spatio-temporal modeling includes Huang and Cressie (1996), Berke (1998), and Mardia, Goodall, Redfern, and Alonso (1998). Handcock and Wallis (1994) take an alternative approach to Bayesian spatio-temporal modeling.

While not covered here, Bayesian analysis of spatial data is popular both because of its flexibility and the fact that it naturally accounts for the effects of estimating covariances, cf. Sect. 4.7. In its simplest form, this involves extending Bayesian standard linear models (see *PA* Sect. 2.9 and Christensen, Johnson, Branscum, & Hanson 2010) by using ideas similar to Chap. 4 and those introduced here to include positive covariances among observations and by incorporating a prior distribution on the additional parameters. Early discussions were given by Kitanidis (1986), Omré and Halvorsen (1989), and Handcock and Stein (1993). Diggle, Tawn, and Moyeed (1998) proposed an extension of generalized linear models to spatial problems along with a Bayesian analysis based on Markov Chain Monte Carlo (MCMC) methods.

Many recent books are devoted to spatial or spatio-temporal data. These include Gelfand, Diggle, Guttorp, Fuentes (2010), Cressie and Wikle (2011), Wikle, Zammit-Mangion, and Cressie (2019), Sherman (2011), and Banerjee, Carlin, and Gelfand (2015). See also Schabenberger and Gotway (2005) and Waller and Gotway (2004), the later focuses on public heath. Hodges (2014) also includes discussion of spatial models. *Many of these emphasize Bayesian methods.*

We begin by discussing the modeling of spatial data in terms of stochastic processes. In Sect. 8.2, linear models and best linear unbiased predictors for spatial data are presented. Best linear unbiased prediction is known in the geostatistics literature as *kriging*. The methods of kriging were developed in France by Matheron (1965, 1969). He was originally inspired by the contributions of D.G. Krige. The French work was performed independently of Goldberger (1962), who first derived general best linear unbiased predictors for linear models. The relationship between prediction based on covariances and prediction based on an alternative measure of variability, the semivariogram, is examined in Sect. 8.3. The role of measurement error is considered in Sect. 8.4. Section 8.5 looks at the effects of estimating the covariances when performing best linear unbiased prediction. Section 8.6 gives models for covariance functions and semivariograms. The special case of spatial lattice data is examined in Sect. 8.7 because of its close relationship to the analysis of time series data observed at regular intervals. Estimation of covariances and the semivariogram is considered in the final section.

8.1 Modeling Spatial Data

Spatial data can be considered as a realization of a stochastic process (*random field*)

$$y(u), \qquad u \in D \subset \mathbf{R}^d.$$

Here, u is a location in D. Most often, d, the dimension of the space, is 1, 2, or 3. (To some extent, time could be added to these problems by merely increasing the dimension.) For every value of u, $y(u)$ is a random variable. To analyze spatial data, we need to model $y(u)$. We begin by assuming that, for any u, $\mathrm{E}[y(u)]$ and $\mathrm{Var}[y(u)]$ exist. It follows that $y(u)$ can be decomposed as

$$y(u) = m(u) + e(u),$$

where $m(u)$ is the fixed mean function of $y(u)$, namely

$$m(u) \equiv \mathrm{E}[y(u)],$$

and $e(u)$ is a stochastic error process with

$$\mathrm{E}[e(u)] = 0.$$

In particular, $e(u) \equiv y(u) - m(u)$. Our approach to modeling $y(u)$ involves modeling both $m(u)$ and $e(u)$.

Begin by assuming a linear structure for $m(u)$. This is known in geostatistics as the *universal kriging* model. In particular, assume that there are, say, p known functions of u, say, $x_1(u), x_2(u), \ldots, x_p(u)$ so that the mean function satisfies

$$m(u) = \sum_{j=1}^{p} \beta_j x_j(u)$$

for some fixed unknown parameters $\beta_1, \ldots, \beta_p$. A special case of the universal kriging model is the *ordinary kriging* model

$$m(u) = \mu$$

for an unknown parameter μ.

In practice, the $x_j(\cdot)$s are often just functions of the location coordinates (cf. Chap. 1), but they can be any variables that are available for every location. Satellite data often provide a variety of interesting predictor functions $x_j(u)$.

A mathematically simpler, but typically unrealistic, model is to assume that $m(u)$ is known. *Simple kriging* is the special case

$$m(u) = \mu_0,$$

where μ_0 is a known value. The case with $m(u)$ known will not be treated further. Simple modifications of the procedures outlined later provide data analysis for the case with $m(u)$ known.

In modeling the error process $e(u)$, we will be primarily interested in its second-order (second moment) properties. The *covariance function* is our primary concern,

$$\sigma(u,w) \equiv \text{Cov}[e(u), e(w)].$$

Note also that

$$\sigma(u,w) = \text{Cov}[y(u), y(w)].$$

and

$$\sigma(u,w) = \sigma(w,u).$$

Often, the covariance function is modeled in terms of an unknown parameter vector θ. In that case, write $\sigma(u,w;\theta)$. Some of the common assumptions made about $e(u)$ are that it is (1) second-order stationary, (2) strictly stationary, (3) intrinsically stationary, (4) increment stationary, or (5) isotropic. These terms are defined in the next subsection.

8.1.1 Stationarity

We restrict attention to processes for which means and variances exist. A process $y(u)$ is said to be *strictly stationary* if for any value k, any locations $u_1, \ldots, u_k$, any (Borel) sets $C_1, \ldots, C_k$, and any vector $h \in \mathbf{R}^d$, we have

$$\Pr[y(u_1) \in C_1, \ldots, y(u_k) \in C_k] = \Pr[y(u_1 + h) \in C_1, \ldots, y(u_k + h) \in C_k]. \quad (8.1.1)$$

The process is stationary in the sense that the joint distribution of the process evaluated at any set of points is not changed if all of the points are relocated in the same way. In particular,

$$m(u) = m(u + h)$$

for any vector h, so $m(u)$ must be a constant, namely

$$m(u) = \mu. \quad (8.1.2)$$

Also, for two locations u and w and any vector h,

$$\sigma(u,w) = \sigma(u + h, w + h).$$

In particular, let $h = -w$ so

$$\sigma(u,w) = \sigma(u - w, 0)$$

and the covariance function can be thought of as a function of $u - w$ alone. To indicate this, let $h = u - w$ and write

$$\sigma(u,w) = \sigma(u-w)$$
$$= \sigma(h). \qquad (8.1.3)$$

Since $\sigma(u,w) = \sigma(w,u)$, for a stationary process,

$$\sigma(u-w) = \sigma(w-u),$$

and

$$\sigma(h) = \sigma(-h).$$

Note that for any location u, equation (8.1.3) implies that

$$\sigma(0) = \sigma(u,u) = \mathrm{Var}[y(u)].$$

If $y(u)$ is strictly stationary and the joint distribution of all the random variables in (8.1.1) is multivariate normal for any k, then the process is called a *Gaussian process*. To specify a particular Gaussian process it is enough to specify (8.1.2) and (8.1.3).

As we have discussed, when variances exist property (8.1.1) implies properties (8.1.2) and (8.1.3). A *second-order* (weak) *stationary process* is any process that satisfies (8.1.2) and (8.1.3). A second-order stationary process may be strictly stationary in the sense that (8.1.1) holds, but it need not be. Any second-order stationary process has a corresponding strictly stationary Gaussian process.

We define a process to be *increment stationary* if it satisfies (8.1.2) and, for any integer k, locations $u_1,\dots,u_k$, sets $C_1,\dots,C_{k-1}$, and vector h,

$$\Pr[y(u_2)-y(u_1) \in C_1,\dots,y(u_k)-y(u_{k-1}) \in C_{k-1}] =$$
$$\Pr[y(u_2+h)-y(u_1+h) \in C_1,\dots,y(u_k+h)-y(u_{k-1}+h) \in C_{k-1}].$$

It is immediate that stationary processes are increment stationary, but the converse need not be true. Brownian motion (see Breiman 1968) is a well-known process with $u \in [0,\infty)$ that is increment stationary but not stationary.

In the geostatistics literature, second-order properties are not typically characterized using the covariance function. Instead, they are represented by either the variogram or the semivariogram. These functions are similar to the covariance function but are more appropriate for use with increment stationary processes because they are defined directly on the increments. For a process satisfying (8.1.2), the *semivariogram* is defined as

$$\gamma(u,w) \equiv \frac{1}{2}\mathrm{E}[y(u)-y(w)]^2$$
$$= \frac{1}{2}\mathrm{Var}[y(u)-y(w)]$$
$$= \frac{1}{2}\{\mathrm{Var}[y(u)] + \mathrm{Var}[y(w)] - 2\mathrm{Cov}[y(w),y(u)]\}$$
$$= \frac{1}{2}[\sigma(u,u) + \sigma(w,w) - 2\sigma(w,u)].$$

The *variogram* is twice the semivariogram. Clearly, the two functions contain similar information. The variogram has a more natural definition but, as will be seen later, for second-order stationary processes the semivariogram has advantages. Our discussion will use the semivariogram exclusively.

For an increment stationary process, $\gamma(u,w) = \gamma(u+h, w+h)$ for any h. Letting $h = -w$,

$$\gamma(u,w) = \gamma(u-w, 0),$$

and we write

$$\gamma(u,w) = \gamma(u-w). \tag{8.1.4}$$

A process is said to be *intrinsically stationary* if it satisfies both (8.1.4) and (8.1.2). Intrinsic stationary processes need not be increment stationary, but increment stationary processes are intrinsically stationary. The relationship between increment stationarity and intrinsic stationarity is similar to the relationship between stationarity and second-order stationarity. Just as stationary processes are increment stationary, second-order stationary processes are intrinsically stationary. To whit, with second-order stationarity

$$\begin{aligned}
\gamma(u,w) &= \frac{1}{2}\left[\sigma(u,u) + \sigma(w,w) - 2\sigma(w,u)\right]. \\
&= \frac{1}{2}\left[\sigma(0) + \sigma(0) - 2\sigma(w-u)\right] \\
&= \sigma(0) - \sigma(w-u) \\
&= \sigma(0) - \sigma(u-w).
\end{aligned}$$

This is a function of $u - w$, so second-order stationary processes are also intrinsically stationary. In particular, if $h = u - w$,

$$\gamma(h) = \sigma(0) - \sigma(h). \tag{8.1.5}$$

If variances do not exist, a process can be increment stationary without being intrinsically stationary.

It has been argued that methods based on the semivariogram are preferable to methods based on the covariance function because the semivariogram can exist in cases where the covariance function does not (e.g., processes with infinite variances). I have never heard of a measuring device that allows infinitely large observations, so I cannot imagine a need for modeling data with infinite variances. A more interesting rationale for preferring models based on intrinsic stationarity, rather than second-order stationarity, is that second-order stationarity implies a constant variance for all observations, whereas intrinsic stationarity allows different variances. However, the real goal is prediction and, as we will see in Sect. 8.3, predictions based on intrinsic stationarity are identical to predictions based on second-order stationarity.

Another generalization of stationarity is contained in the ideas of *generalized covariance functions* and *intrinsic random functions of order k*. Just as an intrinsically stationary process with its stationary semivariogram can be used to model a process with a nonstationary covariance function, an intrinsic random function of order k with its stationary generalized covariance function can be used to model a process with a nonstationary covariance function. Knowledge of the generalized covariance function does not completely specify the covariance structure of a process, but it can be used to obtain best linear unbiased predictions for universal kriging models. The equivalence of predictions based on the covariance function and the generalized covariance function can be established by modifying the results given in Sect. 8.3; see Christensen (1990, 1993). Generalized covariance functions and intrinsic random functions were originally introduced by Matheron (1973) to avoid problems encountered with using residuals to estimate second-order properties in universal kriging. For data analysis, the fundamental idea is the same as in REML estimation, see Sect. 4.3, but the methods are not necessarily based on likelihood analysis. Delfiner (1976) gives an introduction to these topics; they will not be discussed further.

Often the second-order properties of a process can be assumed to depend only on the distance between two points and not on the direction between them. A second-order stationary process is *isotropic* if

$$\sigma(u-w) = \sigma(\|u-w\|).$$

An intrinsically stationary process is isotropic if

$$\gamma(u-w) = \gamma(\|u-w\|).$$

A process that is not isotropic is said to be *anisotropic*.

Finally, a characterization of processes that leads to computational advantages is separability. Let $h' = (h_1, \ldots, h_d)$. A second-order stationary process is said to be *separable* if its covariance function can be written as

$$\sigma(h) = \prod_{i=1}^{d} \sigma_i(h_i)$$

for some one-dimensional covariance functions $\sigma_1(\cdot), \ldots, \sigma_d(\cdot)$. A similar property for the semivariogram defines separability for intrinsically stationary processes. See Zimmerman (1989) and Zimmerman and Harville (1990) for applications of separability. More generally, if we partition h as $h' = (\tilde{h}_1, \ldots, \tilde{h}_r)$, we can define separability as

$$\sigma(h) = \prod_{k=1}^{r} \tilde{\sigma}_k(\tilde{h}_k).$$

For example, if time is used as a dimension, the time covariance may be multiplied by the spatial covariance to get the spatio-temporal covariance. In three dimensions characterized by longitude, latitude, and depth, sometimes depth is taken as a separable covariance term.

In this subsection, concepts of stationarity have been discussed for an arbitrary process $y(u)$ for which variances exist. In the remainder of the chapter, these variants of stationarity will be incorporated into diverse models for the error process $e(u)$. In the universal kriging model, $E[y(u)]$ is not constant, so the process $y(u)$ fails to satisfy any of the definitions related to stationarity. On the other hand, the error process has a constant mean of zero, so it may be modeled with some sort of stationarity assumption. As will be seen in the next two sections, the actual process of best linear unbiased prediction does not require any version of stationarity. Nonetheless, modeling the covariance structure is vital to the analysis of spatial data, and stationarity assumptions are an important aspect of covariance modeling.

It seems to have been the case that most people either put a lot of effort into modeling $m(\cdot)$ or $\sigma(\cdot, \cdot)$ but not both. People who did ordinary kriging seemed to put a lot of work into modeling the covariance (or semivariogram) function. People who did sophisticated modeling of the mean function seemed to worry less about the covariance function they used.

8.2 Best Linear Unbiased Prediction of Spatial Data: Kriging

The purpose of this section is to illustrate the relationship between kriging and best linear unbiased prediction. The object of kriging is the prediction of unobserved spatial random variables based on the values of observed random variables. There is little to do except establish that we are working with a linear model. Given that fact, the BLUP is well-known; see Sect. 4.1. Kriging is often presented as either point kriging or block kriging. We give a detailed discussion of point kriging and mention the variations needed for block kriging.

Assume that the universal kriging model

$$m(u) = \sum_{j=1}^{p} \beta_j x_j(u)$$

holds, that observations have been taken at locations $u_1, \ldots, u_n$, and that we wish to predict the value of $y(u_0)$. Set the following notation:

$$y_i = y(u_i), \quad i = 0, \ldots, n,$$
$$Y = (y_1, \ldots, y_n)',$$
$$x_{ij} = x_j(u_i), \quad i = 0, \ldots, n,$$
$$x_i' = (x_{i1}, \ldots, x_{ip}), \quad i = 0, \ldots, n,$$

$$X = \begin{bmatrix} x_1' \\ \vdots \\ x_n' \end{bmatrix},$$

$$\beta = (\beta_1, \ldots, \beta_p)',$$
$$e_i = e(u_i), \quad i = 0, \ldots, n,$$
$$e = (e_1, \ldots, e_n)',$$

and
$$\sigma_{ij} = \sigma(u_i, u_j), \quad i,j = 0, \ldots, n.$$

The universal kriging model, as applied to the observations, can be written as

$$Y = X\beta + e, \quad E(e) = 0, \quad \text{Cov}(e) = \Sigma, \tag{8.2.1}$$

where
$$\Sigma = [\sigma_{ij}], \quad i,j = 1, \ldots, n.$$

Generally, Σ is nonsingular and the functions $x_j(u)$ and locations are taken so that X has full column rank with $J \in C(X)$. Under the full column rank assumption, β is estimable, so, for any location u_0, $x_0'\beta$ is estimable. Let

$$\Sigma_{Y0} \equiv \begin{bmatrix} \sigma_{10} \\ \vdots \\ \sigma_{n0} \end{bmatrix}.$$

Applying the results of Sect. 4.1, the best linear unbiased predictor of y_0 is

$$\hat{y}_0 = x_0'\hat{\beta} + \delta'(Y - X\hat{\beta}), \tag{8.2.2}$$

where
$$\hat{\beta} = (X'\Sigma^{-1}X)^{-1}X'\Sigma^{-1}Y$$

and
$$\delta = \Sigma^{-1}\Sigma_{Y0}.$$

As in Sect. 4.1, we can write
$$\hat{y}_0 = b'Y,$$

where

$$b' = x_0'(X'\Sigma^{-1}X)^{-1}X'\Sigma^{-1} + \delta'\left(I - X(X'\Sigma^{-1}X)^{-1}X'\Sigma^{-1}\right). \tag{8.2.3}$$

The mean squared prediction error (prediction variance) is

$$\text{Var}(y_0 - \hat{y}_0) = \sigma_{00} - \Sigma_{Y0}'\Sigma^{-1}\Sigma_{Y0} + [x_0 - X'\delta]'(X'\Sigma^{-1}X)^{-1}[x_0 - X'\delta] \tag{8.2.4}$$
$$= \sigma_{00} - 2b'\Sigma_{Y0} + b'\Sigma b.$$

Often, the object of kriging is to produce a two- or three-dimensional map of the variable $y(u)$ over the region D. To do this, a large number of predictions for various locations u_0 are needed. Rewrite (2) as

$$\hat{y}_0 = x_0'\hat{\beta} + \Sigma_{Y0}'\Sigma^{-1}(Y - X\hat{\beta}).$$

Given $\hat{\beta}$ and $\Sigma^{-1}(Y - X\hat{\beta})$, computation of different predictions is very inexpensive. It requires only the computation of inner products between x_0 and $\hat{\beta}$ and between Σ_{Y0} and $\Sigma^{-1}(Y - X\hat{\beta})$ for the various values of x_0 and Σ_{Y0}.

Kitanidis (1986) gives a Bayesian analysis of the universal kriging model. In particular, he relates Bayes point predictions to the best linear unbiased predictors and the variance of the predictive distribution to the prediction variance. Cressie (1986, 1993) presents prediction methods based on exploratory data analysis. Computations are discussed in the more modern books mentioned earlier.

8.2.1 Block Kriging

So far we have assumed that our observations have been taken at point locations $u_1, \ldots, u_n$. In fact, it is physically impossible to take observations at points. An observation taken at u_i must actually be an observation on some neighborhood of u_i. If the neighborhood is small, the approximation to point observations should be good. However, in many applications (particularly mining), the neighborhood is sufficiently large that the properties of the neighborhood need to be incorporated into the analysis. In the geostatistical literature, the neighborhoods are referred to as blocks. The theory presented earlier carries through with almost no change. We need only to redefine the terms of the linear model.

Let B_i be the block associated with the ith observation. Let $|B_i|$ be the volume (area) of the block. The observations and value to be predicted are now

$$y_i = \frac{1}{|B_i|} \int_{B_i} y(u) du$$

for $i = 0, 1, \ldots, n$. The elements of the model matrix X are

$$x_{ij} = \frac{1}{|B_i|} \int_{B_i} x_j(u) du.$$

Note that under the universal kriging model

$$
\begin{aligned}
\mathrm{E}(y_i) &= \frac{1}{|B_i|} \int_{B_i} \sum_{j=1}^{p} \beta_j x_j(u) du \\
&= \sum_{j=1}^{p} \beta_j \left\{ \frac{1}{|B_i|} \int_{B_i} x_j(u) du \right\} \\
&= \sum_{j=1}^{p} \beta_j x_{ij}.
\end{aligned}
$$

Thus, the y_is follow a linear model.

Note that $y_i - \mathrm{E}(y_i) = (1/|B_i|) \int_{B_i} e(u) du$, so the covariance of two observations is

$$\mathrm{Cov}(y_i, y_j) = \mathrm{E} \left\{ \frac{1}{|B_i|} \frac{1}{|B_j|} \int_{B_i} e(u) du \int_{B_j} e(v) dv \right\}$$

$$= \frac{1}{|B_i|}\frac{1}{|B_j|}\int_{B_i}\int_{B_j} E[e(u)e(v)]dudv$$

$$= \frac{1}{|B_i|}\frac{1}{|B_j|}\int_{B_i}\int_{B_j} \sigma(u,v)dudv.$$

Given this linear model with known covariance structure, the BLUP can be obtained as usual.

8.2.2 Gaussian Process Regression

If one chooses to think of $u \in \mathbf{R}^d$ as a vector of predictor variables in a regression, rather than as a vector of locations, the ideas discussed in this chapter are sometimes referred to as *Gaussian process regression*. Many discussions of Gaussian process regression restrict themselves to the ordinary kriging model. In the machine learning community the most common reference seems to be Rasmussen and Williams (2006).

8.3 Prediction Based on the Semivariogram: Geostatistical Kriging

In much of the geostatistics literature, kriging is presented as a procedure that uses the semivariogram of the error process to determine optimal predictions. In general, given the semivariogram, one cannot reproduce the covariance function, so it is not clear that kriging can be performed using only the semivariogram. A special case in which the semivariogram and the covariance function are equivalent is that of a second-order stationary process with

$$\lim_{\|u\|\to\infty} \sigma(u) = 0. \tag{8.3.1}$$

In this case, by (8.1.5), $\gamma(h) = \sigma(0) - \sigma(h)$, so

$$\lim_{\|u\|\to\infty} \gamma(u) = \sigma(0)$$

and

$$\sigma(h) = \lim_{\|u\|\to\infty} \gamma(u) - \gamma(h).$$

But generally, for intrinsically stationary processes and for second-order stationary processes that do not have property (8.3.1), we cannot expect to reproduce the covariance function from the semivariogram.

In this section, we give a condition under which the BLUP can be found as a function of the semivariogram. Interestingly, the condition does not involve the variability of the error process; the condition is essentially that the linear model (8.2.1) contains an intercept and that for predicting $y(u_0)$, the mean $m(u_0) = x'_0 \beta$ also contains an intercept. In practice, this condition can be specified as

$$x_1(u) = 1 \qquad \text{all } u.$$

More generally, we can assume that

$$J_{n+1} \in C\left(\begin{bmatrix} X \\ x'_0 \end{bmatrix}\right),$$

or equivalently that, for some vector d,

$$J = Xd$$

and

$$1 = x'_0 d.$$

Our proof is based on the result of *PA-V* Exercise 6.8c (Christensen 2011, Exercise 12.1c) that if $b'Y$ is the BLUP, there exists a vector, that we will call λ, such that

$$\begin{bmatrix} \Sigma & X \\ X' & 0 \end{bmatrix} \begin{bmatrix} b \\ \lambda \end{bmatrix} = \begin{bmatrix} \Sigma_{Y0} \\ x_0 \end{bmatrix}. \tag{8.3.2}$$

There are two simple ways to prove this. First, b' as given in (8.2.3) can be substituted into (8.3.2) and a vector λ can be identified that satisfies the equality. Alternatively, if the inverses exist, one can establish that

$$\begin{bmatrix} \Sigma & X \\ X' & 0 \end{bmatrix}^{-1} = \begin{bmatrix} \Sigma^{-1}(I-A) & \Sigma^{-1}X(X'\Sigma^{-1}X)^{-1} \\ (X'\Sigma^{-1}X)^{-1}X'\Sigma^{-1} & -(X'\Sigma^{-1}X)^{-1} \end{bmatrix},$$

where $A = X(X'\Sigma^{-1}X)^{-1}X'\Sigma^{-1}$; see *PA* Exercise B.21. The solution of (8.3.2) is

$$\begin{bmatrix} b \\ \lambda \end{bmatrix} = \begin{bmatrix} \Sigma & X \\ X' & 0 \end{bmatrix}^{-1} \begin{bmatrix} \Sigma_{Y0} \\ x_0 \end{bmatrix},$$

and substituting for the inverse matrix it is easily seen that b is the same as (8.2.3). In any case, if (8.3.2) has a unique solution, b must be as in (8.2.3).

Having established (8.3.2), we show that b satisfies a similar equation that involves only the semivariogram. If this equation has a unique solution, the solution must determine the BLUP.

In particular, define semivariogram matrices similar to Σ and Σ_{Y0}, say

$$\Gamma = [\gamma(u_i, u_j)]_{n \times n}$$

and

$$\Gamma_{Y0} = [\gamma(u_i, u_0)]_{n \times 1}.$$

The equation analogous to (8.3.2) is

$$\begin{bmatrix} -\Gamma & X \\ X' & 0 \end{bmatrix} \begin{bmatrix} b \\ \xi \end{bmatrix} = \begin{bmatrix} -\Gamma_{Y0} \\ x_0 \end{bmatrix}. \tag{8.3.3}$$

We now establish the equivalence of Eqs. (8.3.2) and (8.3.3).

Proposition 8.3.1. Suppose there exists a vector d such that $J = Xd$ and $1 = x_0' d$. Any solution $[b', \xi']'$ to (8.3.3) determines a solution $[b', \lambda']'$ to (8.3.2) and conversely.

PROOF. Let $\sigma_y' = [\sigma_{11}, \ldots, \sigma_{nn}]$. It is easily seen that

$$\Sigma = \frac{1}{2}[\sigma_y J' + J\sigma_y'] - \Gamma \tag{8.3.4}$$

and

$$\Sigma_{Y0} = \frac{1}{2}[\sigma_y + \sigma_{00} J] - \Gamma_{Y0}. \tag{8.3.5}$$

Both Eqs. (8.3.2) and (8.3.3) imply that $X'b = x_0$, so a solution to either equation satisfies

$$J'b = d'X'b = d'x_0 = 1.$$

Suppose $[b', \xi']'$ is a solution to (8.3.3). We need to show that this determines a solution $[b', \lambda']'$ for (8.3.2). From Eq. (8.3.3)

$$-\Gamma b + X\xi = -\Gamma_{Y0}.$$

Substituting for $-\Gamma$ and $-\Gamma_{Y0}$ from (8.3.4) and (8.3.5) gives

$$\Sigma b - \frac{1}{2}\sigma_y J'b - \frac{1}{2}J\sigma_y'b + X\xi = \Sigma_{Y0} - \frac{1}{2}\sigma_y - \frac{1}{2}\sigma_{00}J.$$

Using the fact that $J'b = 1$ and rearranging terms gives

$$\Sigma b - \frac{1}{2}J\sigma_y'b + X\xi + \frac{1}{2}\sigma_{00}J = \Sigma_{Y0}.$$

However, $J = Xd$, so

$$\Sigma b + X\left(-\frac{1}{2}d\sigma_y'b + \xi + \frac{1}{2}\sigma_{00}d\right) = \Sigma_{Y0}.$$

Pick $\lambda = -\frac{1}{2}d\sigma_y'b + \xi + \frac{1}{2}\sigma_{00}d$, and $[b', \lambda']'$ is a solution to (8.3.2).

To show that any solution $[b', \lambda']'$ for (8.3.2) yields a solution to (8.3.3), use a similar argument. In particular, use (8.3.4) and (8.3.5) to substitute for Σ and Σ_{Y0} in (8.3.2). □

Based on Proposition 8.3.1, if (8.3.2) and (8.3.3) have unique solutions, then a solution to (8.3.3) determines a solution to (8.3.2), which determines the BLUP. The traditional method for finding the BLUP in geostatistics is to find a solution to (8.3.3). We have established mild conditions for the validity of that approach.

The prediction variance can be written in terms of the semivariogram. By (8.2.4), the fact that $J'b = 1$, and Eqs. (8.3.4) and (8.3.5),

$$
\begin{aligned}
E(y_0 - \hat{y}_0)^2 &= \sigma_{00} - 2b'\Sigma_{Y0} + b'\Sigma b \\
&= \sigma_{00} - 2b'\left\{\frac{1}{2}\sigma_y + \frac{1}{2}\sigma_{00}J - \Gamma_{Y0}\right\} + b'\left\{\frac{1}{2}\sigma_y J' + \frac{1}{2}J\sigma_y' - \Gamma\right\}b \\
&= \sigma_{00} - b'\sigma_y - \sigma_{00}b'J + 2b'\Gamma_{Y0} + \frac{1}{2}b'\sigma_y J'b + \frac{1}{2}b'J\sigma_y'b - b'\Gamma b \\
&= 2b'\Gamma_{Y0} - b'\Gamma b.
\end{aligned}
\tag{8.3.6}
$$

8.4 Measurement Error and the Nugget Effect

One aspect of best linear (unbiased) prediction that is unappealing for some purposes is the fact, illustrated in *PA-V* Exercise 6.8b (Christensen 2011, Exercise 12.1b), that the predictor of a point that has been observed is just the point itself (i.e., $\hat{y}_i = y_i$). This phenomenon occurs because in predicting $y_0 \equiv y_i$,

$$
\Sigma_{Y0} = \begin{bmatrix} \sigma_{1i} \\ \vdots \\ \sigma_{ni} \end{bmatrix},
$$

which is just the ith column of Σ. Clearly, a solution to $\Sigma\delta = \Sigma_{Y0}$ is given by the vector $\delta = (0, \ldots, 0, 1, 0, \ldots, 0)'$, where the 1 is in the ith place, i.e., δ is the ith column of I_n. Thus,

$$
\begin{aligned}
\hat{y}_i &= x_i'\hat{\beta} + \delta'(Y - X\hat{\beta}) \\
&= x_i'\hat{\beta} + (y_i - x_i'\hat{\beta}) \\
&= y_i.
\end{aligned}
$$

Given our current model, this is only reasonable. We have available only one realization of the stochastic process, so there is only one value observable at u_i and we have observed it. If we wish to predict at u_i, the predictor must be y_i.

For many purposes, more smoothing is desired in the predictor. This can be accomplished by imagining that subsequent measurements taken at u_i could be different from our original observation y_i. For this to happen, our observations must be subject to measurement error. Measurement errors are generally modeled as being uncorrelated with constant variance (white noise). In particular, we assume that observations follow the model

$$y(u) = m(u) + e(u) + e_M(u),$$

where all terms are defined as in Sect. 8.1 except that $e_M(u)$ is a second-order stationary measurement error process, namely

$$E[e_M(u)] = 0,$$
$$\text{Var}[e_M(u)] = \sigma_M^2,$$
$$\text{Cov}[e_M(u), e_M(w)] = 0 \qquad u \neq w,$$

and

$$\text{Cov}[e_M(u), e(w)] = 0 \qquad \text{any } u, w.$$

Define covariance functions $\sigma_e(u,w)$ for $e(u)$, $\sigma_M(u,w)$ for $e_M(u)$ and $\sigma_\varepsilon(u,w)$ for $\varepsilon(u) \equiv e(u) + e_M(u)$. Note that $\sigma_\varepsilon(u,w) = \sigma_e(u,w) + \sigma_M(u,w)$ and $\sigma_M(u,w) = \sigma_M^2 \delta_{u,w}$, where $\delta_{u,w}$ is the Kronecker delta: 1 if $u = w$ and 0 otherwise.

Letting $e_M = (e_M(u_1), \ldots, e_M(u_n))'$, the spatial linear model is

$$Y = X\beta + \varepsilon,$$

where $\varepsilon = e + e_M$, $E(\varepsilon) = 0$, and writing $\text{Cov}(\varepsilon) = V$ gives

$$V = \Sigma + \sigma_M^2 I.$$

With measurement error, the covariance matrix of Y is now V but the covariance between Y and a future observation at u_i is still Σ_{Y0}. The covariance is based entirely on $e(u)$. The measurement error process $e_M(u)$ contributes nothing. The prediction is

$$\hat{y}_i = x_i'\hat{\beta} + \Sigma_{Y0}'V^{-1}(Y - X\hat{\beta}),$$

where

$$\hat{\beta} = (X'V^{-1}X)^{-1}X'V^{-1}Y.$$

Typically, the prediction does not simplify to y_i.

A well-known concept in geostatistics is that of the *nugget effect*. The terminology comes from mining in that a spatial process might not be smooth because one finds nuggets of gold etc. A nugget effect is said to occur for an intrinsically stationary process if

$$\lim_{\|h\| \to 0} \gamma(h) = \gamma_0 > 0.$$

Note that by definition, $\gamma(0) = 0$ (see Sect. 8.1). The existence of a nugget effect is in some sense equivalent to measurement error.

First, we show that adding measurement error to any intrinsically stationary process creates a nugget effect. Assume an arbitrary measurement error model. The total error process is

$$\varepsilon(u) = e(u) + e_M(u).$$

The measurement error process $e_M(u)$ is second-order stationary by definition and it is easy to see that its semivariogram is $\gamma_M(h) = \sigma_M^2(1 - \delta_{h,0})$. Assume that

$e(u)$ is intrinsically stationary. Because the measurement error is assumed to be uncorrelated with $e(u)$, it is not hard to show that $\varepsilon(u)$ must also be intrinsically stationary with

$$\gamma_\varepsilon(h) = \gamma_e(h) + \gamma_M(h).$$

If $\lim_{\|h\|\to 0} \gamma_e(h)$ exists, it must be nonnegative, and

$$\lim_{\|h\|\to 0} \gamma_\varepsilon(h) = \lim_{\|h\|\to 0} \gamma_e(h) + \lim_{\|h\|\to 0} \gamma_M(h) = \lim_{\|h\|\to 0} \gamma_e(h) + \sigma_M^2 > 0.$$

Conversely, if $\varepsilon(u)$ is an intrinsically stationary error process with a nugget effect, then there exists an intrinsically stationary error process $e(u)$ and an uncorrelated measurement error process $e_M(u)$ such that the process $\eta(u) = e(u) + e_M(u)$ has the same semivariogram as $\varepsilon(u)$. Take $\lim_{\|h\|\to 0} \gamma_\varepsilon(h) \equiv \gamma_0$ and simply define $e(u)$ to be a second-order stationary process with

$$\sigma_e(u,w) = \begin{cases} \sigma_\varepsilon(u,w) & u \neq w \\ \sigma_\varepsilon(u,u) - \gamma_0 & u = w \end{cases}$$

and $e_M(h)$ to be the measurement error process with

$$\sigma_M(h) = \begin{cases} 0 & h \neq 0 \\ \gamma_0 & h = 0 \end{cases}.$$

Because the measurement error process is taken to be uncorrelated with $e(h)$ and both are intrinsically stationary, $\eta(u)$ is intrinsically stationary with

$$\gamma_\eta(h) = \gamma_e(h) + \gamma_M(h) = \gamma_\varepsilon(h).$$

What remains to be shown is that $\sigma_e(u,w)$ is a valid covariance function. To do that, one needs to show that for any n and locations u_i, the matrix $\Sigma_e \equiv [\sigma_e(u_i, u_j)]$ is nonnegative definite.

For prediction, the difference between measurement error and a nugget effect without measurement error exists only in Σ_{Y0} and then only when $y_0 = y_i$ for some $i = 1, \ldots, n$. With measurement error, a new observation at location u_i can be different from y_i. The covariance between Y and a new observation at u_i is

$$\Sigma_{YiM} = \begin{bmatrix} \sigma_e(u_1, u_i) \\ \vdots \\ \sigma_e(u_n, u_i) \end{bmatrix}.$$

With a pure nugget effect, the only observation has already been taken, so

$$\Sigma_{YiN} = \begin{bmatrix} \sigma_\varepsilon(u_1, u_i) \\ \vdots \\ \sigma_\varepsilon(u_n, u_i) \end{bmatrix} = \Sigma_{YiM} + \sigma_M^2 \delta,$$

where $\delta' = (0, \ldots, 0, 1, 0, \ldots, 0)$ with the 1 in the ith place. For predictions not at a data point, either model gives

$$\Sigma_{Y0} = \begin{bmatrix} \sigma_e(u_1, u_0) \\ \vdots \\ \sigma_e(u_n, u_0) \end{bmatrix}.$$

If $\sigma_e(u, w)$ is continuous, as $u_0 \to u_i$, the measurement error model gives continuous predictions because $\Sigma_{Y0} \to \Sigma_{YiM}$. In the pure nugget effect model, Σ_{Y0} does not converge to Σ_{YiM}, so the predictions are discontinuous at the data points. The discontinuities exist so that data points will be predicted as themselves.

With only one observation at each location, the effects of measurement error and the nugget effect are statistically indistinguishable. Nonetheless, measurement error and the nugget effect are distinct concepts. The idea of a nugget effect is that the spatial process is fundamentally discontinuous. (One stumbles upon nuggets of gold, but gold is not continuously spread around the surface of the earth.) There is nothing one can do about such discontinuity. The corresponding spatial predictions are smooth everywhere except at the observed data locations, where the predicted value is the actual observation. In a sense, having a nugget effect simply means that the spatial correlation is weaker because no matter how close two locations are, the covariance between them remains smaller than the spatial variance, so there is no assurance that the observations will be close to each other. On the other hand, the effects of measurement error can be reduced and modeled by repeated sampling. As we take more observations at a given location, the variability should be entirely due to measurement error. We can then estimate both the mean value at that location, which should be the realization of the spatial process, and the variance at that location, which is the measurement error variance.

8.5 The Effect of Estimated Covariances on Prediction

As the universal kriging model simply defines a spatial linear model, all of the results of Chap. 4 continue to apply. In particular, the covariance matrix Σ is rarely known, so must be estimated. Thus, under conditions presented in Sect. 4.7, point estimates of spatial mean values and point predictions at individual locations are unbiased and naive estimates of their variability often seriously underestimate the true variability. This is an excellent motivation for performing Bayesian analysis, which gives similar estimates and predictions without the tendency to underestimate variability.

Since predictions based on the semivariogram are identical to predictions based on the covariance function, such predictions also share the properties discussed in Chap. 4 including the tendency towards naively underestimating prediction error.

Cressie (1988) observes that to ensure consistency of estimators in ordinary kriging one must assume not only that the process is strictly stationary but also *ergodic*;

see Adler (1981). Recall that strict stationarity (and the existence of second moments) implies second-order, increment, and intrinsic stationarities. None of these three imply stationarity. For Gaussian processes, ergodicity is implied by

$$\lim_{\|h\|\to\infty} \sigma(h) = 0.$$

Because ordinary kriging is a special case of universal kriging, even stronger assumptions may be necessary to ensure consistency in the more general model. In particular, some assumptions about the asymptotic behavior of the model matrix X are probably needed.

It should be emphasized that we are not discussing consistency of predictors. Even if the joint distributions (first and second moments) were known, the best predictor (best linear predictor) would not give perfect predictions. Best linear unbiased estimates depend on estimating the mean function. Consistency is concerned with the estimated mean converging to the true mean so that the BLUP converges to the BLP. When the covariances are also estimated, we obtain only an estimated BLUP. In this case again, consistency refers to the estimated BLUP converging to the BLP. Stein (1988, 1999) discusses asymptotically efficient prediction. Diamond and Armstrong (1983) indicate that prediction is reasonably robust to the choice of different covariance functions. See also Zimmerman and Cressie (1992).

8.6 Models for Covariance Functions and Semivariograms

In practice, the covariance function $\sigma(u,w)$ is rarely known. To obtain an estimate of the covariance matrix of Y (i.e., Σ), some method of estimating $\sigma(u,w)$ is needed. Recalling that Σ is an $n \times n$ matrix with $n(n+1)/2$ distinct elements, there is little hope of estimating Σ from the n observations in Y without making assumptions about Σ or, equivalently, assumptions about $\sigma(\cdot,\cdot)$. In particular, we assume that the covariance function depends on a vector of parameters θ. Write the covariance function as

$$\sigma(u,w;\theta),$$

which is a known function for given θ. We can also write

$$\mathrm{Cov}(Y) = \Sigma(\theta) = [\sigma(u_i,u_j;\theta)]$$

and

$$\mathrm{Cov}(Y,y_0) = \Sigma_{Y0}(\theta) = [\sigma(u_i,u_0;\theta)].$$

Alternatively, we can assume that the semivariogram depends on a vector of parameters θ and write

$$\gamma(u,w;\theta),$$

$$\Gamma(\theta) = [\gamma(u_i,u_j;\theta)],$$

and
$$\Gamma_{Y0}(\theta) = [\gamma(u_i, u_0; \theta)].$$

In this section, we consider some of the standard models for $\sigma(u, w; \theta)$ and $\gamma(u, w; \theta)$. In Sect. 8.8, methods of estimating θ are discussed. Section 11.6 discusses nonparameteric covariance functions for functional data. They can also be applied to spatial data.

8.6.1 The Linear Covariance Model

This model assumes that $\sigma(u, w; \theta)$ is linear in the components of θ. In particular, write $\theta = (\theta_1, \ldots, \theta_s)'$ and, for $k = 1, \ldots, s$, let $\sigma_k(u, w)$ be a known covariance function. The linear covariance model is

$$\sigma(u, w; \theta) = \sum_{k=1}^{s} \theta_k \sigma_k(u, w).$$

To ensure that $\sigma(u, w; \theta)$ is a legitimate covariance function, we assume that $\theta_k \geq 0$ for all k. Writing the $n \times n$ matrix

$$\Sigma_k = [\sigma_k(u_i, u_j)],$$

we have

$$\Sigma(\theta) = \sum_{k=1}^{s} \theta_k \Sigma_k.$$

Because each σ_k is a covariance function, the matrices Σ_k are nonnegative definite. Frequently one takes, $\Sigma_1 = I_n$. Clearly, this is the linear covariance structure of Sect. 4.4 with $V_0 = 0$ and $V_k = \Sigma_k$.

The linear semivariogram model is defined similarly,

$$\gamma(u, w; \theta) = \sum_{k=0}^{r} \theta_k \gamma_k(u, w),$$

for known semivariograms $\gamma_k(\cdot, \cdot)$.

A commonly used linear semivariogram is the isotropic function

$$\gamma(\|h\|; \theta) = \theta_0 + \theta_1 \|h\|, \tag{8.6.1}$$

where θ_0 and θ_1 are nonnegative and $\theta_0 = \sigma_M^2$. In fact, this is often referred to as *the* linear semivariogram model. This semivariogram cannot correspond to a second-order stationary process because

$$\lim_{\|h\| \to \infty} \{\theta_0 + \theta_1 \|h\|\} = \infty.$$

Recall that for a second-order stationary process $\gamma(\|h\|) = \sigma(0) - \sigma(\|h\|)$ and, by Cauchy–Schwartz, $|\sigma(\|h\|)| \leq \sigma(0)$; thus, the variance would have to be infinite. Brownian motion is a process in $\mathbf{R}$ that has a linear semivariogram. The linear semi-variogram model would seem to be most appropriate for data that have a logical origin (e.g., data collected around a smelter) and that resemble a random walk in that the variability increases as one gets further from the origin.

There is a temptation to modify (8.6.1) so that

$$\lim_{\|h\| \to \infty} \gamma(\|h\|; \theta) \neq \infty.$$

The idea is that the linear semivariogram may be a reasonable approximation up to a point but that the variability of real data would not go on increasing indefinitely. It is sometimes suggested that the function

$$g(\|h\|; \theta) = \begin{cases} \theta_0 + \theta_1 \|h\| & \|h\| \leq \theta_2 \\ \theta_0 + \theta_1 \theta_2 & \|h\| > \theta_2 \end{cases}$$

could be used. This would correspond to a second-order stationary process with measurement error $\sigma_M^2 = \theta_0$ and a variance for each observation of $K = \theta_0 + \theta_1 \theta_2$. Unfortunately, the corresponding "covariance" function

$$s(\|h\|; \theta) = K - g(\|h\|; \theta)$$

is not a legitimate covariance function. It is not nonnegative definite. One can find locations $u_1, \ldots, u_k$ such that the $k \times k$ matrix $[s(\|u_i - u_j\|; \theta)]$ is not nonnegative definite. Moreover, $g(\cdot; \theta)$ does not satisfy a property similar to nonnegative definiteness that is necessary for all semivariograms; see Journel and Hüijbregts (1978).

8.6.2 Nonlinear Isotropic Covariance Models

We now present some of the standard isotropic covariance models that are nonlinear. In the next subsection, some methods of dealing with nonisotropic (anisotropic) covariances will be considered.

The *spherical covariance function* is

$$\sigma(\|h\|; \theta) = \begin{cases} \theta_1 \left[1 - \frac{3\|h\|}{2\theta_2} + \frac{\|h\|^3}{2\theta_2^3} \right] & 0 < \|h\| \leq \theta_2 \\ \theta_0 + \theta_1 & \|h\| = 0 \\ 0 & \|h\| > \theta_2 \end{cases}$$

for θ_0, θ_1, θ_2 nonnegative. This covariance function arises naturally on $\mathbf{R}^3$ (see Matern 1986, Section 3.2) and also defines a covariance function on $\mathbf{R}^2$. The measurement error variance is $\sigma_M^2 = \theta_0$. The total variance is $\theta_0 + \theta_1$. The *range* of a covariance function is the distance after which observations become uncorrelated.

For the spherical model, observations more than θ_2 units apart are uncorrelated, so the range is θ_2.

Another class of covariance functions is

$$\sigma(\|h\|;\theta) = \begin{cases} \theta_1 \exp[-(\theta_2\|h\|)^v] & \|h\| > 0 \\ \theta_0 + \theta_1 & \|h\| = 0 \end{cases}$$

for θ_0, θ_1, θ_2 nonnegative and $0 < v \le 2$. The measurement error variance is θ_0, the total variance is $\theta_0 + \theta_1$, and the range is infinite. For $v = 1$, this is called the *exponential covariance function*, and for $v = 2$ it is called the *Gaussian covariance function*. While the range is infinite, correlations decrease very rapidly as $\|h\|$ increases. Of course, this phenomenon depends on the values of v and θ_2. For $v = 1$, the covariance structure is very similar to an *AR(1)* process, see Exercise 8.2. Using this covariance function for $v < 2$, the spatial process $e(u)$ is continuous but not differentiable. For $v = 2$, $e(u)$ is infinitely differentiable. Moreover, with $v = 2$ and $\theta_0 = 0$, the covariance matrix Σ is often nearly singular, hence making the resulting analysis quite unreliable. The use of $v = 2$ seems to be almost universally frowned upon.

Whittle (1954) has shown that a covariance function that depends on $K_1(\cdot)$, the first-order modified Bessel function of the second kind, arises naturally in $\mathbf{R}^2$. In particular, for $\theta_1, \theta_2 > 0$, the function is

$$\sigma(\|h\|;\theta) = \theta_1\|h\|\theta_2 K_1(\|h\|\theta_2).$$

This can be modified by adding a measurement error of variance θ_0 when $\|h\| = 0$. Whittle (1963) considers more general functions

$$\sigma(\|h\|;\theta) = [\theta_1/2^{v-1}\Gamma(v)](\theta_2\|h\|)^v K_v(\theta_2\|h\|),$$

where $v > 0$ and $K_v(\cdot)$ is the v order modified Bessel function of the second kind. Ripley (1981, p. 56) gives some graphs of these functions and mentions that, for $v = 1/2$, one gets the exponential model (without measurement error). Also, as $v \to \infty$, it approaches the Gaussian model. This class of covariance functions is often called the Matern class and seems to be increasingly popular—in part because it makes $e(u)$ finitely differentiable.

In both of the last two families, v can either be fixed or it can be treated as a parameter.

8.6.3 Modeling Anisotropic Covariance Functions

Anisotropic covariance functions are simply covariance functions that are not isotropic. We mention only two possible approaches to modeling such functions. Suppose that $h = (h_1, h_2, h_3)'$ and that we suspect the variability in the direction $(0, 0, 1)'$ is causing the anisotropicity. (Isn't anisotropicity a wonderful word?) For example, h_1 and h_2 could determine a surface location (e.g., longitude and latitude), while h_3 determines depth. For fixed h_3, variability might very well be isotropic in h_1 and h_2; however, the variability in depth may not behave like that in the other two directions.

Ripley (1981) suggests modifying isotropic models. Rather than using $\sigma(\|h\|)$, where $\|h\| = \sqrt{h_1^2 + h_2^2 + h_3^2}$, use $\sigma(\sqrt{h_1^2 + h_2^2 + \lambda h_3^2})$, where λ is an additional parameter to be estimated. For example, the exponential model becomes

$$\sigma(h; \theta, \lambda) = \begin{cases} \theta_1 \exp\left(-\theta_2 \sqrt{h_1^2 + h_2^2 + \lambda h_3^2}\right) & \|h\| > 0 \\ \theta_0 + \theta_1 & \|h\| = 0 \end{cases}.$$

This is a special case of the elliptical covariance functions discussed by Matern (1986). Elliptical covariance functions are isotropic functions $\sigma(\cdot)$ evaluated at $\sqrt{h'Ah}$. Here, A can be taken as any nonnegative definite matrix and may involve additional parameters.

Hüijbregts (1975) suggests adding different isotropic models, for example,

$$\sigma(h; \theta) = \sigma_1(\|h\|; \theta_1) + \sigma_2(|h_3|; \theta_2),$$

where $\sigma_1(\cdot; \theta_1)$ is an isotropic covariance on the entire vector h that depends on a parameter vector θ_1. Similarly, $\sigma_2(\cdot; \theta_2)$ is isotropic in the component h_3 and depends on the parameter vector θ_2.

8.6.4 Nonlinear Semivariograms

In geostatistics, if the semivariogram $\gamma(\cdot)$ has the property that

$$\lim_{\|h\| \to \infty} \gamma(h) = \gamma_\infty < \infty,$$

then γ_∞ is called the *sill* of the semivariogram. Any semivariogram with a sill can be obtained from a second-order stationary process with the property that

$$\lim_{\|h\| \to \infty} \sigma(h) = 0.$$

In particular, the stationary covariance function is

$$\sigma(0) = \gamma_\infty$$
$$\sigma(h) = \gamma_\infty - \gamma(h).$$

Conversely, any stationary covariance function with

$$\lim_{\|h\| \to \infty} \sigma(h) = 0 \tag{8.6.2}$$

determines a semivariogram with a sill. This follows from the fact that for second-order stationary processes

$$\gamma(h) = \sigma(0) - \sigma(h). \tag{8.6.3}$$

All of the nonlinear covariance functions we have considered satisfy (8.6.2). It is a simple matter to convert them to semivariogram models using (8.6.3).

8.7 Models for Spatial Lattice Data

When there exist only a finite (or countable) number of locations at which data would be collected, models for the covariances can be created to exploit that special structure. Such data are referred to as lattice data. When the locations occur only on a regular grid, they are referred to as regular lattice data. Throughout this section, we will assume normal errors, thus our model is

$$Y = X\beta + e, \quad e \sim N(0, \Sigma). \tag{8.7.1}$$

We consider three covariance models: spatial covariance selection models, spatial autoregression models, and spatial autoregressive moving average models. In all three cases, parameter estimation is through maximum likelihood. Predictions are made using empirical BLUPs.

8.7.1 Spatial Covariance Selection Models

Covariance selection is a graphical modeling device for specifying conditional independence between normally distributed random variables. Whittaker (1990, Chapter 6) and Edwards (2000, Chapter 3) discuss covariance selection in the context of *graphical Gaussian models*. Christensen (1997, Chapter 5) discusses graphical models for count data. The key point is that, for normal data, 0s in Σ^{-1} determine conditional independencies among the random variables. (Graphical representations help one interpret the conditional independencies.) In particular, an off-diagonal element of Σ^{-1} is zero if and only if the partial covariance between the two corresponding random variables given all of the remaining random variables is zero. For multivariate normal data, the partial covariance is zero (i.e., $\sigma_{ij \cdot k \neq i, j} = 0$), if and only if the random variables y_i and y_j are conditionally independent given all of the other random variables y_k with $k \neq i, j$; see Whittaker (1990, Chapter 5), Edwards (2000, Section 3.1), or Exercise B.3 and *PA* Sect. 6.5 and Exercise B.21.

The idea in spatial covariance selection is to model Σ^{-1} directly by incorporating nonzero off-diagonal parameters only for locations that are close to one another. Often, the model is parameterized as

$$\Sigma^{-1} = \frac{1}{\sigma^2}(I - C),$$

where σ^2 is a scalar parameter and the parameter matrix $C = [c_{ij}]$ has $c_{ii} \equiv 0$. C must be a symmetric matrix such that $I - C$ is positive definite. Typically, $c_{ij} \equiv 0$ unless the locations u_i and u_j are close to one another.

EXAMPLE 8.7.1.   For a regular lattice, we reindex the locations u_i, $i = 1, \ldots, n$ into $u_{g,h}$, $g = 1, \ldots, G$, $h = 1, \ldots, H$ with the idea that all the locations fall on a regular grid. We can then posit models for C such as $0 = c_{(g,h),(g',h')}$ except

$$\theta_{0,1} = c_{(g,h),(g,h-1)} = c_{(g,h),(g,h+1)},$$
$$\theta_{1,0} = c_{(g,h),(g-1,h)} = c_{(g,h),(g+1,h)},$$
$$\theta_{1,1} = c_{(g,h),(g-1,h-1)} = c_{(g,h),(g+1,h-1)}$$
$$= c_{(g,h),(g-1,h+1)} = c_{(g,h),(g+1,h+1)}.$$

Using a general result from graphical models, this particular model can be interpreted as saying that an observation $y_{g,h}$ is independent of all the other data given the values of the eight observations that immediately surround it. (This includes two horizontal neighbors, two vertical neighbors, and four diagonal neighbors.) The model reduces the covariance matrix to a four-parameter family (including σ^2). The θ parameters model the spatial correlation. Note that some care must be taken about how to model covariances at the edges of the lattice.   □

For nonregular lattices, one approach is to impose a regular lattice over the map of the nonregular lattice and then associate each point in the nonregular lattice with the closest point in the regular lattice.

See Cressie (1993, Section 7.2) for details of computing maximum likelihood estimates for covariance selection models.

8.7.2 Spatial Autoregression Models

One approach to producing a spatial autoregression is to model the error vector in (8.7.1). Let $\xi \sim N(0, \sigma^2 I)$ and for a parameter matrix Φ, define e in (8.7.1) through

$$(I - \Phi)e = \xi.$$

This is similar in spirit to the time series $AR(p)$ model (7.2.2).

EXAMPLE 8.7.2.   For a regular lattice, we could posit a model

$$e_{g,h} = \phi_{0,1}e_{g,h+1} + \phi_{0,-1}e_{g,h-1} + \phi_{1,0}e_{g+1,h} + \phi_{-1,0}e_{g-1,h} + \xi_{g,h},$$

or equivalently

$$e_{g,h} - \phi_{0,1}e_{g,h+1} - \phi_{0,-1}e_{g,h-1} - \phi_{1,0}e_{g+1,h} - \phi_{-1,0}e_{g-1,h} = \xi_{g,h}.$$

Obviously, more complicated models involving more locations can be built as well as models that simplify the parameters (e.g., taking $\phi_{0,1} = \phi_{0,-1}$). The matrix Φ is determined by how the entries $e_{g,h}$ are listed in the vector e as well as the $\phi_{r,s}$ parameters and how edge effects are handled. Typically, many of the entries in Φ will be zero.   □

Of course, $e = (I - \Phi)^{-1}\xi$, so

$$\Sigma = \sigma^2(I - \Phi)^{-1}(I - \Phi)^{-1\prime}$$

and

$$\Sigma^{-1} = \frac{1}{\sigma^2}(I - \Phi)'(I - \Phi).$$

As with covariance selection models, 0s in Σ^{-1} imply conditional independencies. In Example 8.7.2, y_{gh} is independent of all the other observations given the values of the four neighbors $y_{g,h+1}, y_{g,h-1}, y_{g+1,h}$, and $y_{g-1,h}$, and Σ is a five-parameter family. Note that in Example 8.7.1, if we set $\theta_{1,1} = 0$, we get the same conditional independence relationship with a three parameter family for Σ.

Exercise 8.1. Throughout, assume in Example 8.7.1 that $\theta_{1,1} = 0$. Establish whether setting $\phi_{0,1} = \phi_{0,-1}$ and $\phi_{1,0} = \phi_{-1,0}$ in Example 8.7.2 gives the covariance selection model of Example 8.7.1. Establish whether generalizing the parameters in Example 8.7.1 into $c_{(g,h),(g,h-1)} = \theta_{0,-1}$, $c_{(g,h),(g,h+1)} = \theta_{0,1}$, $c_{(g,h),(g-1,h)} = \theta_{-1,0}$, $c_{(g,h),(g+1,h)} = \theta_{1,0}$ gives the autoregressive model of Example 8.7.2.

8.7.3 Spatial Autoregressive Moving Average Models

As with the spatial autoregression model, let $\xi \sim N(0, \sigma^2 I)$. Now, define a spatial autoregressive moving average model for e through

$$(I - \Phi)e = (I - \Theta)\xi.$$

This is similar in spirit to the time series $ARMA(p,q)$ model (7.2.11). For example, with a regular lattice, we could posit a model

$$e_{g,h} = \phi_{0,1}e_{g,h+1} + \phi_{0,-1}e_{g,h-1} + \phi_{1,0}e_{g+1,h} + \phi_{-1,0}e_{g-1,h}$$
$$+ \xi_{g,h} - \theta_{0,1}\xi_{g,h+1} - \theta_{0,-1}\xi_{g,h-1} - \theta_{1,0}\xi_{g+1,h} - \theta_{-1,0}\xi_{g-1,h}.$$

The covariance matrix is

$$\Sigma = \sigma^2(I - \Phi)^{-1}(I - \Theta)(I - \Theta)'(I - \Phi)^{-1\prime}.$$

Computations for the maximum likelihood and REML estimates seem to be complicated.

Covariance selection and spatial autoregression models are both naturally specified using the inverse covariance matrix. In Chap. 4 the likelihood and REML equations were developed for a parameterization like $\Sigma(\theta)$, and used $\mathbf{d}_{\theta_j}\Sigma(\theta)$. However, the use of an inverse covariance parameterization was also discussed and equations were given.

Exercise 8.2. Show that the covariance function of an $AR(1)$ time series process is a special case of the exponential covariance function on $\mathbf{R}^1$.

8.8 Estimation of Covariance Functions and Semivariograms

The spatial linear model

$$Y = X\beta + e, \quad \mathrm{E}(e) = 0, \quad \mathrm{Cov}(e) = \Sigma(\theta), \tag{8.8.1}$$

lends itself immediately to the methods discussed in Chap. 4 for estimation of the covariance parameters. In this model any measurement error has been built into $\Sigma(\theta)$. Unlike the mixed models of Chap. 5, there is very little simplification to be obtained from the specific models considered for spatial data except that the linear covariance model has the linear covariance structure of Sect. 4.4.

Assuming normality and writing $\theta = (\theta_1, \ldots, \theta_s)'$, the likelihood equations are

$$\mathrm{tr}\left\{\Sigma^{-1}(\theta)[\mathbf{d}_{\theta_j}\Sigma(\theta)]\right\} = [Y - X\hat{\beta}(\theta)]'\Sigma^{-1}(\theta)[\mathbf{d}_{\theta_j}\Sigma(\theta)]\Sigma^{-1}(\theta)[Y - X\hat{\beta}(\theta)]$$

$$= Y'[I - A(\theta)]'\Sigma^{-1}(\theta)[\mathbf{d}_{\theta_j}\Sigma(\theta)]\Sigma^{-1}(\theta)[I - A(\theta)]Y$$

for $j = 1, \ldots, s$. The REML equations are

$$\mathrm{tr}\left\{\Sigma^{-1}(\theta)[I - A(\theta)][\mathbf{d}_{\theta_j}\Sigma(\theta)]\right\}$$

$$= Y'[I - A(\theta)]'\Sigma^{-1}(\theta)[\mathbf{d}_{\theta_j}\Sigma(\theta)]\Sigma^{-1}(\theta)[I - A(\theta)]Y,$$

$j = 1, \ldots, s$,

Based on Eq. (8.3.4), when $J \in C(X)$ and $C(B) = C(X)^{\perp}$, for any semivariogram model we have

$$B'\Sigma(\theta)B = B'\Gamma(\theta)B.$$

It is clear from their derivations that estimates of θ from REML or, when appropriate, MINQUE are the same when based on either a covariance function $\sigma(u, w; \theta)$ or a corresponding semivariogram $\gamma(u, w; \theta)$. To apply the REML or MINQUE equations, replace all occurrences of $\Sigma^{-1}(\theta)[I - A(\theta)]$ with $B[B'\Gamma(\theta)B]^{-1}B'$ where B is chosen as a full rank matrix. Maximum likelihood is ill-defined based on the semi-variogram alone.

Exercise 8.3.   Using Eq. (8.3.4), show that if $J \in C(X)$, then

$$\text{tr}\left\{\Sigma^{-1}(\theta)[I - A(\theta)][\mathbf{d}_{\theta_j}\Sigma(\theta)]\right\} = \text{tr}\left\{\Sigma^{-1}(\theta)[I - A(0)][\mathbf{d}_{\theta_j}\Gamma(\theta)]\right\}$$

and

$$[I - A(\theta)]'\Sigma^{-1}(\theta)[\mathbf{d}_{\theta_j}\Sigma(\theta)]\Sigma^{-1}(\theta)[I - A(\theta)]$$
$$= [I - A(\theta)]'\Sigma^{-1}(\theta)[\mathbf{d}_{\theta_j}\Gamma(\theta)]\Sigma^{-1}(\theta)[I - A(\theta)].$$

Unfortunately, these facts cannot be used directly to solve the REML or MINQUE equations because $\Gamma(\theta)$ does not determine either $\Sigma^{-1}(\theta)$ or $A(\theta)$, even though $\Gamma(\theta)$ and B determine $[I - A(\theta)]'\Sigma^{-1}(\theta) = \Sigma^{-1}(\theta)[I - A(\theta)]$.

The geostatistics literature also includes various ad hoc model fitting procedures; see Cressie (1985). These are usually based on empirical estimates $\tilde{\sigma}(u_i, u_j)$ of the elements of $\Sigma(\theta)$ combined with an ad hoc method of choosing $\hat{\theta}$ so that the values $\sigma(u_i, u_j; \hat{\theta})$ are in some sense close to the values $\tilde{\sigma}(u_i, u_j)$. These are addressed briefly in the last subsection.

Of course, the observed data are Y, so estimation must be based on Y and the model for Y. It is a subtle point, but worth mentioning, that we must estimate the parameter θ in $\Sigma(\theta)$ rather than the parameter θ in $\sigma(u, w; \theta)$. Of course, given an estimate $\tilde{\theta}$, not only does $\Sigma(\tilde{\theta})$ estimate the covariance matrix but also $\sigma(u, w; \tilde{\theta})$ estimates the covariance function. Nonetheless, our observations only give direct information about $\Sigma(\theta)$.

8.8.1 Nonlinear Covariance Functions

For nonlinear covariance functions, Chap. 4 applies but there are not many simplifications available. The matrix

$$\mathbf{d}_{\theta_i}\Sigma(\theta) = [\mathbf{d}_{\theta_i}\sigma(u_j, u_k; \theta)]$$

depends on the particular covariance model being used. For example, assuming the isotropic exponential model without measurement error ($\theta_0 = 0$) gives

$$\sigma(u, w; \theta) = \theta_1 e^{-\theta_2\|u-w\|}.$$

Differentiation yields

$$\mathbf{d}_{\theta_1}\sigma(u, w; \theta) = e^{-\theta_2\|u-w\|},$$

thus defining $\mathbf{d}_{\theta_1}\Sigma(\theta)$. Also,

$$\mathbf{d}_{\theta_2}\sigma(u,w;\theta) = -\theta_1\|u-w\|e^{-\theta_2\|u-w\|},$$

which defines $\mathbf{d}_{\theta_2}\Sigma(\theta)$.

The standard covariance models are discontinuous at $\|h\| = 0$ when measurement error occurs. This might give one doubts about whether the methods for obtaining MLEs and REML estimates can be executed. There is no problem. Derivatives are taken with respect to the θ_i's, and all of the standard models are continuous in θ. If measurement error is incorporated, rewrite the model as

$$Y = X\beta + e, \quad E(e) = 0, \quad \text{Cov}(e) = V(\theta),$$

$$V(\theta) = \theta_0 I_n + \Sigma(\theta_*)$$

where $\theta' = (\theta_0, \theta_*') = (\theta_0, \theta_1, \ldots, \theta_s)'$. Now $[\mathbf{d}_{\theta_j}V(\theta)] = [\mathbf{d}_{\theta_j}\Sigma(\theta)]$, $j = 1, \ldots, s$ but there is one more equation to consider, one that involves $[\mathbf{d}_{\theta_0}V(\theta)] = I_n$. The derivatives of $\Sigma(\theta_*)$ remain the same as without measurement error.

The maximum likelihood approach for spatial data was apparently first proposed by Kitanidis (1983) for linear covariance functions. Mardia and Marshal (1984) independently proposed using MLEs for general covariance models. Kitanidis and Lane (1985) also extended Kitanidis (1983) to general covariance functions. All of these articles discuss computational procedures. In their article on analyzing field-plot experiments, Zimmerman and Harville (1990) present a nice general discussion of maximum likelihood and residual maximum likelihood methods. They also point out that results in Zimmerman (1989) can be used to reduce the computational burden when many of the standard covariance models are used. Warnes and Ripley (1987) have pointed out that the likelihood function is often multimodal and that care must be taken to obtain the global rather than some local maximum of the likelihood function; see also Mardia and Watkins (1989). (Multimodality of the [restricted] likelihood is another good reason for doing a Bayesian analysis. A Bayesian analysis explores the likelihood rather than focusing on its maximum.) As always, high correlations between the parameter estimates can cause instability in the estimates.

8.8.2 Linear Covariance Functions

With

$$\Sigma(\theta) = \sum_{k=1}^{s} \theta_k \Sigma_k$$

the likelihood equations become

$$\sum_{k=1}^{s} \theta_k \text{tr}\left[\Sigma_k \Sigma^{-1}(\theta)\Sigma_j \Sigma^{-1}(\theta)\right] = Y'[I - A(\theta)]'\Sigma^{-1}(\theta)\Sigma_j\Sigma^{-1}(\theta)[I - A(\theta)]Y$$

$j = 1, \ldots, s$. The REML equations become

$$\sum_{k=1}^{s} \theta_k \text{tr}\left\{\Sigma_k [I - A(\theta)]' \Sigma^{-1}(\theta) \Sigma_j \Sigma^{-1}(\theta) [I - A(\theta)]\right\}$$
$$= Y'[I - A(\theta)]' \Sigma^{-1}(\theta) \Sigma_j \Sigma^{-1}(\theta) [I - A(\theta)] Y$$

$j = 1, \ldots, s$. The MINQUE equations are

$$\sum_{k=1}^{s} \theta_k \text{tr}\left\{\Sigma_k [I - A(w)]' \Sigma^{-1}(w) \Sigma_j \Sigma^{-1}(w) [I - A(w)]\right\}$$
$$= Y'[I - A(w)]' \Sigma^{-1}(w) \Sigma_j \Sigma^{-1}(w) [I - A(w)] Y$$

$j = 1, \ldots, s$.

Because the Σ_ks are all nonnegative definite, they can be written as $\Sigma_k = Z_k Z_k'$ for some matrix Z_k and as a result, the spatial model can be rewritten as a variance component model. The theory for linear covariance structures of Sect. 4.4 preempts the value of this observation. The equivalence of the linear covariance function model and variance component model was apparently first recognized by Kitanidis (1983, 1985). Marshall and Mardia (1985) also proposed MINQUE estimation. Stein (1987) gives asymptotic efficiency and consistency results for MINQUE estimates.

8.8.3 Traditional Geostatistical Estimation

The traditional approach to covariance function or semivariogram estimation (see Journel & Hüijbregts 1978 or David 1977) is to obtain an "empirical" estimate and to fit a model to the empirical estimate. We concentrate on fitting covariance functions and discuss fitting semivariograms at the end of the subsection. We begin by discussing empirical estimation. In order to have enough data to perform estimation, we assume second-order stationarity (i.e., $\sigma(u, w) = \sigma(u - w)$). The empirical estimate is nonparametric in the sense that estimates are not based on any covariance model with a small number of parameters. To begin the procedure, choose a nonnegative definite weighting matrix, say Σ_0, and fit the model

$$Y = X\beta + e, \quad \text{E}(e) = 0, \quad \text{Cov}(e) = \Sigma_0$$

to obtain residuals

$$\hat{e}_0 = (I - A_0)Y = Y - X\hat{\beta}_0,$$

where

$$A_0 = X(X' \Sigma_0^{-1} X)^- X' \Sigma_0^{-1}$$

and

$$X\hat{\beta}_0 = A_0 Y.$$

These residuals are the basis of empirical estimates of the covariance function. For any vector h, there is a finite number N_h of pairs of observations y_i and y_j for which $u_i - u_j = h$. For each of these pairs, list the corresponding residual pairs, say $(\hat{e}_{0i}, \hat{e}_{0i(h)})$, $i = 1, \ldots, N_h$. If $N_h \geq 1$, the traditional empirical estimator is

$$\hat{\sigma}(h) = \hat{\sigma}(-h) = \frac{1}{N_h} \sum_{i=1}^{N_h} \hat{e}_i \hat{e}_{i(h)} .$$

If N_h is zero, no empirical estimate is possible because no data have been collected with $u_i - u_j = h$. Often, any vector $u_i - u_j$ in a neighborhood of h is included in the computation. With a finite number of observations, there will be only a finite number of vectors, say $h(1), h(2), \ldots, h(q)$, that have $N_{h(k)} > 0$. In practice, if $N_{h(k)}$ is not substantially greater than 1, we may not wish to include $h(k)$ as a vector for which the covariance function will be estimated.

Given a parametric stationary covariance function, say $\sigma(h; \theta)$, a least squares estimate of θ can be obtained by minimizing

$$\sum_{i=1}^{q} \{\hat{\sigma}[h(i)] - \sigma[h(i); \theta]\}^2 .$$

Weighted least squares estimates can also be computed. If the covariances or asymptotic covariances of $\hat{\sigma}[h(i)]$ and $\hat{\sigma}[h(j)]$ can be computed, say $\mathrm{Cov}\{\hat{\sigma}[h(i)], \hat{\sigma}[h(j)]\} = v_{ij}$, write

$$V = [v_{ij}]$$

and choose θ to minimize

$$S'V^{-1}S,$$

where $S' = \{\hat{\sigma}[h(1)] - \sigma[h(1); \theta], \ldots, \hat{\sigma}[h(q)] - \sigma[h(q); \theta]\}$. In many cases, the covariances will be small relative to the variances, so a reasonable estimate can be obtained by minimizing

$$\sum_{i=1}^{q} \{\hat{\sigma}[h(i)] - \sigma[h(i); \theta]\}^2 / v_{ii} .$$

The Gauss-Newton procedure described in Sect. 7.4.1 can be used to find the estimate, say $\hat{\theta}_0$, using any of these criteria.

The best fit of the linear model and hence the best residuals is obtained by taking $\Sigma_0 = \Sigma$. Because Σ is unknown, it is reasonable to use $\hat{\theta}_0$ to estimate it. Let

$$\Sigma_1 = [\sigma(u_i, u_j; \hat{\theta}_0)],$$

and find residuals

$$\hat{e}_1 = (I - A_1)Y,$$

where
$$A_1 = X(X'\Sigma_1^{-1}X)^- X'\Sigma_1^{-1}.$$

These residuals lead to pairs $(\hat{e}_{1i}, \hat{e}_{1i(h)})$, $i = 1,\ldots,N_h$ and estimates $\hat{\sigma}_1(h) = (1/N_h)\sum_{i=1}^{N_h} \hat{e}_{1i}\hat{e}_{1i(h)}$. The estimates can then be used to obtain $\hat{\theta}_1$ and define

$$\hat{\Sigma}_2 = [\sigma(u_i,u_j;\hat{\theta}_1)].$$

This procedure can be iterated in the hope that the sequence $\hat{\theta}_t$ converges to some value $\hat{\theta}$. Armstrong (1984) presents criticisms of this method.

The idea of using weighted least squares as a criterion for fitting semivariograms was first presented by Cressie (1985). The preceeding presentation is a covariance function version of Cressie's ideas. Cressie's discussion was restricted to the ordinary kriging model. For this model, he computed the necessary variances and covariances. Cressie also suggested using robust empirical semivariogram estimates, in particular those proposed by Hawkins and Cressie (1984). Again, Cressie computed the variances and covariances necessary for weighted least squares. The traditional empirical semivariogram estimator in ordinary kriging is

$$\hat{\gamma}(h) = \frac{1}{2N_h}\sum_{i=1}^{N_h}(y_i - y_{i(h)})^2,$$

where the pairs $(y_i, y_{i(h)})$ are the N_h pairs whose locations differ by h. The robust estimates from Hawkins and Cressie (1984) are

$$2\tilde{\gamma}(h) = \left[\frac{1}{N_h}\sum_{i=1}^{N_h}|y_i - y_{i(h)}|^{1/2}\right]^4 \Big/ [0.457 + 0.494/N_h]$$

and

$$2\tilde{\gamma}(h) = [\text{median}\{|y_i - y_{i(h)}|^{1/2}\}]^4 \Big/ B_h,$$

where B_h is a bias correction factor.

Methods other than least squares and weighted least squares are often used to fit covariance functions $\sigma(h;\theta)$ and semivariograms $\gamma(h;\theta)$ to their empirical counterparts. Various methods have been devised for particular covariance and semivariogram models. Models have also frequently been fit by visual inspection.

If an isotropic covariance function or semivariogram is assumed, the empirical estimates change slightly. For covariance functions,

$$\hat{\sigma}(\|h\|) = \frac{1}{N_h}\sum_{i=1}^{N_h}\hat{e}_i\hat{e}_{i(\|h\|)},$$

where the pairs $(\hat{e}_i, \hat{e}_{i(\|h\|)})$, $i = 1,\ldots,N_h$ are all residual pairs with locations separated by the distance $\|h\|$. For the semivariogram in ordinary kriging,

$$2\hat{\gamma}(\|h\|) = \frac{1}{N_h} \sum_{i=1}^{N_h} [y_i - y_{i(\|h\|)}]^2.$$

The pairs $(y_i, y_{i(\|h\|)})$, $i = 1, \ldots, N_h$ consist of all observations with locations separated by $\|h\|$.

Zimmerman and Zimmerman (1991) present results from a Monte Carlo experiment comparing various techniques of estimating the variogram in ordinary kriging. Cressie (1989) gives a very complete illustration of traditional methods for semivariogram estimation in ordinary kriging.

References

Adler, R. J. (1981). *The geometry of random fields*. New York: Wiley.

Armstrong, M. (1984). Problems with universal kriging. *Journal of the International Association for Mathematical Geology, 16*, 101–108.

Banerjee, S., Carlin, B. P., Gelfand, A. E. (2015). *Hierarchical modeling and analysis for spatial data* (2nd ed.). Boca Raton, FL: Chapman & Hall/CRC.

Berke, O. (1998). On spatio-temporal prediction for on-line monitoring data. *Communications in Statistics, Series A, 27*, 2343–2369.

Breiman, L. (1968). *Probability*. Reading, MA: Addison-Wesley.

Christensen, R. (1990). The equivalence of predictions from universal kriging and intrinsic random function kriging. *Mathematical Geology, 22*, 655–664.

Christensen, R. (1993). Quadratic covariance estimation and equivalence of predictions. *Mathematical Geology, 25*, 541–558.

Christensen, R. (1997). *Log-linear models and logistic regression* (2nd ed.). New York: Springer.

Christensen, R. (2011). *Plane answers to complex questions: The theory of linear models* (4th ed.). New York: Springer.

Christensen, R., Johnson, W., Branscum, A., & Hanson, T. E. (2010). *Bayesian ideas and data analysis: An introduction for scientists and statisticians*. Boca Raton, FL: Chapman and Hall/CRC Press.

Cliff, A., & Ord, J. K. (1981). *Spatial processes: Models and applications*. London: Pion.

Cressie, N. (1985). Fitting variogram models by weighted least squares. *Journal of the International Association for Mathematical Geology, 17*, 563–586.

Cressie, N. (1986). Kriging nonstationary data. *Journal of the American Statistical Association, 81*, 625–634.

Cressie, N. (1988). Spatial prediction and ordinary kriging. *Mathematical Geology, 20*, 405–421.

Cressie, N. (1989). Geostatistics. *The American Statistician, 43*, 197–202.

Cressie, N. A. C. (1993). *Statistics for spatial data* (Revised ed.). New York: Wiley.

Cressie, N. A. C., & Wikle, C. K. (2011). *Statistics for spatio-temporal data*. New York: Wiley.

David, M. (1977). *Geostatistical ore reserve estimations*. New York: Elsevier.

Delfiner, P. (1976). Linear estimation of nonstationary spatial phenomena. In M. Guarascia, M. David, & C. Hüijbregts (Eds.) *Advanced geostatistics in the mining industry*. Dordrecht: Reidel.

Diamond, P., & Armstrong, M. (1983). Robustness of variograms and conditioning of kriging matrices. *Journal of the International Association for Mathematical Geology, 16*, 809–822.

Diggle, P. J., Tawn, J. A., & Moyeed, R. A. (1998). Model-based geostatistics. *Applied Statistics, 47*, 299–326.

Edwards, D. (2000). *Introduction to graphical modeling* (2nd ed.). Berlin: Springer.

Gelfand, A. E., Diggle, P., Guttorp, P., Fuentes, M. (2010). *Handbook of spatial statistics*. Boca Raton, FL: Chapman & Hall/CRC.

Goldberger, A. S. (1962). Best linear unbiased prediction in the generalized linear regression model. *Journal of the American Statistical Association, 57*, 369–375.

Handcock, M. S., & Stein, M. L. (1993). A Bayesian analysis of kriging. *Technometrics, 35*, 403–410.

Handcock, M. S., & Wallis, J. R. (1994). An approach to statistical spatial-temporal modeling of meterological fields (with discussion). *Journal of the American Statistical Association, 89*, 368–378.

Hawkins, D. M., & Cressie, N. (1984). Robust kriging – a proposal. *Journal of the International Association for Mathematical Geology, 16*, 3–18.

Hodges, J. S. (2014). *Richly parameterized linear models: Additive, time series, and spatial models using random effects*. Boca Raton, FL: Chapman and Hall/CRC.

Huang, H.-C., & Cressie, N. (1996). Spatio-temporal prediciton of snow water equivalent using the Kalman filter. *Computational Statistics and Data Analysis, 22*, 159–175.

Hüijbregts, C. J. (1975). Regionalized variables and quantitative analysis of spatial data. In J. C. Davis & M. J. McCullagh (Eds.), *Display and analysis of spatial data*. New York: Wiley.

Isaaks, E. H., & Srivastava, R. M. (1989). *An introduction to applied geostatistics*. Oxford: Oxford University Press.

Journel, A. G., & Hüijbregts, Ch. J. (1978). *Mining geostatistics*. New York: Academic Press.

Kitanidis, P. K. (1983). Statistical estimation of polynomial generalized covariance functions and hydrologic applications. *Water Resources Research, 19*, 909–921.

Kitanidis, P. K. (1985). Minimum-variance unbiased quadratic estimation of covariances of regionalized variables. *Journal of the International Association for Mathematical Geology, 17*, 195–208.

Kitanidis, P. K. (1986). Parameter uncertainty in estimation of spatial functions: Bayesian analysis. *Water Resources Research, 22*, 499–507.

Kitanidis, P. K. (1997). *Introduction to geostatistics: Applications to hydrogeology*. Cambridge: Cambridge University Press.

Kitanidis, P. K., & Lane, R. W. (1985). Maximum likelihood parameter estimation of hydrologic spatial processes by the Gauss–Newton method. *Journal of Hydrology, 79*, 53–71.

Mardia, K. V., Goodall, C., Redfern, E. J., & Alonso, F. J. (1998). The kriged Kalman filter (with discussion). *Test, 7,* 217–252.

Mardia, K. V., & Watkins, A. J. (1989). On multimodality of the likelihood in the spatial linear model. *Biometrika, 76,* 289–295.

Mardia, K. V., & Marshal, R. J. (1984). Maximum likelihood estimation of models for residual covariance in spatial regression. *Biometrika, 71,* 135–146.

Marshall, R. J., & Mardia, K. V. (1985). Minimum norm quadratic estimation of components of spatial covariance. *Journal of the International Association for Mathematical Geology, 17,* 517–525.

Matern, B. (1986). *Spatial variation* (2nd ed.). New York: Springer.

Matheron, G. (1965). Les variable regionalisees et leur estimation: Masson, Paris, xxxp.

Matheron, G. (1969). Le krigeage universal: Fascicule 1, Cahiers du CMMM., 82p.

Matheron, G. (1973). The intrinsic random functions and their applications. *Advances in Applied Probability, 5,* 439–468.

Omré, H., & Halvorsen, K. B. (1989). The Bayesian bridge between simple and universal kriging. *Mathematical Geology, 21,* 767–786.

Rasmussen, C. E., & Williams, C. K. I. (2006). *Gaussian processes for machine learning.* Cambridge: MIT Press.

Ripley, B. D. (1981). *Spatial statistics.* New York: Wiley.

Schabenberger, O., & Gotway, C. A. (2005). *Statistical methods for spatial data analysis.* Boca Raton, FL: Chapman & Hall/CRC.

Sherman, M. (2011). *Spatial statistics and spatio-temporal data: Covariance functions and directional properties.* New York: Wiley.

Stein, M. L. (1987). Minimum norm quadratic estimation of spatial variograms. *Journal of the American Statistical Association, 82,* 765–772.

Stein, M. L. (1988). Asymptotically efficient prediction of a random field with misspecified covariance function. *The Annals of Statistics, 16,* 55–64.

Stein, M. L. (1999). *Interpolation of spatial data: Some theory for kriging.* New York: Springer.

Waller, L. A., & Gotway, C. A. (2004). *Applied spatial statistics for public health data.* New York: Wiley.

Warnes, J. J., & Ripley, B. D. (1987). Problems with likelihood estimation of covariance functions of spatial Gaussian processes. *Biometrika, 74,* 640–642.

Whittaker, J. (1990). *Graphical models in applied multivariate statistics.* New York: Wiley.

Whittle, P. (1954). On stationary processes in the plane. *Biometrika, 41,* 434–449.

Whittle, P. (1963). Stochastic processes in several dimensions. *Bulletin of the International Statistical Institute, 40*(1), 974–994.

Wikle, C. K., Zammit-Mangion, A., Cressie, N. (2019). *Spatio-temporal statistics with R.* Boca Raton: Chapman and Hall/CRC.

Zimmerman, D. L. (1989). Computationally exploitable structure of covariance matrices and generalized covariance matrices in spatial models. *Journal of Statistical Computation and Simulation, 32,* 1–15.

Zimmerman, D. L., & Cressie, N. (1992). On the stability of the geostatistical method. *Mathematical Geology, 24*, 45–59.

Zimmerman, D. L., & Harville, D. A. (1990). A random field approach to the analysis of field-plot experiments and other spatial experiments. *Biometrics, 47*, 223–239.

Zimmerman, D. L., & Zimmerman, M. B. (1991). A Monte Carlo comparison of spatial variogram estimators and kriging predictors. *Technometrics, 33*, 77–91.

Chapter 9
Multivariate Linear Models: General

Abstract This chapter introduces the basic theory for linear models with more than one dependent variable.

Chapters 9 through 14 examine topics in multivariate analysis. Specifically, they discuss the theory of multivariate linear models, applications of multivariate linear models, generalized multivariate linear models and associated longitudinal models, discriminant analysis (with binary regression), and principal components (with factor analysis and classical multidimensional scaling). The basic ideas behind these subjects are closely related to linear model theory. A multivariate linear model is a collection of linear models with more than one dependent variable that share the same model matrix X. A *standard multivariate linear model* has identical covariances among dependent variables for every individual but with data from different individuals uncorrelated. Generalized multivariate linear models extend the standard multivariate linear model by positing a mean structure that relates the dependent variables. Discriminant analysis is closely related to both Mahalanobis distance (*PA-V* Sect. 12.1 or Christensen 2011, Sect. 13.1) and multivariate one-way analysis of variance. Principal components, cf. *PA-V* Sect. 14.2.1 (Christensen 2011, Subsection 15.2.1), are user-constructed variables that are best linear predictors of the original data. Factor analysis has ties to both multivariate linear models and principal components.

These six chapters are introductory in nature. The discussions benefit from the advantage of being based on linear model theory. They suffer from the disadvantage of being relatively brief. More detailed discussions are available in numerous other sources, such as Anderson (2003), Arnold (1981), Dillon and Goldstein (1984), Eaton (1983), Gnanadesikan (1977), Johnson and Wichern (2007), Mardia, Kent and Bibby (1979), Morrison (2004), Muirhead (1982), Press (1982), and Seber (1984). More recent contributions to this literature include Everitt and Hothorn (2011), Rencher and Christensen (2012), and Marden (2015).

As mentioned earlier, the distinction between standard multivariate linear models and standard (univariate) linear models is simply that multivariate linear models involve more than one dependent variable. For multivariate data, let the dependent

© Springer Nature Switzerland AG 2019
R. Christensen, *Advanced Linear Modeling*, Springer Texts in Statistics,
https://doi.org/10.1007/978-3-030-29164-8_9

variables be $y_1, \ldots, y_q$. The idea is that all q random variables will be observed on each of n individuals. The standard assumption is that the random variables have some unknown covariance matrix Σ that is the same for all individuals but different individuals are uncorrelated. If n observations are taken on each dependent variable, we have $y_{i1}, \ldots, y_{iq}$, $i = 1, \ldots, n$. Let $Y_1 = [y_{11}, \ldots, y_{n1}]'$ and, in general, $Y_h = [y_{1h}, \ldots, y_{nh}]'$, $h = 1, \ldots, q$. For each h, the vector Y_h is the vector of n responses on the variable y_h and can be used as the response vector for a linear model. For $h = 1, \ldots, q$, write the linear model

$$Y_h = X\beta_h + e_h, \quad \mathrm{E}(e_h) = 0, \quad \mathrm{Cov}(e_h) = \sigma_{hh}I, \tag{9.0.1}$$

where X is a known $n \times p$ matrix that is the same for all dependent variables (it depends on the individuals but not on the variable being measured), but β_h and the error vector $e_h = [e_{1h}, \ldots, e_{nh}]'$ are peculiar to the dependent variable. Here we are using σ_{hh} (rather than σ_h^2) to denote the variance associated with y_h. The assumed covariance structure determines the covariances between the Y_h vectors, see Exercise 9.1.

The multivariate linear model consists of fitting the q linear models simultaneously. Write the matrices

$$Y_{n \times q} = [Y_1, \ldots, Y_q], \quad B_{p \times q} = [\beta_1, \ldots, \beta_q], \quad e_{n \times q} = [e_1, \ldots, e_q].$$

The multivariate linear model is

$$Y = XB + e. \tag{9.0.2}$$

The key to the analysis of the standard multivariate linear model is the random nature of the $n \times q$ error matrix $e = [e_{ih}]$. At a minimum, we assume that $\mathrm{E}(e) = 0$ and that different individuals i are uncorrelated,

$$\mathrm{Cov}(e_{ih}, e_{i'h'}) = \begin{cases} \sigma_{hh'} & \text{if } i = i' \\ 0 & \text{if } i \neq i' \end{cases}.$$

Let

$$\delta_{ii'} = \begin{cases} 1 & \text{if } i = i' \\ 0 & \text{if } i \neq i', \end{cases}$$

then the covariances can be written simply as

$$\mathrm{Cov}(e_{ih}, e_{i'h'}) = \sigma_{hh'}\delta_{ii'}.$$

To construct tests and confidence regions, we assume that the e_{ij}s have a multivariate normal distribution with the previously indicated mean and covariances. Note that this covariance structure implies that the error vector in model 9.0.1 has $\mathrm{Cov}(e_h) = \sigma_{hh}I$, as indicated previously.

Exercise 9.1. For any two columns of Y, say Y_r and Y_s, show that $\mathrm{Cov}(Y_r, Y_s) = \sigma_{rs}I$.

An alternative but equivalent way to state the standard multivariate linear model is by examining the rows of model (9.0.2). Write

$$Y = \begin{bmatrix} y_1' \\ \vdots \\ y_n' \end{bmatrix}, \quad X = \begin{bmatrix} x_1' \\ \vdots \\ x_n' \end{bmatrix}, \quad \text{and} \quad e = \begin{bmatrix} \varepsilon_1' \\ \vdots \\ \varepsilon_n' \end{bmatrix}.$$

The standard multivariate linear model is also

$$y_i' = x_i' B + \varepsilon_i', \tag{9.0.3}$$

$i = 1, \ldots, n$. The error vector ε_i has the properties

$$E(\varepsilon_i) = 0, \quad \text{Cov}(\varepsilon_i) = \Sigma_{q \times q} = [\sigma_{hh'}],$$

and, for $i \neq j$,

$$\text{Cov}(\varepsilon_i, \varepsilon_j) = 0.$$

To construct tests and confidence regions, the vectors ε_i are assumed to have independent multivariate normal distributions.

To reiterate, the multivariate model (9.0.2) holds if and only if the q univariate models (9.0.1) hold simultaneously and the multivariate model holds if and only if the n models (9.0.3) hold simultaneously. All of these models have errors with mean zero and all models determine the same covariance structure. The unknown covariance parameters are the unique parameters in Σ. We assume throughout that Σ is positive definite.

9.1 The Univariate Model

The key to estimation in the multivariate linear model is rewriting the model as a univariate linear model. The multivariate model is

$$Y = XB + e, \quad E(e) = 0, \quad \text{Cov}(e_{ih}, e_{i'h'}) = \sigma_{hh'} \delta_{ii'}. \tag{9.1.1}$$

Observing that

$$XB = [X\beta_1, \ldots, X\beta_1],$$

the multivariate model holds if and only if each of the q univariate linear models hold. To have all q models hold simultaneously is to have

$$\begin{bmatrix} Y_1 \\ Y_2 \\ \vdots \\ Y_q \end{bmatrix} = \begin{bmatrix} X\beta_1 \\ X\beta_2 \\ \vdots \\ X\beta_q \end{bmatrix} + \begin{bmatrix} e_1 \\ e_2 \\ \vdots \\ e_q \end{bmatrix}$$

which in turn can be rewritten as a single (univariate) linear model

$$
\begin{bmatrix} Y_1 \\ Y_2 \\ \vdots \\ Y_q \end{bmatrix} = \begin{bmatrix} X & 0 & \cdots & 0 \\ 0 & X & & \vdots \\ \vdots & & \ddots & 0 \\ 0 & 0 & & X \end{bmatrix} \begin{bmatrix} \beta_1 \\ \beta_2 \\ \vdots \\ \beta_q \end{bmatrix} + \begin{bmatrix} e_1 \\ e_2 \\ \vdots \\ e_q \end{bmatrix}, \tag{9.1.2}
$$

where the error vector has mean zero and the covariance matrix is

$$
\begin{bmatrix} \sigma_{11} I_n & \sigma_{12} I_n & \cdots & \sigma_{1q} I_n \\ \sigma_{12} I_n & \sigma_{22} I_n & \cdots & \sigma_{2q} I_n \\ \vdots & \vdots & \ddots & \vdots \\ \sigma_{1q} I_n & \sigma_{2q} I_n & \cdots & \sigma_{qq} I_n \end{bmatrix}. \tag{9.1.3}
$$

The Vec operator stacks the columns of a matrix, it is clear that we have rewritten model (9.1.1) as

$$
\mathrm{Vec}(Y) = \mathrm{Vec}(XB) + \mathrm{Vec}(e)
$$

and that model (9.1.2) goes a step further in writing $\mathrm{Vec}(XB)$ as the product of the model matrix in (9.1.2) times the parameter vector $\mathrm{Vec}(B)$. *Vec operators, Kronecker products and their algebra are reviewed in Appendix A.2* The model matrix in (9.1.2) is the Kronecker product $[I_q \otimes X]$. Model (9.1.2) can now be rewritten as

$$
\mathrm{Vec}(Y) = [I_q \otimes X]\mathrm{Vec}(B) + \mathrm{Vec}(e). \tag{9.1.4}
$$

The first two moments of $\mathrm{Vec}(e)$ are

$$
E[\mathrm{Vec}(e)] = 0
$$

and, rewriting (9.1.3) as a Kronecker product,

$$
\mathrm{Cov}[\mathrm{Vec}(e)] = [\Sigma \otimes I_n]. \tag{9.1.5}
$$

9.2 BLUEs

For estimation, the nice thing about the standard multivariate linear model (9.1.1) is that least squares estimates are optimal. In particular, it will be shown that optimal estimation is based on

$$
\hat{Y} = X\hat{B} = MY,
$$

where $M = X(X'X)^- X'$ is, as always, the perpendicular projection operator onto the column space of X, $C(X)$. This is a simple generalization of the univariate linear model results of *PA* Chap. 2. To show that least squares estimates are best linear unbiased estimates (BLUEs) in model (9.1.4) with the covariance in (9.1.5), apply *PA* Proposition 2.7.5 or *PA* Theorem 10.4.5, both of which state that for a univariate linear model $Y_{n\times 1} = X\beta + e$, $E(e) = 0$, $Cov(e) = \sigma^2 V$, least squares estimates are BLUEs if $C(VX) \subset C(X)$.

The model matrix in (9.1.4) is $[I_q \otimes X]$. The covariance matrix is $[\Sigma \otimes I_n]$. We need to show that $C([\Sigma \otimes I_n][I_q \otimes X]) \subset C([I_q \otimes X])$. In particular,

$$[\Sigma \otimes I_n][I_q \otimes X] = [\Sigma \otimes X]$$

$$= \begin{bmatrix} \sigma_{11}X & \cdots & \sigma_{1q}X \\ \vdots & & \vdots \\ \sigma_{1q}X & \cdots & \sigma_{qq}X \end{bmatrix}$$

$$= [I_q \otimes X][\Sigma \otimes I_p].$$

Recalling that $C(RS) \subset C(R)$ for any conformable matrices R and S, it is clear that

$$C([\Sigma \otimes I_n][I_q \otimes X]) = C([I_q \otimes X][\Sigma \otimes I_p]) \subset C([I_q \otimes X]).$$

Applying PA Proposition 2.7.5 or PA Theorem 10.4.5 establishes that least squares estimates are best linear unbiased estimates for model (9.1.4) with the covariance in (9.1.5).

To find least squares estimates in model (9.1.4), we need the perpendicular projection operator onto $C([I_q \otimes X])$. The ppo is

$$\mathcal{M} = [I_q \otimes X]\left([I_q \otimes X]'[I_q \otimes X]\right)^{-}[I_q \otimes X]'.$$

Because $[A \otimes B]' = [A' \otimes B']$, we have

$$[I_q \otimes X]'[I_q \otimes X] = [I_q \otimes X'][I_q \otimes X]$$
$$= [I_q \otimes X'X].$$

It is easily seen from the definition of a generalized inverse that

$$\left([I_q \otimes X'X]\right)^{-} = [I_q \otimes (X'X)^{-}].$$

It follows that

$$\mathcal{M} = [I_q \otimes X][I_q \otimes (X'X)^{-}][I_q \otimes X]'$$
$$= [I_q \otimes X(X'X)^{-}X']$$
$$= [I_q \otimes M].$$

By PA Theorem 2.2.1, in a univariate linear model $Y_{n\times 1} = X\beta + e$, least squares estimates $\hat{\beta}$ satisfy $X\hat{\beta} = MY_{n\times 1}$; thus, for the univariate linear model (9.1.4), least squares estimates of $\text{Vec}(B)$, say $\text{Vec}(\hat{B})$, satisfy

$$[I_q \otimes X]\text{Vec}(\hat{B}) = [I_q \otimes M]\text{Vec}(Y);$$

that is,

$$
\begin{bmatrix} X\hat{\beta}_1 \\ \vdots \\ X\hat{\beta}_q \end{bmatrix} = \begin{bmatrix} MY_1 \\ \vdots \\ MY_q \end{bmatrix}.
$$

In terms of the multivariate linear model (9.1.1), this is equivalent to

$$
X\hat{B} = MY.
$$

9.3 Unbiased Estimation of Σ

The usual unbiased estimate of Σ depends only on the mean and covariance assumptions that we have made. The result looks like the result for a standard univariate linear model, indeed the standard univariate linear model is just the special case where $q = 1$. An unbiased estimate of Σ is

$$
S \equiv \frac{1}{n - r(X)} Y'(I - M)Y.
$$

The proof that S is unbiased is similar in spirit to the proof of *PA* Theorem 1.3.2, a result that gives the expected value of a quadratic form, cf. *PA* Exercise 1.10a. To establish the unbiasedness, consider the i, j element of $Y'(I - M)Y$.

$$
\begin{aligned}
\mathrm{E}[Y_i'(I - M)Y_j] &= \mathrm{E}[(Y_i - X\beta_i)'(I - M)(Y_j - X\beta_j)] \\
&= \mathrm{E}\{\mathrm{tr}[(Y_i - X\beta_i)'(I - M)(Y_j - X\beta_j)]\} \\
&= \mathrm{E}\{\mathrm{tr}[(I - M)(Y_j - X\beta_j)(Y_i - X\beta_i)']\} \\
&= \mathrm{tr}\{\mathrm{E}[(I - M)(Y_j - X\beta_j)(Y_i - X\beta_i)']\} \\
&= \mathrm{tr}\{(I - M)\mathrm{E}[(Y_j - X\beta_j)(Y_i - X\beta_i)']\} \\
&= \mathrm{tr}\{(I - M)\mathrm{Cov}(Y_j, Y_i)\} \\
&= \mathrm{tr}\{(I - M)\sigma_{ji}I\} \\
&= \sigma_{ij}[n - r(X)].
\end{aligned}
$$

It follows that each element of S is an unbiased estimate of the corresponding element of Σ.

EXAMPLE 9.3.1.   *Partial Correlation Coefficients.*
Partial correlations were discussed in *PA* Sect. 6.5. Suppose we have n observations on two dependent variables y_1, y_2 and $p - 1$ independent variables $x_1, \ldots, x_{p-1}$. Write

$$
Y = \begin{bmatrix} y_{11} & y_{12} \\ \vdots & \vdots \\ y_{n1} & y_{n2} \end{bmatrix} = [Y_1, Y_2]
$$

and

$$Z = \begin{bmatrix} x_{11} & \cdots & x_{1\,p-1} \\ \vdots & & \vdots \\ x_{n1} & \cdots & x_{n\,p-1} \end{bmatrix}.$$

Write a multivariate linear model as

$$Y = [J, Z]B + e,$$

where J is an $n \times 1$ vector of 1's. As discussed earlier, the unbiased estimate of Σ is $S = [s_{ij}]$, where

$$S = Y'(I - M)Y / [n - r(X)]$$
$$= \frac{1}{n - r(X)} \begin{bmatrix} Y_1'(I-M)Y_1 & Y_1'(I-M)Y_2 \\ Y_2'(I-M)Y_1 & Y_2'(I-M)Y_2 \end{bmatrix}.$$

From *PA* Sect. 6.5, the sample partial correlation coefficient is

$$r_{y \cdot x} = \frac{Y_1'(I-M)Y_2}{[Y_1'(I-M)Y_1 \ Y_2'(I-M)Y_2]^{1/2}}$$
$$= \frac{s_{12}}{\sqrt{s_{11}s_{22}}}.$$

The sample partial correlation coefficient is just the sample correlation coefficient as estimated in a multivariate linear model in which the effects of the x variables have been eliminated. □

We assumed that Σ is positive definite and it is clear that S and $E \equiv Y'(I-M)Y$ are nonnegative definite. We will often need to know that S and E are also nonsingular, hence positive definite. S and E are random matrices, so it is by no means clear that they are nonsingular. Fortunately, the following theorem often establishes that S and E are nonsingular with probability one.

Theorem 9.3.2. Let Y be an $n \times q$ random matrix and let A be a fixed symmetric $n \times n$ matrix. If the joint distribution of the nq elements of Y admits a density function with respect to Lesbesgue measure on $\mathbf{R}^{nq}$, then

$$\Pr\{r(Y'AY) = \min[q, r(A)]\} = 1.$$

PROOF. See Okamoto (1973). □

If the density condition of Theorem 9.3.2 holds, with probability one,

$$r(E) = r[Y'(I-M)Y]$$
$$= \min[q, r(I-M)]$$
$$= \min[q, n - r(X)].$$

We will henceforth assume that the density conditions holds and that the number of observations n is large enough so that

$$q \leq n - r(X).$$

In this case, with probability one, E is a nonnegative definite $q \times q$ matrix of rank q, hence E is almost surely positive definite. In the discussion to follow, we will simply treat E, and S, as positive definite.

In particular, for maximum likelihood estimation and for testing in multivariate linear models we assume that the rows of Y are independent with

$$y_i \sim N(B'x_i, \Sigma).$$

This satisfies the density condition of Theorem 9.3.2.

9.4 Maximum Likelihood Estimates

Write the matrices Y and X using their component rows,

$$Y = \begin{bmatrix} y_1' \\ \vdots \\ y_n' \end{bmatrix} \quad \text{and} \quad X = \begin{bmatrix} x_1' \\ \vdots \\ x_n' \end{bmatrix}.$$

We assume that the rows of Y are independent and $y_i \sim N(B'x_i, \Sigma)$ with Σ positive definite. The likelihood function for Y is

$$L(B, \Sigma) = \prod_{i=1}^{n} (2\pi)^{-q/2} |\Sigma|^{-1/2} \exp\left[-(y_i - B'x_i)'\Sigma^{-1}(y_i - B'x_i)/2\right],$$

where $|\Sigma|$ denotes the determinant of Σ. The log of the likelihood function is

$$\ell(B, \Sigma) = -\frac{nq}{2} \log(2\pi) - \frac{n}{2} \log(|\Sigma|) - \frac{1}{2} \sum_{i=1}^{n} (y_i - B'x_i)'\Sigma^{-1}(y_i - B'x_i).$$

Maximum likelihood estimates (MLEs) maximize these functions.

Consider the multivariate model as the univariate model (9.1.2). As with any other univariate linear model, if the nonsingular covariance matrix is fixed, then the MLE of $[I_q \otimes X]\text{Vec}(B)$ is the same as the BLUE. As seen in Sect. 9.2, least squares estimates are BLUEs. The least squares estimate of XB does not depend on the covariance matrix; hence, for any value of Σ, $X\hat{B} = MY$ maximizes the likelihood function. It remains only to find the MLE of Σ.

The log-likelihood, and thus the likelihood, are maximized for any Σ by substituting a least squares estimate for B. Write $\hat{B} = (X'X)^{-}X'Y$. We need to maximize

$$\ell(\hat{B}, \Sigma) = -\frac{nq}{2} \log(2\pi) - \frac{n}{2} \log(|\Sigma|)$$
$$-\frac{1}{2} \sum_{i=1}^{n} [y_i - Y'X(X'X)^- x_i]' \Sigma^{-1} [y_i - Y'X(X'X)^- x_i]$$

subject to the constraint that Σ is positive definite. The last term on the right-hand side can be simplified. Define the $n \times 1$ vector

$$\rho_i = (0, \ldots, 0, 1, 0, \ldots, 0)'$$

with the 1 in the ith place.

$$\sum_{i=1}^{n} [y_i - Y'X(X'X)^- x_i]' \Sigma^{-1} [y_i - Y'X(X'X)^- x_i]$$
$$= \sum_{i=1}^{n} [y_i' - x_i'(X'X)^- X'Y] \Sigma^{-1} [y_i - Y'X(X'X)^- x_i]$$
$$= \sum_{i=1}^{n} \rho_i' [Y - X(X'X)^- X'Y] \Sigma^{-1} [Y' - Y'X(X'X)^- X'] \rho_i$$
$$= \sum_{i=1}^{n} \rho_i' (I - M)Y \Sigma^{-1} Y'(I - M) \rho_i$$
$$= \text{tr}[(I - M)Y \Sigma^{-1} Y'(I - M)]$$
$$= \text{tr}[\Sigma^{-1} Y'(I - M)Y].$$

Thus, our problem is to maximize

$$\ell(\hat{B}, \Sigma) = -\frac{nq}{2} \log(2\pi) - \frac{n}{2} \log(|\Sigma|) - \frac{1}{2} \text{tr}[\Sigma^{-1} Y'(I - M)Y]. \qquad (9.4.1)$$

We will find the maximizing value by setting all the partial derivatives (with respect to the σ_{ij}'s) equal to zero. To find the partial derivatives, we need the derivative results in Appendix A.1. In particular, we need

$$\mathbf{d}_{\sigma_{ij}} \log |\Sigma| = \text{tr}[\Sigma^{-1} \mathbf{d}_{\sigma_{ij}} \Sigma] \qquad (9.4.2)$$
$$= \text{tr}[\Sigma^{-1} T_{ij}],$$

where the symmetric $q \times q$ matrix T_{ij} has ones in row i column j and row j column i and zeros elsewhere. We need

$$\mathbf{d}_{\sigma_{ij}} \Sigma^{-1} = -\Sigma^{-1} [\mathbf{d}_{\sigma_{ij}} \Sigma] \Sigma^{-1}$$
$$= -\Sigma^{-1} T_{ij} \Sigma^{-1}.$$

Finally, we need a result involving the derivative of a trace. Using the chain rule,

$$
\begin{aligned}
\mathbf{d}_{\sigma_{ij}}\mathrm{tr}[\Sigma^{-1}Y'(I-M)Y] &= \mathrm{tr}[\mathbf{d}_{\sigma_{ij}}\{\Sigma^{-1}Y'(I-M)Y\}] \\
&= \mathrm{tr}[\{\mathbf{d}_{\sigma_{ij}}\Sigma^{-1}\}Y'(I-M)Y] \qquad (9.4.3) \\
&= \mathrm{tr}[-\Sigma^{-1}T_{ij}\Sigma^{-1}Y'(I-M)Y].
\end{aligned}
$$

Applying (9.4.2) and (9.4.3) to (9.4.1), we get

$$
\mathbf{d}_{\sigma_{ij}}\ell(\hat{B},\Sigma) = -\frac{n}{2}\mathrm{tr}[\Sigma^{-1}T_{ij}] + \frac{1}{2}\mathrm{tr}[\Sigma^{-1}T_{ij}\Sigma^{-1}Y'(I-M)Y].
$$

Setting the partial derivatives equal to zero leads to finding a positive definite matrix Σ that solves

$$
\mathrm{tr}[\Sigma^{-1}T_{ij}] = \mathrm{tr}[\Sigma^{-1}T_{ij}\Sigma^{-1}Y'(I-M)Y/n] \qquad (9.4.4)
$$

for all i and j.

Let $\hat{\Sigma} = \frac{1}{n}Y'(I-M)Y$. $\hat{\Sigma}$ is positive definite whenever S and E are. $\hat{\Sigma}$ provides our solution to the likelihood equations. Substituting $\hat{\Sigma}$ for Σ in (4) gives

$$
\begin{aligned}
\mathrm{tr}[\hat{\Sigma}^{-1}T_{ij}] &= \mathrm{tr}[\hat{\Sigma}^{-1}T_{ij}\hat{\Sigma}^{-1}Y'(I-M)Y/n] \\
&= \mathrm{tr}[\hat{\Sigma}^{-1}T_{ij}].
\end{aligned}
$$

Obviously, this holds for all i and j. (Readers should at least ponder Exercise 9.10.10.)

Of course we have not actually shown that $\hat{\Sigma}$ maximizes the likelihood, only that it is a critical point. As usual, if $q=1$, this reduces to the standard univariate result.

9.5 Hypotheses and Statistics

Consider testing the multivariate model

$$
Y = XB + e \qquad (9.5.1)
$$

against a reduced model

$$
Y = X_0\Gamma + e, \qquad (9.5.2)
$$

where $C(X_0) \subset C(X)$ and the elements of e are multivariate normal. The covariance matrix $[\Sigma \otimes I_n]$ from the univariate model (4.1.2) is unknown, so standard univariate methods of testing do not apply. Let $M_0 = X_0(X_0'X_0)^-X_0'$ be the perpendicular projection operator onto $C(X_0)$. Multivariate tests of model (9.5.2) versus model (9.5.1) are based on the *hypothesis statistic*

$$
H \equiv Y'(M - M_0)Y
$$

and the *error statistic*

$$
E \equiv Y'(I - M)Y.
$$

These statistics look identical to the sums of squares used in standard univariate linear models. The difference is that the univariate sums of squares are scalars, while in multivariate models these statistics are matrices. The matrices have diagonals that consist of sums of squares for the various dependent variables and off-diagonals that are sums of cross products of the different dependent variables.

For univariate models, the test statistic is proportional to the scalar $Y'(M - M_0)Y[Y'(I - M)Y]^{-1}$. For multivariate models, the test statistic is often taken as a function of the matrix $Y'(M - M_0)Y[Y'(I - M)Y]^{-1}$ or some closely related matrix. For multivariate models, there is no one test statistic that is the universal standard for performing tests. Various test statistics are discussed in the next section.

The existence of multiple test statistics is not that unusual. Even in a univariate balanced one-way ANOVA, the range test (that underlies Tukey's HSD multiple comparison procedure) is a reasonable alternative to the F test for no group differences. Ott's Analysis of Means is similarly based on an alternative to the F test. See Christensen (1996, Chapter 6.)

The procedure for testing an hypothesis about an estimable parametric function, say

$$H_0: \Lambda' B = 0 \tag{9.5.3}$$

where $\Lambda' = P'X$, follows from model testing exactly as in *PA* Sect. 3.3. The test is based on the hypothesis statistic

$$H \equiv Y' M_{MP} Y$$
$$= (\Lambda'\hat{B})'[\Lambda'(X'X)^-\Lambda]^-(\Lambda'\hat{B})$$

and the error statistic

$$E \equiv Y'(I - M)Y .$$

The projection operator in H is $M_{MP} = MP(P'MP)^-P'M$. Its use follows from the fact that $\Lambda'B = 0$ puts a constraint on the model that requires $E(Y_h) \in C[(I - M_{MP})X] = C(M - M_{MP})$ for each h. Thus, the reduced model can be written

$$Y = (M - M_{MP})\Gamma + e,$$

and the hypothesis statistic is

$$Y'[M - (M - M_{MP})]Y = Y'M_{MP}Y .$$

Just as in *PA* Sects. 3.1 and 3.2, both the reduced model hypothesis and the parametric function hypothesis can be generalized. The reduced model can be generalized to

$$Y = X_0\Gamma + XQ + e, \tag{9.5.4}$$

where XQ is a known $n \times q$ matrix. As in *PA* Sect. 3.2, model (9.5.1) is rewritten as

$$(Y - XQ) = XB_* + e \tag{9.5.5}$$

for some appropriate reparameterization B_*. Model (9.5.4) is rewritten as

$$(Y - XQ) = X_0 \Gamma + e. \tag{9.5.6}$$

The test of (9.5.4) versus (9.5.1) is performed by testing (9.5.6) against (9.5.5). The hypothesis statistic for the test is

$$H \equiv (Y - XQ)'(M - M_0)(Y - XQ).$$

The error statistic is

$$E \equiv (Y - XQ)'(I - M)(Y - XQ)$$
$$= Y'(I - M)Y.$$

Similarly, for a known matrix $\tilde{W}$, a test can be performed of

$$H_0: \Lambda'B = \tilde{W}$$

where $\Lambda' = P'X$ and the equation $\Lambda'B = \tilde{W}$ has at least one solution. Let Q be a known solution $\Lambda'Q = \tilde{W}$; then, the hypothesis statistic is

$$H \equiv (Y - XQ)'M_{MP}(Y - XQ)$$
$$= (\Lambda'\hat{B} - \tilde{W})'[\Lambda'(X'X)^- \Lambda]^-(\Lambda'\hat{B} - \tilde{W})$$

and the error statistic is

$$E \equiv (Y - XQ)'(I - M)(Y - XQ)$$
$$= Y'(I - M)Y.$$

An interesting variation on the hypothesis $H_0: \Lambda'B = 0$ is

$$H_0: \Lambda'B\xi = 0, \tag{9.5.7}$$

where ξ can be any $q \times 1$ vector and again $\Lambda' = P'X$. To test (9.5.7), transform model (9.5.1) into

$$Y\xi = XB\xi + e\xi.$$

This is a standard univariate linear model with dependent variable vector $Y\xi$, parameter vector $B\xi$, and error vector $e\xi \sim N(0, \xi'\Sigma\xi I_n)$. It is easily seen that the least squares estimate of $B\xi$ in the univariate model is $\hat{B}\xi$. From univariate theory, a test for (9.5.7) is based on the noncentral F distribution, in particular

$$\frac{(\Lambda'\hat{B}\xi)'[\Lambda'(X'X)^- \Lambda]^-(\Lambda'\hat{B}\xi)/r(\Lambda)}{\xi'Y'(I - M)Y\xi/[n - r(X)]} \sim F[r(\Lambda), n - r(X), \pi],$$

where

$$\pi = \xi'B'\Lambda[\Lambda'(X'X)^- \Lambda]^- \Lambda'B\xi/2\xi'\Sigma\xi.$$

This test can also be generalized in several ways. For example, let W be a known $q \times r$ matrix with $r(W) = r < q$. A test of

$$H_0 : \Lambda' B W = 0 \qquad (9.5.8)$$

can be performed by examining the transformed multivariate linear model

$$YW = XBW + eW .$$

Here, the dependent variable matrix is YW, the error matrix is eW, and the parameter matrix is BW. The test of (9.5.8) follows precisely the form of (9.5.3). The hypothesis statistic is

$$H_* \equiv W'Y'M_{MP}YW = (\Lambda'\hat{B}W)'(\Lambda'(X'X)^-\Lambda)^-(\Lambda'\hat{B}W)$$

and the error statistic is

$$E_* \equiv W'Y'(I - M)YW .$$

It was convenient to assume that W has full column rank. If B can be any $p \times q$ matrix and W has full column rank, then BW can be any $p \times r$ matrix. Thus, BW can serve as the parameter matrix for a multivariate linear model. If W does not have full column rank, then YW has linear dependencies, $\mathrm{Cov}(W'\varepsilon_i) = W'\Sigma W$ is singular, and BW is not an arbitrary $p \times r$ matrix. None of these problems is an insurmountable difficulty for conducting the analysis of the transformed model, but proving that the analysis works for a nonfull rank W is more trouble than it is worth.

Exercise 9.2. Show that under the multivariate linear model

$$\mathrm{E}\left[Y'(M - M_0)Y\right] = r(M - M_0)\Sigma + B'X'(M - M_0)XB.$$

9.6 Test Statistics

Various functions of the hypothesis and error statistics have been proposed as test statistics. Four of the more commonly used are discussed here. A complete survey will not be attempted, and an exhaustive treatment of the related distribution theory will certainly not be given.

In this section, we consider testing only reduced models such as (9.5.2) or parametric hypotheses such as (9.5.3). Hence,

$$E = Y'(I - M)Y$$

and, according to the context, either

$$H = Y'(M - M_0)Y$$

or

$$H = Y'M_{MP}Y.$$

Adjustments for other hypotheses are easily made.

The test statistics discussed in this section are all functions of H and E. Under normality, the null distributions of these statistics depend on H and E only through the fact that they have independent central *Wishart distributions*.

Definition 9.6.1.   Let $w_1, w_2, \cdots, w_n$ be independent $N(\mu_i, \Sigma)$; then

$$W = \sum_{i=1}^{n} w_i w_i'$$

has a noncentral Wishart distribution with n degrees of freedom, covariance matrix Σ, and noncentrality parameter matrix Q, where

$$Q = \frac{1}{2}\Sigma^{-1}\sum_{i=1}^{n}\mu_i\mu_i'.$$

If $Q = 0$, the distribution is a central Wishart. In general, write

$$W \sim W(n, \Sigma, Q).$$

Under the full model and assuming normal distributions, H and E have independent Wishart distributions. In particular,

$$E \sim W\big(n - r(X), \Sigma, 0\big)$$

and

$$H \sim W\left(r(X) - r(X_0), \Sigma, \frac{1}{2}\Sigma^{-1}B'X'(M - M_0)XB\right).$$

The reduced model is true if and only if

$$H \sim W\big(r(X) - r(X_0), \Sigma, 0\big).$$

Exercise 9.3.

(a) Use Definition 9.6.1 to show that E and H have the distributions indicated earlier.

(b) Show that H and E are independent.

(c) Show that MY and E are independent.

Hint: For (b), show that $(I - M)Y$ and $(M - M_0)Y$ are independent. This exercise assumes that you know basic properties of multivariate normal distributions.

The covariance matrix of a Wishart was previously introduced in the discussion of MIVQUE in Chap. 4. See Muirhead (1982) for a detailed discussion. Additional information is available from the online supplemental material associated with Tarpey, Ogden, Petkova, and Christensen (2015) and at http://stat.unm.edu/~fletcher/Wishart.pdf.

A more traditional approach to the distribution theory of multivariate linear models would be to define the distributions to be used later (i.e., U, ϕ_{max}, T^2, and V) as various functions of random matrices with independent central Wishart distributions. One could then show that the corresponding functions of H and E have these distributions. For the present purposes, we are only interested in these distributions because they are interesting functions of the hypothesis and error statistics. There seems to be little point in defining the distributions as anything other than the appropriate functions of H and E.

We begin by considering the *likelihood ratio test statistic*. This is simply the maximum of the likelihood under H_0 divided by the overall maximum of the likelihood. The overall maximum of the likelihood function is obtained at the MLEs, so substituting $\hat{\Sigma}$ for Σ in (9.4.1) gives the maximum value of the log-likelihood as

$$\ell(\hat{B}, \hat{\Sigma}) = -\frac{nq}{2}\log(2\pi) - \frac{n}{2}\log(|\hat{\Sigma}|) - \frac{1}{2}\mathrm{tr}[\hat{\Sigma}^{-1}n\hat{\Sigma}]$$
$$= -\frac{nq}{2}\log(2\pi) - \frac{n}{2}\log(|\hat{\Sigma}|) - \frac{nq}{2},$$

where, again, $|\hat{\Sigma}|$ is the determinant of $\hat{\Sigma}$. The maximum value of the likelihood function is

$$L(\hat{B}, \hat{\Sigma}) = 2\pi^{-nq/2}|\hat{\Sigma}|^{-n/2}e^{-nq/2}.$$

Similarly, if we assume that the reduced model (9.5.2) is true, then MLEs are $\hat{\Gamma} = (X_0'X_0)^- X_0'Y$ and $\hat{\Sigma}_H \equiv Y'(I - M_0)Y/n$. The maximum value of the likelihood function under the assumption that H_0 is true is

$$L(\hat{\Gamma}, \hat{\Sigma}_H) = 2\pi^{-nq/2}|\hat{\Sigma}_H|^{-n/2}e^{-nq/2}.$$

The likelihood ratio test statistic is

$$\frac{L(\hat{\Gamma}, \hat{\Sigma}_H)}{L(\hat{B}, \hat{\Sigma})} = \frac{|\hat{\Sigma}_H|^{-n/2}}{|\hat{\Sigma}|^{-n/2}}$$
$$= \left[\frac{|\hat{\Sigma}|}{|\hat{\Sigma}_H|}\right]^{n/2}.$$

The null hypothesis is rejected if the maximum value of the likelihood under H_0 is too much smaller than the overall maximum (i.e., if the likelihood ratio test statistic is too small). Because the function $f(x) = x^{2/n}$ is strictly increasing, the likelihood ratio test is equivalent to rejecting H_0 when

$$U = f([|\hat{\Sigma}|/|\hat{\Sigma}_H|]^{n/2})$$
$$= \frac{|\hat{\Sigma}|}{|\hat{\Sigma}_H|}$$

is too small. Noting that $\hat{\Sigma} = E/n$ and $\hat{\Sigma}_H = (E+H)/n$, we see that

$$U = \frac{|E|}{|E+H|}.$$

Multiplying the numerator and denominator of U by $|E^{-1}|$, we get a slightly different form,

$$U = \frac{|I|}{|I+HE^{-1}|} = |I+HE^{-1}|^{-1}.$$

When H_0 is true, U has some distribution, say

$$U \sim U(q,d,n-r(X)),$$

where d is either $r(X) - r(X_0)$ or $r(\Lambda)$ depending on which kind of hypothesis is being tested. An α-level test is rejected if the observed value of U is smaller than the α percentile of the $U(q,d,n-r(X))$ distribution.

The likelihood ratio test statistic (LRTS) is often referred to as *Wilks's* Λ. The symbol Λ is used here for other purposes, so to minimize internal confusion, the LRTS is denoted U. In reading applications and computer output, the reader needs to remember that references to Wilks's Λ are references to the LRTS.

Rao (1951) has established the following approximate distribution for U when H_0 is true. Let

$$r = r(X),$$
$$d = r(X) - r(X_0) = r(\Lambda),$$
$$s = \frac{qd}{2} + 1,$$
$$f = (n-r) + d - \frac{1}{2}(d+q+1),$$

and

$$t = \begin{cases} [(q^2d^2 - 4)/(q^2 + d^2 - 5)]^{\frac{1}{2}} & \text{if } \min(q,d) \geq 2 \\ 1 & \text{if } \min(q,d) = 1 \end{cases};$$

then, it is approximately true that

$$\frac{1 - U^{1/t}}{U^{1/t}} \frac{ft - s}{qd} \sim F(qd, ft - s).$$

The test is rejected for large values of $(1 - U^{1/t})/U^{1/t}$. If $\min(q,d)$ is 1 or 2, the distribution is exact. For properties and tables of the U distribution, see a text on multivariate analysis, such as Seber (1984, p. 413).

An alternative to the likelihood ratio test statistic can be generated by *Roy's union-intersection principle* (see Roy 1953). His method is based on the fact that the reduced model (9.5.2) holds if and only if, for every $q \times 1$ vector ξ, the univariate linear model

$$Y\xi = X_0\Gamma\xi + e\xi \ , \ e\xi \sim N(0,(\xi'\Sigma\xi)I) \tag{9.6.1}$$

holds. If we denote $H_{0\xi}$ as the hypothesis that model (9.6.1) holds and H_0 as the hypothesis that model (9.5.2) holds, then

$$H_0 = \bigcap_{\text{all } \xi} H_{0\xi} \ .$$

The intersection is appropriate because the equivalence requires the validity of *every* model of the form (9.6.1).

Similarly, the full model (9.5.1) holds if and only if every model

$$Y\xi = XB\xi + e\xi \tag{9.6.2}$$

holds. The alternative hypothesis H_A that (9.5.1) holds but (9.5.2) does not is simply that, for some ξ, the univariate model (9.6.2) holds but model (9.6.1) does not. Let $H_{A\xi}$ be the univariate alternative. Because H_A requires only one of the $H_{A\xi}$ to be true, we have

$$H_A = \bigcup_{\text{all } \xi} H_{A\xi} \ .$$

This reformulation of H_A and H_0 in terms of unions and intersections is the genesis of the name "union-intersection principle." (Merely mentioning an alternative hypothesis makes me feel like an apostate.)

For testing model (9.6.1) against model (9.6.2), there is a standard univariate F statistic available, $F_\xi = [\xi'Y'(M - M_0)Y\xi/(r(X) - r(X_0))]/[\xi'Y'(I - M)Y\xi/(n - r(X))]$. The union-intersection principle test statistic is simply

$$\psi_{\text{max}} \equiv \sup_\xi\{F_\xi\} \ .$$

The test is rejected if ψ_{max} is too large. In practice, an alternative but equivalent test statistic is used. Because

$$F_\xi = \frac{n - r(X)}{r(X) - r(X_0)} \frac{\xi'H\xi}{\xi'E\xi} \ ,$$

it is equivalent to reject H_0 if

$$\phi_{\text{max}} \equiv \sup_\xi \left[\frac{\xi'H\xi}{\xi'E\xi} \right]$$

is too large.

We will show that ϕ_{max} is distributed as the largest eigenvalue of HE^{-1}. The argument is based on the following result.

Lemma 9.6.2. If E is a positive definite matrix, then there exists a nonsingular matrix $E^{1/2}$ such that

$$E = E^{1/2}E^{1/2}$$

and

$$E^{-1/2}EE^{-1/2} = I.$$

PROOF. As in the proof of *PA* Theorem B.22, write

$$E = PD(\lambda_i)P',$$

where P is an orthonormal matrix, $D(\lambda_i)$ is diagonal, and the λ_i's are all positive. Pick

$$E^{1/2} = PD(\sqrt{\lambda_i})P',$$

and the results are easily shown. □

Note that any vector ξ can be written as $E^{-1/2}\rho$ for some vector ρ, and any vector ρ determines a vector ξ, so

$$\phi_{max} = \sup_{\rho} \left[\frac{\rho'E^{-1/2}HE^{-1/2}\rho}{\rho'\rho} \right].$$

Writing $E^{-1/2}HE^{-1/2}$ as $P_0D(\phi_i)P_0'$, where P_0 is an orthonormal matrix and the ϕ_i's are the eigenvalues of $E^{-1/2}HE^{-1/2}$, we can replace ρ with $v = P_0\rho$, giving

$$\phi_{max} = \sup_{v} \left[\frac{v'D(\phi_i)v}{v'v} \right].$$

Because

$$v'D(\phi_i)v/v'v = \sum_{i=1}^{q} v_i^2\phi_i \bigg/ \sum_{i=1}^{q} v_i^2$$

is a weighted average of the ϕ_i's, the maximum value is attained when all of the weight is placed on the largest eigenvalue. Thus,

$$\phi_{max} = \max_{i} \phi_i.$$

Finally, we need to establish that $E^{-1/2}HE^{-1/2}$ and HE^{-1} have the same eigenvalues. Let w be an eigenvector for $E^{-1/2}HE^{-1/2}$ corresponding to the eigenvalue ϕ. Because $E^{-1/2}HE^{-1/2}w = \phi w$, we have

$$E^{1/2}E^{-1/2}HE^{-1/2}E^{-1/2}E^{1/2}w = \phi E^{1/2}w$$

and

$$HE^{-1}E^{1/2}w = \phi E^{1/2}w.$$

Clearly, $E^{1/2}w$ is an eigenvector of HE^{-1} corresponding to ϕ. Thus, we have shown that ϕ_{max} is the largest eigenvalue of HE^{-1}.

Rather than using ϕ_{max}, tests are often performed using θ_{max}, where θ_{max} is the maximum eigenvalue of $H(E+H)^{-1}$. It is a simple matter to see that the eigenvalues of $H(E+H)^{-1}$ and HE^{-1} are related by the equation $\theta = \phi/(1+\phi)$. Thus, θ_{max} is a one-to-one increasing transformation of ϕ_{max} and the test based on θ_{max} is equivalent to the one based on ϕ_{max}. Tables of the distribution of θ_{max} under H_0 were worked out by Heck (1960). Derivations and tables for the distribution of ϕ_{max} or θ_{max} under the null hypothesis can be found in many standard multivariate analysis books, such as Seber (1984, Section 2.5).

Although the likelihood ratio and union-intersection test statistics are probably the best known, several other test statistics have also been proposed. Two of these are the *Lawley-Hotelling trace*

$$T^2 = [n - r(X)]\,\text{tr}[HE^{-1}] = \text{tr}[HS^{-1}]$$

and *Pillai's trace*

$$V = \text{tr}[H(E+H)^{-1}].$$

Exercise 9.4. Show that θ is an eigenvalue of $H(E+H)^{-1}$ if and only if ϕ is an eigenvalue of HE^{-1}, where $\theta = \phi/(1+\phi)$. Show that

$$U = \prod_{h=1}^{q} 1/(1+\phi_i), \quad T^2 = \sum_{h=1}^{q} \phi_i/[n - r(X)], \quad V = \sum_{h=1}^{q} \phi_i/(1+\phi_i).$$

Hints: Show that the eigenvalues of AB^{-1} and $B^{-1}A$ are the same. Use the fact that

$$I = (E+H)^{-1}H + (E+H)^{-1}E.$$

Though the exact distributions of T^2 and V require special tables, their distributions when H_0 is true can be approximated using standard tables. Let $d = r(X) - r(X_0) = r(\Lambda)$ and let $n - r(X) = n - r$. A very good approximation to T^2 has been suggested by McKeon (1974). He advocates using

$$GT^2 \sim F(qd, D),$$

where

$$D = 4 + \frac{qd+2}{B-1},$$

$$B = \frac{(n-r+d-q-1)(n-r-1)}{(n-r-q-3)(n-r-q)},$$

and

$$G = (qd)^{-1}\left[\frac{D}{D-2}\right]\left[\frac{n-r-q-1}{n-r}\right].$$

This gives the exact distribution for $\min(q,d) - 1$. In particular, if $d = 1$,

$$B = \frac{n-r-1}{n-r-q-3}, \quad D = n-r-q+1, \quad G = \frac{1}{n-r}\frac{n-r-q+1}{q}.$$

It follows that when $r(H) = 1$,

$$F \equiv \frac{1}{dfE}\frac{dfE-q+1}{q}T^2 \sim F(q, dfE-q+1).$$

Note that, for large n, the term $B-1$ gets very small so that D, the denominator degrees of freedom, gets large. In particular, for large n, the distribution of GT^2 is approximately that of a $\chi^2(qd)$ divided by its degrees of freedom. This result is equivalent to the standard asymptotic approximation

$$T^2 \sim \chi^2(qd).$$

The null distribution of Pillai's trace can be approximated by

$$\frac{n-r-q+s}{|q-d|+s}\frac{V}{s-V} \sim F(s[|q-d|+s], s[n-r-q+s]),$$

where

$$s = \min(q, d).$$

Asymptotically,

$$(n-r)V \sim \chi^2(qd).$$

Seber (1984, p. 414) compares all four test statistics. He also provides tables and other details about the distributions under H_0 (see Seber 1984, Section 2.5). Kres (1983) is another good source for tables related to multivariate linear models.

For standard linear models, the univariate F test should be about 1 if the null model is correct, regardless of whether the data are multivariate normal. Similarly, although formal tests based on these four statistics depend on the assumption of multivariate normality, these are reasonable test statistics even without the normality assumption as long as the assumed mean and covariance structures are reasonable. To evaluate the test statistics intuitively, we need some idea of the values of the test statistics when H_0 is true. As we have seen, even without the assumption of normality

$$E(S) = E(E/[n-r(X)]) = \Sigma,$$

and from Exercise 9.3, under H_0,

$$E(H/r(M-M_0)) = \Sigma.$$

Using these equalities as crude approximations, it follows that if H_0 is true,

$$U \doteq \left| I + \left(\frac{r(X) - r(X_0)}{n - r(X)}\right) I \right|^{-1}$$

$$= \left[\frac{n - r(X)}{n - r(X_0)}\right]^q,$$

$$\phi_{\max} \doteq \frac{r(X) - r(X_0)}{n - r(X)},$$

$$\theta_{\max} \doteq \frac{n - r(X)}{n - r(X_0)},$$

$$T^2 \doteq \mathrm{tr}(I)\,[r(X) - r(X_0)]$$
$$= q\,[r(X) - r(X_0)],$$

and

$$V \doteq \mathrm{tr}(I)\,[r(X) - r(X_0)]\,/\,[n - r(X_0)]$$
$$= q\,[r(X) - r(X_0)]\,/\,[n - r(X_0)].$$

These *comparison values* can be very useful in exploring the data. If the observed value of U is much smaller than $([n - r(X)]\,/\,[n - r(X_0)])^q$ or if the observed values of the other test statistics are much larger than the comparison values, the null hypothesis is called in question. In a formal test, the null distribution of the test statistic is used to quantify the meaning of the word "much" in the previous sentence.

An important parameter in the distributions of the test statistics is the rank of HE^{-1}. With probability one, E is nonsingular, so

$$r\left(HE^{-1}\right) = r(H).$$

By definition,

$$H = Y'(M - M_0)Y$$

is a $q \times q$ matrix. Because $r(M - M_0) = r(X) - r(X_0)$, Theorem 9.3.2 implies that if Y has a density, $r(HE^{-1}) = r(H) = \min[q, r(X) - r(X_0)]$ with probability one.

9.6.1 Equivalence of Test Statistics

Whenever H is a rank one matrix, all four of the test statistics discussed are equivalent. First, note that E is nonsingular (with probability one), so HE^{-1} has rank one. A rank one matrix has only one nonzero eigenvalue, hence the maximum eigenvalue equals the sum of the eigenvalues. In other symbols,

$$\phi_{\max} = \mathrm{tr}[HE^{-1}],$$

or equivalently

$$\phi_{\max} = \frac{1}{n - r(X)}T^2.$$

Thus, the union-intersection test statistic is equivalent to Hotelling's T^2.
The likelihood ratio test uses the statistic

$$U = |I + HE^{-1}|^{-1}.$$

It is easy to see that the eigenvalues of $I + HE^{-1}$ are just one plus the eigenvalues of HE^{-1}. The determinant is the product of the eigenvalues $\prod_{h=1}^{q}(1 + \phi_i)$, and HE^{-1} has only one nonzero eigenvalue, so

$$|I + HE^{-1}| = 1 + \phi_{max}$$

and

$$U = (1 + \phi_{max})^{-1}.$$

The statistic U is a strictly decreasing function of ϕ_{max}, thus the likelihood ratio test is equivalent to the other tests.

The equivalence of Pillai's trace to Hotelling's T^2 is established in Exercise 9.5.

Exercise 9.5.   Use *PA-V* Proposition 12.5.1 (Christensen 2011, Proposition 13.5.1) to show that the test based on Pillai's trace is equivalent to the test based on Hotelling's T^2. The proposition states that

$$(A + a'b)^{-1} = A^{-1} - A^{-1}a'(I + bA^{-1}a')^{-1}bA^{-1}.$$

9.7 Prediction and Confidence Regions

Suppose we wish to predict the value of a new observation vector y_0' with $E(y_0') = x_0'B$, where $x_0'B$ is estimable (i.e., $x_0'B = \rho_0'XB$ for some vector ρ_0). It is natural to assume that y_0 is generated by the same process as Y, thus $Cov(y_0) = \Sigma$ and y_0 is independent of Y. As discussed in Chap. 4, it follows that the best linear unbiased predictor of y_0 is $\hat{y}_0 \equiv \hat{B}'x_0$.

A prediction region for y_0 can be based on the distribution of $y_0 - \hat{y}_0$. Clearly, $E(y_0 - \hat{y}_0) = 0$ and, recalling Exercise 9.1,

$$
\begin{aligned}
Cov(y_0 - \hat{y}_0) &= \Sigma + Cov(\hat{y}_0) \\
&= \Sigma + Cov(Y'M\rho_0) \\
&= \Sigma + Cov\left(\begin{bmatrix} Y_1'M\rho_0 \\ \vdots \\ Y_q'M\rho_0 \end{bmatrix}\right) \\
&= \Sigma + \rho_0'M\rho_0 \Sigma \\
&= \left(1 + x_0'(X'X)^- x_0\right)\Sigma.
\end{aligned}
$$

If y_0 and Y are multivariate normal, then $y_0 - \hat{y}_0$ is also normal, independent of E, and

$$\left(1 + x_0'(X'X)^- x_0\right)^{-1} (y_0 - \hat{y}_0)(y_0 - \hat{y}_0)' \sim W(1, \Sigma, 0).$$

It follows that

$$\frac{\text{tr}\left[(y_0 - \hat{y}_0)(y_0 - \hat{y}_0)'S^{-1}\right]}{1 + x_0'(X'X)^- x_0},$$

or equivalently

$$\frac{(y_0 - \hat{y}_0)'S^{-1}(y_0 - \hat{y}_0)}{1 + x_0'(X'X)^- x_0},$$

has the same distribution as the null distribution of the Lawley-Hotelling T^2 statistic with $d = 1$. From our discussion of McKeon's approximation with $dfE \equiv n - r(X)$, the exact distribution is

$$\frac{(y_0 - \hat{y}_0)'S^{-1}(y_0 - \hat{y}_0)}{(dfE)\left(1 + x_0'(X'X)^- x_0\right)} \frac{dfE + 1 - q}{q} \sim F(q, dfE + 1 - q, 0).$$

A $(1 - \alpha)100\%$ prediction ellipsoid consists of all y_0 vectors that satisfy

$$(y_0 - \hat{y}_0)'S^{-1}(y_0 - \hat{y}_0) \leq$$
$$F(1 - \alpha, q, dfE + 1 - q)\frac{q}{dfE + 1 - q}(dfE)\left(1 + x_0'(X'X)^- x_0\right).$$

The methods illustrated here can also be used to obtain confidence ellipsoids. Examples of such ellipsoids are given in Sects. 10.1 and 10.2.

9.8 Multiple Testing Methods

A number of multiple comparison methods for univariate linear models were discussed in *PA* Chap. 5. *PA* also discusses general issues of multiple testing in more detail than we do here. Three of the methods discussed in *PA* are easily adapted for use with multivariate linear models.

Consider a collection of single degree of freedom estimable null hypotheses $H_{0j} : \lambda_j'B = 0$, $j = 1, \ldots, s$. For example, the hypotheses could be contrasts in a one-way ANOVA or interaction contrasts in a balanced two-way ANOVA. Define $\Lambda \equiv [\lambda_1, \ldots, \lambda_s]$. Then $\Lambda'B = 0$, for example, could be for testing equality of treatment groups in a one-way ANOVA or for testing no interaction in a balanced two-way ANOVA. The relevant λs are determined by X, and are appropriate for any q. Clearly, $\Lambda'B = 0$ if and only if $\lambda_j'B = 0$, $j = 1, \ldots, s$. Moreover, both of these conditions are equivalent to $\lambda'B = 0$ for every $\lambda \in C(\Lambda)$. Equivalently, there exists a $\lambda \in C(\Lambda)$ such that $\lambda'B \neq 0$ if and only if $\Lambda'B \neq 0$ which, since the λ_js are a spanning set for $C(\Lambda)$, occurs if and only if there exists a j such that $\lambda_j'B \neq 0$. Indeed, we do not even need to define $\Lambda = [\lambda_1, \ldots, \lambda_s]$ as long as

$$C(\Lambda) = \text{span}\{\lambda_1, \ldots, \lambda_s\}.$$

Henceforth, *any reference within this section to* $\lambda'B$ *presumes that* $\lambda \in C(\Lambda)$.

For testing $\Lambda'B = 0$, $H_\Lambda \equiv Y'M_{MP}Y$. For testing $\lambda'B = 0$ with $\lambda \in C(\Lambda)$ the hypothesis matrix is $H_\lambda \equiv Y'M_{M\rho}Y$ where $H_\Lambda - H_\lambda$ is nonnegative definite. Here P and ρ are defined by $\Lambda' = P'X$ and $\lambda' = \rho'X$ and have the property that $M\rho \in C(MP)$. Finally, define $H_j \equiv H_{\lambda_j}$ and more generally we may replace a subscript λ_j with j in this section. For our four test statistics based on H_Λ, H_λ, and E,

$$U_\Lambda \leq U_\lambda,$$
$$\phi_{max,\Lambda} \geq \phi_{max,\lambda}.$$
$$T_\Lambda^2 \geq T_\lambda^2,$$
$$V_\Lambda \geq V_\lambda.$$

See Exercise 9.6 for the validity of these statements. Unlike univariate models, when $q \geq 2$ only ϕ_{max} typically has a λ for which equality can always be achieved.

Let a generic α level test of $\Lambda'B = 0$ be to reject when

$$Q_\Lambda > Q[1 - \alpha, q, r(\Lambda), n - r(X)]$$

and a generic α level test of $\lambda'B = 0$ be to reject when

$$Q_\lambda > Q[1 - \alpha, q, 1, n - r(X)].$$

Specifically, Q can be any of ϕ_{max}, T^2, V, or $1/U$.

The *Least Significant Difference (LSD)* method often (I think incorrectly) associated with Fisher, amounts to

1. Check the truth of

$$Q_\Lambda > Q[1 - \alpha, q, r(\Lambda), n - r(X)]$$

2. If the inequality is not true, quit and go home. None of the $\lambda'B$s are declared significant.
3. If the inequality is true, declare $\lambda'B$ significantly different from 0 if and only if

$$Q_\lambda > Q[1 - \alpha, q, 1, n - r(X)].$$

Note that step 3 is just the usual α level Q test for $\lambda'B = 0$. Collective experience indicates that the method works reasonably well if you restrict yourself in step 3 to testing only a prespecified finite collection $\lambda_j'B = 0$, $j = 1, \ldots, s$. Experience indicates that this method performs very badly, in the sense that it rejects too many hypotheses $\lambda'B = 0$ that are true, if you use it to test arbitrarily chosen $\lambda'B$.

Scheffé's method amounts to

1. Check the truth of

$$Q_\Lambda > Q[1 - \alpha, q, r(\Lambda), n - r(X)]. \tag{9.8.1}$$

2. If the inequality is not true, quit and go home. None of the $\lambda'B$s are declared significant.

3. If the inequality is true, declare $\lambda'B$ significantly different from 0 if and only if

$$Q_\lambda > Q[1 - \alpha, q, r(\Lambda), n - r(X)].$$

The basis of Scheffé's method is replacing Q_Λ in (9.8.1) with Q_λ. The motivation is based on the earlier discussed fact that

$$Q_\Lambda \geq Q_\lambda, \tag{9.8.2}$$

so it is impossible to reject $\lambda'B = 0$ unless you have previously rejected $\Lambda'B = 0$. In fact, you can skip steps 1 and 2 of the algorithm because they take care of themselves. Unfortunately, unlike Scheffé univariate test, except for $Q = \phi_{max}$, there does not seem to exist a $\lambda \in C(\Lambda)$ for which equality occurs in (9.8.2), so there is no assurance in step 3 that there exists a $\lambda'B = 0$ that would be rejected, even though, based on step 1, we are confident that there exist $\lambda'B \neq 0$. In particular (and this agrees with the univariate case), there is no assurance that any of the specific $\lambda_j'B = 0$ hypotheses will be rejected, even though we are reasonably confident that they cannot all be true.

Scheffé's procedure makes it very difficult to reject a $\lambda'B = 0$. Collective experience indicates that this method performs well if you use it to test arbitrary $\lambda'B$ (including ones suggested by the data) but is too conservative if your interest is only in testing a prespecified finite collection $\lambda_j'B = 0$, $j = 1, \ldots, s$. Too conservative means not finding enough of the $\lambda_j'B$s that are different from 0.

Bonferroni's method simply involves declaring $\lambda_j'B$ significantly different from 0 if and only if

$$Q_{\lambda_j} > Q[1 - \alpha_j, q, 1, n - r(X)],$$

where the α_js add up to α. Typically one takes $\alpha_j = \alpha/s$.

When considering λ_js that define orthogonal constraints, i.e., λ_js for which $\rho_j'M\rho_k = 0$ when $j \neq k$, $T_\Lambda^2 = T_1^2 + \cdots + T_s^2$, but no similar decompositions seem to work for the other statistics.

For any $q \times r$ matrix W, all of these ideas apply immediately to testing H_{0j} : $\lambda_j'BW = 0$, $j = 1, \ldots, s$, $H_0 : \Lambda'BW = 0$, and $H_0 : \lambda'BW = 0$, $\lambda \in C(\Lambda)$. You just apply the ideas to the multivariate linear model $YW = XBW + eW$.

Exercise 9.6. Let A, A_1, A_2, and $A_1 - A_2$ be nonnegative definite and let B be positive definite.

1. Provide a simple induction proof that for nonnegative ϕ_hs, $\prod_{h=1}^{q}(1 + \phi_h) \geq 1 + \prod_{h=1}^{q}\phi_h$.

2. Show that

$$|A + B| \geq |A||I + A^{-1/2}BA^{-1/2}| \geq |A|[1 + |A^{-1/2}BA^{-1/2}|] \geq |A|.$$

3. Show that

$$|I + E^{-1/2} H_\Lambda E^{-1/2}| =$$
$$|I + E^{-1/2} H_\lambda E^{-1/2} + E^{-1/2} [H_\Lambda - H_\lambda] E^{-1/2}| \geq |I + E^{-1/2} H_\lambda E^{-1/2}|$$

so that $U_\Lambda \leq U_\lambda$.

4. Show that $\xi' Y' M_{MP} Y \xi \geq \xi' Y' M_{M\rho} Y \xi$ and that $\phi_{\max,\Lambda} \geq \phi_{\max,\lambda}$.
5. Show that $T_\Lambda^2 \geq T_\lambda^2$.
6. Show that $\text{tr}[(A_2 + B)A_1] \geq \text{tr}[A_2(A_1 + B)]$.
7. Show that $\text{tr}[A_1(A_1 + B)^{-1}] \geq \text{tr}[(A_2 + B)^{-1}A_2]$.
8. Show that $V_\Lambda \geq V_\lambda$.
9. Extend the multiple testing ideas to a collection of tests involving a known matrix $\tilde{W}$ with $C(\tilde{W}) \subset C(\Lambda')$ and hypotheses determined by $H_0: \Lambda'B = \tilde{W}$.

Perhaps a more interesting multiple testing problem examines null hypotheses $H_{0jk}: \lambda_j' B \xi_k = 0$, $j = 1, \ldots, s$, $k = 1, \ldots, r$. The Bonferroni idea would be to test each hypothesis at the α/rs level. Alternatively, similar to Λ, write $W = [\xi_1, \ldots, \xi_r]$ so that the multiple testing problem is related to testing $\Lambda'BW = 0$ but also to $\lambda'B\xi$ whenever $\lambda \in C(\Lambda)$ and $\xi \in C(W)$. An alternative to Bonferroni is to test $\Lambda'BW = 0$ first and not declare any $\lambda'B\xi \neq 0$ unless the first test is rejected. Note that

$$\Lambda'BW = 0$$

iff

$$\lambda'B\xi = 0, \quad \forall \lambda \in C(\Lambda), \xi \in C(W)$$

iff

$$\lambda_j'B\xi_k = 0, \quad j = 1, \ldots, s, k = 1, \ldots, r.$$

Thus

$$\Lambda'BW \neq 0$$

iff

$$\exists \lambda \in C(\Lambda), \xi \in C(W) \text{ such that } \lambda'B\xi \neq 0$$

iff

$$\exists j, k \text{ such that } \lambda_j'B\xi_k \neq 0.$$

The fact that no $\lambda'B\xi = 0$ is rejected unless $\Lambda'BW = 0$ is rejected immediately controls the weak experimentwise error rate. If $\Lambda'BW = 0$ is rejected, LSD then just tests each $H_{0jk}: \lambda_j'B\xi_k = 0$ at the α level. (Recall LSD does not work well for arbitrary $\lambda'B\xi = 0$.) Of course, if you perform a test of $\Lambda'BW = 0$ first, and quit if it is not rejected, any scheme for declaring some of $\lambda'B\xi = 0$ significant has controlled the weak experimentwise error rate. It is not clear how to apply the Scheffé idea except for Roy's test.

Roy's method of test construction for

$$H_0: \Lambda'BW = 0$$

is to maximize the F statistics for the univariate hypotheses

$$H_0: \Lambda'BW\zeta = 0.$$

In particular, this is

$$\sup_{\zeta} \frac{(\Lambda'\hat{B}W\zeta)' [\Lambda'(X'X)^-\Lambda]^{-1} (\Lambda'\hat{B}W\zeta) \Big/ r(\Lambda)}{\zeta'W'SW\zeta} = \frac{n-r(X)}{r(\Lambda)}\phi_{\max},$$

where $\phi_{\max}$ is the maximum eigenvalue of $(W'Y'M_{MP}YW)[W'Y'(I-M)YW]^{-1}$. It immediately follows that

$$\frac{(\Lambda'\hat{B}W\zeta)' [\Lambda'(X'X)^-\Lambda]^{-1} (\Lambda'\hat{B}W\zeta) \Big/ r(\Lambda)}{\zeta'W'SW\zeta} \leq \frac{n-r(X)}{r(\Lambda)}\phi_{\max}, \qquad (9.8.3)$$

with equality achieved for some ζ (cf., Sect. 9.6).

Moreover, from our knowledge of univariate tests (cf., *PA* Sect. 5.1)

$$(\lambda'\hat{B}W\zeta)' [\lambda'(X'X)^-\lambda]^{-1} (\lambda'\hat{B}W\zeta) = \zeta'W'Y'M_{MP}YW\zeta$$
$$\leq \zeta'W'Y'M_{MP}YW\zeta = (\Lambda'\hat{B}W\zeta)' [\Lambda'(X'X)^-\Lambda]^{-1} (\Lambda'\hat{B}W\zeta)$$

and there exists a λ that achieves equality, namely $\lambda' = \rho'X$ with $\rho = M_{MP}YW\zeta$.

Taking $\xi = W\zeta$, it follows that the multiple comparison procedure that rejects $H_0: \lambda'B\xi = 0$ if and only if

$$\frac{(\lambda'\hat{B}\xi)^2 \Big/ r(\Lambda)}{[\lambda'(X'X)^-\lambda]\xi S\xi} > \frac{n-r(X)}{r(\Lambda)}\phi_{\max}[1-\alpha, r, r(\Lambda), n-r(X)]$$

has an experimentwise error rate no greater than α when applied to testing any or all hypotheses of the form $\lambda'B\xi = 0$ and there exists a pair λ and ξ that will reject whenever the test of $\Lambda'BW = 0$ is rejected, so the experimentwise error rate for testing all hypotheses of the form $H_0: \lambda'BW\xi = 0$ is precisely α.

For testing something less than all hypotheses, such as a finite number H_{0jk}: $\lambda'_jB\xi_k = 0$, this method, first developed in Roy and Bose (1953), has an experimentwise error rate that is actually something less that α. Typically for the methods discussed in this section the actual experimentwise error rate is something less that α.

As with Scheffé's method, the Roy–Bose procedure for controlling the simultaneous error rate of multiple tests can be adapted to providing simultaneous confidence intervals. With confidence coefficient $(1-\alpha)100\%$, the intervals

$$\lambda'\hat{B}\xi \pm \sqrt{\xi'E\xi[\lambda'(X'X)^-\lambda]\phi_{\max}[1-\alpha, q, r(\Lambda), n-r(X)]}$$

contain all parameters of the form $\lambda'B\xi$.

Exercise 9.7. Show that the constraint imposed on a univariate linear model by any hypothesis H_0: $\zeta'\Lambda'\beta = 0$ is contained in the constraint subspace determined by the estimable hypothesis H_0: $\Lambda'\beta = 0$. Hint: Review *PA* Sect. 3.3.

9.9 Multivariate Time Series and Spatial Models

Model (9.1.1) is easily generalized into a model that has common spatial-temporal relationships between the n observation units but an arbitrary correlation matrix associated with q measurements on each unit. Again, write the multivariate model $Y = XB + e$, $E(e) = 0$ as

$$\text{Vec}(Y) = [I_q \otimes X]\text{Vec}(B) + \text{Vec}(e), \qquad E[\text{Vec}(e)] = 0 \qquad (9.9.1)$$

but now we incorporate a common spatial-temporal relationship among the observational units by defining

$$\text{Cov}[\text{Vec}(e)] = [\Sigma \otimes V(\phi)], \qquad (9.9.2)$$

where $V(\phi)$ can incorporate the correlation structure of any of our standard time series or spatial models.

When $V(\phi)$ is known it is not difficult to show that the BLUEs for this model are the individual generalized least squares estimates discussed in *PA* Sect. 2.7, i.e.,

$$X\hat{B} = AY; \qquad A \equiv X\left[X'V^{-1}(\phi)X\right]^{-}X'V^{-1}(\phi) \qquad (9.9.3)$$

and that an unbiased estimate of Σ is

$$S(\phi) \equiv \frac{1}{n - r(X)}Y'(I - A)'V^{-1}(\phi)(I - A)Y.$$

Unfortunately, ϕ is never known, so these are not really estimates. Of course maximum likelihood and REML can be used to estimate ϕ and Σ, but depending on how $V(\phi)$ is defined, there may be identifiability issues with the parameters, e.g. you might want to make $V(\phi)$ a correlation matrix rather than a covariance matrix.

Exercise 9.8. Prove that for $V(\phi)$ known, solutions to (9.9.3) are BLUEs and that S is unbiased for Σ.

9.9.1 A Tensor Model

Consider a sequence of spatial multivariate linear models

$$\mathcal{Y}_t = XB_t + e_t \qquad t = 1, \ldots, T,$$

where each $\mathcal{Y}_t$ is an $n \times q$ matrix of observations. For each t turn the multivariate model into a univariate linear model

$$\text{Vec}(\mathcal{Y}_t) = [I_q \otimes X]\text{Vec}(B_t) + \text{Vec}(e_t), \quad \text{E}[\text{Vec}(e_t)] = 0, \quad \text{Cov}[\text{Vec}(e_t)] = [\Sigma \otimes V].$$

Combine the univariate versions of the multivariate models into another spatial multivariate linear model,

$$[\text{Vec}(\mathcal{Y}_1) \cdots \text{Vec}(\mathcal{Y}_T)] = [I_q \otimes X][\text{Vec}(B_1) \cdots \text{Vec}(B_T)] + [\text{Vec}(e_1) \cdots \text{Vec}(e_T)]$$

wherein the observation matrix is now $nq \times T$ and the rows of this matrix are assumed to have a covariance structure that is a multiple of some matrix Ψ. We can again turn this into a univariate linear model

$$\begin{bmatrix} \text{Vec}(\mathcal{Y}_1) \\ \vdots \\ \text{Vec}(\mathcal{Y}_T) \end{bmatrix} = [I_T \otimes I_q \otimes X] \begin{bmatrix} \text{Vec}(B_1) \\ \vdots \\ \text{Vec}(B_T) \end{bmatrix} + \begin{bmatrix} \text{Vec}(e_1) \\ \vdots \\ \text{Vec}(e_T) \end{bmatrix} \qquad (9.9.4)$$

wherein the covariance matrix is

$$[\Psi \otimes \Sigma \otimes V].$$

If we think of $\mathcal{Y}$ as a three-dimensional matrix with subscripts iht, $i = 1, \ldots, n$, $h = 1, \ldots, q, t = 1, \ldots, T$ we can define $\text{Vec}(\mathcal{Y})$ as the dependent variable vector in model (9.9.4). This is essentially a three-dimensional tensor notation and similar to the notation used for ANOVA models in PA.

9.10 Additional Exercises

Exercise 9.10.1.   Show that if $W \sim W(n, \Sigma, 0)$, then

$$AWA' \sim W(n, A\Sigma A', 0).$$

Exercise 9.10.2.   Show that if $W_1, \cdots W_r$ are independent with $W_i \sim W(n_i, \Sigma, 0)$, then

$$\sum_{i=1}^{r} W_i \sim W\left(\sum_{i=1}^{r} n_i, \Sigma, 0\right).$$

Exercise 9.10.3.   Show that if $W \sim W(n, \Sigma, 0)$, then

$$\frac{\lambda' W \lambda}{\lambda' \Sigma \lambda} \sim \chi^2(n,0).$$

Exercise 9.10.4.   For $i = 1, 2, 3$, let $y_i \sim N(\mu + (i-2)\xi, \Sigma)$, where Σ is known and y_1, y_2, and y_3 are independent. Find the maximum likelihood estimates of μ and ξ.

Exercise 9.10.5.   Based on the multivariate linear model

$$Y = XB + e, \; \mathrm{E}(e) = 0, \; \mathrm{Cov}(\varepsilon_i, \varepsilon_j) = \delta_{ij} \Sigma,$$

find a 99% prediction interval for $y_0' \xi$, where y_0 is an independent observation that is distributed $N(B'x_0, \Sigma)$.

Exercise 9.10.6.   Let $y_1, y_2, \cdots, y_n$ be i.i.d. $N(X\beta, \Sigma)$, where $\Sigma_{n \times n}$ is unknown. Show that the maximum likelihood estimate of $X\beta$ is

$$X\hat{\beta} = X(X'E^{-1}X)^{-}X'E^{-1}\bar{y}_{..}.$$

Exercise 9.10.7.   Consider the multivariate linear model $Y = XB + e$ and the parametric function $\Lambda' BW$, where W is a $q \times r$ matrix of rank r. Find simultaneous confidence intervals for all parameters of the form $\zeta' \Lambda' BW \xi$.

Exercise 9.10.8.   Use Lemma 9.6.2 to show that if A is nonnegative definite and B is positive definite, then AB^{-1} is nonnegative definite. Hint: Show that the nonnegative definite matrix $B^{-1/2}AB^{-1/2}$ has the same eigenvalues as AB^{-1}.

Exercise 9.10.9.   Rewrite the multivariate linear model in terms of $\mathrm{Vec}(Y')$. Write it similarly to both (9.1.2) with covariance (9.1.3) and using the Vec operator with Kronecker products as in (9.1.4) with (9.1.5).

Exercise 9.10.10.   In the multivariate linear model $Y = XB + e$ the likelihood equations reduce to

$$\mathrm{tr}\left\{ \Sigma^{-1}[d_{\theta_j}\Sigma] \right\} = \mathrm{tr}\left\{ \Sigma^{-1}[d_{\theta_j}\Sigma]\Sigma^{-1}\hat{\Sigma} \right\},$$

for $j = 1, \ldots, s$ where $\hat{\Sigma} \equiv Y'(I-M)Y/n$ and $\theta_1, \theta_2, \ldots, \theta_s$ from Chap. 4 correspond to σ_{gh} for $g \leq h$ so that $s = q(q+1)/2$. We recognize that $\Sigma = \hat{\Sigma}$ provides a solution to the likelihood equations. Now suppose that $\Sigma \equiv \sigma^2[(1-\rho)I_1 + \rho J_q^q]$, so that $s = 2$ and $(\theta_1, \theta_2) = (\sigma^2, \rho)$. It is clear that $\Sigma = \hat{\Sigma}$ still provides a solution to the likelihood equations, why is $\hat{\Sigma}$ no longer the MLE? (The argument has almost nothing to do with the particular parametric form that we chose for Σ.)

Exercise 9.10.11. Use the univariate linear model form of the multivariate linear model and the likelihood equations from Sect. 4.3 to obtain the MLEs for the multivariate linear model.

References

Anderson, T. W. (2003). *An introduction to multivariate statistical analysis* (3rd ed.). New York: Wiley.

Arnold, S. F. (1981). *The theory of linear models and multivariate analysis*. New York: Wiley.

Christensen, R. (1996). *Analysis of variance, design, and regression: Applied statistical methods*. London: Chapman and Hall.

Christensen, R. (2011). *Plane answers to complex questions: The theory of linear models* (4th ed.). New York: Springer.

Dillon, W. R., & Goldstein, M. (1984). *Multivariate analysis: Methods and applications*. New York: Wiley.

Eaton, M. L. (1983). *Multivariate statistics: A vector space approach*. New York: Wiley.

Everitt, B., & Hothorn, T. (2011). *An introduction to applied multivariate analysis with R*. New York: Springer.

Gnanadesikan, R. (1977). *Methods for statistical data analysis of multivariate observations*. New York: Wiley.

Heck, D. L. (1960). Charts of some upper percentage points of the distribution of the largest characteristic root. *Annals of Mathematical Statistics, 31*, 625–642.

Johnson, R. A., & Wichern, D. W. (2007). *Applied multivariate statistical analysis* (6th ed.). Englewood Cliffs, NJ: Prentice-Hall

Kres, H. (1983). *Statistical tables for multivariate analysis*. New York: Springer.

McKeon, J. J. (1974). F approximations to the distribution of Hotelling's T_0^2. *Biometrika, 61*, 381–383.

Marden, J. I. (2015). *Multivariate statistics: Old school*. http://stat.istics.net/Multivariate

Mardia, K. V., Kent, J. T., & Bibby, J. M. (1979). *Multivariate analysis*. New York: Academic.

Morrison, D. F. (2004). *Multivariate statistical methods* (4th ed.). Pacific Grove, CA: Duxbury Press.

Muirhead, R. J. (1982). *Aspects of multivariate statistical theory*. New York: Wiley.

Okamoto, M. (1973). Distinctness of the eigenvalues of a quadratic form in a multivariate sample. *Annals of Statistics, 1*, 763–765.

Press, S. J. (1982). *Applied multivariate analysis: Using Bayesian and frequentist methods of inference* (2nd ed.). Malabar, FL: R.E. Krieger (Latest reprinting, Dover Press, 2005).

Rao, C. R. (1951). An asymptotic expansion of the distribution of Wilks' criterion. *Bulletin of the International Statistical Institute, 33*, 177–180.

Rencher, A. C., & Christensen, W. F. (2012). *Methods of multivariate analysis* (3rd ed.). New York: Wiley.

Roy, S. N. (1953). On a heuristic method of test construction and its use in multivariate analysis. *Annals of Mathematical Statistics, 24*, 220–238.

Roy, S. N., & Bose, R. C. (1953). Simultaneous confidence interval estimation. *Annals of Mathematical Statistics, 24*, 513–536.

Seber, G. A. F. (1984). *Multivariate observations*. New York: Wiley.

Tarpey, T., Ogden, R. T., Petkova, E., & Christensen, R. (2015). A paradoxical result in estimating regression coefficients. *The American Statistician, 68*, 271–276.

Chapter 10
Multivariate Linear Models: Applications

Abstract This chapter applies the results of Chap. 9 to the one-sample, two-sample, and one-way ANOVA problems. A major tool in MANOVA is profile analysis, which is analogous to performing a split-plot analysis. Profile analysis leads us to the consideration of generalized multivariate linear models (growth curve models, GMANOVA models). Finally, we consider testing for whether a subset of the dependent variables actually provides us with additional information over and above the variables not considered in the subset. In Chap. 12 testing for additional information is seen as an important tool in linear discriminant analysis.

10.1 One-Sample Problems

The multivariate one-sample problem has the same linear structure as the univariate one-sample problem that was explored in *PA* Exercises 2.3 and 3.3. Let $y_1, \ldots, y_n$ be i.i.d. $N(\mu, \Sigma)$, where μ is $q \times 1$ and Σ is $q \times q$. Write $y_i' = (y_{i1}, \ldots, y_{iq})$,

$$Y = \begin{bmatrix} y_1' \\ \vdots \\ y_n' \end{bmatrix},$$

and

$$Y = J\mu' + e,$$

where again J is an $n \times 1$ vector of 1s. The ppo onto $C(J)$ is $\frac{1}{n}J_n^n$ so by the Fundamental Theorem of Least Squares Estimation, least squares estimates satisfy

$$J\hat{\mu}' = \frac{1}{n}J_n^n Y = J\bar{y}'.,$$

where $\bar{y}'. = \frac{1}{n}\sum_{i=1}^n y_i' = (\bar{y}._1, \ldots, \bar{y}._q)$. The sample mean $\bar{y}.$ is also the MLE of μ. The MLE of Σ is

© Springer Nature Switzerland AG 2019
R. Christensen, *Advanced Linear Modeling*, Springer Texts in Statistics,
https://doi.org/10.1007/978-3-030-29164-8_10

389

$$\hat{\Sigma} = \frac{n-1}{n}S = \frac{1}{n}Y'\left(I - \frac{1}{n}J_n^n\right)Y$$

$$= \frac{1}{n}\left[\left(I - \frac{1}{n}J_n^n\right)Y\right]'\left[\left(I - \frac{1}{n}J_n^n\right)Y\right]$$

$$= \frac{1}{n}[(y_1 - \bar{y}.), \dots, (y_n - \bar{y}.)]\begin{bmatrix}(y_1 - \bar{y}.)' \\ \vdots \\ (y_n - \bar{y}.)'\end{bmatrix}$$

$$= \frac{1}{n}\sum_{i=1}^{n}(y_i - \bar{y}.)(y_i - \bar{y}.)'.$$

In particular, writing $\hat{\Sigma} = [\hat{\sigma}_{jk}]$, we have

$$\hat{\sigma}_{jk} = \frac{1}{n}Y_j'\left(I - \frac{1}{n}J_n^n\right)Y_k$$

$$= \frac{1}{n}\left[\left(I - \frac{1}{n}J_n^n\right)Y_j\right]'\left[\left(I - \frac{1}{n}J_n^n\right)Y_k\right]$$

$$= \frac{1}{n}\sum_{i=1}^{n}(y_{ij} - \bar{y}._j)(y_{ik} - \bar{y}._k).$$

The standard unbiased estimate of the covariance is $s_{jk} = \hat{\sigma}_{jk}[n/(n-1)]$.

To test the hypothesis $H_0 : \mu = \mu_0$ we need to recognize that, because the one-sample model is a regression model, μ is estimable, and Λ' is just the scalar 1. The hypothesis and error statistics are

$$H = (\Lambda'\hat{B} - W)'[\Lambda'(X'X)^-\Lambda]^-(\Lambda'\hat{B} - W)$$
$$= (\bar{y}. - \mu_0')'[1(J'J)^{-1}1]^{-1}(\bar{y}. - \mu_0')$$
$$= n(\bar{y}. - \mu_0)(\bar{y}. - \mu_0)'$$

and

$$E = (n-1)S.$$

In the one-sample problem, the Lawley-Hotelling trace is the famous Hotelling T^2 statistic.

$$T^2 = (n-1)\operatorname{tr}[HE^{-1}]$$
$$= \operatorname{tr}[n(\bar{y}. - \mu_0)(\bar{y}. - \mu_0)'S^{-1}]$$
$$= n\operatorname{tr}[(\bar{y}. - \mu_0)'S^{-1}(\bar{y}. - \mu_0)]$$
$$= n(\bar{y}. - \mu_0)'S^{-1}(\bar{y}. - \mu_0).$$

The reason the test statistic simplifies so nicely is because H is a rank one matrix, which also implies that the other three test statistics are equivalent. Under H_0,

$$\frac{T^2}{(n-1)}\frac{n-q}{q} \sim F(q, n-q).$$

This follows from the exact part of McKeon's approximation, see also Seber (1984, Section 2.4). The test is rejected for large values of T^2. There is no need for any unusual tables to perform the test.

Arguments similar to those given in Sect. 9.7 for the development of the prediction region yield a $(1 - \alpha)100\%$ confidence ellipsoid for μ consisting of all μ vectors that satisfy

$$(\mu - \bar{y}.)' S^{-1} (\mu - \bar{y}.) \leq F(1 - \alpha, q, n - q) \frac{q}{n - q} \frac{n - 1}{n}.$$

EXAMPLE 10.1.1. Mosteller and Tukey (1977) consider data from *The Coleman Report* on the relationships between several variables and mean verbal test scores for sixth graders at twenty schools in the New England and Mid-Atlantic regions of the United States. The data are also given in Christensen (2015, Table 6.14). In this example, we consider only two variables, x_2, the percentage of sixth-graders' fathers employed in white-collar jobs and x_5, one-half of the sixth-graders' mothers' mean number of years of schooling. The data are given in Table 10.1.

Table 10.1 Coleman report data

x_2	$\sqrt{x_2}$	x_5	x_2	$\sqrt{x_2}$	x_5
28.87	5.37	6.19	12.20	3.49	5.62
20.10	4.48	5.17	22.55	4.75	5.34
69.05	8.31	7.04	14.30	3.78	5.80
65.40	8.09	7.10	31.79	5.64	6.19
29.59	5.44	6.15	11.60	3.41	5.62
44.82	6.69	6.41	68.47	8.27	6.94
77.37	8.80	6.86	42.64	6.53	6.33
24.67	4.97	5.78	16.70	4.09	6.01
65.01	8.06	6.51	86.27	9.29	7.51
9.99	3.16	5.57	76.73	8.76	6.96

In particular, we will test the null hypothesis that the percentage of white-collar fathers is 50% and that the mothers are on average high-school graduates (i.e., have 12 years of schooling). To perform a formal test of this hypothesis, we need to establish that the data have a multivariate normal distribution. A normal plot (see PA-V Sect. 12.2 or Christensen 2011, Sect. 13.2) of x_2 is given in Fig. 10.1a. It is not very encouraging. It has a very noticeable shoulder at the low end. It also has a gap and a flat spot in the middle. The Shapiro–Francia statistic for the plot is $W' = 0.904$, which has a P value of about 0.05. Figure 10.1b contains a normal plot of $\sqrt{x_2}$. This is much better behaved and has a W' value of 0.933. The variance stabilizing transformation $\text{Arcsin}(\sqrt{x_2/100})$ was also considered, but its behavior was actually worse than that of $\sqrt{x_2}$. The normal plot for x_5 in Fig. 10.1c has a minor shoulder near the top but a W' value of 0.976; we leave x_5 untransformed. Figure 10.2 contains a plot of $\sqrt{x_2}$ versus x_5. This should look elliptical. Except for the smallest value of x_5, it is not too bad.

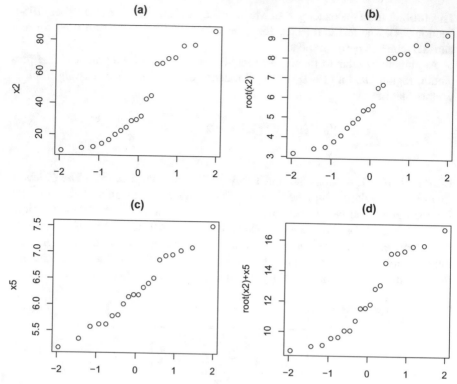

Fig. 10.1 Normal plots of x_2, $\sqrt{x_2}$, x_5, and $\sqrt{x_2} + x_5$

To test the hypothesis, we need the statistics

$$S = \begin{bmatrix} 4.288 & 1.244 \\ 1.244 & 0.428 \end{bmatrix}$$

and

$$\bar{y}'_{.} = (6.069, 6.255).$$

The hypothesized mean is

$$\mu'_0 = (\sqrt{50}, 12/2),$$

and Hotelling's statistic is

$$T^2 = 20(\bar{y}_{.} - \mu_0)'S^{-1}(\bar{y}_{.} - \mu_0) = 93.45.$$

The comparison value for T^2 is 2 so T^2 seems to be highly significant. To perform a formal test based on multivariate normality, compute

$$\frac{T^2}{n-1} \frac{n-q}{q} = \frac{93.45}{19} \frac{18}{2} = 44.27$$

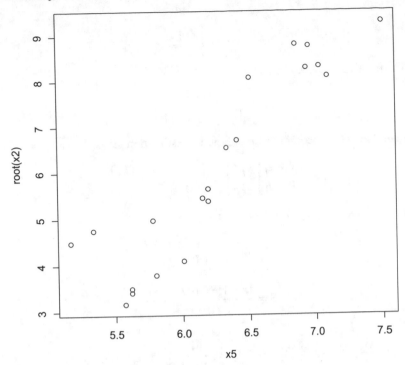

Fig. 10.2 Scatter plot of $\sqrt{x_2}$, x_5

and compare it to an $F(2, 18)$ distribution. Again, the result is highly significant.

We have not yet sufficiently analyzed the question of multivariate normality. Not only must the marginal distributions of a multivariate normal be normal but all linear combinations of the variables must also be normal. Figure 10.1d contains a normal plot of $\sqrt{x_2} + x_5$. While this has a not unrespectable W' value of 0.925, the plot looks horrible. The normal plot for $\sqrt{x_2} - x_5$ is also disturbing. Nevertheless, the null model is so clearly untrue that the lack of multivariate normality is probably not crucial. □

10.2 Two-Sample Problems

PA Exercises 2.2 and 3.2 examine the univariate two-sample problem. The multivariate two-sample problem has a similar linear structure. Let $y_{11}, \ldots, y_{1r}$ be i.i.d. $N_q(\mu_1, \Sigma)$, let $y_{21}, \ldots, y_{2t}$ be i.i.d. $N_q(\mu_2, \Sigma)$, and let the two samples be independent. Write

$$\tilde{Y}_1 = \begin{bmatrix} y_{11}' \\ \vdots \\ y_{1r}' \end{bmatrix}, \quad \tilde{Y}_2 = \begin{bmatrix} y_{21}' \\ \vdots \\ y_{2t}' \end{bmatrix}$$

and

$$Y = \begin{bmatrix} \tilde{Y}_1 \\ \tilde{Y}_2 \end{bmatrix}.$$

The multivariate linear model is

$$Y = \begin{bmatrix} J_r & 0 \\ 0 & J_t \end{bmatrix} \begin{bmatrix} \mu_1' \\ \mu_2' \end{bmatrix} + e.$$

The Fundamental Theorem of Least Squares Estimation gives

$$\begin{bmatrix} J_r \hat{\mu}_1' \\ J_t \hat{\mu}_2' \end{bmatrix} = \begin{bmatrix} J_r & 0 \\ 0 & J_t \end{bmatrix} \begin{bmatrix} \hat{\mu}_1' \\ \hat{\mu}_2' \end{bmatrix} = \begin{bmatrix} \frac{1}{r} J_r^r & 0 \\ 0 & \frac{1}{t} J_t^t \end{bmatrix} \begin{bmatrix} \tilde{Y}_1 \\ \tilde{Y}_2 \end{bmatrix} = \begin{bmatrix} J_r \bar{y}_1' \\ J_t \bar{y}_2' \end{bmatrix},$$

so $\hat{\mu}_1 = \bar{y}_1.$ and $\hat{\mu}_2 = \bar{y}_2..$ It follows that

$$(I - M)Y = \begin{bmatrix} \tilde{Y}_1 - J_r \bar{y}_1' \\ \tilde{Y}_2 - J_t \bar{y}_2' \end{bmatrix},$$

hence

$$S = \frac{(r-1)S_1 + (t-1)S_2}{r+t-2}, \tag{10.2.1}$$

where

$$S_1 = \frac{1}{r-1} [\tilde{Y}_1 - J_r \bar{y}_1']' [\tilde{Y}_1 - J_r \bar{y}_1'] = \frac{1}{r-1} \sum_{j=1}^{r} (y_{1j} - \bar{y}_1.)(y_{1j} - \bar{y}_1.)' \tag{10.2.2}$$

and

$$S_2 = \frac{1}{t-1} [\tilde{Y}_2 - J_t \bar{y}_2']' [\tilde{Y}_2 - J_t \bar{y}_2'] = \frac{1}{t-1} \sum_{j=1}^{t} (y_{2j} - \bar{y}_2.)(y_{2j} - \bar{y}_2.)'. \tag{10.2.3}$$

Exercise 10.1.   Prove formulas (10.2.1)–(10.2.3).

To test $\mu_1 = \mu_2$, write the hypothesis as $H_0 : \mu_1 - \mu_2 = 0$. The null hypothesis is equivalent to assuming a reduced model

$$Y = J_n \mu' + e,$$

where $n = r + t$ and μ is a $q \times 1$ vector. The parametric form of the hypothesis statistic is

$$H = (\bar{y}_1. - \bar{y}_2.) \left[(1, -1) \begin{pmatrix} r & 0 \\ 0 & t \end{pmatrix}^{-1} \begin{pmatrix} 1 \\ -1 \end{pmatrix} \right]^{-1} (\bar{y}_1. - \bar{y}_2.)'$$

$$= \left(\frac{1}{r} + \frac{1}{t} \right)^{-1} (\bar{y}_1. - \bar{y}_2.)(\bar{y}_1. - \bar{y}_2.)'.$$

The error statistic is

$$E = (r + t - 2)S.$$

H is again a rank one matrix, so the four test statistics are equivalent. In particular, the Lawley–Hotelling trace is

$$T^2 = (r+t-2)\mathrm{tr}[HE^{-1}]$$
$$= \left(\frac{1}{r}+\frac{1}{t}\right)^{-1}(\bar{y}_{1\cdot}-\bar{y}_{2\cdot})'S^{-1}(\bar{y}_{1\cdot}-\bar{y}_{2\cdot})$$
$$= \left(\frac{rt}{r+t}\right)(\bar{y}_{1\cdot}-\bar{y}_{2\cdot})'S^{-1}(\bar{y}_{1\cdot}-\bar{y}_{2\cdot}).$$

This is Hotelling's T^2 for the two-sample problem. Under H_0,

$$\frac{T^2}{r+t-2}\frac{r+t-q-1}{q} \sim F(q,r+t-q-1).$$

Arguments similar to those given in Sect. 9.7 for the development of the prediction interval yield a $(1-\alpha)100\%$ confidence ellipsoid consisting of all $\mu_1 - \mu_2$ vectors that satisfy

$$\left[(\mu_1-\mu_2)-(\bar{y}_{1\cdot}-\bar{y}_{2\cdot})\right]'S^{-1}\left[(\mu_1-\mu_2)-(\bar{y}_{1\cdot}-\bar{y}_{2\cdot})\right]$$
$$\leq F(1-\alpha,q,r+t-q-1)\frac{q(r+t-2)}{r+t-q-1}\left(\frac{1}{r}+\frac{1}{t}\right).$$
$$(10.2.4)$$

Exercise 10.2.   Prove that the confidence ellipsoid for $\mu_1 - \mu_2$ given in (10.2.4) is correct.

EXAMPLE 10.2.1.   Lubischew (1962) presents data on four characteristics of male flea-beetles from two species within the genus *Haltica*. The four characteristics are y_1, the distance, in microns, from the posterior border of the prothorax to the transverse groove; y_2, the length, in 0.01 mm units, of the elytra; y_3, the length, in microns, of the second antennal joint; and y_4, the length, in microns, of the third antennal joint. While various plots of the data do not look too bad, there is reason to question the assumptions of multivariate normality and equal covariance matrices for the two groups. For the purpose of this example, we will ignore such problems. Example 12.2.2 includes a discussion of difficulties involved with the assumptions of multivariate normality and equality of covariance matrices using similar data also from Lubischew (1962).

There are $r = 19$ and $t = 20$ observations on the two species. The mean vectors are

$$\bar{y}_{1\cdot} = \begin{bmatrix} 194.47 \\ 267.05 \\ 137.37 \\ 185.95 \end{bmatrix}, \quad \bar{y}_{2\cdot} = \begin{bmatrix} 179.55 \\ 290.80 \\ 157.20 \\ 209.25 \end{bmatrix}.$$

The individual estimated covariance matrices are

$$S_1 = \begin{bmatrix} 187.596 & 176.863 & 48.371 & 113.582 \\ 176.863 & 345.386 & 75.980 & 118.781 \\ 48.371 & 75.980 & 66.357 & 16.243 \\ 113.582 & 118.781 & 16.243 & 239.941 \end{bmatrix}$$

and

$$S_2 = \begin{bmatrix} 101.839 & 128.063 & 36.989 & 32.592 \\ 128.063 & 389.010 & 165.358 & 94.368 \\ 36.989 & 165.358 & 167.537 & 66.526 \\ 32.592 & 94.368 & 66.526 & 177.882 \end{bmatrix}.$$

The pooled estimate of the covariance matrix is

$$S = \begin{bmatrix} 143.559 & 151.803 & 42.527 & 71.993 \\ 151.803 & 367.788 & 121.877 & 106.245 \\ 42.527 & 121.877 & 118.314 & 42.064 \\ 71.993 & 106.245 & 42.064 & 208.073 \end{bmatrix}.$$

To test the hypothesis $H_0: \mu_1 - \mu_2 = 0$, we need the statistics

$$(\bar{y}_{1.} - \bar{y}_{2.})' = (14.92, -23.75, -19.83, -23.30)$$

and

$$S^{-1} = \begin{bmatrix} 0.0132580 & -0.0053492 & 0.0015135 & -0.0021618 \\ -0.0053492 & 0.0066679 & -0.0047338 & -0.0005969 \\ 0.0015135 & -0.0047338 & 0.0130491 & -0.0007445 \\ -0.0021618 & -0.0005969 & -0.0007445 & 0.0060093 \end{bmatrix}.$$

These yield

$$(\bar{y}_{1.} - \bar{y}_{2.})' S^{-1} (\bar{y}_{1.} - \bar{y}_{2.}) = 13.70$$

and Hotelling's test statistic

$$T^2 = \left(\frac{19 \cdot 20}{19 + 20} \right) 13.70 = 133.5.$$

The test statistic is huge, so one can feel reasonably confident that the populations are different in spite of any doubts about the validity of the assumptions. Standardizing T^2 so that it can be compared to an F distribution gives

$$\frac{T^2}{19 + 20 - 2} \frac{19 + 20 - 4 - 1}{4} = 30.66.$$

The corresponding reference distribution is an $F(4, 34)$. ☐

10.3 One-Way Analysis of Variance and Profile Analysis

Consider a multivariate one-way analysis of variance

$$y'_{ij} = \mu'_i + \varepsilon'_{ij}, \tag{10.3.1}$$

where $i = 1,\dots,a$, $j = 1,\dots,N_i$, the ε_{ij}s are independent $N(0,\Sigma)$ random vectors, and $\mu_i = (\mu_{i1},\cdots,\mu_{iq})'$. One can do the standard test of $H_0: \mu_1 = \cdots = \mu_a$ by testing the MANOVA model against the reduced model

$$y'_{ij} = \mu' + \varepsilon'_{ij}. \tag{10.3.2}$$

Write the dependent variable matrix as

$$Y = [y_{ij,h}],$$

where the pair (i, j) denote a row of Y and $h = 1,\dots,q$ denotes a column. Let X be the model matrix for the one-way MANOVA. As in PA Chap. 4, $X = [x_{ij,k}]$, where $x_{ij,k} = \delta_{ik}$, $k = 1,\dots,a$. This is just the standard model matrix for a univariate one-way ANOVA that is parameterized without a grand mean. Let

$$B = \begin{bmatrix} \mu'_1 \\ \vdots \\ \mu'_a \end{bmatrix}$$

and, in conformance with $Y = [y_{ij,h}]$, write a matrix with ε'_{ij} as the ij row, say

$$e = [\varepsilon'_{ij}].$$

The full model (10.3.1) is

$$Y = XB + e,$$

and the reduced model (10.3.2) is

$$Y = J\mu' + e.$$

Because the linear structure of X is just that of a one-way ANOVA, the analysis is similar to that of PA Chap. 4. Consider the estimation of Σ. The error matrix is

$$E = Y'(I - M)Y = \left[Y'_h(I - M)Y_k \right],$$

where, from PA Chap. 4, the elements of this matrix are

$$Y'_h(I - M)Y_k = [(I - M)Y_h]' [(I - M)Y_k]$$
$$= \sum_{i=1}^{a} \sum_{j=1}^{N_i} \left(y_{ij,h} - \bar{y}_{i\cdot,h} \right) \left(y_{ij,k} - \bar{y}_{i\cdot,k} \right).$$

It follows easily that

$$E = Y'(I - M)Y = \sum_{i=1}^{a} \sum_{j=1}^{N_i} (y_{ij} - \bar{y}_{i\cdot})(y_{ij} - \bar{y}_{i\cdot})'.$$

Thinking of the observations as separate samples and using the results of Sect. 10.1, define

$$S_i = \sum_{j=1}^{N_i} (y_{ij} - \bar{y}_{i\cdot})(y_{ij} - \bar{y}_{i\cdot})'/(N_i - 1).$$

From the fact that

$$S = E/(n - a),$$

we can write S as a weighted average of the S_is,

$$S = \sum_{i=1}^{a} \left(\frac{N_i - 1}{n - a} \right) S_i,$$

with $n \equiv N_1 + \cdots + N_a$. This is the usual pooled estimate of Σ.

Estimation of the parameter μ_{ih} is performed exactly as in a one-way ANOVA based on the model

$$Y_h = X\beta_h + e_h,$$

where $\beta_h = (\mu_{1h}, \cdots, \mu_{ah})'$. The estimate is $\hat{\mu}_{ih} = \bar{y}_{i\cdot,h}$ which implies that

$$\hat{\mu}_i = \bar{y}_{i\cdot}$$

for $i = 1, \ldots, a$ and

$$\hat{B} = \begin{bmatrix} \bar{y}_{1\cdot}' \\ \vdots \\ \bar{y}_{a\cdot}' \end{bmatrix}.$$

Finally, the hypothesis matrix for testing the reduced model of no treatment effects, i.e., $H_0: \mu_1 = \cdots = \mu_a$, is

$$H = Y' \left(M - \frac{1}{n} J_n^n \right) Y = \sum_{i=1}^{a} N_i (\bar{y}_{i\cdot} - \bar{y}_{\cdot\cdot})(\bar{y}_{i\cdot} - \bar{y}_{\cdot\cdot})'.$$

EXAMPLE 10.3.1.   *One-Way Analysis of Variance with Repeated Measures.*
A study was conducted to examine the effects of two drugs on heart rates. Thirty women were randomly divided into three groups of ten. An injection was given to each person. Depending on their group, women received either a placebo, drug A, or drug B. Repeated measurements of their heart rates were taken beginning at 2 min after the injection and at 5 min intervals thereafter. Four measurements were taken on each individual. The data are given in Table 10.2.

Clearly, observations taken over time on the same individual are correlated. We can consider the heart rate measurements taken at the four times to be four depen-

Table 10.2 Heart rate data

		DRUG										
		Placebo				A				B			
TIME	1	2	3	4	1	2	3	4	1	2	3	4
SUBJECT												
1	80	77	73	69	81	81	82	82	76	83	85	79
2	64	66	68	71	82	83	80	81	75	81	85	73
3	75	73	73	69	81	77	80	80	75	82	80	77
4	72	70	74	73	84	86	85	85	68	73	72	69
5	74	74	71	67	88	90	88	86	78	87	86	77
6	71	71	72	70	83	82	86	85	81	85	81	74
7	76	78	74	71	85	83	87	86	67	73	75	66
8	73	68	64	64	81	85	86	85	68	73	73	66
9	76	73	74	76	87	89	87	82	68	75	79	69
10	77	78	77	73	77	75	73	77	73	78	80	70

dent variables. This is a completely randomized design, so a multivariate one-way analysis of variance is appropriate. The treatments are the two drugs and the placebo. The multivariate model can be written as

$$y'_{ij} = \mu'_i + \varepsilon'_{ij},$$

where $i = 1, 2, 3$ and $j = 1, 2, \ldots, 10$. Because $q = 4$,

$$y'_{ij} = (y_{ij1}, y_{ij2}, y_{ij3}, y_{ij4})$$

and

$$\mu'_i = (\mu_{i1}, \mu_{i2}, \mu_{i3}, \mu_{i4}) .$$

The least squares estimate of B is

$$\hat{B} = \begin{bmatrix} \hat{\mu}'_1 \\ \hat{\mu}'_2 \\ \hat{\mu}'_3 \end{bmatrix} = \begin{bmatrix} \bar{y}'_1. \\ \bar{y}'_2. \\ \bar{y}'_3. \end{bmatrix} = \begin{bmatrix} \bar{y}_{1\cdot1} & \bar{y}_{1\cdot2} & \bar{y}_{1\cdot3} & \bar{y}_{1\cdot4} \\ \bar{y}_{2\cdot1} & \bar{y}_{2\cdot2} & \bar{y}_{2\cdot3} & \bar{y}_{2\cdot4} \\ \bar{y}_{3\cdot1} & \bar{y}_{3\cdot2} & \bar{y}_{3\cdot3} & \bar{y}_{3\cdot4} \end{bmatrix}$$

$$= \begin{bmatrix} 73.8 & 72.8 & 72.0 & 70.3 \\ 82.9 & 83.1 & 83.4 & 82.9 \\ 72.9 & 79.0 & 79.6 & 72.0 \end{bmatrix} .$$

The unbiased estimates of the covariance matrix computed within each treatment group are

$$S_1 = \begin{bmatrix} 18.62 & 15.07 & 8.22 & 0.73 \\ 15.07 & 17.07 & 11.11 & 3.07 \\ 8.22 & 11.11 & 13.33 & 8.89 \\ 0.73 & 3.07 & 8.89 & 11.34 \end{bmatrix},$$

$$S_2 = \begin{bmatrix} 10.54 & 13.57 & 12.93 & 6.77 \\ 13.57 & 22.54 & 18.62 & 10.12 \\ 12.93 & 18.62 & 21.82 & 12.82 \\ 6.77 & 10.12 & 12.82 & 8.99 \end{bmatrix},$$

and

$$S_3 = \begin{bmatrix} 24.10 & 25.11 & 19.18 & 18.89 \\ 25.11 & 28.22 & 23.11 & 22.33 \\ 19.18 & 23.11 & 24.93 & 19.00 \\ 18.89 & 22.33 & 19.00 & 22.00 \end{bmatrix}.$$

In general, a weighted average of these gives S. Because the sample sizes for the drugs are all equal, the weights are equal and a simple average gives

$$S = \begin{bmatrix} 17.76 & 17.91 & 13.44 & 8.80 \\ 17.91 & 22.61 & 17.61 & 11.84 \\ 13.44 & 17.61 & 20.03 & 13.57 \\ 8.80 & 11.84 & 13.57 & 14.11 \end{bmatrix}.$$

The error matrix is

$$E = (30-3)S = \begin{bmatrix} 479.4 & 483.7 & 363.0 & 237.5 \\ 483.7 & 610.5 & 475.6 & 319.7 \\ 363.0 & 475.6 & 540.8 & 366.4 \\ 237.5 & 319.7 & 366.4 & 381.0 \end{bmatrix}.$$

The correlation matrix is

$$R = \begin{bmatrix} 1.000 & 0.894 & 0.713 & 0.556 \\ 0.894 & 1.000 & 0.828 & 0.663 \\ 0.713 & 0.828 & 1.000 & 0.807 \\ 0.556 & 0.663 & 0.807 & 1.000 \end{bmatrix}.$$

It consists of the correlations between all the variables, e.g.,

$$r_{12} = 0.894 = \frac{17.91}{\sqrt{17.76}\sqrt{22.61}}.$$

The reason for treating these data as a multivariate one-way ANOVA was our initial claim that the observations made on an individual are correlated. This certainly seems to be borne out by the large off-diagonal elements of the correlation matrix. In fact, for normal data, we could test whether the correlations are zero. In PA Sect. 6.5, a t test for partial correlations was presented. While these are not partial correlations in the usual sense, the correlations are pooled over three groups so these correlations represent a special case of partial correlations. A test of H_0: $\rho_{34} = 0$ can be based on comparing $\sqrt{30-4}(0.807)/\sqrt{1-0.807^2} = 6.97$ to a $t(30-4)$ distribution. The correlation is highly significant.

Of course, there are methods available for analyzing correlated data other than the multivariate linear model. One alternative is to consider a split plot model, where individual women are whole plots, drugs are whole plot treatments, and the four times are subplot treatments. (There are no blocks in the whole plots.) As discussed in *PA* Chap. 11, the split plot model assumes that observations in different whole plots are uncorrelated while all observations in the same whole plot have a common positive correlation. In the multivariate one-way, the covariance matrix for observations on the same person is denoted by Σ. The split plot model assumes that

$$\Sigma = \sigma^2 \begin{bmatrix} 1 & \rho & \cdots & \rho \\ \rho & 1 & \cdots & \rho \\ \vdots & \vdots & \ddots & \vdots \\ \rho & \rho & \cdots & 1 \end{bmatrix}. \tag{10.3.3}$$

If this assumption is correct, the split plot model is more appropriate than the multivariate one-way ANOVA. Although the correlation matrix R does not seem very supportive of the split plot model assumption, a detailed comparison of the two analyses will be made in Example 10.3.3.

Returning to the multivariate one-way analysis of these data, we might wish to test for differences in the treatment means. If we fit the reduced model

$$y'_{ij} = \mu' + e'_{ij},$$

we obtain

$$S_0 = \begin{bmatrix} 37.64 & 31.52 & 28.02 & 33.72 \\ 31.52 & 39.60 & 37.10 & 32.29 \\ 28.02 & 37.10 & 41.89 & 35.39 \\ 33.72 & 32.29 & 35.39 & 45.37 \end{bmatrix}$$

and

$$E_0 = (30-1)S_0 = \begin{bmatrix} 1091.5 & 914.2 & 812.7 & 977.9 \\ 914.2 & 1148.3 & 1076.0 & 936.4 \\ 812.7 & 1075.9 & 1214.7 & 1026.3 \\ 977.9 & 936.4 & 1026.3 & 1315.9 \end{bmatrix}.$$

The hypothesis matrix can be computed as

$$H = Y'(M - M_0)Y = Y'(I - M_0)Y - Y'(I - M)Y$$
$$= E_0 - E$$
$$= \begin{bmatrix} 612.1 & 430.5 & 449.7 & 740.4 \\ 430.5 & 537.8 & 600.4 & 616.7 \\ 449.7 & 600.4 & 673.9 & 659.9 \\ 740.4 & 616.7 & 659.9 & 934.9 \end{bmatrix}.$$

If there are no differences in the drug means, then H divided by its degrees of freedom should estimate Σ. Computing this gives

$$\frac{1}{2}H = \begin{bmatrix} 306 & 215 & 225 & 370 \\ 215 & 269 & 300 & 308 \\ 225 & 300 & 337 & 330 \\ 370 & 308 & 330 & 467 \end{bmatrix}.$$

Even though this estimate has only two degrees of freedom, it is clear that this is not estimating the same thing that S is estimating.

If the data have a multivariate normal distribution, formal tests can be performed. The various test statistics, comparison values, and $\alpha = 0.01$ normal theory critical values are as follows.

Statistic	Observed value	Comparison value	Critical value
U	0.0628	0.75	0.440
ϕ_{max}	5.52	0.07	Less than 1.19
T^2	188.0	8	26.66
V	1.44	0.276	Less than 0.725

In particular,

$$T^2 = 187.99,$$

which is much larger than the intuitive comparison value $q[r(X) - r(X_0)] = 4(2) = 8$. Comparing T^2 to the exact small sample distribution, McKeon's approximate distribution, or the asymptotic χ^2 distribution with $q[r(X) - r(X_0)] = 8$ degrees of freedom leads to the clear conclusion that the drugs have different multivariate means.

The validity of multivariate linear model tests depends on the data having a multivariate normal distribution. In particular, each dependent variable must be normal and any linear combinations of the variables must also be normal. As in *PA-V* Sect. 12.2 (Christensen 2011, Section 13.2), the normality of data can be evaluated using normal plots. If the sample size for each drug were large, it would be appropriate to check for normality within the treatment groups. Since it is difficult to draw distributional conclusions using only ten observations, the residual matrix $\hat{e} = (I - M)Y$ was used. In general, the standardized residuals are more appropriate for normal plots, but for this model all cases have the same leverage, so plotting the residuals would be equivalent.

Normal plots of the standardized residuals were made for each dependent variable (Fig. 10.3) and also for one linear combination of the residuals (Fig. 10.4). The linear combination was the residuals for the sum of the four variables. All of the normal plots looked reasonably linear. The Shapiro–Francia statistic W' was also computed for each plot (see *PA-V* Sect. 12.2 or Christensen 2011, Sect. 13.2). The results are as follows.

Variable	W'
Time 1	0.986
Time 2	0.974
Time 3	0.949
Time 4	0.980
Sum of variables	0.960

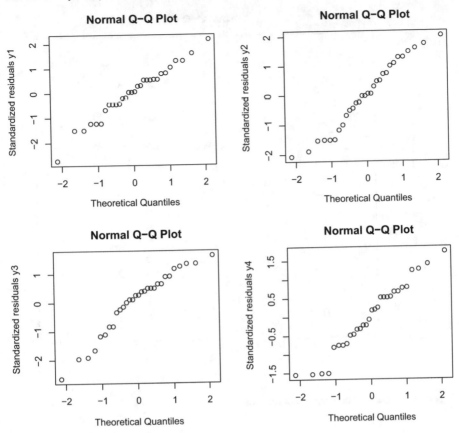

Fig. 10.3 Heart rate data: Normal plots

Comparing these values to those tabled in Christensen (2015) gives no cause for concern. Of course, to do a proper check for multivariate normality, one should inspect the normal plot for every linear combination of the columns of the residual matrix. Unfortunately, that can be rather time-consuming. The data analyst must usually be satisfied with evaluating some finite number of linear combinations.

Another assumption of the multivariate linear model is that the covariance matrix is the same for every observation vector y_i. As in *PA-V* Sect. 12.4, (Christensen 2011, Section 13.4), residual plots can be performed, checking for constant variance in each of the dependent variables. Residual plots for linear combinations of the columns of the residual matrix should also display constant variance. Figure 10.5 plots the standardized residuals from each variable against their predicted values. Nothing too untoward appears.

In a one-way ANOVA, the assumption of a constant covariance matrix can also be checked by comparing the estimated covariance matrices for each of the

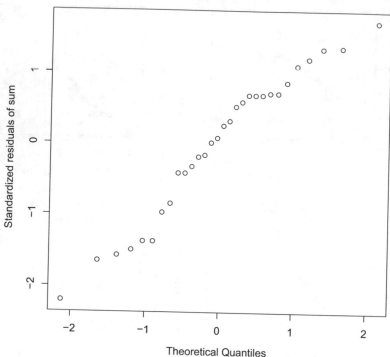

Fig. 10.4 Heart rate data: normal plot for sum

individual groups. The sample covariance matrices S_1, S_2, and S_3 were given earlier. Although they display some differences that seem to be fairly substantial (the estimated covariances between Time 1 and Time 4 vary from 0.73 to 18.89), taken as a whole, the covariance matrices are reasonably consistent. In fact, Bartlett's modification to the likelihood ratio test for equality of the covariance matrices gives a P value greater than 0.05 (based on a $\chi^2(20)$ approximation). This in spite of the fact that the test is so notoriously sensitive to nonnormality that it is rarely used.

A visual approach to evaluating the individual covariance matrices can be based on plotting pairs of variables. The 10 pairs of observations at times 1 and 2 can be plotted for each of the three drugs. Each of these three plots should be roughly elliptical and the ellipses should have the same orientation in two-dimensional space. A similar set of $a = 3$ plots can be made for each of the $\binom{q}{2} = 6$ pairs of dependent variables. Given the difficulties of evaluating plots based on only ten observations, there seems to be no reason to doubt the assumption of equal covariance matrices for this example. In particular, the orientations of the three plots in each set are consistent.   □

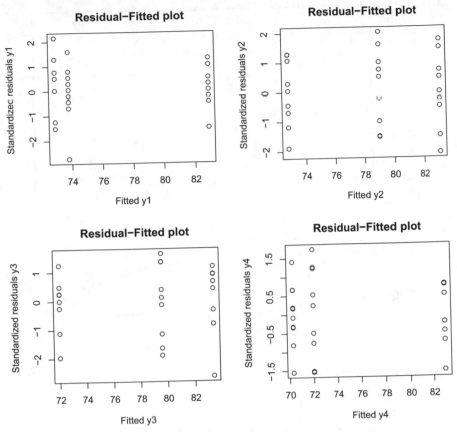

Fig. 10.5 Heart rate data: standardized residual-fitted values plots

10.3.1 Profile Analysis

Profile analysis seeks to examine possible similarities between the μ_i vectors. The name derives from the plots of the a theoretical curves defined by (h, μ_{ih}), $h = 1, \ldots, q$. These curves are referred to as profiles. Because the μ_is are not available, the estimated profiles $(h, \hat{\mu}_{ih})$ provide a valuable visual display. (The q points available for each i are connected by line segments to obtain the a different profiles.)

EXAMPLE 10.3.2. Figure 10.6 contains a plot of the estimated profiles from Example 10.3.1. □

Three questions are commonly asked in profile analysis. First, whether the curves are parallel. Second, whether the curves have the same average level. The average level for each curve is defined as the average over different dependent variables. The third question is whether the average curve is horizontal. The average curve is

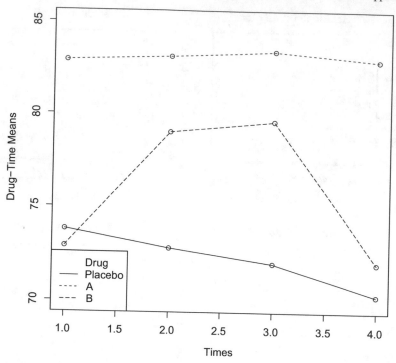

Fig. 10.6 Heart rate profiles

obtained by averaging over groups. Note that the hypotheses involving averages are of particular interest when the curves are parallel. In this case, if the average levels are the same, the profiles are the same, and if the average profile is horizontal, all profiles are horizontal.

To test these questions, it is convenient to define, for a general value r, the $(r-1) \times r$ matrix

$$\Lambda'_r = [J_{r-1}, -I_{r-1}].$$

Note that

$$\Lambda'_a B = \begin{bmatrix} \mu'_1 - \mu'_2 \\ \mu'_1 - \mu'_3 \\ \vdots \\ \mu'_1 - \mu'_a \end{bmatrix},$$

so $H_0 : \Lambda'_a B = 0$ is precisely $H_0 : \mu_1 = \mu_2 = \cdots = \mu_a$.

The hypothesis that the profiles are parallel is that for every possible choice of i and i'

$$\mu_{i1} - \mu_{i'1} = \mu_{i2} - \mu_{i'2} = \cdots = \mu_{iq} - \mu_{i'q}.$$

Equivalently, the hypothesis is that

$$\mu_{11} - \mu_{i1} = \cdots = \mu_{1q} - \mu_{iq} \quad \text{for} \quad i = 2, \ldots, a,$$

that is, the other profiles are all parallel to the group 1 profile. Finally, the hypothesis can be written $(\mu_{11} - \mu_{i1}) - (\mu_{12} - \mu_{i2}) = 0$, $(\mu_{11} - \mu_{i1}) - (\mu_{13} - \mu_{i3}) = 0$, $\ldots$, $(\mu_{11} - \mu_{i1}) - (\mu_{1q} - \mu_{iq}) = 0$ for $i = 2, \ldots, a$. With $\Lambda_a' B$ illustrated earlier, it is not difficult to see that the profiles are parallel if and only if $\Lambda_a' B \Lambda_q = 0$. The test of parallel profiles is $H_0 : \Lambda_a' B \Lambda_q = 0$. This is a standard multivariate hypothesis and yields the hypothesis statistic

$$H_* = (\Lambda_a' \hat{B} \Lambda_q)' [\Lambda_a' (X'X)^{-1} \Lambda_a]^{-1} (\Lambda_a' \hat{B} \Lambda_q).$$

For our one-way ANOVA model matrix, $X'X$ is $\text{Diag}(N_1, \ldots, N_a)$ so it is easily inverted. The hypothesis statistic is compared to

$$E_* = \Lambda_q' E \Lambda_q.$$

To test whether the average levels of the curves are the same, we need the average level for each curve. For the ith curve, the average level is $\bar{\mu}_{i\cdot} = \frac{1}{q} \sum_{h=1}^{q} \mu_{ih}$. The hypothesis $\bar{\mu}_{1\cdot} = \cdots = \bar{\mu}_{a\cdot}$ can be written as $\bar{\mu}_{1\cdot} - \bar{\mu}_{2\cdot} = 0$, $\bar{\mu}_{1\cdot} - \bar{\mu}_{3\cdot} = 0$, $\ldots$, $\bar{\mu}_{1\cdot} - \bar{\mu}_{a\cdot} = 0$. Equivalently, we can look for equality of the curve totals, where the ith curve total is $\mu_{i\cdot} = \mu_i' J_q = \sum_{h=1}^{q} \mu_{ih}$. Recalling the form of $\Lambda_a' B$, it is easily seen that the null hypothesis is $H_0 : \Lambda_a' B J_q = 0$ and

$$H_* = (\Lambda_a' \hat{B} J_q)' [\Lambda_a' (X'X)^{-1} \Lambda_a]^{-1} (\Lambda_a' \hat{B} J_q).$$

The error statistic is

$$E_* = J_q' E J_q.$$

The third hypothesis is that the average curve is horizontal. The average curve is based on $\bar{\mu}' = \frac{1}{a} J_a' B = \frac{1}{a} \sum_{i=1}^{a} \mu_i'$. Write $\bar{\mu}' = (\bar{\mu}_{\cdot 1}, \ldots, \bar{\mu}_{\cdot q})$. Testing whether the curve is horizontal amounts to testing that $\bar{\mu}_{\cdot 1} - \bar{\mu}_{\cdot 2} = 0$, $\ldots$, $\bar{\mu}_{\cdot 1} - \bar{\mu}_{\cdot q} = 0$. Equivalently, we can test whether the total curve is horizontal by examining $\mu' = J_a' B$. Clearly, the test that the total curve is horizontal is $H_0 : J_a' B \Lambda_q = 0$ with

$$H_* = (J_a' \hat{B} \Lambda_q)' [J_a' (X'X)^{-1} J_a]^{-1} (J_a' \hat{B} \Lambda_q)$$

and

$$E_* = \Lambda_q' E \Lambda_q.$$

Finally, another word about combining the hypotheses. The curves are both parallel and have the same average level if and only if the curves are identical (i.e., $\mu_1 = \cdots = \mu_a$). In other notation, $\Lambda_a' B \Lambda_q = 0$ and $\Lambda_a' B J_q = 0$ if and only if $\Lambda_a' B = 0$. Also, the curves are parallel and the average curve is horizontal if and only if all of the curves are horizontal. Putting it another way, $\Lambda_a' B \Lambda_q = 0$ and $J_a' B \Lambda_q = 0$ if and only if $B \Lambda_q = 0$. The hypothesis that $B \Lambda_q = 0$ is simply that all of the curves are horizontal.

10.3.2 Comparison with Split Plot Analysis

EXAMPLE 10.3.3.   We again consider the heart rate data of Example 10.3.1. It was mentioned earlier that using a split plot model is an alternative method of analyzing these data. The nature of profile analysis will be clearer if we contrast profile analysis with the more familiar split plot analysis. We begin with the split plot analysis; see the theoretical discussion in *PA* Chap. 11, the balanced examples in Christensen (1996, Chapter 12), and the unbalanced examples in Christensen (2015, Chapter 19).

The split plot model for this experimental design is

$$y_{ijk} = \mu + \delta_i + \eta_{ij} + \tau_k + (\delta\tau)_{ik} + e_{ijk},$$

$i = 1, 2, 3;\ j = 1, \ldots, 10;\ k = 1, 2, 3, 4$. Here δ_i indicates the drug effect, τ_k indicates the time effect, η_{ij} indicates a random error for the ij individual, and e_{ijk} is a random error specific to the observation on the ij individual at the kth time period. As usual, we assume $\mathrm{Var}(\eta_{ij}) = \sigma_w^2$, $\mathrm{Var}(e_{ijk}) = \sigma_s^2$, and that all of the random errors have zero covariance with all of the other random errors. For the construction of tests and confidence regions, it is assumed that the errors have a joint multivariate normal distribution. To construct the ANOVA table for a balanced split plot model, it is often convenient to compute the sums of squares treating the data as a complete factorial experiment and then combine terms to get the correct table. Considering the data as a complete factorial, there are three treatments, Drugs (D), Times (T), and Individuals (I). The corresponding ANOVA table is

Source	df	SS	MS
Drugs	2	2438.5	1219.2
Individuals	9	404.3	44.93
$D \times I$	18	1221.5	67.86
Times	3	222.3	74.10
$T \times D$	6	320.1	53.36
$T \times I$	27	64.0	2.37
$T \times D \times I$	54	321.9	5.96

Since individuals do not constitute whole plot blocks, the whole plot error is found by pooling the Individuals and the $D \times I$ terms. The subplot error contains the $T \times I$ and $T \times D \times I$ terms. The correct split plot ANOVA table is

Source	df	SS	MS	F
Drugs	2	2438.5	1219.2	20.25
Whole plot error	27	1625.8	60.22	
Times	3	222.3	74.10	15.56
$T \times D$	6	320.1	53.36	11.20
Subplot error	81	385.9	4.76	

All of the effects are highly significant. The next step in the split plot analysis might be to examine contrasts in the interactions. This involves looking at contrasts in the 3×4 table of means $\bar{y}_{i \cdot k}$.

The three basic tests in profile analysis are analogous to the tests for whole plot treatments, subplot treatments, and interaction. The test for parallelism is equivalent to testing for interaction. The test of whether the average levels of the curves are the same is equivalent to testing for whole plot treatments. The test of whether the average curve is horizontal is equivalent to testing for subplot treatments. Profile analysis is based on examining the structure of

$$\hat{B} = \begin{bmatrix} 73.8 & 72.8 & 72.0 & 70.3 \\ 82.9 & 83.1 & 83.4 & 82.9 \\ 72.9 & 79.0 & 79.6 & 72.0 \end{bmatrix}.$$

This is the 3×4 table of means, $\bar{y}_{i \cdot k}$. The split plot analysis examines the structure of exactly the same means table. While not all interaction contrasts have obvious multivariate tests, any interaction contrast that can be written as $\lambda' B \xi$ can also be tested in the multivariate model.

Depending on the software one has available, to test the hypothesis of parallel profiles $H_0: \Lambda'_3 B \Lambda_4 = 0$, it may be convenient to do a one-way ANOVA on a new set of variables $(y^*_{ij1}, y^*_{ij2}, y^*_{ij3}) = (y_{ij1} - y_{ij2}, y_{ij1} - y_{ij3}, y_{ij1} - y_{ij4}) = y'_{ij} \Lambda_4$. The one-way MANOVA on the transformed data yields

$$S_* = \Lambda'_4 S \Lambda_4 = \begin{bmatrix} 4.54 & 4.01 & 2.89 \\ 4.01 & 10.90 & 9.09 \\ 2.89 & 9.09 & 14.27 \end{bmatrix}$$

and

$$E_* = \Lambda'_4 E \Lambda_4 = \begin{bmatrix} 122.5 & 108.3 & 77.9 \\ 108.3 & 294.2 & 245.3 \\ 77.9 & 245.3 & 385.4 \end{bmatrix}. \tag{10.3.4}$$

Fitting the model of no drug effects yields.

$$S_{0*} = \Lambda'_4 S_0 \Lambda_4 = \begin{bmatrix} 14.19 & 15.19 & 4.68 \\ 15.19 & 23.48 & 11.28 \\ 4.68 & 11.28 & 15.57 \end{bmatrix}$$

and

$$E_{0*} = \Lambda'_4 E_0 \Lambda_4 = \begin{bmatrix} 411.4 & 440.6 & 135.7 \\ 440.6 & 680.8 & 327.2 \\ 135.7 & 327.2 & 451.5 \end{bmatrix}.$$

The hypothesis statistic is

$$H_* = E_{0*} - E_* = \begin{bmatrix} 288.9 & 332.3 & 57.8 \\ 332.3 & 386.6 & 81.9 \\ 57.8 & 81.9 & 66.1 \end{bmatrix}.$$

Dividing by the degrees of freedom gives

$$\frac{1}{2}H_* = \begin{bmatrix} 144.5 & 166.1 & 28.8 \\ 166.1 & 193.3 & 40.8 \\ 28.8 & 40.8 & 33.1 \end{bmatrix}.$$

It is clear that $\frac{1}{2}H_*$ and S_* are not estimating the same thing. The test statistics, comparison values, and normal theory critical values for an $\alpha = 0.05$ test follow.

Statistic	Observed value	Comparison value	Critical value
U	0.204	0.807	0.510
ϕ_{max}	3.22	0.07	less than 0.886
T^2	91.4	6	22.08
V	0.902	0.207	less than 0.569

All of the values are far from their comparison values and are in the critical regions of the tests. The hypothesis that the profiles are parallel is rejected. In other words, the relationship between the time means depends on which drug you look at (i.e., there is interaction).

To explore further the lack of parallelism, consider the orthogonal contrasts in drugs determined by $\lambda_1' = (1, -1, 0)$ and $\lambda_2' = (-1, -1, 2)$. The test of $H_0: \lambda_1' B \Lambda_4 = 0$ examines whether the placebo curve is parallel to the drug A curve. $H_0: \lambda_2' B \Lambda_4 = 0$ hypothesizes that the drug B curve is parallel to the average of the others. Define the hypothesis matrices

$$H_i = (\lambda_i' \hat{B} \Lambda_4)' \left[\lambda_i' (X'X)^{-1} \lambda_i \right]^{-1} (\lambda_i' \hat{B} \Lambda_4).$$

Thus,

$$H_1 = \begin{bmatrix} 1.2 \\ 2.3 \\ 3.5 \end{bmatrix} (5)[1.2, 2.3, 3.5]$$

$$= \begin{bmatrix} 7.20 & 13.80 & 21.00 \\ 13.80 & 26.42 & 40.25 \\ 21.00 & 40.25 & 61.25 \end{bmatrix}$$

and

$$H_2 = \begin{bmatrix} -13 \\ -14.7 \\ -1.7 \end{bmatrix} (5/3)[-13, -14.7, -1.7]$$

$$= \begin{bmatrix} 281.67 & 318.50 & 36.83 \\ 318.50 & 360.15 & 41.65 \\ 36.83 & 41.65 & 4.82 \end{bmatrix}.$$

Because the contrasts λ_i are orthogonal

$$H_* = H_1 + H_2.$$

The test statistics, comparison values, and $\alpha = 0.01$ critical values follow.

Statistic	$\lambda_1' B \Lambda_4 = 0$	$\lambda_2' B \Lambda_4 = 0$	Comp. value	Crit. value
U	0.853	0.237	0.900	0.617
ϕ_{max}	0.17	3.21	0.07	0.56
T^2	4.65	86.75	3.00	15.16
V	0.147	0.763	0.103	0.359

Clearly, the predominant cause of nonparallelism is due to the fact that drug B is not parallel to the placebo and drug A. The placebo and drug A appear to be reasonably parallel.

A problem with doing formal tests for the orthogonal contrasts considered earlier is that they were chosen after looking at the sample profiles in Fig. 10.6. This invalidates the distributions used. Since $H_* = H_1 + H_2$, we have

$$T_*^2 = T_1^2 + T_2^2,$$

where $T_*^2 = \text{tr}(H_* S_*^{-1})$, $T_1^2 = \text{tr}(H_1 S_*^{-1})$, and $T_2^2 = \text{tr}(H_2 S_*^{-1})$. Moreover,

$$T_i^2 \le T_*^2.$$

If we use the Scheffé idea and reject H_0: $\lambda_i' B \Lambda_4 = 0$ only when T_i^2 is greater than the critical value appropriate for an α-level test based on T_*^2, then the experimentwise error rate is no greater than α (and probably much less). Since $T_2^2 = 86.75$ for the second contrast and the critical value appropriate for T_*^2 is 22.08, the profile for drug B and the average profile for the placebo and drug A display a significant lack of parallelism, regardless of the fact that the contrast was chosen after examining the data.

Although not every contrast in the drug-time interaction can be written as $\lambda' B \xi$ for some drug contrast vector λ and time contrast vector ξ, these are often the most interpretable contrasts. Clearly, any such contrast can be tested in the multivariate model. As just illustrated, orthogonal contrasts in the drugs are a useful tool for balanced data, especially in relation to T^2. Because of the correlation between times, orthogonal contrasts are of little interest relative to the times. For example, it is natural in the split plot model to examine orthogonal polynomial contrasts in the times. Unfortunately, the relationship between orthogonal polynomial contrasts and polynomial regression depends on the validity of the least squares analysis. Using the standard polynomial contrasts is not particularly appropriate for the multivariate model. Fitting a polynomial in the times requires the use of a growth curve model (see Sect. 10.4).

In univariate ANOVA, when there is interaction, the tests of main effects are difficult to interpret. This is precisely because they involve averaging over the interactions. In profile analysis, when the profiles are not parallel, the hypotheses that involve averages are also difficult to interpret. For example, if the curves are not parallel, testing whether the average curve is horizontal does not seem too interesting. If this null hypothesis were true, it would tell us very little. On the other hand, if the profiles are parallel and the average profile is horizontal, then every profile must be horizontal. *Because there is clear evidence that the profiles are not parallel, the other two standard tests in profile analysis are of little interest.* In spite of this fact, we now illustrate their computation and interpretation.

The test for equality of the average levels, H_0: $\Lambda_3' BJ_4 = 0$, can be performed by doing a univariate one-way ANOVA on the dependent variable constructed by adding together the four time variables. The test corresponds to testing for drug main effects in the split plot model. The ANOVA table is as follows.

Source	df	SS	MS	F
Drugs	2	9754	4877	20.25
Error	27	6503	241	
Total	29	16,257		

The F test is highly significant, thus indicating that the drugs affect the average level of the curves. In other words, averaging over times, there are differences in the drugs. Note that this is precisely the split plot model test for Drug main effects. The sums of squares for Drugs and Error are exactly four times those reported earlier for the split plot model. This is a result of the fact that $q = 4$.

The test of H_0: $J_3' B\Lambda_4 = 0$ is a test of whether the average curve is horizontal. It tests for differences in Times averaging over Drugs. Using the dependent variables formed to test for parallelism, we get

$$\hat{B}\Lambda_4 = \begin{bmatrix} 1.0 & 1.8 & 3.5 \\ -0.2 & -0.5 & 0.0 \\ -6.1 & -6.7 & 0.9 \end{bmatrix}$$

and

$$J_3' \hat{B}\Lambda_4 = (-5.3, -5.4, 4.4) .$$

From standard ANOVA theory,

$$(X'X) = \begin{bmatrix} 10 & 0 & 0 \\ 0 & 10 & 0 \\ 0 & 0 & 10 \end{bmatrix}$$

and

$$\left[J_3'(X'X)^{-1}J_3 \right]^{-1} = \frac{10}{3} .$$

It follows that

$$H_* = \frac{10}{3} \begin{bmatrix} -5.3 \\ -5.4 \\ 4.4 \end{bmatrix} [-5.3, -5.4, 4.4]$$

$$= \begin{bmatrix} 93.6 & 95.4 & -77.7 \\ 95.4 & 97.2 & -79.2 \\ -77.7 & -79.2 & 64.5 \end{bmatrix} .$$

Clearly, this matrix has rank one, hence one degree of freedom. The error matrix E_* is the same as used to test parallelism. It is given in Eq. (10.3.4). An ad hoc evaluation of the null hypothesis compares $H_*/1$ with S_*. The matrices are vastly different. Formal tests confirm the conclusion that the average curve is not horizontal. In particular,

$$T^2 = 56.3893, \qquad F = 17.4, \qquad F(0.999, 3, 25) = 7.451. \qquad \Box$$

It is interesting to examine the relationship between the split plot model test for Time main effects and the multivariate test for average time effects. With one degree of freedom for H_*, the standard multivariate test statistics are equivalent. In particular,

$$T^2 = [n - r(X)] \operatorname{tr} \left[E_*^{-1} H_* \right]$$
$$= \operatorname{tr} \left[(\Lambda_4' S \Lambda_4)^{-1} H_* \right].$$

If we require that S have the structure of a split plot covariance matrix, write

$$S = \hat{\sigma}^2 \begin{bmatrix} 1 & \hat{\rho} & \hat{\rho} & \hat{\rho} \\ \hat{\rho} & 1 & \hat{\rho} & \hat{\rho} \\ \hat{\rho} & \hat{\rho} & 1 & \hat{\rho} \\ \hat{\rho} & \hat{\rho} & \hat{\rho} & 1 \end{bmatrix}$$
$$= \hat{\sigma}^2 (1 - \hat{\rho}) I_4 + \hat{\sigma}^2 \hat{\rho} J_4^4,$$

then it is easily seen that

$$\Lambda_4' S \Lambda_4 = \hat{\sigma}^2 (1 - \hat{\rho}) \Lambda_4' \Lambda_4$$

so

$$T^2 = \operatorname{tr} \left[(\Lambda_4' \Lambda_4)^{-1} H_* \right] / \left[\hat{\sigma}^2 (1 - \hat{\rho}) \right].$$

Recall from *PA* Chap. 11 that MS(Subplot Error) is the unbiased estimate of $\sigma^2 (1 - \rho)$. Also note that $\operatorname{tr} \left[(\Lambda_4' \Lambda_4)^{-1} H_* \right] = \text{SS(Times)}$. Thus, restricting the form of Σ leads naturally to

$$T^2 = \text{SS(Times)/MS(Subplot Error)}.$$

The split plot F statistic for testing main effects in Times is just $T^2/3$. However, by imposing additional structure on Σ, one gains degrees of freedom for the denominator of the F statistic, so one ends up comparing $F = 15.5$ to an $F(3, 81)$ distribution rather than the multivariate linear model comparison of $F = 17.4$ to an $F(3, 25)$ distribution.

A similar argument relates the T^2 for parallelism to the F statistic for interaction.

10.3.3 Computations

There are a number of ways that one can approach the computational problems involved in one-way MANOVA and profile analysis. The three main computational approaches are through use of flexible interactive statistics packages, structured programs, and matrix manipulation programs. R and MINITAB are both flexible interactive statistical packages and matrix manipulation programs. SAS is a structured program with matrix manipulation. Code for the MANOVA and split plot analyses in Christensen (2015) is on my website for that book.

The primary computation involved is finding covariance matrices: one for each treatment, a pooled covariance matrix, and one ignoring treatments. These can be found using a flexible interactive package such as MINITAB that includes some matrix commands. More structured programs require less thought but more reading. For example, BMDP 4V automatically provides the univariate split plot analysis, the multivariate profile analysis, and compromise tests based on the Greenhouse and Geisser (1959) and Huynh and Feldt (1976) adjustments to the degrees of freedom of the univariate tests. The additional reading is needed to identify such things as the definition of Λ used in profile analysis. (The definition given earlier was just one of an infinite number of equally good choices. While the choice will not affect the standard test statistics, the matrices H_* and E_* do depend on Λ.) Other things that need to be identified in structured programs are the exact definitions of the test statistics. For example, in BMDP 4V, the TRACE statistic is $\mathrm{tr}[HE^{-1}]$ rather than T^2 or V and MXROOT is θ_{max} rather than ϕ_{max}. Of course, with a good matrix-manipulation package, such as MATLAB or R, you can do any of these computations directly.

10.3.4 Covariance Matrix Modeling

The multivariate model makes no assumptions about the form of Σ. The split plot model assumes Σ has the form of Eq. (10.3.3). Increasingly, other forms are being used. For example, with the heart rate data, one might use

$$
\Sigma = \sigma^2 \begin{bmatrix} 1 & \phi & \phi^2 & \phi^3 \\ \phi & 1 & \phi & \phi^2 \\ \phi^2 & \phi & 1 & \phi \\ \phi^3 & \phi^2 & \phi & 1 \end{bmatrix}.
$$

This model would be appropriate if, for each ij, the $\varepsilon_{ij,h}$s $h = 1, 2, 3, 4$ follow the same $AR(1)$ time series process (see Sect. 7.2). When using covariance models other than those appropriate for the multivariate linear model or the split plot model, exact tests and BLUEs are typically not available. Typically, maximum likelihood estimation is used along with large sample likelihood ratio tests for these covariance models. The results of Chap. 4 all apply because multivariate linear models can be written as linear models. See also the discussion in Sect. 11.5.2 and the references given near its end.

□

10.4 Growth Curves for One-Way MANOVA

A natural extension of profile analysis is to develop models for the profiles of the various groups. For the one-way MANOVA model $Y = XB + e$ with the cell means parameterization

$$B = \begin{bmatrix} \mu_1' \\ \vdots \\ \mu_a' \end{bmatrix},$$

the q observations in each row of Y might be taken at times $t_1, \ldots, t_q$. We might then incorporate a growth curve model which posits that the components of μ_i, when plotted against time, form the parabola $\mu_{ih} = \gamma_{i0} + \gamma_{i1} t_h + \gamma_{i2} t_h^2$. Write this in matrix form as

$$\mu_i = \begin{bmatrix} \mu_{i1} \\ \vdots \\ \mu_{iq} \end{bmatrix} = \begin{bmatrix} 1 & t_1 & t_1^2 \\ \vdots & \vdots & \vdots \\ 1 & t_q & t_q^2 \end{bmatrix} \begin{bmatrix} \gamma_{i0} \\ \gamma_{i1} \\ \gamma_{i2} \end{bmatrix}$$

$$= Z\gamma_i .$$

Note that the coefficients of the parabola are allowed to vary with the group i but that the model matrix of the growth curve is the same for each group.

In general, for a fixed $q \times r$ matrix Z with $r(Z) = r < q$, we assume a linear growth curve model for each μ_i, say

$$\mu_i = Z\gamma_i .$$

Incorporating the models for the growth curves into the one-way MANOVA gives

$$B = \begin{bmatrix} \mu_1' \\ \vdots \\ \mu_a' \end{bmatrix} = \begin{bmatrix} \gamma_1' Z' \\ \vdots \\ \gamma_a' Z' \end{bmatrix} = \Gamma Z',$$

where

$$\Gamma = \begin{bmatrix} \gamma_1' \\ \vdots \\ \gamma_a' \end{bmatrix} .$$

The complete multivariate growth curve model is

$$Y = X\Gamma Z' + e. \tag{10.4.1}$$

In the next chapter we examine the analysis of models having this general form. Such models are usually called growth curve models or generalized multivariate analysis of variance (GMANOVA) models. We refer to them as *generalized multivariate linear models*.

10.5 Testing for Additional Information

In some multivariate linear model problems, it is of interest to examine whether all the information about XB can be obtained from a subset of the dependent variables. If so, the other variables provide no additional information beyond that available in the subset. Let y be the vector of dependent variables, and partition it as $y' = (y_1', y_2')$. Our interest is in determining whether y_1 contains all the useful information or whether there is useful information in y_2.

Assume a multivariate linear model

$$y_i' = x_i' B + \varepsilon_i'$$

$i = 1, \ldots, n$, with $\varepsilon_i \sim N(0, \Sigma)$, and for $i \neq j$, ε_i and ε_j independent. Partition each y_i as $y_i' = (y_{i1}', y_{i2}')$ and partition B and Σ in conformance with the y_is to give

$$(y_{i1}', y_{i2}') = x_i'[B_1, B_2] + (\varepsilon_{i1}', \varepsilon_{i2}'),$$

where

$$\text{Cov}\left(\begin{bmatrix} \varepsilon_{i1} \\ \varepsilon_{i2} \end{bmatrix}\right) = \begin{bmatrix} \Sigma_{11} & \Sigma_{12} \\ \Sigma_{21} & \Sigma_{22} \end{bmatrix}.$$

Let

$$Y_j = \begin{bmatrix} y_{1j}' \\ \vdots \\ y_{nj}' \end{bmatrix}$$

$j = 1, 2$. In matrix form, the linear model is

$$[Y_1, Y_2] = X[B_1, B_2] + [e_1, e_2].$$

The procedure for evaluating additional information is based on the conditional distribution of y_{i2} given y_{i1} and is similar to the analysis of covariance method used in the next chapter for generalized multivariate linear models. The conditional distribution is normal with mean equal to the best linear predictor of y_{i2} based on y_{i1} and covariance matrix equal to the covariance of the prediction error. Thus,

$$y_{i2} \mid y_{i1} \sim N\left(B_2' x_i + \Sigma_{21} \Sigma_{11}^{-1}(y_{i1} - B_1' x_i), \Sigma_{22} - \Sigma_{21} \Sigma_{11}^{-1} \Sigma_{12}\right).$$

$i = 1, \ldots, n$. Conditioning on Y_1, the random vectors $y_{i2}, i = 1, \ldots n$ are still independent, so the conditional distribution also defines a multivariate linear model. Let

$$e_{2 \cdot 1} = \begin{bmatrix} \xi_1' \\ \vdots \\ \xi_n' \end{bmatrix},$$

where the ξ_is are i.i.d. $N(0, \Sigma_{22} - \Sigma_{21} \Sigma_{11}^{-1} \Sigma_{12})$. Then,

$$Y_2 = XB_2 + (Y_1 - XB_1)\Sigma_{11}^{-1}\Sigma_{12} + e_{2\cdot1}$$
$$= X(B_2 - B_1\Sigma_{11}^{-1}\Sigma_{12}) + Y_1(\Sigma_{11}^{-1}\Sigma_{12}) + e_{2\cdot1} \qquad (10.5.1)$$
$$\equiv X\Delta + Y_1\Gamma + e_{2\cdot1} \,.$$

Our interest is in whether Y_2 contains any additional information about XB beyond that available in Y_1. If $X\Delta = 0$, the conditional distribution of Y_2 depends only on Σ; XB is not involved. Thus, Y_2 provides no additional information on XB. In particular, if $\Delta \equiv B_2 - B_1\Sigma_{11}^{-1}\Sigma_{12} = 0$, then knowledge of B_1 determines XB. The linear structure involved in the components $Y_2 = XB_2 + e_2$ is determined entirely through the dependence between Y_1 and Y_2.

EXAMPLE 10.5.1.   Suppose $Y = XB + e$ is a multivariate one-way analysis of variance. The matrix X merely identifies the treatment group for each observation. If the conditional distribution of Y_2 does not involve X, then the conditional distribution does not depend on the treatment groups. It follows that for the purpose of modeling treatment means, there is no additional information in Y_2.   □

The actual test procedure is straightforward. The hypothesis H_0: $X\Delta = 0$ can be tested by comparing the fit of model (10.5.1) to that of

$$Y_2 = Y_1\Gamma + e_{2\cdot1} \,. \qquad (10.5.2)$$

Alternatively, if Δ is estimable, analysis of covariance methods can be used to obtain $\hat{\Delta}$ and a test can be based on the estimate. In practice, the idea that Δ could really be zero is rather far-fetched. The important thing is that the test provides a way of evaluating whether Y_2 is worth bothering about. If the reduced model (10.5.2) fits nearly as well as (10.5.1), then the $X\Delta$ structure is not really needed to explain Y_2. If Y_2 has almost no relationship to $C(X)$ except through Y_1, an analysis of Y_1 should provide essentially all the information on the relationship between (Y_1, Y_2) and $C(X)$.

EXAMPLE 10.5.2.   Typically, the method of testing for additional information is of most value when the number of dependent variables q is quite large. It provides a method of *reducing the dimensionality* of the problem. The current example is restricted to demonstrating some elementary computations using the heart rate data of Example 10.3.1. In particular, we test whether the first two measurements add any information beyond that contained in the last two measurements.

The error matrix for the test can be obtained from a multivariate analysis of covariance (ACOVA). The components of the ACOVA are available in the matrix E reported in Example 10.3.1. The error matrix for the analysis of covariance model is

$$E = \begin{bmatrix} 479.5 & 483.7 \\ 483.7 & 610.5 \end{bmatrix} - \begin{bmatrix} 363.0 & 237.5 \\ 475.6 & 319.7 \end{bmatrix} \begin{bmatrix} 540.8 & 366.4 \\ 366.4 & 381.0 \end{bmatrix}^{-1} \begin{bmatrix} 363.0 & 475.6 \\ 237.5 & 319.7 \end{bmatrix}$$
$$- \begin{bmatrix} 235.21 & 164.30 \\ 164.30 & 192.19 \end{bmatrix}$$

The reduced model involves only the covariates and not the analysis of variance structure. To obtain the reduced model error matrix E_0, one can regress y_1 on y_3, y_4 and y_2 on y_3, y_4 without including intercepts. The diagonal elements of E_0 are the sums of the squared residuals. The off-diagonal element is the sum of the crossproducts of the two sets of residuals. These computations yield

$$E_0 = \begin{bmatrix} 422.747 & 205.842 \\ 205.842 & 206.801 \end{bmatrix}.$$

The hypothesis matrix is

$$H = E_0 - E = \begin{bmatrix} 187.539 & 41.538 \\ 41.538 & 14.610 \end{bmatrix}$$

and

$$HE^{-1} = \begin{bmatrix} 1.60460 & -1.15564 \\ 0.30659 & -0.18608 \end{bmatrix}.$$

Under the null hypothesis, HE^{-1} should approximate a scalar multiple of the identity matrix. The observed matrix HE^{-1} is nothing like that. In particular, $T^2 = 25 \operatorname{tr}[HE^{-1}] = 35.46$, while the comparison value for T^2, as introduced in Chap. 9, is only 6. Formal tests of $H_0: \Delta = 0$ are highly significant. The first two time measurements are *not* extraneous.

□

The analysis of additional information can be generalized to examine whether Y_2 contains additional information for specific purposes. For example, if X is the model matrix for a one-way ANOVA, we can test whether Y_2 contains additional information for the purpose of distinguishing treatment groups. (This is different from modeling treatment means.) Rather than testing model (10.5.1) against the reduced model (10.5.2), test model (10.5.1) against the analysis of covariance model without treatment groups,

$$Y_2 = J\mu' + Y_1\Gamma + e_{2.1}. \tag{10.5.3}$$

By assuming a parameterization of model (10.5.1) in which

$$X = [J, X_*]$$

and

$$B = \begin{bmatrix} B_{10} & B_{20} \\ B_{11} & B_{21} \end{bmatrix},$$

it is easily seen that this is a test of whether Y_2 contains any additional information about the part of the model that distinguishes treatment groups,

$$X_*[B_{11}, B_{21}]. \tag{10.5.4}$$

It is of interest to note that if Y_2 consists of only one column, the test of model (10.5.3) versus model (10.5.1) is a standard univariate F test. In particular, sequential (stepwise) evaluation of the individual variables is quite simple. Moreover, the F test can be constructed from the diagonal elements of E^{-1} and $(H+E)^{-1}$ that correspond to Y_2; here, E and H are defined for testing the multivariate linear models $Y = XB + e$ and $Y = J\mu' + e$. This provides a simple way to compute all of the F statistics needed for a stepwise evaluation. Variable selection methods in discriminant analysis are based on testing the additional information available for distinguishing treatment groups; see Sect. 12.6.

Exercise 10.3. Find the F test in terms of E^{-1} and $(H+E)^{-1}$.

Exercise 10.4. Test whether the first two time variates in the heart rate data are needed for distinguishing the treatment groups.

10.6 Additional Exercises

Exercise 10.6.1. Jolicoeur and Mosimann (1960) give data on the length, width, and height of painted turtle shells. The carapace dimensions of 24 females and 24 males are given in Table 10.3. Use Hotelling's T^2 statistic to test whether there is a sex difference in shell dimensions. Is there a significant sex difference between any of the individual dimensions? Use plots to check the validity of your assumptions.

Table 10.3 Carapace dimensions

Female			Female			Male			Male		
Length	Width	Height	Length	Width	Height	Length	Width	Height	Length	Width	Height
98	81	38	138	98	51	93	74	37	116	90	43
103	84	38	138	99	51	94	78	35	117	90	41
103	86	42	141	105	53	96	80	35	117	91	41
105	86	42	147	108	57	101	84	39	119	93	41
109	88	44	149	107	55	102	85	38	120	89	40
123	92	50	153	107	56	103	81	37	120	93	44
123	95	46	155	115	63	104	83	39	121	95	42
133	99	51	155	117	60	106	83	39	125	93	45
133	102	51	158	115	62	107	82	38	127	96	45
133	102	51	159	118	63	112	89	40	128	95	45
134	100	48	162	124	61	113	88	40	131	95	46
136	102	49	177	132	67	114	86	40	135	106	47

Exercise 10.6.2. Smith, Gnanadesikan, and Hughes (1962) provide data on characteristics of the urine of young men. The men are categorized into four groups based on their degree of obesity. The four variables given in Table 10.4 consist of a covariate $x = 10^3((\text{specific gravity}) - 1)$ and three dependent variables: $y_1 = $ pigment creatinine, $y_2 = $ chloride, and $y_3 = $ chlorine.

(a) Do a one-way MANOVA on these data ignoring the covariate.

(b) Do a one-way multivariate ACOVA (MACOVA) on these data.

Table 10.4 Excretory characteristics

Group I				Group II				Group III				Group IV			
x	y_1	y_2	y_3	x	y_1	y_2	y_3	x	y_1	y_2	y_3	x	y_1	y_2	y_3
24	17.6	5.15	7.5	31	18.1	9.00	14.5	18	17.0	4.55	1.9	32	12.5	2.90	22.5
32	13.4	5.75	7.1	23	19.7	5.30	12.5	10	12.5	2.65	0.7	25	8.7	3.00	19.5
17	20.3	4.35	2.3	32	16.9	9.85	8.0	33	21.5	6.50	8.3	28	9.4	3.40	1.3
30	22.3	7.55	4.0	20	23.7	3.60	4.9	25	22.2	4.85	9.3	27	15.0	5.40	20.0
30	20.5	8.50	2.0	18	19.2	4.05	0.2	35	13.0	8.75	13.0	23	12.9	4.45	1.0
27	18.5	10.25	2.0	23	18.0	4.40	3.6	33	13.0	5.20	18.3	25	12.1	4.30	5.0
25	12.1	5.95	16.8	31	14.8	7.15	12.0	31	10.9	4.75	10.5	26	13.2	5.00	3.0
30	12.0	6.30	14.5	28	15.6	7.25	5.2	34	12.0	5.85	14.5	34	11.5	3.40	5.1
28	10.1	5.45	0.9	21	16.2	5.30	10.2	16	22.8	2.85	3.3				
24	14.7	3.75	2.0	20	14.1	3.10	8.5	31	16.5	6.55	6.3				
26	14.8	5.10	0.4	15	17.5	2.40	9.6	28	18.4	6.60	4.9				
27	14.4	4.05	3.8	26	14.1	4.25	6.9								
				24	19.1	5.80	4.7								
				16	22.5	1.55	3.5								

Exercise 10.6.3. Analyze the repeated measures data given by Danford, Hughes, and McNee (1960) in *Biometrics* on pages 562 and 563.

Exercise 10.6.4. Box (1950) gives data on the weights of three groups of rats. One group was given thyroxin in their drinking water, one thiouracil, and the third group was a control. Weights are measured in grams at weekly intervals. The data are given in Table 10.5.

(a) Perform a multivariate one-way analysis of variance.

(b) Evaluate the validity of the assumptions.

(c) My 8-year-old son told me that the true means for the control group are $(60, 80, 100, 120, 140)'$. Use Hotelling's T^2 to test the validity of his claim.

(d) Do a profile analysis.

Table 10.5 Rat weights

Control					Thiouracil				
Time 0	Time 1	Time 2	Time 3	Time 4	Time 0	Time 1	Time 2	Time 3	Time 4
57	86	114	139	172	61	86	109	120	129
60	93	123	146	177	59	80	101	111	122
52	77	111	144	185	53	79	100	106	133
49	67	100	129	164	59	88	100	111	122
56	81	104	121	151	51	75	101	123	140
46	70	102	131	153	51	75	92	100	119
51	71	94	110	141	56	78	95	103	108
63	91	112	130	154	58	69	93	114	138
49	67	90	112	140	46	61	78	90	107
57	82	110	139	169	53	72	89	104	122

Thyroxin									
Time 0	Time 1	Time 2	Time 3	Time 4					
59	85	121	156	191					
54	71	90	110	138					
56	75	108	151	189					
59	85	116	148	177					
57	72	97	120	144					
52	73	97	116	140					
52	70	105	138	171					

At the end of the next chapter you will be asked to perform a growth curve analysis on these data.

References

Box, G. E. P. (1950). Problems in the analysis of growth and wear curves. *Biometrics, 6*, 362–389.

Christensen, R. (1996). *Analysis of variance, design, and regression: Applied statistical methods.* London: Chapman and Hall.

Christensen, R. (2011). *Plane answers to complex questions: The theory of linear models* (4th ed.). New York: Springer.

Christensen, R. (2015). *Analysis of variance, design, and regression: Linear modeling for unbalanced data* (2nd ed.). Boca Raton, FL: Chapman and Hall/CRC Press.

Danford, M. B., Hughes, H. M., & McNee, R. C. (1960). On the analysis of repeated-measurements experiments. *Biometrics, 16*, 547–565.

Greenhouse, S. W., & Geisser, S. (1959). On methods in the analysis of profile data. *Psychometrika, 24*, 95–112.

Huynh, H., & Feldt, L. S. (1976). Estimation of the Box correction for degrees of freedom from sample data in randomized block and split-plot designs. *Journal of Educational Statistics, 1*, 69–82.

Jolicoeur, P., & Mosimann, J. E. (1960). Size and shape variation on the painted turtle: A principal component analysis. *Growth, 24*, 339–354.

Lubischew, A. A. (1962). On the use of discriminant functions in taxonomy. *Biometrics, 18*, 455–477.

Mosteller, F., & Tukey, J. W. (1977). *Data analysis and regression*. Reading, MA: Addison-Wesley.

Seber, G. A. F. (1984). *Multivariate observations*. New York: Wiley.

Smith, H., Gnanadesikan, R., & Hughes, J. B. (1962). Multivariate analysis of variance (MANOVA). *Biometrics, 18*, 22–41.

Chapter 11
Generalized Multivariate Linear Models and Longitudinal Data

Abstract This chapter further develops the growth curve models introduced in Chap. 10. This broader class of models is shown to be intimately related to split plot models. We also illustrate how the models can be modified to deal with missing data which is a common problem in longitudinal studies. The models are also related to functional data analysis.

11.1 Generalized Multivariate Linear Models

In Sect. 10.4 we introduced the growth curve model as a modification of the one-way MANOVA model in order to relate the mean structure across the dependent variables. The same idea can be applied to arbitrary multivariate linear models. Consider the multivariate linear model

$$Y = XB + e$$

with the mean and covariance properties discussed in Chap. 9. If we assume that $B = \Gamma Z'$ for some known $q \times r$ matrix Z and an unknown $p \times r$ matrix of parameters Γ, we get

$$Y = X\Gamma Z' + e. \tag{11.1.1}$$

This is variously called a *growth curve model*, a *generalized multivariate analysis of variance (GMANOVA)* model, or a *generalized multivariate linear model (GMLM)*.

Just as multivariate linear models are a special class of univariate linear models, the generalized multivariate linear model can also be written as a univariate linear model. Unfortunately, if $Z \neq I_q$, the generalized multivariate linear model is not a member of the class that we have been calling multivariate linear models. Multivariate linear models are not allowed to specify structure between the means of the different variables. The results on optimal estimation and the distribution theory based on the Wishart distribution depend on applying the same projection operator to each dependent variable vector Y_h. If, as in the growth curve model, one specifies

© Springer Nature Switzerland AG 2019
R. Christensen, *Advanced Linear Modeling*, Springer Texts in Statistics,
https://doi.org/10.1007/978-3-030-29164-8_11

a linear structure for the means across the dependent variables, then the best estimate for any parameter will incorporate information available from the other variables. For example, in a multivariate linear model estimation of $\rho'X\beta_h$ comes from analyzing the Y_h variable separately, e.g., $\rho'X\hat{\beta}_h = \rho'MY_h$, but in a GMANOVA model the estimate typically depends on the information in the other columns of Y. This is enough to invalidate the usual estimation results and distribution theory. Some alternative form of analysis must be developed. When $C(Z) = C(I_q)$, the generalized multivariate linear model is equivalent to a multivariate linear model.

Rewrite model (11.1.1) as

$$\text{Vec}(Y) = \text{Vec}(X\Gamma Z') + \text{Vec}(e)$$

and then as the linear model

$$\text{Vec}(Y) = [Z \otimes X]\text{Vec}(\Gamma) + \text{Vec}(e). \qquad (11.1.2)$$

We continue to assume that

$$E(e) = 0 \quad \text{and} \quad \text{Cov}[\text{Vec}(e)] = [\Sigma \otimes I_n].$$

For known Σ, optimal estimates of Γ are generalized least squares estimates that satisfy

$$[Z \otimes X]\text{Vec}(\hat{\Gamma}) = \mathscr{A}\text{Vec}(Y)$$

where $\mathscr{A}$ is the appropriate oblique projection operator. It is not hard to see that

$$\mathscr{A} = [A_Z \otimes M]$$

where M is the ppo onto $C(X)$ and $A_Z \equiv Z[Z'\Sigma^{-1}Z]^- Z'\Sigma^{-1}$ is an oblique projection operator onto $C(Z)$. Thus, the optimal estimate of Γ, for known Σ, satisfies

$$[Z \otimes X]\text{Vec}(\hat{\Gamma}) = [A_Z \otimes M]\text{Vec}(Y) \qquad (11.1.3)$$

or

$$X\hat{\Gamma}Z' = MYA_Z'.$$

Exercise 11.1.   Prove that Eq. (11.1.3) gives the BLUEs when Σ is known.

In particular, when the inverses exist we can write generalized least squares estimates for each i, say,

$$\breve{\gamma}_i \equiv [Z'\Sigma^{-1}Z]^{-1}Z'\Sigma^{-1}y_i$$

so that we can write an $n \times r$ matrix

$$\tilde{Y}_1 \equiv \begin{bmatrix} \breve{\gamma}_1' \\ \vdots \\ \breve{\gamma}_n' \end{bmatrix} = Y\Sigma^{-1}Z[Z'\Sigma^{-1}Z]^{-1}$$

Multiplying the GMLM (11.1.1) on the right by $\Sigma^{-1}Z[Z'\Sigma^{-1}Z]^{-1}$ gives the standard multivariate linear model

$$\tilde{Y}_1 = X\Gamma + \tilde{e}_1 \tag{11.1.4}$$

and we could then find $\hat{\Gamma}$ by fitting the multivariate linear model (11.1.4) using least squares, i.e., $X\hat{\Gamma} = M\tilde{Y}_1$.

Of course, A_Z and $\hat{\Gamma}$ depend on knowing Σ, which is an unrealistic assumption. As discussed in Chap. 4, an appropriate procedure is to estimate Σ with, say, $\tilde{\Sigma}$ and use that to compute empirical estimates $\hat{\Gamma}$ using $\tilde{A}_Z$. The standard procedures seem to be based on observing that the linear model (11.1.2) is a reduced model relative to the standard multivariate linear model

$$\text{Vec}(Y) = [I_q \otimes X]\text{Vec}(B) + \text{Vec}(e). \tag{11.1.5}$$

We can use model (11.1.5) to estimate the covariance parameters. From the multivariate linear model we know that an unbiased estimate of Σ is $S \equiv Y'(I-M)Y/[n-r(X)]$. Identifying $S \equiv \tilde{\Sigma}$ leads to an empirical estimate of Γ. Relative to model (11.1.2), S should be a residual type estimate in the sense of Sect. 4.7, so it should lead to the unbiasedness and variance estimation properties discussed in that section. In fact, Khatri (1966) showed that the maximum likelihood estimates, $\hat{\Gamma}_{ML}$ and $\hat{\Sigma}$, are determined by

$$X\hat{\Gamma}_{ML}Z' = MY\tilde{A}'_Z$$

where $\tilde{A}_Z \equiv Z(Z'S^{-1}Z)^- Z'S^{-1}$ and

$$n\hat{\Sigma} = (Y - X\hat{\Gamma}_{ML}Z')'(Y - X\hat{\Gamma}_{ML}Z') = Y'(I-M)Y + (I-\tilde{A})Y'MY(I-\tilde{A})'.$$

Model (11.1.1) must not be confused with the transformation of $Y = XB + e$ considered after Eq. (9.5.8):

$$YW = XBW + eW. \tag{11.1.6}$$

Model (11.1.1) has $\text{E}(Y) = X\Gamma Z'$ whereas model (11.1.6) has $\text{E}(YW) = XBW$. The parameter matrices Γ and B seem interchangeable but in fact B is analogous to $\Gamma Z'$. The known $q \times r$ matrices Z and W seem interchangeable but they are used differently. Model (11.1.1) examines $\text{E}(Y)$ whereas model (11.1.6) examines $\text{E}(YW)$. Model (11.1.6) is still a standard multivariate linear model, unlike the GMLM (11.1.1). Model (11.1.6) reduces the dimensionality of everything from q to r whereas model (11.1.1) has q dimensions for the dependent variable but only r for the parameter matrix. In the transformed model, B is $p \times q$, whereas in model (11.1.1), Γ is $p \times r$. Because B is completely unknown, in model (11.1.6) BW remains completely unknown, which allows us to treat model (11.1.6) as a standard multivariate linear model. These distinctions make all the difference in being able to derive optimal estimates and closed-form small sample distributions, something we can do for model (11.1.6) but not for model (11.1.1).

In the next two sections we discuss methods for analyzing generalized multivariate linear models. Both involve transforming the generalized multivariate linear model into a standard multivariate linear model. Section 11.4 discusses a class of covariance matrices for which least squares estimates are optimal, thus enabling an optimal analysis. Section 11.5 discusses the application of generalized multivariate linear models to longitudinal data and extends those results to data in which some of the q dependent variables are unobserved. Section 11.6 discusses the relationship between generalized multivariate linear models and generalized split plot models.

11.2 Generalized Least Squares Analysis

The problem with Eq. (11.1.3) is that we do not know Σ. One way to analyze growth curve models is, similar in spirit to MINQUE in Chap. 4, just to pick a weighting matrix G to use in place of Σ^{-1}. The most common choice is $G = I_q$. In this ad hoc approach, when the inverses exist, we multiply

$$Y = X\Gamma Z' + e$$

on the right by $GZ[Z'GZ]^{-1}$ to obtain

$$YGZ[Z'GZ]^{-1} = X\Gamma Z'GZ[Z'GZ]^{-1} + eGZ[Z'GZ]^{-1}.$$

Upon simplification we get the standard multivariate linear model

$$YGZ[Z'GZ]^{-1} = X\Gamma + eGZ[Z'GZ]^{-1}.$$

In particular, we can write generalized least squares estimates for each i, say,

$$\breve{\gamma}_i \equiv [Z'GZ]^{-1}Z'Gy_i$$

so that we can write an $n \times r$ matrix

$$\tilde{Y}_1 \equiv \begin{bmatrix} \breve{\gamma}_1' \\ \vdots \\ \breve{\gamma}_n' \end{bmatrix} = YGZ[Z'GZ]^{-1}$$

We can then find $\hat{\Gamma}_G$ by fitting the multivariate linear model

$$\tilde{Y}_1 = X\Gamma + \tilde{e}_1 \tag{11.2.1}$$

using least squares. If X is the model matrix for a one-way ANOVA, the rows of $\hat{\Gamma}_G$ consist of the sample means of the $\breve{\gamma}_i$s from each group.

While $\hat{\Gamma}_G$ is the best linear unbiased estimate of Γ based on $\tilde{Y}_1$, it is *not* the best linear unbiased estimate of Γ based on Y. This $\hat{\Gamma}_G$ loses information due to $\tilde{Y}_1$ being based on suboptimal estimators. None of the estimates $\breve{\gamma}_i$ are optimal because they are based on G rather than Σ^{-1}. As discussed in Sect. 11.4, useful models for Σ exist in which the ordinary least squares estimates obtained with $G = I$ do give optimal estimation. Such models for Σ provide good justification for the analysis discussed here with $G = I$.

Tests of hypotheses for model (11.2.1) are performed as for any other standard multivariate linear model. The most commonly performed test for a multivariate linear model is against the model that replaces X with J_n. It tests whether the regression coefficients of the growth curves are the same for every individual. One advantage of analyzing model (11.2.1) for some fixed choice of G is that the standard test statistics have known distributions under H_0. Unfortunately, likelihood ratio tests for the original model (11.1.1) have thus far proved to have intractable distributions.

Exercise 11.2.   Show that if Z does not have full column rank that it suffices to pick any $(Z'GZ)^-$ but that one should restrict hypotheses being tested to be of the form $\Lambda'\Gamma W$ where $\Lambda'\Gamma$ is estimable in the sense that $\Lambda' = P'X$ for some matrix P but where ΓW is also estimable in the sense that $W = Z'W_0$ for some W_0. What does it mean for $\Lambda'\Gamma W$ to be estimable in model (11.1.2)? (In a standard multivariate linear model there are no restrictions on W nor are there any in a GMLM when Z has full rank.)

EXAMPLE 11.2.1.   Once again, consider the heart rate data of Example 10.3.1. We have seen that standard analysis of variance procedures can be used to examine relationships between drugs. Unfortunately, standard procedures are less applicable for comparing times because observations over time are correlated. If some contrast is of particular interest, it can be examined, but such tools as orthogonal contrasts and tabled polynomial contrasts do not retain their attractive properties. In lieu of using orthogonal polynomial contrasts, we fit a polynomial growth curve model. Specifically, we assume a quadratic growth curve model. Recall that heart rates are measured at 2, 7, 12, and 17 min after the injection. The model for the time means is

$$
\begin{bmatrix} \mu_{i1} \\ \mu_{i2} \\ \mu_{i3} \\ \mu_{i4} \end{bmatrix} = \begin{bmatrix} 1 & 2 & 4 \\ 1 & 7 & 49 \\ 1 & 12 & 144 \\ 1 & 17 & 289 \end{bmatrix} \begin{bmatrix} \gamma_{i0} \\ \gamma_{i1} \\ \gamma_{i2} \end{bmatrix}
$$

$$
= Z\gamma_i,
$$

$i = 1, 2, 3$.

The simplest choice of G is $G = I$. (This choice of G actually *is* equivalent to using a column of ones and tabled polynomial contrasts for W in model (11.1.6).) Let

$$
W_1 = Z(Z'Z)^{-1}
$$
$$
= \begin{bmatrix} 1.41 & -0.25 & 0.01 \\ -0.15 & 0.17 & -0.01 \\ -0.53 & 0.21 & -0.01 \\ 0.27 & -0.13 & 0.01 \end{bmatrix}.
$$

The transformed dependent variable matrix in model (11.2.1) becomes

$$\tilde{Y}_1 \equiv YW_1,$$

so we fit the multivariate one-way MANOVA

$$\tilde{Y}_1 = X\Gamma + \tilde{e}_1 .$$

The estimates obtained are

$$\hat{\Gamma} = \begin{bmatrix} 73.96 & -0.093 & -0.007 \\ 82.60 & 0.139 & -0.007 \\ 68.19 & 2.561 & -0.137 \end{bmatrix}$$

and

$$\tilde{S}_1 = \begin{bmatrix} 19.447 & -0.701 & -0.0025 \\ -0.701 & 0.51570 & -0.02485 \\ -0.0025 & -0.02485 & 0.001379 \end{bmatrix}.$$

Computing the usual test statistics for no treatment effects in the multivariate one-way ANOVA model, we find substantial differences among the three injections. For example, $T^2 = 184.81$, which is huge when compared to the asymptotic null distribution $\chi^2(6)$.

Write

$$\tilde{Y}_1 = [\tilde{Y}_{10}, \tilde{Y}_{11}, \tilde{Y}_{12}] .$$

The vector $\tilde{Y}_{10} = (\tilde{\gamma}_{10}, \ldots, \tilde{\gamma}_{n0})'$ is used for inferences about intercepts. The vector $\tilde{Y}_{11} = (\tilde{\gamma}_{11}, \ldots, \tilde{\gamma}_{n1})'$ is used for slopes; $\tilde{Y}_{12}$ is used to analyze the coefficients of the quadratic terms in the polynomials. Each of these vectors can be examined in a univariate one-way ANOVA. Pairs of them can be examined in a multivariate one-way.

The coefficient of the quadratic term for drug B is γ_{32}. The least squares estimate $\hat{\gamma}_{32} = -0.137$ has a standard error of 0.01174 and a t statistic of -11.67. The quadratic term for drug B is clearly important. All of this can be obtained from the univariate one-way ANOVA on $\tilde{Y}_{12}$.

Similarly, we can test whether parabolas are needed to model the placebo and drug A. This is a test of $H_0 \colon \gamma_{12} = \gamma_{22} = 0$ and depends only on $\tilde{Y}_{12}$. Again, the test is just a univariate ANOVA test. The reduced model is

$$\tilde{y}_{12,ij} = \delta_{i3} \gamma_{i2} + \tilde{e}_{12,ij},$$

where δ_{i3} is 1 if $i = 3$ and zero otherwise. From the matrix $\tilde{S}_1$ reported earlier, we find that $\text{SSE}(Full) = (27)(0.001379) = 0.03723$. For the reduced model, $\text{SSE}(Red.) = 0.03821$, so the F statistic is

$$F = \frac{[0.03821 - 0.03723]/2}{0.001379} = 0.355$$

and there is no evidence of the need for a parabola in either the placebo or drug A.

We can also test for equality of the slopes in the placebo and drug A. Because our interest is in the slopes, the dependent variable is $\tilde{Y}_{11}$. The reduced model is again a one-way ANOVA, but now the placebo and drug A are considered as the same treatment. The error sums of squares for the full and reduced models take the values

$$SSE(Full) = 13.924,$$

$$SSE(Red.) = 14.193.$$

The F statistic is

$$F = \frac{14.193 - 13.924}{[13.924/27]} = 0.522,$$

so there is no evidence of a difference between the placebo and drug A. We can also test whether both slopes are zero. The new reduced model is $\tilde{y}_{11,ij} = \delta_{i3}\gamma_{i1} + \tilde{e}_{11,ij}$ with

$$SSE(Red.) = 14.203$$

and

$$F = \frac{[14.203 - 13.924]/2}{0.5157} = 0.271.$$

There is no evidence of a nonzero slope for the placebo and drug A.

The procedure for testing whether the placebo and drug A act the same can be applied to $\tilde{Y}_{10}$ to determine if there is evidence of a difference in intercepts between the placebo and drug A. If no difference is found, we would have found no evidence of any difference between the placebo and drug A on any of the dependent variables. The statistics are

$$SSE(Full) = 525.07,$$

$$SSE(Red.) = 898.83,$$

$$F = \frac{898.83 - 525.07}{19.45} = 19.22.$$

Comparing this to $F(0.995, 1, 27) = 9.34$ or even to the Scheffé critical point $2F(0.995, 2, 27) = 12.98$ establishes that there is a significant difference at the 0.005 level between the intercepts for the placebo and drug A.

The heart rates under the placebo seem to be relatively constant at about $\hat{\gamma}_{10} \doteq 74$ beats per minute. Over the course of the 17 min experiment, drug A yields approximately constant heart rates at $\hat{\gamma}_{20} \doteq 82.5$ beats per minute. Over the course of the experiment, heart rates for drug B can be approximated by the parabola $68.19 + 2.561t - 0.137t^2$. Clearly, this is only an approximation. It is unlikely that heart rates would really become negative after thirty-three and a half minutes.

This entire analysis is essentially an exercise in quantifying and evaluating the visual impressions given by Fig. 10.6. For example, the downward trend seen in the placebo profile is not statistically significant from the current data and analysis. Both $\hat{\gamma}_{11}$ and $\hat{\gamma}_{12}$ are negative, so the downward trend is being modeled, but neither coefficient is significant, so we do not have firm evidence of a downward trend.

Although we only tested that these coefficients were zero in combination with the corresponding values for drug A, the same results occur for the individual tests. □

Exercise 11.3.

(a) Using $\tilde{Y}_{11}$ and $\tilde{Y}_{12}$ as dependent variables, do a multivariate test of $\gamma_{11} = \gamma_{12} = 0$.
(b) Perform an analysis similar to the one just given using a split plot model. Compare the results to those of the growth curve model.

11.3 MACOVA Analysis

The use of model (11.2.1) to analyze growth curves was apparently first proposed by Potthoff and Roy (1964). An extension of this method based on multivariate analysis of covariance (MACOVA) and normality was proposed by Rao (1965, 1966, 1967) and Khatri (1966). Let Z have full rank, $W_1 = GZ(Z'GZ)^{-1}$ and let W_2 be a full column rank matrix with $C(W_2) = C(Z)^{\perp}$. As established later in Exercise 11.4, with these choices the matrix $W = [W_1, W_2]$ is nonsingular. Because W is a nonsingular matrix, the generalized multivariate linear model (11.1.1) is equivalent to

$$YW = X\Gamma Z'W + eW. \tag{11.3.1}$$

Write $\tilde{Y}_1 = YW_1$ and $\tilde{Y}_2 = YW_2$. Note that $\tilde{Y}_1$ is precisely the dependent variable matrix in model (11.2.1). Using the definition of W, (11.3.1) can be rewritten as

$$[\tilde{Y}_1, \tilde{Y}_2] = [X\Gamma, 0] + [eW_1, eW_2],$$

which is similar to a multivariate linear model. As will be seen later, under normality, the conditional distribution of $\tilde{Y}_1$ given $\tilde{Y}_2$ is determined by the multivariate linear model

$$\tilde{Y}_1 = X\Gamma + \tilde{Y}_2\Psi + \tilde{e}_{1\cdot2}. \tag{11.3.2}$$

In particular, this is a multivariate analysis of covariance model. Estimates and quadratic forms for tests are derived as in *PA* Chap. 9, so

$$\hat{\Psi} = [\tilde{Y}_2'(I-M)\tilde{Y}_2]^{-1}\tilde{Y}_2'(I-M)\tilde{Y}_1,$$
$$X\hat{\Gamma} = M(\tilde{Y}_1 - \tilde{Y}_2\hat{\Psi}),$$

and, with $\tilde{\Sigma}_{1\cdot2}$ denoting the conditional covariance matrix of a row of $\tilde{Y}_1$, the estimate of the covariance matrix is based on

$$[n - r(X, \tilde{Y}_2)]\tilde{S}_{1\cdot2} = \tilde{Y}_1'\{(I-M) - (I-M)\tilde{Y}_2[\tilde{Y}_2'(I-M)\tilde{Y}_2]^{-1}\tilde{Y}_2'(I-M)\}\tilde{Y}_1.$$

In general, Γ need not be estimable because X need not have full column rank. To test that the growth curves are the same for everybody (in one-way MANOVA

$H_0\colon \gamma_1 = \gamma_2 = \cdots = \gamma_a$), simply test the full model (11.3.2) against the reduced model

$$\tilde{Y}_1 = J\mu' + \tilde{Y}_2\Psi + \tilde{e}_{1\cdot2}.$$

One substantial advantage of the Rao-Khatri modification is that, for inferences about Γ, the method does not depend on the specific choice of the matrix W. This can be shown for all aspects of the problem; we illustrate only that $X\hat{\Gamma}$ does not depend on G. Let $E \equiv Y'(I - M)Y = [n - r(X)]S$.

$$\begin{aligned}
X\hat{\Gamma} &= M(\tilde{Y}_1 - \tilde{Y}_2\hat{\Psi}) \\
&= M(YW_1 - YW_2\hat{\Psi}) \\
&= MY(W_1 - W_2[\tilde{Y}_2'(I - M)\tilde{Y}_2]^{-1}\tilde{Y}_2'(I - M)\tilde{Y}_1) \\
&= MY(W_1 - W_2[W_2'EW_2]^{-1}W_2'EW_1) \\
&= MY(I - A_2)W_1 ,
\end{aligned}$$

where $A_2 \equiv W_2(W_2'EW_2)^{-1}W_2'E$. Because $W_2'Z = 0$, Lemma 4.3.1 implies that $(I - A_2)$ is the oblique projection operator onto $C(E^{-1}Z)$ along $C(W_2)$. By Lemma 4.3.2, $E^{-1}Z(Z'E^{-1}Z)^{-1}Z'$ is the projection operator onto $C(E^{-1}Z)$ along $C(W_2)$, so

$$\begin{aligned}
X\hat{\Gamma} &= MYE^{-1}Z(Z'E^{-1}Z)^{-1}Z'W_1 \\
&= MYE^{-1}Z(Z'E^{-1}Z)^{-1}Z'GZ(Z'GZ)^{-1} \\
&= MYE^{-1}Z(Z'E^{-1}Z)^{-1} \\
&= MYS^{-1}Z(Z'S^{-1}Z)^{-1} ,
\end{aligned}$$

which does not depend on W. Moreover, this is precisely the intuitively appealing estimator based on taking $\Sigma = S$ in model (11.1.4) or $G^{-1} = S$ in model (11.2.1).

It remains to show that (11.3.2) is a valid multivariate linear model. A typical row of Y is y_i' so a row of YW is $y_i'W = [y_i'W_1, y_i'W_2]$. Define $[\tilde{y}_{1i}', \tilde{y}_{2i}'] = [y_i'W_1, y_i'W_2]$. We will find the conditional distribution of $\tilde{y}_{1i}$ given $\tilde{y}_{2i}$. Because $[\tilde{y}_{1i}', \tilde{y}_{2i}']$ is multivariate normal, the conditional distribution is also normal. The conditional mean is the best linear predictor of $\tilde{y}_{1i}$ based on $\tilde{y}_{2i}$. The conditional covariance matrix is the prediction covariance matrix. (These concepts were discussed in Chap. 4 and PA Chap. 6.) The parameters of the conditional distribution are simple functions of the parameters of the joint distribution. We begin by finding the mean and covariance matrix of the joint distribution. The rows of X can be written as x_i', so $E(y_i') = x_i'\Gamma Z'$ and, as usual, $Cov(y_i) = \Sigma$. It follows that

$$\begin{aligned}
E[\tilde{y}_{1i}', \tilde{y}_{2i}'] &= E(y_i'W) = x_i'\Gamma Z'W \\
&= [x_i'\Gamma ZW_1, x_i'\Gamma Z'W_2] \\
&= [x_i'\Gamma, 0],
\end{aligned}$$

where the last equality follows from the choice of W. Also,

$$\text{Cov}\left(\begin{bmatrix} \tilde{y}_{1i} \\ \tilde{y}_{2i} \end{bmatrix}\right) = \text{Cov}(W'y_i) = W'\Sigma W$$

$$= \begin{bmatrix} W_1'\Sigma W_1 & W_1'\Sigma W_2 \\ W_2'\Sigma W_1 & W_2'\Sigma W_2 \end{bmatrix}.$$

From the theory of best linear predictors,

$$\tilde{y}_{1i}|\tilde{y}_{2i} \sim N(\Gamma'x_i + \Psi'\tilde{y}_{2i}, \tilde{\Sigma}_{1\cdot2}), \qquad (11.3.3)$$

where $\Psi' \equiv W_1'\Sigma W_2(W_2'\Sigma W_2)^{-1}$ and

$$\tilde{\Sigma}_{1\cdot2} \equiv W_1'\Sigma W_1 - W_1'\Sigma W_2(W_2'\Sigma W_2)^{-1}W_2'\Sigma W_1.$$

Because the y_is are independent, the $W'y_i$s are also independent. In particular, $\tilde{y}_{1i}$ is independent of $\tilde{y}_{2i'}$ whenever $i \neq i'$. Thus, the distribution in (11.3.3) is also the distribution of $\tilde{y}_{1i}$ given $\tilde{Y}_2$. Moreover, because y_i and $y_{i'}$ are independent for $i \neq i'$, the random vector $\tilde{y}_{1i}$ given $\tilde{y}_{2i}$ is independent of any other observations, say $\tilde{y}_{1i'}$ given $\tilde{y}_{2i'}$. Because both $\tilde{y}_{2i}$ and $\tilde{y}_{2i'}$ can be replaced by $\tilde{Y}_2$ in the conditioning, it follows that, given $\tilde{Y}_2$, the $\tilde{y}_{1i}$s are i.i.d. with the distribution in (11.3.3). This is precisely the definition of the multivariate linear model (11.3.2). Because (11.3.2) is a conditional normal theory multivariate linear model, the estimates $\hat{\Psi}$ and $X\hat{\Gamma}$ are conditional maximum likelihood estimates. In fact, since $X\hat{\Gamma}$ does not depend on Σ yet maximizes the conditional likelihood for any Σ, and since the marginal distribution of $\tilde{Y}_2$ does not depend on Γ, the conditional maximum likelihood estimate of $X\Gamma$ is also the unconditional MLE.

EXAMPLE 11.3.1.   We now reanalyze the heart rate data using the analysis of covariance growth curve model. Recall from Example 11.2.1 that

$$Z = \begin{bmatrix} 1 & 2 & 4 \\ 1 & 7 & 49 \\ 1 & 12 & 144 \\ 1 & 17 & 289 \end{bmatrix},$$

and we took

$$W_1 = Z(Z'Z)^{-1}.$$

The measurements are taken at equally spaced time intervals, so a standard table of polynomial contrasts can be used to obtain the matrix W_2. The cubic contrast coefficients with four equally spaced time periods is $(-1, 3, -3, 1)$ and W_2 can be taken as

$$W_2 = \begin{bmatrix} -1 \\ 3 \\ -3 \\ 1 \end{bmatrix}.$$

Computing $\tilde{Y}_1 = YW_1$ and $\tilde{Y}_2 = YW_2$ and fitting the multivariate one-way analysis of covariance model

$$\tilde{Y}_1 = X\Gamma + \tilde{Y}_2\Psi + \tilde{e}_{1\cdot2}$$

gives

$$\hat{\Gamma} = \begin{bmatrix} 74.179 & -0.1209 & -0.00630 \\ 82.785 & 0.1162 & -0.00640 \\ 68.730 & 2.4926 & -0.13528 \end{bmatrix},$$

$$\hat{\Psi} = [0.1998, -0.02533, 0.000635],$$

and the error statistic

$$\tilde{E} \equiv \tilde{E}_{1\cdot2} = \begin{bmatrix} 477.88 & -12.95 & -0.2176 \\ -12.95 & 13.17 & -0.6519 \\ -0.2176 & -0.6519 & 0.0368 \end{bmatrix}$$

with 26 degrees of freedom. Again, X is the full rank model matrix associated with the cell means parameterization, so Γ is estimable.

The reduced model for no differences between drugs in the regression coefficients is

$$\tilde{Y}_1 = J\mu' + \tilde{Y}_2\Psi + \tilde{e}_{1\cdot2},$$

with an error statistic of

$$\tilde{E}_0 = \begin{bmatrix} 1470.31 & -162.16 & 8.01 \\ -162.16 & 54.29 & -2.76 \\ 8.01 & -2.76 & 0.14 \end{bmatrix}.$$

For testing the reduced model, the hypothesis statistic is

$$\tilde{H} = \tilde{E}_0 - \tilde{E} = \begin{bmatrix} 992.43 & -149.21 & 8.23 \\ -149.21 & 41.12 & -2.11 \\ 8.23 & -2.11 & 0.11 \end{bmatrix}.$$

The standard test statistics are $U = 0.064$, $\phi_{max} = 5.43$, $T^2 = 185.3$, $V = 1.434$. There are clear differences due to drugs in the coefficients of the parabolas.

We can now repeat the detailed analysis of the parabolas given in Example 11.3.1. The full and reduced models used in the analysis are all the same except that all now include the covariate matrix $\tilde{Y}_2$. The specific models used for various tests were discussed in the earlier example.

In looking at the coefficients of the quadratic terms,

$$\hat{\gamma}_{32} = -0.13528,$$

with

$$SE(\hat{\gamma}_{32}) = 0.01225$$

and

$$t_{obs} = -11.04.$$

These results can be obtained by fitting the univariate linear model $\tilde{Y}_{12} = X\gamma_2 + \tilde{Y}_2\psi_2 + e_2$, where $\tilde{Y}_{12}$ is the third component of $\tilde{Y}_1 = [\tilde{Y}_{10}, \tilde{Y}_{11}, \tilde{Y}_{12}]$. The analogous results in Example 11.5.1 were obtained by analyzing the model without the covariate $\tilde{Y}_2$. This analogy holds between all the models of Example 11.5.1 and the models needed here, so only summary statistics are given in the remainder of this example.

For testing $H_0: \gamma_{12} = \gamma_{22} = 0$, the error sums of squares for the full and reduced models are

$$SSE(Full) = 0.036753; \qquad SSE(Red.) = 0.037550$$

and the F statistic is

$$F = 0.282,$$

which is not significant.

For looking at equality of the slopes in the placebo and drug A,

$$SSE(Full) = 13.166; \qquad SSE(Red.) = 13.447$$

and

$$F = 0.555.$$

For testing whether the slopes are both zero, to three decimal places we again happen to have

$$SSE(Red.) = 13.447,$$

but the numerator has two degrees of freedom, so the F statistic is half as large:

$$F = 0.278.$$

Neither test is significant.

To test for differences in intercepts between the placebo and drug A

$$SSE(Full) = 477.88; \qquad SSE(Red.) = 848.13$$

and

$$F = 20.14.$$

Clearly, the placebo and drug A have different intercepts.

As in Example 11.5.1, we have found that heart rates are fairly constant for the placebo and drug A. The rates are approximately 74 and 83 beats per minute. Drug B again follows a parabola. The coefficients are remarkably close in the two analyses.

□

Rao (1965) has suggested that the analysis of covariance procedure be used only with columns of $\tilde{Y}_2$ that are important contributors to model (11.3.2). Formal tests can be made for the various columns, but these depend on the choice of G.

EXAMPLE 11.3.2.   To test whether $\tilde{Y}_2$ contributes additional information to the model using the heart rate data with $G = I$, the error matrix from Example 11.2.1

is the appropriate $\tilde{E}_0$, i.e., $\tilde{E}_0 = [n - r(X)]\tilde{S}_1$ from Example 11.2.1, and the error matrix from Example 11.3.1 is $\tilde{E}$. Subtracting these gives

$$\tilde{H} = \begin{bmatrix} 47.18844 & -5.98145 & 0.15006 \\ -5.98145 & 0.75819 & -0.01902 \\ 0.15006 & -0.01902 & 0.00048 \end{bmatrix},$$

and multiplying by E^{-1} leads to

$$T^2 = 4.45.$$

The comparison value for T^2 is 3, so there is no overwhelming evidence of the importance of $\tilde{Y}_2$. The hypothesis matrix H has only one degree of freedom, so McKeon's approximate distribution for T^2 is exact. The formal test is based on comparing

$$\frac{T^2}{dfE}\frac{dfE+1-q}{q} = \frac{4.45}{26}\frac{24}{3} = 1.37$$

to an $F(q, dfE + 1 - q)$ distribution. The result is far from significant. This is consistent with the fact that the results of our two analyses of these data are not very different. □

Exercise 11.4.

(a) Show that the estimate of the covariance matrix $\tilde{\Sigma}_{1\cdot 2}$ does not depend on the choice of the matrix G in W_1.
(b) Show that W is nonsingular if and only if $C(W_1) \cap C(W_2) = \{0\}$.
(c) Show that $C(W_1) \cap C(W_2) = \{0\}$.

Hint: Since G is positive definite, so is G^{-1}. Recall that $v = 0$ if and only if $v'Gv = 0$.

11.4 Rao's Simple Covariance Structure

Rao's *Simple Covariance Structure (SCS)* is a class of covariance matrices Σ for which least squares estimates are optimal. Under a SCS, the least squares analysis illustrated in Sect. 11.2 is optimal. Under SCS the MACOVA analysis in Sect. 11.3 becomes unnecessary but that analysis provides a useful test of whether the SCS is an appropriate model for the data. Finally, the optimality of least squares also suggests a way of testing a reduced model $X\Gamma_0 Z_0'$ in lieu of the full model $X\Gamma Z'$.

As in the previous section, let $C(W_2) = C(Z)^\perp$. Rao's SCS is a model for the covariance matrix in which

$$\Sigma = Z\tilde{D}Z' + W_2\Theta W_2',$$

see Geisser (1970). In the general SCS model, parts or all of $\tilde{D}$ and Θ can be unknown parameters. In particular, the covariance models

$$\Sigma = \sigma_s^2 I_q + \sigma_w^2 J_q^q \quad \text{and} \quad \Sigma = \sigma_0^2 I_q + ZDZ'$$

are special cases of the SCS. The first of these is known variously as the *intraclass correlation* structure, the *compound symmetry* structure, and the *equal covariance* structure. For this to be of SCS we need $J_q \in C(Z)$. The second covariance structure is just a mixed model but is often called the *random regression coefficients* structure and most often involves an unstructured covariance matrix D. Most often, compound symmetry is a special case of random regression coefficients. In particular, compound symmetry is determined by adding a random (intercept) effect for each individual. In the growth curve context, adding a random slope and intercept for each individual gives a random regression coefficient structure.

Exercise 11.5.   Show that compound symmetry and random regression coefficients are special cases of SCS. Hint: Write $I = M_Z + (I - M_Z)$.

With SCS and using *PA* Proposition 2.7.5, because $C(\Sigma Z) \subset C(Z)$ we have $A_Z = M_Z$ and least squares estimates become optimal in model (11.1.1). Potthoff and Roy's (1964) method for analyzing generalized multivariate linear models in Sect. 11.2 is typically suboptimal because it is based on suboptimal estimates of the parameters in Γ, but knowing that least squares estimates are optimal provides a strong justification for using the least squares methods illustrated in Sect. 11.2.

The analysis of the multivariate linear model

$$\tilde{Y}_1 = X\Gamma + \tilde{e}_1,\tag{11.4.1}$$

is based on having an unstructured covariance matrix for the rows of $\tilde{Y}_1 \equiv YZ(Z'Z)^{-1}$. If we assume that $\Sigma = \sigma_s^2 I_q + \sigma_w^2 J_q^q$ with $J_q \in C(Z)$, it is possible to show that the univariate version of the GMLM (11.1.2) is a generalized split plot model as discussed in *PA* Chap. 11 and Sect. 11.6, and therefore can be analyzed with the more powerful methods available for split plot models rather than with multivariate methods for model (11.4.1).

On the other hand, if we assume that $\Sigma = \sigma^2 I + ZDZ'$ where D is an unknown and unstructured covariance matrix, the covariance matrix associated with model (11.4.1) is

$$(Z'Z)^{-1}Z'\Sigma Z(Z'Z)^{-1} = \sigma^2(Z'Z)^{-1} + D.$$

We will see later that unbiased estimates of the parameters σ^2 and D exist. But observe that if D is an unstructured covariance matrix, there is no usable structure in the covariance of model (11.4.1) that could improve estimation or inference in that model.

In the general SCS case,

$$(Z'Z)^{-1}Z'\Sigma Z(Z'Z)^{-1} = \tilde{D}$$

and regardless of whether $\tilde{D}$ contains known structure, the analysis based on the multivariate linear model (11.4.1) remains valid, even if it may be suboptimal when $\tilde{D}$ is structured.

11.4.1 Reduced Models in Z

While standard multivariate linear model theory applied to model (11.4.1) allows testing hypotheses involving Γ, another approach to testing reduced models relative to the $Z\Gamma'$ structure is available through the device of testing for additional information as in Sect. 10.5. The details of testing were described there. Here we illustrate how to partition the data into two parts so as to allow testing of whether one part contains useful information.

Suppose $Z = [Z_0, Z_1]$ and we want to test whether the reduced model

$$Y = X\Gamma_0 Z_0' + e$$

provides an adequate fit. Using ideas from analysis of covariance, without loss of generality we can replace Z in model (11.1.1) with $[Z_0, (I - M_{Z0})Z_1]$ where $M_{Z0} = Z_0(Z_0'Z_0)^{-1}Z_0'$. Then, instead of examining model (11.1.1), we examine the equivalent growth curve model

$$Y = X[\Gamma_0, \quad \Gamma_1]\begin{bmatrix} Z_0' \\ Z_1'(I - M_{Z0}) \end{bmatrix} + e = X\Gamma_0 Z_0' + X\Gamma_1 Z_1'(I - M_{Z0}) + e. \quad (11.4.2)$$

With SCS, instead of transforming model (11.1.1) into model (11.4.1) we transform model (11.4.2) into

$$[\tilde{Y}_{10}, \tilde{Y}_{11}] \equiv [YZ_0(Z_0'Z_0)^{-1}, Y(I - M_{Z0})Z_1\{Z_1'(I - M_{Z0})Z_1\}^{-1}] = X[\Gamma_0, \quad \Gamma_1] + \tilde{e}. \quad (11.4.3)$$

We can now test the efficacy of $X\Gamma_1 Z_1'(I - M_{Z0})$ in model (11.4.2) by using the standard test from Sect. 10.5 of whether $\tilde{Y}_{11}$ adds additional information beyond that provided by $\tilde{Y}_{10}$ in model (11.4.3).

11.4.2 Unbiased Covariance Parameter Estimation

As mentioned earlier and demonstrated in Sect. 11.6, with $\Sigma = \sigma_s^2 I_q + \sigma_w^2 J_q^q$ and $J_q \in C(Z)$, the growth curve model is a generalized split plot model, so unbiased estimates of σ_s^2 and σ_w^2 are available from that theory.

For general SCS, we mentioned earlier that

$$(Z'Z)^{-1}Z'\Sigma Z(Z'Z)^{-1} = \tilde{D}.$$

Model (11.4.1) then provides an unbiased estimate of an unstructured matrix $\tilde{D}$ via

$$(Z'Z)^{-1}Z'Y'(I-M)YZ(Z'Z)^{-1}/[n-r(X)].$$

Of course if $\tilde{D}$ contains structure, this estimate will not be particularly efficient. For example, under the assumptions associated with compound symmetry,

$$(Z'Z)^{-1}Z'\Sigma Z(Z'Z)^{-1} = \sigma_s^2(Z'Z)^{-1} + \sigma_w^2 bb'$$

for a vector b with $J_q = Zb$. We can get better estimates of this matrix by estimating the individual parameters from the generalized split plot model. Incidentally, it is not hard to see that an unbiased estimate of Θ is available from

$$(W_2'W_2)^{-1}W_2'Y'(I-M)YW_2(W_2'W_2)^{-1}/[n-r(X)].$$

We devote most of our attention to looking at unbiased estimation of the parameters in $\Sigma = \sigma^2 I_q + ZDZ'$. In particular, we show how to write a growth curve model with this covariance structure as a mixed model as in Chap. 5. For the purpose of finding unbiased covariance parameter estimates, we assume a slightly more general covariance structure $\Sigma = \sigma^2 I_q + (ZU)D(U'Z')$ where D is an unstructured covariance matrix but U is a known matrix. Incorporating U allows the possibility of involving, say, quadratic growth curves or linear growth curves with harmonic effects in Z, but restricting the random effects for individuals to be only, say, random slopes and intercepts.

On an individual level the mixed model generating such a covariance structure is

$$y_i = Z\Gamma'x_i + ZU\xi_i + \varepsilon_i, \qquad \text{Cov}(\xi_i) = D, \qquad \text{Cov}(\varepsilon_i) = \sigma^2 I_q,$$

which, as advertised, gives $\Sigma = \sigma^2 I_q + ZUDU'Z'$ under the usual assumptions. An unbiased estimate of

$$\sigma^2(U'Z'ZU)^{-1} + D$$

is

$$(U'Z'ZU)^{-1}U'Z'Y'(I-M)YZU(U'Z'ZU)^{-1}/[n-r(X)].$$

An unbiased estimate of σ^2 will then provide us with an unbiased estimate of D. The mixed model for the complete data is

$$\text{Vec}(Y) = [Z \otimes X]\text{Vec}(\Gamma) + [ZU \otimes I_n]\text{Vec}(\Xi) + \text{Vec}(e),$$

$$\text{Cov}[\text{Vec}(\Xi)] = [D \otimes I_n], \quad \text{Cov}[\text{Vec}(e)] = [\sigma^2 I_m \otimes I_n],$$

$$\text{E}[\text{Vec}(\Xi)] = 0, \quad \text{Cov}[\text{Vec}(\Xi), \text{Vec}(e)] = 0, \quad \text{E}[\text{Vec}(e)] = 0.$$

From Henderson's method 3, an unbiased estimate of σ^2 is available from

$$\text{Vec}(Y)'(I-\mathscr{P})\text{Vec}(Y)/[n-r(\mathscr{P})],$$

where $\mathscr{P}$ is the ppo onto $C([Z \otimes X], [ZU \otimes I_n])$. In the special case that $U = I_r$, this estimate is just the mean square for subplot error in the generalized split plot model associated with model (11.1.1).

11.4.3 Testing the SCS Assumption

The virtue of the MACOVA method in Sect. 11.3 is that it automatically incorporates an estimate of Σ into the process of estimating Γ while retaining the ability to conduct inferences using results known for multivariate linear models. With the SCS, least squares estimates are optimal so there is no need to estimate Σ in order to obtain a good estimate of Γ. However, the MACOVA model still provides a valuable test, namely a test of the validity of the SCS that we have assumed for Σ.

For the SCS, the MACOVA method uses $W_1 = Z(Z'Z)^{-1}$ and takes W_2 to be a full column rank matrix with $C(W_2) = C(Z)^{\perp}$. With these choices the matrix $W = [W_1, W_2]$ is nonsingular so the GMLM (11.1.1) is equivalent to

$$YW = X\Gamma ZW + e.$$

Exploiting the partitioning in W we get

$$[\tilde{Y}_1, \tilde{Y}_2] \equiv YW = [X\Gamma, 0] + e.$$

This leads to the ACOVA model

$$\tilde{Y}_1 = X\Gamma + \tilde{Y}_2 \Psi + e_{1.2}$$

where

$$\Psi = W_1' \Sigma W_2 (W_2' \Sigma W_2)^{-1}.$$

However, under SCS,

$$W_1' \Sigma W_2 = (Z'Z)^{-1} Z' [Z\tilde{D}Z' + W_2 \Theta W_2'] W_2 = 0.$$

Thus, rejecting a test of $H_0 : \Psi = 0$, provides evidence that the SCS is incorrect. In particular, if one assumes either compound symmetry or random regression coefficients, rejecting this test is evidence that those models are incorrect. However, it is possible for those models to be incorrect even when SCS is true.

Example 11.3.2 provides a test for whether the heart rate data are consistent with a quadratic growth curve model with a SCS.

11.5 Longitudinal Data

Models for the analysis of longitudinal (*repeated measures*) data are primarily concerned with the behavior of individuals over time. Both profile analysis and growth curve models can address the issue of behavior over time. In this section, we examine the structure of generalized multivariate linear models in more detail and generalize that structure to unbalanced data. We begin with the full data case. For simplicity we discuss modifications of the one-way MANOVA model.

11.5.1 Full Data

Consider the growth curve model (10.4.1) based on a one-way MANOVA. On an individual level, the model is a q dimensional linear model

$$y_{ij} = Z\gamma_i + \varepsilon_{ij}, \quad \varepsilon_{ij} \sim N(0, \Sigma),$$

with $i = 1, \ldots, a$, $j = 1, \ldots, N_i$ and different y_{ij}s independent. This model is appropriate when we have samples of individuals from each of a groups with exactly q observations on each individual, observations that are taken at exactly the same times. Typically, the columns of Z are functions of time. For growth curves, these functions of time can be polynomials as illustrated in Sect. 10.4, or other functions used for curve fitting like sines and cosines.

To create an overall linear model from the growth curve model (11.1.1) we find it convenient to transpose the model and write

$$\mathrm{Vec}(Y') = \mathrm{Vec}(Z\Gamma'X') + \mathrm{Vec}(e'), \quad \mathrm{E}[\mathrm{Vec}(e')] = 0_{nq}, \quad \mathrm{Cov}[\mathrm{Vec}(e')] = [I_n \otimes \Sigma].$$

Using result 8 from Appendix A.2, this can be written as the linear model

$$\mathrm{Vec}(Y') = [X \otimes Z]\mathrm{Vec}(\Gamma') + \mathrm{Vec}(e'), \quad \mathrm{E}[\mathrm{Vec}(e')] = 0_{nq}, \quad \mathrm{Cov}[\mathrm{Vec}(e')] = [I_n \otimes \Sigma].$$

With X a matrix of group indicators having N_i observations in group i, the overall linear model is

$$
\begin{bmatrix} y_{11} \\ \vdots \\ y_{1N_1} \\ y_{21} \\ \vdots \\ y_{2N_2} \\ \vdots \\ y_{a1} \\ \vdots \\ y_{aN_a} \end{bmatrix} = \begin{bmatrix} Z & 0 & & 0 \\ \vdots & \vdots & & \vdots \\ Z & 0 & & 0 \\ 0 & Z & & 0 \\ \vdots & \vdots & & \vdots \\ 0 & Z & & 0 \\ & & \ddots & \\ 0 & 0 & & Z \\ \vdots & \vdots & & \vdots \\ 0 & 0 & & Z \end{bmatrix} \begin{bmatrix} \gamma_1 \\ \gamma_2 \\ \vdots \\ \gamma_a \end{bmatrix} + \begin{bmatrix} \varepsilon_{11} \\ \vdots \\ \varepsilon_{1N_1} \\ \varepsilon_{21} \\ \vdots \\ \varepsilon_{2N_2} \\ \vdots \\ \varepsilon_{a1} \\ \vdots \\ \varepsilon_{aN_a} \end{bmatrix},
$$

This is a generalized multivariate linear model. None of the entries in this model is a scalar.

To explore the model further, partition

$$
Z = [J_q, Z_*] \quad \text{and} \quad \gamma_i = \begin{bmatrix} \gamma_{i0} \\ \gamma_{i*} \end{bmatrix}.
$$

A model without group-time interaction would have $\gamma_{1*} = \cdots = \gamma_{a*} \equiv \gamma_*$. The corresponding model is

$$
\begin{bmatrix} y_{11} \\ \vdots \\ y_{1N_1} \\ \vdots \\ y_{a1} \\ \vdots \\ y_{aN_a} \end{bmatrix} = \begin{bmatrix} J_q & & 0 & Z_* \\ \vdots & & 0 & Z_* \\ J_q & & 0 & Z_* \\ & \ddots & & \vdots \\ 0 & & J_q & Z_* \\ & & \vdots & \vdots \\ 0 & & J_q & Z_* \end{bmatrix} \begin{bmatrix} \gamma_{10} \\ \vdots \\ \gamma_{a0} \\ \gamma_* \end{bmatrix} + \begin{bmatrix} \varepsilon_{11} \\ \vdots \\ \varepsilon_{1N_1} \\ \vdots \\ \varepsilon_{a1} \\ \vdots \\ \varepsilon_{aN_a} \end{bmatrix}.
$$

This does not seem to be a generalized multivariate linear model but it can be written as $y_{ij} = J_q \gamma_{i0} + Z_* \gamma_* + \varepsilon_{ij}$, $i = 1, \ldots, a$, $j = 1, \ldots, N_i$.

If there is no interaction, we can consider a model without group effects,

$$
\begin{bmatrix} y_{11} \\ \vdots \\ y_{aN_a} \end{bmatrix} = \begin{bmatrix} J_q & Z_* \\ \vdots & \vdots \\ J_q & Z_* \end{bmatrix} \begin{bmatrix} \gamma_0 \\ \gamma_* \end{bmatrix} + \begin{bmatrix} \varepsilon_{11} \\ \vdots \\ \varepsilon_{aN_a} \end{bmatrix}.
$$

Equivalently, we can write this as $y_{ij} = Z\gamma + \varepsilon_{ij}$, $i = 1, \ldots, a$, $j = 1, \ldots, N_i$ or as $\text{Vec}(Y') = \text{Vec}(Z\gamma J_n') + \text{Vec}(e')$.

If there is no interaction, we can also consider a model without time effects:

$$
\begin{bmatrix} y_{11} \\ \vdots \\ y_{1N_1} \\ \vdots \\ y_{a1} \\ \vdots \\ y_{aN_a} \end{bmatrix} = \begin{bmatrix} J_q & & 0 \\ \vdots & & \vdots \\ J_q & & 0 \\ \vdots & \ddots & \vdots \\ 0 & & J_q \\ \vdots & & \vdots \\ 0 & & J_q \end{bmatrix} \begin{bmatrix} \gamma_{10} \\ \vdots \\ \gamma_{a0} \end{bmatrix} + \begin{bmatrix} \varepsilon_{11} \\ \vdots \\ \varepsilon_{1N_1} \\ \vdots \\ \varepsilon_{a1} \\ \vdots \\ \varepsilon_{aN_a} \end{bmatrix}.
$$

This model can be written as $y_{ij} = J_q \gamma_{i0} + \varepsilon_{ij}$, $i = 1,\dots,a$, $j = 1,\dots,N_i$ or $\mathrm{Vec}(Y') = \mathrm{Vec}(J_q \gamma'_0 X') + \mathrm{Vec}(e')$.

Obviously, when the groups have factorial structure, higher order ANOVA models for the groups can be incorporated into the analysis.

An additional complication allows s covariates to be measured for each individual. If w_{ij} is an s vector of covariates for the ij individual, say, initial systolic and diastolic blood pressures and cholesterol, we can also incorporate such terms into the overall model as

$$
\begin{bmatrix} y_{11} \\ \vdots \\ y_{1N_1} \\ y_{21} \\ \vdots \\ y_{2N_2} \\ \vdots \\ y_{a1} \\ \vdots \\ y_{aN_a} \end{bmatrix} = \begin{bmatrix} Z & 0 & & 0 & J_q w'_{11} \\ \vdots & \vdots & & \vdots & \vdots \\ Z & 0 & & 0 & J_q w'_{1N_1} \\ 0 & Z & & 0 & J_q w'_{21} \\ \vdots & \vdots & & \vdots & \vdots \\ 0 & Z & & 0 & J_q w'_{2N_2} \\ & & \ddots & & \vdots \\ 0 & 0 & & Z & J_q w'_{a1} \\ \vdots & \vdots & & \vdots & \vdots \\ 0 & 0 & & Z & J_q w'_{aN_a} \end{bmatrix} \begin{bmatrix} \gamma_1 \\ \gamma_2 \\ \vdots \\ \gamma_a \\ \xi \end{bmatrix} + \begin{bmatrix} \varepsilon_{11} \\ \vdots \\ \varepsilon_{1N_1} \\ \varepsilon_{21} \\ \vdots \\ \varepsilon_{2N_2} \\ \vdots \\ \varepsilon_{a1} \\ \vdots \\ \varepsilon_{aN_a} \end{bmatrix}.
$$

Defining the $n \times s$ matrix W to have an ij row of w'_{ij}, the model can be written

$$
\mathrm{Vec}(Y') = [X \otimes Z]\mathrm{Vec}(\Gamma') + [W \otimes J_q]\xi + \mathrm{Vec}(e'),
$$

$$
\mathrm{E}[\mathrm{Vec}(e')] = 0_{nq}, \qquad \mathrm{Cov}[\mathrm{Vec}(e')] = [I_n \otimes \Sigma].
$$

Similarly, covariates can be incorporated into the no interaction model, the time effects only model, and the group effects only model. Notice that these covariates are not time dependent, for example, blood pressure is only allowed to be measured once for each person, not measured at each time. It is not hard to see how this model could be modified to allow the covariate effects to depend on the group i. However, when the data come from a randomized experiment, models that incorporate treatment by covariate interaction create problems with causal interpretations of the results, cf. Cox (1958) or Christensen (2015, Section 17.8).

All of these are linear models with a parametric covariance structure so all of them are subject to the results of Chap. 4. For models that are also generalized multivariate linear models, the methods of the previous section provide data analysis. The incorporation of time-dependent covariates into the analysis eliminates any simplifying structure, so models with time-dependent covariates are modeled as if they have incomplete data.

11.5.2 Incomplete Data

With longitudinal data, it is rare to observe every person at every time, but the modeling ideas that apply to complete data also apply to incomplete data.

A natural generalization of the growth curve model is to allow different numbers of observations on different individuals making y_{ij} a $q(ij)$ vector. There are a total of q observations possible, but only $q(ij) \leq q$ are observed. With different numbers of measurements, the times of the measurements depend on the individual; hence we generalize Z to matrices Z_{ij} that depend on the individual. Now, model the data as

$$y_{ij} = Z_{ij}\gamma_i + \varepsilon_{ij}, \quad \varepsilon_{ij} \sim N(0, \Sigma_{ij}(\theta)). \tag{11.5.1}$$

The y_{ij}s remain independent, and $\Sigma_{ij}(\theta)$ is a function of some parameters θ that are the same regardless of the individual. The matrix Z_{ij} is $q(ij) \times r$ for $r \leq q$. It is permissible to have $r > q(ij)$ for some ijs. If $r = q$, typically no additional structure is being imposed related to the times, so $C(Z_{ij}) = C(I_{q(ij)}) = \mathbf{R}^{q(ij)}$, and the model is similar to a two-way ANOVA with interaction, where groups and times are the factors but with missing data and correlated observations. In the original growth curve model, it is hard to imagine Z depending on anything other than the times of measurement because Z must be the same for every individual. However, with Z depending on i and j, we can easily incorporate interesting additional covariates into Z_{ij}. For example, these measurements could include such time-dependent covariates as the weight and blood pressure of an individual taken at each time.

For each ij, the longitudinal model (11.5.1) is simply a linear model with a parametric covariance matrix. The individual models can be combined into a model for the total data

$$
\begin{bmatrix} y_{11} \\ \vdots \\ y_{1N_1} \\ y_{21} \\ \vdots \\ y_{2N_2} \\ \vdots \\ y_{a1} \\ \vdots \\ y_{aN_a} \end{bmatrix}
=
\begin{bmatrix}
Z_{11} & 0 & 0 \\
\vdots & \vdots & \vdots \\
Z_{1N_1} & 0 & 0 \\
0 & Z_{21} & 0 \\
\vdots & \vdots & \vdots \\
0 & Z_{2N_2} & 0 \\
 & & \ddots \\
0 & 0 & Z_{a1} \\
\vdots & \vdots & \vdots \\
0 & 0 & Z_{aN_a}
\end{bmatrix}
\begin{bmatrix} \gamma_1 \\ \gamma_2 \\ \vdots \\ \gamma_a \end{bmatrix}
+
\begin{bmatrix} \varepsilon_{11} \\ \vdots \\ \varepsilon_{1N_1} \\ \varepsilon_{21} \\ \vdots \\ \varepsilon_{2N_2} \\ \vdots \\ \varepsilon_{a1} \\ \vdots \\ \varepsilon_{aN_a} \end{bmatrix},
\tag{11.5.2}
$$

with

$$
E\begin{bmatrix} \varepsilon_{11} \\ \vdots \\ \varepsilon_{aN_a} \end{bmatrix} = 0 \quad \text{and} \quad \text{Cov}\begin{bmatrix} \varepsilon_{11} \\ \vdots \\ \varepsilon_{aN_a} \end{bmatrix} = \text{Blk diag}[\Sigma_{ij}(\theta)].
$$

This is *not* a standard multivariate linear model or generalized multivariate linear model, so the estimation and testing results given earlier do not apply. The analysis depends only on the general results of Chap. 4.

To explore this model further, partition

$$
Z_{ij} = [J_{q(ij)}, \; Z_{ij*}] \quad \text{and} \quad \gamma_i = \begin{bmatrix} \gamma_{i0} \\ \gamma_{i*} \end{bmatrix}.
$$

We now present a model in which the time dependent covariates (including any simple functions of time that are the same for all individuals) do not interact with the treatment groups. A model without group-time interaction would have $\gamma_{1*} = \cdots = \gamma_{a*} \equiv \gamma_*$. The corresponding model is

$$
\begin{bmatrix} y_{11} \\ \vdots \\ y_{1N_1} \\ \vdots \\ y_{a1} \\ \vdots \\ y_{aN_a} \end{bmatrix}
=
\begin{bmatrix}
J_{q(11)} & & 0 & Z_{11*} \\
\vdots & & \vdots & \vdots \\
J_{q(1N_1)} & & 0 & Z_{1N_1*} \\
 & \ddots & & \vdots \\
0 & & J_{q(a1)} & Z_{a1*} \\
\vdots & & \vdots & \vdots \\
0 & & J_{q(aN_a)} & Z_{aN_a*}
\end{bmatrix}
\begin{bmatrix} \gamma_{10} \\ \vdots \\ \gamma_{a0} \\ \gamma_* \end{bmatrix}
+
\begin{bmatrix} \varepsilon_{11} \\ \vdots \\ \varepsilon_{1N_1} \\ \vdots \\ \varepsilon_{a1} \\ \vdots \\ \varepsilon_{aN_a} \end{bmatrix}.
$$

If there is no interaction, we can consider a model without group effects,

$$
\begin{bmatrix} y_{11} \\ \vdots \\ y_{aN_a} \end{bmatrix}
=
\begin{bmatrix} J_{q(11)} & Z_{11*} \\ \vdots & \vdots \\ J_{q(aN_a)} & Z_{aN_a*} \end{bmatrix}
\begin{bmatrix} \gamma_0 \\ \gamma_* \end{bmatrix}
+
\begin{bmatrix} \varepsilon_{11} \\ \vdots \\ \varepsilon_{aN_a} \end{bmatrix},
$$

or, equivalently, $y_{ij} = Z_{ij}\gamma + \varepsilon_{ij}$. If there is no interaction, we can also consider a model without time effects:

$$
\begin{bmatrix} y_{11} \\ \vdots \\ y_{1N_1} \\ \vdots \\ y_{a1} \\ \vdots \\ y_{aN_a} \end{bmatrix}
=
\begin{bmatrix}
J_{q(11)} & & 0 \\
\vdots & & \vdots \\
J_{q(1N_1)} & & 0 \\
& \ddots & \\
0 & & J_{q(a1)} \\
& & \vdots \\
0 & & J_{q(aN_a)}
\end{bmatrix}
\begin{bmatrix} \gamma_{10} \\ \vdots \\ \gamma_{a0} \end{bmatrix}
+
\begin{bmatrix} \varepsilon_{11} \\ \vdots \\ \varepsilon_{1N_1} \\ \vdots \\ \varepsilon_{a1} \\ \vdots \\ \varepsilon_{aN_a} \end{bmatrix}.
$$

Obviously, when the groups have factorial structure, higher order ANOVA models for the groups can be incorporated into the analysis.

An additional complication allows some time dependent covariates to interact with groups while others do not. Let W_{ij} be a $q(ij) \times s$ matrix that contains time dependent covariates for each person, e.g., systolic and diastolic blood pressures measured at each time. We can incorporate such noninteracting terms into the overall model as

$$
\begin{bmatrix} y_{11} \\ \vdots \\ y_{1N_1} \\ y_{21} \\ \vdots \\ y_{2N_2} \\ \vdots \\ y_{a1} \\ \vdots \\ y_{aN_a} \end{bmatrix}
=
\begin{bmatrix}
Z_{11} & 0 & & 0 & W_{11} \\
\vdots & \vdots & & \vdots & \vdots \\
Z_{1N_1} & 0 & & 0 & W_{1N_1} \\
0 & Z_{21} & & 0 & W_{21} \\
\vdots & \vdots & & \vdots & \vdots \\
0 & Z_{2N_2} & & 0 & W_{2N_2} \\
& & \ddots & & \\
0 & 0 & & Z_{a1} & W_{a1} \\
\vdots & \vdots & & \vdots & \vdots \\
0 & 0 & & Z_{aN_a} & W_{aN_a}
\end{bmatrix}
\begin{bmatrix} \gamma_1 \\ \gamma_2 \\ \vdots \\ \gamma_a \\ \xi \end{bmatrix}
+
\begin{bmatrix} \varepsilon_{11} \\ \vdots \\ \varepsilon_{1N_1} \\ \varepsilon_{21} \\ \vdots \\ \varepsilon_{2N_2} \\ \vdots \\ \varepsilon_{a1} \\ \vdots \\ \varepsilon_{aN_a} \end{bmatrix},
$$

In this model we have presupposed that the effects of the time dependent covariates in W_{ij} are the same for each treatment group. If we allowed the effects to be different in each treatment group, we would reincorporate the W_{ij} predictors back into the Z_{ij} matrices and have the original model (11.5.2). The useful thing about this model is that it allows us to model some time dependent variables as interacting with treatments and some as having no interaction. For example, we could incorporate cholesterol into the Z_{ij} matrices and the two blood pressures into the W_{ij}s. Such a model has no blood pressure by group interaction but does have a group by cholesterol interaction.

There are many complications that can be incorporated. We could allow different intercepts for each individual or even different regressions. This might involve redefining Z_{ij} as a $q(ij) \times r_{ij}$ matrix with different definitions of the parameter vector.

To construct standard errors and tests of reduced models, use the general ideas discussed in Chap. 4, specifically in Sects. 4.1.2 and 4.7.

The biggest modeling concern is what to use for the covariance matrix $\Sigma_{ij}(\theta)$. One possibility is to take $\Sigma_{ij}(\theta)$ as a submatrix of a $q \times q$ positive definite matrix of parameters Σ. The submatrix is determined by which $q(ij)$ of the q possible times are observed in y_{ij}. Another possibility is to use a mixed model

$$y_{ij} = Z_{ij}\gamma_i + W_{ij}b_{ij} + \xi_{ij}, \quad b_{ij} \sim N(0, D), \quad \xi_{ij} \sim N(0, \sigma^2 I_{q(ij)}),$$

with the b_{ij}s and ξ_{ij}s all independent. In this model,

$$\Sigma_{ij}(\theta) = W_{ij}DW_{ij}' + \sigma^2 I_{q(ij)},$$

where D is a matrix of parameters, all of which need to be estimated along with σ^2. Alternatively, one could let D depend on some smaller number of parameters. Note that the size of D does not depend on ij, so the number of columns in W_{ij} must not depend on ij. In particular, $W_{ij} = J_{q(ij)}$ leads to a random effect for each individual and W_{ij} can also model random slopes and intercepts for each individual.

Other models for $\Sigma_{ij}(\theta)$ can be taken from time domain analysis (see Sect. 7.2), spatial data analysis (see Sect. 8.6), the next section on functional data, or linear combinations of the various possibilities. In particular, one can think of time as a one-dimensional space making the spatial data models immediately applicable.

Once a covariance structure is determined, the model can be fitted by maximum likelihood or a combination of REML and empirical generalized least squares. While these models look more complicated than spacial data models, they are in fact simpler. For each individual ij, the longitudinal model (11.5.1) is fundamentally similar to the universal kriging models considered in Chap. 8 and to the mixed models considered in Chap. 5. However, this data structure should be the fondest wish of anyone doing either spatial data or mixed models. Longitudinal data provide the luxury of having independent replications providing extra information on the common parameters. In particular, the methods for fitting spatial data models discussed in Chap. 8 all apply to these models.

For more extensive discussions of longitudinal data analysis see Hand and Crowder (1996), Diggle, Heagerty, Liang, and Zeger (2002), or Fitzmaurice, Laird, and Ware (2011).

11.6 Functional Data Analysis

Functional data analysis is really just longitudinal data analysis when q and the various $q(ij)$s are large. Unlike the previous section where we focused on one-way MANOVA type data, we here use h to identify individuals.

Consider two linear models for functional responses on each of n independent subjects, $h = 1, \ldots, n$. First, a model that concerns itself with relationships between subjects,

$$y_h(t) = x_h'\beta(t) + \varepsilon_h(t), \tag{11.6.1}$$

where $y_h(t)$ is the observed functional response, x_h is a known p vector, $\beta(t)$ is a fixed unknown p dimensional parameter function and $\varepsilon_h(t)$ is an unobservable random error function with mean 0 and covariance function $\sigma(t,\tilde{t})$.

Second is a model that treats all subjects the same but concerns itself with relationships over time

$$y_h(t) = \gamma'z(t) + \varepsilon_h(t), \tag{11.6.2}$$

where $z(t)$ is a known r dimensional function with $r \leq q$ and γ is an unknown parameter vector common to all subjects.

Of course nobody has yet figured out how to actually record an infinite amount of data, so of necessity one must evaluate the functions at a finite number of times. For simplicity in introducing concepts, assume that all subjects are observed at the same times t_j, $j = 1,\ldots,q$. Models (11.6.1) and (11.6.2) now correspond to standard longitudinal models. Restricting attention to the observed times, define

$$y_h = \begin{bmatrix} y_h(t_1) \\ \vdots \\ y_h(t_q) \end{bmatrix}, \quad Y_{n\times q} = \begin{bmatrix} y_1' \\ \vdots \\ y_n' \end{bmatrix}, \quad \varepsilon_h = \begin{bmatrix} \varepsilon_h(t_1) \\ \vdots \\ \varepsilon_h(t_q) \end{bmatrix}, \quad e = \begin{bmatrix} \varepsilon_1' \\ \vdots \\ \varepsilon_n' \end{bmatrix},$$

and

$$\beta_j = \beta(t_j), \quad B_{p\times q} = [\beta_1 \quad \cdots \quad \beta_q].$$

This allows us to write model (11.6.1) as the multivariate linear model

$$Y = XB + e. \tag{11.6.3}$$

If we further define

$$Z_{q\times r} = \begin{bmatrix} z(t_1)' \\ \vdots \\ z(t_q)' \end{bmatrix}$$

we can write model (11.6.2) as a one sample GMLM

$$Y = J\gamma'Z' + e \tag{11.6.4}$$

where J is an n vector of ones. Moreover, we can combine the modeling ideas of (11.6.1) and (11.6.2) into a GMLM

$$Y = X\Gamma Z' + e \tag{11.6.5}$$

or

$$\text{Vec}(Y) = [Z \otimes X]\text{Vec}(\Gamma) + \text{Vec}(e).$$

In all of these models,

$$\text{Cov}(y_h) = \Sigma = [\sigma_{jj'}], \quad \sigma_{jj'} = \sigma(t_j, t_{j'})$$

so that

$$\text{Cov}[\text{Vec}(Y)] = [\Sigma \otimes I_n].$$

The fundamental assumption that subjects are independent is reflected by the I_n term of the covariance matrix. In the multivariate model (11.6.3), a standard assumption is that $n \geq q + r(X)$. This provides an almost surely nonsingular (nonparametric) estimate of the covariance matrix Σ.

As we did for longitudinal models, rewrite the GMLM as

$$\text{Vec}(Y') = [X \otimes Z]\text{Vec}(\Gamma') + \text{Vec}(e'), \quad \text{Cov}[\text{Vec}(Y')] = [I_n \otimes \Sigma]$$

or

$$y_h = \sum_{j=1}^{p} x_{hj} Z\gamma_j + \varepsilon_h, \quad i = 1, \ldots, n. \tag{11.6.6}$$

In particular, if X corresponds to a one-way ANOVA, model (11.6.6) reduces to

$$y_h = Z\gamma_j + \varepsilon_h,$$

where γ_j is the regression parameter vector unique to the group that contains subject h. One advantage of this reexpression is that the covariance matrix becomes block diagonal.

A primary complication to these models viewed either as functional data or other longitudinal data is that frequently the observation times depend on the subject so that we observe subject h at t_{hj}, $j = 1, \ldots, q(h)$. Without loss of generality, we consider all the t_{hj}s as contained in the set $\{t_j | j = 1, \ldots, q\}$ so that $q(h) \leq q$. This leads to redefining

$$y_h = \begin{bmatrix} y_h(t_{h1}) \\ \vdots \\ y_h(t_{hq(h)}) \end{bmatrix} = \begin{bmatrix} y_{h1} \\ \vdots \\ y_{hq(h)} \end{bmatrix},$$

with a similar definition of ε_h. Let W_h be a $q(h) \times q$ matrix of indicators defined by

$$\begin{bmatrix} t_{h1} \\ \vdots \\ t_{hq(h)} \end{bmatrix} = W_h \begin{bmatrix} t_1 \\ \vdots \\ t_q \end{bmatrix}$$

and define

$$Z_h = W_h Z$$

so that Z_h is a $q(h) \times r$ matrix. The linear model (11.6.6) becomes

$$y_h = \sum_{j=1}^{p} x_{hj} Z_h \gamma_j + \varepsilon_h, \quad i = 1, \ldots, n. \tag{11.6.7}$$

For model (11.6.7), denote

$$\text{Cov}(y_h) = \Sigma_h = W_h \Sigma W_h'$$

which is also determined by applying $\sigma(t,\tilde{t})$ to the observation times associated with subject h, that is,

$$\Sigma_h = [\sigma_{h,jj'}], \quad \sigma_{h,jj'} = \sigma(t_{hj}, t_{hj'}).$$

A unique feature of functional data is that if one is willing to think of t as continuous, one should be prepared to deal with a large number of time observations, in particular, $q > n$. To make progress in that situation, one must assume some parametric family for $\sigma(t,\tilde{t})$. Wolfinger (1996) discussed a variety of covariance parameterizations for longitudinal models. Allowing for heteroscedasticity immediately involves q parameters. The most general models that Wolfinger considered were the antedependence, Toeplitz, and first order factor analytic models. For modeling covariances, the first two incorporate $q - 1$ additional parameters and the last adds q parameters. Thus these covariance models involve about $2q$ parameters, whereas an unstructured covariance matrix involves $q(q+1)/2$ free parameters. With q large, there is considerable room for developing "nonparametric" covariance estimates with more than $2q$ but fewer than $q(q+1)/2$ parameters.

One should also note the trade-off between fixed effects modeling of time effects and modeling the covariance structure. This trade-off is common wisdom in spatial data analysis. Typically, the more one invests in developing a complicated model for fixed effects, the less one needs to invest in developing a complicated covariance model. Relative to the models discussed here, a complicated fixed effects model for time corresponds to having Z_h matrices with a large number of columns r.

The covariance function $\sigma(t,\tilde{t})$ is a positive definite function. The theory of RKHSs relies on a result by Mercer (1909) that provides sufficient conditions on $\sigma(t,\tilde{t})$ for being able to write

$$\sigma(t,\tilde{t}) = \sum_{k=1}^{\infty} \theta_k \phi_k(t) \phi_k(\tilde{t})$$

for some functions $\phi_k(t)$ and nonnegative θ_ks. RKHS theory assumes that $\sigma(t,\tilde{t})$ is known and avoids working with the functions $\phi_k(t)$. Similar to Chap. 1, we instead use a collection of known spanning functions for the $\phi_k(t)$s and estimate the θ_ks. Of course, we cannot handle an infinite sum so we use the approximation

$$\sigma(t,\tilde{t}) \doteq \sum_{k=1}^{s} \theta_k \phi_k(t) \phi_k(\tilde{t}). \tag{11.6.8}$$

For our purposes, we would choose s between $2q$ and $q(q+1)/2$.

When applying (11.6.8)

$$\Sigma_h = [\sigma_{h,jj'}], \quad \sigma_{h,jj'} = \sigma(t_{hj}, t_{hj'}) = \sum_{k=1}^{s} \theta_k \phi_k(t_{hj}) \phi_k(t_{hj'}).$$

Incorporating this covariance structure into the models is quite simple via random effects. First define

$$\Phi_{hk} = [\phi_k(t_{h1}) \quad \cdots \quad \phi_k(t_{hq(h)})]', \quad \text{and} \quad \Phi_h = [\Phi_{h1} \quad \cdots \quad \Phi_{hs}].$$

In any of the subjectwise models simply replace the term ε_h with a term $\Phi_h b_h$ where b_h is a random s vector having mean 0, uncorrelated components, and variances $\theta_1, \ldots, \theta_s$. Thus, in model (11.6.7) fit

$$y_h = \sum_{j=1}^{p} x_{hj} Z_h \gamma_j + \Phi_h b_h.$$

Technically, this is very similar to the random coefficients models discussed by Wolfinger (1996). He discussed using both first order and second order polynomials, so $s = 1, 2$. In addition, Wolfinger retained ε_h but assumed that $\text{Cov}(\varepsilon_h) = \sigma^2 I_{q(h)}$, thus making a model

$$y_h = \sum_{j=1}^{p} x_{hj} Z_h \gamma_j + \Phi_h b_h + \varepsilon_h.$$

I can see no reason not to include such an error term. Relative to the trade-off between fixed effects and covariance modeling, using Z_hs with large r allow s to be smaller.

The use of sines and cosines for the basis functions is particularly appealing because of their close connection with frequency domain time series analysis. Let t_q be the last time at which any observation was taken, then

$$\phi_{2k-1}(t) = \cos\left(2\pi \frac{k}{t_q} t\right), \quad \text{and} \quad \phi_{2k}(t) = \sin\left(2\pi \frac{k}{t_q} t\right).$$

Note that if we assume $\theta_{2k-1} = \theta_{2k}$ for all k, both $\sigma(t, \tilde{t})$ and its approximation will be homoscedastic.

Wolfinger also examined a factor analysis model, cf. Sect. 14.4, with

$$\Sigma = \begin{bmatrix} \theta_{01} & & \\ & \ddots & \\ & & \theta_{0q} \end{bmatrix} + \begin{bmatrix} \theta_{11} \\ \vdots \\ \theta_{1q} \end{bmatrix} \begin{bmatrix} \theta_{11} \\ \vdots \\ \theta_{1q} \end{bmatrix}'.$$

He mentions that this can be generalized to

$$\Sigma = \begin{bmatrix} \theta_{01} & & \\ & \ddots & \\ & & \theta_{0q} \end{bmatrix} + \sum_{k=1}^{s} \begin{bmatrix} \theta_{k1} \\ \vdots \\ \theta_{kq} \end{bmatrix} \begin{bmatrix} \theta_{k1} \\ \vdots \\ \theta_{kq} \end{bmatrix}'.$$

This covariance model includes $q + qs - s(s-1)/2$ free parameters.

If the functional data involve function domains with more than one dimension, the situation is analogous to fitting a spatial model to each individual. For example, if $t \in \mathbf{R}^2$ so that $t = (t_1, t_2)'$, one can use two sets of functions ϕ_k and ψ_k and model

$$\sigma(t, \tilde{t}) \doteq \sum_{k=1}^{s} \theta_k \phi_k(t_1) \psi_k(t_2) \phi_k(\tilde{t}_1) \psi_k(\tilde{t}_2).$$

or

$$\sigma(t, \tilde{t}) \doteq \sum_{k=1}^{s_1} \sum_{k'=1}^{s_2} \theta_{kk'} \phi_k(t_1) \psi_{k'}(t_2) \phi_k(\tilde{t}_1) \psi_{k'}(\tilde{t}_2).$$

11.7 Generalized Split Plot Models

Christensen (1987) introduced generalized split plot (GSP) models as a broad family of linear models that display the salient characteristics of split plot models. For example, least squares estimates are optimal, exact F tests are available for whole plot effects where they are tested against a whole-plot error term, and subplot effects (including subplot treatment by whole-plot treatment interactions) have exact F tests based on a subplot error term.

In this section we relate generalized multivariate linear models (GMLMs) to GSP models following, with modifications, the notation of *PA* Chap. 11. Let $y_i = (y_{i1}, \ldots, y_{im})'$ be the vector of observations associated with whole-plot i. A fundamental assumption of split plot models is that

$$\mathrm{Cov}(y_i) = \sigma_s^2 I_m + \sigma_w^2 J_m^m$$

where σ_s^2 is the subplot variance and σ_w^2 is the whole-plot variance. Different y_is are assumed to be uncorrelated and for testing they are collectively assumed to be multivariate normal.

We denote the number of whole plots by $\tilde{n}$ and, with m subplots in each whole plot, the total number of observations is $n = \tilde{n}m$. GSP models do not require any form of balance among the whole-plots, but there is little hope of developing a simple whole-plot analysis without the assumption that the subplots are balanced. The whole plot analysis is based on averaging over the observations in the subplots, and if those averages are not balanced, the whole-plot means are no longer comparable to one another. Christensen (2015, Subsection 19.2.1) exploits results in Christensen (1984) to provide an analysis of split plot data with missing subplots but this is done at the expense of abandoning the whole-plot analysis.

In addition to subplot balance, GSP models assume a particular structure to the model matrix. Let $\mathscr{Y}$ be the n vector of all observations. Write the GSP model as

$$\mathscr{Y} = \mathscr{X}_* \delta + \mathscr{X}_2 \gamma + \mathbf{e},$$

$$\mathrm{E}(\mathbf{e}) = 0, \quad \mathrm{Cov}(\mathbf{e}) = \sigma_s^2 I_n + \sigma_w^2 \mathscr{X}_1 \mathscr{X}_1' = \mathrm{Cov}(\mathscr{Y})$$

where $\mathscr{X}_1$ is a matrix whose columns are indicator variables for the whole plots. Note that the ppo onto $C(\mathscr{X}_1)$, is

$$\mathscr{M}_1 \equiv \frac{1}{m} \mathscr{X}_1 \mathscr{X}_1'.$$

The conditions required of GSP models are that

$$C(\mathscr{X}_*) \subset C(\mathscr{X}_1) \quad \text{and} \quad C(\mathscr{X}_*, \mathscr{X}_2) = C[\mathscr{X}_*, (I - \mathscr{M}_1)\mathscr{X}_2].$$

To get this second condition satisfied, it is enough to show that

$$C(\mathscr{M}_1 \mathscr{X}_2) \subset C(\mathscr{X}_*).$$

The ppo onto $C(\mathscr{X}_*, \mathscr{X}_2)$ is $\mathscr{M}_* + \mathscr{M}_2$ where $\mathscr{M}_*$ is the ppo onto $C(\mathscr{X}_*)$ and $\mathscr{M}_2$ is the ppo onto $C[(I - \mathscr{M}_1)\mathscr{X}_2]$ and the two ppos are orthogonal. The sum of squares for whole plot error is $\mathscr{Y}'(\mathscr{M}_1 - \mathscr{M}_*)\mathscr{Y}$ and the sum of squares for subplot error is $\mathscr{Y}'(I - \mathscr{M}_1 - \mathscr{M}_2)\mathscr{Y}$.

11.7.1 GMLMs Are GSP Models

We now demonstrate how to associate the GMLM matrix $[Z \otimes X]$ from (11.1.2) with the model matrix for a GSP. In this association, $q \equiv m$, that is, the number of dependent variables in the GMLM is precisely the number of subplot units. While $n = \tilde{n}m$ in a GSP is the total number of observations, the number of rows in Y and X of the GMLM is $\tilde{n}$. (In all other discussions we have called this number n.) The matrix X determines the model for the whole-plot analysis, e.g., in a split plot experimental design, X is determined by the whole-plot design. We assume that we can partition Z as

$$Z = [J_m, Z_2].$$

Let $\mathscr{Y} \equiv \mathrm{Vec}(Y)$ with

$$\mathscr{X}_1 = [J_m \otimes I_{\tilde{n}}], \quad \mathscr{X}_* = [J_m \otimes X], \quad \mathscr{X}_2 = [Z_2 \otimes X].$$

This makes the first column of Γ into δ, the whole plot effect vector. Clearly,

$$C(\mathscr{X}_*, \mathscr{X}_2) = C([J_m \otimes X], [Z_2 \otimes X]) = C([Z \otimes X]).$$

Defining M_2 as the ppo onto $C\left[\left(I_m - \frac{1}{m}J_m^m\right)Z_2\right]$, we get

$$\mathscr{M}_1 = \left[\frac{1}{m}J_m^m \otimes I_{\tilde{n}}\right], \qquad \mathscr{M}_* = \left[\frac{1}{m}J_m^m \otimes M\right], \qquad \mathscr{M}_2 = [M_2 \otimes M]$$

and also, by direct computation,

$$\mathcal{M}_* + \mathcal{M}_2 = [M_Z \otimes M].$$

To prove that model (11.1.2) satisfies the model matrix conditions for a GSP model, we need to show that $C(\mathcal{X}_*, \mathcal{X}_2) = C[\mathcal{X}_*, (I - \mathcal{M}_1)\mathcal{X}_2]$, which is equivalent to showing that $C(\mathcal{M}_1 \mathcal{X}_2) \subset C(\mathcal{X}_*)$. In the GMLM,

$$\mathcal{M}_1 \mathcal{X}_2 = \left[\frac{1}{m} J_m^m \otimes I_{\tilde{n}}\right][Z_2 \otimes X] = \left[\frac{1}{m} J_m^m Z_2 \otimes X\right].$$

It follows that

$$C(\mathcal{M}_1 \mathcal{X}_2) = C\left(\left[\frac{1}{m} J_m^m Z_2 \otimes X\right]\right) \subset C([J_m \otimes X]) = C(\mathcal{X}_*).$$

GMLMs are not as flexible as GSP models. For example, the standard split plot model involves whole-plot blocks, whole-plot treatments, subplot treatments, and whole-plot treatment by subplot treatment interaction. A GMLM that contained all of those terms would also have a whole-plot blocks by subplot treatment interaction.

11.8 Additional Exercises

Exercise 11.8.1. Box (1950) gives data on the weights of three groups of rats. One group was given thyroxin in their drinking water, one thiouracil, and the third group was a control. Weights are measured in grams at weekly intervals. The data were given in Table 10.5. Analyze the data using the analysis of covariance method for growth curve analysis.

Exercise 11.8.2. Box (1950) presents data on the weight loss of a fabric due to abrasion. Two fillers were used in three proportions. Some of the fabric was given a surface treatment. Weight loss was recorded after 1000, 2000, and 3000 revolutions of a machine designed to test abrasion resistance. The data are given in Table 11.1. Perform a multivariate analysis of variance, a profile analysis, and fit a simple linear growth curve model. Check your assumptions.

Table 11.1 Abrasion resistance

Surface		Proportions								
		25%			50%			75%		
Treatment	Fill	1000	2000	3000	1000	2000	3000	1000	2000	3000
Yes	A	194	192	141	233	217	171	265	252	207
	B	208	188	165	241	222	201	269	283	191
	A	239	127	90	224	123	79	243	117	100
	B	187	105	85	243	123	110	226	125	75
No	A	155	169	151	198	187	176	235	225	166
	B	173	152	141	177	196	167	229	270	183
	A	137	82	77	129	94	78	155	76	91
	B	160	82	83	98	89	48	132	105	67

References

Box, G. E. P. (1950). Problems in the analysis of growth and wear curves. *Biometrics, 6*, 362–389.

Christensen, R. (1984). A note on ordinary least squares methods for two-stage sampling. *Journal of the American Statistical Association, 79*, 720–721.

Christensen, R. (1987). The analysis of two-stage sampling data by ordinary least squares. *Journal of the American Statistical Association, 82*, 492–498.

Christensen, R. (2015). *Analysis of variance, design, and regression: Linear modeling for unbalanced data* (2nd ed.). Boca Raton, FL: Chapman and Hall/CRC Press.

Cox, D. R. (1958). *Planning of experiments*. New York: Wiley.

Diggle, P. J., Heagerty, P., Liang, K.-Y., & Zeger, S. L. (2002). *Analysis of longitudinal data* (2nd ed.). Oxford: Oxford University Press.

Fitzmaurice, G. M., Laird, N. M., & Ware, J. H. (2011). *Applied longitudinal analysis* (2nd ed.). New York: Wiley.

Geisser, S. (1970). Bayesian analysis of growth curves, *Sankhya Series A, 32*, 53–64.

Hand, D. J., & Crowder, M. J. (1996). *Practical longitudinal data analysis*. London: Chapman and Hall.

Khatri, C. G. (1966). A note on a MANOVA model applied to problems in growth curves. *Annals of the Institute of Statistical Mathematics, 18*, 75–86.

Mercer, J. (1909) Functions of positive and negative type and their connection with the theory of integral equations. *Philophical Transactions of the Royal Society of London, A, 209*, 415–446.

Potthoff, R. F., & Roy, S. N. (1964). A generalized multivariate analysis of variance model useful especially for growth curve problems. *Biometrika, 51*, 313–326.

Rao, C. R. (1965). The theory of least squares when the parameters are stochastic and its application to the analysis of growth curves. *Biometrika, 52,* 447–458.

Rao, C. R. (1966). Covariance adjustment and related problems in multivariate analysis. In P. R. Krishnaiah (Ed.), *Multivariate analysis - II.* New York: Academic Press.

Rao, C. R. (1967). Least squares theory using an estimated dispersion matrix and its application to measurement signals. In *Proceedings of the Fifth Berkeley Symposium on Mathematical Statistics and Probability* (vol. 1, pp. 355–372).

Wolfinger, R. D. (1996). Heterogeneous variance-covariance structures for repeated measures. *Journal of Agricultural, Biological, and Environmental Statistics, 1,* 205–230.

Chapter 12
Discrimination and Allocation

Abstract This chapter discusses discrimination and allocation. Regression data are commonly the result of sampling a population, taking two or more measurements on each individual sampled, and then examining how those variables relate to one another. Discrimination problems have a very different sampling scheme. In discrimination problems data are obtained from multiple groups and we seek efficient means of telling the groups apart, i.e., discriminating between them. Discrimination is closely related to one-way multivariate analysis of variance in that we seek to tell groups apart. One-way multivariate analysis of variance addresses the question of whether the groups are different whereas discriminant analysis seeks to specify how the groups are different. Allocation is the problem of assigning new individuals to their appropriate group. Allocation procedures have immediate application to diagnosing medical conditions.

Consider the eight populations of people determined by all combinations of sex (male, female) and age (adult, adolescent, child, infant). These are commonly used distinctions, but the populations are not clearly defined. It is not obvious when infants become children, when children become adolescents, nor when adolescents become adults. On the other hand, most people can clearly be identified as members of one of these eight groups. It might be of interest to see whether one can *discriminate* among these populations on the basis of, say, various aspects of their blood chemistry. The discrimination problem is sometimes referred to as the problem of *separation*. Another potentially interesting problem is trying to predict the population of a new individual given only the information on their blood chemistry. The problem of predicting the population of a new case is referred to as the problem of *allocation*. Other names for this problem are *identification* and *classification*.

EXAMPLE 12.0.1.   Aitchison and Dunsmore (1975) present data on Cushing's syndrome, a medical condition characterized by overproduction of cortisol by the adrenal cortex. Twenty-one individuals were identified as belonging to one of three types: *adenoma*, *bilateral hyperplasia*, and *carcinoma*. The amounts of tetrahydrocortisone and pregnanetriol excreted in the urine were measured. The data are given

© Springer Nature Switzerland AG 2019

R. Christensen, *Advanced Linear Modeling*, Springer Texts in Statistics,

https://doi.org/10.1007/978-3-030-29164-8_12

in Table 12.1. We wish to discriminate among the three types of Cushing's syndrome based on the urinary excretion data. A quick glance at Table 12.1 establishes that none of the data groups seems to be from a bivariate normal distribution. The pregnanetriol value for case 4 looks out of place because it is the only value that is nonzero in the hundredths place but, from looking at similar data, it is apparently a valid observation. Following Aitchison and Dunsmore (1975), the analysis is performed on the logarithms of the data. The log data are plotted in Fig. 12.1. Ripley (1996, Section 2.4) discusses these data in the context of predictive allocation.   □

Table 12.1 Cushing's syndrome data

Case	Type	TETRA	PREG	Case	Type	TETRA	PREG
1	A	3.1	11.70	12	B	15.4	3.60
2	A	3.0	1.30	13	B	7.7	1.60
3	A	1.9	0.10	14	B	6.5	0.40
4	A	3.8	0.04	15	B	5.7	0.40
5	A	4.1	1.10	16	B	13.6	1.60
6	A	1.9	0.40	17	C	10.2	6.40
7	B	8.3	1.00	18	C	9.2	7.90
8	B	3.8	0.20	19	C	9.6	3.10
9	B	3.9	0.60	20	C	53.8	2.50
10	B	7.8	1.20	21	C	15.8	7.60
11	B	9.1	0.60				

Most books on multivariate analysis contain extensive discussions of discrimination and allocation. The author can particularly recommend the treatments in Anderson (2003), Johnson and Wichern (2007), and Seber (1984). In addition, Hand (1981) and Lachenbruch (1975) have written monographs on the subject. As is so often the case in statistics, the first modern treatment of these problems was by Sir Ronald A. Fisher; see Fisher (1936, 1938). The discussion in this chapter is closely related to methods associated with the multivariate normal distribution. One alternative approach is based on logistic regression; see also Christensen (1997), Press (1982), Press and Wilson (1978), or Seber (1984). There are also a variety of nonparametric methods available. More recently methods such as support vector machines and other "machine learning" tools have been applied to this problem, cf. Hastie, Tibshirani, and Friedman (2016). Logistic regression and discrimination, other similar generalized linear models, and support vector machines are examined in the next chapter.

Discrimination seems to be a purely descriptive endeavor. The observations are vectors in $\mathbf{R}^q$. All observations come from known populations. Discriminate analysis uses the observations to partition $\mathbf{R}^q$ into regions, each uniquely associated with a particular population. Given a partition, it is easy to allocate future observations. An observation y is allocated to population r if y falls into the region of $\mathbf{R}^q$ associated with the rth population. The difficulty lies in developing a rational approach to partitioning $\mathbf{R}^q$.

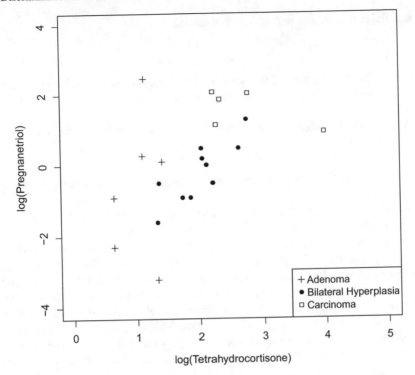

Fig. 12.1 Cushing's syndrome data

Just as a solution to the discrimination problem implicitly determines an allocation rule, a solution to the allocation problem implicitly solves the discrimination problem. The set of all y values to be allocated to population r determines the region associated with population r.

Our discussion will be centered on the allocation problem. We present allocation rules based on Mahalanobis distance, maximum likelihood, and Bayes theorem. An advantage of the Mahalanobis distance method is that it is based solely on the means and covariances of the population distributions. The other methods require knowledge of the entire distribution in the form of a density. Not surprisingly, the Mahalanobis, the maximum likelihood, and the Bayes rules are similar for normally distributed populations.

In general, consider the situation in which there are t populations and q variables $y_1, \ldots, y_q$ with which to discriminate among them. In particular, if $y = (y_1, \ldots, y_q)'$ is an observation from the ith population, we assume that either the mean and covariance matrix of the population are known, say

$$E(y) = \mu_i$$

and

$$\text{Cov}(y) = \Sigma_i,$$

or that the density of the population distribution, say

$$f(y|i),$$

is known. In practice, neither the density, the mean, nor the covariance matrix will be known for any population. These must be estimated using data from the various populations.

An important special case is where the covariance matrix is the same for all populations, say

$$\Sigma \equiv \Sigma_1 = \cdots = \Sigma_t.$$

In this case, samples from the populations constitute data for a standard one-way MANOVA.

Section 12.1 deals with the general allocation problem. Section 12.2 examines quadratic discriminant analysis (QDA). It applies the general ideas by estimating parameters and densities. Section 12.3 examines linear discriminant analysis (LDA). It is the special case of equal covariance matrices. Section 12.4 introduces ideas of cross-validation for estimating error rates. Section 12.5 contains some general discussion. Sections 12.6 and 12.7 examine the relationship between MANOVA and LDA and in particular discuss a method for selecting variables in LDA and introduce discrimination coordinates that are useful in visualizing the discrimination procedure. Finally, Sect. 12.8 discusses a broader idea of linear discrimination that includes both LDA and QDA and relates it to the ideas explored in the next chapter.

In recent years it has become quite common to refer to binary regression problems as classification problems. In olden times, classification was used as a synonym for what we are doing here. Discrimination tries to tell two or more groups apart. Binary regression also tries to tell two groups apart and generalizations of it seek to tell more than two groups apart. The big difference between binary regression and discrimination is in the nature of the data. In binary regression you sample a population and try to predict the subgroup in which a new observation will fall. In discrimination you separately sample from the subgroups but still try to predict where a new observation from the population will fall. The key difference is that with discrimination data the researcher determines the sample size from each group, whereas in binary regression the number observed from each group is itself a binomial random variable determined by the overall sample size and the population probability for that group. With discrimination data, allocating a new observation to a group requires knowledge from outside the data collection about the population probabilities of the groups.

It is tempting to use binary regression methods to analyze discrimination data but such methods always need to be adjusted for the group probabilities. Regression methods will incorrectly treat the group sample size proportions as estimates of the group probabilities, when in fact the group sample size proportions are determined by the researcher. In fact a frequent reason for collecting discrimination data is to obtain more information on groups that occur infrequently in the general population, e.g., to better diagnose rare diseases. Binary generalized linear models and related regression/discrimination methods are examined in the next chapter.

12.1 The General Allocation Problem

In this section, we discuss allocation rules based on Mahalanobis distance, maximum likelihood, and Bayes theorem. These rules are based on populations with either known means and covariances or known densities.

12.1.1 Mahalanobis Distance

As discussed in *PA-V* Sect. 12.1 (Christensen 2011, Section 13.1), the Mahalanobis distance

$$D^2 = (y - \mu)' \Sigma^{-1} (y - \mu)$$

is a frequently used measure of how far a random vector is from the center of its distribution. In the allocation problem, we have a random vector y and t possible distributions from which it could arise. A reasonable allocation procedure is to assign y to the population that minimizes the observed Mahalanobis distance. In other words, allocate y to population r if

$$(y - \mu_r)' \Sigma_r^{-1} (y - \mu_r) = \min_i (y - \mu_i)' \Sigma_i^{-1} (y - \mu_i). \qquad (12.1.1)$$

12.1.2 Maximum Likelihood

If the densities $f(y|i)$ are known for each population, the population index i is the only unknown parameter. Given an observation y, the likelihood function is

$$L(i) \equiv f(y|i),$$

which is defined for $i = 1, \ldots, t$. The maximum likelihood allocation rule assigns y to population r if

$$L(r) = \max_i L(i),$$

or equivalently if

$$f(y|r) = \max_i f(y|i).$$

If the observations have a multivariate normal distribution, the maximum likelihood rule is very similar to the Mahalanobis distance rule. From *PA* Sect. 1.3, the likelihoods (densities) are

$$L(i) = f(y|i) = (2\pi)^{-q/2} |\Sigma_i|^{-1/2} \exp[-(y - \mu_i)' \Sigma_i^{-1} (y - \mu_i)/2],$$

$i = 1, \ldots, t$, where $|\Sigma_i|$ is the determinant of the covariance matrix. The logarithm is a monotone increasing function, so maximizing the log-likelihood is equivalent to maximizing the likelihood. The log likelihood is

$$\ell(i) \equiv \log[L(i)] = -\frac{q}{2}\log(2\pi) - \frac{1}{2}\log(|\Sigma_i|) - \frac{1}{2}(y - \mu_i)'\Sigma_i^{-1}(y - \mu_i).$$

If we drop the constant term $-\frac{q}{2}\log(2\pi)$ and minimize twice the negative of the log-likelihood rather than maximizing the log-likelihood, we see that the maximum likelihood rule for normally distributed populations is: assign y to population r if

$$\log(|\Sigma_r|) + (y - \mu_r)'\Sigma_r^{-1}(y - \mu_r) = \min_i \left\{ \log(|\Sigma_i|) + (y - \mu_i)'\Sigma_i^{-1}(y - \mu_i) \right\}.$$

(12.1.2)

The only difference between the maximum likelihood rule and the Mahalanobis rule is the inclusion of the term $\log(|\Sigma_i|)$ which does not depend on y, the case to be allocated. Both the Mahalanobis rule and the maximum likelihood rule involve quadratic functions of y. Methods related to these rules are often referred to as *quadratic discrimination* methods. (As discussed in Chap. 1, quadratic functions of the predictor variables define linear functions of an expanded set of predictor variables. More on this in the chapter's last section and in Exercise 13.4.)

EXAMPLE 12.1.1.   The simplest case of allocation is assigning a new observation y to one of two normal populations with the same variance. The top panel of Fig. 12.2 contains normal densities with variance 1. The one on the left has mean 2; the one on the right has mean 5. The solid dot is the point at which the two densities are equal. For y to the right of the dot, the maximum likelihood allocation is to the mean 5 population. To the left of the dot, the maximum likelihood allocation is to the mean 2 population. The Mahalanobis distance in this one dimensional problem is $|y - \mu_i|/1$. The black dot is also the solution to $y - 2 = y - 5$, so y values to the left of the dot are closer to the mean 2 population and those to the right are closer to the mean 5 population.

The bottom panel of Fig. 12.2 is far more complicated because it involves unequal variances. The population on the left is $N(2, 1)$ whereas the population on the right is now $N(5, 9)$. The two squares are where the densities from the two distributions are equal. To the left of the left square and to the right of the right square, the $N(5, 9)$ has a higher density, so a y in those regions would be assigned to that population. Between the two squares, the $N(2, 1)$ has a higher density, so a y between the squares is assigned to the mean 2 population. The squared Mahalanobis distances are $(y - 2)^2/1$ and $(y - 5)^2/9$. Setting these equal gives the two black dots. Again, y is assigned to the mean 2 population if and only if y is between the black dots. The normal theory maximum likelihood and Mahalanobis methods are similar but distinct.

□

12.1.3 Bayesian Methods

We will discuss two procedures for Bayesian allocation. One is an intuitive rule. The other is a formal procedure based on costs of misclassification. It will also be shown that the intuitive rule can be arrived at by the formal procedure.

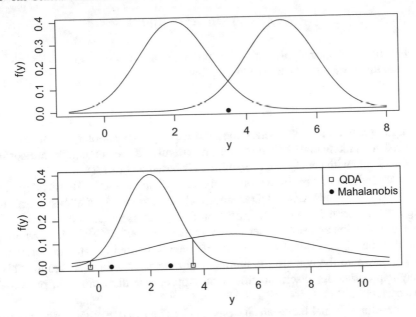

Fig. 12.2 One dimensional normal discrimination

Bayesian allocation methods presuppose that for each population i there exists a prior probability, say $\pi(i)$, that the new observation y comes from that population. Typically, these prior probabilities are arrived at either from previous knowledge of the problem or through the use of the maximum entropy principle (see Berger 1985, Section 3.4). The maximum entropy principle dictates that the $\pi(i)$s should be chosen to minimize the amount of information contained in them. This is achieved by selecting

$$\pi(i) = 1/t, \tag{12.1.3}$$

$i = 1, \ldots, t$.

Given the prior probabilities and the data y, the posterior probability that y came from population i can be computed using Bayes theorem (see Berger 1985, Section 4.2 or Christensen, Johnson, Branscum, & Hanson 2010, Chapter 2). The posterior probability is

$$\pi(i|y) = f(y|i)\pi(i) \Big/ \sum_{j=1}^{t} f(y|j)\pi(j). \tag{12.1.4}$$

A simple intuitive allocation rule is to assign y to the population with the highest posterior probability. In other words, assign y to population r if

$$\pi(r|y) = \max_i \pi(i|y).$$

The denominator in (12.1.4) does not depend on i, so the allocation rule is equivalent to choosing r such that

$$f(y|r)\pi(r) = \max_i \{f(y|i)\pi(i)\}.$$

In the important special case in which the $\pi(i)$ are all equal, this corresponds to maximizing $f(y|i)$; that is, choosing r so that

$$f(y|r) = \max_i f(y|i).$$

Thus, for equal initial probabilities, the intuitive Bayes allocation rule is the same as the maximum likelihood allocation rule. In particular, if the populations are normal, the Bayes rule with equal prior probabilities is based on (12.1.2).

Most methods of discrimination are based on estimating the density functions $f(y|i)$. These include LDA, QDA, and such nonparametric methods as nearest neighbors and kernels (in the sense of Sect. 1.7.2). As discussed in the next chapter, logistic regression and some other binary regression methods give direct estimates of $\pi(i|y)$, but those estimates are based on having implicitly estimated $\pi(i)$ with $N_i/\sum_j N_j$, where the N_js are the numbers of observations in each group. This is rarely appropriate for discrimination data. Appropriate discrimination procedures must correct for this.

To develop a formal Bayesian analysis requires knowledge of the costs of correct classification (allocation) and of misclassification. Let $c(j|i)$ be the cost of classifying y into population j when y is in fact from population i. Note that $c(i|i)$ is the cost of correct classification. For any $j \neq i$, $c(j|i)$ is a cost for misclassification. Typically, we would take $c(i|i) \leq c(j|i)$ for all j.

The expected cost of classifying y into population i (given the data) is simply

$$C(i|y) = \sum_{j=1}^{t} c(i|j)\pi(j|y).$$

The formal Bayes allocation rule is to assign y to population r if

$$C(r|y) = \min_i C(i|y).$$

Now, consider the special case where

$$c(i|i) = 0 \tag{12.1.5}$$

and

$$c(j|i) = c \quad \text{for} \quad j \neq i. \tag{12.1.6}$$

This cost structure leads to the intuitive Bayes rule discussed earlier. Clearly, the formal Bayes rule is equivalent to choosing r such that

$$c - C(r|y) = \max_i \{c - C(i|y)\}.$$

Note two things. First, by the definition of probability,

$$\sum_{j=1}^{t} \pi(j|y) = 1.$$

Second, from the assumed cost structure,

$$C(i|y) = c \sum_{j \neq i} \pi(j|y).$$

Using these two facts,

$$c - C(i|y) = c \sum_{j=1}^{t} \pi(j|y) - c \sum_{j \neq i} \pi(j|y)$$

$$= c \, \pi(i|y).$$

Thus, the formal Bayes rule allocates y to population r if

$$c \, \pi(r|y) = \max_{i} c \, \pi(i|y),$$

or equivalently if

$$\pi(r|y) = \max_{i} \pi(i|y).$$

This is precisely the intuitive rule given earlier.

Combining our Bayesian results we see that assuming equal prior probabilities as in (12.1.3) and the classification cost structure of (12.1.5) and (12.1.6), the formal Bayes rule is identical to the maximum likelihood rule. In particular, for normally distributed populations, the Bayes rule is given by (12.1.2).

12.2 Estimated Allocation and QDA

One serious problem with the allocation rules of the previous section is that typically the moments and the densities are unknown. In (12.1.1) and (12.1.2) typically the values μ_i and Σ_i are unknown. In practice, allocation is often based on estimated means and covariances or estimated densities.

We assume that a random sample of $N_i \equiv N(i)$ observations is available from the ith population. The jth observation from the ith population is denoted $y_{ij} \equiv (y_{ij,1}, \ldots, y_{ij,q})'$. Note that

$$E(y_{ij}) = \mu_i$$

and

$$\text{Cov}(y_{ij}) = \Sigma_i.$$

It is important to recognize that in the special case of equal covariance matrices, the data follow a multivariate one-way ANOVA model with t groups. It will be convenient to write a matrix that contains the ith sample,

$$Y_i \equiv \begin{bmatrix} y'_{i1} \\ \vdots \\ y'_{iN(i)} \end{bmatrix}.$$

The estimated Mahalanobis distance rule is that an observation y is allocated to population r if

$$(y - \bar{y}_{r\cdot})' S_r^{-1} (y - \bar{y}_{r\cdot}) = \min_i (y - \bar{y}_{i\cdot})' S_i^{-1} (y - \bar{y}_{i\cdot}),$$

where

$$S_i = \sum_{j=1}^{N_i} (y_{ij} - \bar{y}_{i\cdot})(y_{ij} - \bar{y}_{i\cdot})' \Big/ (N_i - 1) = Y_i' \left[I - \frac{1}{N_i} J_{N(i)}^{N(i)} \right] Y_i \Big/ (N_i - 1).$$

An estimated maximum likelihood allocation rule is to assign y to population r if

$$\hat{f}(y|r) = \max_i \hat{f}(y|i).$$

If q is not too large, the estimate $\hat{f}(y|i)$ can be estimated nonparametrically using nearest neighbors or kernels (similar to those used in Sect. 1.7.3) or, if $\hat{f}(y|i)$ depends on parameters θ_i, $\hat{f}(y|i)$ can be obtained by estimating the parameters, which is what we do for multivariate normals.

For multivariate normal densities, an estimated maximum likelihood allocation rule is to assign y to population r if

$$\log(|S_r|) + (y - \bar{y}_{r\cdot})' S_r^{-1} (y - \bar{y}_{r\cdot}) = \min_i \{ \log(|S_i|) + (y - \bar{y}_{i\cdot})' S_i^{-1} (y - \bar{y}_{i\cdot}) \}.$$

Application of this estimated normal theory allocation rule is often referred to as *quadratic discriminant analysis (QDA)*.

Although the QDA allocation decision is based a quadratic function of the predictor variables in y, the quadratic function in y can also be written as a linear combination of the y predictor variables, their squares, and their crossproducts. In one sense, QDA is not linear discrimination but in another sense it is. (In terms of the "training" data y_{ij}, it is neither quadratic nor linear but rather a very complicated function.) This issue is discussed further in Sect. 12.8.

With estimated parameters, the Bayes allocation rule is really only a quasi-Bayesian rule. The allocation is Bayesian, but the estimation of $f(y|i)$ is not. Geisser's (1971) suggestion of using the Bayesian predictive distribution as an estimate of $f(y|i)$ has been shown to be optimal under frequentist criteria by Aitchison (1975), Murray (1977), and Levy and Perng (1986). It also provides the optimal Bayesian allocation. In particular, for normal data, treating the maximum likelihood estimates as the mean and covariance matrix of the normal gives an inferior estimate for the distribution of new observations. The appropriate distribution (for "noninformative" priors) is a multivariate t with the same location vector and a covariance matrix that is a multiple of the MLE. See Geisser (1977) for a discussion of these issues in relation to discrimination.

In any case, plugging in estimates of μ_i and Σ_i requires that good estimates be available. Friedman (1989) has proposed an alternative estimation technique for use with small samples.

Finally, in examining the data, it is generally not enough to look just at the results of the allocation. Typically, one is interested not only in the population to which a case is allocated but also in the clarity of the allocation. It is desirable to know whether a case y is clearly in one population or whether it could have come from two or more populations. The posterior probabilities from the Bayesian method address these questions in the simplest fashion. Similar information can be gathered from examining the entire likelihood function or the entire set of Mahalanobis distances.

We now illustrate QDA on the Cushing's syndrome data. Clarity of allocation is addressed later in Example 12.5.1. An alternative analysis of these data based on assuming equal covariance matrices for the groups is presented in the next section.

EXAMPLE 12.2.1.   *Cushing' Syndrome: Quadratic Discrimination Analysis.*
Performing a QDA on the Cushing's syndrome data of Example 12.0.1 begins with estimating the means and covariance matrices for the three populations. These are as follows.

Variable	Group means			Grand mean
	a	b	c	
log(Tet)	1.0433	2.0073	2.7097	1.8991
log(Preg)	−0.60342	−0.20604	1.5998	0.11038

Covariance matrix for adenoma
	log(Tet)	log(Preg)
log(Tet)	0.1107	0.1239
log(Preg)	0.1239	4.0891

Covariance matrix for bilateral hyperplasia
	log(Tet)	log(Preg)
log(Tet)	0.2119	0.3241
log(Preg)	0.3241	0.7203

Covariance matrix for carcinoma
	log(Tet)	log(Preg)
log(Tet)	0.5552	−0.2422
log(Preg)	−0.2422	0.2885

The results of QDA are given in Table 12.2. Group probabilities are computed in two ways. First, the data are used to estimate the parameters of the normal densities and the estimated densities are plugged into Bayes theorem. Second, the probabilities are estimated from cross-validation, the details of which will be discussed in

Sect. 12.4. The analyses are based on *equal costs and prior probabilities*; hence, they are also maximum likelihood allocations. Using the resubstitution method, only two cases are misallocated: 9 and 12. Under cross-validation, cases 1, 8, 19, and 20 are also misclassified.

□

Table 12.2 Quadratic discrimination analysis

Allocated to group	Resubstitution True group			Cross-validation True group		
	a	b	c	a	b	c
a	6	1	0	5	2	0
b	0	8	0	0	7	2
c	0	1	5	1	1	3

Case	Group	Resubstitution Probability			Cross-validation Probability		
		a	b	c	a	b	c
1 **	a	0.61	0.00	0.39	0.24	0.00	0.76
2	a	1.00	0.00	0.00	1.00	0.00	0.00
3	a	0.91	0.09	0.00	0.81	0.19	0.00
4	a	1.00	0.00	0.00	0.96	0.04	0.00
5	a	0.92	0.08	0.00	0.87	0.13	0.00
6	a	1.00	0.00	0.00	1.00	0.00	0.00
7	b	0.00	1.00	0.00	0.00	1.00	0.00
8 **	b	0.42	0.58	0.00	0.62	0.38	0.00
9 **	b	0.61	0.39	0.00	0.93	0.07	0.00
10	b	0.00	1.00	0.00	0.00	1.00	0.00
11	b	0.00	1.00	0.00	0.00	1.00	0.00
12 **	b	0.00	0.28	0.72	0.00	0.12	0.88
13	b	0.01	0.99	0.00	0.01	0.98	0.01
14	b	0.02	0.98	0.00	0.03	0.97	0.00
15	b	0.04	0.96	0.00	0.05	0.95	0.00
16	b	0.00	0.96	0.04	0.00	0.94	0.06
17	c	0.00	0.01	0.99	0.00	0.01	0.99
18	c	0.00	0.00	1.00	0.00	0.00	1.00
19 **	c	0.00	0.40	0.60	0.00	1.00	0.00
20 **	c	0.00	0.00	1.00	0.00	1.00	0.00
21	c	0.00	0.05	0.95	0.00	0.08	0.92

12.3 Linear Discrimination Analysis: LDA

If the populations are all multivariate normal with the same covariance matrix, say $\Sigma \equiv \Sigma_1 = \cdots = \Sigma_t$, then the Mahalanobis distance rule (12.1.1) and the maximum-likelihood/Bayes allocation rule (12.1.2) are identical. The maximum-likelihood/Bayes allocation rule assigns y to the population r that satisfies

$$\log(|\Sigma|) + (y - \mu_r)'\Sigma^{-1}(y - \mu_r) = \min_i \{\log(|\Sigma|) + (y - \mu_i)'\Sigma^{-1}(y - \mu_i)\}.$$

However, the term $\log(|\Sigma|)$ is the same for all populations, so the rule is equivalent to choosing the population r that satisfies

$$(y - \mu_r)'\Sigma^{-1}(y - \mu_r) = \min_i (y - \mu_i)'\Sigma^{-1}(y - \mu_i).$$

This is precisely the Mahalanobis distance rule.

In practice, estimates must be substituted for Σ and the μ_is. With equal covariance matrices the data fit the multivariate one-way ANOVA model of Sect. 10.3, so the standard estimates $\bar{y}_{i\cdot}$, $i = 1, \ldots, t$ and $S = E/(n-t)$ are reasonable. The allocation rule is: assign y to population r if

$$(y - \bar{y}_{r\cdot})'S^{-1}(y - \bar{y}_{r\cdot}) = \min_i (y - \bar{y}_{i\cdot})'S^{-1}(y - \bar{y}_{i\cdot}).$$

Recall that in a one-way ANOVA, the estimated covariance matrix is a weighted average of the individual estimates, namely

$$S = \sum_{i=1}^{t} (N_i - 1)S_i/(n-t).$$

Although $(y - \bar{y}_{i\cdot})'S^{-1}(y - \bar{y}_{i\cdot})$ is a quadratic function of y, the allocation only depends on a linear function of y. Note that

$$(y - \bar{y}_{i\cdot})'S^{-1}(y - \bar{y}_{i\cdot}) = y'S^{-1}y - 2\bar{y}_{i\cdot}'S^{-1}y + \bar{y}_{i\cdot}'S^{-1}\bar{y}_{i\cdot}.$$

The term $y'S^{-1}y$ is the same for all populations. Subtracting this constant and dividing by -2, the allocation rule can be rewritten as: assign y to population r if

$$y'S^{-1}\bar{y}_{r\cdot} - \frac{1}{2}\bar{y}_{r\cdot}'S^{-1}\bar{y}_{r\cdot} = \max_i \left\{ y'S^{-1}\bar{y}_{i\cdot} - \frac{1}{2}\bar{y}_{i\cdot}'S^{-1}\bar{y}_{i\cdot} \right\}.$$

This is based on a linear function of y, so methods related to this allocation rule are often referred to as (traditional) *linear discriminant analysis (LDA)*.

EXAMPLE 12.3.1.   *Cushing's Syndrome data.*
LDA results from estimating normal densities with equal covariance matrices. In this example we also *assume equal prior probabilities and costs* so the Bayesian allocation corresponds to a maximum likelihood allocation. All results are based on pooling the covariance matrices from Example 12.2.1 into

Pooled Covariance Matrix

	log(Tet)	log(Preg)
log(Tet)	0.2601	0.1427
log(Preg)	0.1427	1.5601.

The results of the linear discriminant analysis are summarized in Table 12.3. Based on resubstitution, four cases are misallocated, all from group b. Based on leave-one-out cross-validation, three additional cases are misallocated. Cases 8 and 9 are consistently classified as belonging to group a, and cases 12 and 16 are classified as c. In addition, when they are left out of the fitting process, cases 1 and 4 are allocated to groups c and b, respectively, while case 19 is misallocated as b. It is interesting to note that linear discrimination has a hard time deciding whether case 19 belongs to group b or c. A more detailed discussion of these cases is given later.

□

Consider $t = 2$ populations. If $q = 2$ we can easily plot the data from each population as we did in Fig. 12.1. Linear discrimination seeks to find a line that best separates the two clouds of data points (as is illustrated in the next chapter). If $q = 3$, while it is harder to do, we can still plot the data from each population. Linear discrimination now seeks to find a plane in three dimensions that best separates the two clouds of data points. For $q > 3$, visualization becomes problematic but linear discrimination uses hyperplanes, i.e., translated (shifted) vector spaces (affine spaces) of dimension $q - 1$, to separate the populations. In q dimensions, quadratic discrimination uses quadratic, rather than linear functions to separate the data. (But quadratics are linear functions in higher dimensional spaces.)

In particular, if there are just two groups, it is easy to see that the traditional linear discrimination rule assigns y to group 1 if

$$y'S^{-1}\bar{y}_{1.} - \frac{1}{2}\bar{y}'_{1.}S^{-1}\bar{y}_{1.} > y'S^{-1}\bar{y}_{2.} - \frac{1}{2}\bar{y}'_{2.}S^{-1}\bar{y}_{2.},$$

which occurs if and only if

$$y'S^{-1}(\bar{y}_{1.} - \bar{y}_{2.}) > \frac{1}{2}\bar{y}'_{1.}S^{-1}\bar{y}_{1.} - \frac{1}{2}\bar{y}'_{2.}S^{-1}\bar{y}_{2.},$$

iff

$$y'S^{-1}(\bar{y}_{1.} - \bar{y}_{2.}) > \frac{1}{2}(\bar{y}_{1.} + \bar{y}_{2.})'S^{-1}(\bar{y}_{1.} - \bar{y}_{2.}),$$

iff, with $\hat{\mu} \equiv (\bar{y}_{1.} + \bar{y}_{2.})/2$,

$$(y - \hat{\mu})'S^{-1}(\bar{y}_{1.} - \bar{y}_{2.}) > 0. \tag{12.3.1}$$

The separating line (or plane or hyperplane) in a plot consists of the y values with $(y - \hat{\mu})'S^{-1}(\bar{y}_{1.} - \bar{y}_{2.}) = 0$ or $y'S^{-1}(\bar{y}_{1.} - \bar{y}_{2.}) = \hat{\mu}'S^{-1}(\bar{y}_{1.} - \bar{y}_{2.})$.

As discussed in the next chapter, any linear function of y can be used to form a two group discrimination rule, say, assign y to a group based on whether $y'\beta_* > \beta_0$. The trick is to find a good vector β_* and a good constant β_0. LDA uses obviously good estimates of what are the optimal choices for β_* and β_0 when the data are multivariate normal with equal covariance matrices.

Table 12.3 Linear discrimination

Allocated to group	Resubstitution True group			Cross-validation True group		
	a	b	c	a	b	c
a	6	2	0	4	2	0
b	0	6	0	1	6	1
c	0	2	5	1	2	4

Case	Group	Resubstitution Probability			Cross-validation Probability		
		a	b	c	a	b	c
1 **	a	0.80	0.14	0.05	0.17	0.30	0.54
2	a	0.83	0.16	0.01	0.80	0.19	0.01
3	a	0.96	0.04	0.00	0.94	0.06	0.00
4 **	a	0.61	0.39	0.00	0.12	0.88	0.00
5	a	0.60	0.37	0.03	0.56	0.41	0.03
6	a	0.96	0.04	0.00	0.96	0.04	0.00
7	b	0.08	0.71	0.21	0.09	0.69	0.22
8 **	b	0.64	0.35	0.00	0.75	0.25	0.00
9 **	b	0.64	0.35	0.01	0.69	0.30	0.01
10	b	0.10	0.69	0.22	0.11	0.67	0.23
11	b	0.06	0.77	0.17	0.07	0.75	0.19
12 **	b	0.00	0.20	0.80	0.00	0.12	0.88
13	b	0.10	0.64	0.26	0.11	0.62	0.27
14	b	0.19	0.75	0.06	0.21	0.73	0.06
15	b	0.29	0.68	0.04	0.31	0.65	0.04
16 **	b	0.01	0.42	0.58	0.01	0.36	0.63
17	c	0.02	0.26	0.73	0.02	0.31	0.67
18	c	0.02	0.26	0.72	0.04	0.36	0.61
19 **	c	0.03	0.43	0.53	0.03	0.49	0.47
20	c	0.00	0.02	0.98	0.00	0.14	0.86
21	c	0.00	0.10	0.89	0.00	0.12	0.88

12.4 Cross-Validation

It is of considerable interest to be able to evaluate the performance of allocation rules. Depending on the populations involved, there is generally some level of mis-classification that is unavoidable. (If the populations are easy to tell apart, why are you worrying about them?) If the distributions that determine the allocation rules are known, one can simply classify random samples from the various populations to see how often the data are misclassified. This provides simple yet valid estimates of the error rates. Unfortunately, things are rarely that straightforward. In practice, the data available are used to estimate the distributions of the populations. If the same data are also used to estimate the error rates, a bias is introduced. Typically, this double dipping in the data overestimates the performance of the allocation rules. The method of estimating error rates by reclassifying the data used to construct the classification rules is often called the *resubstitution method*.

To avoid the bias of the resubstitution method, *cross-validation* is often used, cf. Geisser (1977) and Lachenbruch (1975). Cross-validation often involves leaving out one data point, estimating the allocation rule from the remaining data, and then classifying the deleted case using the estimated rule. Every data point is left out in turn. Error rates are estimated by the proportions of misclassified cases. This version of cross-validation is also known as the *jackknife*. (The jackknife was originally a tool for reducing bias in location estimates.) The computation of the cross-validation error rates can be simplified by the use of updating formulae similar to those discussed in *PA-V* Sect. 12.5 (Christensen 2011, Section 13.5).

While resubstitution underestimates error rates, cross-validation may tend to overestimate them. In standard linear models, if one thinks of the variance σ^2 as the error rate, resubstitution is analogous to estimating the variance with the *naive estimate SSE/n* (also the normal theory MLE) whereas leave-one-out cross-validation is analogous to estimating the variance with $PRESS/n$ where PRESS is the *predicted residual sum of squares* discussed in *PA-V* Sect. 12.5 (Christensen 2011, Section 13.5). Using Jensen's inequality it is not too difficult to show that

$$\mathrm{E}\left[\frac{SSE}{n}\right] < \mathrm{E}\left[\frac{SSE}{n-r(X)}\right] = \sigma^2 \le \mathrm{E}\left[\frac{PRESS}{n}\right],$$

and, indeed, that on average the cross-validation estimate $PRESS/n$ over estimates σ^2 by at least as much as the naive (resubstitution) method underestimates σ^2. While this is not really an issue in linear models, because we know how to find an unbiased estimate of the error in linear models, this result calls in question the idea of blindly using leave-one-out cross-validation to estimate error rates in allocation problems and logistic regression. In fact, since the over-estimation and the under-estimation seem to have similar orders of magnitude in standard linear models, one might consider averaging the two estimates.

For large data sets, K group cross-validation seems to be more popular than leave-one-out cross-validation. This involves (1) randomly dividing the data into K groups and (2) fitting the model on $K-1$ of those groups. Evaluate the error by (3) using this fitted model to allocate the data for the one omitted group, and (4) comparing these allocations to the true group memberships for the omitted group. This is done K times, where each group is omitted one time. The overall estimates of error are averages from the K different estimates. This approach requires quite a bit of data to be effective. Leave-one-out cross-validation uses $K = n$ but it seems popular to pick K considerably smaller than n and, indeed, this seems likely to reduce the bias problem of the previous paragraph.

When I first wrote this book (and contrary to the previous discussion), leave-one-out cross-validation was considered to have less bias than the resubstitution method but typically a considerably larger variance, cf. more recently (Hastie et al. 2016, Section 7.10). Hastie et al. also suggest that smaller values of K like $K = 5$ should have less variability but possibly more bias.

If the number of observations is much larger than the number of parameters to be estimated, resubstitution is often adequate for estimating the error rates. When the number of parameters is large relative to the number of observations, the bias becomes unacceptably high. Under normal theory, the parameters involved are simply the means and covariances for the populations. Thus, the key issue is the number of variables used in the discrimination relative to the number of observations. (While we have not discussed nonparametric discrimination, in the context of this discussion nonparametric methods should be considered as highly parametric methods.)

The bootstrap has also been suggested as a tool for estimating error rates. It often has both small bias and small variance, but it is computationally intensive and handles large biases poorly. The interested reader is referred to Efron (1983) and the report of the Panel on Discriminant Analysis, Classification, and Clustering in *Statistical Science* (1989).

12.5 Discussion

EXAMPLE 12.5.1. *Cushing's Syndrome Data.*
Careful inspection of Table 12.1 and Fig. 12.1 sheds light on both the LDA and QDA procedures. From Fig. 12.1, there seems to be almost no evidence that the covariance matrices are equal. Adenoma displays large variability in log(pregnanetriol), very small variability in log(tetrahydrocortisone), and almost no correlation between the variables. Carcinoma is almost the opposite. It has large variability in log(Tet) and small variability in log(Preg). Carcinoma seems to have a negative correlation. Bilateral hyperplasia displays substantial variability in both variables, with a positive correlation. These conclusions are also visible from the estimated covariance matrices. Given that the covariance structure seems to differ from group to group, linear discrimination does surprisingly well when evaluated by resubstitution. Recall that linear discriminant analysis has been found to be rather robust. Of course, quadratic discrimination does a much better job for these data.

The fact that the assessments based on cross-validation are much worse than those based on resubstitution is due largely to the existence of influential observations. The mean of group c and especially the covariance structure of group c are dominated by the large value of log(Tet) for case 20. Case 20 is not misclassified by the LDA because its effect on the covariance structure is minimized by the pooling of covariance estimates over groups. In cross-validated QDA, its effect on the covariance of group c is eliminated, so case 20 seems more consistent with group b. The large log(Preg) value of case 1 is also highly influential. With case 1 dropped out and case 20 included, case 1 is more consistent with carcinoma than with adenoma. The reason that cases 8 and 9 are misclassified is simply that they tend to be consistent with group a, see Fig. 12.1. In examining Table 12.1, a certain symmetry can be seen involving cases 12 and 19. Because of case 19, when case 12 is unassigned it looks more like group c than its original group. Similarly, because of case 12, when case 19 is unassigned it looks more like group b than group c under quadratic discrimination. Case 19 is essentially a toss-up under LDA. Cases 4 and 16 are misclassified under LDA because they involve very unusual data. Case 4 has an extremely small pregnanetriol value, and case 16 has a very large tetrahydrocortisone value for being part of the bilateral hyperplasia group.

In a data set this small, it seems unreasonable to drop influential observations. If we cannot believe the data, there is little hope of being able to arrive at a reasonable analysis. If further data bear out the covariance tendencies visible in Fig. 12.1, the better analysis is provided by quadratic discrimination. It must be acknowledged that the error rates obtained by resubstitution are unreliable. They are generally biased toward underestimating the true error rates and may be particularly bad for these data. QDA simply provides a good description of the data. There is probably insufficient data to produce really good predictions.   □

The methods discussed explicitly in this chapter are all related to the normal distribution. If the true distributions $f(y|i)$ are elliptically symmetric, both the quadratic and linear methods work well. Moreover, the linear discrimination method is generally quite robust; it even seems to work quite well for discrete data. See Lachenbruch, Sneeringeer, and Revo (1973), Lachenbruch (1975), and Hand (1983) for details.

The gold standard for discrimination seems to be, depending on one's philosophical bent, maximum likelihood or Bayesian discrimination. But they are only the gold standard if you know what the distributions are. If you know the densities, those are the only functions of the data that need concern you.

Linear and quadratic discrimination for nonnormal data can be based on Mahalanobis distances rather than on densities. Since they are not based on densities, they are ad hoc methods. Many of the binary regression methods discussed in the next chapter provide direct estimates of $\pi(i|y)$ that are (typically) inappropriate for discrimination data but from which appropriate density estimates can be inferred. Often the regression methods implicitly or explicitly perform discrimination in higher dimensions. Instead of linear or quadratic discrimination on the basis of, say, $y = (y_1, y_2, y_3)'$, they discriminate on the basis of some extended vector, for example, $\tilde{y} = (y_1, y_2, y_3, y_1^2, y_2^2, y_3^2, y_1 y_2, y_1 y_3, y_2 y_3)'$. *If you know the densities, there is little point in expanding the dimensionality, because the density is the only relevant function of the data.* But if you do not know the densities, expanding the dimensionality can be very useful. In particular, support vector machines typically use expanded data. Of course, one could also perform traditional linear or quadratic discrimination on the new $\tilde{y}$ and I suspect that, when practical, linear and quadratic discrimination on $\tilde{y}$ will often be competitive with the newer methods. Personally, I am more comfortable using expanded data in logistic (or log-linear) discrimination than in LDA or QDA.

For $\tilde{y}$ as given above, any linear discrimination method based on $\tilde{y}'\beta_*$ is equivalent to a quadratic discrimination based on y. This is not to say that LDA applied to $\tilde{y}$ is QDA, but merely that $\tilde{y}'\beta_*$ is always a quadratic function of y. If you know that the data y are normal, QDA on y is pretty nearly optimal. (For known mean vectors and covariance matrices it is optimal.) And if the data are normal with equal covariance matrices, those optimal quadratic discriminate functions reduce to linear functions of y. But if y is normal, $\tilde{y}$ is certainly *not* normal and applying traditional LDA methods to $\tilde{y}$ is unlikely to agree with QDA. Nonetheless, LDA on $\tilde{y}$ is some form of quadratic discrimination.

12.6 Stepwise LDA

One interesting problem in linear allocation is the choice of variables. Including variables that have no ability to discriminate among populations can only muddy the issues involved. By analogy with multiple regression, one might expect to find advantages to allocation procedures based solely on variables with high discrimina tory power. In multiple regression, methods for eliminating independent variables are either directly based on, or closely related to, testing whether exclusion of the variables hurts the regression model. In other words, a test is performed of whether, given the included variables, the excluded variables contain any additional informa- tion for prediction. In discrimination and allocation, methods for eliminating dis- criminatory variables such as *stepwise discrimination* are based on testing whether, given the included variables, the excluded variables contain any additional informa- tion for discrimination among the populations. We have noted that LDA is closely related to the multivariate one-way ANOVA model. Tests of additional information can be performed as in Sect. 10.5. In particular, they are typically performed by testing a one-way ACOVA model such as (10.5.1) against the no treatment effects ACOVA model (10.5.3).

EXAMPLE 12.2.2. We now illustrate the process of stepwise discrimination us- ing data given by Lubischew (1962). He considered the problem of discriminating among three populations of flea-beetles within the genus *Chaetocnema*. Six vari- ables were given: y_1, the width, in microns, of the first joint of the first tarsus, y_2, the same measurement for the second joint, y_3, the maximum width, in microns, of the aedeagus in the fore part, y_4, the front angle, in units of $7.5°$, of the aedeagus, y_5, the maximum width of the head, in 0.01 mm units, between the external edges of the eyes, and y_6, the width of the aedeagus from the side, in microns. In addi- tion, Lubischew mentions that $r_{12} \equiv y_1/y_2$ is very good for discriminating between one of the species and the other two. The vector of dependent variables is taken as $y' = (y_1, y_2, y_3, y_4, y_5, y_6, r_{12})$. Stepwise discrimination is carried out by testing for additional information in the one-way MANOVA.

Evaluating the assumptions of a one-way MANOVA with three groups and seven dependent variables is a daunting task. There are three 7×7 covariance matrices that should be roughly similar. To wit, there are $\binom{7}{2} = 21$ bivariate scatter plots to check for elliptical patterns. If the capability exists for the user, there are $\binom{7}{3} = 35$ three-dimensional plots to check. There are $3(7) = 21$ normal plots to evaluate the marginal distributions and at least some linear combinations of the variables should be evaluated for normality. Of course, if y_1 and y_2 are multivariate normal, the constructed variable r_{12} cannot be. However, it may be close enough for our purposes.

If the assumptions break down, it is difficult to know how to proceed. After any transformation, everything needs to be reevaluated, with no guarantee that things will have improved. It seems like the best bet for a transformation is some model- based system similar to the Box and Cox (1964) method (see Andrews, Gnanade- sikan, & Warner 1971).

For the most part, in this example, we will cross our fingers and hope for the best. In other words, we will rely on the robustness of the procedure. While it is certainly true that the P values used in stepwise discriminant analysis should typically not be taken at face value (this is true for almost any statistical modeling technique), the P values can be viewed as simply a one-to-one transformation of the test statistics. Thus, decisions based on P values are based on the relative sizes of comparable test statistics. The test statistics are reasonable even without the assumption of multivariate normality so, from this point of view, multivariate normality is not a crucial issue.

The assumption of equal covariance matrices is a stickier issue. From *PA* Sect. 3.2, univariate test statistics are based on the squared length of the vector of differences between the optimal fitted values under the full model and the optimal fitted values under the reduced model. If the reduced model is (nearly) correct, the difference vector should be near zero (but subject to random variation that affects its length). For the multivariate linear model, the hypothesis statistic H consists of the squared length for each variable and inner products between distinct variables. Small matrices H (if they are not *too* small) are consistent with the reduced model, while large matrices H are inconsistent with it. Even with unequal covariance matrices for the groups, the least squares estimates are reasonable, so valid conclusions can be based on the size of H. In fact, with unequal covariance matrices, least squares estimates are still optimal for the full one-way MANOVA model. To see this, write the model as a univariate linear model and, as in Sect. 9.2, use Theorem 10.4.5 from *PA*. The problem with unequal covariance matrices is twofold. First, it is difficult to evaluate explicitly what we mean by large and small matrices H. Second, the inner products (and thus the lengths) are being evaluated with the Euclidean inner product, which is less than optimal.

Although the properties of formal tests can be greatly affected by the invalidity of the MANOVA assumptions, crude but valid evaluations can still be made based on the test statistics. This is often the most that we have any right to expect from multivariate procedures. For univariate models, Scheffé (1959, Chapter 10) gives an excellent discussion of the effects of invalid assumptions on formal tests.

The three species of flea-beetles considered will be referred to as simply A, B, and C and indexed as 1, 2, and 3, respectively. There are 21 observations on species A with

$$\bar{y}_{1.}' = (183.1, 129.6, 51.2, 146.2, 14.1, 104.9, 1.41)$$

and

$$S_1 = \begin{bmatrix} 147.5 & 66.64 & 18.53 & 15.08 & -5.21 & 14.21 & 0.406 \\ 66.64 & 51.25 & 11.55 & 2.48 & -1.81 & 3.09 & -0.044 \\ 18.53 & 11.55 & 4.99 & 5.85 & -0.524 & 5.49 & 0.017 \\ 15.08 & 2.48 & 5.85 & 31.66 & -0.969 & 15.63 & 0.090 \\ -5.21 & -1.81 & -0.524 & -0.969 & 0.791 & -1.99 & -0.021 \\ 14.21 & 3.09 & 5.49 & 15.63 & -1.99 & 38.23 & 0.078 \\ 0.406 & -0.044 & 0.017 & 0.090 & -0.021 & 0.078 & 0.0036 \end{bmatrix}.$$

Species B has 31 observations with

$$\bar{y}'_{2\cdot} = (201.0, 119.3, 48.9, 124.6, 14.3, 81.0, 1.69)$$

and

$$S_2 = \begin{bmatrix} 222.1 & 63.40 & 22.60 & 30.37 & 4.37 & 29.47 & 0.926 \\ 63.40 & 44.16 & 7.91 & 11.82 & 0.337 & 11.47 & -0.100 \\ 22.60 & 7.91 & 5.52 & 5.69 & 0.005 & 4.23 & 0.075 \\ 30.37 & 11.82 & 5.69 & 21.37 & -0.327 & 11.70 & 0.088 \\ 4.37 & 0.337 & 0.005 & -0.327 & 1.21 & 1.27 & 0.029 \\ 29.47 & 11.47 & 4.23 & 11.70 & 1.27 & 79.73 & 0.085 \\ 0.926 & -0.100 & 0.075 & 0.088 & 0.029 & 0.085 & 0.009 \end{bmatrix}.$$

For species C, there are 22 observations with

$$\bar{y}'_{3\cdot} = (138.2, 125.1, 51.6, 138.3, 10.1, 106.6, 1.11)$$

and

$$S_3 = \begin{bmatrix} 87.33 & 44.55 & 20.53 & 19.17 & -0.736 & 15.29 & 0.301 \\ 44.55 & 73.04 & 15.71 & 14.02 & -0.390 & 21.23 & -0.267 \\ 20.53 & 15.71 & 8.06 & 8.21 & -0.294 & 4.97 & 0.027 \\ 19.17 & 14.02 & 8.21 & 2.16 & -0.502 & 7.93 & 0.027 \\ 0.736 & -0.390 & -0.294 & -0.502 & 0.944 & 0.277 & -0.002 \\ 15.29 & 21.23 & 4.97 & 7.93 & 0.277 & 34.25 & -0.061 \\ 0.301 & -0.267 & 0.027 & 0.027 & -0.002 & -0.061 & 0.0046 \end{bmatrix}.$$

The pooled estimate of the covariance is a weighted average of S_1, S_2, and S_3, with approximately 50% more weight on S_2 than on the other estimates.

Although, typically, backward elimination is to be preferred to forward selection in stepwise procedures, it is illustrative to demonstrate forward selection on these data. We will begin by making a very rigorous requirement for inclusion: variables will be included if the P value for adding them is 0.01 or less.

The first step in forward selection consists of performing the univariate one-way ANOVA F tests for each variable.

Step 1: Statistics for entry, $df = 2, 71$		
Variable	F_{obs}	$\Pr[F > F_{obs}]$
y_1	160.339	0.0001
y_2	12.499	0.0001
y_3	9.659	0.0002
y_4	134.353	0.0001
y_5	129.633	0.0001
y_6	101.314	0.0001
r_{12}	351.292	0.0001

The P values are all sufficiently small to warrant inclusion of the variables. By far the largest F statistic, and thus the smallest P value, is for r_{12}, so this is the first variable included for use in discrimination. Note that r_{12} is the variable constructed by Lubischew.

The second and all subsequent steps of the procedure involve performing a one-way analysis of covariance for each variable not yet included. For the second step, the sole covariate is r_{12}, and a test is made for treatment effects in the analysis of covariance model. For the dependent variables y_1 through y_6, the results are as follows.

Step 2: Statistics for entry, $df = 2, 70$

Variable	F_{obs}	$Pr[F > F_{obs}]$
y_1	9.904	0.0002
y_2	8.642	0.0004
y_3	6.386	0.0028
y_4	87.926	0.0001
y_5	30.549	0.0001
y_6	28.679	0.0001

The largest F statistic is for y_4, and the corresponding P value is less than 0.01, so y_4 is included for discrimination.

At the third step, both r_{12} and y_4 are used as covariates in a one-way analysis of covariance. Again, the F tests for treatment differences are performed.

Step 3: Statistics for entry, $df = 2, 69$

Variable	F_{obs}	$Pr[F > F_{obs}]$
y_1	2.773	0.0694
y_2	3.281	0.0436
y_3	6.962	0.0018
y_5	24.779	0.0001
y_6	3.340	0.0412

Variable y_5 is included for discrimination. Note the large difference between the F statistic for y_5 and that for the other variables. There is an order-of-magnitude difference between the abilities of the r_{12}, y_4, and y_5 to discriminate and the abilities of the other variables. Considering the questionable validity of formal tests, this is an important point. It should also be mentioned that this conclusion is based on one sequence of models. There is a possibility that other sequences would lead to different conclusions about the relative importance of the variables. In fact, it would be desirable to check all models or, better yet, have an algorithm to identify the best models.

Step 4 simply adds weight to our conclusions of the previous paragraph. In performing the analysis of covariance with three covariates, none of the variables considered have the very large F statistics seen earlier.

Step 4: Statistics for entry, $df = 2,68$		
Variable F_{obs}		$\Pr[F > F_{obs}]$
y_1	1.985	0.1453
y_2	2.567	0.0842
y_3	3.455	0.0372
y_6	3.359	0.0406

Any rule that terminates forward selection when all P values exceed 0.0371 will stop the selection process at Step 4. In particular, our stringent stopping rule based on P values of 0.01 terminates here.

In practice, it is much more common to use a stopping rule based on P values of 0.05, 0.10, or 0.15. By any of these rules, we would add variable y_3 and continue checking variables. This leads to Step 5 and the corresponding F statistics.

Step 5: Statistics for entry, $df = 2,67$		
Variable F_{obs}		$\Pr[F > F_{obs}]$
y_1	7.040	0.0017
y_2	8.836	0.0004
y_6	3.392	0.0395

Surprisingly, adding y_3 has changed things dramatically. While the F statistic for y_6 is essentially unchanged, the F values for y_1 and y_2 have more than tripled. Of course, we are still not seeing the huge F statistics that were encountered earlier, but apparently one can discriminate much better with y_3 and either y_1 or y_2 than would be expected from the performance of any of these variables individually. This is precisely the sort of thing that is very easily missed by forward selection procedures and one of the main reasons why they are considered to be poor methods for model selection. Forward selection does have advantages. In particular, it is cheap and it is able to accommodate huge numbers of variables.

The stepwise procedure finishes off with two final steps. Variable y_2 was added in the previous step. The results from Step 6 are as follows.

Step 6: Statistics for entry, $df = 2,66$		
Variable F_{obs}		$\Pr[F > F_{obs}]$
y_1	0.827	0.4418
y_6	3.758	0.0285

Variable y_6 is added if our stopping rule is not extremely stringent. This leaves just y_1 to be evaluated.

Step 7: Statistics for entry, $df = 2,65$		
Variable F_{obs}		$\Pr[F > F_{obs}]$
y_1	0.907	0.4088

By any standard y_1 would not be included. Of course, r_{12} is the ratio of y_1 and y_2, so it is not surprising that there is no need for all three variables. A forward selection procedure that does not include r_{12} would simply include all of the variables.

We have learned that r_{12}, by itself, is a powerful discriminator. The variables r_{12}, y_4, and y_5, when taken together, have major discriminatory powers. Variable y_3, taken together with either y_1 or y_2 and the previous three variables, may provide substantial help in discrimination.

Finally, y_6 may also contribute to distinguishing among the populations. Most of these conclusions are visible from Table 12.4 that summarizes the results of the forward selection.

Table 12.4 Summary of forward selection

Step	Variable entered	F_{obs}	$\Pr[F > F_{obs}]$
1	r_{12}	351.292	0.0001
2	y_4	87.926	0.0001
3	y_5	24.779	0.0001
4	y_3	3.455	0.0372
5	y_2	8.836	0.0004
6	y_6	3.758	0.0285

It is also of interest to see the results of a multivariate analysis of variance for all of the variables included at each step. For example, after Step 3, variables r_{12}, y_4, and y_5 were included for discrimination. The likelihood ratio test statistic for no group effects in the one-way MANOVA is $U = 0.0152$. This is a very small, hence very significant, number. Table 12.5 lists the results of such tests for each step in the process. Based on their P values, all of the variables added had substantial discriminatory power. Thus, it is not surprising that the U statistics in Table 12.5 decrease as each variable is added.

In practice, decisions about the practical discriminatory power of variables should not rest solely on the P values. After all, the P values are often unreliable. Other methods, such as the graphical methods presented in the next section, should be used in determining the practical usefulness of results based on multivariate normal distribution theory.

□

Table 12.5 Forward stepwise discrimination: MANOVA tests

Step	Variable entered	LRTS U_{obs}	$\Pr[U < U_{obs}]$
1	r_{12}	0.09178070	0.0001
2	y_4	0.02613227	0.0001
3	y_5	0.01520881	0.0001
4	y_3	0.01380601	0.0001
5	y_2	0.01092445	0.0001
6	y_6	0.00980745	0.0001

12.7 Linear Discrimination Coordinates

As mentioned earlier, one is typically interested in the clarity of classification. This can be investigated by examining the posterior probabilities, the entire likelihood function, or the entire set of Mahalanobis distances. It is done by computing the allocation measures for each element of the data set. The allocation measure can be estimated either by the entire data set or the data set having deleted the case currently being allocated. To many people, the second, cross-validatory, approach is more appealing.

An alternative approach to examining the clarity of discrimination is through the use of *linear discrimination coordinates*. This approach derives from the work of Fisher (1938) and Rao (1948, 1952). It consists of redefining the coordinate system in $\mathbf{R}^q$ in such a way that the different treatment groups in the one-way ANOVA have, in some sense, maximum separation in each coordinate. The clarity of discrimination can then be examined visually by inspecting one-, two-, or three-dimensional plots of the data. In these plots, cases are identified by their populations. If the new coordinate system is effective, observations from the same population should be clustered together and distinct populations should be well-separated.

It is standard practice to redefine the coordinate system by taking linear combinations of the original variables. It is also standard practice to define the new coordinate system sequentially. In particular, the first coordinate is chosen to maximize the separation between the groups. The second coordinate maximizes the separation between the groups given that the second linear combination is uncorrelated with the first. The third maximizes the separation given that the linear combination is uncorrelated with the first two. Subsequent coordinates are defined similarly. In the following discussion, we assume a constant covariance matrix for the t groups. It remains to define what precisely is meant by "maximum separation of the groups."

Recall that with equal covariance matrices, the data available in a discriminant analysis fit a multivariate one-way ANOVA,

$$Y = XB + e.$$

Thus,

$$E = \sum_{i=1}^{t} \sum_{j=1}^{N_i} (y_{ij} - \bar{y}_{i\cdot})(y_{ij} - \bar{y}_{i\cdot})'$$

and

$$H = \sum_{i=1}^{t} N_i (\bar{y}_{i\cdot} - \bar{y}_{\cdot\cdot})(\bar{y}_{i\cdot} - \bar{y}_{\cdot\cdot})'.$$

Also, define

$$H_* = \sum_{i=1}^{t} (\bar{y}_{i\cdot} - \bar{y}_{\cdot\cdot})(\bar{y}_{i\cdot} - \bar{y}_{\cdot\cdot})'.$$

The linear discrimination coordinates are based on E and either H or H_*. We will examine the use of H in detail. Some comments will also be made on the motivation for using H_*.

For any vector $y = (y_1, \ldots, y_q)'$, the first linear discrimination coordinate is defined by

$$y'a_1,$$

where the vector a_1 is chosen so that the univariate one-way ANOVA model

$$(Ya_1) = X(Ba_1) + (ea_1)$$

has the largest possible F statistic for testing equality of group effects. Intuitively, the linear combination of the variables that maximizes the F statistic must have the greatest separation between groups. The degrees of freedom are not affected by the choice of a_1, so we need to find a_1 that maximizes

$$\frac{(Ya_1)' \left(M - \frac{1}{n}J_n^n\right)(Ya_1)}{(Ya_1)'(I - M)(Ya_1)},$$

or equivalently

$$\frac{a_1'Ha_1}{a_1'Ea_1}.$$

A one-dimensional plot of the n elements of Ya_1 shows the maximum separation between groups that can be achieved in a one-dimensional plot.

The second linear discrimination coordinate is

$$y'a_2$$

such that

$$\frac{a_2'Ha_2}{a_2'Ea_2}$$

is maximized subject to the constraint that, for any i and j, the estimated covariance between $y_{ij}'a_1$ and $y_{ij}'a_2$ is zero. The covariance condition can be rewritten as

$$a_1'Sa_2 = 0,$$

or equivalently as

$$a_1'Ea_2 = 0.$$

Another way of thinking of this condition is that a_1 and a_2 are orthogonal in the inner product space defined using the matrix E.

A one-dimensional plot of Ya_2 illustrates visually the separation in the groups. Even more productively, the n ordered pairs that are the rows of $Y(a_1, a_2)$ can be plotted to illustrate the discrimination achieved by the first two linear discrimination coordinates.

For $h = 3, \ldots, r(H)$ the hth linear discriminant coordinate is

$$y' a_h,$$

where

$$a_h' H a_h / a_h' E a_h$$

is maximized subject to the covariance condition

$$a_h' E a_i = 0 \qquad i = 1, 2, \ldots, h-1.$$

Note that, using the inner product for $\mathbf{R}^q$ based on E, this defines an orthogonal system of coordinates (i.e., $a_1, \ldots, a_q$ define an orthogonal basis for $\mathbf{R}^q$ using the inner product defined by E).

Unfortunately, the discrimination coordinates are not uniquely defined. Given a vector a_h, any scalar multiple of a_h also satisfies the requirements listed earlier. One way to avoid the nonuniqueness is to impose another condition. The most commonly used extra condition is that $a_h' E a_h = 1$, so that $a_1, \ldots, a_q$ is an orthonormal basis for $\mathbf{R}^q$ under the inner product defined by E. Alas, even this does not quite solve the uniqueness problem because $-a_h$ has the same properties as a_h.

Before going into the details of actually finding the linear discrimination coordinates, we illustrate their use. It will be established later that the linear discrimination coordinate vectors a_i, $i = 1, \ldots, q$ are eigenvectors of $E^{-1}H$. Moreover, the appropriate metric for examining variables transformed into the linear discrimination coordinates is the standard Euclidean metric. This allows simple visual inspection of the transformed data. Writing $A = [a_1, \ldots, a_q]$, the mapping Y into YA gives the data matrix in the linear discrimination coordinates.

EXAMPLE 12.6.1.   Consider again the heart rate data of Example 10.3.1. The data structure needed for development of linear discrimination coordinates is the same as for a one-way MANOVA. We have already examined these data for multivariate normality and equal covariance matrices. The data seem to satisfy the assumptions.

The linear discrimination coordinates are defined by a matrix of eigenvectors of $E^{-1}H$. One is

$$A = \begin{bmatrix} 0.739 & 0.382 & 0.581 & 0.158 \\ -0.586 & -0.323 & -0.741 & 0.543 \\ -0.353 & -0.234 & 0.792 & -0.375 \\ 0.627 & -0.184 & -0.531 & -0.218 \end{bmatrix}.$$

Recall that E and H were given in Example 10.3.1. The columns of A define four new data vectors Ya_1, Ya_2, Ya_3, and Ya_4 but remember that eigenvectors are not uniquely defined. Different software often give different eigenvectors but (when the eigenvalues are unique) they only vary by a scale factor, so the differences typically do not matter (unless you are trying to reproduce existing results). If we perform an analysis of variance on each variable, we get F statistics for discriminating between groups. All have 2 degrees of freedom in the numerator and 27 in the denominator.

Variable	F
Ya_1	74.52
Ya_2	19.47
Ya_3	0.0
Ya_4	0.0

As advertised, the F statistics are nonincreasing. The first two F statistics clearly establish that there are group differences in the first two coordinates. The last two F statistics are zero because with three groups there are 2 degrees of freedom for treatments and H is a 4×4 matrix of rank 2. Only two of the linear discrimination coordinates can have positive F statistics. This issue is discussed in more detail in the next subsection.

The big advantage of linear discrimination coordinates is that they allow us to plot the data in ways that let us visualize the separation in the groups. Figure 12.3 shows two plots that display the first discrimination coordinate values for each population. The software placed the populations in different positions. Note that the degree of separation is substantial and about the same for all three groups. The edges of the middle group are close to the edges of the other groups. The placebo has one observation that is consistent with drug A.

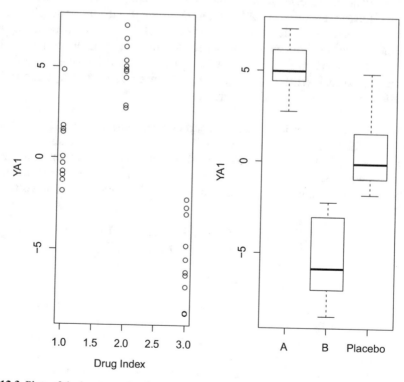

Fig. 12.3 Plots of the heart rate data in the first linear discrimination coordinate

Figure 12.4 is similar to Fig. 12.3 except that it plots the data in the second discrimination coordinate. Note that in the second coordinate it is very difficult to distinguish between drugs A and B. The placebo is separated from the other groups, but there is more overlap around the edges than was present in the first coordinate.

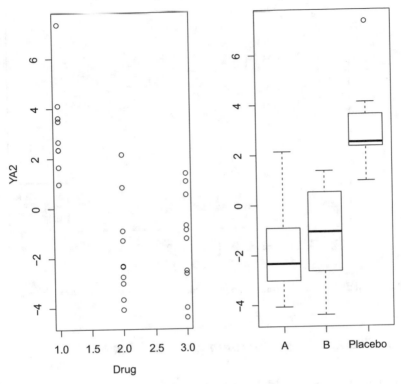

Fig. 12.4 Plots of the heart rate data in the second discrimination coordinate

Figure 12.5 is a scatter plot of the data in the first two discrimination coordinates. Together, the separation is much clearer than in either of the individual coordinates. There is still one observation from drug A that is difficult to distinguish from the placebo group but, other than that, the groups are very well-separated. That the one observation from drug A is similar to the placebo is a conclusion based on the Euclidean distance of the point from the centers of the groups for drug A and the placebo. It is not clear that Euclidean distances are appropriate, but that will be shown in the next subsection.   ☐

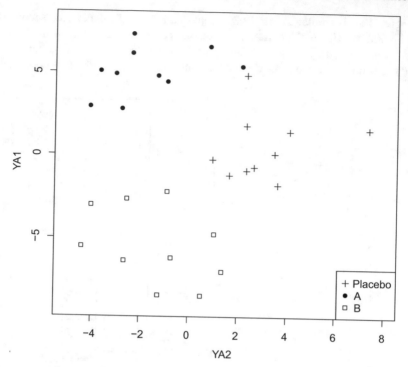

Fig. 12.5 Scatter plot of the heart rate data in the first two linear discrimination coordinates

12.7.1 Finding Linear Discrimination Coordinates

The vectors $a_1, \ldots, a_q$ can be taken as eigenvectors of the matrix $E^{-1}H$. Before showing this we prove a result similar to Theorem B.15 in *PA*. Theorem B.15 states that, given any symmetric matrix, say W, there exists an orthonormal basis for $\mathbf{R}^n$ consisting of eigenvectors of W. Lemma 12.7.2 states that, relative to an inner product on $\mathbf{R}^q$ defined by an arbitrary positive definite matrix E, there exists an orthonormal basis for $\mathbf{R}^q$ consisting of eigenvectors of $E^{-1}H$, where H is an arbitrary symmetric matrix. Note that although we have continued to use the symbols H and E and our immediate interest is in application of these results to the specific matrices H and E defined earlier, the result does not depend on the choice of these matrices except as indicated in the lemma. The following series of results will also be used in Chap. 14 to derive principal components. In Chap. 14, E and H will not be the error and hypothesis matrices from a multivariate linear model.

Lemma 12.7.2.   Let E be any $q \times q$ positive definite matrix and let H be any symmetric $q \times q$ matrix. Then there exists a $q \times q$ diagonal matrix Λ and a matrix A such that

$$E^{-1}HA = A\Lambda \quad \text{and} \quad A'EA = I.$$

Observe that the columns of A must be eigenvectors and the elements of Λ must be eigenvalues of $E^{-1}H$.

PROOF.   Define $E^{1/2}$ as in Lemma 9.6.2. By PA Theorem B.15 there exists B such that

$$E^{-1/2}HE^{-1/2}B = B\Lambda \tag{12.7.1}$$

with

$$I = BB' = B'B.$$

Let $A = E^{-1/2}B$ then multiplying (12.7.1) on the left by $E^{-1/2}$ gives

$$E^{-1}HA = A\Lambda$$

and

$$A'EA = B'E^{-1/2}EE^{-1/2}B = B'B = I. \qquad \square$$

We will later need the following result.

Corollary 12.7.3.   If A is $q \times q$, $E^{-1}HA = A\Lambda$, and $A'EA = I$ then

$$E^{-1} = AA'.$$

PROOF.   If $A'EA = I$ and both E and A are $q \times q$, then A must be nonsingular. It follows that

$$I = (A'EA)^{-1} = A^{-1}E^{-1}(A')^{-1}.$$

Multiplying on the left by A and on the right by A' gives

$$AA' = AA^{-1}E^{-1}(A')^{-1}A' = E^{-1}. \qquad \square$$

The argument that the linear discrimination coordinates can be taken as eigenvectors of $E^{-1}H$ has similarities to the proof of Lemma 12.7.2 and also to the argument in Sect. 9.6 that relates Roy's ϕ_{max} test statistic to the maximum eigenvalue of HE^{-1}. Once again, the argument does not depend on the specific choices of H and E.

Proposition 12.7.4.   Let E be any $q \times q$ positive definite matrix and let H be any $q \times q$ symmetric matrix. The vectors $a_1, \ldots, a_q$ satisfy

$$\frac{a_1'Ha_1}{a_1'Ea_1} = \sup_a \frac{a'Ha}{a'Ea},$$

and, for $i > 1$,

$$a_i'Ea_j = 0, \quad j = 1, \ldots, i-1,$$

and

$$\frac{d_i'Ha_i}{d_i'Ea_i} = \sup_a \left\{ \frac{a'Ha}{a'Ea} \middle| a'Ea_j = 0, \quad j = 1,\ldots,i-1 \right\}$$

if and only if for $i = 1,\ldots,q$, a_i is an eigenvector of $E^{-1}H$ corresponding to an eigenvalue ϕ_i, where $\phi_1 \geq \cdots \geq \phi_q$ and where, for $i > 1$, $a_i'Ea_j = 0$, $j = 1,\ldots,i-1$.

PROOF. The idea of the proof is to transform the difficult problem of looking at $a'Ha/a'Ea$ to a simpler problem of looking at $c'\Lambda c$, where Λ is a diagonal matrix. The conditions $a'Ea_j = 0$ are transformed into conditions $c'c_j = 0$. The proof is by induction. The inductive step requires the assumption of an induction hypothesis that is more detailed than one might expect. We begin by examining the transformation.

Define $E^{1/2}$ as in Lemma 9.6.2. Write

$$\Lambda = D(\phi_1,\ldots,\phi_q),$$

and pick B and A as in the proof of Lemma 12.7.2. Note that

$$E^{-1/2}HE^{-1/2} = B\Lambda B',$$

and recall that

$$E^{-1}HA = A\Lambda.$$

For any vector a, define

$$b = E^{1/2}a$$

and

$$c = B'b = B'E^{1/2}a.$$

The transformation from a to c is invertible, namely

$$a = E^{-1/2}Bc$$
$$= Ac.$$

Observe that

$$\frac{a'Ha}{a'Ea} = \frac{b'E^{-1/2}HE^{-1/2}b}{b'b} = \frac{b'B\Lambda B'b}{b'b} = \frac{b'B\Lambda B'b}{b'BB'b} = \frac{c'\Lambda c}{c'c}.$$

Moreover, if a and a_* are two vectors, by the choice of B

$$a'Ea_* = b'b_* = b'BB'b_* = c'c_*.$$

Thus, properties related to $a'Ha/a'Ea$ and $a'Ea_*$ can be examined by investigating $c'\Lambda c$ and $c'c_*$.

Finally, we need to establish that a is an eigenvector of $E^{-1}H$ corresponding to some value ϕ if and only if c is an eigenvector of Λ corresponding to ϕ. First, if $E^{-1}Ha = a\phi$, then $\Lambda c = A^{-1}(A\Lambda)c = A^{-1}(E^{-1}HA)c = A^{-1}E^{-1}Ha = A^{-1}a\phi = A^{-1}Ac\phi = c\phi$. Conversely, if $\Lambda c = c\phi$, then $E^{-1}Ha = E^{-1}HAc = A\Lambda c = Ac\phi = a\phi$.

We begin by showing that $a'_1 Ha_1/a'_1 Ea_1$ maximizes $a'Ha/a'Ea$ if and only if a_1 is an eigenvector of $E^{-1}H$ corresponding to ϕ_1. A vector a_1 maximizes $a'Ha/a'Ea$ if and only if the corresponding vector c_1 maximizes

$$\frac{c'\Lambda c}{c'c} = \sum_{j=1}^{q} c_j^2 \phi_j \Big/ \sum_{j=1}^{q} c_j^2.$$

Because this is a weighted average of the ϕ_js and the ϕ_js are ordered from largest to smallest, the maximum value is ϕ_1. Given the previously demonstrated relationship between eigenvectors of $E^{-1}H$ and Λ, the proof is complete if we show that c_1 maximizes $c'\Lambda c/c'c$ if and only if c_1 is an eigenvector of Λ corresponding to ϕ_1. If c_1 is an eigenvector of Λ corresponding to ϕ_1, that is, if

$$\Lambda c_1 = c_1 \phi_1,$$

then

$$c'_1 \Lambda c_1/c'_1 c_1 = \phi_1,$$

so c_1 maximizes $c'\Lambda c/c'c$. Conversely, suppose $\phi_1 = \cdots = \phi_s > \phi_{s+1}$; then, to maximize $c'\Lambda c/c'c$, a vector $c = (k_1, \ldots, k_q)$ must have $k_{s+1} = \cdots = k_q = 0$. Such vectors clearly satisfy $\Lambda c = c\phi_1$.

We complete the proof by induction. Suppose that for $i = 1, \ldots, h-1$

$$\frac{a'_i Ha_i}{a'_i Ea_i} = \sup_a \left\{ \frac{a'Ha}{a'Ea} \Big| a'Ea_j = 0 \quad j = 1, \ldots, i-1 \right\}$$

with

$$a'_i Ea_j = 0 \quad j = 1, \ldots, i-1$$

if and only if a_i is an eigenvector of $E^{-1}H$ corresponding to ϕ_i with $a'_i Ha_j = 0$, $j = 1, \ldots, i-1$. We need to prove that

$$\frac{a'_h Ha_h}{a'_h Ea_h} = \sup_a \left\{ \frac{a'Ha}{a'Ea} \Big| a'Ea_j = 0 \quad j = 1, \ldots, h-1 \right\}$$

with

$$a'_h Ea_j = 0 \quad j = 1, \ldots, h-1$$

if and only if a_h is an eigenvector of $E^{-1}H$ corresponding to ϕ_h with $a'_h Ea_j = 0$, $j = 1, \ldots, h-1$. From the equivalence established earlier, a_h maximizes $a'Ha/a'Ea$ subject to the conditions if and only if the corresponding vector c_h maximizes $c'\Lambda c/c'c$ subject to the conditions $c'c_j = 0$, $j = 1, \ldots, h-1$ and a_h is an eigenvector with $a'_h Ea_j = 0$ if and only if c_h is an eigenvector with $c'_h c_j = 0$.

Suppose $\phi_r = \cdots = \phi_h = \cdots = \phi_s$ with either $r = 1$ or $\phi_{r-1} > \phi_r$ and either $s = q$ or $\phi_s > \phi_{s+1}$. Write $c = (k_1, \ldots, k_q)'$ and $c_i = (k_{i1}, \ldots, k_{iq})'$. Think of the element in the jth row of a vector as corresponding to ϕ_j. As additional parts of the induction hypothesis, assume that the terms in c_i corresponding to eigenvalues less than ϕ_i must be zero and that if c is orthogonal to the c_is the terms in c corresponding to eigenvalues greater than ϕ_h must be zero. Specifically, for $i = 1, \ldots, h-1$ if $\phi_j < \phi_i$, then $k_{ij} = 0$, and that if $c'c_i = 0$, $i = 1, \ldots, h-1$ and $\phi_j > \phi_h$, then $k_j = 0$. Note that the first of these conditions holds for c_1 and that the second condition also holds because it does not apply to c_1.

The second of the assumptions implies that for c orthogonal to $c_1, \ldots, c_{h-1}$,

$$c'\Lambda c / c'c = \sum_{j=r}^{q} k_j^2 \phi_j \Big/ \sum_{j=r}^{q} k_j^2.$$

This is a weighted average of the values ϕ_j, $j = r, \ldots, q$. The maximum value is $\phi_r = \phi_h$. As before, c_h will attain the maximum if and only if c_h is an eigenvector. If c_h is an eigenvector (i.e., $\Lambda c_h = c_h \phi_h$), then the maximum is attained. Conversely, a maximum is only attained if $k_{s+1} = \cdots = k_q = 0$. Thus, an orthogonal maximizing vector c_h must have $k_{h1} = \cdots = k_{hr-1} = k_{hs+1} = \cdots = k_{hq} = 0$. Clearly, any such vector satisfies $\Lambda c = c\phi_h$ and thus is an eigenvector of ϕ_h. In particular, c_h can be any eigenvector that satisfies $c_h' c_j = 0$, $j = 1, \ldots, h-1$.

To complete the proof, we need to prove that our additional induction hypotheses hold for $i = h$. We have already established that $k_{h,s+1} = \cdots = k_{hq} = 0$, which is precisely the condition that if $\phi_j < \phi_h$, then $k_{hj} = 0$. We also need to show that if $c'c_i = 0$, $i = 1, \ldots, h$, and $\phi_j > \phi_{h+1}$, then $k_j = 0$. Equivalently, we need to show that if $h < s$, $k_1 = \cdots = k_{r-1} = 0$ and if $h = s$, $k_1 = \cdots = k_s = 0$. If $h < s$, there is nothing to prove; the result follows from the induction hypothesis. For $h = s$, the two induction hypotheses and the argument in the previous paragraph give: (1) for $i = r, \ldots, s$, $k_{is+1} = \cdots = k_{iq} = 0$ and (2) $k_1 = \cdots = k_{r-1} = 0$. Writing $d_i = (k_{ir}, \ldots, k_{is})$ and $d = (k_r, \ldots, k_s)$, we see that for $i = r, \ldots, s$, $c'c_i = 0$ if and only if $d'd_i = 0$. In particular, $d_r, d_{r+1}, \ldots, d_s$ is an orthogonal basis for $\mathbf{R}^{s-r+1}$. Any other vector d that is orthogonal to $d_r, \ldots, d_s$ must be the zero vector. Thus, if $c'c_i = 0$, $i = r, \ldots, s$, then $k_r = \cdots = k_s = 0$. This is precisely what we needed to prove. $\square$

12.7.2 Using Linear Discrimination Coordinates

In the previous subsection we established a practical procedure for finding linear discriminant coordinates and for transforming the original data into the discrimination coordinates. Simply find A such that

$$E^{-1}HA = A\Lambda,$$

where $\Lambda = D(\phi_1, \ldots, \phi_q)$ and

$$A'EA = I.$$

This is not difficult to do using a good matrix manipulation computer program. The transformed data are

$$Z = YA.$$

As illustrated in Example 12.7.1, the n rows of Z can be plotted in a variety of ways to examine the efficacy of the different coordinates for discrimination. The columns of Z corresponding to the largest eigenvalues show the clearest discrimination because they maximize the one-way ANOVA F test. The estimated covariance matrix of a transformed vector $A'y$ is

$$\widehat{\text{Cov}}(A'y) = A'SA = \frac{1}{n-t}A'EA = \frac{1}{n-t}I,$$

so the Euclidean metric is appropriate for evaluating relationships between data points. This is important in that it allows intuitive evaluation of plots.

With t groups, there are at most $t-1$ coordinates that are valuable for discrimination. If $t > q$, this is not of much interest, but if $t \le q$, this means that some coordinates have no discriminatory power. Recall that

$$H = Y'\left(M - \frac{1}{n}J_n^n\right)Y$$

and that $r\left(M - \frac{1}{n}J_n^n\right) = t-1$. It follows that for the $q \times q$ matrix H

$$r(H) \le \min(q, t-1).$$

Any choice of linear discrimination coordinates corresponds to a set of eigenvectors for $E^{-1}H$. Write $A = (a_1, \ldots, a_q)$ and partition A as $A = (A_*, A_0)$, where the columns of A_* correspond to the nonzero eigenvalues of $E^{-1}H$ and the columns of A_0 correspond to the eigenvalue zero. If a_h is a column of A_0, then, as will be seen in later,

$$a_h'Ha_h / a_h'Ea_h = 0. \tag{12.7.2}$$

In other words, if the data are transformed to the hth discrimination coordinate (i.e., Ya_h), then a one-way ANOVA applied to the transformed data gives an F statistic of zero for testing differences between groups. Thus, the coordinate is useless in discrimination. This result (12.7.2) is an immediate consequence of the following lemma.

Lemma 12.7.5. $HA_0 = 0.$

PROOF.   The columns of A_0 are eigenvectors of $E^{-1}H$ corresponding to the eigenvalue 0. Hence, $E^{-1}HA_0 = 0$. Multiplying on the left by E gives the result.   □

12.7.3 Relationship to Mahalanobis Distance Allocation

Sometimes, the linear discrimination coordinates are used to allocate new observations. Generally, this is done when the coordinates are chosen as in Lemma 12.7.2. Suppose the first s coordinates are to be used. Let

$$A_s = [a_1, \ldots, a_s].$$

The standard allocation rule is to assign y to population r if

$$(y - \bar{y}_{r.})'[A_s A_s'](y - \bar{y}_{r.}) = \min_i (y - \bar{y}_{i.})'[A_s A_s'](y - \bar{y}_{i.}). \tag{12.7.3}$$

Note that

$$(y - \bar{y}_{i.})'[A_s A_s'](y - \bar{y}_{i.}) = \sum_{j=1}^{s} [(y - \bar{y}_{i.})' a_j]^2,$$

so the allocation is based on the squared values of the first s linear discrimination coordinates of the vector $(y - \bar{y}_{i.})$. Once again, we see that the Euclidean metric is appropriate for the transformed variables. If the coordinates are not chosen as in Lemma 12.7.2 (i.e., if the vectors $a_1, \ldots, a_q$ are not an orthonormal set in the appropriate inner product), then taking a simple sum of squares is not appropriate. Thus, the restriction on the choice of coordinates is imposed.

This allocation procedure is closely related to the Mahalanobis distance method. If $s = q$, then $A_s = A$ and by Corollary 12.7.3, $AA' = E^{-1}$. The allocation rule is based on the (squared) distances

$$(y - \bar{y}_{i.})' E^{-1} (y - \bar{y}_{i.}).$$

The estimated covariance matrix is $S^{-1} = (n-t)E^{-1}$, so these distances are simply a constant multiple of the estimated Mahalanobis distances used in Sect. 12.2. The distances differ by merely a constant multiple; therefore, the allocation rules are identical.

Recall that there are at most $\min(q, t-1)$ useful linear discrimination coordinates. Eigenvectors of $E^{-1}H$ that correspond to the eigenvalue zero are not useful for discrimination. It makes little sense to choose s greater than the number of nonzero eigenvalues of $E^{-1}H$. For reasons of collinearity (see *PA-V* Chap. 12 or Christensen 2011, Chapter 13), it might make sense to choose s less than the number of nonzero eigenvalues of $E^{-1}H$ when some of those nonzero eigenvalues are very close to zero. In fact, choosing the number of linear discrimination coordinates is reminiscent of choosing the number of principal components to use in principal component regression. (Recall the similarities between variable selection in discrimination and variable selection in regression.)

We now show that if s is chosen to be exactly the number of nonzero eigenvalues (i.e., $s = r(E^{-1}H)$), then the linear discrimination coordinate allocation rule based on (12.7.3) is precisely the same as the Mahalanobis distance rule. If $r(E^{-1}H) = q$, then $A_s = A$ and we have already shown the result. We need consider only the case where $r(E^{-1}H) \leq t - 1 < q$. Using notation from earlier in the section, $A = [A_*, A_0]$. With $s = r(E^{-1}H)$, we have $A_s = A_*$. Before proving the equivalence of allocation rules, we need the following lemma.

Lemma 12.7.6. If $t - 1 < q$, then for any $i = 1, \ldots, t$,

$$A_0'(\bar{y}_{i\cdot} - \bar{y}_{\cdot\cdot}) = 0.$$

PROOF. Recall that

$$H = \sum_{j=1}^{t} N_j(\bar{y}_{j\cdot} - \bar{y}_{\cdot\cdot})(\bar{y}_{j\cdot} - \bar{y}_{\cdot\cdot})'$$

$$= [\sqrt{N_1}(\bar{y}_{1\cdot} - \bar{y}_{\cdot\cdot}), \ldots, \sqrt{N_t}(\bar{y}_{t\cdot} - \bar{y}_{\cdot\cdot})] \begin{bmatrix} \sqrt{N_1}(\bar{y}_{1\cdot} - \bar{y}_{\cdot\cdot})' \\ \vdots \\ \sqrt{N_t}(\bar{y}_{t\cdot} - \bar{y}_{\cdot\cdot})' \end{bmatrix}.$$

By Proposition B.51 in *PA*,

$$C(H) = C([\sqrt{N_1}(\bar{y}_{1\cdot} - \bar{y}_{\cdot\cdot}), \ldots, \sqrt{N_t}(\bar{y}_{t\cdot} - \bar{y}_{\cdot\cdot})]).$$

It follows that for $i = 1, \ldots, t$,

$$\bar{y}_{i\cdot} - \bar{y}_{\cdot\cdot} \in C(H),$$

and for some vector d,

$$\bar{y}_{i\cdot} - \bar{y}_{\cdot\cdot} = Hd.$$

By Lemma 12.7.5,

$$0 = A_0'Hd = A_0'(\bar{y}_{i\cdot} - \bar{y}_{\cdot\cdot}). \qquad \square$$

The equivalence of the allocation rules is established in the following proposition.

Proposition 12.7.7. If $t - 1 < q$, then for any y,

$$(y - \bar{y}_{i\cdot})'E^{-1}(y - \bar{y}_{i\cdot}) = (y - \bar{y}_{i\cdot})'[A_*A_*'](y - \bar{y}_{i\cdot}) + (y - \bar{y}_{\cdot\cdot})'[A_0A_0'](y - \bar{y}_{\cdot\cdot}), \quad (12.7.4)$$

and the Mahalanobis distance allocation rule is identical to the linear discrimination coordinate allocation rule with $s = r(E^{-1}H)$.

PROOF. We begin by arguing that if Eq. (12.7.4) holds, the two allocation rules are identical. As mentioned earlier, the Mahalanobis rule minimizes $(y - \bar{y}_{i\cdot})'E^{-1}(y - \bar{y}_{i\cdot})$ with respect to i. The term $(y - \bar{y}_{\cdot\cdot})'[A_0A_0'](y - \bar{y}_{\cdot\cdot})$, on the right of (12.7.4), does not depend on i, so the Mahalanobis rule minimizes $(y - \bar{y}_{i\cdot})'[A_*A_*'](y - \bar{y}_{i\cdot})$ with respect to i. However, this is simply the linear discrimination coordinate rule for $s = r(E^{-1}H)$.

We now prove Eq. (12.7.4). From Corollary 12.7.3 and the partition of A,

$$E^{-1} = AA' = A_* A_*' + A_0 A_0',$$

so

$$(y - \bar{y}_{i\cdot})' E^{-1}(y - \bar{y}_{i\cdot}) = (y - \bar{y}_{i\cdot})' A_* A_*'(y - \bar{y}_{i\cdot}) + (y - \bar{y}_{i\cdot})' A_0 A_0'(y - \bar{y}_{i\cdot}).$$

It suffices to show that

$$(y - \bar{y}_{i\cdot})' A_0 A_0'(y - \bar{y}_{i\cdot}) = (y - \bar{y}_{\cdot\cdot})' A_0 A_0'(y - \bar{y}_{\cdot\cdot})$$

or

$$A_0'(y - \bar{y}_{i\cdot}) = A_0'(y - \bar{y}_{\cdot\cdot}).$$

Clearly,

$$\begin{aligned} A_0'(y - \bar{y}_{i\cdot}) &= A_0'(y - \bar{y}_{\cdot\cdot} - \bar{y}_{i\cdot} + \bar{y}_{\cdot\cdot}) \\ &= A_0'(y - \bar{y}_{\cdot\cdot}) - A_0'(\bar{y}_{i\cdot} - \bar{y}_{\cdot\cdot}). \end{aligned}$$

By Lemma 12.7.6, $A_0'(\bar{y}_{i\cdot} - \bar{y}_{\cdot\cdot}) = 0$ and the proof is complete. □

12.7.4 Alternate Choice of Linear Discrimination Coordinates

Our motivation for the choice of linear discrimination coordinates has been based entirely on maximizing analysis of variance F statistics. An alternative motivation, based on population rather than sample values, leads to slightly different results. Consider a linear combination of the dependent variable vector y, say $a'y$. It follows that $\mathrm{Var}(a'y) = a'\Sigma a$ and, depending on its population, $\mathrm{E}(a'y) = a'\mu_i$. Define

$$\Omega = \sum_{i=1}^{t} (\mu_i - \bar{\mu}_\cdot)(\mu_i - \bar{\mu}_\cdot)'.$$

The value

$$\frac{a'\Omega a}{a'\Sigma a}$$

can be viewed as a measure of the variability between the population means $a'\mu_i$ relative to the variance of $a'y$. Choosing a to maximize this measure may be a reasonable way to choose linear discrimination coordinates. Both Ω and Σ are unknown parameters and must be estimated. The covariance matrix can be estimated with S, and Ω can be estimated with

$$H_* = \sum_{i=1}^{t} (\bar{y}_{i\cdot} - \bar{y}_{\cdot\cdot})(\bar{y}_{i\cdot} - \bar{y}_{\cdot\cdot})'.$$

The sample version of $a'\Omega a / a'\Sigma a$ is

$$\frac{a'H_* a}{a'Sa}.$$

The error statistic E is a constant multiple of S, so it is equivalent to work with

$$\frac{a'H_* a}{a'Ea}.$$

The subsequent development of linear discrimination coordinates follows as in our discussion based on $a'Ha / a'Ea$.

12.8 Linear Discrimination

In linear model theory we consider a vector of predictor variables x and linear models $x'\beta$. The key feature of a linear model is that $x'\beta$ is a linear function of the predictor variables, whatever the predictor variables may be. The predictor variables are *not* restricted to be a set of measurements originally taken on a collection of observational units. Chapter 1 examined the large variety of transformations that can be applied to the original measurements that make linear models far more flexible.

In this chapter, our predictor variables have been denoted y, rather than x. We have examined the traditional linear and quadratic discrimination methods LDA and QDA. We have pointed out that both of these methods are linear in the sense that they involve linear combinations of predictor variables, it is just that quadratic discrimination includes squares and cross-products of the original measurements as additional predictors. *The key aspect of LDA and QDA is not that they involve linear or quadratic functions of the original measurements but that the methods assume that the original data have a multivariate normal distribution and involve estimating appropriate normal densities.* If the data really are multivariate normal, no other procedure will give much of an improvement on LDA or QDA. If the data are not multivariate normal, nor easily transformed to multivariate normal, alternative discrimination procedures should be able to improve on them.

Obviously one *could* apply LDA to a y vector that includes not only the original measurements but also squares and cross-products (or other transformations) and LDA would probably give reasonable results, even though such a y vector could not possibly have a multivariate normal distribution. In the next chapter we focus on linear discrimination methods that do not assume multivariate normality. These methods include both logistic discrimination and support vector machines (SVMs). All such methods admit as predictor variables, transformations of the original measurements.

EXAMPLE 12.8.1.   Figure 12.6 contains data of a form that have often been used
to sell support vector machines because neither linear nor quadratic discriminant
analysis can distinguish the two populations whereas SVMs separate them easily.
Such a claim is comparing apples with oranges. It is true that the most naive forms
of LDA and QDA cannot separate them. But the most naive form of SVMs cannot
separate them either. If you transform the data into polar coordinates, you get the
data representation in Fig. 12.7. It is trivial to separate the data in Fig. 12.7 with a
vertical line. The difference between LDA and SVMs is that they may pick different
vertical lines to do the separation. And in this case, how much do you really care
which vertical line you use?

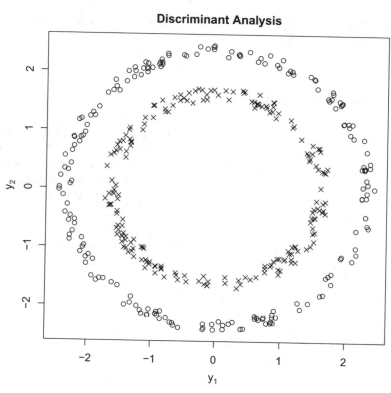

Fig. 12.6 Doughnut data

Obviously, no line is going to separate the data in Fig. 12.6. To separate the
groups with a line, you have to transform the data. The main difference is that computer programs for SVMs have simple transformation methods built into them by
allowing the specification of an appropriate reproducing kernel. Logistic regression
programs could also allow the specification of an appropriate reproducing kernel,
but typically they do not. For the more traditional methods, like LDA and QDA,
it seems that the transformations need to be specified explicitly. (Even though you

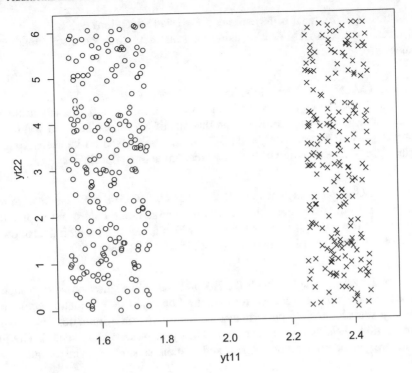

Fig. 12.7 Doughnut data in polar coordinates

can apply LDA and QDA to transformed data, it is rarely a good idea unless the transformation is designed to make the data more multivariate normal.) □

12.9 Additional Exercises

Exercise 12.9.1. Consider the data of Example 10.3.1. Suppose a person has heart rate measurements of $y = (84, 82, 80, 69)'$.

(a) Using normal theory linear discrimination, what is the estimated maximum likelihood allocation for this person?

(b) Using normal theory quadratic discrimination, what is the estimated maximum likelihood allocation for this person?

(c) If the two drugs have equal prior probabilities but the placebo is twice as probable as the drugs, what is the estimated maximum posterior probability allocation?

(d) Suppose the costs of correct classification are zero, the costs of misclassification depend only on the true population, and the cost of misclassifying an individual who actually has taken drug B is twice that of the other populations. Using the prior probabilities of (c), what is the estimated Bayesian allocation?

(e) What is the optimal allocation using only the first two linear discrimination coordinates?

Exercise 12.9.2. In the motion picture *Diary of a Mad Turtle* the main character, played by Richard Benjamin Kingsley, claims to be able to tell a female turtle by a quick glance at her carapace. Based on the data of Exercise 10.6.1, do you believe that it is possible to accurately identify a turtle's sex based on its shell? Explain. Include graphical evaluation of the linear discrimination coordinates.

Exercise 12.9.3. Using the data of Exercise 10.6.3, do a stepwise discriminant analysis to distinguish among the thyroxin, thiouracil, and control rat populations based on their weights at various times. To which group is a rat with the following series of weights most likely to belong: $(56, 75, 104, 114, 138)$?

Exercise 12.9.4. Lachenbruch (1975) presents information on four groups of junior technical college students from greater London. The information consists of summary statistics for the performance of the groups on arithmetic, English, and form relations tests that were given in the last year of secondary school. The four groups are Engineering, Building, Art, and Commerce students. The sample means are:

	Engineering	Building	Art	Commerce
Arithmetic (y_1)	27.88	20.65	15.01	24.38
English (y_2)	98.36	85.43	80.31	94.94
Form Relations (y_3)	33.60	31.51	32.01	26.69
Sample Size	404	400	258	286

The pooled estimate of the covariance matrix is

$$S_p = \begin{bmatrix} 55.58 & 33.77 & 11.66 \\ 33.77 & 360.04 & 14.53 \\ 11.66 & 14.53 & 69.21 \end{bmatrix}.$$

What advice could you give to a student planning to go to a junior technical college who just achieved scores of $(22, 90, 31)'$?

Exercise 12.9.5. Suppose the concern in Exercise 12.9.4 is minimizing the cost to society of allocating students to the various programs of study. The great bureaucrat in the sky, who works on the top floor of the tallest building in Whitehall, has determined that the costs of classification are as follows:

Cost	Optimal Study Program			
	Engineering	Building	Art	Commerce
Allocated Engineering	1	2	8	2
Study Building	4	2	7	3
Program Art	8	7	4	4
Commerce	4	3	5	2

Evaluate the program of study that the bureaucrat thinks is appropriate for the student from Exercise 12.9.4.

Exercise 12.9.6. Show that the Mahalanobis distance is invariant under affine transformations $z = Ay + b$ of the random vector y when A is nonsingular.

Exercise 12.9.7. Let y be an observation from one of two normal populations that have means of μ_1 and μ_2 and common covariance matrix Σ. Define $\lambda' = (\mu_1 - \mu_2)'\Sigma^{-1}$.
(a) Show that, under linear discrimination, y is allocated to population 1 if and only if

$$\lambda'y - \lambda'\frac{1}{2}(\mu_1 + \mu_2) > 0.$$

(b) Show that if y is from population 1,

$$E(\lambda'y) - \lambda'\frac{1}{2}(\mu_1 + \mu_2) > 0$$

and if y is from population 2,

$$E(\lambda'y) - \lambda'\frac{1}{2}(\mu_1 + \mu_2) < 0.$$

Exercise 12.9.8. Consider a two group allocation problem in which the prior probabilities are $\pi(1) = \pi(2) = 0.5$ and the sampling distributions are exponential, namely

$$f(y|i) = \theta_i e^{-\theta_i y}, \quad y \geq 0.$$

Find the optimal allocation rule. Assume a cost structure where $c(i|j)$ is zero for $i = j$ and one otherwise. The *total probability of misclassification* for an allocation rule is precisely the Bayes risk of the allocation rule under this cost structure. Let $\delta(y)$ be an allocation rule. The frequentist risk for the true population j is $R(j, \delta) = \int c(\delta(y)|j)f(y|j)dy$ and the Bayes risk is $r(p, \delta) = \sum_{j=1}^{t} R(j, \delta)\pi(j)$. See Berger (1985, Section 1.3) for more on risk functions. Find the total probability of misclassification for the optimal rule.

Exercise 12.9.9.   Suppose that the distributions for two populations are bivariate normal with the same covariance matrix. For $\pi(1) = \pi(2) = 0.5$, find the value of the correlation coefficient that minimizes the total probability of misclassification. The total probability of misclassification is defined in Exercise 12.9.8.

References

Aitchison, J. (1975). Goodness of prediction fit. *Biometrika, 62*, 547–554.

Aitchison, J., & Dunsmore, I. R. (1975). *Statistical prediction analysis*. Cambridge: Cambridge University Press.

Anderson, T. W. (2003). *An introduction to multivariate statistical analysis* (3rd ed.). New York: Wiley.

Andrews, D. F., Gnanadesikan, R., & Warner, J. L. (1971). Transformations of multivariate data. *Biometrics, 27*, 825–840.

Berger, J. O. (1985). *Statistical decision theory and bayesian analysis*. New York: Springer.

Box, G. E. P. & Cox, D. R. (1964). An analysis of transformations (with discussion). *Journal of the Royal Statistical Society, Series B, 26*, 211–246.

Christensen, R. (1997). *Log-linear models and logistic regression* (2nd ed.). New York: Springer.

Christensen, R. (2011). *Plane answers to complex questions: The theory of linear models* (4th ed.). New York: Springer.

Christensen, R., Johnson, W., Branscum, A., & Hanson, T. E. (2010). *Bayesian ideas and data analysis: An introduction for scientists and statisticians*. Boca Raton: Chapman and Hall/CRC Press.

Efron, B. (1983). Estimating the error rate of a prediction rule: Improvement on cross-validation. *Journal of the American Statistical Association, 78*, 316–331.

Fisher, R. A. (1936). The use of multiple measurements in taxonomic problems. *Annals of Eugenics, 7*, 179–188.

Fisher, R. A. (1938). The statistical utilization of multiple measurements. *Annals of Eugenics, 8*, 376–386.

Friedman, J. H. (1989). Regularized discriminant analysis. *Journal of the American Statistical Association, 84*, 165–175.

Geisser, S. (1971). The inferential use of predictive distributions. In V. P. Godambe & D. A. Sprott (Eds.), *Foundations of statistical inference*. Toronto: Holt, Rinehart, and Winston.

Geisser, S. (1977). Discrimination, allocatory and separatory, linear aspects. In J. Van Ryzin (Ed.), *Classification and clustering*. New York: Academic Press.

Hand, D. J. (1981). *Discrimination and classification*. New York: Wiley.

Hand, D. J. (1983). A comparison of two methods of discriminant analysis applied to binary data. *Biometrics, 39*, 683–694.

Hastie, T., Tibshirani, R., & Friedman, J. (2016). *The elements of statistical learning: Data mining, inference, and prediction* (2nd ed.). New York: Springer.

Johnson, R. A., & Wichern, D. W. (2007). *Applied multivariate statistical analysis* (6th ed.). Englewood Cliffs: Prentice–Hall.

Lachenbruch, P. A. (1975). *Discriminate analysis*. New York: Hafner Press.

Lachenbruch, P. A., Sneeringeer, C., & Revo, L. T. (1973). Robustness of the linear and quadratic discriminant function to certain types of non-normality. *Communications in Statistics, 1*, 39–57.

Levy, M. S., & Perng, S. K. (1986). An optimal prediction function for the normal linear model. *Journal of the American Statistical Association, 81*, 196–198.

Lubischew, A. A. (1962). On the use of discriminant functions in taxonomy. *Biometrics, 18*, 455–477.

Murray, G. D. (1977). A note on the estimation of probability density functions. *Biometrika, 64*, 150–152.

Panel on Discriminant Analysis, Classification, and Clustering. (1989). Discriminant analysis and clustering. *Statistical Science, 4*, 34–69.

Press, S. J. (1982). *Applied multivariate analysis: Using Bayesian and frequentist methods of inference* (2nd ed.). Malabar: R.E. Krieger. (Latest reprinting, Dover Press, 2005).

Press, S. J., & Wilson, S. (1978). Choosing between logistic regression and discriminant analysis. *Journal of the American Statistical Association, 73*, 699–705.

Rao, C. R. (1948). The utilization of multiple measurements in problems of biological classification. *Journal of the Royal Statistical Society, Series B, 10*, 159–203.

Rao, C. R. (1952). *Advanced statistical methods in biometric research*. New York: Wiley.

Ripley, B. D. (1996). *Pattern recognition and neural networks*. Cambridge: Cambridge University Press.

Scheffé, H. (1959). *The analysis of variance*. New York: Wiley.

Seber, G. A. F. (1984). *Multivariate observations*. New York: Wiley.

Chapter 13
Binary Discrimination and Regression

Abstract This chapter examines support vector machines. To do that properly it includes background on binomial regression and discrimination. Much of the technical material on support vector machines is relegated to Appendix A.

Traditional discrimination was examined in Chap. 12. Binary discrimination is the special case in which the number of groups is $t = 2$. Obviously LDA and QDA apply to the special case but the purpose of this chapter is to explore some alternatives to LDA and QDA. Specifically we examine logistic discrimination, other forms of generalized linear model discrimination, and the application of Support Vector Machines (SVMs). We cannot really extend our discussion of discrimination without first examining logistic regression and similar methods. Binary regression problems are of great interest in their own right. The special case of logistic linear regression models is treated in more detail in many places including Christensen (1997, 2015). Those references also include discussions about generalizing logistic discrimination to log-linear model discrimination for $t \geq 2$. In fact, they contain no specific discussion of logistic discrimination.

The general prediction problem is discussed in *PA* Sect. 6.3. It considers the problem of predicting a random variable y from, say, a $d - 1$ dimensional random vector $\mathbf{x}$. We now focus on the special case in which the dependent variable y takes only the values 0 and 1. (We will use p to denote the probability of a 1, so we use d in labeling the number of predictor variables.) 0–1 random variables are called *Bernoulli random variables*, which suggests that we should call this Bernoulli prediction. Boolean prediction also seems like a reasonable name. But the gang all seems to like "binary" so, in an attempt to fit-in, I am sticking with that.

Under squared error prediction loss, the best predictor is

$$p(\mathbf{x}) \equiv \mathrm{E}(y|\mathbf{x}),$$

which is also the conditional probability of getting a 1 (success). Under *Hamming prediction loss* the best predictor is 0 when $p(\mathbf{x}) < 0.5$ and 1 when $p(\mathbf{x}) > 0.5$. (Hamming loss is 0 if the prediction equals y and 1 otherwise.) In both cases it is incumbent upon us to obtain a good estimate of $p(\mathbf{x})$.

© Springer Nature Switzerland AG 2019
R. Christensen, *Advanced Linear Modeling*, Springer Texts in Statistics,
https://doi.org/10.1007/978-3-030-29164-8_13

503

Regression analysis and discrimination involve different data collection schemes. Regression collects independent observations from the joint distribution of $(y, \mathbf{x}')$. Aldrich (2005) suggests that it was Fisher who first argued that regression estimation should condition on the predictor variables $\mathbf{x}$. The first part of this chapter examines how to estimate $p(\mathbf{x})$ from the conditional distribution of y given $\mathbf{x}$. In this part we need only assume that the observations are conditionally independent and that $y \sim \text{Bin}[1, p(\mathbf{x})] \equiv \text{Bern}[p(\mathbf{x})]$. The last section of the chapter considers estimates of $p(\mathbf{x})$ derived from sampling the conditional distribution of $\mathbf{x}$ given y. This is the binary ($t = 2$) version of the discrimination problem considered in Chap. 12. The middle of the chapter examines Hamming prediction without explicitly estimating $p(\mathbf{x})$.

Throughout we will explicitly incorporate ideas on penalized estimates similar to Chap. 2. Implicit throughout is that the models can exploit the nonparametric linear structures discussed in Chap. 1.

We begin with the binomial regression problem that most generalized linear model computer programs are written to handle.

13.1 Binomial Regression

Suppose there are a number of independent observations with $y_h \sim \text{Bin}[1, p(\mathbf{x}_h)]$. Often such data get reported only as the total number of successes for each vector of predictor variables. In such cases, we implicitly reindex the original data as

$$(y_{ij}, \mathbf{x}_i'), \quad i = 1, \ldots, n, \ j = 1, \ldots, N_i$$

so that the reported data are

$$(y_{i\cdot}, \mathbf{x}_i'), \quad i = 1, \ldots, n, \quad \text{where } y_{i\cdot} \equiv \sum_{j=1}^{N_i} y_{ij}.$$

We now have independent binomial random variables

$$N_i \bar{y}_{i\cdot} \equiv y_{i\cdot} \sim \text{Bin}[N_i, p(\mathbf{x}_i)]; \quad i = 1, \ldots, n,$$

where the binomial proportions $\bar{y}_{i\cdot}$ are between 0 and 1. It is common practice to write binomial generalized linear model computer programs using the binomial proportions as the input data and specifying the N_is as weights. Obviously such programs can also handle the original binary data $(y_h, \mathbf{x}_h')$ by writing $h = 1, \ldots, n$ but with $N_h = 1$ for all h. *In conformance with such programs, we write*

$$y_i \equiv \bar{y}_{i\cdot}$$

for the rest of this section.

The likelihood function for independent data with $N_i y_i \sim \text{Bin}[N_i, p(\mathbf{x}_i)]$ is

$$L[p(\cdot)] \equiv \prod_{i=1}^{n} \binom{N_i}{N_i y_i} [p(\mathbf{x}_i)]^{N_i y_i} [1 - p(\mathbf{x}_i)]^{N_i - N_i y_i}.$$

The *deviance* is defined as -2 times the log-likelihood so

$$
\begin{aligned}
&D[p(\cdot)] \\
&\equiv -2\sum_{i=1}^{n}\{N_i y_i \log[p(\mathbf{x}_i)] + (N_i - N_i y_i)\log[1 - p(\mathbf{x}_i)]\} - 2\sum_{i=1}^{n}\log\left[\binom{N_i}{N_i y_i}\right] \\
&= \sum_{i=1}^{n} -2N_i\{y_i\log[p(\mathbf{x}_i)] + (1 - y_i)\log[1 - p(\mathbf{x}_i)]\} - 2\sum_{i=1}^{n}\log\left[\binom{N_i}{N_i y_i}\right].
\end{aligned}
$$

A maximum likelihood estimate of $p(\cdot)$ maximizes the likelihood or, equivalently, minimizes the deviance. To simplify notation denote the constant term in the deviance

$$
K \equiv -2\sum_{i=1}^{n}\log\left[\binom{N_i}{N_i y_i}\right].
$$

The constant term has no effect on estimation. For binary regression models in which $N_i \equiv 1$ so that y_i is 1 or 0, the constant term in the deviance vanishes and only one of the two terms in the braces actually applies. Either y_i or $1 - y_i$ has to be zero, so one of the terms in the braces always gets multiplied by 0.

If the function $p(\mathbf{x})$ is known except for some unknown parameter vector θ, write $p(\mathbf{x}; \theta)$. The maximum likelihood estimate of θ maximizes the parameterized likelihood

$$
L(\theta) \equiv \prod_{i=1}^{n}\binom{N_i}{N_i y_i}[p(\mathbf{x}_i; \theta)]^{N_i y_i}[1 - p(\mathbf{x}_i; \theta)]^{N_i - N_i y_i}
$$

or minimizes the parameterized deviance

$$
D(\theta) \equiv \sum_{i=1}^{n} -2N_i\{y_i\log[p(\mathbf{x}_i; \theta)] + (1 - y_i)\log[1 - p(\mathbf{x}_i; \theta)]\} + K.
$$

Henceforth, we take x to be a d vector that includes all explanatory variables. In most cases $x' = (1, \mathbf{x}')$.

Binomial generalized linear models typically specify that the conditional probability is a known function of $x'\beta$. In particular,

$$
p(\mathbf{x}) \equiv p(x) = F(x'\beta)
$$

for some known cumulative distribution function (cdf) F for which the inverse function F^{-1} exists. F is a cdf so that the real valued term $x'\beta$ is transformed into a number between 0 and 1. The inverse function is called a *link* function and is used to isolate the linear structure $x'\beta$ of the model, i.e.,

$$
F^{-1}[p(x)] = x'\beta.
$$

The most common choices for F are the standard versions of the *logistic*, normal, and *Gumbel (minimum)* distributions. With $\Phi(\cdot)$ denoting the cdf for a $N(0, 1)$ random variable,

$$p(x) = F(x'\beta) = \begin{cases} e^{x'\beta} \Big/ \left[1 + e^{x'\beta}\right] & \text{Logistic} \\ \Phi(x'\beta) & \text{Normal} \\ 1 - \exp\left[-e^{x'\beta}\right] & \text{Gumbel.} \end{cases}$$

Most often the procedures are referred to by the names of the inverse functions rather than the names of the original cdfs:

$$x'\beta = F^{-1}[p(x)] = \begin{cases} \log\{p(x)/[1-p(x)]\} & \text{Logit} \\ \Phi^{-1}[p(x)] & \text{Probit} \\ \log\{-\log[1-p(x)]\} & \text{Complementary log-log.} \end{cases}$$

In the case of logit/logistic models, logit often refers to ANOVA type models and logistic is often used for regression models. I use the terms interchangeably but prefer calling them logit models. (In this chapter, all references to "generalized linear models" refer to the subclass of binomial generalized linear models defined using an inverse cdf link.)

In any case, the likelihood function for such data is

$$L_F(\beta) \equiv \prod_{i=1}^{n} \binom{N_i}{N_i y_i} [F(x_i'\beta)]^{N_i y_i} [1 - F(x_i'\beta)]^{N_i - N_i y_i}$$

and the deviance is

$$D_F(\beta) \equiv \sum_{i=1}^{n} -2N_i \left\{ y_i \log\left[F(x_i'\beta)\right] + (1 - y_i) \log\left[1 - F(x_i'\beta)\right] \right\} + K. \quad (13.1.1)$$

As always, the constant term K in the deviance is irrelevant to the estimation of β.

Note that minimum deviance (maximum likelihood) estimation fits into the pattern discussed in *PA-V* Sect. 13.6 of estimating β by defining weights $w_i > 0$ and a loss function $\mathcal{L}(y, u)$ and minimizing

$$\sum_{i=1}^{n} w_i \mathcal{L}(y_i, x_i'\beta).$$

Here the weights are $w_i = N_i$ and the loss function is

$$\mathcal{L}(y, u) = -2\{y \log[F(u)] + (1 - y) \log[1 - F(u)]\}.$$

In later sections we will see that the loss function is easier to interpret for binary data and that *support vector machines* use a very similar procedure when estimating β.

In analogy to penalized least squares estimates, we can form *penalized minimum deviance (penalized maximum likelihood)* estimates that minimize

$$D_F(\beta) + k \mathcal{P}(\beta).$$

Typically, we use the same penalty functions $\mathscr{P}(\beta)$ as discussed in Chap. 2 for penalized least squares. For a multiple regression with $x_i'\beta \equiv \beta_0 + \sum_{j=1}^{d-1} \beta_j x_{ij}$, write $\beta_* \equiv (\beta_1, \ldots, \beta_{d-1})'$. As with standard regression, we typically would not penalize the intercept. By choosing

$$\mathscr{P}_L(\beta) \equiv \sum_{j=1}^{d-1} |\beta_j| = \|\beta_*\|_1$$

we get lasso binomial regression. By choosing

$$\mathscr{P}_R(\beta) \equiv \beta_*'\beta_* = \sum_{j=1}^{d-1} \beta_j^2$$

we get one form of ridge binomial regression. Elastic net binomial regression is obtained by taking

$$\mathscr{P}_E(\beta) \equiv k_1 \mathscr{P}_R(\beta) + k_2 \mathscr{P}_L(\beta).$$

As mentioned in Chap. 2, these penalty functions penalize each coefficient the same amount, so typically one would *standardize the predictor variables to a common length* before applying such a penalty. (The penalization ideas apply to all generalized linear models, not just these binomial generalized linear models, and are fundamental to support vector machines.)

As in Chap. 2, it is a simple matter to generalize the penalized estimation ideas to a partitioned model,

$$F^{-1}[p(x_i, z_i)] = x_i'\beta + z_i'\gamma$$

where x_i and z_i are known, β and γ are unknown parameters and where we only penalize γ.

13.1.1 Data Augmentation Ridge Regression

For linear ridge regression we established in Sect. 2.2 that the ridge estimates could be obtained by fitting an augmented linear model. We now define an analogous augmented binomial regression model and infer the penalty function that it implicitly incorporates. (The penalty is not the traditional ridge penalty.) Although ridge regression requires no assumption of normality, the analogies between standard regression and binomial regression will be clearer making it.

Model (2.2.5) is an augmented, partitioned linear model that provides generalized ridge estimates. By specifying $Q = I$, model (2.2.5) provides standard ridge estimates. With $Q = I$, model (2.2.5) treats the augmented observations 0 as observations on independent random variables $\tilde{y}_j$, $j = 1, \ldots, s$ with the distribution $\tilde{y}_j \sim N(\gamma_j, \sigma^2/k)$. The model involves finding the simplest form of generalized least squares estimates: weighted least squares. The vector of weights becomes $w = [J_n', kJ_s']'$. Relatively few computer programs for linear models incorporate the ability to perform generalized least squares but I cannot remember ever using a regression program that did not include the capability for weighted least squares.

Data augmentation binomial ridge regression takes s augmenting observations as $\tilde{y}_j = F(0)$ and treats them as independent with $k\tilde{y}_j \sim \text{Bin}[k, F(\gamma_j)]$. For logit and probit models $\tilde{y}_j = 0.5$. To analyze such data you need software that is coded in enough generality that it permits analysis on binomials with non-integer numbers of trials. An augmented observation $\tilde{y}_j = F(0)$ comes from a case with probability $F(\gamma_j)$ so it forces γ_j towards 0. The parameter k determines how many Bernoulli trials $\tilde{y}_j$ corresponds to, so it determines how much γ_j gets forced towards 0. These augmented data define the same augmented model matrix as in (2.2.5). Model (2.2.5) augments the data Y with a string of 0s but instead we augment Y into $[Y', F(0)J_s']'$. The weight vector we need for the augmented binomial model is exactly the same as the weight vector for model (2.2.5).

The penalty associated with this procedure is defined by what the augmenting observations add to the deviance function. Ignoring the constant term that the augmented data add to the deviance, it is not hard to see that the penalty function, say, $\mathscr{P}_{R2}(\gamma)$ is defined via

$$k\mathscr{P}_{R2}(\gamma) \equiv k \sum_{j=1}^{s} -2\left\{ F(0) \log[F(\gamma_j)] + [1 - F(0)] \log[1 - F(\gamma_j)] \right\}.$$

13.2 Binary Prediction

Henceforth we use binary data ($N_i = 1$) to make binary predictions. In the machine learning community, binary prediction is called *classification*, cf. Hastie, Tibshirani, and Friedman (2016). I cannot overemphasize that when performing this activity, there is an important distinction to be made over how the data were collected. In regression problems, a sample is taken from a population and individuals randomly fall into a group: 0 or 1. In the discrimination problems considered in Chap. 12, data are randomly sampled from each group separately. The term "classification" has traditionally been associated with discrimination problems, but maintaining that distinction is surely a losing battle.

Exercise 13.1.   How does sampling from fixed groups, as associated with ANOVA models, fit into the regression/discrimination distinction?

One of the beauties of using the binomial/binary generalized linear models in Sect. 13.1 for binary prediction of regression data is that they provide estimated probabilities for belonging in the two groups. Having good estimates of $p(x) \equiv E(y|x)$ helps in making good predictions for any prediction loss: squared error, absolute error, even Hamming. However, under Hamming prediction loss, all that matters is, for any x, estimating whether $p(x) > 0.5$ or $p(x) < 0.5$. Hamming loss only cares whether cases get assigned to the correct group.

The estimate of $p(x)$ leads to a linear prediction rule. A *linear prediction rule* amounts to defining a hyperplane of x vectors and predicting that points on one side of the hyperplane will be a 1 and points on the other side will be a 0. In Sect. 13.4 we consider a wider class of linear prediction rules that merely assign cases to groups without actually estimating the probability function. The motivation for this will be by analogy to the estimation methods for generalized linear models discussed in Sect. 13.3. These linear prediction rules include the support vector machines considered in Sect. 13.5 and they seem to implicitly assume that the data are regression data rather than discrimination data. Section 13.6 looks at how to estimate $p(x)$ from the best predictor associated with a specific loss function.

EXAMPLE 13.2.1.   Consider again the Cushing's syndrome data from Table 12.1, To illustrate binary prediction, we restrict our attention to the 15 cases that are bilateral hyperplasia or carcinoma. Again, the analysis is performed on the logarithms of the predictor variables. Figure 13.1 plots the points and includes three linear prediction rules: logistic regression, probit regression and a support vector machine. Points above a line are classified as carcinoma and points below a line are identified as bilateral hyperplasia. To anthropomorphize, the generalized linear models seem to care more about not getting any point too badly wrong. The SVM almost seems like if it cannot get that one bilateral point correctly classified, it doesn't care how far it is from the line. (Indeed, even if the SVM could get that bilateral point correctly classified, if ignoring it will get the fitted line far enough away from all the other points, the SVM would still largely ignore the misclassified point. For more on this see the artificially simple example in the online R commands document.) Christensen (1997, Section 4.7) and Ripley (1996, Section 2.4) discuss the complete three group data.

It is not clear whether the Cushing Syndrome data are regression data or discrimination data. If regression data, someone would have sampled 21 Cushing's Syndrome patients who fell into the categories: 6 adenoma, 10 bilateral hyperplasia, 5 carcinoma. If discrimination data, someone decided to sample 6 adenoma patients, 10 bilateral hyperplasia patients, and 5 carcinoma patients. We assumed the latter in Chap. 12. For discrimination data, the generalized linear model methods of the next section require the adjustments discussed at the end of the chapter before they will make proper predictions. Linear prediction methods that are not closely associated with estimating $p(x)$ have shakier justifications when used for discrimination data because they do not lend themselves to the adjustments that are clearly needed for generalized linear models. The discrimination methods of Chap. 12 *are* closely associated with estimating $p(x)$ but they do it indirectly by estimating the density $f(y|i)$ of the predictor variables given the group.   □

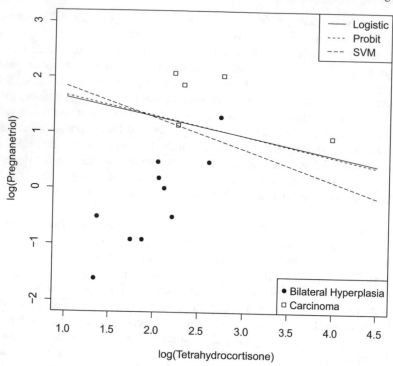

Fig. 13.1 Logistic regression, probit regression, and an SVM: Cushing's syndrome data (subset)

13.3 Binary Generalized Linear Model Estimation

For binary data the deviance in (13.1.1) reduces to

$$D_F(\beta) \equiv \sum_{i=1}^{n} -2 \left\{ y_i \log \left[F(x_i'\beta) \right] + (1 - y_i) \log \left[1 - F(x_i'\beta) \right] \right\}. \tag{13.3.1}$$

Again, minimum deviance (maximum likelihood) estimation fits into the pattern of estimating β by defining a loss function $\mathscr{L}(y, u)$ and weights $w_i > 0$ and then minimizing

$$\sum_{i=1}^{n} w_i \mathscr{L}(y_i, x_i'\beta).$$

For binary generalized linear models the weights are all 1 and, as before, the loss function is

$$\mathscr{L}_F(y, u) = -2 \left\{ y \log \left[F(u) \right] + (1 - y) \log \left[1 - F(u) \right] \right\}$$

but now, because of the binary nature of the data, we can write

$$\mathscr{L}_F(y, u) = \begin{cases} -2 \log \left[F(u) \right] & \text{if } y = 1 \\ -2 \log \left[1 - F(u) \right] & \text{if } y = 0 \, . \end{cases} \tag{13.3.2}$$

The logit and probit loss functions are plotted in Fig. 13.2. The loss functions are quite similar, as were the prediction lines in Fig. 13.1.

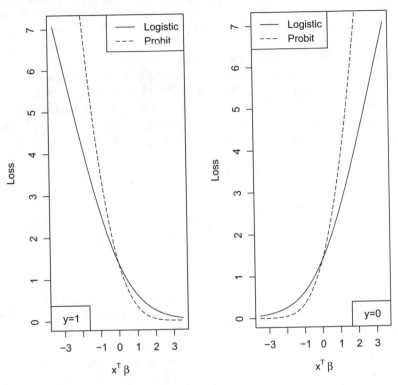

Fig. 13.2 Binary logistic regression and probit regression loss functions

A penalized minimum deviance (penalized maximum likelihood) estimate is defined as in Sect. 13.1. It can be viewed as minimizing

$$\sum_{i=1}^{n} \mathscr{L}_F(y_i, x_i'\beta) + k\mathscr{P}(\beta).$$

(The artificial example in the R code document includes a data augmentation ridge fit that is reasonably similar to the default SVM.)

13.4 Linear Prediction Rules

We now examine linear prediction rules in detail. First, that generalized linear models lead to linear prediction rules and then, that similar ideas can produce linear prediction rules without an explicit probability model. In Sect. 13.6 we will try to relate such rules back to probability models.

For the generalized linear models, the optimal Hamming loss predictor is 1 when $F(x'\beta) > 0.5$ and 0 when $F(x'\beta) < 0.5$. These conditions are equivalent to predicting 1 when $x'\beta > F^{-1}(0.5)$ and 0 when $x'\beta < F^{-1}(0.5)$. Thus the hyperplane of x vectors that satisfy $x'\beta = F^{-1}(0.5)$ implicitly defines a linear prediction rule which is the optimal Hamming rule for the generalized linear model.

With $\hat{\beta}$ the minimum deviance estimate and $F(x'\hat{\beta})$ the estimated probability for group 1, the logistic and probit lines in Fig. 13.1 were constructed by setting $F(x'\hat{\beta}) = 0.5$, i.e., $x'\hat{\beta} = F^{-1}(0.5) = 0$. (The last equality only holds when 0 is a median of F and always holds when F is symmetric about 0.) When $x'\hat{\beta} > 0$, the logistic and probit models have $F(x'\hat{\beta}) > 0.5$. When $x'\hat{\beta} < 0$, they have $F(x'\hat{\beta}) < 0.5$.

In a regression setting we typically have

$$x'\beta \equiv \beta_0 + \sum_{j=1}^{d-1} \beta_j x_j = \beta_0 + \mathbf{x}'\beta_*$$

where

$$\mathbf{x}' \equiv (x_1, \ldots, x_{d-1}).$$

The hyperplane $x'\hat{\beta} = F^{-1}(0.5)$ is the same creature as $\left[\beta_0 - F^{-1}(0.5)\right] + \mathbf{x}'\beta_* = 0$. As a function of the predictor variables in $\mathbf{x}$, the orientation of the hyperplane is determined by β_*. Hyperplanes with β_* vectors that are multiples of one another are parallel in $\mathbf{R}^{d-1}$.

Any hyperplane $x'\beta = 0$ can be used to predict binary outcomes using the rule: *if $x'\beta > 0$ the case is predicted as group 1 and if $x'\beta < 0$ the case is predicted as group 0.* Using the regression notation that means: group 1 if $-\beta_0 < \mathbf{x}'\beta_*$ and group 0 if $-\beta_0 > \mathbf{x}'\beta_*$. To use a hyperplane $x'\beta = C$ is simply to redefine β_0 as $\beta_0 - C$. If η is a nonzero scalar multiple of β, the vectors x with $x'\beta = 0$ are precisely the same as the vectors x with $x'\eta = 0$, so β and η define the same linear predictor. (Although if the constant of proportionality is negative, the codes for the two groups will be reversed.)

Exercise 13.2.   Show that $|x'\beta|$ is $\|\beta\|$ times the perpendicular distance from x to the prediction hyperplane (subspace) $\{x|x'\beta = 0\}$ by finding the perpendicular distance using $M_\beta \equiv \beta(\beta'\beta)^{-1}\beta'$.

Since any β determines a binary predictor, we can certainly pick one by minimizing

$$\sum_{i=1}^{n} \mathscr{L}(y_i, x_i'\beta) + k\mathscr{P}(\beta),$$

for any loss function $\mathscr{L}$, for any penalty function $\mathscr{P}$, and any tuning parameter k. The question is, "Will it be any good?" Certainly if we use the loss function associated with minimizing the deviance of a generalized linear model having $F(0) = 0.5$ and either take $k = 0$ or any reasonable penalty function with k small, the linear predictor will be reasonable.

Since the numerical value of $x'\beta$ essentially measures the distance of x from the hyperplane defined by $x'\beta = 0$, it should provide a measure of how clearly a case belongs to a group. Ideally, we would like to know $p(x)$. For a general differentiable loss function, i.e. one not associated with a generalized linear model, we will probably need to rely on Eq. (13.6.2) to estimate probabilities. In the next section we will see that SVMs use a loss function that looks reasonable, but not one that is consistent with a generalized linear model nor is it differentiable, so there is no obvious method of turning an estimated SVM into group probabilities.

Like all linear models, the linearity of a linear prediction rule is linearity in the unknown regression coefficients, not in the predictor variables.

EXAMPLE 13.4.2. In Fig. 13.1, the linear structure used for determining the logit and probit linear prediction rules was

$$x'\beta = \beta_0 + \beta_1 TL + \beta_2 PL.$$

where TL and PL are the logs of the tetrahydrocortisone and pregnanetriol scores. Figure 13.3 illustrates the use of the quadratic model

$$x'\beta = \beta_{00} + \beta_{10}TL + \beta_{01}PL + \beta_{20}TL^2 + \beta_{02}PL^2 + \beta_{11}TL \times PL.$$

The logistic and probit linear predictors $0 = x'\hat{\beta}$ take the form of parabolas when plotted in two (rather than 5) dimensions. Even more than in Fig. 13.1, their linear predictors are almost on top of one another. Unlike Fig. 13.1, the parabolas completely separate the carcinoma cases from the bilateral hyperplacia cases. (More on this later.) Figure 13.3 also illustrates a quadratic support vector machine. Only one of the two SVM parabolic curves appears on this plot and the one that appears, over the range of this plot, is almost a straight line. The SVM fails to separate the two groups of observations. More details on the SVM linear predictor are given in the next section. □

Exercise 13.3. Use the algebra of Vec operators to show that the quadratic model in Example 13.4.2 can be written as

$$\beta_{00} + \beta_{10}TL + \beta_{01}PL + \beta_{20}TL^2 + \beta_{02}PL^2 + \beta_{11}TL \times PL = \begin{bmatrix} 1 \\ TL \\ PL \end{bmatrix}' B \begin{bmatrix} 1 \\ TL \\ PL \end{bmatrix}$$

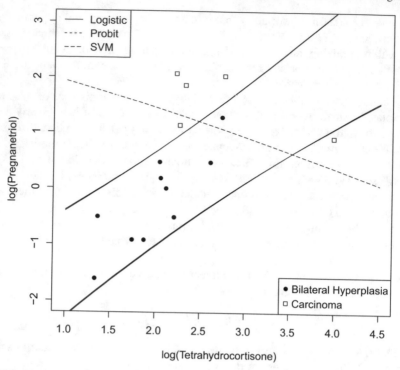

Fig. 13.3 Quadratic model logistic regression, probit regression (indistinguishable from logistic), and an SVM: Cushing's syndrome data (subset)

for a symmetric matrix B and that this is also a linear function of the predictor row vector $(1, TL, PL)' \otimes (1, TL, PL)'$. (The B matrix will be the subject of some linear (equality) constraints.)

13.4.1 Loss Functions

We have discussed the loss functions associated with binomial/binary generalized linear models. The loss function for support vector machines is discussed in the next section. The use of squared error loss is related to the normal theory discriminant analysis of Chap. 12 and is also discussed in the next subsection. In the machine learning community the use of squared error loss together with a penalty function is sometimes called the *proximal support vector machine*. Another loss function that gets used as an approximation to AdaBoost is

$$\mathscr{L}_{Ada}(y, u) = \begin{cases} e^{-u} & \text{if } y = 1 \\ e^{u} & \text{if } y = 0 \,. \end{cases}$$

13.4.2 Least Squares Binary Prediction

As defined in this section, both LDA and QDA provide estimated linear predictors. In particular, there is a relationship between fitting a least squares regression model to binary y data and LDA, cf. Williams (1959). The algebra involved in the demonstration is tedious. (I have about four pages of formulae with no explanations of what I am doing.) Suffice it to say that if the number of successes equals the number of failures, LDA agrees with least squares regression on 0–1 data where a case is assigned to group 1 if and only if its predicted (fitted) value is greater than 0.5. Moreover, if the number of successes and failures are not equal, there exists a cutoff point for the least squares predicted values (typically different from 0.5) that will give the same predictions as LDA.

13.5 Support Vector Machines

Support vector machines are linear predictors that pick $\beta = (\beta_0, \beta'_*)'$ by minimizing

$$\sum_{i=1}^{n} \mathscr{L}_S(y_i, x'_i\beta) + k\mathscr{P}_R(\beta).$$

The penalty function is the standard ridge regression penalty, $\mathscr{P}_R(\beta) \equiv \beta'_*\beta_*$, but most importantly the loss function is

$$\mathscr{L}_S(y, u) = \begin{cases} (1-u)_+ & \text{if } y = 1 \\ (1+u)_+ & \text{if } y = 0. \end{cases}$$

(Recall that $a_+ = a$ if a is positive and $a_+ = 0$ is a is not positive.) Figure 13.4 plots the logit and SVM loss functions. The SVM loss function is certainly reasonable.

EXAMPLE 13.5.2. In Figs. 13.1 and 13.3, the SVM curves presented were the default linear and quadratic fits for unscaled predictor variables from the R library e1071's program svm. Figure 13.5 is similar to Fig. 13.3 but plots the SVM when the tuning parameter k associated with the penalty function has been reduced by a factor of 100. The new fitted SVM is much more like the maximum likelihood fits and it separates the two classes. □

As discussed in PA-V Sect. 13.6, least squares is all about minimizing a squared error loss function. Similarly, ridge and lasso regression problems are dominated by the problem of minimizing the squared error loss function subject to quadratic and linear inequality constraints. Somewhat ironically, programs for finding SVM estimates seem to focus on minimizing the quadratic penalty function subject to

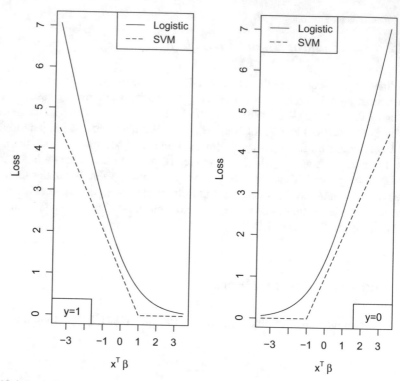

Fig. 13.4 Logistic regression and SVM loss functions

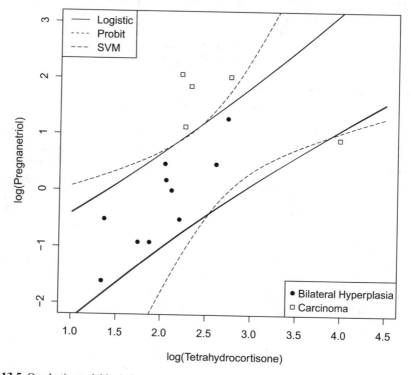

Fig. 13.5 Quadratic model logistic regression, probit regression, and an SVM with reduced tuning parameter: Cushing's syndrome data (subset)

the constraints imposed by needing to minimize the loss function. The issue is less about whether the loss is more important than the penalty function and more about the highest order polynomial involved in the minimization. Appendix A.3 discusses the general problem of minimizing quadratic functions subject to linear inequality constraints and a subsection applies the general results to the SVM problem. In this chapter, we merely cite the most important of those results. Hastie et al. (2016), Zhu (2008), and Moguerza and Muñoz (2006) all present introductions to SVMs.

13.5.1 Probability Estimation

The value $|x'\beta|$, which is $\|\beta\|$ times the perpendicular distance from x to the prediction hyperplane, should measure the assuredness of a classification. The bigger the value, the more sure we should be of the classification. Unfortunately, for SVMs this does not obviously convert to a classification probability. First, the loss function associated with SVMs is similar to the logit and probit losses, so SVMs might be generalized linear models for some F. If so, that would give us a way to associate probabilities with the SVM predictor. We will show that SVMs cannot be generalized linear models. Second, the general equation for determining probabilities from a best predictor given in Eq. (13.6.2) does not apply because the SVM loss function is not differentiable everywhere.

From (13.3.2) it is easy to see that a generalized linear model has

$$1 = F(u) + [1 - F(u)] = \exp[-\mathcal{L}_F(1,u)/2] + \exp[-\mathcal{L}_F(0,u)/2].$$

Making a similar computation for SVMs,

$$\exp[-\mathcal{L}_S(1,u)/2] + \exp[-\mathcal{L}_S(0,u)/2] = e^{-(1-u)_+/2} + e^{-(1+u)_+/2}.$$

Evaluating this at $u = \pm 1$ gives $1 + e^{-1}$, which is not equal to 1 but even more importantly, evaluating this at $u = 0$ gives a different value, $2e^{-1/2}$, so there is no hope of rescaling the SVM loss function into one that corresponds to a generalized linear model.

13.5.2 Parameter Estimation

Finding the SVM parameter estimates is generally performed by turning the estimation problem into a quadratic optimization problem, cf. Appendix A.3.

Write our binary data in vector form as

$$Y \equiv \begin{bmatrix} Y_1 \\ Y_0 \end{bmatrix}$$

where N_1 successes are in $Y_1 \equiv J_{N_1}$ and $N_0 = n - N_1$ failures are in $Y_0 \equiv 0_{N_0}^1$. For this discussion only

$$J_1 \equiv J_{N_1}; \qquad J_0 \equiv J_{N_0}.$$

Similarly write the model matrix, which includes an intercept predictor, as

$$X \equiv \begin{bmatrix} X_1 \\ X_0 \end{bmatrix} \equiv \begin{bmatrix} J_1 & \mathbf{X}_1 \\ J_0 & \mathbf{X}_0 \end{bmatrix}.$$

Any n vector v may be written

$$v = \begin{bmatrix} v_1 \\ v_0 \end{bmatrix}$$

in conformance with Y_1 and Y_0.

Support vector machines pick $\beta = (\beta_0, \beta_*')'$ by minimizing

$$\sum_{h=1}^{n} \mathscr{L}_S(y_h, x_h'\beta) + k\beta_*'\beta_* \tag{13.5.1}$$

where

$$\mathscr{L}_S(y,u) = \begin{cases} (1-u)_+ & \text{if } y = 1 \\ (1+u)_+ & \text{if } y = 0 . \end{cases}$$

The innovative idea is to introduce slack variables $\xi = (\Xi_1, \ldots, \Xi_n)'$ that serve as upper bounds for the contributions to the loss function. Because the loss function is nonnegative, the slack variables are also. Minimizing the sum of the slack variables, because they are unknown upper bounds, amounts to minimizing the sum of the losses, so minimizing (13.5.1) is equivalent to finding

$$\inf_{\beta,\xi} \left(k\beta_*'\beta_* + \xi'J \right) \tag{13.5.2}$$

subject to

$$\mathscr{L}_S(y_h, x_h'\beta) \leq \Xi_h, \quad h = 1, \ldots, n. \tag{13.5.3}$$

To establish this as a quadratic optimization problem, we need to replace the loss function constraints (13.5.3) with linear constraints. When $y_h = 1$ the loss is $\mathscr{L}_S(1, x_h'\beta) = (1 - x_h'\beta)_+ = \max\{0, 1 - x_h'\beta\}$. We place two linear constraints on the slack variables to force them to be upper bounds for the loss function: $0 \leq \Xi_h$ and $1 - x_h'\beta \leq \Xi_h$, which gives us $\mathscr{L}_S(1, x_h'\beta) \leq \Xi_h$. In matrix form these constraints are

$$0_{N_1}^1 \leq \xi_1; \qquad J_1 - X_1\beta \leq \xi_1 \tag{13.5.3a}$$

where *an inequality applied to a matrix is understood to apply elementwise*. Similarly for $y_h = 0$ impose $0 \leq \Xi_h$ and $1 + x_h'\beta \leq \Xi_h$ or

$$0_{N_0}^1 \leq \xi_0; \qquad J_0 + X_0\beta \leq \xi_0. \tag{13.5.3b}$$

In total there are $2n$ linear inequality constraints being imposed on the criterion function (13.5.2).

In matrix notation, rewrite the penalized loss function in standard form for quadratic optimization as

$$k\beta_*'\beta_* + \xi'J = \frac{1}{2}\begin{bmatrix} \beta_0 \\ \beta_* \\ \xi \end{bmatrix}' \begin{bmatrix} 0 & 0 & 0 \\ 0 & 2kI & 0 \\ 0 & 0 & 0 \end{bmatrix} \begin{bmatrix} \beta_0 \\ \beta_* \\ \xi \end{bmatrix} + \begin{bmatrix} 0 \\ 0 \\ J_n \end{bmatrix}' \begin{bmatrix} \beta_0 \\ \beta_* \\ \xi \end{bmatrix}, \qquad (13.5.4)$$

which is to be minimized subject to the constraints (13.5.3) rewritten in standard form as

$$\begin{bmatrix} -X_1 & -I & 0 \\ X_0 & 0 & -I \\ 0 & -I & 0 \\ 0 & 0 & -I \end{bmatrix} \begin{bmatrix} \beta \\ \xi_1 \\ \xi_0 \end{bmatrix} \leq \begin{bmatrix} -J_1 \\ -J_0 \\ 0 \\ 0 \end{bmatrix}$$

or

$$\begin{bmatrix} -J_1 & -X_1 & -I & 0 \\ J_0 & X_0 & 0 & -I \\ 0 & 0 & -I & 0 \\ 0 & 0 & 0 & -I \end{bmatrix} \begin{bmatrix} \beta_0 \\ \beta_* \\ \xi_1 \\ \xi_0 \end{bmatrix} \leq \begin{bmatrix} -J_1 \\ -J_0 \\ 0 \\ 0 \end{bmatrix}.$$

In the SVM literature the criterion function (13.5.4) often gets multiplied by $\tilde{C} \equiv 1/2k$ which results in minor changes to the results, cf. Exercise A.2.

As discussed in Appendix A.3, typically one finds an n vector λ_1 (*not* an N_1 vector like v_1 in $v' = (v_1', v_0')'$) that *maximizes* the dual criterion

$$\frac{-1}{2k}\lambda_1' \begin{bmatrix} X_1X_1' & -X_1X_0' \\ -X_0X_1' & X_0X_0' \end{bmatrix} \lambda_1 + \lambda_1' J_n.$$

(Appendix A.3 involves another n vector λ_2 because there are $2n$ linear inequality constraints.) The dual criterion has λ_1 subject to the constraints

$$-J_1'\lambda_{11} + J_0'\lambda_{10} = 0$$

and

$$0 \leq \lambda_1 \leq 1.$$

Actual solutions β and ξ need to incorporate the well-known KKT conditions.

Appendix A.3 establishes that

$$\hat{\beta}_* = \frac{1}{2k}\left(X_1'\lambda_{11} - X_0'\lambda_{10}\right).$$

Often many of the λ_1 values are zero, so it makes sense to report only the values of λ_1 that are nonzero and report the corresponding rows of X_1 and $-X_0$. The computer programs I have seen do something equivalent; they report the nonzero coefficients of $\frac{1}{2k}(\lambda'_{11}, -\lambda'_{10})$ and report the corresponding rows of $\mathbf{X}_1$ and $\mathbf{X}_0$ as *support vectors*. Typically, they make you figure out $\hat{\beta}_*$.

As discussed in Appendix A.3, if $0 < \lambda_{1h} < 1$, depending on whether y_h is 1 or 0, we must have $\hat{\beta}_0 + \mathbf{x}'_h\hat{\beta}_* = 1$ or $\hat{\beta}_0 + \mathbf{x}'_h\hat{\beta}_* = -1$, respectively. Changing notation a bit, think about $\lambda'_1 = (\lambda'_{11}, \lambda'_{10})$. If y_{1j} denotes an element of Y_1 with $0 < \lambda_{11j} < 1$, then $\hat{\beta}_0 = 1 - \mathbf{x}'_{1j}\hat{\beta}_*$ and similarly, when y_{0j} has $0 < \lambda_{10j} < 1$, $\hat{\beta}_0 = -1 - \mathbf{x}'_{0j}\hat{\beta}_*$. If you have the correct λ_1, and thus the correct $\hat{\beta}_*$, all of these cases should give the same $\hat{\beta}_0$.

It may seem curious that finding $\hat{\beta}_0$ is so directly tied to cases with $x'\beta = \pm 1$, but remember that any multiple of $\hat{\beta}$ defines the same hyperplane, so we have merely chosen a multiple that defines $\hat{\beta}_0$ in terms of being 1 unit away from an appropriate value of $\mathbf{x}'_h\hat{\beta}_*$.

Computer programs often report $-\hat{\beta}_0$ rather than $\hat{\beta}_0$. Surprisingly, computer programs, and even published works, often make some fuss about how to obtain $\hat{\beta}_0$ from the various cases that have $0 < \lambda_{1h} < 1$. Indeed, when fitting the linear (as opposed to quadratic) model to the Cushing's Syndrome data, the svm program for R reports three vectors with $0 < \lambda_{1h} < 1$. These imply the values $\hat{\beta}_0 = 3.101790, 3.102093, 3.101790$. Your guess is as good as mine for why the middle one is slightly different. The program reports $-\hat{\beta}_0 = -3.101891$, which is the average of the three.

13.5.2.1 The Kernel Trick

As discussed in Chap. 1, you could just replace X with $\tilde{R}$ and proceed exactly as before. However, the SVM methodology admits a more particular approach to using kernels. Every time you evaluate $\mathbf{x}'\mathbf{x}_h$ in the discussion, you could replace it with an evaluation of $R(\mathbf{x}, \mathbf{x}_h)$. The primary change that ensues is that instead of evaluating

$$
\begin{aligned}
\mathbf{x}'\hat{\beta}_* &= \mathbf{x}'\left[\frac{1}{2k}\left(\mathbf{X}'_1\lambda_{11} - \mathbf{X}'_0\lambda_{10}\right)\right] \\
&= \mathbf{x}'\left[\frac{1}{2k}\left(\sum_{j=1}^{N_1}\mathbf{x}_{1j}\lambda_{11j} - \sum_{j=1}^{N_0}\mathbf{x}_{0j}\lambda_{10j}\right)\right] \\
&= \frac{1}{2k}\left(\sum_{j=1}^{N_1}\mathbf{x}'\mathbf{x}_{1j}\lambda_{11j} - \sum_{j=1}^{N_0}\mathbf{x}'\mathbf{x}_{0j}\lambda_{10j}\right),
\end{aligned}
$$

you evaluate

$$
\frac{1}{2k}\left(\sum_{j=1}^{N_1}R(\mathbf{x}, \mathbf{x}_{1j})\lambda_{11j} - \sum_{j=1}^{N_0}R(\mathbf{x}, \mathbf{x}_{0j})\lambda_{10j}\right).
$$

Again, this computation is simplified by dropping terms with $\lambda_{1ij} = 0$.

Exercise 13.4. Apply cubic logistic regression along with a cubic kernel SVM to the Cushing's syndrome data. Provide a two-dimensional plot of the separating curves.

13.5.3 Advantages of SVMs

To be honest, the whole point of writing this chapter was to address support vector machines. I had planned a subsection listing the advantages of SVMs but, after studying SVMs, I no longer see any advantages. I once thought that their ability to involve the kernel trick was an advantage. But we established in Sect. 1.8.2 that $C(\Phi) = C(\tilde{R})$, so the kernel trick applies to any linear structure $X\beta$, whether it is applied to regular linear models, generalized linear models, proportional hazard models, or anything else. (I once gave a talk where I jokingly referred to $C(X) = C(XX')$ as the Fundamental Theorem of RKHSs for Statistics.) The other advantage I imagined for SVMs was computational, because the vector λ_1 is often nonzero on only a relatively small subset of the data involving "support vectors." But in my review of the literature (which was far from complete but more extensive than the references I have given) I did not notice any such claims being made for SVMs; no more does the theory in Appendix A.3 suggest to me any such advantage. In fact, I found some discussion of the need to deal with the computational problems that SVMs have with big data (something that would be unlikely to arise if the computational complexity was being driven by a relatively small number of support vectors). This is *not* to say that SVMs don't give reasonable answers; they do. I am just not aware of any advantages they have over using logistic regression with the kernel trick and penalized estimation.

13.5.4 Separating Hyper-Hogwash

SVMs are often sold as finding the optimal hyperplane that has all the data from one group above the hyperplane and all the data from the other group below the hyperplane. The "optimal" hyperplane is defined as the hyperplane that maximizes the distance from the plane to the points on either side that are closest to the hyperplane. While this technical argument is correct, as a reason for using SVMs I think it is quite simply hogwash. I am *not* saying that SVMs are hogwash, only this argument for using them. The optimal separating hyperplane phenomenon is based almost entirely on the fact that SVMs involve minimizing the ridge regression penalty.

- The whole point of binomial generalized linear models is to find good ways of estimating probabilities for the cases that are not obviously from one group or the other. If a separating hyperplane exists, the problem is trivial! All of the MLE probabilities can be pushed arbitrarily close to 1 or 0, cf. Exercise 13.5. *The important question for SVMs (like for all linear predictors) is not how to pick a separating hyperplane but how to pick a hyperplane when separation is not possible!*

In Fig. 13.1, using a linear function of TL and PL, separation was not possible. In Fig. 13.3, using quadratic functions of TL and PL, separation is possible, so the *reported* maximum likelihood logistic and probit fits do that; they separate the cases. In fact, because it is possible to separate the cases, unique maximum likelihood fits to the linear predictors do not exist. The reported curves in Fig. 13.3 for logistic and probit regression are merely those reported when R's glm function stopped iterating. Anderson (1972) argued that any logistic regression program will find you a separating hyperplane when they exist. Essentially, when the program finds a separating hyperplane, that fact establishes that no unique maximum likelihood estimate will exist. Anderson (1972) and Albert and Anderson (1984) show that there are no unique maximum likelihood estimators for separable logistic regression.

- *Finding the **optimal** separating hyperplane is **largely** a waste of time.* Figure 13.5 illustrates three separating hyperplanes in the form of parabolas. What basis is there for picking one separating hyperplane over another one? In terms of maximizing the likelihood, they are all equally good. Why should you think there would be a best separating hyperplane? Do you really need to impose some artificial optimality criterion to find a "best" separating hyperplane? (I admit that maximizing the distance from the separating hyperplane to the closest points on either side is a nice choice, if you think it is worth the trouble to make a choice.)

- *If a separating hyperplane exists, and the procedure does not give you a separating hyperplane, then clearly the procedure is not about finding the optimal separating hyperplane.* Figure 13.3 shows that the default parabola fitted by svm does *not* separate the two groups, even though the logit and probit fitted parabolas do separate the groups. I am not saying that the svm solution is bad, only that it is not finding a separating hyperplane when one clearly exists.

Exercise 13.5. For a generalized linear model with $F(0) = 0.5$, suppose $x'\tilde{\beta} = 0$ defines a separating hyperplane. Show that for $C > 1$, $\hat{\beta} = C\tilde{\beta}$ has at least as high a likelihood because $|x_i'\hat{\beta}|$ is closer to ∞ than $|x_i'\tilde{\beta}|$, hence $F(x_i'\hat{\beta})$ is closer to 0 or 1 than $F(x_i'\tilde{\beta})$.

By letting C go to infinity in the exercise, the associated likelihood will approach 1, which is its supremum. If $\tilde{\beta}$ defines a separating hyperplane, typically *any* sufficiently small modification of $\tilde{\beta}$ will give another separating hyperplane, which will also lead to maximizing the likelihood.

13.6 Best Prediction and Probability Estimation

We began by assuming independent binary data $y_h \sim \text{Bern}[p(x_h)]$ and showed that fitting generalized linear models leads us to minimizing certain loss functions. In the last two sections we have ignored the distributional assumptions and discussed linear predictors based on minimizing different loss functions. We now go back and relate minimization of arbitrary loss functions to best prediction and to estimation of probabilities. Earlier we made the case that good estimation of probabilities was vital to estimating the best predictors for standard predictive loss functions such as squared error and Hamming.

The best predictor $\hat{f}$ for an arbitrary predictive loss function $\mathscr{L}(y, u)$ satisfies

$$E_{y,x}\left\{\mathscr{L}[y, \hat{f}(\mathbf{x})]\right\} = \inf_{f} E_{y,x}\left\{\mathscr{L}[y, f(\mathbf{x})]\right\}. \tag{13.6.1}$$

The best predictor, if it can be found, is found by conditioning on $\mathbf{x}$ and is the number $\hat{u} \equiv \hat{f}(\mathbf{x})$ that achieves

$$E_{y|x}\left[\mathscr{L}(y, \hat{u})\right] = \inf_{u} E_{y|x}\left[\mathscr{L}(y, u)\right].$$

If $\mathscr{L}(y, u)$ is differentiable in u for all y and if the derivative can be taken under the integral of the conditional expectation, cf. Cramér (1946), the best prediction for a fixed $\mathbf{x}$ should occur when

$$0 = \mathbf{d}_u E_{y|x}\left[\mathscr{L}(y, u)\right] = E_{y|x}\left[\mathbf{d}_u \mathscr{L}(y, u)\right].$$

In the special case of binary prediction, this easily becomes

$$0 = p(\mathbf{x})\left[\mathbf{d}_u \mathscr{L}(1, u)\right] + [1 - p(\mathbf{x})]\left[\mathbf{d}_u \mathscr{L}(0, u)\right].$$

Typically, for known $p(\mathbf{x})$, we would solve for $\hat{u} \equiv \hat{f}(\mathbf{x})$ to find the best predictor for the loss function. As alluded to earlier, we can find the best predictor for square error, Hamming, and even absolute error loss functions.

In binary regression, sometimes people solve the equation for $p(\mathbf{x})$,

$$p(\mathbf{x}) = \frac{-\mathbf{d}_u \mathscr{L}[0, u]}{[\mathbf{d}_u \mathscr{L}(1, u)] - [\mathbf{d}_u \mathscr{L}(0, u)]}. \tag{13.6.2}$$

The original idea was to use the conditional distribution of y to find the best predictor (BP) under the loss function. Solving for $p(\mathbf{x})$ is using the BP to find the conditional distribution. It presumes that you know the BP without knowing the conditional distribution. In practice, an estimated predictor $u = \tilde{f}(\mathbf{x})$ is sometimes plugged into (13.6.2) to obtain an estimate of $p(\mathbf{x})$.

Predictive estimators $\tilde{f}$ are often chosen to achieve

$$\inf_{f \in \mathscr{F}} \left\{ \sum_{h=1}^{n} \mathscr{L}[y_h, f(x_h)] + k\mathscr{P}(f) \right\},$$

cf. Sect. 3.5.3. The fact that the sum puts the same weight on each observation pair (y_h, x'_h) is something that is (only?) appropriate when the data come from a simple random sample of some population, e.g., not discrimination data.

Standard limit theorems assure that $(1/n) \sum_{h=1}^{n} \mathscr{L}[y_h, f(x_h)]$ will be a reasonable estimate of $E_{y,x} \{\mathscr{L}[y, f(x)]\}$ and if $k\mathscr{P}(f)/n \to 0$, we should be able to evaluate the effectiveness of f for large samples. But the estimated predictor $\tilde{f}$ is not generally an estimate of the best predictor $\hat{f}$ as defined by (13.6.1), it is an estimate of the best predictor in $\mathscr{F}$. Only if you are willing to assume that the best predictor is in $\mathscr{F}$ does it make sense to use Eq. (13.6.2) to estimate the conditional probabilities. But, as discussed in Chap. 3, that is an assumption that we often make. Zhang, Liu, and Wu (2013) discuss these issues and argue that regularization, i.e., incorporating a penalty function when estimating the predictor $\tilde{f}$, can have deleterious effects on using (13.6.2) for probability estimation.

It seems to be the case that people often define best prediction with one loss function, e.g., squared error or Hamming, but are willing to use a completely different predictive loss function to obtain an estimated predictor $\tilde{f}$ and an estimate of $p(x)$.

The following Exercise establishes that Eq. (13.6.2) can work very well as a method for estimating probabilities but that it can also work very poorly. It depends on the loss function being used.

Exercise 13.6.
(a) Show that, for linear models with squared error loss, Eq. (13.6.2) gives the rather unsatisfactory result $p(x) = x'\beta$. Why is this unsatisfactory?
(b) Show that for a generalized linear model based on a cdf F that is symmetric about 0, Eq. (13.6.2) returns the standard answer $p(x) = F(x'\beta)$.
(c) Show that if a loss function has the properties $\mathscr{L}[0, -u] = \mathscr{L}[1, u]$ and $d_u \mathscr{L}[1, u] < 0$, then Eq. (13.6.2) gives a number between 0 and 1 with $p(0) = 0.5$.
(d) When do the generalized linear models of the previous section have the properties in (c)?

Incidentally, Eq. (13.6.2) and fitting binomial generalized linear models are not the only ways to associate a linear predictor with group probabilities given the predictor variables. In Chap. 12 we used LDA and QDA (which are both linear predictors) to estimate group probabilities via estimation of the sampling distribution $f(y|i)$. Doing that required knowledge of the prevalences (marginal group probabilities) as will be needed in the next section.

13.7 Binary Discrimination

Binary discrimination shares the same predictive goal as binary regression but it involves using a different type of data and therefore requires outside information about the prevalences of the groups within the overall population. Instead of sampling from the joint distribution of the dependent and predictor variables (or the conditional distribution of the dependent variable given the predictor variables), discrimination involves sampling from the conditional distribution of the predictor variables given the dependent variable. If the reader has no previous experience with the difference between logistic regression and logistic discrimination, reading a more elementary treatment such as Christensen (2015, Section 21.9) is advisable.

To be consistent with the notation of Chap. 12, we need to change the binary regression notation. In Chap. 12, y is a q dimensional vector of predictor variables (rather than a $d-1$ vector $\mathbf{x}$) and we will now use z (rather than y) to denote group membership. To predict z from y we need to envision a joint distribution for (z, y'). Call the density (with respect to an appropriate dominating measure) $f(z, y)$. Denote the marginal density (*prevalence*) of z as $\pi(z)$, the conditional density of z given y as $\pi(z|y)$, the marginal density of y as $f(y)$, and the conditional density of y given z as $f(y|z)$. In Chap. 12, z was fixed, not random, and we wrote i in place of z. In binary regression and discrimination we denote the two z groups as 0 and 1. (In Chap. 12, two groups would have been labeled 1 and 2.)

Since we are focused on predicting z, in both regression [sampling from either $f(z, y)$ or $\pi(z|y)$] and discrimination [sampling from $f(y|z)$], our goal is to estimate $\pi(z|y)$. (A sample from the joint distribution $f(z, y)$ can be viewed as a sample from either conditional scheme.) Regression data gives direct information on $\pi(z|y)$. Discrimination data gives direct information on $f(y|z)$ but only indirect information on $\pi(z|y)$. Using Bayes' Theorem with only two groups, the posterior probabilities in (12.1.4) become

$$\pi(1|y) = \frac{f(y|1)\pi(1)}{f(y|1)\pi(1) + f(y|0)\pi(0)}; \qquad \pi(0|y) = 1 - \pi(1|y).$$

If we know the prevalence distribution $\pi(z)$, discrimination data allow us to estimate $f(y|z)$ and, indirectly, $\pi(z|y)$.

Bayes theorem also determines the posterior odds for seeing $z = 1$,

$$O(1|y) \equiv \frac{\pi(1|y)}{\pi(0|y)} = \frac{f(y|1)}{f(y|0)} \frac{\pi(1)}{\pi(0)}. \tag{13.7.1}$$

We will see that binomial regression methods applied to discrimination data are easily adjusted to give appropriate posterior odds. The formulae for logistic regression is particularly nice.

Define the $q+1$ dimensional row vectors $x' \equiv (1, y')$ and $\beta' = (\beta_0, \beta'_*)$ so that $x'\beta = \beta_0 + y'\beta_*$. Further combining the notation of Chap. 12 with the binary regression notation, the probability of seeing an observation in group 1 given the predictors is defined interchangeably as

$$\pi(1|y) \equiv p(y) \equiv p(x).$$

Binary generalized linear models further assume $p(y) = F(x'\beta)$ for a known, invertible cdf F. The densities $f(z,y)$, $f(y|z)$, and $f(y)$ bear *no* relationship to the cdf F used in specifying binary generalized linear models.

Binary regression assumes (conditionally) independent observations

$$z_h \sim \mathrm{Bin}[1, p(y_h)]; \qquad h = 1, \dots, n$$

with the associated likelihood function, cf. Sect. 13.1. The likelihood function for discrimination data is

$$\prod_{h=1}^{n} f(y_h|z_h) \equiv \prod_{i=0}^{1} \prod_{j=1}^{N_i} f(y_{ij}|i),$$

where there are N_i observations y_{ij} on group i. Christensen (1997) argues that the logistic regression likelihood function can be viewed as a *partial likelihood* function for discrimination data. (The argument is actually for the log-linear model that is equivalent to the logistic model.) This allows one to use binary regression methods to estimate the densities $f(y|i)$ associated with discrimination data but it requires a correction for the prevalences implicitly assumed when treating discrimination data as if it were regression data.

Unconditional data constitute a random sample of (z, y') values. Clearly, one can estimate the marginal probabilities (prevalences) of the groups from unconditional data. The obvious estimate of $\pi(1)$ is the number of observed values $z = 1$ divided by the sample size, $N_1/(N_1 + N_0)$. If the ys are preselected and one samples from $z|y$, i.e. the usual regression sampling scheme, there is no statistical basis for estimating $\pi(1)$ without knowing the marginal density $f(y)$. If the y_h were sampled from the appropriate distribution for y, it would make (z_h, y'_h) a random sample from the appropriate joint distribution, and z_h a random sample with the prevalence probabilities, which makes $N_1/(N_1 + N_0)$ the obvious estimate of $\pi(1)$. Any estimation scheme that puts equal weight on the losses associated with each observation is implicitly treating the y_hs as a random sample and using $N_1/(N_1 + N_0)$ as an estimate of $\pi(1)$.

There is no possible way to estimate $\pi(1)$ from discrimination data. A value for $\pi(1)$ has to be obtained from some source outside the data before it becomes possible to estimate $\pi(1|y)$. If we want to use regression estimates computed from discriminant data, we need to correct for the implicit use of $N_1/(N_1 + N_0)$ as an estimate of $\pi(1)$.

In discriminating between two groups, the maximum likelihood allocation can be based on the relative densities (likelihood ratio) $f(y|1)/f(y|0)$. We now derive the likelihood ratio estimate, say, $\hat{f}(y|1)/\hat{f}(y|0)$ from some arbitrary fitted binary regression estimates $\tilde{\pi}(1|y) \equiv \tilde{p}(y) \equiv \tilde{p}(x)$ that incorporate the inappropriate prevalence $\tilde{\pi}(1) = N_1/(N_1 + N_0)$. From these we further obtain an estimate $\hat{\pi}(1|y)$ of $\pi(1|y)$ for an appropriate prior $\pi(1)$.

The binary regression estimated posterior odds are

$$\tilde{O}(y) = \frac{\tilde{\pi}(1|y)}{\tilde{\pi}(0|y)} = \frac{\hat{f}(y|1)}{\hat{f}(y|0)} \frac{\tilde{\pi}(1)}{\tilde{\pi}(0)} = \frac{\hat{f}(y|1)}{\hat{f}(y|0)} \frac{N_1}{N_0}.$$

These induce the estimated relative likelihoods

$$\frac{\hat{f}(y|1)}{\hat{f}(y|0)} = \frac{\tilde{\pi}(1|y)}{\tilde{\pi}(0|y)} \frac{N_0}{N_1},$$

To obtain the actual estimated posterior probabilities $\hat{\pi}(i|y)$ for discrimination using the actual prevalences $\pi(i)$, use the estimated odds

$$\hat{O}(y) \equiv \frac{\hat{\pi}(1|y)}{\hat{\pi}(0|y)} = \frac{\hat{f}(y|1)}{\hat{f}(y|0)} \frac{\pi(1)}{\pi(0)} = \frac{\tilde{\pi}(1|y)}{\tilde{\pi}(0|y)} \frac{N_0}{N_1} \frac{\pi(1)}{\pi(0)}.$$

The discrimination odds $\hat{O}$ give discrimination probabilities $\hat{\pi}$ through $\hat{\pi} = \hat{O}/(1 + \hat{O})$.

When fitting a logistic discrimination, the odds take a particularly nice form:

$$\log\left[\hat{O}(y)\right] \equiv x'\tilde{\beta} - \log\left(\frac{N_1}{N_0}\right) + \log\left(\frac{\pi(1)}{\pi(0)}\right)$$

$$= y'\tilde{\beta}_* + \left[\tilde{\beta}_0 - \log\left(\frac{N_1}{N_0}\right) + \log\left(\frac{\pi(1)}{\pi(0)}\right)\right]$$

$$= x'\hat{\beta}$$

where

$$\hat{\beta}' \equiv \left(\tilde{\beta}_0 - \log\left(\frac{N_1}{N_0}\right) + \log\left(\frac{\pi(1)}{\pi(0)}\right), \tilde{\beta}_*'\right).$$

This leads to

$$\hat{\pi}(1|y) = e^{x'\hat{\beta}} / \left[1 + e^{x'\hat{\beta}}\right].$$

Logistic discrimination defines a hyperplane of y values by $x'\hat{\beta} \equiv \hat{\beta}_0 + y'\hat{\beta}_* = 0$ which corresponds to $0.5 = \hat{\pi}(1|y)$. If $x'\hat{\beta} > 0$, $\hat{\pi}(1|y) > 0.5$. If $x'\hat{\beta} < 0$, $\hat{\pi}(1|y) < 0.5$. The logistic regression model also defines a hyperplane of y values defined by $0.5 = \tilde{p}(x'\tilde{\beta})$ which is equivalent to $0 = x'\tilde{\beta}$ or $-\tilde{\beta}_0 = y'\tilde{\beta}_*$. Because $\hat{\beta}_* = \tilde{\beta}_*$, these hyperplanes are parallel in q dimensions, but unless the prevalences $\pi(i)$ are proportional to the sample sizes N_i, the regression and discrimination hyperplanes are distinct.

There are several ways of generalizing logistic discrimination to handle $t > 2$. Christensen (1997, 2015) focuses on the fact that logit/logistic models are actually log-linear models and that the appropriate log-linear model can easily be generalized to handle more than two populations. In particular, he illustrates for $t = 3$ how to turn estimated odds into allocations. Without probability estimates, SVMs often rely on performing all of the $\binom{t}{2}$ binary discrimination problems and "voting" for a winning allocation. Voting could also be used when probability estimates exist, not that I would do that.

EXAMPLE 13.7.1.   Figures 13.6, 13.7, and 13.8 are discrimination versions of the regression Figs. 13.1, 13.3, and 13.5. Figures 13.6, 13.7, and 13.8 have dropped the probit regression curves and replaced them with LDA and QDA as appropriate. They have also replaced the logistic regression curves with logistic discrimination curves. The discrimination curves are all based on $\pi(1) = 0.5$. The SVM curves are unchanged from the previous plots.

In Fig. 13.1 the logistic and probit regression lines were almost on top of each other. The logistic discrimination line in Fig. 13.6 is parallel but lower than the regression line. This is consistent with the fact that putting equal prior probabilities (prevalences) on the groups makes the carcinoma group more probable relative to the prior probabilities proportional to sample sizes built into the logistic regression line which makes bilateral twice as probable as carcinoma. The LDA line turns out to be nearly parallel to the logistic discrimination line but is even lower. From the limited amount of data, there is no reason to think that the covariance matrices of the two groups are equal; the spreads of the points are not at all similar. Despite this, the LDA does not do a bad job on these data. The SVM is unchanged.

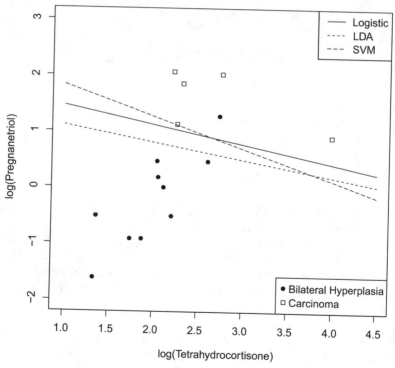

Fig. 13.6 Logistic discrimination, LDA, and an SVM: Cushing's syndrome data (subset)

In Fig. 13.3 the logistic and probit regression parabolas were almost on top of each other. The logistic discrimination parabolas in Fig. 13.7 should have the same shape as the logistic regression parabolas but move closer to the bilateral group. In fact, I cannot *see* any difference between the logistic regression and logistic discrimination parabolas for these data. The QDA parabolas have a radically different shape than the logistic parabolas but, although they do not completely separate the two groups, they do not do a bad job. The SVM is unchanged and the curve that is visible on the plot has a different shape from both other methods. Figure 13.8 is the same as Fig. 13.7 except that it replaces the default SVM with the one from Fig. 13.5 that has a reduced tuning parameter (increased cost). ☐

The code on my website for constructing these figures also contains code for producing tables of estimated posterior probabilities for logistic discrimination that are similar to those presented in Chap. 12 and in Christensen (1997, 2015).

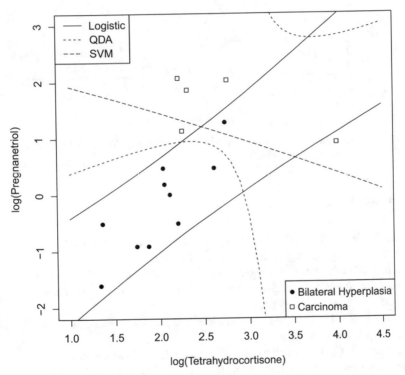

Fig. 13.7 Quadratic model logistic discrimination, QDA, and an SVM: Cushing's syndrome data (subset)

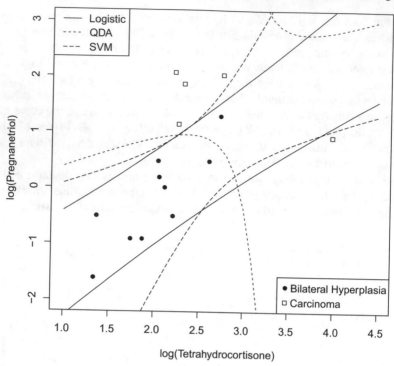

Fig. 13.8 Quadratic model logistic discrimination, QDA, and an SVM with reduced tuning parameter: Cushing's syndrome data (subset)

References

Albert, A., & Anderson, J. A. (1984). On the existence of maximum likelihood estimates in logistic regression models. *Biometrika, 71,* 1–10.

Aldrich, J. (2005). Fisher and regression. *Statistical Science, 20,* 401–417.

Anderson, J. A. (1972). Separate sample logistic discrimination. *Biometrika, 59,* 19–35.

Christensen, R. (1997). *Log-linear models and logistic regression* (2nd ed.). New York: Springer-Verlag.

Christensen, R. (2015). *Analysis of variance, design, and regression: Linear modeling for unbalanced data* (2nd ed.). Boca Raton: Chapman and Hall/CRC Press.

Cramér, H. (1946). *Mathematical methods of statistics.* Princeton: Princeton University Press.

Hastie, T., Tibshirani, R., & Friedman, J. (2016). *The elements of statistical learning: Data mining, inference, and prediction* (2nd ed.). New York: Springer.

Moguerza, J. M., & Muñoz, A. (2006). Support vector machines with applications. *Statistical Science, 21,* 322–336.

Ripley, B. D. (1996). *Pattern recognition and neural networks.* Cambridge: Cambridge University Press.

Williams, E. J. (1959). *Regression analysis*. New York: John Wiley and Sons.

Zhang, C., Liu, Y., & Wu, Z. (2013). On the effect and remedies of shrinkage on classification probability estimation. *The American Statistician, 67,* 134–142.

Zhu, M. (2008). Kernels and ensembles. *The American Statistician, 62,* 97–109.

Chapter 14
Principal Components, Classical Multidimensional Scaling, and Factor Analysis

Abstract This chapter introduces the theory and application of principal components, classical multidimensional scaling, and factor analysis. Principal components seek to effectively summarize high dimensional data as lower dimensional scores. Multidimensional scaling gives a visual representation of points when all we know about the points are the distances separating them. Classical multidimensional scaling is seen to be an application of principal components when the distances are standard Euclidean distances. Principal components and factor analysis are often used for similar purposes but their theoretical background is quite different.

Suppose that observations are available on q variables. When q is quite large it can be very difficult to grasp the relationships among the many variables. It might be convenient if the variables could be reduced to a more manageable number. Clearly, it is easier to work with 4 or 5 variables than with, say, 25. (In the era of big data, perhaps I should be arguing that 400 or 500 variables are easier to work with than 2500.) Of course, one cannot reasonably expect to get a substantial reduction in dimensionality without some loss of information. We want to minimize that loss. Assuming that a reduction in dimensionality is desirable, how can it be performed efficiently? One reasonable method is to choose a small number of linear combinations of the variables based on their ability to reproduce the entire set of variables. In effect, we want to create a few new variables that are best able to predict the original variables. *Principal component analysis (PCA)* finds linear combinations of the original variables that are best linear predictors of the full set of variables. This predictive approach to *dimensionality reduction* seems intuitively reasonable. We emphasize this interpretation of principal component analysis rather than the traditional motivation of finding linear combinations that account for most of the variability in the data. The predictive approach is mentioned in Rao (1973, p. 591). Seber (1984) takes an approach that is essentially predictive. Seber's discussion is derived from Okamoto and Kanazawa (1968). Schervish (1986) gives an explicit derivation in terms of prediction. Other approaches, that are not restricted to linear combinations of the dependent variables, are discussed by Gnanadesikan (1977,

© Springer Nature Switzerland AG 2019
R. Christensen, *Advanced Linear Modeling*, Springer Texts in Statistics,
https://doi.org/10.1007/978-3-030-29164-8_14

Section 2.4) and Li and Chen (1985). Jolliffe (1986) gives a thorough discussion with many examples.

More recently, *independent component analysis (ICA)* has become a popular method of data reduction. Hyvärinen, Karhunen, and Oja (2001) introduce the subject as both a generalization of principal components and as a generalization of factor analysis. (The latter seems more appropriate to me.) The R (package and) program fastICA begins by computing the principal components and obtains the "independent components" from them.

Principal components are similar in spirit to the linear discrimination coordinates discussed in Sect. 12.7. Principal components actually form a new coordinate system for $\mathbf{R}^q$. These coordinates are defined sequentially so that they are mutually orthogonal in an appropriate inner product and have successively less ability to predict the original dependent variables. In practice, only the first few coordinates are used to represent the entire vector of dependent variables.

Section 14.1 presents several alternative derivations for theoretical principal components including both predictive and nonpredictive motivations. The predictive discussion depends heavily on concepts of best linear prediction which are reviewed in Appendix B. Section 14.2 examines the use of sample principal components. Section 14.3 introduces *classical multidimensional scaling (CMDS)*, which seeks to plot the locations of cases when one only knows the distances between the cases. The reason for examining CMDS here is their close relation to PCA. The final section examines *factor analysis*. Although many people consider principal component analysis a special case of factor analysis, in fact their theoretical bases are quite different.

14.1 The Theory of Principal Components

In this section we give several derivations of principal components. First, principal components are derived as a sequence of orthogonal linear combinations of the variable vector y. Each linear combination has maximum capability to predict the full set of variables subject to the condition that each combination is orthogonal to the previous linear combinations. In this sequence, orthogonality is defined using the inner product determined by Σ, the covariance matrix of y. Second, it is shown that the first r principal components have maximum capability to predict y among all sets of r linear combinations of y. Thus, if r linear combinations of y are to be analyzed instead of the full vector y, the first r principal components of y are linear combinations from which y can be most nearly reconstructed. The section closes with a discussion of alternate derivations of principal components and of principal components based on the correlation matrix.

14.1.1 Sequential Prediction

Let $y = (y_1, \ldots, y_q)'$ be a vector in $\mathbf{R}^q$. We define new coordinates $a_1'y, a_2'y, \ldots, a_q'y$ having certain statistical properties. If we think of y as a random vector with

$E(y) = \mu$ and $\text{Cov}(y) = \Sigma$, where Σ is positive definite, then the random variables $a'_1 y, \ldots, a'_q y$ give the random vector y represented in the new coordinate system. The coordinate vectors $a_1, \ldots, a_q$ are to be chosen so that they are orthogonal in the inner product defined by Σ; that is,

$$a'_i \Sigma a_j = 0 \qquad i \neq j.$$

This condition implies that the corresponding random variables are uncorrelated; that is,

$$\text{Cov}(a'_i y, a'_j y) = a'_i \Sigma a_j = 0, \qquad i \neq j.$$

The new coordinates are to provide sequentially optimal predictions of y given the orthogonality conditions. Thus, a_1 is chosen to minimize the expected loss of the best linear predictor (cf. Appendix B)

$$E\{[y - \hat{E}(y|a'y)]'[y - \hat{E}(y|a'y)]\}, \tag{14.1.1}$$

and, for $i > 1$, a_i is chosen to minimize (14.1.1) subject to the condition that $a'_i \Sigma a_j = 0$, $j = 1, \ldots, i - 1$.

If we restate the problem, we can use Proposition 12.7.4 to solve it. Note that

$$\begin{aligned} E\{[y - \hat{E}(y|a'y)]'[y - \hat{E}(y|a'y)]\} &= \text{tr}(E\{[y - \hat{E}(y|a'y)][y - \hat{E}(y|a'y)]'\}) \\ &= \text{tr}\{\text{Cov}[y - \hat{E}(y|a'y)]\}. \end{aligned}$$

Thus, minimizing (14.1.1) is identical to minimizing the trace of the prediction error covariance matrix. Write

$$V_a \equiv \text{Cov}[y - \hat{E}(y|a'y)].$$

Given $a_1, \ldots, a_{r-1}$, we wish to find a_r such that

$$\text{tr}(V_{a_r}) = \inf_a \{\text{tr}(V_a) | a' \Sigma a_i = 0; \ i = 1, \ldots, r - 1\}.$$

It is surprising that eigenvectors $a_1, \ldots, a_q$ of Σ corresponding to the ordered eigenvalues $\phi_1 \geq \cdots \geq \phi_q > 0$ solve this optimal prediction problem. To see this, first note that

$$V_a \equiv \text{Cov}[y - \hat{E}(y|a'y)] = \Sigma - \Sigma a (a' \Sigma a)^{-1} a' \Sigma,$$

so

$$\begin{aligned} \text{tr}(V_a) &= \text{tr}(\Sigma) - \text{tr}[\Sigma a (a' \Sigma a)^{-1} a' \Sigma] \\ &= \text{tr}(\Sigma) - \text{tr}[(a' \Sigma a)^{-1} a' \Sigma \Sigma a] \\ &= \text{tr}(\Sigma) - \frac{a' \Sigma^2 a}{a' \Sigma a}. \end{aligned}$$

Thus, minimizing $\text{tr}(V_a)$ is equivalent to maximizing $a' \Sigma^2 a / a' \Sigma a$.

It suffices to find $a_1, \ldots, a_q$ such that

$$\frac{a_1' \Sigma^2 a_1}{a_1' \Sigma a_1} = \sup_a \frac{a' \Sigma^2 a}{a' \Sigma a},$$

and, for $i = 2, \ldots, q$,

$$a_i' \Sigma a_j = 0, \; j = 1, \ldots, i-1,$$

with

$$\frac{a_i' \Sigma^2 a_i}{a_i' \Sigma a_i} = \sup_a \left\{ \frac{a' \Sigma^2 a}{a' \Sigma a} \bigg| a' \Sigma a_j = 0; \; j = 1, \ldots, i-1 \right\}.$$

By taking $H = \Sigma^2$ and $E = \Sigma$, this is just a special case of the problem solved in Proposition 12.7.4. An optimal vector a_i is an eigenvector of $E^{-1}H = \Sigma^{-1}\Sigma^2 = \Sigma$ that corresponds to the ith largest eigenvalue ϕ_i of Σ.

To evaluate how well each coordinate predicts y, observe that

$$\begin{aligned}
\operatorname{tr}(V_{a_i}) &= \operatorname{tr}(\Sigma) - a_i' \Sigma^2 a_i / a_i' \Sigma a_i \\
&= \sum_{j=1}^{q} \phi_j - \phi_i^2 / \phi_i \\
&= \sum_{j=1}^{q} \phi_j - \phi_i.
\end{aligned}$$

Thus, because the ϕ_is are decreasing, each coordinate does no better at predicting y than the previous components.

14.1.2 Joint Prediction

It is also of interest to evaluate the overall predictive ability of a set of principal components, say $a_1' y, \ldots, a_r' y$, that provide optimal sequential prediction. Using Propositions B.1.5 and B.1.9, a simple inductive proof yields

$$\operatorname{tr}\left\{ \operatorname{Cov}\left[y - \hat{E}(y | a_1' y, \ldots, a_r' y) \right] \right\} = \sum_{j=r+1}^{q} \phi_j. \tag{14.1.2}$$

Together, the first r principal components do a good job of predicting y if the ratio

$$\frac{\operatorname{tr}\left\{ \operatorname{Cov}\left[y - \hat{E}(y | a_1' y, \ldots, a_r' y) \right] \right\}}{\operatorname{tr}\left\{ \operatorname{Cov}[y] \right\}} = \frac{\sum_{j=r+1}^{q} \phi_j}{\sum_{j=1}^{q} \phi_j}$$

is very small. This occurs if $\phi_{r+1}, \ldots, \phi_q$ are all very small relative to $\sum_{j=1}^{q} \phi_j$. While the user must decide how much information can be sacrificed to the goal of dimensionality reduction, Johnson and Wichern (2007, Section 8.2) suggest that for many purposes the principal components form an effective substitute for the original variables when the ratio is 0.2 or less for large q and $r = 1, 2,$ or 3.

Exercise 14.1.   Prove Eq. (14.1.2).

As a matter of fact, $a_1'y, \ldots, a_r'y$ do as good or better at predicting y than any other r linear combinations of y. Let B be a $q \times r$ matrix of rank r. Then, $B'y$ is a vector consisting of r linear combinations of y. The best linear predictor of y based on $B'y$ has

$$\text{Cov}[y - \hat{E}(y|B'y)] = \Sigma - \Sigma B(B'\Sigma B)^{-1}B'\Sigma$$
$$= \Sigma^{1/2}(I - \Sigma^{1/2}B(B'\Sigma B)^{-1}B'\Sigma^{1/2})\Sigma^{1/2}$$
$$= \Sigma^{1/2}(I - M_{\Sigma^{1/2}B})\Sigma^{1/2},$$

where $M_{\Sigma^{1/2}B}$ is the perpendicular projection operator onto the space $C(\Sigma^{1/2}B)$ under the standard Euclidean inner product. It will be shown later (by the reader) that the prediction error satisfies

$$E\{[y - \hat{E}(y|B'y)]'[y - \hat{E}(y|B'y)]\} = \text{tr}\{\text{Cov}[y - \hat{E}(y|B'y)]\}$$
$$= \text{tr}\{\Sigma^{1/2}(I - M_{\Sigma^{1/2}B})\Sigma^{1/2}\}$$
$$= \text{tr}\{(I - M_{\Sigma^{1/2}B})\Sigma\}$$
$$\geq \sum_{j=r+1}^{q} \phi_j. \qquad (14.1.3)$$

By Eq. (14.1.2), the r linear combinations given by the principal components achieve the lower bound, so the first r principal components are not only sequentially optimal predictors but also jointly optimal predictors.

The inequality (14.1.3) can be established using the following lemma.

Lemma 14.2.1.   Let $\phi_1 \geq \cdots \geq \phi_q > 0$ be the eigenvalues of Σ. Let $v_1, \ldots, v_r$ be *any* orthonormal vectors in $\mathbf{R}^q$. Then,

(a) $\displaystyle\sum_{j=q-r+1}^{q} \phi_j \leq \sum_{j=1}^{r} v_j'\Sigma v_j \leq \sum_{j=1}^{r} \phi_j.$

(b) If M is a perpendicular projection operator on $\mathbf{R}^q$ (standard inner product) and $r(M) = r$, then

$$\sum_{j=q-r+1}^{q} \phi_j \leq \text{tr}\{M\Sigma\} \leq \sum_{j=1}^{r} \phi_j.$$

Exercise 14.2.   Prove Lemma 14.2.1.   Hints: For (a), first consider the special case Σ diagonal and use some ideas from the proof of Proposition 12.7.4. Then, use the eigenvector-eigenvalue decomposition $\Sigma = P\Lambda P'$, where $P'P = I_q$ and Λ is diagonal. For (b), use the orthonormal basis decomposition for a perpendicular projection operator (i.e., $M = OO'$ with $O'O = I_r$).

To see that (14.1.3) is a result of Lemma 14.2.1, note that $I - M_{\Sigma^{1/2}B}$ is a perpendicular projection operator of rank $q - r$ and that part (b) of the lemma leads immediately to (14.1.3).

In applications, the correlation between the ith principal component $a'_i y$ and the hth variable y_h is often cited. The correlations for $h = 1, \ldots, q$ are frequently used in trying to develop an interpretation for the ith principal component. The idea is to recognize some common characteristic(s) of the y_hs that have large correlations with $a'_i y$. The principal component is then interpreted as an underlying factor that measures this characteristic. This procedure is really a factor-analytic use of principal components and is open to criticisms similar to those made of factor analysis (see Sect. 14.4). In particular, such interpretations are not only subjective (subjectivity is unavoidable) but they are apparently unverifiable (a much more serious problem). Regardless of the appropriate use of these correlations, they have a simple mathematical form. Note that because a_i is an eigenvector of Σ,

$$\text{Cov}(y, a'_i y) = \Sigma a_i = \phi_i a_i.$$

Thus, for the h component of y,

$$\text{Cov}(y_h, a'_i y) = \phi_i a_{ih},$$

where

$$a'_i = (a_{i1}, \ldots, a_{iq}).$$

In addition,

$$\text{Var}(a'_i y) = a'_i \Sigma a_i = \phi_i a'_i a_i$$

and

$$\text{Var}(y_h) - \sigma_{hh}$$

so

$$\text{Corr}(y_h, a'_i y) = \phi_i a_{ih} / \sqrt{\sigma_{hh} \phi_i a'_i a_i}.$$

Often, the vectors a_i are chosen so that $a'_i a_i = 1$. This generates a simplification in the formula for the correlation.

14.1.3 Other Derivations of Principal Components

Although the methodology of principal components was first proposed by Pearson (1901), principal component analysis in its modern form was originated by Hotelling (1933). It is curious that, although Hotelling was originally interested in a prediction problem (actually a factor analysis problem), he transformed his problem into one of finding vectors $a_1, \ldots, a_r$ of fixed length such that $a_i' y$ has maximum variance subject to the condition that $\mathrm{Cov}(a_i' y, a_j' y) = 0$, $j = 1, \ldots, i-1$. This is the form in which principal component analysis is traditionally presented.

Apparently, it was intuitively clear to Hotelling that his linear combinations $a_i' y$ should have maximum predictive capability. In fact, the solution to Hotelling's problem is identical to the solution just given for the prediction problem. Unfortunately, that fact is not obvious to many of us who do not share Hotelling's keen insight into multivariate analysis. We now prove that this equivalence is true. To begin, we solve a slightly different problem and then show that the solution to Hotelling's problem and the alternative problem are the same.

Suppose we want to find vectors $a_1, \ldots, a_q$ that satisfy three properties: $a_i' a_i = K$ for all i and some constant K, $\mathrm{Var}(a_1' y)$ is maximized, and given that $a_i' a_j = 0$, $j = 1, \ldots, i-1$, $\mathrm{Var}(a_i' y)$ is maximized. The condition that $a_i' a_i = K$ is necessary because multiplying a_i by a constant changes the variance. If we allow different size vectors, we could never attain a maximum variance. Alternatively, we could allow different size vectors, but maximize the standardized variance. In other words, maximizing $\mathrm{Var}(a' y)$ for a fixed size vector is identical to maximizing $\mathrm{Var}(a' y)/a' a = a' \Sigma a / a' a$. Using this equivalence, our problem is to find a_1 such that

$$a_1' \Sigma a_1 / a_1' a_1 = \sup_a a' \Sigma a / a' a$$

and for $i = 2, \ldots, q$ find a_i such that

$$\frac{a_i' \Sigma a_i}{a_i' a_i} = \sup_a \left\{ \frac{a' \Sigma a}{a' a} \,\middle|\, a' a_j = 0; \ j = 1, \ldots, i-1 \right\}$$

and

$$a_i' a_j = 0; \ j = 1, \ldots, i-1.$$

Using Proposition 12.7.4 with $E = I$ and $H = \Sigma$, just as in the sequential prediction problem, the vectors a_i are eigenvectors of Σ with respect to the eigenvalues $\phi_1 \geq \cdots \geq \phi_q$.

Hotelling's problem was slightly different. For $i = 2, \ldots, q$ and $j = 1, \ldots, i-1$, Hotelling wanted $\mathrm{Cov}(a_i' y, a_j' y) = a_i' \Sigma a_j = 0$ instead of $a_i' a_j = 0$. We want to show that the eigenvectors that solve our modified problem also solve Hotelling's problem. We can show this inductively. Clearly, a_1 is the same for either problem. Now suppose the first $i-1$ eigenvectors solve both problems. We need to show that if a_i solves our modified problem it will also solve Hotelling's problem. The key point is that for any vector a and $j = 1, \ldots, i-1$,

$$a' \Sigma a_j = \phi_j a' a_j$$

because a_j is an eigenvector of Σ. It follows that with Σ positive definite,

$$a' \Sigma a_j = 0 \quad \text{if and only if} \quad a' a_j = 0.$$

Thus,

$$\left\{\frac{a'\Sigma a}{a'a}\middle| a'a_j = 0;\ j=1,\ldots,i-1\right\} = \left\{\frac{a'\Sigma a}{a'a}\middle| a'\Sigma a_j = 0;\ j=1,\ldots,i-1\right\},$$

so any a_i that is a solution to our modified problem also satisfies

$$\frac{a_i'\Sigma a_i}{a_i'a_i} = \sup_a \left\{\frac{a'\Sigma a}{a'a}\middle| a'\Sigma a_j = 0;\ j=1,\ldots,i-1\right\},$$

with

$$a_i'\Sigma a_j = 0;\ j=1,\ldots,i-1.$$

This establishes that eigenvectors of Σ provide solutions to Hotelling's problem just

as they do for the modified problem and the prediction problems.

Finally, principal components can be related to ellipsoids. If $y \sim N(\mu,\Sigma)$, the set of points that have constant likelihood (the same value of the density) fall on ellipsoids defined by Σ^{-1}. Figure 14.1 illustrates a density isobar for a

$$N\left(\begin{bmatrix}1\\2\end{bmatrix},\begin{bmatrix}1.0 & 0.9\\0.9 & 2.0\end{bmatrix}\right)$$

random vector. The major and minor axes are denoted a_1 and a_2, respectively. We want to show that axes of the ellipse are determined by the eigenvectors of Σ.

In general, an ellipsoid centered at zero defined by Σ^{-1} is

$$\{a|a'\Sigma^{-1}a = c\},$$

where c is some constant. Let the vectors $a_1,\ldots,a_q$ denote the directions of the axes of the ellipsoid. The principal axis is the longest vector a on the ellipsoid. The subsequent axes are the longest vectors on the ellipsoid that are orthogonal (in the Euclidean inner product) to the previous axes. We want to show that these are determined by the eigenvectors of Σ.

Any vector a can be made to fit on the ellipsoid by standardizing it, i.e., $(\sqrt{c}/\sqrt{a'\Sigma^{-1}a})a$ is always on the ellipsoid because

$$(\sqrt{c}/\sqrt{a'\Sigma^{-1}a})a'\Sigma^{-1}(\sqrt{c}/\sqrt{a'\Sigma^{-1}a})a = c.$$

The principal axis is the longest vector of the form $(\sqrt{c}/\sqrt{a'\Sigma^{-1}a})a$. Because c is a constant, a vector a_1 is in the direction of the principal axis if and only if

$$\frac{a_1'a_1}{a_1'\Sigma^{-1}a_1} = \sup_a \frac{a'a}{a'\Sigma^{-1}a}.$$

Principal Components

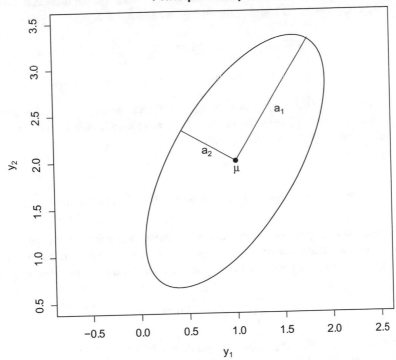

Fig. 14.1 Two-dimensional normal density isobar ($\mu = (1,2)'$, $\sigma_{11} = 1.0$, $\sigma_{12} = 0.9$, $\sigma_{22} = 2.0$) with major and minor axes

Similarly, vectors in the directions of the other axes must satisfy

$$\frac{a_i' a_i}{a_i' \Sigma^{-1} a_i} = \sup_a \left\{ \frac{a'a}{a' \Sigma^{-1} a} \middle| a'a_j = 0; \; j-1,\ldots,i-1 \right\}$$

and

$$a_i' a_j = 0; \; j = 1,\ldots,i-1 .$$

Again, to show that the eigenvectors of Σ provide a solution to this problem, we solve a related problem in which we require

$$\frac{a_i' a_i}{a_i' \Sigma^{-1} a_i} = \sup_a \left\{ \frac{a'a}{a' \Sigma^{-1} a} \middle| a' \Sigma^{-1} a_j = 0; \; j = 1,\ldots,i-1 \right\}$$

and

$$a_i' \Sigma^{-1} a_j = 0 .$$

The related problem is solved using Proposition 12.7.4 with $E = \Sigma^{-1}$ and $H = I$. The eigenvectors of $E^{-1} H = \Sigma$ provide solutions. The equivalence of the solutions to the ellipsoid problem and our related problem is based on the facts that $\Sigma a = \phi a$ if and only if $\frac{1}{\phi} a = \Sigma^{-1} a$ and that for an eigenvector a_j of Σ

$$a' \Sigma^{-1} a_j = 0 \quad \text{if and only if} \quad a' a_j = 0.$$

Finally, the principal axis is actually $(\sqrt{c} / \sqrt{a_1' \Sigma^{-1} a_1}) a_1$ with the other axes, in order, being $(\sqrt{c} / \sqrt{a_i' \Sigma^{-1} a_i}) a_i$. Note that, not only are the original a_is eigenvectors, but the axes are also eigenvectors of Σ, just eigenvectors that have been given the appropriate length to be on the ellipsoid.

14.1.4 Principal Components Based on the Correlation Matrix

In our discussion of principal components, we have used minimization of the trace of the prediction error covariance matrix as a criterion for optimal prediction. This is precisely the sum of the prediction error variances for each component of y, namely,

$$\sum_{h=1}^{q} \text{Var}\left[y_h - \hat{E}(y_h | a'y)\right].$$

If, for example, y is the length, width, and height of a randomly chosen dog house (or turtle shell) and all the measurements are taken in centimeters, this is probably a reasonable prediction error criterion. However, if the length and width are measured in centimeters and the height is measured in kilometers, the height numbers will all be small, so the height variances will be small, so getting good height predictions will be nearly irrelevant to the process of getting good overall predictions as determined by this criterion.

In general, if an individual variable, say y_h, has a very small variance, then any linear combination $a'y$ generates a very small prediction error variance for y_h because the prediction error variance is never greater than the original variance. An optimal linear combination $a'y$ minimizes the sum of all the prediction variances, so prediction based on $a'y$ may lead to incongruities. Variables that happen to be measured on scales with small absolute variability are almost ignored in favor of predicting variables that are measured with large variability. Moreover, because the variance of any individual variable can be made arbitrarily large or small simply by multiplying the variable by a constant, the principal components are subject to the whims of measurement scale.

An obvious way to avoid this problem is to standardize the variance of the individual variables. Suppose $\Sigma = [\sigma_{hh'}]$, and let $D = \text{Diag}(\sigma_{11}, \ldots, \sigma_{qq})$; then, the variable $z = D^{-1/2} y$ has

$$\text{Cov}(z) = D^{-1/2} \Sigma D^{-1/2},$$

which is the correlation matrix for y. Because $\text{Cov}(z)$ is a matrix that has 1's down the main diagonal, the principal components for predicting $z = (z_1, \ldots, z_q)$ will give comparable weight to predicting each of the individual variables z_h. Moreover, because z is a nonsingular linear transformation of y, no information is lost in analyzing z in place of y. The transformation of y to z is simply an implicit way of defining a new criterion for optimal prediction of y. Schervish (1986) gives an explicit derivation of principal components based on the correlation matrix.

14.2 Sample Principal Components

In practice, the covariance matrix Σ is unknown, so the principal components cannot be computed. However, if a sample $y_1, \ldots, y_n$ of observations on y is available, sample principal components can be computed from either the sample covariance matrix

$$S = \sum_{i=1}^{n} (y_i - \bar{y})(y_i - \bar{y})'/(n-1)$$

or the sample correlation matrix

$$R = D^{-1/2}SD^{-1/2},$$

where $S = [s_{ij}]$ and $D = \text{Diag}(s_{11}, \ldots, s_{qq})$. Most often, the correlation matrix seems to be the appropriate choice.

It is convenient to use S or R to define an inner product and to choose $a_1, \ldots, a_q$ as an orthonormal set of eigenvectors corresponding to the eigenvalues $\phi_1 \geq \cdots \geq \phi_q$ of S or R, respectively. Write $A = [a_1, \ldots, a_q]$ and, for $r \leq q$, $A_r = [a_1, \ldots, a_r]$. A vector w, rewritten in the principal component coordinate system, is $A'w$. Write the entire data set as

$$Y = \begin{bmatrix} y_1' \\ \vdots \\ y_n' \end{bmatrix}.$$

Using S, the data in the principal component coordinate system are

$$YA.$$

If principal components are based on the correlation matrix R, the rescaled data are

$$\begin{bmatrix} z_1' \\ \vdots \\ z_n' \end{bmatrix} = Z = YD^{-1/2}$$

and the data in the principal component coordinate system are

$$ZA.$$

The point of principal component analysis is to reduce dimensionality. If the smallest ordered eigenvalues of S, $\phi_{r+1}, \ldots, \phi_q$ are small, a random vector y with covariance matrix S can be predicted well by $A'_r y$. If the entire data set is transformed in this way, a principal component observation matrix is obtained,

$$YA_r = [Ya_1, \ldots, Ya_r],$$ (14.2.1)

where

$$Ya_i = \begin{bmatrix} a'_i y_1 \\ \vdots \\ a'_i y_n \end{bmatrix}.$$

The elements of the vector Ya_i consist of the ith principal component applied to each of the n observation vectors. The principal component observation matrix combines these vectors for each of the first r principal components.

The analysis of the data can be performed on the principal component observations with a minimal loss of information. This includes various plots and formal statistical techniques for the analysis of a sample from one population.

14.2.1 The Sample Prediction Error

A question that arises immediately is just how much information is lost by using r principal components rather than the entire data set. Based on the substitutions $\bar{y}. = \mu$ and $S = \Sigma$, the total error of prediction from using the first r principal components is

$$\sum_{i=1}^{n} [y_i - \hat{E}(y|A'_r y_i)]' [y_i - \hat{E}(y|A'_r y_i)],$$

where

$$\hat{E}(y|A'_r y_i) = \bar{y}. + SA_r (A'_r SA_r)^{-1} A'_r (y_i - \bar{y}.).$$

The total error can be rewritten in terms of the eigenvalues of S, $\phi_1 \geq \cdots \geq \phi_q$. Note that

$$y_i - \hat{E}(y|A'_r y_i) = (y_i - \bar{y}.) - SA_r (A'_r SA_r)^{-1} A'_r (y_i - \bar{y}.)$$
$$= [I - SA_r (A'_r SA_r)^{-1} A'_r](y_i - \bar{y}.).$$

Let $\mathscr{A}_r \equiv A_r (A'_r SA_r)^{-1} A'_r S$; this is an oblique projection operator. In particular, $\mathscr{A}'_r S = S\mathscr{A}_r$. Using this fact and basic results on traces, observe that

$$\sum_{i=1}^{n} [y_i - \hat{E}(y|A'_r y.)]' [y_i - \hat{E}(y|A'_r y_i)]$$

$$= \mathrm{tr} \left\{ \sum_{i=1}^{n} [y_i - \hat{E}(y|A'_r y_i)] [y_i - \hat{E}(y|A'_r y_i)]' \right\}$$

$$
\begin{aligned}
&= \mathrm{tr}\left\{\sum_{i=1}^{n}[I - \mathscr{A}_r'](y_i - \bar{y}.)(y_i - \bar{y}.)'[I - \mathscr{A}_r']'\right\} \\
&= \mathrm{tr}\left\{[I - \mathscr{A}_r']\left[\sum_{i=1}^{n}(y_i - \bar{y}.)(y_i - \bar{y}.)'\right][I - \mathscr{A}_r]\right\} \\
&= \mathrm{tr}\left\{[I - \mathscr{A}_r'][(n-1)S][I - \mathscr{A}_r]\right\} \\
&= \mathrm{tr}\left\{(n-1)\left[S - SA_r(A_r'SA_r)^{-1}A_r'S\right]\right\} \\
&= (n-1)\mathrm{tr}\left\{S - SA_r(A_r'SA_r)^{-1}A_r'S\right\}.
\end{aligned}
$$

From the subsection on joint prediction,

$$
\mathrm{tr}\{S - SA_r(A_r'SA_r)^{-1}A_r'S\} = \mathrm{tr}\{\mathrm{Cov}[y - \hat{E}(y|A_r'y)]\},
$$

where $\mathrm{Cov}(y) = S$. By Eq. (14.2.2),

$$
\sum_{i=1}^{n}\left[y_i - \hat{E}(y|A_r'y_i)\right]'\left[y_i - \hat{E}(y|A_r'y_i)\right] = (n-1)\sum_{j=r+1}^{q}\phi_j.
$$

To evaluate the quality of prediction, the total error of prediction using r components can be compared to the maximum possible prediction error. The maximum possible prediction error can be viewed as using zero principal components. The maximum is

$$
\begin{aligned}
\sum_{i=1}^{n}(y_i - \bar{y}.)'(y_i - \bar{y}.) &= \mathrm{tr}\left\{\sum_{i=1}^{n}(y_i - \bar{y}.)(y_i - \bar{y}.)'\right\} \\
&= (n-1)\mathrm{tr}\{S\} \\
&= (n-1)\sum_{j=1}^{q}\phi_j.
\end{aligned}
$$

The value

$$
100\sum_{j=r+1}^{q}\phi_j \bigg/ \sum_{j=1}^{q}\phi_j
$$

is the percentage of the maximum prediction error left unexplained by $\hat{E}(y_i|A_r'y_i)$, $i = 1,\ldots,n$. Alternatively,

$$
100\sum_{j=1}^{r}\phi_j \bigg/ \sum_{j=1}^{q}\phi_j
$$

is the percentage of the maximum prediction error accounted for by $A_r'y$.

14.2.2 Using Principal Components

Principal components are designed to reduce dimensionality. They provide a number $r < q$ of linear combinations $a_i'y$ that maximize the ability to linearly predict the original random q-vector y. Thus they are appropriate to use when you are taking a random sample of ys.

Although the analysis of data can be performed on the principal component observations with a minimal loss of information, why accept any loss of information? Two possibilities come to mind. First, if q is very large, an analysis of all q variables may be untenable. If one must reduce the dimensionality before any work can proceed, principal components are a reasonable place to begin. However, it should be kept in mind that principal components are based on linear combinations of y and linear predictors of y. If the important structure in the data is nonlinear, principal components can totally miss that structure.

A second reason for giving up information is when you do not trust all of the information. In prediction theory the underlying idea is that a vector (y, x') would be randomly sampled from some population and we would seek to predict y based on x. Principal component regression (cf. *PA-V*, Chap. 13, or Christensen 2011, Chapter 15) can be used to reduce the dimensionality of x. But *PA* argues that using the principal components are an effective way to treat collinearity, even if you only have a sample from $y|x$. The main idea was that, with errors in the model matrix, directions corresponding to small eigenvalues are untrustworthy. In the present context, we might say that any statistical relationships depending on linear combinations that do not provide substantial power of prediction are questionable.

As a general principle, to reduce the dimensionality of data successfully you need to know ahead of time that it can be reduced. Whether it can be reduced depends on the goal of the analysis. The work involved in figuring out whether data reduction can be accomplished often negates the value of doing it. The situation is similar to that associated with Simpson's paradox in contingency table analysis (see Christensen 1997, Section 3.1; Christensen 2014). Valid inferences cannot always be obtained from a collapsed contingency table. To know whether valid inferences can be obtained, one needs to analyze the full table first. Having analyzed the full table, there may be little point in collapsing to a smaller-dimensional table. Often, it would be convenient to reduce a data set using principal components and then do a MANOVA on the reduced data. Unfortunately, about the only way to find out if that approach is reasonable is to examine the results of a MANOVA on the entire data set.

Principal components are well designed for data reduction within a given population. If there are samples available from several populations with the same covariance matrix, then the optimal data reduction will be the same for every group and can be estimated using the pooled covariance matrix. Note that this essentially requires doing a one-way MANOVA prior to the principal component analysis. If an initial MANOVA is required, you may wonder why one would bother to reduce the data having already done a significant analysis on the unreduced set.

In particular, my friend Ed Bedrick has point out that if y is sampled from more than one population, reducing the dimensionality, without having first accounted for the different populations, can cause you to loose the ability to distinguish the populations. As illustrated in Fig. 14.2, with two normal populations having means μ_1 and μ_2, if the vector $\mu_1 - \mu_2$ is orthogonal to the eigenvectors that you are using to define your principal components, then there may be no information in your principal components capable of distinguishing the populations. In fact, even if $\mu_1 - \mu_2$ is merely close to $C(a_1, \ldots, a_r)^\perp$ you may lose most of the information for distinguishing the populations. Basically, the only way to tell that this is not happening is to do the one-way MANOVA prior to doing the principal component analysis. Again, there may be no point in doing principal components after doing MANOVA. Jolliffe (1986, Section 9.1) discusses this problem in more detail.

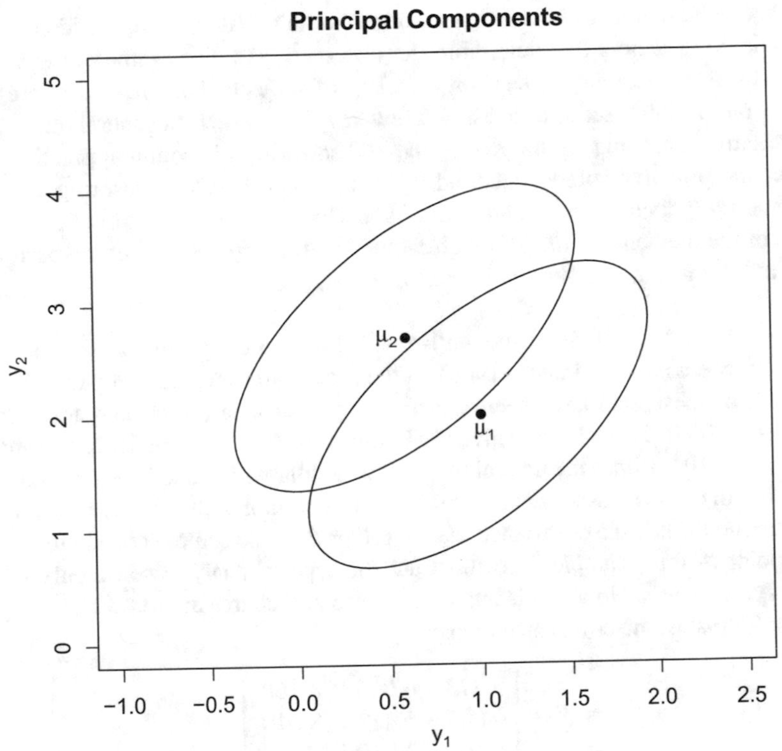

Principal Components

Fig. 14.2 Two populations indistinguishable in the first principal component

Exercise 14.3. Consider two q-vectors y_j with $E(y_j) = \mu_i$ and $Cov(y_j) = \Sigma$ and z independent of the y_js with $z \sim \text{Bern}(p)$. Define the mixture random vector $y \equiv zy_1 + (1-z)y_2$. Assume that $a_1, \ldots, a_q$ are eigenvectors of Σ associated with eigenvalues $\phi_1 > \quad > \phi_q > 0$.

1. Show that $\text{Cov}(y) = \Sigma + p(1-p)(\mu_1 - \mu_2)(\mu_1 - \mu_2)'$.
2. Show that if $\mu_1 - \mu_2$ is an eigenvector for ϕ_k, then the a_is are all eigenvectors of $\text{Cov}(y)$ with corresponding eigenvalues ϕ_i for $i \neq k$ and the eigenvalue $\phi_k + p(1-p)\|\mu_1 - \mu_2\|^2$ corresponding to a_k.
3. Show that if $\mu_1 - \mu_2$ is proportional to a_k and $\phi_k + p(1-p)\|\mu_1 - \mu_2\|^2 < \phi_r$, then the first r principal components of y_i agree with the first r principal components of y and that for $i = 1, \ldots, r$, $E(a_i'y_1) = E(a_i'y_2)$.
4. With $\mu_1 - \mu_2$ proportional to a_k, what happens to the first r principal components of y when $\phi_k + p(1-p)\|\mu_1 - \mu_2\|^2 > \phi_r$?
5. Show that if $(\mu_1 - \mu_2) \in C(a_{r+1}, \ldots, a_q)$ and $\phi_{r-1} + p(1-p)\|\mu_1 - \mu_2\|^2 < \phi_r$, then the first r principal components of y_i agree with the first r principal components of y.

Data reduction is also closely related to a more nebulous idea, the identification of underlying factors that determine the observed data. For example, the vector y may consist of a battery of tests on a variety of subjects. One may seek to explain scores on the entire set of tests using a few key factors such as general intelligence, quantitative reasoning, verbal reasoning, and so forth. It is common practice to examine the principal components and try to interpret them as measuring some sort of underlying factor. Such interpretations are based on examination of the relative sizes of the elements of a_i. Although factor identification is commonly performed, it is, at least in some circles, quite controversial.

EXAMPLE 14.2.1.   One of the well-traveled data sets in multivariate analysis is from Jolicoeur and Mosimann (1960) on the shell (carapace) sizes of painted turtles. Aspects of these data have been examined by Morrison (2004) and Johnson and Wichern (2007). The data were given in Exercise 10.6.1 and Table 10.3. The analysis is based on $10^{3/2}$ times the natural logs of the height, width, and length of the shells. Because all of the measurements are taken on a common scale, it may be reasonable to examine the sample covariance matrix rather than the sample correlation matrix. The point of this example is to illustrate the type of analysis commonly used in identifying factors. No claim is made that these procedures are reasonable.

For 24 males, the covariance matrix is

$$S = \begin{bmatrix} 6.773 & 6.005 & 8.160 \\ 6.005 & 6.417 & 8.019 \\ 8.160 & 8.019 & 11.072 \end{bmatrix}.$$

The eigenvalues and corresponding eigenvectors for S are as follows.

ϕ_i	23.303	0.598	0.360
	a_1	a_2	a_3
$10^{3/2}\ln(\text{height})$	0.523	0.788	−0.324
$10^{3/2}\ln(\text{width})$	0.510	−0.594	−0.622
$10^{3/2}\ln(\text{length})$	0.683	−0.159	0.713

Recall that eigenvectors are not uniquely defined. Eigenvectors of a matrix B corresponding to ϕ (along with the zero vector) constitute the null space of $B - \phi I$. Often, the null space has rank 1, in which case every eigenvector is a multiple of every other eigenvector. If we standardize the eigenvectors of S so that each has a maximum element of 1, we get the following eigenvectors.

	a_1	a_2	a_3
$10^{3/2} \ln(\text{height})$	0.764	1	−0.451
$10^{3/2} \ln(\text{width})$	0.747	−0.748	−0.876
$10^{3/2} \ln(\text{length})$	1	−0.205	1
ϕ	23.30	0.60	0.36

The first principal component accounts for $100(23.30)/(23.30 + 0.60 + 0.36) = 96\%$ of the predictive capability (variance) of the variables. The first two components account for $100(23.30 + 0.60)/(24.26) = 98.5\%$ of the predictive capability (variance) of the variables. All the elements of a_1 are positive and approximately equal, so $a_1'y$ can be interpreted as a measure of overall size. The elements of a_2 are a large positive value for $10^{3/2} \ln(\text{height})$, a large negative value for $10^{3/2} \ln(\text{width})$, and a small value for $10^{3/2} \ln(\text{length})$. The component $a_2'y$ can be interpreted as a comparison of the $\ln(\text{height})$ and the $\ln(\text{width})$. Finally, if one considers the value $a_{31} = -0.451$ small relative to $a_{32} = -0.876$ and $a_{33} = 1$, one can interpret $a_3'y$ as a comparison of width versus length.

Interpretations such as these necessarily involve rounding values to make them more interpretable. The interpretations just given are actually appropriate for the three linear combinations of y, $b_1'y$, $b_2'y$, and $b_3'y$ that follow.

	b_1	b_2	b_3
$10^{3/2} \ln(\text{height})$	1	1	0
$10^{3/2} \ln(\text{width})$	1	−1	−1
$10^{3/2} \ln(\text{length})$	1	0	1

The first interpreted component is

$$b_1'y = 10^{3/2} \ln[(\text{height})(\text{width})(\text{length})]$$
$$= 10^{3/2} \ln[\text{volume}],$$

where the volume is that of a box. It is interesting to note that in this particular example, the first principal component can be interpreted without changing the coefficients of a_1.

$$a_1'y = 10^{3/2}[0.764 \ln(\text{height}) + 0.747 \ln(\text{width}) + \ln(\text{length})]$$
$$= 10^{3/2} \ln[(\text{height})^{0.764}(\text{width})^{0.747}(\text{length})].$$

The component $a_1'y$ can be thought of as measuring the log volume with adjustments made for the fact that painted turtle shells are somewhat curved and thus not a perfect box. Because the first principal component accounts for 96% of the predictive capability, to a very large extent, if you know this pseudovolume measurement, you know the height, length, and width.

In this example, we have sought to interpret the elements of the vectors a_i. Alternatively, one could base interpretations on estimates of the correlations $\text{Corr}(y_h, a_i'y)$ that were discussed in Sect. 14.2. The estimates of $\text{Corr}(y_h, a_1'y)$ are very uniform, so they also suggest that a_1 is an overall size factor.   □

Linear combinations $b_i'y$ that are determined by the effort to interpret principal components will be called *interpreted components*. Although it does not seem to be common practice, it is interesting to examine how well interpreted components predict the original data and compare that to how well the corresponding principal components predict the original data. As long as the interpreted components are linearly independent, a full set of q components will predict the original data perfectly. *Any* nonsingular transformation of y will predict y perfectly because it amounts to simply changing the coordinate system. If we restrict attention to r components, we know from the theoretical results on joint prediction that the interpreted components can predict no better than the actual principal components. In general, to evaluate the predictive capability of r interpreted components, write $B_r = [b_1, \ldots, b_r]$ and compute

$$\sum_{i=1}^{n} [y_i - \hat{E}(y|B_r'y_i)]'[y_i - \hat{E}(y|B_r'y_i)] = (n-1)\text{tr}\{S - SB_r(B_r'SB_r)^{-1}B_r'S\}.$$

One hundred times this value divided by $(n-1)\sum_{i=1}^{q}\phi_i = (n-1)\text{tr}(S)$ gives the percentage of the predictive error unaccounted for by the r interpreted components. If this is not much greater than the corresponding percentage for the first r principal components, the interpretations are to some extent validated.

EXAMPLE 14.2.2. Using the first two interpreted components from Example 14.2.1,

$$\text{tr}[SB_2(B_2'SB_2)^{-1}B_2'S] = 23.88$$

and

$$\frac{100\text{tr}[S - SB_2(B_2'SB_2)^{-1}B_2'S]}{\text{tr}[S]} = \frac{100(24.26 - 23.88)}{24.26}$$
$$= \frac{100(0.38)}{24.26}$$
$$= 1.6.$$

Using the first two principal components,

$$\frac{100\text{tr}[S - SA_2(A_2'SA_2)^{-1}A_2'S]}{\text{tr}[S]} = \frac{100\sum_{j=1}^{3}\phi_j - \sum_{j=1}^{2}\phi_j}{\sum_{j=1}^{3}\phi_j}$$

$$= \frac{100(0.36)}{24.26}$$

$$= 1.5.$$

Thus, in this example, there is almost no loss of predictive capability by using the two interpreted components rather than the first two principal components. □

There is one aspect of principal component analysis that is often overlooked. It is possible that the most interesting components are those that have the least predictive power. Such components are taking on very similar values for all cases in the sample. It may be that these components can be used to characterize the population. Jolliffe (1986) has a fairly extensive discussion of uses for the *last few* principal components.

EXAMPLE 14.2.3.   The smallest eigenvalue of S is 0.36 and corresponds to the linear combination

$$a_3'y = 10^{3/2}[-0.451\ln(\text{height}) - 0.876\ln(\text{width}) + \ln(\text{length})].$$

This linear combination accounts for only 1.5% of the variability in the data. It is essentially a constant. All male painted turtles in the sample have about the same value for this combination. The linear combination is a comparison of the ln-length with the ln-width and ln-height. This might be considered as a measurement of the general shape of the carapace. One would certainly be very suspicious of any new data that were supposedly the shell dimensions of a male painted turtle but which had a substantially different value of $a_3'y$. On the other hand, this should not be thought of as a discrimination tool except in the sense of identifying whether data are or are not consistent with the male painted turtle data. We have no evidence that other species of turtles will produce substantially different values of $a_3'y$. □

14.3 Classical Multidimensional Scaling

Multidimensional Scaling starts with a matrix containing the squared distances between a set of objects and produces a plot of the objects that reflects those distances. There are a number of methods for doing this but we restrict our attention to *Classical Multidimensional Scaling (CMDS)* because it reproduces the (mean corrected) sample principal component scores from only the squared distance matrix.

Consider a matrix $\mathscr{D}$ that consists of the squared distances between n objects. To obtain an r dimensional graphical representation of the objects, find eigenvectors $a_1, \ldots, a_r$ of $[I - (1/n)JJ']\mathscr{D}[I - (1/n)JJ']$ corresponding to its r largest eigenvalues. Create the matrix

$$A_r \equiv [a_1, \cdots, a_r] \equiv \begin{bmatrix} \mathbf{a}_1' \\ \vdots \\ \mathbf{a}_n' \end{bmatrix}$$

and in r dimensions, plot the n vectors $\mathbf{a}_i$ to represent the n objects.

EXAMPLE 14.3.1.   I computed the distances between the 21 observations in the Cushing's syndrome data of Table 12.1 and applied CMDS to the squared distances. The result appears in Fig. 14.3. The plot is just a recentering and rotation of the data appearing in Fig. 12.1. (The data are rotated about 45° counterclockwise.) This occurs because, as we will show, the two-dimensional CMDS method is essentially just plotting the first two principal components of the data. Because the original data were two dimensional, the first two principal components contain all the information in the data, so we just get a recentered, rotated plot of the data.   □

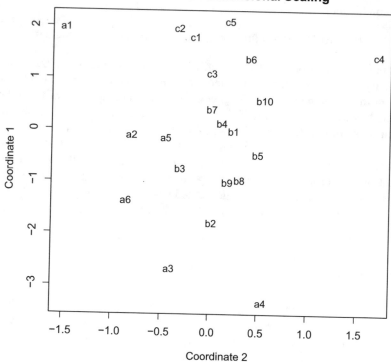

Classical Multidimensional Scaling

Fig. 14.3 Classical multidimensional scaling: Cushing syndrome data

Starting with a data matrix

$$Y = \begin{bmatrix} y_1' \\ \vdots \\ y_n' \end{bmatrix}$$

that contains observations on n objects, we construct the squared Euclidian distance matrix $\mathscr{D}$ for which the ij element d_{ij} is the squared distance between y_i and y_j. In particular,

$$d_{ij} = (y_i - y_j)'(y_i - y_j).$$

We then establish that we can find the mean corrected sample principal components directly from $\mathscr{D}$. CMDS consists of plotting the mean corrected sample principal components.

Multiplying out the squared distances gives

$$d_{ij} \equiv (y_i - y_j)'(y_i - y_j) = y_i' y_i + y_j' y_j - 2 y_i' y_j.$$

The squared distances are functions of the inner products and all of the inner products are given by

$$YY' = \begin{bmatrix} y_1' y_1 & y_1' y_2 & \cdots & y_1' y_n \\ y_2' y_1 & y_2' y_2 & \cdots & y_2' y_n \\ \vdots & \vdots & \ddots & \vdots \\ y_n' y_1 & y_n' y_2 & \cdots & y_n' y_n \end{bmatrix}.$$

Create a vector consisting of the squared lengths of the vectors of observations,

$$\mathbf{d} \equiv (y_1' y_1, y_2' y_2, \cdots, y_n' y_n)'.$$

It is not hard to see that the squared Euclidean distance matrix is

$$\mathscr{D} = \mathbf{d}J' + J\mathbf{d}' - 2YY'. \tag{14.3.1}$$

To relate this to principal components, we need to relate $\mathscr{D}$ to the sample covariance matrix

$$S \equiv \frac{1}{n-1} Y'[I - (1/n)JJ']Y = \frac{1}{n-1} \left\{ [I - (1/n)JJ']Y \right\}' \left\{ [I - (1/n)JJ']Y \right\}.$$

The mean of the row vectors y_i' is

$$\bar{y}_. = (1/n)J'Y.$$

If we recenter the data, that does not change their orientation in space, it only shifts the data to being centered at 0. The centered data matrix is

$$Y - J\bar{y}_. = [I - (1/n)JJ']Y.$$

From (14.2.1) the principal component scores are YA_r where A_r has columns that are eigenvectors of S corresponding to the r largest eigenvalues. The mean corrected principal component scores are

$$(Y - J\bar{y}')A_r = [I - (1/n)JJ']YA_r.$$

The key mathematical fact in relating squared distances to principal components is that if λ and b are an eigenvalue and eigenvector for $B'B$, then λ and Bb are an eigenvalue and eigenvector for BB'. In particular, if ϕ and a are an eigenvalue and eigenvector of $(n-1)S$, then ϕ and

$$\left\{[I - (1/n)JJ']Y\right\}a \tag{14.3.2}$$

are an eigenvalue and eigenvector of

$$\left\{[I - (1/n)JJ']Y\right\}\left\{[I - (1/n)JJ']Y\right\}' = [I - (1/n)JJ']YY'[I - (1/n)JJ']. \tag{14.3.3}$$

The formula in (14.3.2) is just a mean corrected principal component.

We now show that (14.3.3) can be obtained directly from the squared distance matrix. Using (14.3.1) write

$$[I - (1/n)JJ']\mathscr{D}[I - (1/n)JJ'] = [I - (1/n)JJ']\left(\mathbf{d}J' + J\mathbf{d}' - 2YY'\right)[I - (1/n)JJ']$$
$$= [I - (1/n)JJ']\left(-2YY'\right)[I - (1/n)JJ']$$

Thus

$$[I - (1/n)JJ']\left(YY'\right)[I - (1/n)JJ'] = \frac{-1}{2}[I - (1/n)JJ']\mathscr{D}[I - (1/n)JJ'].$$

Since S and $(n-1)S$ have the same eigenvectors and the same ordering of the eigenvalues, we can compute the mean corrected principal components from the squared distance matrix.

While Example 14.3.1 is informative about what CMDS is doing, in practice multidimensional scaling is often used in situations where only a measure of distance between objects is available; not the raw data from which the distances were computed.

EXAMPLE 14.3.2.   Lawley and Maxwell (1971) and Johnson and Wichern (2007) examine data on the examination scores of 220 male students. The dependent variable vector consists of test scores on (Gaelic, English, history, arithmetic, algebra, geometry). The correlation matrix is

$$R = \begin{bmatrix} 1.000 & 0.439 & 0.410 & 0.288 & 0.329 & 0.248 \\ 0.439 & 1.000 & 0.351 & 0.354 & 0.320 & 0.329 \\ 0.410 & 0.351 & 1.000 & 0.164 & 0.190 & 0.181 \\ 0.288 & 0.354 & 0.164 & 1.000 & 0.595 & 0.470 \\ 0.329 & 0.320 & 0.190 & 0.595 & 1.000 & 0.464 \\ 0.248 & 0.329 & 0.181 & 0.470 & 0.464 & 1.000 \end{bmatrix}.$$

We are going to treat the 6 tests as objects and use the correlation matrix as a measure of similarity between the objects. In particular, we used

$$\mathscr{D} = 1 - R$$

as our squared distance measure with results displayed in the top panel of Fig. 14.4. The bottom panel of Fig. 14.4 contains the CMDS representation when the distance is measured as 1 minus the squared correlation between the variables. (This does *not* involve multiplying the matrix R times itself.) □

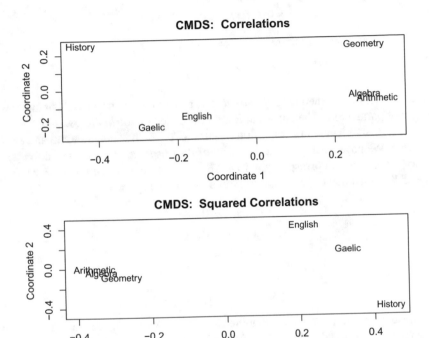

Fig. 14.4 Classical multidimensional scaling: examination data

14.4 Factor Analysis

Principal components are often used in an attempt to identify factors underlying the observed data. There is also a formal modeling procedure called *factor analysis* that is used to address this issue. The model looks similar to a multivariate linear model but several of the assumptions are changed and, most importantly, you don't get to see the matrix of predictor variables. It is assumed that the observation vectors y_i

are uncorrelated and have $E(y_i) = \mu$ and $Cov(y_i) = \Sigma$. For n observations, the factor analysis model is

$$y_i' = \mu' + x_i'B + \varepsilon_i', \quad i = 1, \ldots, n$$

or

$$Y = J\mu' + XB + e, \tag{14.4.1}$$

where Y is $n \times q$ and X is $n \times r$. Most of the usual multivariate linear model assumptions about the rows of e are made,

$$E(\varepsilon_i) = 0,$$
$$Cov(\varepsilon_i, \varepsilon_j) = 0 \quad i \neq j,$$

and for Ψ nonnegative definite

$$Cov(\varepsilon_i) \equiv \Psi,$$

except Ψ is now assumed to be *diagonal*. The matrix B remains a fixed but unknown matrix of parameters. The entries in B are now called *factor loadings*.

The primary change in assumptions relates to X which is now random and *unobservable*. We have assumed that the mean of the observation vector on individual i is $E(y_i') = \mu'$, so the corresponding row $x_i'B$ better be random with mean zero. Each row of X is assumed to be an unobservable random vector with

$$E(x_i) = 0,$$
$$Cov(x_i, x_j) = 0, \quad i \neq j,$$
$$Cov(x_i) = I_r,$$

and

$$Cov(x_i, \varepsilon_j) = 0, \quad \text{any } i, j.$$

The idea behind the model is that the elements of X consist of r underlying *common factors*. For fixed individual i, the different observations, $y_{ih}, h = 1, \ldots, q$ are different linear combinations of the r (random) common factors for that individual $x_{ik}, k = 1, \ldots, r$ plus some (white noise) error $\varepsilon_{ih}, h = 1, \ldots, q$. The errors for the individual are allowed to have different variances but they are assumed to be uncorrelated.

Specifically, every individual i has observations y_{ih} that involve the same linear combinations (loadings) of the common factors, $\sum_{k=1}^{r} \beta_{hk} x_{ik}$, where the β_{hk}s do not depend on the individual i but change depending on h, so the linear combination depends on which variable y_{ih} is being considered for individual i. Of course different individuals i have different realizations of the random common factors x_{ik}. So, while all the common factors affect each column of Y, different linear combinations of the common factors apply to different columns. The kth column of X consists of n realizations of the kth common factor. The row k of B therefore tells how the kth factor is incorporated into all of observations. It serves as the basis for trying to interpret what the kth factor contributes. The hth column of B are the coefficients that generate the hth dependent variable y_{ih}.

For the model to be of interest, the number of factors r should be less than the number of variables q. Based on this model, $y_i = \mu + B' x_i + \varepsilon_i$, $i = 1, \ldots, n$, so

$$\begin{aligned} \text{Cov}(y_i) &= \text{Cov}(B' x_i + \varepsilon_i) \\ &= B'B + \Psi. \end{aligned}$$

In most of the discussion to follow, we will work directly with the matrix $B'B$. It is convenient to have a notation for this matrix. Write

$$\Lambda \equiv B'B.$$

The matrix Λ is characterized by two properties: (1) Λ is nonnegative definite and (2) $r(\Lambda) = r$. Recalling our initial assumption that $\text{Cov}(y_i) = \Sigma$, the factor analysis model has imposed the restriction that

$$\Sigma = \Lambda + \Psi. \tag{14.4.2}$$

Clearly, one cannot have $r(\Lambda) = r > q$. It is equally clear that if $r = q$, one can always find matrices Λ and Ψ that satisfy (14.4.2). Just choose $\Lambda = \Sigma$ and $\Psi = 0$. If $r < q$, Eq. (14.4.2) may place a real restriction on Σ, see Exercise 14.4b.

In practice, Σ is unknown and estimated by S, so one seeks matrices $\hat{B}$ and $\hat{\Psi}$ such that

$$S \doteq \hat{B}'\hat{B} + \hat{\Psi}.$$

The interesting questions now become (a) how many factors r are needed to get a good approximation and (b) which matrices $\hat{B}$ and $\hat{\Psi}$ give good approximations. The first question is certainly amenable to analysis. Clearly, $r = q$ will always work, so there must be ways to decide when $r < q$ is doing an adequate job. The second question ends up being tricky. The problem is that if U is any orthonormal matrix, $U\hat{B}$ works just as well as $\hat{B}$ because

$$\hat{B}'\hat{B} = (U\hat{B})'(U\hat{B}).$$

14.4.1 Additional Terminology and Applications

As may already be obvious, factor analysis uses quite a bit of unusual terminology.

The elements of a row of e are called *unique* or *specific factors*. These are uncorrelated random variables that are added to the linear combinations of common factors to generate the observations. They are distinct random variables for each distinct observation (i.e., they are specific to the observation). The ith diagonal element of $\Psi \equiv \text{Cov}(\varepsilon_i)$ is called the *uniqueness*, *specificity*, or *specific variance* of the ith variable.

The diagonal elements of Λ are called *communalities*. Writing $\Lambda = [\lambda_{ij}]$, the communality of the ith variable is generally denoted

$$h_i^2 \equiv \lambda_{ii}.$$

Note that if $B = [\beta_{ij}]$,

$$h_i^2 = \sum_{k=1}^{r} \beta_{ki}^2.$$

The total variance is

$$\text{tr}[\Sigma] = \text{tr}[\Lambda] + \text{tr}[\Psi].$$

The *total communality* is

$$v \equiv \text{tr}[\Lambda] = \sum_{i=1}^{q} h_i^2 = \sum_{i=1}^{q} \sum_{k=1}^{r} \beta_{ki}^2.$$

The matrix

$$\Lambda = \Sigma - \Psi = \text{Cov}(B'x)$$

is called the *reduced covariance matrix*, for obvious reasons. Often the observations are standardized so that Σ is actually a correlation matrix. If this has been done, Λ is sometimes called the reduced correlation matrix (even though it need not be a correlation matrix).

In practice, factor analysis is used primarily to obtain estimates of B. One then tries to interpret the estimated factor loadings in some way that makes sense relative to the subject matter of the data. As is discussed later, this is a fairly controversial procedure. One of the reasons for the controversy is that B is not uniquely defined. Given any orthonormal $r \times r$ matrix U, write $X_0 = XU'$ and $B_0 = UB$; then,

$$XB = XU'UB = X_0 B_0,$$

where X_0 again satisfies the assumptions made about X. Unlike standard linear models, X is not observed, so there is no way to tell X and X_0 apart. There is also no way to tell B and B_0 apart. Actually, this indeterminacy is used in factor analysis to increase the interpretability of B. This will be discussed again later. At the moment, we examine ways in which the matrix B is interpreted.

One of the key points in interpreting B is recognizing that it is the rows of B that are important and not the columns. A column of B is used to explain one dependent variable. A row of B consists of all of the coefficients that affect a single common factor. The q elements in the jth row of B represent the contributions made by the jth common factor to the q dependent variables. Traditionally, if a factor has all of its large loadings with the same sign, the subject matter specialist tries to identify some common attribute of the dependent variables that correspond to the high loadings. This common attribute is then considered to be the underlying factor. A *bipolar* factor involves high loadings that are both positive and negative; the user identifies common attributes for both the group of dependent variables with positive signs and the group with negative signs. The underlying factor is taken to be one that causes individuals who are high on some scores to be low on other scores. The following example involves estimated factor loadings. Estimation is discussed in the following two subsections.

EXAMPLE 14.4.1. Again consider the correlation matrix

$$R = \begin{bmatrix} 1.000 & 0.439 & 0.410 & 0.288 & 0.329 & 0.248 \\ 0.439 & 1.000 & 0.351 & 0.354 & 0.320 & 0.329 \\ 0.410 & 0.351 & 1.000 & 0.164 & 0.190 & 0.181 \\ 0.288 & 0.354 & 0.164 & 1.000 & 0.595 & 0.470 \\ 0.329 & 0.320 & 0.190 & 0.595 & 1.000 & 0.464 \\ 0.248 & 0.329 & 0.181 & 0.470 & 0.464 & 1.000 \end{bmatrix}$$

from Lawley and Maxwell (1971) and Johnson and Wichern (2007) that was obtained from (Gaelic, English, history, arithmetic, algebra, geometry) examination scores on 220 male students.

For $r = 2$, maximum likelihood estimation gives one choice of estimates,

$$\hat{B} = \begin{bmatrix} 0.553 & 0.568 & 0.392 & 0.740 & 0.724 & 0.595 \\ 0.429 & 0.288 & 0.450 & -0.273 & -0.211 & -0.132 \end{bmatrix}$$

and

$$(\hat{\psi}_1, \ldots, \hat{\psi}_6) = (0.510, 0.594, 0.644, 0.377, 0.431, 0.628).$$

Factor interpretation involves looking at the rows of $\hat{B}$ and trying to interpret them. Write

$$\hat{B} = \begin{bmatrix} \hat{b}'_1 \\ \hat{b}'_2 \end{bmatrix}.$$

All of the elements of $\hat{b}_1$ are large and fairly substantial. This suggests that the first factor is a factor that indicates general intelligence. The second factor is bipolar, with positive scores on math subjects and negative scores on nonmath subjects. The second factor might be classified as some sort of math–nonmath factor. This example will be examined again later with a slightly different slant. □

Rather than taking the factor analysis model as a serious model for the behavior of data, it may be more appropriate to view factor analysis as a data analytic procedure that seeks to discover structure in the covariance matrix and may *suggest* the presence of underlying factors. My son Fletcher (not to be confused with my imaginary son Basil) has convinced me that if you have a previous idea of the important factors it may be a worthwhile exercise to see whether the data are capable of being contorted into consistency with those previous factors.

14.4.2 Maximum Likelihood Theory

Maximum likelihood theory can be used for both estimation and testing. Maximum likelihood factor analysis is based on assuming that the random vectors in the factor model have a joint multivariate normal distribution and rewriting the factor analysis model as a standard multivariate linear model. (By contrast, ICA allows at most one of the factors to have a normal distribution.) To do this, the random terms are pooled together as, say

$$\xi_i = B'x_i + \varepsilon_i$$

and

$$\xi = XB + e.$$

With $\Lambda = B'B$, the factor analysis model is a special case of the one-sample model of Sect. 10.1,

$$Y = J\mu' + \xi, \tag{14.4.3}$$

where

$$E(\xi_i) = 0,$$
$$\text{Cov}(\xi_i, \xi_j) = 0 \qquad i \neq j,$$

and

$$\text{Cov}(\xi_i) = \Lambda + \Psi,$$

with Ψ diagonal, Λ nonnegative definite, and $r(\Lambda) = r$.

In Sect. 10.1, the assumption was simply that

$$\text{Cov}(\xi_i) = \Sigma.$$

The new model places the restriction on Σ that

$$\Sigma = \Lambda + \Psi, \tag{14.4.4}$$

where $r(\Lambda) = r$ and Ψ is diagonal. For ξ_is with a joint multivariate normal distribution, the likelihood function for an arbitrary Σ was discussed in Chap. 9. Clearly, the likelihood can be maximized subject to the restrictions that $\Sigma = \Lambda + \Psi$, Λ is nonnegative definite, $r(\Lambda) = r$, and Ψ is diagonal. However Seber (1984, Exercise 5.4) argues that even these parameters are not identifiable without additional restrictions.

Because $\Lambda + \Psi$ is just a particular choice of Σ, as in Chap. 9 the maximum likelihood estimate of μ is always the least squares estimate, $\hat{\mu} = \bar{y}_{..}$. This simplifies the maximization problem. Unfortunately, with the additional restrictions on Σ, closed-form estimates of the covariance matrix are no longer available. Computational methods for finding MLEs are discussed in Lawley and Maxwell (1971) and Jöreskog (1975). They can be quite difficult.

Exercise 14.4.

(a) Show that the maximization problem reduces to finding a rank r matrix $\hat{\Lambda}$ and a diagonal matrix $\hat{\Psi}$ that minimize

$$\log(|\Lambda + \Psi|) + \text{tr}\{(\Lambda + \Psi)^{-1}\hat{\Sigma}_q\},$$

where $\hat{\Sigma}_q = \dfrac{n-1}{n}S$ and $\hat{\Lambda}$ is nonnegative definite.

(b) Show that any positive definite Σ can be written with a factor analysis structure having $r = q - 1$. Hint: Look at $\Sigma = PD(\phi_i)P' = P[D(\phi_i) - \phi_q I + \phi_q I]P'$

One advantage of the maximum likelihood method is that standard asymptotic results apply. Maximum likelihood estimates are asymptotically normal. Minus two times the likelihood ratio test statistic is asymptotically chi-squared under the null hypothesis. See Geweke and Singleton (1980) for a discussion of sample size requirements for the asymptotic test.

Of specific interest are tests for examining the rank of Λ. If $r < s$, the restriction $r(\Lambda) = r$ is more stringent than the restriction $r(\Lambda) = s$. To test $H_0 : r(\Lambda) = r$ versus $H_A : r(\Lambda) = s$, one can use the likelihood ratio test statistic. This is just the maximum value of the likelihood under $r(\Lambda) = r$ divided by the maximum value of the likelihood under $r(\Lambda) = s$. Under H_0, -2 times the log of this ratio has an asymptotic chi-squared distribution. The degrees of freedom are the difference in the number of independent parameters for the models with $r(\Lambda) = s$ and $r(\Lambda) = r$. If we denote

$$\hat{\Sigma}_r = \hat{\Lambda} + \hat{\Psi}$$

when $r(\Lambda) = r$ with a similar notation for $r(\Lambda) = s$, -2 times the log of the likelihood ratio test statistic is easily shown to be

$$n\left[\ln\left(\frac{|\hat{\Sigma}_r|}{|\hat{\Sigma}_s|}\right) + \text{tr}\{(\hat{\Sigma}_r^{-1} - \hat{\Sigma}_s^{-1})\hat{\Sigma}_q\}\right],$$

where again

$$\hat{\Sigma}_q = \frac{n-1}{n}S.$$

As will be seen later, the degrees of freedom for the test are usually

if $s = q, q-1$	$df = q(q+1)/2 - q - [qr - r(r-1)/2]$,
if $s < q-1$	$df = [qs - s(s-1)/2] - [qr - r(r-1)/2].$

The formula for degrees of freedom is derived from the number of independent parameters in each model. If $r(\Lambda) \equiv r = q, q-1$, the covariance matrix Σ is unrestricted. The independent parameters are the q elements of μ and the $q(q+1)/2$ distinct elements of Σ. Recall that because Σ is symmetric, not all of its elements are distinct. Thus, for $r = q, q-1$, the model has

$$q + q(q+1)/2$$

degrees of freedom.

Counting the degrees of freedom when $r(\Lambda) = r < q - 1$ is a bit more compli-cated. The model involves the restriction

$$\Sigma = \Lambda + \Psi,$$

where Ψ is diagonal and Λ is nonnegative definite with rank r. Clearly, Ψ has q independent parameters, the diagonal elements. Because Λ is of rank r, then $q - r$ columns out of the $q \times q$ matrix are linear combinations of the other r columns. Thus, the independent parameters are at most the elements of these r columns. There are qr of these parameters. However, Λ is also symmetric. All of the parameters above the diagonal are redundant. In the first r columns there are $1 + 2 + \cdots + (r - 1) = r(r-1)/2$ of these redundant values. Thus, Λ has at most $qr - r(r-1)/2$ parameters. Finally, μ again involves q independent parameters. Adding the number of independent parameters in μ, Ψ, and Λ gives the maximum model degrees of freedom as

$$q + q + [qr - r(r-1)/2].$$

Taking differences in model degrees of freedom gives the test degrees of freedom indicated earlier. However, if r and q are both large, the number of factor model parameters can exceed the number of parameters for the unrestricted covariance matrix model, so an unrestricted covariance matrix should be used. In particular, if $r = q - 1$, the number of factor model parameters always exceeds the number of parameters in the unrestricted covariance matrix. Fortunately, r is usually taken to be small.

Thus far in the discussion, we have ignored B in favor of $\Lambda = B'B$. Given a function of Λ, say $B = f(\Lambda)$, and the maximum likelihood estimate $\hat{\Lambda}$, the MLE of B is $\hat{B} = f(\hat{\Lambda})$. The problem is in defining the function f. There are an uncountably infinite number of ways to define f. If f defines B and U is an orthonormal matrix, then

$$f_1(\Lambda) = U f(\Lambda)$$

is just as good a definition of B because $\Lambda = B'B = B'U'UB$. As mentioned, this indeterminacy is used to make the results more interpretable. The matrix B is rede-fined until the user gets a pleasing $\hat{B}$. The procedure starts with any $\hat{B}$ and then $\hat{B}$ is rotated (multiplied by an orthonormal matrix) until $\hat{B}$ seems to be interpretable to the user. In fact, there are some standard rotations (e.g., varimax and quartimax), that are often used to increase interpretability. For a more complete discussion of rotations see Williams (1979).

Often, in an effort to make B well defined, it is taken to be $D\left(\sqrt{\phi_1}, \ldots, \sqrt{\phi_r}\right) A'_r$, where $A_r = [a_1, \ldots, a_r]$ with a_i an eigenvector of Λ with length one corresponding to a positive eigenvalue ϕ_i. To accomplish the goal one would need to address the issues of eigenvalues not being unique and the nonidentifiability of Λ.

EXAMPLE 14.4.2.   For $r = 2$, the orthonormal matrices U used in rotations are 2×2 matrices. Thus, the effects of orthogonal rotations can be plotted. The plots consist of q points, one for each dependent variable. Each point consists of the two values in each column of $\hat{B}$. Figure 14.5 gives a plot of the unrotated factor loadings presented in Example 14.4.1 for the examination score data. The points labeled 1 through 6 indicate the corresponding dependent variable $h = 1, \ldots, 6$. Two commonly used rotations are the varimax rotation and the quartimax rotation (see Exercise 14.5.10). The *varimax* rotation for these data is

$$\hat{B}_V = \begin{bmatrix} 0.235 & 0.323 & 0.088 & 0.771 & 0.724 & 0.572 \\ 0.659 & 0.549 & 0.590 & 0.170 & 0.213 & 0.210 \end{bmatrix},$$

and the *quartimax* rotation is

$$\hat{B}_Q = \begin{bmatrix} 0.260 & 0.344 & 0.111 & 0.777 & 0.731 & 0.580 \\ 0.650 & 0.536 & 0.587 & 0.139 & 0.184 & 0.188 \end{bmatrix}.$$

A plot of the varimax factor loadings is presented in Fig. 14.6. It is a substantial counterclockwise rotation about the origin (0,0) of Fig. 14.5. A plot (not given) of the quartimax loadings is a very slight clockwise rotation of the varimax loadings. Rather than isolating a general intelligence factor and a bipolar factor as seen in the unrotated factors, these both identify factors that can be interpreted as one for mathematics ability and one for nonmathematics ability. ($\hat{B}_V$ has changed slightly since *ALM-II* with different software.) □

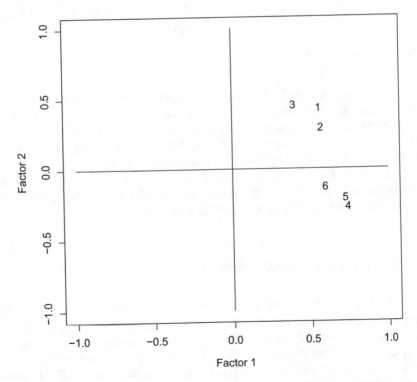

Fig. 14.5 Unrotated factor loadings

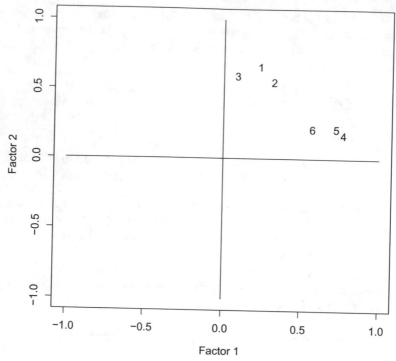

Fig. 14.6 Varimax factor loadings

The factor analysis model for maximum likelihood assumes that the matrix of common factors X has rows consisting of independent observations from a multivariate normal distribution with mean zero and covariance matrix I_r. While X is not observable, it is possible to predict the rows of X. In Exercise 14.5, it will be seen that $\hat{E}(x_i|Y) = B(\Lambda + \Psi)^{-1}(y_i - \mu)$. Thus, estimated best linear predictors of the x_is can be obtained. These *factor scores* are frequently used to check the assumption of multivariate normality. Bivariate plots can be examined for elliptical shapes and outliers. Univariate plots can be checked for normality.

14.4.3 Principal Factor Estimation

It would be nice to have a method for estimating the parameters of model (14.4.3) that did not depend on the assumption of normality. Thurstone (1931) and Thompson (1934) have proposed *principal (axes) factor estimation* as such a method. The parameters to be estimated are μ, Λ, and Ψ. As mentioned earlier, model (14.4.3) is just a standard multivariate linear model with a peculiar choice for Σ. The results of Sect. 9.2 imply that $\bar{y}.$ is the best linear unbiased estimate of μ.

It remains to estimate Λ and Ψ. If Ψ is known, estimation of Λ is easy. Using Eq. (14.4.4)

$$\Lambda = \Sigma - \Psi,$$

where Λ is assumed to be nonnegative definite of rank r. If it were not for the rank condition, a natural estimate would be

$$\tilde{\Lambda} \equiv S - \Psi.$$

Incorporating the rank condition, one natural way to proceed is to choose a nonnegative definite matrix of rank r, say $\hat{\Lambda}$, that minimizes, say,

$$\mathrm{tr}\{(S - \Psi) - \Lambda\}.$$

Although other functions of $(S - \Psi) - \Lambda$ might be reasonable, the trace is a convenient choice because we have already solved a version of this problem.

Let $\phi_1 \geq \cdots \geq \phi_q$ be the eigenvalues of S, let $a_1, \ldots, a_q$ be the corresponding eigenvectors, let $A_r = [a_1, \ldots, a_r]$, and let G be a $q \times r$ matrix of rank r. In our discussion of principal components, we established that

$$\mathrm{tr}[S - SA_r(A_r'SA_r)^{-1}A_r'S] = \min_G \mathrm{tr}[S - SG(G'SG)^{-1}G'S].$$

Clearly, $SG(G'SG)^{-1}G'S$ is nonnegative definite of rank r. If we consider the problem of estimating Σ when $r(\Sigma) = r$ and restrict ourselves to the class of estimates $SG(G'SG)^{-1}G'S$, then the matrix $SA_r(A_r'SA_r)^{-1}A_r'S$ is an optimal rank r estimate of S.

Applying this result in the factor analysis problem gives an optimal estimate

$$\hat{\Lambda} = \tilde{\Lambda}A_r(A_r'\tilde{\Lambda}A_r)^{-1}A_r'\tilde{\Lambda},$$

where A_r consists of eigenvectors of $\tilde{\Lambda} = S - \Psi$. If we choose the eigenvectors so that $A_r'A_r = I_r$, $\hat{\Lambda}$ simplifies to

$$\hat{\Lambda} = A_r D(\phi_1, \ldots, \phi_r)A_r',$$

where $\phi_1, \ldots, \phi_r$ are the r largest eigenvalues of $\tilde{\Lambda}$. An obvious estimate of B is

$$\hat{B} = D(\sqrt{\phi_1}, \ldots, \sqrt{\phi_r})A_r'.$$

Of course, any rotation of $\hat{B}$ is an equally appropriate estimate.

All of this assumes that Ψ is known. In practice, one makes an initial guess Ψ_0 that leads to initial estimates $\tilde{\Lambda}_0 = S - \Psi_0$ and $\hat{\Lambda}_0$. Having computed $\hat{\Lambda}_0$, compute Ψ_1 from the diagonal elements of $S - \hat{\Lambda}_0$ and repeat the process to obtain $\hat{\Lambda}_1$. This iterative procedure can be repeated until convergence. A common choice for $\Psi_0 = D(\psi_{i0})$ is

$$\psi_{i0} = 1/s^{ii},$$

where s^{ii} is the ith diagonal element of S^{-1}.

Another common choice for Ψ_0 is taking $\psi_{i0} = 0$ for all i. This choice yields $\tilde{\Lambda} = S$, and the rows of $\hat{B} = D(\sqrt{\phi_i})A'_r$ are eigenvectors of S. These are the same vectors as used to determine principal components. In fact, principal components are often used to address questions about underlying factors. The difference is that in a principal component analysis the elements of the eigenvector determine a linear combination of the dependent variable y. In the factor analysis model, the elements of an eigenvector, say a_1, are the q coefficients applied to the first hypothetical factor. Although factor interpretations are based on these q values, in the factor analysis model, data are generated using the r values $a_{1h}, \ldots, a_{rh}$ taken *across* eigenvectors.

Some of the problems with principal factor estimation are that r is assumed to be known, there are no tests available for the value of r, and the matrix $S - \Psi$ may not be nonnegative definite.

In our examination of principal components, we found that the eigenvectors of Σ provided solutions to several different problems: sequential prediction, joint prediction, sequential variance maximization, and geometrical interpretation. The principal factor estimation method can also be motivated by a sequential optimization problem, see Gnanadesikan (1977).

14.4.4 Computing

For the previous edition, I think I did this analysis in an old version of Minitab. I repeated the analysis in Minitab 18, in R's factanal and in the R library psych's program fa. R's factanal fits using only maximum likelihood, Minitab allows both methods, and fa allows these two and several more. Between these various programs, the third digit of the loadings often differed by 1.

As for rotations, R's factanal allows none and varimax as well as promax which is a transformation but not a rotation. Minitab allows none, varimax, quartimax, equimax, and a family of rotations called orthomax. Psych's fa allows none, varimax, quartimax, equamax, and four others as well as promax and six more transformations that are not rotations.

14.4.5 Discussion

There is no question that model (14.4.3) is a reasonable model. There is considerable controversy about whether the factor analysis model (14.4.1) has any meaning beyond that of model (14.4.3).

Factor analysis is a frequently used methodology. Obviously, its users like it. Users like to rotate the estimated factor loadings $\hat{B}$ and interpret their results. On the other hand, many people, often of a more theoretical bent, are deeply disturbed by the indeterminacy of the factors and the factor loadings. Many people claim it is impossible to understand the nature of the underlying factors and the basis of their interpretation. Personally, I have always tried to straddle this particular fence. There are people on both sides that I respect. (OK, as I have gotten older, I've fallen on the "disturbed by the indeterminacy" side of the fence.)

An important criterion for evaluating models is that if a model is useful it should be useful for making predictions about future observables. The maximum likelihood model (14.4.3), like all linear models, satisfies this criterion. The prediction of a new case would be $\bar{y}_{\cdot\cdot}$. The peculiar covariance matrix of model (14.4.3) plays a key role in predicting the unobserved elements of a new case when some of the elements have been observed.

The factor analysis model (14.4.1) looks like it is more than the corresponding linear model. The interpretation of factor loadings depends on (14.4.1) being more than the linear model. If the factor analysis model really is more than the linear model, it should provide predictions that are distinct from the linear model. When the factor analysis model is correct, these predictions should be better than the linear model predictions.

Unfortunately, the factor analysis model does not seem to lend itself to prediction except through the corresponding linear model. One can predict the factor vectors x_i (assuming that μ, B, and Ψ are known), but this does not affect prediction of y_i.

Exercise 14.5. Show that
(a) $\hat{E}(x_i|Y) = \hat{E}(x_i|y_i) = B(\Lambda + \Psi)^{-1}(y_i - \mu)$.
(b) $\hat{E}(\mu + B'x_i + \varepsilon_i|y_i) = y_i$.
Do not use the fact that $\hat{E}(y_i|y_i) = y_i$.

Though the factor analysis model may not hold up to careful scrutiny, it does not follow that the data-analytic method known as factor analysis is a worthless endeavor. Rather than thinking of factor analysis as a theoretical method of estimating the loadings on some unspecified factors, it may be better to think of it as a data-analytic method for identifying structure in the covariance matrix. As a data-analytic method, it is neither surprising nor disconcerting that different people (using different rotations) obtain different results. It is more important whether, in practice, users working on similar problems often obtain similar results.

The factor analysis model is one motivation for this method of data analysis. We now will present a slightly different view. We begin by decomposing the covariance matrix into the sum of r different covariance matrices plus Ψ. In other words, write

$$B = \begin{bmatrix} b'_1 \\ \vdots \\ b'_r \end{bmatrix}$$

and

$$\Lambda_i = b_i b'_i .$$

Thus,

$$\begin{aligned}
\Sigma &= \Lambda + \Psi \\
&= B'B + \Psi \\
&= \sum_{i=1}^{r} b_i b_i' + \Psi \\
&= \sum_{i=1}^{r} \Lambda_i + \Psi.
\end{aligned}$$

We can think of y as being a random observation vector and Λ_i as being the covariance matrix for some factor, say w_i, where $y = \mu + \sum_{i=1}^{r} w_i + \varepsilon$ with $\text{Cov}(w_i, w_j) = 0$, $\text{Cov}(\varepsilon) = \Psi$, and $\text{Cov}(w_i, \varepsilon) = 0$. In the usual factor analysis model with factors $x = (x_1, \dots, x_r)'$ and $B' = [b_1, \dots, b_r]$, we have $w_i = x_i b_i$. The question then becomes what kind of underlying factor w_i would generate a covariance matrix such as Λ_i. The advantage to this point of view is that attention is directed towards explaining the observable correlations. In traditional factor analysis, attention is directed towards estimating the ill-defined factor loadings. Of course, the end result is the same.

Just as the matrix B is not unique, neither is the decomposition

$$\Lambda = \sum_{i=1}^{r} \Lambda_i.$$

In practice, one would rotate B to make the Λ_is more interpretable. Moreover, as will be seen later, one need not actually compute Λ_i to discover its important structure. The key features of Λ_i are obvious from examination of b_i.

EXAMPLE 14.4.3.   Using $\hat{B}$ from Example 14.4.1

$$\hat{\Lambda}_1 = \hat{b}_1 \hat{b}_1' = \begin{bmatrix}
0.31 & 0.31 & 0.22 & 0.41 & 0.40 & 0.33 \\
0.31 & 0.32 & 0.22 & 0.42 & 0.41 & 0.34 \\
0.22 & 0.22 & 0.15 & 0.29 & 0.28 & 0.23 \\
0.41 & 0.42 & 0.29 & 0.55 & 0.54 & 0.44 \\
0.40 & 0.41 & 0.28 & 0.44 & 0.52 & 0.43 \\
0.33 & 0.34 & 0.23 & 0.44 & 0.43 & 0.35
\end{bmatrix}.$$

All of the variances and covariances are uniformly high because all of the elements of b_1 are uniformly high. The factor w_1 must be some kind of overall measure—call it general intelligence.

The examination of the second covariance matrix

$$\hat{\Lambda}_2 = \begin{bmatrix}
0.18 & 0.12 & 0.19 & -0.12 & -0.09 & -0.07 \\
0.12 & 0.08 & 0.13 & -0.08 & -0.06 & -0.04 \\
0.19 & 0.13 & 0.20 & -0.12 & -0.09 & -0.06 \\
-0.12 & -0.08 & -0.12 & 0.07 & 0.06 & 0.04 \\
-0.09 & -0.06 & -0.09 & 0.06 & 0.04 & 0.03 \\
-0.07 & -0.04 & -0.06 & 0.04 & 0.03 & 0.02
\end{bmatrix}$$

is trickier. The factor w_2 has two parts; there is positive correlation among the first three variables: Gaelic, English, and history. There is positive correlation among the last three variables: arithmetic, algebra, and geometry. However, the first three variables are negatively correlated with the last three variables. Thus, w_2 can be interpreted as a math factor and a nonmath factor that are negatively correlated.

A totally different approach to dealing with Λ_2 is to decide that any variable with a *variance* less than, say, 0.09 is essentially constant. This leads to

$$\tilde{\Lambda}_2 = \begin{bmatrix} 0.18 & 0 & 0.19 & 0 & 0 & 0 \\ 0 & 0 & 0 & 0 & 0 & 0 \\ 0.19 & 0 & 0.20 & 0 & 0 & 0 \\ 0 & 0 & 0 & 0 & 0 & 0 \\ 0 & 0 & 0 & 0 & 0 & 0 \\ 0 & 0 & 0 & 0 & 0 & 0 \end{bmatrix}.$$

Thus, the second factor puts weight on only Gaelic and history. The second factor would then be interpreted as some attribute that only Gaelic and history have in common.

Either analysis of Λ_2 can be arrived at by direct examination of

$$b_2' = (-0.429, -0.288, -0.450, 0.273, 0.211, 0.132).$$

The pattern of positives and negatives determines the corresponding pattern in Λ_2. Similarly, the requirement that a variance be greater than 0.09 to be considered nonzero corresponds to a variable having an absolute factor loading greater than 0.3. Only Gaelic and history have factor loadings with absolute values greater than 0.3. Both are negative, so Λ_i will display the positive correlation between them. □

The examination of underlying factors is, of necessity, a very slippery enterprize. The parts of factor analysis that are consistent with traditional ideas of modeling are estimation of Λ and Ψ and the determination of the rank of Λ. The rest is pure data analysis. It is impossible to prove that underlying factors actually exist. The argument in factor analysis is that if these factors existed they could help explain the data.

Factor analysis can only suggest that certain factors might exist. The appropriate question is not whether they really exist but whether their existence is a useful idea. For example, does the idea of a factor for general intelligence help people to understand the nature of test scores. A more stringent test of usefulness is whether the idea of a general intelligence factor leads to accurate predictions about future observable events. Recall from Exercise 14.5 that one can predict factor scores, so those predictions can be used as a tool in making predictions about future observables for the individuals in the study.

An interesting if unrelated example of these criteria for usefulness involves the force of gravity. For most of us, it is impossible to prove that such a force exists. However, the idea of this force allows one to both explain and predict the behavior of physical objects. The fact that accurate predictions can be made does not prove that gravity exists. If an idea explains current data in an intelligible manner and/or allows accurate prediction, it is a useful idea. For example, the usefulness of Newton's laws of motion cannot be disregarded just because they break down for speeds approaching that of light.

14.5 Additional Exercises

Exercise 14.5.1.

(a) Find the vector b that minimizes

$$\sum_{i=1}^{q} \left[y_i - \mu_i - b'(x - \mu_x) \right]^2.$$

(b) For given weights w_i, $i = 1, \ldots, q$, find the vector b that minimizes

$$\sum_{i=1}^{q} w_i^2 \left[y_i - \mu_i - b'(x - \mu_x) \right]^2.$$

(c) Find the vectors b_i that minimize

$$\sum_{i=1}^{q} w_i^2 \left[y_i - \mu_i - b_i'(x - \mu_x) \right]^2.$$

Exercise 14.5.2.   In a population of large industrial corporations, the covariance matrix for $y_1 = assets/10^6$ and $y_2 = net\ income/10^6$ is

$$\Sigma = \begin{bmatrix} 75 & 5 \\ 5 & 1 \end{bmatrix}.$$

(a) Determine the principal components.

(b) What proportion of the total prediction variance is explained by $a_1'y$?
(c) Interpret $a_1'y$.
(d) Repeat (a), (b), and (c) for principal components based on the correlation matrix.

Exercise 14.5.3.   What are the principal components associated with

$$\Sigma = \begin{bmatrix} 5 & 0 & 0 & 0 \\ 0 & 3 & 0 & 0 \\ 0 & 0 & 3 & 0 \\ 0 & 0 & 0 & 2 \end{bmatrix} ?$$

Discuss the problem of reducing the variables to a two-dimensional space.

Exercise 14.5.4. Let $v_1 = (2,1,1,0)'$, $v_2 = (0,1,-1,0)'$, $v_3 = (0,0,0,2)'$, and

$$\Sigma = \sum_{i=1}^{3} v_i v_i'.$$

(a) Find the principal components of Σ.

(b) What is the predictive variance of each principal component? What percentage of the maximum prediction error is accounted for by the first two principal components?

(c) Interpret the principal components.

(d) What are the correlations between the principal components and the original variables?

Exercise 14.5.5. Do a principal components analysis of the female turtle carapace data of Exercise 10.6.1.

Exercise 14.5.6. The data in Table 14.1 are a subset of the Chapman data reported by Dixon and Massey (1983). It contains the age, systolic blood pressure, diastolic blood pressure, cholesterol, height, and weight for a group of men in the Los Angeles Heart Study. Do a principal components analysis of the data.

Table 14.1 Chapman data

Age	Sbp	Dbp	Chol	Ht	Wt	Age	Sbp	Dbp	Chol	Ht	Wt
44	124	80	254	70	190	37	110	70	312	71	170
35	110	70	240	73	216	33	132	90	302	69	161
41	114	80	279	68	178	41	112	80	394	69	167
31	100	80	284	68	149	38	114	70	358	69	198
61	190	110	315	68	182	52	100	78	336	70	162
61	130	88	250	70	185	31	114	80	251	71	150
44	130	94	298	68	161	44	110	80	322	68	196
58	110	74	384	67	175	31	108	70	281	67	130
52	120	80	310	66	144	40	110	74	336	68	166
52	120	80	337	67	130	36	110	80	314	73	178
52	130	80	367	69	162	42	136	82	383	69	187
40	120	90	273	68	175	28	124	82	360	67	148
49	130	75	273	66	155	40	120	85	369	71	180
34	120	80	314	74	156	40	150	100	333	70	172
37	115	70	243	65	151	35	100	70	253	68	141
63	140	90	341	74	168	32	120	80	268	68	176
28	138	80	245	70	185	31	110	80	257	71	154
40	115	82	302	69	225	52	130	90	474	69	145
51	148	110	302	69	247	45	110	80	391	69	159
33	120	70	386	66	146	39	106	80	248	67	181

Exercise 14.5.7.   Assume a two-factor model with

$$\Sigma = \begin{bmatrix} 0.15 & 0.00 & 0.05 \\ 0.00 & 0.20 & -0.01 \\ 0.05 & -0.01 & 0.05 \end{bmatrix}$$

and

$$B = \begin{bmatrix} 0.3 & 0.2 & 0.1 \\ 0.2 & -0.3 & 0.1 \end{bmatrix}.$$

What is Ψ? What are the communalities?

Exercise 14.5.8.   Using the vectors v_1 and v_2 from Exercise 14.5.4, let

$$\Lambda = v_1 v_1' + v_2 v_2'.$$

Give the eigenvector solution for B and another set of loadings that generates Λ.

Exercise 14.5.9.   Given that

$$\Sigma = \begin{bmatrix} 1.00 & 0.30 & 0.09 \\ 0.30 & 1.00 & 0.30 \\ 0.09 & 0.30 & 1.00 \end{bmatrix}$$

and

$$\Psi = D(0.1, 0.2, 0.3),$$

find Λ and two choices of B.

Exercise 14.5.10.   Find definitions for the well-known factor loading matrix rotations varimax, direct quartimin, quartimax, equamax, and orthoblique. What is each rotation specifically designed to accomplish? Apply each rotation to the covariance matrices of Exercise 14.5.9.

Exercise 14.5.11.   Do a factor analysis of the female turtle carapace data of Exercise 10.6.1. Include tests for the numbers of factors and examine various factor-loading rotations.

Exercise 14.5.12.   Do a factor analysis of the Chapman data discussed in Exercise 14.5.6.

Exercise 14.5.13.   Show the following determinant equality.

$$|\Psi + BB'| = |I + B'\Psi^{-1}B||\Psi|.$$

Exercise 14.5.14. Find the likelihood ratio test for

$$H_0 : \Sigma = \sigma^2 \left[(1 - \rho)I + \rho JJ' \right]$$

against the general alternative.

References

Christensen, R. (1997). *Log-linear models and logistic regression* (2nd ed.). New York: Springer.

Christensen, R. (2011). *Plane answers to complex questions: The theory of linear models* (4th ed.). New York: Springer.

Christensen, R. (2014). Comment. *The American Statistician, 68*, 13–17.

Dixon, W. J., & Massey Jr., F. J. (1983). *Introduction to statistical analysis*. New York: McGraw-Hill.

Geweke, J. F., & Singleton, K. J. (1980). Interpreting the likelihood ratio statistic in factor models when sample size is small. *Journal of the American Statistical Association, 75*, 133–137.

Gnanadesikan, R. (1977). *Methods for statistical data analysis of multivariate observations*. New York: Wiley.

Hotelling, H. (1933). Analysis of a complex of statistical variables into principal components. *Journal of Educational Psychology, 24*, 417-441, 498–520.

Hyvärinen, A., Karhunen, J. & Oja, E. (2001). *Independent component analysis*. New York: Wiley.

Johnson, R. A., & Wichern, D. W. (2007). *Applied multivariate statistical analysis* (6th ed.). Englewood Cliffs: Prentice–Hall.

Jolicoeur, P., & Mosimann, J. E. (1960). Size and shape variation on the painted turtle: A principal component analysis. *Growth, 24*, 339–354.

Jolliffe, I. T. (1986). *Principal component analysis*. New York: Springer.

Jöreskog, K. G. (1975). Factor analysis by least squares and maximum likelihood. In K. Enslein, A. Ralston, & H. S. Wilf (Eds.) *Statistical methods for digital computers*. New York: Wiley.

Lawley, D. N., & Maxwell, A. E. (1971). *Factor analysis as a statistical methodology* (2nd ed.). New York: American Elsevier.

Li, G. & Chen, Z. (1985). Projection pursuit approach to robust dispersion matrices and principal components: Primary theory and Monte Carlo. *Journal of the American Statistical Association, 80*, 759–766.

Morrison, D. F. (2004). *Multivariate statistical methods* (4th ed.). Pacific Grove, CA: Duxbury Press.

Okamoto, M. & Kanazawa, M. (1968). Minimization of eigenvalues of a matrix and optimality of principal components. *Annals of Mathematical Statistics, 39*, 859–863.

Pearson, K. (1901). On lines and planes of closest fit to systems of points in space. *Philosophical Magazine, 6*(2), 559–572.

Rao, C. R. (1973). *Linear statistical inference and its applications* (2nd ed.). New York: Wiley.

Schervish, M. J. (1986). *A predictive derivation of principal components.* Technical Report 378, Department of Statistics, Carnegie-Mellon University, Pittsburgh, PA.

Seber, G. A. F. (1984). *Multivariate observations.* New York: Wiley.

Thompson, G. H. (1934). Hotelling's method modified to give Spearman's g. *Journal of Educational Psychology, 25,* 366–374.

Thurstone, L. L. (1931). Multiple factor analysis. *Psychological Review, 38,* 406–427.

Williams, J. S. (1979). A synthetic basis for comprehensive factor-analysis theory. *Biometrics, 35,* 719–733.

Appendix A
Mathematical Background

Abstract This appendix reviews two areas of mathematics that arise in dealing with dependent data: differentiation and the use of Vec operators with Kronecker products. It also examines quadratic optimization problems and their connection to support vector machines.

A.1 Differentiation

In *PA* I managed to avoid using very much calculus. There is no avoiding calculus when dealing with dependent data.

If F is a function from $\mathbf{R}^s$ into $\mathbf{R}^t$ with $F(x) = [f_1(x), \ldots, f_t(x)]'$, then the derivative of F at c is the $t \times s$ matrix of partial derivatives,

$$\mathbf{d}_x F(x)|_{x=c} \equiv [\partial f_i(x)/\partial x_j|_{x=c}].$$

When the context is clear, we often use simpler notations such as

$$\mathbf{d}_x F(x)|_{x=c} \equiv \mathbf{d}_x F(c) \equiv \mathbf{d}F(c).$$

Critical points are points c where $\mathbf{d}F(c) = 0$.

A first order Taylor's expansion of F around the point c is

$$F(x) \doteq F(c) + [\mathbf{d}F(c)](x - c)$$

or, to be more mathematically precise,

$$F(x) = F(c) + [\mathbf{d}F(c)](x - c) + o(\|x - c\|),$$

where $\|x - c\|^2 = (x - c)'(x - c)$ and for scalars $a_n \to 0$, $o(a_n)$ has the property that $o(a_n)/a_n \to 0$.

© Springer Nature Switzerland AG 2019
R. Christensen, *Advanced Linear Modeling*, Springer Texts in Statistics,
https://doi.org/10.1007/978-3-030-29164-8

In fact, the first order Taylor's expansion is essentially the mathematical definition of a derivative. The technical definition of a derivative, if it exists, is that it is some $t \times s$ matrix $dF(c)$ such that for any $\varepsilon > 0$, there exists a $\delta > 0$ for which any x with $\|x - c\| < \delta$ has

$$\|F(x) \quad F(c) - [\mathbf{d}F(c)](x - c)\| < \varepsilon \|x - c\|.$$

In other words, the linear function of x defined by $[\mathbf{d}F(c)](x - c)$ is a good approximation to the curved function $F(x) - F(c)$ in neighborhoods of c. Under suitable conditions, the matrix $\mathbf{d}F(c)$ is unique and equals the matrix of partial derivatives of F evaluated at the vector c.

In particular, if g maps $\mathbf{R}^s$ into $\mathbf{R}$, then $\mathbf{d}g(x)$ is a $1 \times s$ *row* vector. In this case we can also define the second derivative matrix of g at c as

$$\mathbf{d}^2_{xx}g(c) \equiv \mathbf{d}^2_{xx}g(x)|_{x=c} \equiv \mathbf{d}_x \left\{ [\mathbf{d}_x g(x)]' \right\}|_{x=c} = [\partial^2 g(x)/\partial x_i \partial x_j|_{x=c}],$$

which is an $s \times s$ matrix. Taylor's second order expansion can be written

$$g(x) \doteq g(c) + [\mathbf{d}g(c)](x - c) + (x - c)'[\mathbf{d}^2 g(c)](x - c)/2$$

or, to be more mathematically precise,

$$g(x) = g(c) + [\mathbf{d}g(c)](x - c) + (x - c)'[\mathbf{d}^2 g(c)](x - c)/2 + o(\|x - c\|^2).$$

First and second order Taylor's expansions are fundamental to the models used in response surface methodology, cf. http://www.stat.unm.edu/~fletcher/TopicsInDesign or *ALM-II*.

The chain rule can be written as a matrix product. If $f : \mathbf{R}^s \to \mathbf{R}^t$ and $g : \mathbf{R}^t \to \mathbf{R}^n$, then the composite function is defined by

$$(g \circ f)(x) \equiv g[f(x)]$$

and its derivative is an $n \times s$ matrix that satisfies

$$\mathbf{d}(g \circ f)(c) = [\mathbf{d}_v g(v)|_{v=f(c)}][\mathbf{d}_x f(x)|_{x=c}] \equiv \mathbf{d}g[f(c)]\mathbf{d}f(c).$$

Since we are examining linear models, we will be particularly interested in the derivatives of linear functions and quadratic functions.

Proposition A.1.   Let A be a fixed $t \times s$ matrix with $t = s$ in part (b).
(a) $\mathbf{d}_x[Ax] = A$.
(b) $\mathbf{d}_x[x'Ax] = 2x'A$.

PROOF.   (a) is proven by writing each element of Ax as a sum and taking partial derivatives. (b) is proven by writing $x'Ax$ as a double sum and taking partial derivatives. $\square$

Let $A(u) = [a_{ij}(u)]$ be an $t \times s$ matrix that is a function of a scalar u. We define the derivative of $A(u)$ with respect to u as the matrix of derivatives for its individual entries, i.e.,

$$\mathbf{d}_u A(u) \equiv [\mathbf{d}_u a_{ij}(u)].$$

Functions like $A(u)$, from $\mathbf{R}$ into a matrix, do not fit into our definition of derivatives, however $A(u)x$ and $x'A(u)x$ are functions of u from $\mathbf{R}$ into $\mathbf{R}^t$ and $\mathbf{R}$, respectively, and it is easy to see that

$$\mathbf{d}_u[A(u)x] = [\mathbf{d}_u A(u)]x \quad \text{and that} \quad \mathbf{d}_u[x'A(u)x] = x'[\mathbf{d}_u A(u)]x.$$

We now present some useful rules for matrix derivatives.

Proposition A.2.   For (c) and (d), $t = s$.

(a) A form of the product rule holds for conformable matrices $A(u)$ and $B(u)$,

$$\mathbf{d}_u[A(u)B(u)] = [\mathbf{d}_u A(u)]B(u) + A(u)[\mathbf{d}_u B(u)].$$

(b) When B and C are fixed matrices of conformable sizes,

$$\mathbf{d}_u[CA(u)B] = C[\mathbf{d}_u A(u)]B.$$

(c) The derivative of an inverse is

$$\mathbf{d}_u A^{-1}(u) = -A^{-1}(u)[\mathbf{d}_u A(u)]A^{-1}(u).$$

(d) The derivative of a trace is

$$\mathbf{d}_u\{\mathrm{tr}[A(u)]\} = \mathrm{tr}[\mathbf{d}_u A(u)].$$

(e) If $V(u)$ is positive definite for all u,

$$\mathbf{d}_u \log\{\det[V(u)]\} = \mathrm{tr}\{V^{-1}(u)[\mathbf{d}_u V(u)]\}.$$

The notations $\det[V]$ and $|V|$ are used interchangeably to indicate the determinant.

PROOF.   See Exercise A.1 for a proof of the proposition.   □

Exercise A.1.   Prove Proposition A.2. Hints: For (a), consider $A(u)B(u)$ elementwise. For (b), use (a) twice. For (c), use (a) and the fact that $0 = \mathbf{d}_u I = \mathbf{d}[A(u)A^{-1}(u)]$. For (d) use the fact that the trace is a linear function of the diagonal elements. For (e), write $V = P\mathrm{Diag}(\phi_i)P'$ and show that both sides equal

$$\sum_{i=1}^q \mathbf{d}_u \phi_i(u) \frac{1}{\phi_i(u)}.$$

For the right-hand side, use (a) and the fact that $0 = \mathbf{d}_u I = \mathbf{d}_u PP'$.

A.2 Vec Operators and Kronecker Products

Kronecker products and Vec operators are extremely useful in multivariate analysis and some approaches to variance component estimation. They are also often used in writing balanced ANOVA models. We present their definitions and basic algebraic properties.

We begin with definitions. Let A be an $r \times c$ matrix. Write $A = [A_1, A_2, \ldots, A_c]$, where A_i is the ith column of A. Then the *Vec* operator stacks the columns of A into an $rc \times 1$ vector; thus,

$$[\text{Vec}(A)]' = [A_1', A_2', \ldots, A_c'].$$

Let $A = [a_{ij}]$ be an $r \times c$ matrix and $B = [b_{ij}]$ be an $s \times d$ matrix. The *Kronecker product* of A and B, written $A \otimes B$, is an $r \times c$ matrix of $s \times d$ matrices. The matrix in the ith row and jth column is $a_{ij}B$. In total, $A \otimes B$ is an $rs \times cd$ matrix.

EXAMPLE A.2.1.   If

$$A = \begin{bmatrix} 1 & 4 \\ 2 & 5 \end{bmatrix}, \quad B = \begin{bmatrix} 1 & 3 \\ 0 & 4 \end{bmatrix},$$

then

$$\text{Vec}(A) = [1, 2, 4, 5]', \quad \text{Vec}(B) = [1, 0, 3, 4]',$$

and

$$A \otimes B = \begin{bmatrix} 1\begin{pmatrix} 1 & 3 \\ 0 & 4 \end{pmatrix} & 4\begin{pmatrix} 1 & 3 \\ 0 & 4 \end{pmatrix} \\ 2\begin{pmatrix} 1 & 3 \\ 0 & 4 \end{pmatrix} & 5\begin{pmatrix} 1 & 3 \\ 0 & 4 \end{pmatrix} \end{bmatrix} = \begin{bmatrix} 1 & 3 & 4 & 12 \\ 0 & 4 & 0 & 16 \\ 2 & 6 & 5 & 15 \\ 0 & 8 & 0 & 20 \end{bmatrix}. \qquad \square$$

The algebraic properties are as follows:

1. If the matrices are of conformable sizes, $[A \otimes (B+C)] = [A \otimes B] + [A \otimes C]$.
2. If the matrices are of conformable sizes, $[(A+B) \otimes C] = [A \otimes C] + [B \otimes C]$.
3. If a and b are scalars, $ab[A \otimes B] = [aA \otimes bB]$.
4. If the matrices are of conformable sizes, $[A \otimes B][C \otimes D] = [AC \otimes BD]$.
5. The transpose of a Kronecker product matrix is $[A \otimes B]' = [A' \otimes B']$.
6. The generalized inverse of a Kronecker product matrix is $[A \otimes B]^- = [A^- \otimes B^-]$.
7. For two vectors v and w, $\text{Vec}(vw') = w \otimes v$.
8. For a matrix W and conformable matrices A and B, $\text{Vec}(AWB')=[B \otimes A]\text{Vec}(W)$.
9. For conformable matrices A and B, $\text{Vec}(A)'\text{Vec}(B) = \text{tr}(A'B)$.
10. The Vec operator commutes with any matrix operation that is performed elementwise. For example, $E\{\text{Vec}(W)\} = \text{Vec}\{E(W)\}$ when W is a random matrix. Similarly, for conformable matrices A and B and scalar ϕ, $\text{Vec}(A + B) = \text{Vec}(A) + \text{Vec}(B)$ and $\text{Vec}(\phi A) = \phi \text{Vec}(A)$.
11. If A and B are positive definite, then $A \otimes B$ is positive definite.

Most of these are well-known facts and easy to establish. Two of them are somewhat more unusual, and we present proofs.

ITEM 8. We show that for a matrix W and conformable matrices A and B, $\text{Vec}(AWB') = [B \otimes A]\text{Vec}(W)$. First note that if $\text{Vec}(AW) = [I \otimes A]\text{Vec}(W)$ and $\text{Vec}(WB') = [B \otimes I]\text{Vec}(W)$, then $\text{Vec}(AWB') = [I \otimes A]\text{Vec}(WB') = [I \otimes A][B \otimes I]\text{Vec}(W) = [B \otimes A]\text{Vec}(W)$.

To see that $\text{Vec}(AW) = [I \otimes A]\text{Vec}(W)$, let W be $r \times s$ and write W in terms of its columns $W = [w_1, \ldots, w_s]$. Then $AW = [Aw_1, \ldots, Aw_s]$ and $\text{Vec}(AW)$ stacks the columns $Aw_1, \ldots, Aw_s$. On the other hand,

$$[I \otimes A]\text{Vec}(W) = \begin{bmatrix} A & & 0 \\ & \ddots & \\ 0 & & A \end{bmatrix} \begin{bmatrix} w_1 \\ \vdots \\ w_s \end{bmatrix} = \begin{bmatrix} Aw_1 \\ \vdots \\ Aw_s \end{bmatrix}.$$

To see that $\text{Vec}(WB') = [B \otimes I]\text{Vec}(W)$, take W as above and write $B_{m \times s} = [b_{ij}]$ with *rows* $b_1', \ldots, b_m'$. First note that $WB' = [Wb_1, \ldots, Wb_m]$, so $\text{Vec}(WB')$ stacks the columns $Wb_1, \ldots, Wb_m$. Now observe that

$$[B \otimes I_r]\text{Vec}(W) = \begin{bmatrix} b_{11}I_r & \cdots & b_{1s}I_r \\ \vdots & \ddots & \vdots \\ b_{m1}I_r & \cdots & b_{ms}I_r \end{bmatrix} \begin{bmatrix} w_1 \\ \vdots \\ w_s \end{bmatrix} = \begin{bmatrix} Wb_1 \\ \vdots \\ Wb_m \end{bmatrix}.$$

ITEM 11. To see that if $A_{r \times r}$ and $B_{s \times s}$ are positive definite, then $A \otimes B$ is positive definite, consider the eigenvalues and eigenvectors of A and B. Recall that a symmetric matrix is positive definite if and only if all of its eigenvalues are positive. Suppose that $Av = \phi v$ and $Bw = \theta w$. We now show that all of the eigenvalues of $A \otimes B$ are positive. Observe that

$$\begin{aligned} [A \otimes B][v \otimes w] &= [Av \otimes Bw] \\ &= [\phi v \otimes \theta w] \\ &= \phi \theta [v \otimes w]. \end{aligned}$$

This shows that $[v \otimes w]$ is an eigenvector of $[A \otimes B]$ corresponding to the eigenvalue $\phi \theta$. As there are r choices for ϕ and s choices for θ, this accounts for all rs of the eigenvalues in the $rs \times rs$ matrix $[A \otimes B]$. Moreover, ϕ and θ are both positive, so all of the eigenvalues of $[A \otimes B]$ are positive.

A.3 Quadratic Optimization

We give a brief introduction to quadratic optimization. The methods presented here can be used for fitting linear models and linear models subject to linear (equality) constraints but we have not needed them for fitting least squares or even for finding ridge regression estimates. The reason for discussing this topic is that both lasso regression and support vector machines involve minimizing quadratic functions subject to linear inequality constraints.

In general, a quadratic optimization problem seeks a vector η that minimizes

$$\frac{1}{2}\eta'A\eta + b'\eta$$

subject to the one dimensional constraints

$$\tilde{q}_i'\eta = \tilde{d}_i, \quad i = 1,\ldots,s; \qquad \tilde{q}_i'\eta \leq \tilde{d}_i, \quad i = s+1,\ldots,s+t.$$

As always with quadratic forms, without loss of generality we assume that A is symmetric. Rewrite the constraints in matrix notation as

$$Q_1\eta = d_1; \qquad Q_2\eta \leq d_2 \tag{1}$$

where matrix inequalities are understood to apply elementwise. Also write

$$Q = \begin{bmatrix} Q_1 \\ Q_2 \end{bmatrix}; \qquad d = \begin{bmatrix} d_1 \\ d_2 \end{bmatrix}.$$

If $t = 0$, i.e., if there are no inequality constraints, the minimizing values solve the linear system

$$\begin{bmatrix} A & Q_1' \\ Q_1 & 0 \end{bmatrix} \begin{bmatrix} \eta \\ \lambda \end{bmatrix} = \begin{bmatrix} -b \\ d_1 \end{bmatrix}.$$

This is a special case of the more general results that follow.

We now present Karush, Kuhn, and Tucker's (KKT's) four necessary conditions for a solution. First set up the Lagrange function associated with the inequality constraints being equalities:

$$\frac{1}{2}\eta'A\eta + b'\eta + \lambda'(Q\eta - d). \tag{2}$$

If the constraints are all equalities, λ is a vector of Lagrange multipliers. With inequalities they are called KKT multipliers. The idea is that since $(Q\eta - d) \leq 0$, if we take $\lambda \geq 0$, we have $\lambda'(Q\eta - d) \leq 0$. The answer to our problem should occur where we have η and λ achieving

$$\sup_{\lambda \geq 0} \inf_{\eta} \left[\frac{1}{2}\eta'A\eta + b'\eta + \lambda'(Q\eta - d) \right].$$

The requirement that $\lambda \geq 0$ is known as KKT *dual feasibility*.

For any fixed λ the infimum should occur as a solution to

$$0 = \mathbf{d}_\eta \left[\frac{1}{2}\eta'A\eta + b'\eta + \lambda'(Q\eta - d) \right]$$
$$= \eta'A + b' + \lambda'Q. \tag{3}$$

This is the KKT *stationarity condition*. For a solution of (3) to exist, we must have

$$b + Q'\lambda \in C(A). \tag{4}$$

We refer to this as the *KKT column space condition* (although it is not one of the four KKT conditions).

Equality (3) implies that

$$0 = \eta'A\eta + b'\eta + \lambda'Q\eta.$$

This allows us to simplify (2) into

$$\frac{-1}{2}\eta'A\eta - \lambda'd \tag{5}$$

which we want to maximize relative to $\lambda \geq 0$. Moreover, equation (3) also allows us to solve for η,

$$\eta' = -\left(b' + \lambda'Q\right)A^-; \qquad \eta = -A^-\left(b + Q'\lambda\right).$$

The choice of generalized inverse for A is crucial to satisfying KKT's conditions other than stationarity.

Having solved for η, we can find (5) as a function of λ alone,

$$\frac{-1}{2}\left(b' + \lambda'Q\right)A^-AA^-\left(b + Q'\lambda\right) - \lambda'd$$

or, due to (4),

$$\frac{-1}{2}\left(b' + \lambda'Q\right)A^-\left(b + Q'\lambda\right) - \lambda'd. \tag{6}$$

The maximizing value of the criterion function does not depend on the choice of generalized inverse, even though an appropriate solution for η that satisfies all of the KKT conditions does.

The KKT *primal feasibility* condition is simply that any solution η must satisfy the original constraints in (1). In terms of λ that means picking λ and A^- so that

$$-QA^-\left(b + Q'\lambda\right) - d \stackrel{=}{\leq} 0,$$

where it is understood that some rows of the vectors are equalities and some are inequalities.

The final KKT condition is *complementary slackness* which means that

$$0 = Diag(\lambda)[Q\eta - d] = -Diag(\lambda)\left[A^-\left(b + Q'\lambda\right) - d\right].$$

This condition is vacuous for the equality constraints for which $Q_1\eta - d_1 = 0$.

If solving the stationarity condition gives λ elements that do not satisfy dual feasibility or complementary slackness, the solution is not a minimizer and some corrective action must be taken.

In (2) we did *not* set the derivative with respect to λ equal to 0. Setting the derivative with respect to λ equal to 0 gives both $Q_1\eta = d_1$ and $Q_2\eta = d_2$. Given (1), it is appropriate to set $Q_1\eta = d_1$ but it is not appropriate to set $Q_2\eta = d_2$.

A.3.1 Application to Support Vector Machines

This discussion relies on the notation of Section 13.5. Any n vector v may be written as an N_1 vector and an N_0 vector, $v' = (v'_1, v'_0)$. For this discussion only, denote $J_1 \equiv J_{N_1}$ and $J_0 \equiv J_{N_0}$. The data are written as

$$Y = \begin{bmatrix} Y_1 \\ Y_0 \end{bmatrix} = \begin{bmatrix} J_1 \\ 0 \end{bmatrix}; \qquad X = \begin{bmatrix} X_1 \\ X_0 \end{bmatrix} = \begin{bmatrix} J_1 & \mathbf{X}_1 \\ J_0 & \mathbf{X}_0 \end{bmatrix}.$$

The criterion function (13.5.4) is

$$\frac{1}{2}\eta'A\eta + b'\eta = \frac{1}{2}\begin{bmatrix} \beta_0 \\ \beta_* \\ \xi \end{bmatrix}' \begin{bmatrix} 0 & 0 & 0 \\ 0 & 2kI & 0 \\ 0 & 0 & 0 \end{bmatrix} \begin{bmatrix} \beta_0 \\ \beta_* \\ \xi \end{bmatrix} + \begin{bmatrix} 0 \\ 0 \\ J_n \end{bmatrix}' \begin{bmatrix} \beta_0 \\ \beta_* \\ \xi \end{bmatrix}. \tag{7}$$

For SVMs we only have inequality constraints:

$$Q\eta - d = \begin{bmatrix} -X_1 & -I & 0 \\ X_0 & 0 & -I \\ 0 & -I & 0 \\ 0 & 0 & -I \end{bmatrix} \begin{bmatrix} \beta \\ \xi_1 \\ \xi_0 \end{bmatrix} - \begin{bmatrix} -J_1 \\ -J_0 \\ 0 \\ 0 \end{bmatrix}$$

$$= \begin{bmatrix} -J_1 & -\mathbf{X}_1 & -I & 0 \\ J_0 & \mathbf{X}_0 & 0 & -I \\ 0 & 0 & -I & 0 \\ 0 & 0 & 0 & -I \end{bmatrix} \begin{bmatrix} \beta_0 \\ \beta_* \\ \xi_1 \\ \xi_0 \end{bmatrix} - \begin{bmatrix} -J_1 \\ -J_0 \\ 0 \\ 0 \end{bmatrix} \leq 0. \tag{8}$$

For stationary solutions to exist we need the KKT column space condition (4) to hold, i.e.,

$$\begin{bmatrix} 0 \\ 0 \\ J_1 \\ J_0 \end{bmatrix} + \begin{bmatrix} -J'_1 & J'_0 & 0 & 0 \\ -\mathbf{X}'_1 & \mathbf{X}'_0 & 0 & 0 \\ -I & 0 & -I & 0 \\ 0 & -I & 0 & -I \end{bmatrix} \begin{bmatrix} \lambda_{11} \\ \lambda_{10} \\ \lambda_{21} \\ \lambda_{20} \end{bmatrix}$$

$$= \begin{bmatrix} 0 \\ 0 \\ J_n \end{bmatrix} + \begin{bmatrix} -J'_1\lambda_{11} + J'_0\lambda_{10} \\ -\mathbf{X}'_1\lambda_{11} + \mathbf{X}'_0\lambda_{10} \\ -\lambda_1 - \lambda_2 \end{bmatrix} = b + Q'\lambda \in C(A) = C\left(\begin{bmatrix} 0 \\ I \\ 0 \end{bmatrix}\right).$$

This requires that

$$-J'_1\lambda_{11} + J'_0\lambda_{10} = 0$$

and $J - \lambda_1 - \lambda_2 = 0$ or

$$\lambda_2 = J - \lambda_1.$$

The first of these displayed equalities is a standard SVM condition. The second eliminates our need to be concerned with λ_2. Moreover, the dual feasibility condition $\lambda \geq 0$ implies that both $\lambda_1 \geq 0$ and $J - \lambda_1 \geq 0$ so

$$0 \leq \lambda_1 \leq 1,$$

which is another standard SVM condition. These two conditions will be referred to as the *column space conditions*.

A.3.1.1 Computing β_*

Using the column space conditions,

$$b + Q'\lambda = \begin{bmatrix} 0 \\ -\mathbf{X}_1'\lambda_{11} + \mathbf{X}_0'\lambda_{10} \\ 0 \end{bmatrix},$$

so

$$\eta = -A^-(b + Q'\lambda) = A^- \begin{bmatrix} 0 \\ \mathbf{X}_1'\lambda_{11} - \mathbf{X}_0'\lambda_{10} \\ 0 \end{bmatrix}$$

In particular, because $AA^-A = A$, $(b + Q'\lambda) \in C(A)$, and the specific form of A,

$$\beta_* = \begin{bmatrix} 0 \\ I \\ 0 \end{bmatrix}' \eta = \begin{bmatrix} 0 \\ I \\ 0 \end{bmatrix}' \frac{1}{2k} A\eta$$

$$= \frac{1}{2k} \begin{bmatrix} 0 \\ I \\ 0 \end{bmatrix}' AA^- \begin{bmatrix} 0 \\ \mathbf{X}_1'\lambda_{11} - \mathbf{X}_0'\lambda_{10} \\ 0 \end{bmatrix}$$

$$= \frac{1}{2k} \left(\mathbf{X}_1'\lambda_{11} - \mathbf{X}_0'\lambda_{10} \right).$$

As discussed in Section 13.5, often many of the λ_1 values are 0 which is reflected in how computer programs report their results.

A.3.1.2 The Dual Criterion

We now evaluate the criterion function (7) in terms of λ by using the solution for η provided by the KKT stationarity and column space conditions. Maximizing this function is often known as the *dual problem*. Since $(b + Q'\lambda) \in C(A)$, in (6) the choice of A^- does not matter when evaluating (7) so we take the simplest generalized inverse.

$$A^- = \begin{bmatrix} 0 & 0 & 0 \\ 0 & (1/2k)I & 0 \\ 0 & 0 & 0 \end{bmatrix}.$$

Substitution gives

$$\frac{-1}{2}(b+Q'\lambda)'A^-(b+Q'\lambda)-\lambda'd$$

$$= \begin{bmatrix} 0 \\ \mathbf{X}_1'\lambda_{11} - \mathbf{X}_0'\lambda_{10} \\ 0 \end{bmatrix}' \begin{bmatrix} 0 & 0 & 0 \\ 0 & (1/2k)I & 0 \\ 0 & 0 & 0 \end{bmatrix} \begin{bmatrix} 0 \\ \mathbf{X}_1'\lambda_{11} - \mathbf{X}_0'\lambda_{10} \\ 0 \end{bmatrix} - \begin{bmatrix} \lambda_1 \\ \lambda_2 \end{bmatrix}' \begin{bmatrix} -J_n \\ 0 \end{bmatrix}$$

$$= \frac{-1}{2k}\lambda_1' \begin{bmatrix} \mathbf{X}_1\mathbf{X}_1' & -\mathbf{X}_1\mathbf{X}_0' \\ -\mathbf{X}_0\mathbf{X}_1' & \mathbf{X}_0\mathbf{X}_0' \end{bmatrix} \lambda_1 + \lambda_1'J_n.$$

The constraints on λ are typically cited as being the dual feasibility condition $\lambda \geq 0$ and the column space conditions. Together these require $0 \leq \lambda_{1i} \leq 1$ and $J_1'\lambda_{11} - J_0'\lambda_{10} = 0$.

Exercise A.2.   Show that if you multiply the original criterion function (7) by $1/2k \equiv \tilde{C}$, the dual criterion function remains unchanged but the column space conditions change to $0 \leq \lambda_1 \leq \tilde{C}$. How does this change the formula for β_*? This form of the problem involving a "cost" $\tilde{C}$, rather than a tuning parameter k for the penalty, seems quite popular.

An actual solution η needs to incorporate the two other KKT conditions: primal feasibility and complementary slackness. Primal feasibility is that the inequality constraints in (8) continue to hold. Complementary slackness is the key to determining β_0.

A.3.1.3 Computing β_0

Complementary slackness is $0 = D(\lambda)[Q\eta - d]$. Together with a column space condition, it requires

$$0 = \begin{bmatrix} D(\lambda_1) & 0 \\ 0 & D(J-\lambda_1) \end{bmatrix} \left(\begin{bmatrix} -J_1 & -\mathbf{X}_1 & -I & 0 \\ J_0 & \mathbf{X}_0 & 0 & -I \\ 0 & 0 & -I & 0 \\ 0 & 0 & 0 & -I \end{bmatrix} \begin{bmatrix} \beta_0 \\ \beta_* \\ \xi \end{bmatrix} - \begin{bmatrix} -J_1 \\ -J_0 \\ 0 \\ 0 \end{bmatrix} \right).$$

If $0 < \lambda_{1h} < 1$, the bottom half of the equation implies $\xi_h = 0$. Using that fact in the top half of the equation, depending on whether y_h is 1 or zero, we must have $\beta_0 + \mathbf{x}_h'\beta_* = 1$ or $\beta_0 + \mathbf{x}_h'\beta_* = -1$. Changing notation a bit, think about $\lambda_1' = (\lambda_{11}', \lambda_{10}')$. If y_{1j} denotes an element of Y_1 with $0 < \lambda_{11j} < 1$, then $\beta_0 = 1 - \mathbf{x}_{1j}'\beta_*$ and similarly for y_{0j} an element of Y_0 with $0 < \lambda_{10j} < 1$, $\beta_0 = -1 - \mathbf{x}_{1j}'\beta_*$.

It may seem curious that β_0 should be so directly tied to the arbitrary number 1, but remember that any multiple of β defines the same hyperplane, so we have just chosen a multiple that defines β_0 in terms of being 1 unit away from an appropriate value of $\mathbf{x}_h'\beta_*$.

As alluded to in Section 13.5, computer programs often report a λ_1 vector that does not satisfy all of the equalities implied in the paragraph above. As a result, computer programs, and even published works, make some ado about how to obtain β_0 from the various cases that have $0 < \lambda_{1h} < 1$.

Appendix B
Best Linear Predictors

Abstract Best linear prediction is a key concept in the development of many topics in statistics. It is introduced in Chap. 6 of *PA* as an alternative to traditional regression theory. In this volume, it is introduced in Chap. 4 and used in the discussions of time series analysis, the Kalman filter, kriging, principal components, and factor analysis.

B.1 Properties of Best Linear Predictors

We need to establish general properties of best linear predictors that are analogous to results for conditional expectations. Let $y = (y_1, \ldots, y_q)'$ and $x = (x_1, \ldots, x_{p-1})'$. Denote

$$E(y) = \mu_y \qquad\qquad E(x) = \mu_x$$
$$\text{Cov}(y) = V_{yy} \qquad\qquad \text{Cov}(x) = V_{xx}$$

and

$$\text{Cov}(y, x) = V_{yx} = V'_{xy}.$$

The best linear predictor (BLP) of y is defined to be the linear function $f(x)$ that minimizes

$$E\left\{[y - f(x)]'\, [y - f(x)]\right\}.$$

The best linear predictor, also called the linear expectation, is

$$\hat{E}(y|x) \equiv \mu_y + B'(x - \mu_x),$$

where $B_{(p-1)\times q}$ is a solution

$$V_{xx}B = V_{xy}.$$

In general, the linear expectation $\hat{E}(y|x)$ is neither the conditional expectation of y given x nor an estimate of the conditional expectation; it is a different concept. The conditional expectation $E(y|x)$ is the best predictor of y based on x. $\hat{E}(y|x)$ is

© Springer Nature Switzerland AG 2019
R. Christensen, *Advanced Linear Modeling*, Springer Texts in Statistics,
https://doi.org/10.1007/978-3-030-29164-8

the best *linear* predictor. Conditional expectations require knowledge of the entire multivariate distribution. Linear expectations depend only on the mean vector and covariance matrix. For some families of multivariate distributions, of which the multivariate normal is the best known, the linear expectation and the conditional expectation happen to be the same. This is similar to the fact that for multivariate normals best linear unbiased estimates are also best within the broader class of (possibly nonlinear) unbiased estimates. Linear expectations have a number of properties that are similar to those of conditional expectations. Many of these properties will be explored in the current section. The notation $\hat{E}(y|x)$ for the linear expectation has been used for at least sixty years, see Doob (1953).

Similar to *PA* Section 6.3, $\hat{E}(y|x)$ is a function of x, $E(y) = E[\hat{E}(y|x)]$, and the *prediction error covariance matrix* is

$$\text{Cov}[y - \hat{E}(y|x)]$$
$$= E\left\{[(y - \mu_y) - V_{yx}V_{xx}^-(x - \mu_x)][(y - \mu_y) - V_{yx}V_{xx}^-(x - \mu_x)]'\right\}$$
$$= V_{yy} - V_{yx}V_{xx}^-V_{xy}.$$

In particular, the *partial correlation* between two components of y given x is defined as the correlation between those components obtained from the prediction error covariance matrix. To simplify the discussion, it is assumed in the following that appropriate inverses exist. In particular,

$$B = V_{xx}^{-1}V_{xy},$$

so that the linear expectation is unique. First, we establish that linear expectation is a linear operator.

Proposition B.1.1.   Let A be an $r \times q$ matrix and let a be an $r \times 1$ vector. The best linear predictor of $Ay + a$ based on x is

$$\hat{E}(Ay + a|x) = A\hat{E}(y|x) + a.$$

Exercise B.1.   Prove Proposition B.1.1.

If we predict a random variable y from a set of random variables that includes y, then the prediction is just y. It is convenient to state this result in terms of the x vector.

Proposition B.1.2.   For $x = (x_1, \ldots, x_{p-1})'$,

$$\hat{E}(x_i|x) = x_i.$$

Exercise B.2.   Prove Proposition B.1.2. Hint: By definition, $\hat{E}(x_i|x) = \mu_{xi} + B'(x - \mu_x)$, where $V_{xx}B = V_{xx_i}$. In this case, V_{xx_i} is the ith column of V_{xx}.

Propositions B.1.1 and B.1.2 lead to the following corollary.

Corollary B.1.3.   If $\beta \in \mathbf{R}^{p-1}$, then

$$\hat{E}(x'\beta|x) = x'\beta .$$

The next result is that a nonsingular affine transformation of the predictors does not change the linear expectation.

Proposition B.1.4.   Let A be a $(p-1) \times (p-1)$ nonsingular matrix and let a be a vector in $\mathbf{R}^{p-1}$; then,

$$\hat{E}(y|Ax + a) = \hat{E}(y|x).$$

PROOF.   Note   that $\text{Cov}(y, Ax + a) = \text{Cov}(y, Ax), \text{Cov}(Ax + a) = \text{Cov}(Ax)$ and $(Ax + a) - \text{E}(Ax + a) = A(x - \mu_x)$. Then,

$$\begin{aligned}
\hat{E}(y|Ax + a) &= \mu_y + \text{Cov}(y, Ax)[\text{Cov}(Ax)]^{-1}A(x - \mu_x) \\
&= \mu_y + V_{yx}A'[AV_{xx}A']^{-1}A(x - \mu_x) \\
&= \mu_y + V_{yx}A'A'^{-1}V_{xx}^{-1}A^{-1}A(x - \mu_x) \\
&= \mu_y + V_{yx}V_{xx}^{-1}(x - \mu_x) \\
&= \hat{E}(y|x) .
\end{aligned}$$ □

The next proposition involves predictors that are uncorrelated with the random vector to be predicted.

Proposition B.1.5.   If $\text{Cov}(y, x) = 0$, then

$$\hat{E}(y|x) = \mu_y .$$

PROOF. $\hat{E}(y|x) = \mu_y + B'(x - \mu_x)$, where $V_{xx}B = V_{xy}$. If $V_{xy} = 0$, the matrix $B = 0$ is the solution, thus giving the result.   □

Again, all of these results are analogous to results for conditional expectations; see *PA* Appendix D. In Proposition B.1.5, the condition $\text{Cov}(y, x) = 0$ is analogous to the idea, for conditional expectations, that y and x are independent. In Proposition B.1.4, the idea that A is nonsingular in the transformation $Ax + a$ corresponds to taking an invertible transformation of the conditioning variable. Because of these analogies, any proofs that depend only on the five results given earlier have corresponding proofs for conditional expectations. This observation generalizes results from best linear predictors to best predictors. As mentioned in *PA* Section 6.3 and earlier in this section, the best predictor of y based on x is $E(y|x)$. The reason for not using best predictors is that they require knowledge of the joint distribution of the random variables. Best linear predictors require knowledge only of the first and second moments. For Gaussian processes, best linear predictors are also best predictors.

Exercise B.3.   Assume that

$$\begin{bmatrix} y \\ x \end{bmatrix} \sim N\left(\begin{bmatrix} \mu_y \\ \mu_x \end{bmatrix}, \begin{bmatrix} V_{yy} & V_{yx} \\ V_{xy} & V_{xx} \end{bmatrix} \right)$$

and that the covariance matrix is nonsingular, so that a density exists for the joint distribution. Show that

$$y|x \sim N\{\hat{E}(y|x), \mathrm{Cov}[y - \hat{E}(y|x)]\}.$$

In applications to the time domain models of Chapter 7, it is occasionally convenient to allow x to be an infinite vector $x = (x_1, x_2, \ldots)'$. For infinite vectors, the condition $V_{xx}\beta = V_{xy}$ can be written rigorously as $\sum_{j=1}^{\infty} \sigma_{ij}\beta_j = \sigma_{iy}$ for $i = 1, 2, \ldots$, where $V_{xx} = [\sigma_{ij}]$ and $V_{xy} = [\sigma_{iy}]$. It is not difficult to see that all of the preceeding propositions continue to hold.

Finally, to derive the Kalman filter and later the joint prediction properties of principal components, some additional results are required. The first involves linear expectations based on a predictor consisting of two vectors with zero correlation. (I am not aware of any corresponding result for conditional expectations given two independent vectors.)

To handle three vectors simultaneously, we need some additional notation. Consider a partition of y, say $y' = (y_1', y_2')$. Denote

$$\mathrm{Cov}(y) = \mathrm{Cov}\begin{pmatrix} y_1 \\ y_2 \end{pmatrix} = \begin{bmatrix} V_{11} & V_{12} \\ V_{21} & V_{22} \end{bmatrix}$$

and

$$\mathrm{Cov}(y_i, x) = V_{ix} \qquad i = 1, 2.$$

Also, let

$$E(y_i) = \mu_i \qquad i = 1, 2.$$

The main result follows.

Proposition B.1.6.   If $\mathrm{Cov}(y_1, x) = 0$, then

$$\hat{E}(y_2|y_1, x) = \hat{E}(y_2|x) + \hat{E}(y_2|y_1) - \mu_2.$$

PROOF.   By definition,

$$\hat{E}(y_2|y_1,x) = \mu_2 + (V_{21}, V_{2x}) \begin{bmatrix} V_{11} & V_{1x} \\ V_{x1} & V_{xx} \end{bmatrix}^{-1} \begin{pmatrix} y_1 - \mu_1 \\ x - \mu_x \end{pmatrix} . \tag{1}$$

By assumption, $V_{1x} = 0$, and it follows that

$$\begin{bmatrix} V_{11} & V_{1x} \\ V_{x1} & V_{xx} \end{bmatrix}^{-1} = \begin{bmatrix} V_{11} & 0 \\ 0 & V_{xx} \end{bmatrix}^{-1} = \begin{bmatrix} V_{11}^{-1} & 0 \\ 0 & V_{xx}^{-1} \end{bmatrix} . \tag{2}$$

Substituting (2) into (1) gives

$$\begin{aligned} \hat{E}(y_2|y_1,x) &= \mu_2 + V_{21}V_{11}^{-1}(y_1 - \mu_1) + V_{2x}V_{xx}^{-1}(x - \mu_x) \\ &= \hat{E}(y_2|y_1) + \hat{E}(y_2|x) - \mu_2 . \end{aligned}$$

$\square$

To simplify notation, let

$$e(y_1|x) \equiv y_1 - \hat{E}(y_1|x) .$$

This is the prediction error from predicting y_1 using x. Our primary application of Proposition B.1.6 will be through the following lemma.

Lemma B.1.7.

$$\text{Cov}\left[e(y_1|x), x\right] = 0 .$$

PROOF.

$$\begin{aligned} \text{Cov}\left[e(y_1|x), x\right] &= \text{Cov}[y_1 - \hat{E}(y_1|x), x] \\ &= \text{Cov}[(y_1 - \mu_1) - V_{1x}V_{xx}^{-1}(x - \mu_x), x] \\ &= \text{Cov}(y_1 - \mu_1, x) - V_{1x}V_{xx}^{-1}\text{Cov}(x - \mu_x, x) \\ &= V_{1x} - V_{1x}V_{xx}^{-1}V_{xx} \\ &= 0 . \end{aligned}$$

$\square$

Lemma B.1.7 leads to the key result.

Proposition B.1.8.

$$\hat{E}(y_2|y_1,x) = \hat{E}(y_2|x) + \text{Cov}\left[y_2, e(y_1|x)\right]\{\text{Cov}\left[e(y_1|x)\right]\}^{-1} e(y_1|x) .$$

PROOF.   First, note that $\hat{E}(y_1|x)$ is an affine transformation of x. Thus, we can write

$$\begin{bmatrix} y_1 - \hat{E}(y_1|x) \\ x \end{bmatrix} = \begin{bmatrix} I & B \\ 0 & I \end{bmatrix} \begin{bmatrix} y_1 \\ x \end{bmatrix} - \begin{bmatrix} -\mu_1 \\ 0 \end{bmatrix}$$

for some matrix B. The upper triangular matrix

$$\begin{bmatrix} I & B \\ 0 & I \end{bmatrix}$$

is nonsingular, so by Proposition B.1.4,

$$\hat{E}(y_2|y_1,x) = \hat{E}\left[y_2|y_1 - \hat{E}(y_1|x),x\right] = \hat{E}\left[y_2|e(y_1|x),x\right].$$

The result follows from Proposition B.1.6, Lemma B.1.7, and observing that by definition

$$\hat{E}\left[y_2|e(y_1|x)\right] = \mu_2 + \text{Cov}\left[y_2,e(y_1|x)\right]\left[\text{Cov}e(y_1|x)\right]^{-1}e(y_1|x),$$

so subtracting μ_2 from both sides gives

$$\hat{E}\left[y_2|e(y_1|x)\right] - \mu_2 = \text{Cov}\left[y_2,e(y_1|x)\right]\left[\text{Cov}e(y_1|x)\right]^{-1}e(y_1|x). \qquad \square$$

Note that if x is vacuous, Proposition B.1.8 is just the standard definition of $\hat{E}(y_2|y_1)$.

The prediction error covariance matrix can be written in a form analogous to results in *PA* Chapter 9.

Proposition B.1.9.

$$\text{Cov}\left[y_2 - \hat{E}(y_2|y_1,x)\right] = \text{Cov}\left[y_2 - \hat{E}(y_2|x)\right]$$
$$- \text{Cov}\left[y_2,e(y_1|x)\right]\left\{\text{Cov}\left[e(y_1|x)\right]\right\}^{-1}\text{Cov}\left[e(y_1|x),y_2\right].$$

PROOF.   Let $e \equiv e(y_1|x)$. By Lemma B.1.7, $\text{Cov}(x,e) = 0$. Moreover, because $\hat{E}(y_2|x)$ is a linear function of x, $\text{Cov}(\hat{E}(y_2|x),e) = 0$ and $\text{Cov}(y_2 - \hat{E}(y_2|x),e) = \text{Cov}(y_2,e) - \text{Cov}(\hat{E}(y_2|x),e) = \text{Cov}(y_2,e)$. Using Proposition B.1.8 and this fact about covariances gives

$$\text{Cov}[y_2 - \hat{E}(y_2|y_1,x)]$$
$$= \text{Cov}\left\{[y_2 - \hat{E}(y_2|x)] - \text{Cov}(y_2,e)[\text{Cov}(e)]^{-1}e\right\}$$
$$= \text{Cov}\left[y_2 - \hat{E}(y_2|x)\right] + \text{Cov}(y_2,e)[\text{Cov}(e)]^{-1}\text{Cov}(e,y_2)$$
$$- \text{Cov}\left[y_2 - \hat{E}(y_2|x),e\right][\text{Cov}(e)]^{-1}\text{Cov}(e,y_2)$$
$$- \text{Cov}(y_2,e)[\text{Cov}(e)]^{-1}\text{Cov}\left[e,y_2 - \hat{E}(y_2|x)\right]$$
$$= \text{Cov}\left[y_2 - \hat{E}(y_2|x)\right] + \text{Cov}(y_2,e)[\text{Cov}(e)]^{-1}\text{Cov}(e,y_2)$$
$$- 2\text{Cov}(y_2,e)[\text{Cov}(e)]^{-1}\text{Cov}(e,y_2)$$
$$= \text{Cov}[y_2 - \hat{E}(y_2|x)] - \text{Cov}(y_2,e)[\text{Cov}(e)]^{-1}\text{Cov}(e,y_2),$$

which proves the result.

$\square$

Proposition B.1.10.

$$\mathrm{Cov}[y, y - \hat{E}(y|x)] = \mathrm{Cov}[y - \hat{E}(y|x)].$$

PROOF.

$$\mathrm{Cov}[y - \hat{E}(y|x)] = V_{yy} - V_{yx}V_{xx}^{-1}V_{xy},$$
$$\mathrm{Cov}[y, y - \hat{E}(y|x)] = \mathrm{Cov}[y] - \mathrm{Cov}[y, \hat{E}(y|x)]$$
$$= V_{yy} - \mathrm{Cov}[y, \mu_y + V_{yx}V_{xx}^{-1}(x - \mu_x)]$$
$$= V_{yy} - V_{yx}V_{xx}^{-1}V_{xy}. \qquad \square$$

Exercise B.4. For A nonsingular, let

$$A = \begin{bmatrix} A_{11} & A_{12} \\ A_{21} & A_{22} \end{bmatrix},$$

and let $A_{1 \cdot 2} = A_{11} - A_{12}A_{22}^{-1}A_{21}$. Show that if all inverses exist,

$$A^{-1} = \begin{bmatrix} A_{1 \cdot 2}^{-1} & -A_{1 \cdot 2}^{-1}A_{12}A_{22}^{-1} \\ -A_{22}^{-1}A_{21}A_{1 \cdot 2}^{-1} & A_{22}^{-1} + A_{22}^{-1}A_{21}A_{1 \cdot 2}^{-1}A_{12}A_{22}^{-1} \end{bmatrix}$$

and that

$$A_{22}^{-1} + A_{22}^{-1}A_{21}A_{1 \cdot 2}^{-1}A_{12}A_{22}^{-1} = \left[A_{22} - A_{21}A_{11}^{-1}A_{12} \right]^{-1}.$$

Exercise B.5. Let $Y = (y_1, \ldots, y_n)'$ and $Y_k = (y_{n+1}, \ldots, y_{n+k})'$. Show that

$$\hat{E}\left[y_{n+k+1} | Y \right] = \hat{E}\left[\hat{E}(y_{n+k+1} | Y, Y_k) | Y \right].$$

Hint: Use the definition of $\hat{E}(\cdot | \cdot)$, Proposition B.1.1, and results on the inverse of a partitioned matrix.

B.2 Irrelevance of Units in Best Multivariate Prediction

Suppose you want to predict the size of a dog house, $y = $ (height,length,width)$'$, from the size of the dog x. It would make sense to measure the components of y in some common unit, like centimeters. If we do, our standard measure of prediction error $[y - f(x)]'[y - f(x)]$ is measured in square centimeters. If, however, we measured height in kilometers but length and width in centimeters the numbers in y_1 would be 5 orders of magnitude smaller than those for y_2 and y_3. You might think that would affect the best predictor or best linear predictor. In fact, based on the prediction criterion, we would expect it to be less important to predict y_1 well, because the squared error that it would contribute to the overall prediction criterion will probably be tiny compared to the squared prediction errors associated with y_2 and y_3.

It would be reasonable, when the components of y involve different units, to minimize the prediction criterion

$$E\left\{[y-f(x)]'V_{yy}^{-1}[y-f(x)]\right\}.$$

One nice thing about this prediction criterion is that it is unitless. The units associated with V_{yy}^{-1} cancel with the units associated with y and $f(x)$.

Quite remarkably, none of these considerations have any affect on the best predictor or the best linear predictor. Write $V_{yy}^{-1} = Q'Q$ where Q is nonsingular. Then

$$E\left\{[y-f(x)]'V_{yy}^{-1}[y-f(x)]\right\}$$
$$= E\left(\{Q[y-f(x)]\}'\{Q[y-f(x)]\}\right) = E\left\{[Qy-\tilde{f}(x)]'[Qy-\tilde{f}(x)]\right\}$$

where $\tilde{f}(x) \equiv Qf(x)$. Because Q is invertible, it follows that prediction of y under the new V_{yy}^{-1} criterion is equivalent to prediction of Qy under the old criterion. In particular, the best predictor $m_V(x)$ under the new criterion will be Q^{-1} times the best predictor of Qy under the old criterion. However, the best predictor of Qy under the old criterion is $E(Qy|x)$ so the best predictor of y under the new criterion is $Q^{-1}E(Qy|x) = Q^{-1}QE(y|x) = E(y|x)$ which was the best predictor of y under the old criterion. Similar arguments establish that the best linear predictor under the new criterion is also the same as the best linear predictor under the old criterion. Moreover, the same arguments hold if we replace V_{yy}^{-1} in the new criterion with any positive definite matrix.

Appendix C
Residual Maximum Likelihood

Abstract Residuals have singular distributions, so they do not have density/likelihood functions. This appendix addresses issues of how to maximize a likelihood function that does not exist.

C.1 Maximum Likelihood Estimation for Singular Normal Distributions

Maximum likelihood estimation involves maximizing the joint density of the observations over the possible values of the parameters. This assumes the existence of a density. A density cannot exist if the covariance matrix of the observations is singular, as is the case with residuals. We consider an approach that allows maximum likelihood estimation and show some uniqueness properties of the approach.

Suppose Y is a random vector in $\mathbf{R}^n$, $\mathrm{E}(Y) = \mu$, $\mathrm{Cov}(Y) = V$. If $r(V) = t < n$, then, as seen in *PA* Lemma 1.3.5, $\Pr[(Y - \mu) \in C(V)] = 1$ and $Y - \mu$ is restricted to an t-dimensional subspace of $\mathbf{R}^n$. It is this restriction of $Y - \mu$ to a subspace of $\mathbf{R}^n$ (with Lesbesgue measure zero) that causes the nonexistence of the density. We seek a linear transformation from $\mathbf{R}^n$ to $\mathbf{R}^t$ that will admit a density for the transformed random vector. The linear transformation should not lose any information and the MLEs should, in some sense, be the unique MLEs.

Suppose we pick an $n \times t$ matrix Q with $C(Q) = C(V)$. $Q'Y$ together with a nonrandom function of Y can reconstruct Y with probability 1. Let M_V be the perpendicular projection operator onto $C(V)$, then $Y = M_V Y + (I - M_V)Y$. $M_V Y = Q(Q'Q)^{-1}Q'Y$ is a function of $Q'Y$ while $(I - M_V)Y = (I - M_V)\mu$ with probability 1, because $\mathrm{Cov}[(I - M_V)Y] = (I - M_V)V(I - M_V) = 0$.

We would also like to see that $\mathrm{Cov}(Q'Y) = Q'VQ$ is nonsingular, so that a density can exist. Since $C(Q) = C(V)$ and V is symmetric, $V = QTQ'$ for some symmetric matrix T. If T is nonsingular, then $Q'VQ = (Q'Q)T(Q'Q)$ is nonsingular, because both T and $Q'Q$ are nonsingular. We now show that T is nonsingular. Suppose T is singular. Then there exists $d \in \mathbf{R}^t$, $d \neq 0$, so that $Td = 0$. Since $r(Q) = t$, there exists

© Springer Nature Switzerland AG 2019
R. Christensen, *Advanced Linear Modeling*, Springer Texts in Statistics,
https://doi.org/10.1007/978-3-030-29164-8

$b \neq 0$ such that $d = Q'b$. Because $Q'b = Q'M_V b$, we can assume that $b \in C(V)$. Now, $Vb = QTQ'b = QTd = 0$. However, for $b \in C(V)$, $Vb = 0$ can only happen if $b = 0$, which is a contradiction.

As mentioned earlier, there is little hope of estimating the entire matrix V. A more manageable problem is to assume that V is a function of a parameter vector θ. It should also be clear that in order to transform to a nonsingular random variable, we will need to know $C(V)$. This forces us to assume that $C(V(\theta))$ does not depend on θ.

Suppose now that $Y \sim N(\mu, V(\theta))$; then $Q'Y \sim N(Q'\mu, Q'V(\theta)Q)$. The density of $Q'Y$ is

$$f(Q'Y|\mu, \theta)$$
$$= \frac{1}{(2\pi)^{\frac{r}{2}}} \frac{1}{|Q'V(\theta)Q|^{\frac{1}{2}}} \exp[-(Q'Y - Q'\mu)'[Q'V(\theta)Q]^{-1}(Q'Y - Q'\mu)/2]$$
$$= (2\pi)^{-r/2}|Q'V(\theta)Q|^{-1/2} \exp[-(Y - \mu)'Q[Q'V(\theta)Q]^{-1}Q'(Y - \mu)/2].$$

The MLEs are obtained by maximizing this with respect to μ and θ. A direct consequence of Proposition C.1.1 below is that maximization of $f(Q'Y)$ does not depend on the choice of Q.

Proposition C.1.1.   If Q and Q_0 are two $n \times t$ matrices of rank t and $C(Q) = C(Q_0)$, then

(1) for some scalar k, $k|Q'VQ| = |Q_0'VQ_0|$;
(2) $Q[Q'VQ]^{-1}Q' = Q_0[Q_0'VQ_0]^{-1}Q_0'$ when the inverses exist.

PROOF.   Since $C(Q) = C(Q_0)$ and both are full rank, $Q_0 = QK$ for some nonsingular K.

(1) $|Q_0'VQ_0| = |K'Q'VQK| = |K|^2|Q'VQ|$. Take $k = |K|^2$.
(2)
$$Q_0[Q_0'VQ_0]^{-1}Q_0' = QK[K'Q'VQK]^{-1}K'Q'$$
$$= QKK^{-1}[Q'VQ]^{-1}(K')^{-1}K'Q'$$
$$= Q[Q'VQ]^{-1}Q'.   \qquad \square$$

Corollary C.1.2. $f(Q'Y|\mu, \theta) = k^{-1/2}f(Q_0'Y|\mu, \theta)$.

C.2 Residual Maximum Likelihood Estimation

Residual maximum likelihood (REML) estimation involves finding maximum likelihood estimates of covariance parameters from the distribution of the residuals. This allows for estimation of the covariance parameters without the complication of the fixed effects. We will see that this procedure is equivalent to the restricted maximum likelihood methods described in Section 4.3.

An apparent problem with this idea is that of defining the residuals when $V(\theta)$ is unknown. We show that any reasonable definition of residuals gives the same answers.

Consider the model

$$Y = X\beta + e, \quad e \sim N(0, V(\theta)),$$

$\theta = (\theta_1, \ldots, \theta_s)'$, As discussed in *PA* Section 10.2, the only reasonable linear unbiased estimates of $X\beta$ are of the form AY, where A is some projection operator onto $C(X)$. The residuals can be defined as $(I-A)Y$. The distribution of the residuals is

$$(I-A)Y \sim N(0, (I-A)V(\theta)(I-A)').$$

$V(\theta)$ is assumed nonsingular, so $C((I-A)V(\theta)(I-A)') = C(I-A)$. Let $r(X) \equiv r$; so $r(I-A) = n - r$. For an $n \times (n-r)$ matrix Q with $C(Q) = C(I-A)$, a residual MLE of θ maximizes

$$f(Q'(I-A)Y|\theta) = (2\pi)^{-(n-s)/2}|Q'(I-A)V(\theta)(I-A)'Q|^{-1/2}$$
$$\times \exp[-Y'(I-A)'Q[Q'(I-A)V(\theta)(I-A)'Q]^{-1}Q'(I-A)Y/2].$$

We will show that this depends on neither A nor Q by showing that $C[(I-A)'Q] = C(X)^{\perp}$ and appealing to Section 1.

Proposition C.2.1. $C((I-A)'Q) = C(X)^{\perp}$.

PROOF.   Clearly, $Q'(I-A)X = 0$, so $C((I-A)'Q) \subset C(X)^{\perp}$. The rank of $C(X)^{\perp}$ is $n - r$, so it is enough to show that the rank of $(I-A)'Q$ is $n - r$. Since $(I-A)'Q$ is an $n \times (n-r)$ matrix it is enough to show that for any $d \in \mathbf{R}^{n-r}$, $(I-A)'Qd = 0$ implies $d = 0$. Since $C(I-A) = C(Q)$, $Qd = (I-A)c$ for some c. If $(I-A)'Qd = 0$, then $c'(I-A)'Qd = d'Q'Qd = 0$; so $Qd = 0$. Since Q is an $n \times (n-r)$ matrix of full column rank, $d = 0$.   □

In Section 4.3, REML is defined as maximum likelihood estimation from $B'Y$, where $r(B) = n - r$ and $B'X = 0$. That definition is equivalent to the one used here. Choose $A = M$. Then $I - M$ is the perpendicular projection operator onto $C(X)^{\perp}$. Take $Q = B$ so that $Q'(I-M)Y = B'Y$.

Reference

Doob, J.L. (1953). *Stochastic Processes*. John Wiley and Sons, New York.

Author Index

© Springer Nature Switzerland AG 2019
R. Christensen, *Advanced Linear Modeling*, Springer Texts in Statistics,
https://doi.org/10.1007/978-3-030-29164-8

Subject Index

© Springer Nature Switzerland AG 2019
R. Christensen, *Advanced Linear Modeling*, Springer Texts in Statistics,
https://doi.org/10.1007/978-3-030-29164-8

Printed in the United States
By Bookmasters